Environmental MICROBIOLOGY

Environmental MICROBIOLOGY

Raina M. Maier

Department of Soil and Water Science
University of Arizona
Tucson, Arizona

Ian L. Pepper

Department of Soil and Water Science
University of Arizona
Tucson, Arizona

Charles P. Gerba

Department of Soil and Water Science
University of Arizona
Tucson, Arizona

ACADEMIC PRESS
An Imprint of Elsevier

San Diego   San Francisco   New York   Boston   London   Sydney   Tokyo

Cover photo credit:: Background photo: Images © 1999 PhotoDisc, Inc.
Insets: left and right, courtesy P. Rusin; center, courtesy K.L. Josephson.

This book is printed on acid-free paper.

Academic Press
An Imprint of Elsevier
525 B Street, Suite 1900, San Diego, California 92101-4495, USA
http://www.academicpress.com

Academic Press
84 Theobald's Road, London WC1X 8RR, UK
http://www.academicpress.com

International Standard Book Number: 0-12-497570-4

PRINTED IN CHINA
03   04   05   06   9   8   7   6   5

Starting a book is one thing, finishing it another. Thanks to my family,
especially Mom, Dad, and Wendy, for their love and support,
without which this book would not have been finished.
I dedicate this book to my daughter, Claire, and to her adventures in the next few years.

Raina M. Maier

I've always been lucky enough to be lucky—lucky to be surrounded by good people.
They include Barbara, Len, Joyce,
and of course Karen and all my graduate students. To them goes the credit for this book.

Ian L. Pepper

This book is dedicated to my parents and my wife for all their support.

Charles P. Gerba

Contents

C H A P T E R

1

Introduction to
Environmental Microbiology

RAINA M. MAIER, IAN L. PEPPER, AND CHARLES P. GERBA

C H A P T E R

2

Microorganisms in the Environment

KELLY A. REYNOLDS AND IAN L. PEPPER

C H A P T E R

3

Bacterial Growth

RAINA M. MAIER

CHAPTER

4

Terrestrial Environments

RAINA M. MAIER AND IAN L. PEPPER

CHAPTER

5

Aeromicrobiology

SCOT E. DOWD AND RAINA M. MAIER

CHAPTER

10

Cultural Methods

KAREN C. JOSEPHSON, CHARLES P. GERBA, AND
IAN L. PEPPER

CHAPTER

11

Physiological Methods

DAVID C. HERMAN AND RAINA M. MAIER

CHAPTER

12

Immunological Methods

SCOT E. DOWD AND RAINA M. MAIER

C H A P T E R

13

Nucleic Acid-Based Methods of Analysis

ELIZABETH M. MARLOWE, KAREN L. JOSEPHSON, AND
IAN L. PEPPER

C H A P T E R

14

Biogeochemical Cycling

RAINA M. MAIER

C H A P T E R

15

Consequences of Biogeochemical Cycles Gone Wild

DAVID C. HERMAN AND RAINA M. MAIER

C H A P T.E R

16

Microorganisms and Organic Pollutants

RAINA M. MAIER

CHAPTER

17

Microorganisms and Metal Pollutants

TIMBERLEY M. ROANE AND IAN L. PEPPER

CHAPTER

18

Beneficial and Pathogenic Microbes in Agriculture

IAN L. PEPPER

CHAPTER

19

Environmentally Transmitted Pathogens

PATRICIA RUSIN, CARLOS E. ENRIQUEZ, DANA JOHNSON,
AND CHARLES P. GERBA

19.1 Environmental Transmitted Pathogens 447
19.2 Bacteria 449
    19.2.1 *Salmonella* 449
    19.2.2 *Escherichia coli* and *Shigella* 450
    19.2.3 *Campylobacter* 451
    19.2.4 *Yersinia* 452
    19.2.5 *Vibrio* 452
    19.2.6 *Helicobacter* 453
    19.2.7 *Legionella* 453
    19.2.8 Opportunistic
        Bacterial Pathogens 455
    19.2.9 Blue–Green Algae 457
19.3 Parasitology 459
    19.3.1 Protozoa 460
    19.3.2 Nematodes 464
    19.3.3 Cestodes (*Taenia saginata*) 466
    19.3.4 Trematodes
        (*Schistosoma mansoni*) 467
    19.3.5 Emerging Pathogens 469
19.4 Viruses 472
    19.4.1 Enteric Viruses 472
    19.4.2 Respiratory Viruses 481
    Questions and Problems 484
    References and Recommended Readings 485

CHAPTER

20

Indicator Microorganisms

CHARLES P. GERBA

20.1 The Concept of Indicator Organisms 491
20.2 Total Coliforms 491
    20.2.1 The Most Probable Number
        (MPN) Test 493
    20.2.2 The Membrane Filter (MF) Test 493
    20.2.3 The Presence–Absence
        (P–A) Test 493
20.3 Fecal Coliforms 496
20.4 Fecal Streptococci 496
20.5 *Clostridium perfringens* 497
20.6 Heterotrophic Plate Count 497
20.7 Bacteriophage 499
20.8 Other Indicator Organisms 500
20.9 Standards and Criteria for Indicators 500
    Questions and Problems 502
    References and Recommended Readings 503

CHAPTER

21

Domestic Wastes and Waste Treatment

CHARLES P. GERBA

21.1 Domestic Wastewater 505
21.2 Modern Wastewater Treatment 508
    21.2.1 Primary Treatment 508
    21.2.2 Secondary Treatment 508
    21.2.3 Tertiary Treatment 512
    21.2.4 Removal of Pathogens by Sewage
        Treatment Processes 512
    21.2.5 Sludge Processing 515
    21.2.6 Pathogen Occurrence and Fate
        in Biosolids 516
    21.2.7 An Example of a Modern Sewage
        Treatment Plant 518
21.3 Oxidation Ponds 520
21.4 Septic Tanks 521
21.5 Land Application of Wastewater 522
21.6 Wetlands and Aquaculture Systems 524
21.7 Solid Waste 528
    21.7.1 Municipal Solid Waste 528
    21.7.2 Modern Sanitary Landfills 528
    21.7.3 Composting of Biosolids and Domestic
        Solid Waste 530
    Questions and Problems 532
    References and Recommended Readings 533

CHAPTER

22

Drinking Water Treatment and Distribution

CHARLES P. GERBA

22.1 Water Treatment Processes 535
22.2 Water Distribution Systems 538
22.3 Assimilable Organic Carbon 539
    Questions and Problems 541
    References and Recommended Readings 541

CHAPTER

23

Disinfection

CHARLES P. GERBA

23.1 Thermal Destruction 543
23.2 Kinetics of Disinfection 544

CHAPTER

24

Risk Assessment

CHARLES P. GERBA

Preface

Historically, environmental microbiology can be traced to studies of municipal waste treatment and disposal. More recently, the area has expanded to the study of soil, water, and air systems, including the interaction of indigenous microbes with organic and inorganic pollutants, as well as the behavior of pathogens introduced into these systems. Further, environmental microbiologists are interested in the discovery and application of new microbes and their products to benefit human health and welfare. While the nature of "environmental microbiology" has changed, there have been few attempts to cover this material in a formal textbook. There have been texts written on "microbial ecology," but it is important to distinguish microbial ecology from environmental microbiology Microbial ecology can be defined as the study of the relationships or interactions of microorganisms between themselves and also their environment, including both abiotic and biotic components. In contrast, environmental microbiology can be defined as the study of the applied effects of microorganisms on the environment and on human activity, health, and welfare. These effects can impact humans directly, as in the case of microbial disease, or indirectly through actions on animals, plants, or ecosystem health in general. These actions can be beneficial, as in the case of nitrogen fixation by *Rhizobium* spp., or detrimental, as in the case of transmission of human bacterial or viral pathogens. In either case, the effects of the microorganisms can often be enhanced or eliminated by human intervention. As such, although the field of microbial ecology is essential to environmental microbiology, the two areas are not synonymous and should be treated differently. In particular, an evaluation of environmental microbiology necessitates an acceptance of the role of human activities in regulating and manipulating microbial activity. Examples of such human activities include municipal

waste treatment, bioaugmentation during remediation of soils, and utilization of biological control agents. Thus, the subjects covered in this textbook are not the same as those covered in traditional texts on microbial ecology. This text has five subject areas presented in a logical progression: (i) foundation chapters to provide an adequate background for the advanced material presented in subsequent chapters; (ii) chapters on microbial environments, including terrestrial, aquatic, and atmospheric; (iii) chapters on detection and quantitation of microbial activity, including cultural, microscopic, physiological, molecular, and immunological approaches; (iv) chapters on the impact of microbial activity on the environment in terms of element cycling and the fate of organic and metal pollutants; and (v) chapters on the transmission, detection, and control of pathogens in the environment.

This textbook is designed for a senior-level undergraduate class or a graduate-level class in environmental microbiology and to serve as a reference for any scientist interested in this field. The overall objectives of the text are to define the important microbes involved in environmental microbiology, the nature of the different environments in which the microbes are situated, and the methodologies used to monitor the microbes and their activities and, finally, to evaluate the effects of these microbes on human activities. This book represents a joint effort led by three authors who have diverse yet complementary backgrounds in environmental microbiology. The authors are close colleagues at the University of Arizona and all have large and active research programs. They have worked together extensively on a variety of practical problems using advanced, interdisciplinary approaches. Examples include biosurfactant-enhanced remediation of contaminated soil, molecular detection of emerging pathogens, transport of microbes and DNA through soil, and microbial risk assessment. Their extensive research programs have provided a number of the examples used in this text to illustrate important learning points. Key contributions to this text were also made by 10 postdoctoral research associates and Ph.D. students who work with the authors at the University of Arizona. This group has worked closely together, resulting in a textbook that has continuity in depth and style and that is state-of-the-art at the time of press.

ACKNOWLEDGMENTS

Artwork and computer graphics: We acknowledge the diligence and creativity of Scot E. Dowd, who spent countless hours developing the artwork and computer graphics for this textbook.

Textbook development: Thanks to Elenor Loya for her patience, good cheer, and endurance throughout the preparation of this textbook. Thanks for proofing go to Christine Stauber, Brandolyn Patterson, Larissa Aumand, and Timberley Roane. Thanks to Timberley Roane for developing the index.

The Authors

All three authors are professors in the Department of Soil, Water and Environmental Science at the University of Arizona.

Raina M. Maier Ph.D., Rutgers University, 1988. Currently, Associate Professor of Environmental Microbiology. Dr. Maier's research is focused on developing a basic understanding of how to evaluate and control microbial activity in disturbed environments. She is known for using an interdisciplinary approach to study the interaction of pollutants with both the biotic and the abiotic environment. Dr. Maier has earned an international reputation for her work on microbial surfactants (biosurfactants). She has studied their role in facilitating biodegradation of organic compounds and been involved in the discoveries that biosurfactants can complex and aid in the removal of metals from contaminated soils and that biosurfactants have potential as biological control agents.

"Environmental microbiology represents one of the relatively unexplored and extremely exciting frontiers of science. So little is yet known about environmental microbes—partially because they quickly become lab rats when taken out of their environment—that the possibilities for new discoveries are limitless."

Ian L. Pepper Ph.D., The Ohio State University, 1975. Currently, Professor of Environmental Microbiology. Dr. Pepper's diverse research interests are reflected in the fact that he is Fellow of the Soil Science Society of America, The American Academy of Microbiology, and the American Society of Agronomy. He is also Director of the National Science Foundation Water Quality Center at the University of Arizona. His pioneering efforts in the use of molecular analyses for the detection of microbial diversity and activity in environmental samples have focused on several areas. These include the detection of bacterial viral and protozoan parasites in soils and waters, as well as the enhancement of remediation of co-contaminated soils through bioaugmentation and gene transfer. Dr. Pepper has also been active in the area of waste utilization, namely biosolids and effluent reuse.

"I was born in Tonypandy, Wales, where the world consisted of coal mines, cricket, soccer, and tea. Now I live in the Sonoran desert—a far cry from those green valleys of my youth. However, both ecosystems are critically affected by microbes and their activities. In fact, all known ecosystems are influenced by environmental microbiology and this is what makes the subject so exciting."

Charles P. Gerba Ph.D., University of Miami, 1973. Currently, Professor of Microbiology. Dr. Gerba is a Fellow of the American Academy of Microbiology and a Member of the EPA Science Advisory Board. As such, he has been influencing national research areas of study within the environmental science arena. He has an international reputation for his methodologies for pathogen detection in water and food, pathogen occurrence in households, and risk assessment.

"My interest in microbiology was sparked by Paul DeKruf's inspiring tales of the scientific achievements of early microbiologists in the book The Microbe Hunters *and my mother's error in giving me a microscope for Christmas instead of the chemistry set I wanted. In my first summer job out of college, I was introduced to environmental microbiology by studying sewage disposal. Later, I examined the fate of viruses in sewage discharged into the ocean. These beginnings led me to an exciting and adventurous career in environmental microbiology where every day brings a new problem to be addressed."*

Contributing Authors

All of the contributing authors to this text are associated with the Department of Soil, Water and Environmental Science at the University of Arizona.

Scot E. Dowd Ph.D. Candidate,
University of Arizona

Carlos E. Enriquez Assistant Research Scientist,
University of Arizona

David C. Herman Postdoctoral Research Associate,
University of Arizona

Dana C. Johnson Ph.D., 1996,
University of Arizona

Karen L. Josephson Research Specialist Senior,
University of Arizona

Elizabeth M. Marlowe Ph.D., 1999,
University of Arizona

Deborah T. Newby Ph.D. Candidate,
University of Arizona

Kelly A. Reynolds Assistant Research Scientist,
University of Arizona

Timberley M. Roane Ph.D. 1999,
University of Arizona

Patricia Rusin Associate Research Scientist,
University of Arizona

1

Introduction to Environmental Microbiology

RAINA M. MAIER   IAN L. PEPPER   CHARLES P. GERBA

1.1 INTRODUCTION

In the 1970s a new area of microbiology emerged and developed into the field of environmental microbiology. The roots of environmental microbiology are widespread but are perhaps most closely related to the field of microbial ecology, which comprises the study of the interaction of microorganisms within an environment, be it air, water, or soil. The primary difference between these two fields is that environmental microbiology is an applied field in which the driving question is, how can we use our understanding of microbes in the environment to benefit society? Thus, these fields are related but not synonymous because they emphasize different viewpoints and address different problems. Consequently, the subjects covered in this textbook are not the same as those covered in traditional texts on microbial ecology. Because environmental microbes can affect so many aspects of life and are easily transported between environments, the field of environmental microbiology interfaces with a number of different subspecialties, including soil, aquatic, and aeromicrobiology, as well as bioremediation, water quality, occupational health and infection control, food safety, and industrial microbiology (Fig. 1.1).

What happened in the 1970s to cause this new field of microbiology to develop? Several events occurred simultaneously that highlighted the need for a better understanding of environmental microorganisms. The first of these events was the emergence of a series of new waterborne and foodborne pathogens that posed a threat to both human and animal health. In the same time frame it became increasingly apparent that, as a result of past waste disposal practices, both surface water and groundwater supplies are frequently contaminated with organic and inorganic chemicals. Finally, the discovery of the structure of DNA in 1953 by James Watson and Francis Crick engendered the development of new technologies based on nucleic acids for measuring and analyzing microbes. The simultaneous impact of these events caused scientists to question the notion that our food and water supplies are safe and also allowed the development of tools to increase our ability to detect and identify microbes and their activities in the environment. Thus, over a relatively short period of time, the new field of environmental microbiology has been established.

1.2 AN HISTORICAL PERSPECTIVE

The initial scientific focus of the field of environmental microbiology was on water quality and the fate of pathogens in the environment in the context of protection of public health. The roots for water quality go back to the turn of the century, when the treatment of water supplies by filtration and disinfection resulted in a dramatic decrease in the incidence of typhoid fever and cholera. Application of these processes throughout the developed world has essentially eliminated waterborne bacterial disease except when treatment failures occur. Until the 1960s, it was thought that threats from waterborne disease had been eliminated. However, case studies began to accumulate

FIGURE 1.1   Environmental microbiology interfaces with many other fields of microbiology

suggesting that other agents, such as viruses and protozoa, were more resistant to disinfection than enteric bacteria. This concern became a reality with the discovery of waterborne outbreaks caused by the protozoan parasite *Giardia* and the Norwalk virus, both of which were found in disinfected drinking water. Water quality continues to be a major focus in environmental microbiology because new waterborne pathogens are discovered all too frequently. For example, in 1993 more than 400,000 people became ill and more than 100 died in Milwaukee, Wisconsin, during a waterborne outbreak caused by the protozoan parasite *Cryptosporidium*. Table 1.1 lists many of the important pathogens that have emerged in the past 30 years. Even more worrisome, recent studies suggest that 10 to 50% of diarrhea-associated illness is caused by waterborne microbial agents we have yet to identify.

Pathogens in our food supply are a second area of immense concern. One of the difficult questions fac-

ing the food industry is that of importation of vegetables and fruits. On the one hand, there is consumer demand for a wide variety and high quality of imported produce; on the other hand, there is increased danger of importing a pathogen with the produce. An example is the protozoan pathogen *Cyclospora*. This organism gained notoriety as the causative agent of a disease outbreak affecting approximately 1000 people in the United States between early May and mid-July 1996. The *Cyclospora* outbreak was due to consumption of raspberries imported from Guatemala. Although the Guatemalan government and the export–import firm that handled the raspberry shipments have cooperated fully, it is still not clear how the raspberries became infected with *Cyclospora*. These examples illustrate the types of questions that are addressed by environmental microbiologists. How do pathogens survive in the environment, and how can they be detected and eliminated rapidly and econom-

TABLE 1.1 Recently Discovered Microbes That Have Had a Significant Impact on Human Health

Agent	Mode of transmission	Disease/Symptoms
Rotavirus	Waterborne	Diarrhea
Legionella	Waterborne	Legionnaire's disease
Escherichia coli 0157:H7	Foodborne Waterborne	Enterohemorrhagic fever, kidney failure
Hepatitis E virus	Waterborne	Hepatitis
Cryptosporidium	Waterborne Foodborne	Diarrhea
Calicivirus	Waterborne Foodborne	Diarrhea
Helicobacter pylori	Foodborne Waterborne	Stomach ulcers
Cyclospora	Foodborne Waterborne	Diarrhea

ically to allow both protection of human health and rapid consumption of fresh produce? Determining the presence and significance of pathogenic organisms in food, water, and air supplies is a challenge that will keep environmental microbiologists occupied in the coming decades.

The important issues in pathogen fate and movement, and detection of microbes in the environment are similar to issues that need to be addressed in the area of microbial interaction with chemicals in the environment. It is natural therefore that as the field of en-

vironmental microbiology has developed, it has expanded to include several other important areas of applied research. These include microbial interaction with chemical pollutants in the environment, and the use of microorganisms for resource production and resource recovery. Chemical pollutants in soil and groundwater have profound effects on human populations, in terms of both potential diseases that the intake of these chemicals can cause, and the economic impact of cleaning up contaminated environments. Table 1.2 shows a partial list of the chemical contami-

TABLE 1.2 Common Organic and Inorganic Contaminants Found in the Environment and the Potential for Cleanup Using Bioremediation[a]

Chemical class	Frequency of occurrence	Status of bioremediation technologies
Gasoline, fuel oil	Very frequent	Established
Polycyclic aromatic hydrocarbons	Common	Emerging
Creosote	Infrequent	Emerging
Alcohols, ketones, esters	Common	Established
Ethers	Common	Emerging
Chlorinated organics	Very frequent	Emerging
Polychlorinated biphenyls (PCBs)	Infrequent	Emerging
Nitroaromatics (TNT)	Common	Emerging
Metals (Cd, Cr, Cu, Hg, Ni, Pb, Zn)	Common	Possible
Nitrate	Common	Emerging

[a] Adapted from National Research Council, 1993. *In situ* bioremediation: When does it work? National Academy Press: Washington, D.C. 1993.

nants that are routinely found in soil and ground-water. The cost of cleanup or remediation of the contaminated sites in the United States alone has been estimated to exceed $1 trillion. Given this price tag, it is increasingly recognized that biological cleanup alternatives, known as bioremediation, may have profound economic advantages over traditional physical and chemical remediation techniques. Once thought unpredictable and ineffective, bioremediation is now a viable alternative for cleanup of many sites. In 1989, the Exxon Valdez spilled almost 11 million gallons of North Slope Crude oil into the Prince William Sound, the largest tanker spill in U.S. history. A combination of physical removal of free product, physical washing of the contaminated shoreline, and bioremediation has been used to slowly restore the Prince William Sound. This was the first large-scale documentation of successful use of bioremediation to clean a contaminated site.

The acceptance of bioremediation as a feasible cleanup alternative has opened the door for environmental microbiologists. In order to apply bioremediation techniques successfully, environmental microbiologists must gain a better understanding of ways to stimulate indigenous organisms to aid in remediation of contamination. A related area of study is how to add microbes to a contaminated site to achieve remediation. In either case, the search goes on for new techniques to better understand microbial behavior in nat-ural habitats, either soil or water. Powerful new tools for detecting and enumerating specific microorganisms have been developed. In particular, the advent of molecular-based techniques such as polymerase chain reaction (PCR), gene probes, gene cloning, reporter genes, and DNA sequencing, to name just a few, is allowing environmental microbiologists a glimpse into what used to be considered a "black box," the environment surrounding microbes in soil and water.

1.3 MODERN ENVIRONMENTAL MICROBIOLOGY

Modern environmental microbiology is more than the study of pathogens and bioremediation. Table 1.3 shows the different areas that constitute this field. Problems addressed by environmental microbiologists include the discovery and identification of new microbes and microbial products that may have practical application for protection of the environment, protection of human health, and commercial application. Examples include: microbial enzymes for use in detergents, food processing, and biotechnology; a variety of metabolites including antibiotics for use in medicine; surfactants that are not only used in cosmetics and food processing, but also have potential for use in remediation. Scientists in search of new microbes are ea-

TABLE 1.3   Scope of Environmental Microbiology

Subject	Microbial issue
Aeromicrobiology	Collection and detection of pathogens or other microbes in aerosols, microbial movement in aerosols
Agriculture, soil microbiology	Biological control, nitrogen fixation, nutrient cycling
Biogeochemistry	Carbon and mineral cycling, control of acid mine drainage, control of loss of fixed nitrogen
Bioremediation	Degradation of organic contaminants, immobilization or removal of inorganic contaminants found in contaminated soil or water environments
Biotechnology	Detection of pathogens or other microbes in the environment, detection of microbial activity in the environment, genetic engineering
Food quality	Detection of pathogens, elimination of pathogens
Resource production	Production of alcohol, single-cell protein
Resource recovery	Microbially mediated recovery of oil and metals
Wastewater treatment	Degradation of waste, reduction of pathogens
Water quality	Removal of organic and inorganic contaminants, detection of pathogens, elimination of pathogens

gerly examining new environments such as the deep sea thermal vents, hot springs, and deep subsurface environments. One challenge is to recover viable microbes from these unique environments. Without an understanding of growth conditions and nutrient requirements, recovery of viable microbes is problematic. Once recovered, these organisms can be characterized and tested to determine whether they produce novel enzymes, antibiotics, or other products that can be harnessed to solve problems. For example, pesticide application continues to increase worldwide and is an important factor in increased crop production. However, as a result of continuous pesticide use, we now find low levels of pesticides in many of our underground drinking water supplies. The question for environmental microbiologists is, can we replace synthetic pesticides with microbially produced pesticides or with specific microbes that will confer protection against plant disease?

Microbes can also be harnessed for resource production, as for example during recovery of metals from ore or recovery of oil, which is also called **microbially enhanced oil recovery (MEOR).** Perhaps the inception of microbially based mining of metal took place in the 18th century in Rio Tinto, Spain, during the extraction of copper from copper ore. Although it was not known at the time, copper solubilization during the extraction process was a result of microbial activity. By 1989 approximately 30% of the copper and uranium mining in the world was done with a microbial process, taking advantage of naturally occurring bacteria such as *Thiobacillus thiooxidans* and *Thiobacillus ferrooxidans*.

An additional area of study for environmental microbiologists is the domain of introduced organisms. At times, microbes are deliberately introduced into the environment for a multitude of purposes (Table 1.4). Many of these are related to enhanced crop production or improved remediation of contaminated sites. Introduced microbes can be naturally occurring or can be genetically engineered to combine and enhance metabolic capabilities. Although **genetically engineered microbes (GEMs)** can be very efficient, their intentional release is currently restricted because of safety concerns. The use of introduced microbes can result in significant benefits provided that they survive long enough to accomplish the desired task. Thus, environmental microbiologists have investigated many of the environmental factors, both biotic and abiotic, that affect survival. Clearly, the concept of distributing modified microbes or augmenting with naturally occurring species has profound implications for protection of the environment and human health. The option of using introduced microbes instead of more traditional approaches also appears to be attractive economically. It is such opportunities that have made the field of environmental microbiology attractive to a broad range of people, including students, researchers, society, and even politicians.

1.4 PURPOSE AND ORGANIZATION OF THIS TEXT

Our purpose in writing this book is to provide a text that can be used in teaching environmental microbiology as well as a general reference book for practitioners in the field of environmental microbiology. This textbook is designed for senior-level undergraduate and graduate classes in environmental microbiology. It is expected that students using this text have a background in general microbiology and a grounding in organic chemistry. The overall objective of the text is to define the important microbes involved in environmental microbiology, the nature of the different possible environments in which the microbes are situated, the methodologies used to monitor the microbes and their activities, and finally the possible effects of these

TABLE 1.4   Role of Introduced Microorganisms in Environmental Microbiology

Introduced microbe	Function
Biocontrol agents, usually bacteria	Reduced seedling and plant disease
Bioremediative agents, usually bacteria	Enhanced remediation of organic- and/or metal-contaminated sites
Genetically engineered microbes	Many potential functions. An example is enhanced frost tolerance by prevention of ice nucleation
Mycorrhizal fungi	Enhanced development of pine seedlings prior to reforestation
Rhizobia	Enhanced nitrogen fixation by legumes

microbes on human activities. This book is organized into five broad sections: (1) a review of basic microbiological concepts; (2) a description of the various microbial environments encountered in the field including soil, water, and the atmosphere; (3) a discussion of methodologies available for detection, enumeration, and identification of microbes and their activities; (4) an examination of microbial activities in the environment and how these activities affect cycling of nutrients as well as how they can be exploited to help solve environmental problems; and (5) a discussion of pathogens in the environment.

2

Microorganisms in the Environment

KELLY A. REYNOLDS   IAN L. PEPPER

Organisms in the environment are diverse in origin and ubiquitous. Microorganisms in the environment are also fundamentally different from laboratory-maintained or clinical isolates of microbes, because they are adapted to harsh and often widely fluctuating environments. In environmental microbiology, we categorize microbes as prokaryotes or eukaryotes, both of which clearly affect human health and welfare and are essential for maintaining life as we know it (Table 2.1). The smallest organisms are the bacteria, which are prokaryotic. Larger and more complex organisms include the eukaryotic fungi, algae, and protozoa. Viruses are unique in that they consist solely of nucleic acids and proteins and are not technically viable living organisms. Although new entities are currently being discovered, such as infectious proteins known as prions, generally all biological entities are characterized by the presence of nucleic acids.

Viruses are significant because of their ability to infect other living organisms and cause disease. Bacteria can also cause infections but are very important in other areas of environmental microbiology including biochemical transformations that affect nutrient cycling, bioremediation, waste disposal, and plant growth. The eukaryotic fungi are involved in these activities as well and are equally important in surface environments, but less so in subsurface environments. Algae affect surface water quality and can also produce microbial toxins, but their overall impact in the environmental microbiology arena is not as significant as that of the bacteria and fungi. Finally, the protozoa are significant sources of pathogens that directly affect human health as evidenced by the outbreak of *Cryptosporidium* contamination of potable water supplies that resulted in over 100 deaths in 1993 in Milwaukee, Wisconsin. They are also key players in surface environments as grazers of bacteria, helping to maintain a steady ecological balance.

In this chapter the basic structure and function of key environmental microbes will be described. We will focus primarily on viruses including phage, bacteria, fungi, and the protists because of their importance and relevance to environmental issues. The objective of this chapter is to enable a reader with a nonmicrobiological background to understand what the characteristics of these microbes are, what their roles in the environment are, and what their requirements are to function successfully in the environment. Special emphasis is placed on the bacteria because of their multiple roles in environmental microbiology. The intent of the chapter is not to establish a treatise on each organism but rather to provide a basic framework of information relevant to subsequent chapters that deal with the effects of these organisms in environmental microbiology. A massive amount of detailed information is available in the literature on each of the types of organisms covered in this chapter.

2.1 VIRUSES

This section provides a general background to viruses found in the environment: their historical discovery, structure, mechanisms of infection, multiplication and transmission. Although most viruses follow the same pattern of replication, there are exceptions. For simplicity, we will outline only general characteristics of the more well-known viruses.

**TABLE 2.1   Scope and Diversity of Microbes Found in
Environmental Microbiology**

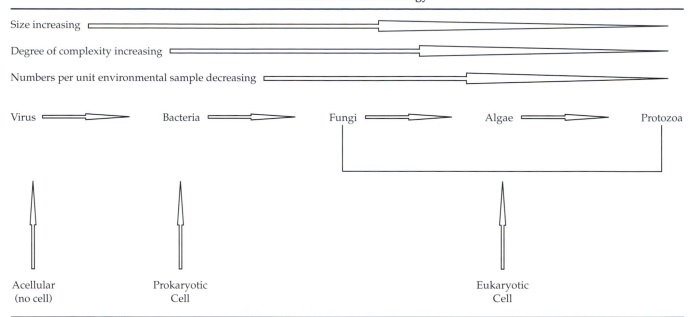

2.1.1 Historical Outbreaks and Discovery

Viruses are defined as small, obligate, intracellular parasites, meaning that they require a host cell for their growth and replication. Although viruses can survive outside a host, their numbers cannot increase without a host. They are generally species specific, infecting either bacteria, plants, or animals, and are initially classified on the basis of the host they infect. Public focus is most often on (1) plant viruses affecting major cash crops such as tobacco, potatoes, and tomatoes; or (2) human viruses causing such ailments as herpes, warts, smallpox, rabies, mumps, measles, meningitis, hepatitis, encephalitis, colds, influenza, diarrhea, yellow fever, dengue fever, and eye infections, just to name a few. Many human viral infections can easily be avoided by reducing exposure factors, such as intravenous or sexual contact, whereas others are caused by viruses present in food, water, soil, and air and are thus more difficult to avoid. Historically, viruses have had a major impact on public health and safety the world over. Most notable is the smallpox virus, documented for centuries and thought to be responsible for the death of millions of Aztecs after the 16th century incursion of Europeans into the New World. Early attempts to define viruses were convoluted because of the focus on bacteria. By the late 19th century, bacteriology developed into a sophisticated science in which many bacterial disease–causing agents were isolated and identified. Although scientists began

to realize that viruses were very different from bacteria, these infectious agents continued to elude identification.

In an attempt to explain their inability to culture the causal agent of smallpox (poxvirus), Ernst Levy and Felix Klemperer, in 1898, wrote that the nutritional requirements of the infectious agent appeared to be different from those of culturable bacteria. Fortunately, lack of absolute identification of viral agents did not prevent scientists from discovering methods of viral disease control. In the late 1700s, Edward Jenner noted that exposure to certain related animal viruses, such as cowpox, extended immunity to the human smallpox virus. By the 1880s, Louis Pasteur was studying rabies. After realizing that the agent could not be grown or detected microscopically, Pasteur began *in vivo* studies, inoculating the brains of dogs and spinal cords of rabbits to isolate pure cultures of the virus. This work was the first evidence of controlled virus culture in animal tissues and soon led to the development of the rabies vaccine using attenuated virus cultures from rabbit spinal cords.

Early studies of viruses as filterable agents (agents that could pass through a bacterial filter) began with the tobacco mosaic virus, which causes mottling of the tobacco leaf. Two men were credited with the discovery of the filterable agent in the early 1900s, Ivanovski and Beijerinck. Their contributions, along with those of Mayer, who conducted some of the earliest research on the disease, led to the concept of the virus. Over the next 50 years, researchers discovered viruses of yellow fever, rabies, polio, measles, mumps, rubella, and more. In 1939, the electron microscope was first used to visualize virus morphology, and *in vitro* cultivation of poliovirus in 1949 led to the evolution of virology as an independent science. With the advent of cultural and microscopic methods, the following decades led to an explosion of information on viruses and their associated diseases. Early investigations focused on economically significant viral diseases of animals and plants, but researchers were soon examining indigenous populations of bacterial viruses, in addition to transmission modes, exposure routes, and the molecular and physical structures of the agents.

2.1.2 Indigenous Virus Populations in the Environment

Although initially studied for potential treatment of bacterial infections in humans, viruses of bacteria (**bacteriophage** or **phage**) have great impact on natural bacterial populations and the transfer of genetic materials between these populations. There are three general types of bacteriophage, classified on the basis of their mode of infection of the host: (1) F-specific or appendage phage, those that infect the host through the sex pili or flagella; (2) capsule phage, those that recognize the outer host layer such as the polysaccharide capsule; and (3) somatic phage, those that enter the host via the cell wall. Discoveries of abundant phage populations (up to 10^8 total counts per milliliter) in natural, unpolluted environments such as rivers, lakes, marine waters, and soil environments throughout the world have caused researchers in microbial ecology to reexamine previous notions of the importance of indigenous viruses. In fact, phage populations are so high in certain habitats that methods of concentration are not needed for detection. Information on the abundance of total viruses in natural environments has been collected by the use of transmission electron microscopy (TEM). Another approach is to filter and directly stain any viruses present with an epifluorescent dye such as DAPI (4,6-diamidino-2-phenylindole) before enumeration with an epifluorescent microscope at 100× magnification. Although helpful for examining total phage counts, both of these methods have the disadvantage of enumerating noninfective virus particles and perhaps overestimating the impact of infective phage on particular bacterial populations.

The recognition that large numbers of environmental viruses exist has led to questions concerning their ecological significance. For example, how do indigenous viruses affect bacterial or other host populations? How do such **phage–host system** interactions affect environmental ecology as a whole, and what factors control the fluctuation of virus populations over time? Using two tools, virus culture and TEM, researchers have found that natural environments contain phage representing wide morphological varieties. This suggests a diversity of phage that corresponds to the wide diversity of bacteria found in these environments. Thus, it seems that phage can exist in any environment capable of supporting a bacterial host.

TEM analysis of depth profiles of marine waters clearly shows a decrease in the amount of viruses as well as bacteria, suggesting a correlation between the two. This is not surprising since, as obligate intracellular parasites, phage are closely related to the host bacterium that supports their life cycle. In fact, in order to isolate a particular phage from the environment, one must first choose the proper phage–host system that is capable of supporting the growth of phage types represented in the particular area being sampled. Once the phage–host system is developed, cultural analysis of phage is simple and rapid compared with cultural techniques for animal viruses. While TEM is problematic if virus counts are low or samples are highly turbid, an obvious disadvantage of using a cultural assay as the only means of detection is that one must first address the task of how to choose the proper phage–host system. This is difficult because some phage–host systems cannot be cultured in the laboratory. In addition, phage isolated from the environment using a phage–host system allow characterization of infective phage for a particular host, but no information is provided for evaluating total phage populations. Here TEM becomes a useful tool because it allows quantitation and identification of phage particles on the basis of number and morphology. Thus, once phage are observed under the microscope, they are generally enumerated and classified into groups on the basis of head size, shape, and the presence and size of a tail.

Survivability and infectivity of viruses depends on a number of environmental and host factors including, temperature, solar radiation, adsorption, heavy metals, salinity, organic chelators, bacterial activity, algae, and protozoa. Therefore as environmental conditions vary, so does the phage distribution. The adsorption (sticking) of viruses to sediments seems to prolong virus survival greatly. As a result, waters with high amounts of particles that sorb viruses protect them from inactivation, allowing more time for a virus to contact a potential host. Another environment conducive to virus sorption is sediment where sediment-dwelling organisms may serve as viral hosts. Recent studies indicate that a bacterial density threshold required before phage can multiply. Under artificial laboratory conditions, this value has been determined to be approximately 10^4 colony-forming units (CFU) for three commonly studied phage–hosts, *Staphylococcus aureus*, *Bacillus subtilis*, and *Escherichia coli*.

Viruses have been implicated as significant participants in the microbial food web. Bratbak *et al.* (1990) observed initial increases in phage populations that correlated with fluctuations in the microbial community brought on by a spring diatom bloom. Phage become active members of the microbial food web by lysing bacterial cells that otherwise might be available for predation by heterotrophic flagellates (Fig. 2.1, see also Chapter 6.3.1.3.2). It is suggested not only that viruses are involved in the parasitism of bacterial populations, but also that they infect a variety of primary producers such as diatoms, cryptophytes, prasinophytes, and chroococcoid cyanobacteria. The question remains, however, what factors control phage populations? The answer may be in the microbial food web, where phage are removed by aggregation and adsorption to nonhost particles. This creates a phage biomass that is available to detritus feeders or that can sink out of the water column.

Little is known about the role phage play in the transfer of genetic information to other organisms within an ecosystem. It is possible that the organic material released by phage may enter the pool of dissolved organic material to be utilized by bacteria or that some of the viral nucleic acid may be returned to bacteria by reinfection. Paul *et al.* (1991) examined the contribution of viral DNA to total dissolved DNA in marine environments. They estimated the viral DNA contents to be an average of 3.7% (range, 0.9 to 12.3%) of the total dissolved DNA in samples from freshwater, estuarine, and offshore oligotrophic environments. From this study it was concluded that viruses themselves do not substantially contribute to dissolved DNA but they may be involved in its production by causing lysis of bacterial and phytoplankton cells. Phage are also involved in transduction, a process by which they multiply in a donor bacterium and carry a piece of genetic material to a recipient bacterium that they subsequently infect. With recent discoveries of large numbers of viruses present in natural environments, more emphasis will undoubtedly be placed on development of methods for their isolation and quantification as an initial step in determining their ecological significance.

No single method, to date, meets all of the criteria for effectively evaluating microbial phage ecology. The development of new molecular methodologies will allow future discovery and study of total phage populations and their interaction with microbial communities regardless of viability or culturability. Used in combination with previously developed cultural and microscopic methodologies, molecular methods hold promise to help further characterize the ecology of phage and their contributions to environmental nutrient cycles.

2.1.3 Distinguishing Characteristics of Viruses and Their Evolution

Although the smallest bacteria approach the size of the largest viruses, as a family the viruses are smaller than most other microorganisms. They were considered "ultramicroscopic" until the advent of the electron microscope, which enabled researchers to study virus morphology visually. Ranging in size from

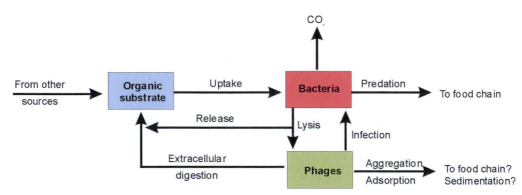

FIGURE 2.1   The role of bacteriophage in the microbial food web. (Adapted with permission from Bratbak *et al.*, 1990.)

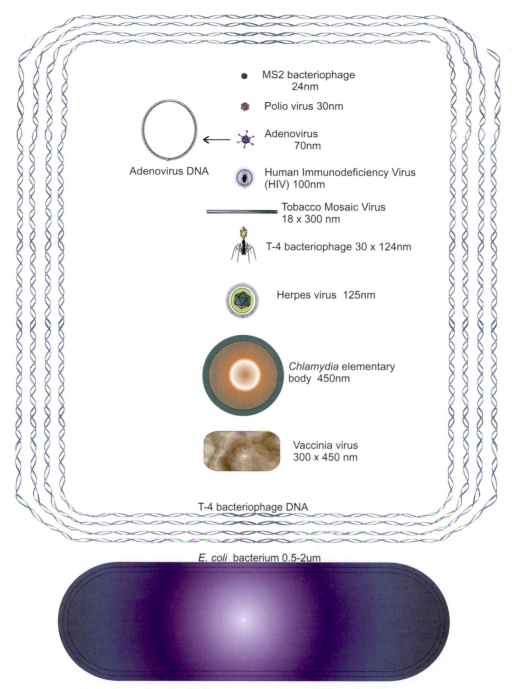

Adenovirus DNA

MS2 bacteriophage
24nm

Polio virus 30nm

Adenovirus
70nm

Human Immunodeficiency Virus
(HIV) 100nm

Tobacco Mosaic Virus
18 x 300 nm

T-4 bacteriophage 30 x 124nm

Herpes virus 125nm

Chlamydia elementary
body 450nm

Vaccinia virus
300 x 450 nm

T-4 bacteriophage DNA

E. coli bacterium 0.5-2um

FIGURE 2.2   Comparative sizes of selected bacteria, viruses, and nucleic acids.

18 nm to several hundred nanometers (Fig. 2.2), their minimal structure is the truly distinctive feature of viruses. **Virus particles** consist of only an inner nucleic acid genome, an outer protein capsid, and sometimes an additional membrane envelope. They are relatively simple in organization. Virus particles may or may not be infectious, while a **virion** is the name given to an infectious virus particle. Unlike other microorganisms, viruses have no ribosomes and do not metabolize. Thus, they cannot grow on nonliving media and are not sensitive to antibiotics. Whereas bacteria contain both DNA and RNA (see Chapter 13), viruses have either one or the other. Unlike any other microorganism, viruses can reproduce themselves from their genome, a single nucleic acid molecule, provided the proper host machinery is available.

How viruses evolved is largely unknown. Several possibilities have been described, with the three most common theories being that viruses are (1) degenerate descendants of larger pathogens, (2) remnants from an ancient precellular environment, or (3) genetic material that has escaped from cells. Viruses are so intertwined with their hosts that it is difficult to imagine their origin from any other source. Of these theories, the most popular is that viruses originated from nucleoproteins released from the host cell. One complication in the theory of virus evolution is the existence of unique viral RNA replication enzymes. Replication of RNA is required for many viruses but is not a normal function in host cells. Tools developed in the field of molecular biology are helping evolutionists to trace genetic origins of viruses. As such, the most recent studies of virus evolution and classification are based on sequencing tools and comparisons of genetic relatedness. Viruses are known to have a high rate of genetic variation, evolving often as a result of mutation. RNA viruses, in particular, have high rates of replication errors, leading to rapid changes in the genetic lineage. It is important to remember that most mutants are eliminated from the growing population by negative selection, but new strains evolve continuously creating the potential for emergence of new pathogens.

Worth mentioning, with respect to theories of virus evolution, is the existence of viroids and prions, not to be confused with virions (Table 2.2). In the past, scientists postulated that viroids, infectious agents of plants, were relatives of conventional viruses. **Viroids** are free RNA sequences able to reproduce in several genetically different hosts. The discovery of viroids adds a new dimension to infectious agent research. This is because viroids differ from viruses in several ways: (1) they exist in *vivo* as free nucleic acids, with no capsid proteins present; (2) they consist solely of RNA that is generally of lower molecular weight than viral RNA; (3) their RNA is not translated although their function and replication are not well understood at this time; and (4) they comprise only two major groups of pathogens. Furthermore, viroids have unique molecular structures and are now thought to

be greatly distanced from viruses with regard to phylogeny. All known viroids are pathogenic only to plants. Agents with similar properties that infect animals have been termed prions. **Prions** (proteinaceous infectious particles), are similar to viroids in that they are smaller than viruses, resistant to heat, unable to elicit an immune response in the host, and are difficult to detect in infected tissues. Prions, in fact, are even smaller than viroids and are not subject to nucleic acid degradative enzymes, suggesting that they are nothing more than proteins. Thus, a protein molecule appears to be all that is needed for prion infectivity. These newly identified agents are responsible for degenerative neurological diseases, such as Creutzfeldt–Jakob disease in humans, scrapie in sheep, and **bovine spongiform encephalopathy (BSE),** also known as "mad cow" disease.

Recent years have seen an increase in viral illness, bringing to question the reasons for emerging viral pathogens. While new virus variants are surely evolving with new environmental health consequences, some viruses are just now being recognized. In fact, a virus may emerge (1) as a previously unrecognized strain; (2) as a newly evolved human variant; (3) as a mutation from a zoonotic (of animal origin) strain expanding the host range; (4) as a result of contact with new conditions, exposing humans to previously unknown strains; (5) from alterations in host immunoconditions; (6) from exposure to vectors in new environments; or (7) from dissemination of disease from a geographically localized subpopulation. Ease of international transportation has also made transcontinental travelers out of viruses. The current era of increased travel, exploration, and growth has greatly increased our exposures to infectious viral agents, thus contributing to "emerging" viral disease.

2.1.4 Structure and Classification

In order to study survival tendencies and methods of concentration, culture, and detection of viruses, physical structure and characteristics must be considered. Early studies of virus structures by biochemist Wendell Stanley utilized x-ray diffraction and crystallography. Using salts to crystallize proteins, Stanley noted that the tobacco mosaic virus structure followed the laws of crystallography to form a crystal-like icosahedron. From this experiment, it was concluded that viruses must have an outer protein capsid structure. Virus structures take on many forms; in fact, there are at least 71 different families of viruses divided into more than 3600 species, all varying in size, structure, and/or chemical characteristics. Basic virus structures include a protomer or capsomer protein coat and in-

TABLE 2.2   **Comparison of Related Terms**

Term	Capsid protein	Nucleic acid	Infectious unit
Viral particles	Yes	Yes	Yes or No
Virions	Yes	Yes	Yes
Viroids	No	Yes	Yes
Prions	Yes	No	Yes

ternal nucleic acid, either RNA or DNA. All viruses have a nucleic acid genome and an outer protein capsid, but some may also have protein and lipid envelopes, glycoprotein spikes, or more complex tail and sheath structures.

2.1.4.1 Surface Characteristics of Viruses

The virus **capsid** is the term for the protein coat. It is composed of a number of protein capsomers held together by noncovalent bonds, surrounding the nucleic acid molecule. Capsids range in size from 18 μm, as in the small parvoviruses of animals, to several hundred nanometers, as in some filamentous plant viruses. Simple virion capsids are constructed from as few as three proteins, whereas complex viral coats are composed of several hundred. This outer coat protects and shields the viral nucleic acid and harbors specific receptor sites for host attachment.

Viral capsids support two kinds of symmetrical organization, helical and icosahedral. Helical viruses resemble a spiral, or helix, with an overall cylindrical shape. Icosahedral viruses have 20 triangular sides but appear spherical when viewed with an electron microscope. Capsid surfaces contain numerous receptor sites and electrical charges resulting from amino acid functional groups. Capsid surfaces also contain hydrophobic groups. Such surface characteristics aid in virus concentration, because oppositely charged filters can be used rather than size entrapment. In neutral solutions, viruses tend to be negatively charged, but as the pH of the suspending solution is lowered, viruses exhibit a conformational shift of their surface proteins that results in an overall positive charge. The pH at which this shift occurs is known as the **isoelectric point.** Being negatively charged in neutral solutions, such as most natural waters, viruses repel surfaces of similar charge but tend to adsorb ions in the aqueous solution to form what is known as an electrical double layer, or **Stern layer** (Fig. 2.3). Beyond the Stern layer, other counter ions that are held farther away, but in the vicinity of the charged particle, create a diffuse layer called the **Gouy layer.** The distance the diffuse second layer extends into the bulk solution determines the force and distance over which particles repel each other. The interactions of the surface of a virus particle with the surrounding environment are important in issues of environmental transport and population exposure.

Some viruses are protected by an outer membranous layer, or **envelope.** Viral envelopes are formed by budding from host cell membranes and thus, they retain part of the host cell outer structure. The lipids of the viral envelope are derived directly from the cell,

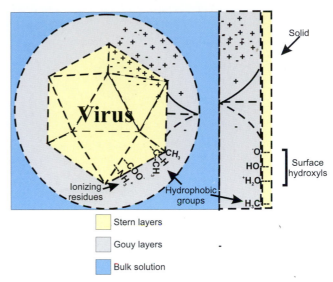

| Stern layers
| Gouy layers
| Bulk solution

FIGURE 2.3 Schematic representation of the surface interactions of viruses. Charge accumulation at the surface of the virus (Stern layer), in addition to charge accumulation in the bulk solution (Gouy layer) and hydrophobic groups, all contribute to attractive forces of viruses.

whereas the proteins in the envelope are virus encoded. Like host cell membranes, virus envelopes are made of lipid bilayers. Lipids are composed of a head group, which is hydrophilic (likes water), and a hydrophobic (dislikes water) tail. The opposite characteristics of the head and tail groups encourage bilayer alignment as in bacterial membranes. Envelope protein structures include glycoprotein spikes or an underlying layer of matrix protein that add rigidity. In enveloped viruses, the added rigidity is necessary for fusion with the host and subsequent infection. Another interesting virus structure is observed in the T-even bacteriophage. These viruses have complicated tail fibers that attach to host cells. A tubular sheath within the tail structure provides a channel for transport and injection of the nucleic acid from the virion structure to the host (Fig. 2.4).

2.1.4.2 Viral Nucleic Acids

Viruses contain either DNA or RNA in the double-stranded or single-stranded form. Viral nucleic acid length corresponds directly to virus size and can vary from 1.7 to over 200 kilobase pairs (kb), coding between 4 and 200 genes. Most viruses contain less than 10 genes. All viruses require **messenger RNA (mRNA)** which is translated to proteins to allow viral replication, but particular virus families go about acquiring mRNA in very different ways. This is because viruses often store their genetic information in unconventional forms such as single-stranded DNA or double-stranded

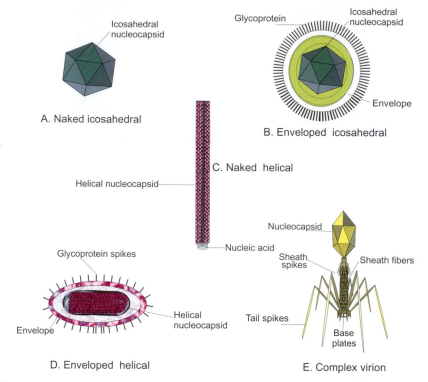

FIGURE 2.4   Simple forms of viruses and their components. The naked icosahedral viruses (A) resemble small crystals: the enveloped icosahedral viruses (B) are made up of icosahedral nucleocapsids surrounded by the envelope: naked helical viruses (C) resemble rods with a fine regular helical pattern in their surface: enveloped helical viruses (D) are helical nucleocapsids surrounded by the envelope: and complex viruses (E) are mixtures of helical and icosahedral and other structural shapes.

RNA. In general, RNA viral genomes tend to be much smaller and code for fewer proteins than DNA genomes. In either case, the goal is to transcribe the original nucleic acid into a readable mRNA sequence for the production of proteins (translation). The three approaches used are (1) mRNA may be transcribed from viral DNA; (2) viral RNA may act directly as the mRNA; or (3) viral RNA may be used to synthesize DNA using a viral enzyme, reverse transcriptase. This DNA is then transcribed to mRNA. These approaches to transcription of mRNA demonstrate one of the unique qualities of viruses, that genetic information can be passed from RNA to DNA as well as from DNA to RNA. Interestingly, more than 70% of all animal and plant infections are due to viruses with RNA genomes.

2.1.5 Infective Nature of Viruses

Outside their hosts, viruses are inert objects, incapable of movement. Thus, these tiny infectious agents require a vehicle for transport. Viruses have many mechanisms of transmission (Fig. 2.5). Transmission routes in the environment include aerosols, vectors, and contamination of food, water, or fomites. Some viruses are known to cross the species barrier from animal to human hosts, whereas others reach humans indirectly using a carrier host or arthropod vector. Virus transmission is also possible via person-to-person exposure, such as through blood, urine, or sexual contact, or by crossing placental barriers to unborn children. Living in close proximity to other hosts greatly increases the chance for continued virus survival.

Once in contact with a potential host, viruses find their way into target cells using specific receptor sites on their capsid or envelope surfaces. This is why viruses of plants do not normally infect humans and vice versa. Once viruses invade host cells and replicate, they can invade neighboring cells to continue the infection process. Infection of a cell by a virus is often but not always debilitating to the cell's regular functions. Thus, viral infection may be asymptomatic or may cause acute, chronic, latent, or slow infections, or may cause cell death.

Latent viruses may be present and even integrated in the host chromosome but do not produce progeny virions unless presented with an appropriate stimulus. Latent infections usually cause an initial disease but then remain dormant and noninfectious, persisting indefinitely, until reactivated. Compromised immunity is the primary reason for reactivation of persistent

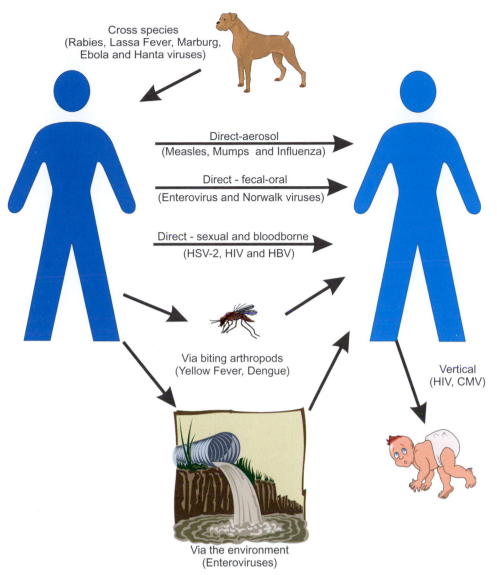

Cross species
(Rabies, Lassa Fever, Marburg,
Ebola and Hanta viruses)

Direct-aerosol
(Measles, Mumps and Influenza)

Direct - fecal-oral
(Enterovirus and Norwalk viruses)

Direct - sexual and bloodborne
(HSV-2, HIV and HBV)

Via biting arthropods
(Yellow Fever, Dengue)

Vertical
(HIV, CMV)

Via the environment
(Enteroviruses)

FIGURE 2.5 A schematic representation of the transmission routes of viral infections in humans with some examples. Direct transmission can include aerosols, sexual contact or fecal–oral contact. The sexually transmitted viruses, human immunodeficiency (HIV) and hepatitis B virus (HBV), may also be transmitted by other processes during which blood is exchanged. Some viruses that are transmitted by fecal–oral contact may also be released into the environment, where contact with contaminated water is a source of infection. Cross-species virus transmission can occur from rodents (e.g., the Hanta viruses and the arenaviruses such as Lassa fever), from monkeys (e.g., African hemorrhagic fever—the Marburg and Ebola viruses), or from domestic animals infected by wild animals (e.g., rabies). There is evidence of some transmission between humans for these infections, but the extent of this interperson transmission is limited. Viruses can be transmitted between humans or between animals and humans via vectors such as arthropods (e.g., yellow fever). Finally there is vertical transmission of virus that occurs between a mother and fetus (e.g., HIV).

virus infections. Animal virus infections may be **acute**, with an episode of illness directly associated with infection, such as with the common cold virus (rhinovirus). Other viruses replicate effectively without killing the host cell and may not lead to any noticeable symptoms. In this case the infection is **asymptomatic.** The host cell may continue to divide and replicate along with the virus. **Oncogenic** viruses do not kill their host cell but do alter its functionality, directing the production of malignant cancer cells. Release of noncytopathogenic viruses via budding allows the cell to continue to survive for prolonged periods during virus replication. Such infections are **chronic,** with no immediate signs of infection, but hosts are still able to shed virus. **Slow** infections gradually increase in effect and progress to a lethal conclusion. Human immunodeficiency virus (HIV) infection is an example of a slow, chronic, latent infection, because some cells have actively replicating and detectable virus while other cells are infected but induced at a later time to produce virus. Viruses that kill the cells in which they replicate are called **cytopathogenic.** Cytopathogenic viruses usually shut down the synthesis of cellular proteins, using structures assembled late in the virus replication cycle that are toxic to host cells. Viral proteins may be inserted into the plasma membrane of host cells, causing loss of osmotic regulation and eventual death.

Some viral infections of bacteria appear to cause no immediate harm to the host cell. Therefore the phage is carried by the host. These carrier hosts, however, may still be sensitive to other phage populations. This condition is known as **lysogeny.** In a lysogenic phage, also known as a **temperate phage,** the nucleic acid is integrated with the chromosome of the host, persisting indefinitely, and is transmitted to host descendants or daughter cells. This stable, noninfectious form of the virus is known as a **prophage.** The phage may remain latent for many generations and then suddenly be mobilized and initiate replication and eventually cause host lysis. Typically, only a portion of temperate phage become lysogenic, while other members of the population remain virulent, multiplying and lysing host cells. Similar to lysogenic bacteriophage, retroviruses integrate their nucleic acid into the animal cell chromosome, producing persistent infections. Such a cycle is typical of herpesvirus infections in humans. In fact, the herpesvirus may be passed from grandparent to grandchild, remaining dormant through two generations. Half of the human population is estimated to be infected by the age of 1 year, and up to 85% of the population is seropositive by puberty. Most latent animal viruses, such as mumps and measles viruses, lack the ability to lyse the host cell or prevent host cell division. Infection is therefore ensured by the production of infected daughter host cells. In latent infections, an equilibrium is reached between host and parasite until a nonspecific stimulus, such as compromised host immunity, evokes active infection.

2.1.6 The Metabolic State of Viruses

Viruses have two stages in their life cycles: 1) the extracellular metabolically inert transmission stage and 2) the intracellular reproductive stage in which host cell machinery is taken over and progeny viruses produced. Viruses are metabolically inert but infectious. Larger than most macromolecules, but smaller than most bacterial cells, they are too complex to be considered macromolecules but they are completely reliant on a cellular host for replication. Whether one consid-

ers viruses dead or alive depends on one's perception of life. Certainly, outside a host system, viruses are nothing more than pieces of protein and nucleic acid. Upon entering the host cell, viruses certainly seem to be alive. They multiply, spread, and remultiply in a continuous cycle. To complicate matters further, viruses not only seem to fluctuate between being "dead" and "alive" but also may exist as infectious and noninfectious units. A complete virus particle consists of a nucleic acid core and an outer capsid; however, a complete virus particle may not be able to initiate disease in a host cell. The condition of the receptors on the surface of the virus particle is critical to the infectious nature of the virus. For example, if a virus or its receptors are damaged in part by disinfectants or natural environmental stress, the virus will not be able to attach to and penetrate the host cell. In any given population of virus particles, it is known that a certain ratio of particles are noninfectious. In laboratory stocks this ratio may be 1 : 100, with 100 inactive particles per infectious unit, but in the environment, where damaging stresses are more probable, the ratio may be even greater. In addition, viruses found in environmental samples may not be adapted to the particular cell culture used, thereby resulting in ratios greater than 1 : 1000. The relevance of this phenomenon will become more obvious in later chapters on the methods used to detect viruses, but, for now, remember that noninfectious viral units are not thought to be a threat to public health.

Viruses that are infectious can sometimes be detected indirectly in laboratory cultures of host cells. As viruses infect layers of host cells in laboratory flasks, clearings or plaques become visible. Each clear zone or plaque is caused by an infectious virus, and as such can be enumerated as **plaque forming units (PFU)** (see Chapter 10.5). Thus, the PFU is a common unit of quantitative measurement for infectious virus concentrations. The ratio of plaques to particles varies with virus type and age, host type and age, and environmental stresses.

2.1.7 Methods of Virus Infection and Replication

The growth cycle of a virus can be described in five steps: (1) adsorption, (2) penetration, (3) replication, (4) maturation, and (5) release (Fig. 2.6). All viruses share a common mechanism of replication at the molecular level, but different viruses replicate at varying rates. For example, bacteriophage often replicate rapidly, in minutes, whereas a typical animal virus replicates in hours to days. All viruses begin infection by adsorption to the host via specific receptors and in-

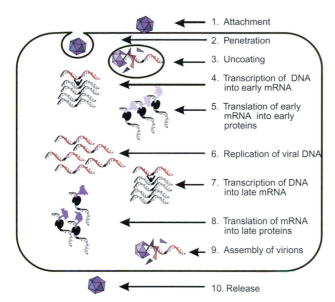

1. Attachment
2. Penetration
3. Uncoating
4. Transcription of DNA into early mRNA
5. Translation of early mRNA into early proteins
6. Replication of viral DNA
7. Transcription of DNA into late mRNA
8. Translation of mRNA into late proteins
9. Assembly of virions
10. Release

FIGURE 2.6 The basic steps of virus multiplication. Representation of an icosahedral DNA virus showing the main steps including adsorption, penetration, replication, maturation, and release.

jection of the nucleic acid or uptake of the total virus particle into the cell. The cycle then goes into what is known as the eclipse phase, a period of time during which no virus particles can be detected because of release and incorporation of the nucleic acid in the host cell machinery. Finally, new viral components are produced, assembled, and released from the host by disruption of the cell or budding at the cell membrane surface. The latter release mechanism is less destructive to the host cell and may support a symbiotic condition between the virus and the host.

2.1.7.1 Adsorption

The first step in the virus multiplication cycle involves attachment of the virus to specific host cells. Virus adsorption takes place by attraction to compatible receptors with antigenic specificity (see Chapter 12) at the surface of the cell membrane. Receptor sites are complex molecules such as lipoproteins or glycoproteins. Each cell has its own receptors, and thus viruses often infect specific species and sites within those species. This is known as the **species barrier.** An example of host and site specificity is that of the human poliovirus, which infects humans and a few species of monkeys. In addition, poliovirus infection in humans is typically isolated in brain, nerve, or intestinal cells. Receptor sites on the virus and the host cell are the most important factor in the specificity and control of a virus infection. This is evidenced by laboratory experiments which show that free viral nucleic acids artificially introduced into an unlikely host may

initiate infection. Receptors may be distributed over the whole surface of the virion, as with poliovirus, or may be isolated at specific sites, such as the glycoprotein spikes or tail fibers of more complex viruses (e.g., influenza virus and T-even phage). Enveloped viruses utilize glycoprotein spikes or penton fibers for specific host attachment, and many bacteriophage utilize tail structures to inject nucleic acid into the host. Most animal viruses lack specialized attachment organs and are thought to have attachment sites distributed over the virion's surface. Cell receptors may be damaged by heat, enzymes, pH, salts, or other stress, thereby preventing virus adsorption and subsequent infection. Likewise, virus receptors may be blocked or damaged to prevent infection of host cells. Temperatures above 55°C to 60°C and repeated freezing and thawing effectively damage receptors of many viruses. In addition, blocking of host receptors protects the cell from infection, and destroying virus attachment proteins renders the virion a noninfectious particle. Enveloped viruses are relatively more resilient but may also be rendered noninfectious by environmental adversities. Given optimal conditions of moisture, pH, and temperature, however, viruses are known to remain infectious for months to years in the environment.

2.1.7.2 Penetration

Animal cells offer few structural defenses to invading viruses, relying mostly on the skin barrier and mucous membranes. Once the virus makes it past the skin and mucous membrane barriers, virions enter hosts by either fusion or **endocytosis.** Endocytosis is a process in which the host cell membrane curves inward and extends appendages to surround the external virus particle. These appendages rejoin to form a pocket or vacuole in the cell. The vacuole eventually merges with the internal contents of the cell, carrying the virus particle with it. Enveloped viruses often shed their envelope during endocytosis, with only the viral capsid and nucleic acid being engulfed into the inner contents of the cell. Virus surfaces may also fuse directly to the plasma membrane of the cell to be released in the cytoplasm. Once virus particles enter a host cell, their nucleic acid must be exposed. Most host cells have cellular enzymes that uncoat the viral nucleic acid by initiation of structural changes in the capsid, but some viruses carry these enzymes with them. Once freed, viral nucleic acid becomes a part of the host cell nucleic acid, instructing it to produce more virus particles.

Following a host infection with a specific virus strain, serum antibodies develop and may persist, giving the host variable immunity to the same virus strain. Viruses, however, are capable of antigenic drift.

Gradual mutations in the gene sequences of virus glycoproteins allow the virus to avoid host antibody responses and reinstate infection in the host. Two viruses that are particularly skilled at such immune system avoidance are HIV and the influenza viruses.

Bacterial cells present more of a barrier to invading viruses. As outlined later in this chapter, bacterial cells usually contain a cell membrane and a rigid cell wall. Many bacteriophage have developed special tail structures to overcome these barriers. The T-even phage (i.e., T2, T4, T6) have polyhedral heads, rodlike tails, and a tail plate with six prongs extending from it. The prongs, or tail fibers, are the first structures to come in contact with the host cell. Once adsorbed to the cell, the rodlike tail may hydrolyze specific surface polysaccharides to create a channel for transport of the viral nucleic acid. The tail sheath contracts and injects the inner nucleic acid into the host. Other bacteriophage do not have tail structures and must find their way through the tough bacterial cell wall using a different mechanism altogether. Some of these phage utilize bacterial pili, attaching and releasing their nucleic acid into the channels of the hairlike structure. Other phage, such as the icosahedral φX174 and the filamentous M13, have protruding spikes that aid in adsorption and injection of the nucleic acid into the host cell.

Viruses of plants have no elaborate mechanisms of injection, presumably because plant cell walls contain cellulose and are very difficult to breach. Instead, plant viruses rely on timely contact with a portion of the plant damaged by insects or weather events. Indeed, most viruses of plants are carried by insects capable of piercing cell walls during feeding. Once a virus is inside the plant cell, transport from cell to cell is easily accomplished via feeding channels, plasmodesmas, or vascular channels.

2.1.7.3 Replication

The diversity of virus replication cycles is very apparent throughout the virus families. Not only do the viral genomes differ, being RNA or DNA, but their structure may be double stranded or single stranded. In addition, the enzymes used to decode the genetic message may originate from the virus or from the host. Following attachment, penetration, and uncoating, a representative replication cycle of a nonenveloped DNA virus includes transcription of early mRNA, translation of early proteins, replication of viral DNA, transcription of late mRNA, translation of late proteins, and finally, assembly and release. Early gene products generally shut down cellular processes and regulate viral genome expression, and late gene products produce virus structural proteins for assembly of

separate components into progeny virions. The ability to transcribe RNA from an RNA template is a phenomenon unique to viruses. The virus itself codes for the enzyme required to perform this task, in contrast to the many viruses that utilize host cell enzymes for transcription. Virus replication is highly varied among different families; however, virus infection generally follows a pattern of eclipse, maturation, and release. During the **eclipse phase**, viruses are intracellular and thus cannot be cultured in laboratory studies. Over time, the viruses take over host cell machinery, mature into whole particles, and are released by either lysis or budding. The eclipse period may last up to 12 hours, depending on the virus type. Once the naked nucleic acid is incorporated in the host cell nucleic acid, virulent viruses turn off cellular synthesis mechanisms and disaggregate cellular polyribosomes to favor the production of viral components. Most DNA viruses replicate in the cell nucleus, and most RNA viruses replicate in the cytoplasm.

2.1.7.4 Maturation and Release

Virus nucleic acid is assembled in capsid proteins following a maturation cycle. Nucleic acid may be packaged in the capsid simultaneously, as each component is synthesized, or sequentially, with nucleic acid production followed by capsid production and then assembly. Structural proteins of simple icosahedral viruses associate spontaneously into capsomers, which then self-assemble into capsids in which the nucleic acid is packaged. Likewise, bacteriophage are often assembled as empty capsids and the nucleic acid packaged later. Small viruses code for only 3–10 proteins but are capable of self-assembly, spontaneously packaging any nucleic acid in the near vicinity. Complex viruses enlist the help of enzymes for assembly. These enzymes are coded for on the viral nucleic acid. In bacteria infected with two closely related virus strains, structural components may be spontaneously mixed during reassembly. This is known as **phenotypic mixing** and may ultimately result in a wider susceptible host range.

To exit the host cell, some viruses use lytic enzymes (lysozymes) to break down the host cell wall. Nonenveloped viruses usually accumulate in the host cytoplasm or nucleus to be released when the cell lyses. Many helical and icosahedral virions mature by budding through host membranes and acquiring an envelope. Enveloped viruses released via budding may acquire their new envelope from the Golgi apparatus, plasma membrane, nuclear membrane, or endoplasmic reticulum of the host cell.

Although diverse in structure and function, all

viruses have a common goal: to reach, infect, and multiply in a host system, as such action is universally vital to their continued survival and evolutionary success.

2.2 BACTERIA

Bacteria are prokaryotic cells and as such are the simplest of microbial cells. In essence they consist of cell protoplasm contained within a retaining structure or cell envelope. They are among the most common and ubiquitous organisms found on earth. For example, even a gram of soil can contain up to 10^{10} bacteria (Paul and Clark, 1989). They can also be found in diverse and extreme environments. The activities and importance of bacteria are evidenced throughout this book which focuses primarily on bacterial activities. A few of their important effects are: biogeochemical processes, nutrient cycling in soils, bioremediation, human and plant diseases, plant–microbe interactions, municipal waste treatment, and the production of important drug agents including antibiotics.

2.2.1 The Bacterial Lifestyle

Molecular phylogeny based on ribosomal RNA genes has shown that there are three domains of living organisms: **Bacteria, Archaea,** and **Eucarya.** Of these domains, the Bacteria and the Archaea are single-celled prokaryotic organisms, whereas the Eucarya are more complex single and multicelled organisms that are eukaryotic in nature. The two prokaryotic domains can be further classified on the basis of their cell wall properties into four types:

Gram negative
Gram positive
Lacking a cell wall—the mycoplasmas
Cell wall lacking peptidoglycan—the Archaea

Gram-negative and gram-positive bacteria are prevalent in soil, water, and subsurface zone materials, with the archaebacteria normally being found in harsh environments. Examples of archaebacteria are the **acidophiles,** which tolerate low-pH environments by exporting H^+ ions from the cell. Archaea tolerant of high salt are known as **halophiles,** and those that tolerate dry habitats are called **xerophiles.**

Regardless of their domain, all four prokaryotic types are characterized by organisms geared toward rapid growth and cell division under favorable conditions. In essence, these microbes consist of a complex cell envelope surrounding cell cytoplasm. The ability to grow and reproduce cells quickly means that these organisms have the ability to adapt quickly to changing environments or environmental stimuli. Thus, mutations that arise in the presence of a stress, such as heavy metals or antibiotics, quickly allow the mutant to become the prevalent organism with resistance to the specific stress.

The modes of nutrition of bacteria vary considerably, as does the necessity for oxygen as a terminal electron acceptor. Bacteria can be free-living in bulk soil, where, because of their high numbers and diversity, substrate is consumed until it becomes limiting. Thus, these free-living bacteria actually exist in a state of limited starvation and become dwarf bacteria of diameter <0.3 μm (Bakken, 1997). The one area in soil where substrate is not limiting in soil is the rhizosphere (see Chapter 18.1.2), where plant photosynthates supply a constant source of available carbon.

Bacteria are extremely numerous and diverse in soils or waters, such that, they appear to be capable of colonizing almost any environment. In the next sections of this chapter the basic structure and function of bacteria is described, illustrating the diverse capabilities of the bacteria. Some of the activities of the Archaea will be described in Chapter 6.4, where examples of extreme environments are examined.

2.2.2 Size and Shape of Bacteria

The first dominant impression of bacteria is that they are small. Of course, small is a relative term, and once an organism is not visible to the naked eye it can be considered as small. In absolute terms, most bacterial cells are about 0.5 to 1 μm in diameter by 1 to 2 μm long. Usually cells do not exist in isolation; instead, they aggregate together to form clumps of cells, which are referred to as bacterial **colonies.** Bacterial colonies are often easily observed by culturing them on petri plates that contain agar with specific nutrients (see Chapter 10.2.1.2). It is important to realize that each bacterial colony that can be seen visually consists of about a million cells. The size of each cell of a particular type of bacterium can be as small as 0.3 μm or as large as 5 μm. The smaller cells known as ultramicrobacteria, can be important because they can be transported more easily through porous media (see Chapter 7.4.1). The small size of the bacterial cell gives the organism a specific advantage over other organisms. The accumulation of nutrients from the cell's environment involves the cell's surface. However, the relationship between the volume and surface area of an object is not constant. Specifically, a smaller sphere has a higher ratio of surface area to volume than does a larger sphere and therefore should have more efficient exchange of nutrients from its environment than the larger sphere. This higher efficiency of a smaller cell manifests itself in the phenomenon of cell **rounding** under environmental stress such as desiccation. During rounding, cells become smaller and also more spherical. It has been estimated that in undisturbed soils at least 50% of the cells may be dwarf-sized with a volume less than 1 μm^3 (Bakken, 1997). The shape of bacterial cells is also variable, as illustrated in Fig. 2.7. Bacteria shaped like spheres are called **cocci** (singular **coccus**), whereas those shaped like cylinders are termed **rods.** Other bacteria are shaped like a spiral with a long **spiral** being known as a **helix** or **helical** cell. Bacteria without a well-defined shape are said to be **pleomorphic.** Many environmental isolates of bacteria are pleomorphic, and, it is often difficult to distinguish between a short rod and a large coccus. A scanning electron micrograph of a typical soil bacterium is shown in Fig. 2.8. The rod-shaped cells have multiple appendages or **flagella** that aid in locomotion. The shape of the cell affects its survival and activity in the environment. **Cocci** have less surface area per volume than rods and thus can survive more severe desiccation than rods or spirals. In addition, being round, they are less distorted upon drying, which again makes them more resistant to desiccation. In

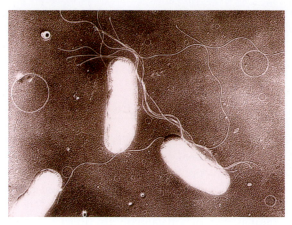

FIGURE 2.8   Scanning electron micrograph of a soil bacterium with multiple flagella.   The circles are detached flagella that have spontaneously assumed the shape of a circle. Reprinted from *Pollution Science,* © 1996, Academic Press, San Diego, CA.

contrast, under low-nutrient conditions, because of their greater ratio of surface area to volume, rods can take up nutrients from dilute solutions more efficiently than cocci. Shape also affects motility, with spiral cells moving with a corkscrew motion that meets with less resistance from surrounding water than rods.

2.2.3 Bacterial Form

Bacteria are structurally and functionally the simplest of the microorganisms and are known as **prokaryotes.** This distinguishes them from more complex organisms such as fungi, which are termed **eukaryotes.** The predominant characteristics of prokaryotic and eukaryotic cells are illustrated in Table 2.3. For comparison, the characteristics of viruses are also included. **Actinomycetes** are prokaryotic organisms that are classified as bacteria but are unique enough to be discussed as an individual group. They are similar to bacteria in that they are prokaryotic. However, morphologically, they resemble fungi because of their elongated cells that branch into filaments or hyphae. These hyphae can be distinguished from fungal hyphae on the basis of size, with actinomycete hyphae being about 2 μm in diameter, compared with the 10-50 μm diameter of fungal hyphae.

The most striking differences between prokaryotes and eukaryotes are that prokaryotes lack: (1) complex internal cell organelles involved in growth, nutrition, or metabolism; (2) a true nucleus; and (3) internal cell membranes. Prokaryotic bacterial cells have evolved with a capacity for rapid growth, metabolism, and reproduction. These attributes allow bacteria to

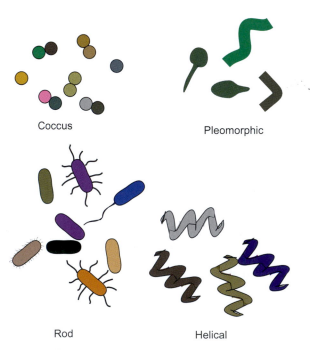

Coccus

Pleomorphic

Rod

Helical

FIGURE 2.7   Typical shapes of representative bacteria.

TABLE 2.3   A General Comparison of Prokaryotic and Eukaryotic Cells and Viruses[a]

Function or structure	Characteristic	Prokaryotic cells	Eukaryotic cells	Viruses[b]
Genetics	Nucleic acids	+	+	+
	Chromosomes	+	+	−
	True nucleus	−	+	−
	Nuclear envelope	−	+	−
Reproduction	Mitosis	−	+	−
	Production of sex cells	+/−	+	−
	Binary fission	+	+	−
Biosynthesis	Independent	+	+	−
	Golgi apparatus	−	+	−
	Endoplasmic reticulum	−	+	−
	Ribosomes	+[c]	+	−
Respiration	Enzymes	+	+	−
	Mitochondria	−	+	−
Photosynthesis	Pigments	+/−	+/−	−
	Chloroplasts	−	+/−	−
Motility/locomotor structures	Flagella	+/−[c]	+/−	−
	Cilia	−	+/−	−
	Cell wall	+[c]	+/−	−
Shape/protection	Capsule	+/−	+/−	−
	Spores	+/−	+/−	−
Complexity of function		+	+	+/−
Size (in general)		0.5−3 μm	2−100 μm	<0.2 μm

Adapted with permission of The McGraw-Hill Companies from Talaro and Talaro (1993) in *Foundations in Microbiology*.

[a] + means that most members of the group exhibit this characteristic; − means that most lack it; +/− means that only some members have it.

[b] Viruses cannot participate in metabolic or genetic activity outside their host cells.

[c] The prokaryotic type is functionally similar to the eukaryotic but structurally unique.

adapt to different environments, often very quickly and with remarkable efficiency. Bacteria such as *Escherichia coli* have the ability to replicate every 20 minutes or less under optimal conditions. The overall form of the bacterial cell is that of a complex cell **envelope** that encloses cell **protoplasm.** Cell **appendages** from the envelope protrude into the environment surrounding the cell. The constituents of each of these cell components are outlined in the following:

PROKARYOTIC CELL

Cell appendages	Cell envelope	Cell protoplasm
Flagella	Glycocalyx	Ribosomes
Pili	Cell wall	Mesosomes
Fimbriae	Cell membrane	Granules
		Nucleoid

2.2.4 Structure of the Bacterial Cell

All bacterial cells have a cell envelope and protoplasm that contains a cell membrane, cell pool, ribo-somes, and a nucleoid. Most also have a cell wall, although a few do not. Other constituents such as flagella or pili are common but not universal in all bacteria. A schematic representation of a typical bacterial cell is shown in Fig. 2.9.

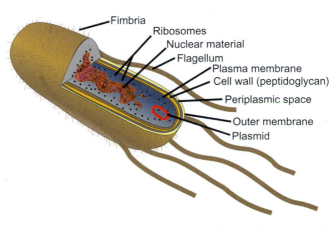

FIGURE 2.9   Schematic representation of a typical bacterial cell.

2.2.4.1 Appendages

Several accessory structures extend from the cell envelope out into the environment surrounding the cell. These appendages are not present in all bacterial types, but they are common, and they typically aid bacteria with motility and attachment to surfaces. The **flagellum** (plural flagella) is a complex appendage that improves the motility of cells by allowing cells to move through an aqueous medium. Almost all spiral bacteria have at least a single flagellum, and many have multiple flagella with variable arrangements (Fig. 2.10). Unusual corkscrew-shaped bacteria called **spirochetes** have modified flagella known as **axial filaments.** These axial filaments are also known as **endoflagella** because they are located within the cell envelope and impart a twisting motion to the cell. Bacteria with a polar arrangement have flagella attached at one or both ends of the cell. Bacteria with a single flagellum at one end only are said to be **monotrichous. Lophotrichous** arrangements consist of multiple flagella emerging from a single end of the bacterium, whereas an **amphitrichous** arrangement consists of multiple flagella at both ends of the cell. In contrast, a **peritrichous** arrangement involves flagella randomly dispersed over the surface of the cell. Unlike the spirilla, very few cocci are flagellated, and rod-shaped cells are intermediate in terms of their degree of flagellation. Overall motility (and therefore flagella) is important in aiding a bacterial cell to move toward nutrients (**positive chemotaxis**) and away from potentially harmful chemicals (**negative chemotaxis**). Chemotaxis

is made possible by the directional rotation, movement, and coordination of the flagella, which results in either a smooth motion of the cell or a tumbling motion. Smooth motion generally results in linear movement of the cell, whereas tumbling results in random motion. It is believed that stimulants inhibit tumbles allowing the cell to move toward a nutrient, while repellents cause multiple tumbles.

Pili and fimbriae are any surface appendages that are not involved in motility. **Fimbriae** (singular fimbria) are numerous short surface appendages that aid the cell in attachment to surfaces. The fimbriae are in part responsible for the microbial colonization of solids such as soil particles as well as the formation of biofilms within pipes. **Pili** (singular pilus) are normally less numerous than fimbriae but are longer. They are found only on gram-negative bacteria and are involved in a mating process between cells known as **conjugation.** In this process the exchange of DNA is facilitated by a pilus forming a connection between two cells. Conjugation in environmental bacteria is important because it enhances microbial diversity, often allowing specific populations to "fit" better in their environmental niche.

2.2.4.2 Cell Envelope

All bacteria are protected from the external environment by a series of layers that are collectively referred to as the **cell envelope.** The envelope for most bacteria consists of the glycocalyx, the cell wall, and the cell membrane, although a few bacteria lack one or both of the first two components. Each component of the envelope performs unique functions, but it is the integrated activity of all three components that ultimately protects the cell. In total, the cell envelope is one of the most complex components of the cell and can account for as much as 50% of the cell mass.

The **glycocalyx** is a coating of macromolecules that surround the cell wall (Fig. 2.11). The coating can be a rather loose informal aggregation of carbohydrates known as a **slime layer** or a more rigid layer of protein that is bonded to the cell wall and is known as a **capsule.** Cells with a capsule produce gummy colonies that are mucoid in nature. Overall, the glycocalyx offers some protection to the cell from water and nutrient loss or other stresses including heavy metal toxicity (Roane and Kellogg, 1996), but it is not essential for the survival of the cell. There are other specialized functions of the glycocalyx associated with attachment and also recognition factors between bacteria and plant roots, such as between nitrogen-fixing rhizobia and leguminous plants (Paul and Clark, 1989).

Beneath the glycocalyx is the cell wall, which deter-

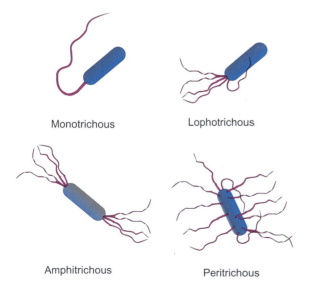

Monotrichous

Lophotrichous

Amphitrichous

Peritrichous

FIGURE 2.10 Arrangement of flagella extending from the cell envelope.

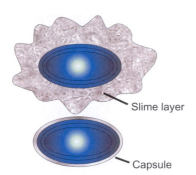

FIGURE 2.11 Two different types of glycocalyces: (1) slime layer; and (2) capsule.

mines the shape of the cell and gives it structural support. The cell wall, in fact, protects the cell from rupturing caused by the intake of water by osmosis. The two basic cell wall types are associated with bacteria are known as **gram positive** and **gram negative,** respectively (Fig. 2.12). In both cell wall types, the rigid protective nature of the wall is due to a macromolecule known as **peptidoglycan.** This polymeric molecule is composed of alternating glycan molecules cross-linked by short peptides. The two glycans are *N*-acetylmuramic acid and *N*-acetylglucosamine. The gram-positive cell wall consists mostly of peptidoglycan plus acidic polysaccharides including teichoic acids. The teichoic acids are negatively charged and are partially responsible for the negative charge of the cell surface as

a whole. As we will see in Chapter 7.2.4, this charge affects the mobility and transport of bacteria in different environments. Some bacteria—the mycoplasmas—lack a cell wall, whereas the Archaea have a modified cell wall that does not contain peptidoglycan.

Beneath the surface of the gram-positive cell wall is the **cell membrane**. The gram-negative cell wall is more complex than the gram-positive wall and contains two membranes; an **outer membrane** as well as an interior cell membrane. Both membranes are composed of phospholipids however, the interior membrane is uniform in structure while the outer leaflet of the outer membrane also contains **lipopolysaccharide (LPS)**. Thus, the outer membrane is sometimes referred to as the LPS layer. The LPS layer is responsible for the antigenic properties of the cell and sometimes exhibits toxic properties. It is known as the endotoxin associated with gram-negative pathogens. Gram-negative cell walls contain a narrow layer of peptidoglycan compared to the thick layer found in gram-positive walls. A further difference is that in gram-negative walls there is an extensive space between the outer membrane and the inner membrane, which is known as the **periplasmic space.** In contrast, in gram-positive walls, the wall and the internal membrane have very little space between them. The periplasmic space contains enzymes important for nutrient uptake. The outer membrane is permeable to small molecules but not to large molecules, including enzymes. Access is provided by special membrane

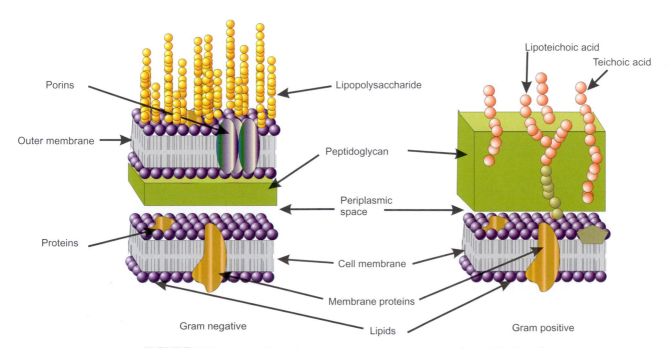

FIGURE 2.12 Comparison of gram-positive and gram-negative bacterial cell walls.

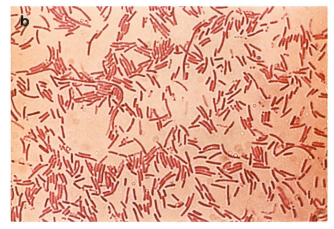

FIGURE 2.13   Typical gram-positive (blue) and gram-negative (red) bacteria identified by the Gram stain.

channels created by porin proteins that completely span the outer membrane.

Overall, the differences in cell wall structure cause differences in the uptake of dyes. It was these differences that allowed Hans Christian Gram in 1884 to differentiate bacteria as gram-positive or gram-negative based on the retention or loss of a crystal violet–iodine complex when treated with an organic solvent, and followed by counterstaining with safranin. Gram-positive organisms stain blue (from the crystal violet), whereas gram-negative organisms stain red (from the safranin) (Fig. 2.13).

Some bacterial groups do not contain these classic cell wall structures or have variations in structure. A few bacteria do not possess a cell wall but instead have more complex cell membranes. For genetic analyses of bacteria, it is normal to remove or rupture the wall, by use of lysozyme for example, prior to cell lysis and DNA extraction. A gram-positive cell without a cell wall is known as a **protoplast.** A gram-negative cell treated with lysozyme loses its peptidoglycan but retains much of its outer membrane and is known as a **spheroplast.**

The interior cell membrane is also known as the **cytoplasmic membrane.** It is a thin, flexible structure, about 5 nm across, that totally surrounds the cell. The general composition of the membrane is that of a lipid bilayer (Fig. 2.14) containing phospholipids (30–40% by mass) and proteins (~60%). The importance of a membrane is that it controls what gets into and out of the cell. Nonpolar small molecules, including water, can pass through the lipid phase of the membrane, but most molecules must be actively transported through protein channels embedded throughout the membrane. Thus, the cell membrane provides a critical bar-

rier between the cell interior and the environment. If the membrane is ruptured, the integrity of the cell is destroyed, its contents leak out, and the cell dies.

The cell membrane is also involved in energy production including respiration and photosynthesis. Specialized forms of energy production including

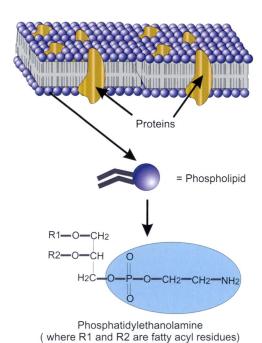

Proteins

= Phospholipid

Phosphatidylethanolamine
(where R1 and R2 are fatty acyl residues)

FIGURE 2.14   Structure of a bacterial cell membrane. The phospholipids contain hydrophobic groups directed inward and hydrophilic groups directed to the exterior of the membrane. Hydrophobic proteins are embedded throughout the membrane and aid in transport of molecules through the membrane.

nitrification and methane oxidation (see Chapter 14.2.4.3.2) also occur close to the membrane.

2.2.4.3 Cell Protoplasm

The cell envelope encloses the cell protoplasm, which consists of metabolites and nutrients in solution. The major constituent of the cell pool is water, which acts as a solvent for carbohydrates, salts, amino acids, and enzymes. Also, within the protoplasm are discrete entities known as the nucleoid, ribosomes, mesosomes, and granules.

The bacterial cell does not have a true nucleus or a nuclear membrane. Instead, the DNA is folded into a dense area of material called the **nucleoid.** The DNA is also known as the chromosome and consists of a single circular molecule of double-stranded DNA (see Chapter 13.1). Whereas the length of a cell may be 2 μm, the total length of the DNA molecule can be as long as 1200 μm. The DNA contains all the genetic material necessary for the growth, metabolism, and survival of the cell. Because each bacterium has certain unique nucleic acid sequences, this allows detection of the organism by nucleic acid based analyses (see Chapter 13). In addition to the chromosome, many bacteria contain smaller accessory pieces of DNA known as **plasmids.** Plasmids normally code for unique characteristics that can aid the cell in some specific circumstances but are not vital for the survival of the cell. Examples of plasmid-encoded functions are heavy metal resistance and the ability to degrade a potentially toxic organic compound. In unique environments, plasmids are critical in aiding bacterial cells function and they are of particular interest because they can be transferred from one bacterial cell to another via horizontal gene transfer (Smit et al., 1991).

Ribosomes are particles responsible for protein synthesis. They are composed of a special type of RNA called ribosomal RNA (60%) and protein (40%). Prokaryotic ribosomes have a function similar to that of eukaryotic ribosomes but different in detail. This difference is expressed in terms of sedimentation constants where bacteria have 70S units that contain subunits of 50S and 30S, whereas eukaryotic ribosomes have 80S units with 40S and 60S subunits. Ribosomes are also important in the identification of bacteria using genetic analyses, because the nucleic acid sequences that code for the ribosomal RNAs are often unique to the bacterial cell type (see Chapter 13.3.3.2).

Bacteria have the ability to store some food reserves intracellularly as **inclusions** or **granules.** Stored nutrients include carbon as glycogen or poly-β-hydroxybu-

tyrate, elemental sulfur, and phosphorus as polyphosphates. Nitrogen is not normally stored as a reserve within the cell. Some bacteria have the ability to precipitate metals intracellularly, which can form the basis of metal resistance (Silver and Phung, 1996). Two other types of inclusions are sometimes found in a cell. These are **gas vesicles,** which are composed of protein and can provide buoyancy for aquatic bacteria, and **endospores.** Endospores are structures produced by a few bacteria such as *Bacillus* or *Clostridium* spp. Spores are dominant structures capable of surviving adverse conditions and hence allow the cell to survive periods where there is a lack of nutrients or other environmental stress.

Viewed as a whole, the bacterial cell has the capacity for rapid periods of growth when conditions are favorable. In addition, because of their ability to reproduce rapidly, bacteria have the potential to adapt to changing conditions through the production of spontaneous mutants that can quickly become the dominant cell type. It is these characteristics plus the extraordinary numbers and diversity of bacteria that make them so important in environmental microbiology.

2.2.5 Plasmid-Chromosome Relationships

The nature, properties, and functions of plasmids have been reviewed by Pepper (1994). Most of the genetic information necessary for a functional bacterial cell is contained in the bacterial chromosome. For many soil organisms this is normally a single circular chromosome consisting of about $1-3 \times 10^6$ nucleic acid base pairs, which code for 1000 or more different genes. The chromosome provides fundamental gene sequences necessary for metabolism and reproduction of the cell. In addition, bacteria may contain extrachromosomal or accessory genetic elements including plasmids. Plasmids are additional DNA sequences that are separate from the chromosome. Normally, plasmids code for genes with cellular functions that are not mandatory for cell growth and division. For example, rhizobia contain symbiotic plasmids that code for nitrogen fixation. However, at times plasmids may be important for bacterial growth, such as when plasmid-coded biodegradative genes are necessary for the utilization of toxic waste organic materials in soil. Plasmids are autonomous in that the plasmid copy number, or number of identical plasmids per cell, is normally independent of the number of chromosome copies. Plasmids are also expendable, meaning that they are an accessory genetic element not essential to the growth of the organism in its normal environment

(Krawiec and Riley, 1990). However, plasmids are nonetheless often critical for the cell to function efficiently in a specific environmental niche. This is due to the phenotypic expression of plasmid genes that confer an advantage to the cell.

Bacterial plasmids can be very small ($\simeq$10 kb), such as the plasmids associated with *Streptomyces*, or can be very large (300–1000 kb), as in the case of rhizobia. Such megaplasmids contain genetic information equivalent to one fifth of the chromosome. In addition, different bacterial species or even strains of species may contain several different plasmids with variable copy numbers. The functions of genes encoded by plasmids are sometimes well defined. However, the functions of many plasmids are not known, and these are referred to as cryptic plasmids. Some plasmids can integrate into the chromosome during replication and function as part of the chromosome. During later replications, this process can be reversed, with the plasmid DNA being excised and allowed to function as a self-replicating entity within the cell. A well-known example of this is the integration and excision of the F plasmid in *E. coli*. In addition, DNA transfers between *Streptomyces* spp. occur in which stable genetic elements in the donor species become plasmids in the recipient strains.

2.2.5.1 Plasmid Nomenclature

- **Copy number.** Low-copy-number plasmids are plasmids that usually exist with one or two copies per cell. High-copy-number plasmids are usually small plasmids (<10 kb) that have 10 to 100 copies per cell.
- **Stringency.** Relaxed plasmids are plasmids whose replication does not depend on the initiation of cell replication. Therefore these plasmids can be amplified (i.e., copy number increased) relative to cell number. Stringent plasmids are dependent on cell replication, and plasmid replication is synchronized with replication of the bacterial chromosome. Thus, stringent plasmids have low copy numbers that cannot be amplified. Since cells growing rapidly may have three or four chromosomes, stringent plasmids can still be present with copy numbers greater than one per cell.
- **Conjugative plasmids.** Conjugative plasmids are self-transmissible and can be transferred from one bacterial cell to another during conjugation between the two cells. The cells can be of the same species or of different species. Conjugative plasmids are usually large and contain transfer genes known as **tra genes**. Nonconjugative plasmids do not contain tra genes and are not self-transmissible. However, some plasmids can transfer to other cells by "mobilization" to other conjugative plasmids, although not all nonconjugative plasmids are mobilizable. In the process of mobilization, transfer of nonconjugative plasmids relies on the tra genes of the conjugative plasmids.

- **Incompatibility.** Incompatible plasmids vary in their ability to coexist within the same cell. Incompatible plasmids cannot exist together and give rise to incompatibility groups (Inc groups). Compatible plasmids belong to different Inc groups and vice versa. Most plasmids exist in only a few closely related species and are known as narrow-host-range plasmids. Certain plasmids, such as those in the Inc P group, are able to exist in a wide variety of bacterial species.

2.2.5.2 Functions of Plasmids

All plasmids contain gene sequences that are necessary for their own replication, and may also contain tra genes necessary for self transmission (Nordstrom *et al.*, 1984). Additional gene sequences code for a variety of phenotypic traits.

- **Cryptic plasmids.** These are plasmids with sequences that encode unknown phenotypic traits with no known function. These are often present as megaplasmids in rhizobia and can be "cured" or deleted only with great difficulty. Other megaplasmids in rhizobia have known functions, including genes required for nodulation and nitrogen fixation. Overall, most bacterial plasmids, regardless of size, are cryptic.
- **Resistance plasmids.** Resistance or R-plasmids contain gene sequences that code for protection of the bacterial cell against specific deleterious substances. Specific plasmids are known that confer resistance to antibiotics, heavy metals, colicins, or bacteriophage.
- **Degradative plasmids.** These plasmids are also known as catabolic plasmids and code for the breakdown of unusual metabolites. These metabolites can be unusual carbon compounds of microbial origin or human origin, including petroleum constituents or pesticides. Many soil bacterial isolates contain degradative plasmids, particularly *Pseudomonas* spp. Such plasmids are now useful in the production of genetically engineered microorganisms (GEMs) used to degrade hazardous wastes contaminating soil or

water. Examples of degradative plasmids include the TOL plasmid involved in the breakdown of toluene and the pJP4 plasmid that contains gene sequences coding for the degradation of 2,4-dichlorophenoxyacetic acid.

- **Plant-interactive plasmids.** There are two well-known interactions involving plasmids and higher plants (see Chapter 18). The first concerns the symbiosis between the bacterium *Rhizobium* and leguminous plants. Typical *Rhizobium* spp. contain Sym (symbiotic) plasmids that code for infection and nodule formation on the plants, as well as gene sequences necessary for biological nitrogen fixation. The second system involves parasitism between *Agrobacterium* spp. and higher plants. Tumor-inducing (Ti) plasmids from the bacterium integrate into the host DNA and induce tumors that result in crown gall disease. This is the best documented case in which prokaryotic DNA is integrated into a eukaryotic genome. This unique property is now being investigated for its use in inserting novel genes into plants.
- **Miscellaneous plasmids.** There are various other kinds of plasmids with unique functions such as RNA metabolism, conjugation, and bacterial cell envelope alteration. Given that chromosomal DNA sequences can and do become integrated into plasmids during plasmid transfer, the potential functions associated with plasmids appear to be unlimited.

2.2.6 Bacterial Metabolism

To survive and proliferate in the environment, bacteria require water, nutrients, a source of energy, and a terminal electron acceptor. We will examine the nutrient requirements and modes of growth of different types of bacteria.

2.2.6.1 Microbial Nutrition

Regardless of their cell type, all bacteria require certain **essential elements** for growth and metabolism (Table 2.4). Nutrients required in relatively large amounts are known as macronutrients and are generally involved in cell structure and metabolism. Elements that are required in smaller amounts are referred to as the **micronutrients** or **trace elements.** Micronutrients are often metals that are needed as catalysts for enzymes.

These essential nutrients are used in the creation of all of the components of the bacterial cell. A typical cell

TABLE 2.4 The Essential Elements Required by All Bacteria

Macroelements	Microelements
Carbon (C)	Manganese (Mn)
Oxygen (O)	Molybdenum (Mo)
Nitrogen (N)	Copper (Cu)
Hydrogen (H)	Cobalt (Co)
Phosphorus (P)	Boron (B)
Sulfur (S)	
Potassium (K)	
Sodium (Na)	
Calcium (Ca)	
Magnesium (Mg)	
Chloride (Cl)	
Iron (Fe)	

composition is shown in Table 2.5. The most prevalent compounds are water, protein, and nucleic acids, with about 97% of the dry weight of the cell being organic (Neidhardt *et al.*, 1990). Most bacteria contain about 5000 different compounds, but the number that most bacteria need to take up from the environment to synthesize these compounds can be as low as eight. *E. coli* require only mineral salts and glucose for survival, whereas other bacteria have requirements for preformed vitamins and other growth factors. **Oligotrophic** organisms are those that are capable of living in nutrient-deficient environments, whereas **copiotrophic** organisms can survive only in nutrient-rich environments.

Autotrophic bacteria derive energy and carbon from inorganic sources. Carbon is obtained via carbon dioxide, whereas energy can be derived from two sources. **Photoautotrophs** are photosynthetic and obtain energy from sunlight. Photosynthetic bacteria such as the cyanobacteria are important in causing eutrophication of surface waters. However, their prevalence in soil and vadose zones is limited because these materials are impermeable to light. **Chemoautotrophs** obtain energy by the oxidation of inorganic substances. Examples of autotrophic modes of nutrition are shown in Table 2.6 of this chapter. Autotrophs are important in the cycling of inorganic nutrients, including the methanogens, which produce methane, and the nitrifiers, which convert ammonia to nitrate (see Chapter 14). Thus, the autotrophs can in subtle ways reduce or amplify potential pollutants.

TABLE 2.5   Overall Macromolecular Composition of an Average *E. coli* B/r cell

Macromolecule	Percentage of total dry weight	Weight per cell ($10^{15} \times$ weight, grams)	Molecular weight	Number of molecules per cell	Different kinds of molecules
Protein	55.0	155.0	4.0×10^4	2,360,000	1050
RNA	20.5	59.0			
23S rRNA		1.0	1.0×10^6	18,700	1
16S rRNA		16.0	5.0×10^5	18,700	1
5S rRNA		1.0	3.9×10^4	18,700	1
Transfer		8.6	2.5×10^4	205,000	60
Messenger		2.4	1.0×10^6	1,380	400
DNA	3.1	9.0	2.5×10^9	2.13	1
Lipid	9.1	26.0	705	22,000,000	4
Lipopolysaccharide	3.4	10.0	4346	1,200,000	1
Petidoglycan	2.5	7.0	$(904)_n$	1	1
Glycogen	2.5	7.0	1.0×10^6	4,360	1
Total macromolecules	96.1	273.0			
Soluble pool	2.9	8.0			
Building blocks		7.0			
Metabolites, vitamins		1.0			
Inorganic ions	1.0	3.0			
Total dry weight	100.0	284.0			
Total dry weight/cell		2.8×10^{-13}g			
Water (at 70% of cell)		6.7×10^{-13}g			
Total weight of one cell		9.5×10^{-13}g			

Adapted with permission from Neidhardt *et al.* (1990).

TABLE 2.6   Biological Generation of Energy

Biological activity requires energy, and all microorganisms generate energy. This energy is subsequently stored as **adenosine triphosphate** (ATP), which can be utilized for growth and metabolism as needed, subject to the second law of thermodynamics.

The Second Law of Thermodynamics

In a chemical reaction, only part of the energy is used to do work. The rest of the energy is lost as entropy.

For any chemical reaction the free energy ΔG is the amount of energy available for work.

For reaction A + B $\rightleftharpoons$ C + D, the thermodynamic equilibrium constant is defined as

$$K_{eq} = \frac{[C]\,[D]}{[A]\,[B]}$$

Case 1: If product formation is favored, the thermodynamic equilibrium constant is positive. That is, if

$$[C]\,[D] > [A]\,[B] \text{ then } K_{eq} > 1$$

Therefore, the logarithm of the thermodynamic equilibrium constant is also positive:

$$K_{eq} > 1$$

For example, if $K_{eq} = 2$,

$$\log K_{eq} = 0.301$$

(continues)

TABLE 2.6 *(Continued)*

Case 2: If product formation is not favored, K_{eq} is negative. That is, if

$$[A]\,[B] > [C]\,[D] \text{ then } K_{eq} < 1$$

and

$$\log K_{eq} < 1$$

For example, if $K_{eq} = 0.2$,

$$\log K_{eq} = -1 + 0.301 = -0.699$$

The relationship between the equilibrium coefficient and the free energy ΔG is given by

$$\Delta G = -RT \log K_{eq}$$

where R is the universal gas constant and T the absolute temperature (K).

If $\underline{\Delta G \text{ is negative}}$, energy is released from the reaction due to a spontaneous reaction. This is because if ΔG is negative, $\log K_{eq}$ must be positive; and therefore $K_{eq} > 1$, which means energy must be released.

If $\underline{\Delta G \text{ is positive}}$, energy is needed to make reaction proceed. This is because if $\underline{\Delta G}$ is positive, $\log K_{eq}$ must be negative; and therefore $K_{eq} < 1$, which means energy must be added to promote the reaction.

Thus we can use ΔG values for any biochemical reaction mediated by microbes to determine whether energy is liberated for work and how much energy is liberated.

Soil organisms can generate energy via several mechanisms, which can be divided into two main categories.

1. **Photosynthesis**

$$2H_2O + CO_2 \xrightarrow{\text{light}} CH_2O + O_2 + H_2O$$
$$\text{biomass}$$
$$\Delta G \simeq +115 \text{ kcal mol}^{-1}$$

For this reaction energy supplied by sunlight is necessary. The fixed organic carbon is then used to generate energy via respiration. Examples of soil organisms that undergo photosynthesis are *Rhodospirillum*, *Chromatium*, and *Chlorobium*.

2. **Respiration**

   (a) <u>Aerobic heterotrophic respiration</u>: Many organisms undergo aerobic, heterotrophic respiration, for example, *Pseudomonas* and *Bacillus*.

   $$C_6 H_{12} O_6 + 6O_2 \rightarrow 6CO_2 + 6H_2O$$
   $$\Delta G = -686 \text{ kcal mol}^{-1}$$

   (b) <u>Aerobic autotrophic respiration</u>: The reactions carried out by *Nitrosomonas* and Nitrobacter are known as **nitrification**:

   $$NH_3 + 1\tfrac{1}{2}O_2 \rightarrow HNO_2 + H_2O \qquad (\textit{Nitrosomonas})$$
   $$\Delta G = -66 \text{ kcal mol}^{-1}$$

   $$KNO_2 + \tfrac{1}{2}O_2 \rightarrow KNO_3 \qquad (\textit{Nitrobacter})$$
   $$\Delta G = -17.5 \text{ kcal mol}^{-1}$$

   The following two reactions are examples of **sulfur oxidation**:

   $$2H_2S + O_2 \rightarrow 2H_2O + 2S \qquad (\textit{Beggiatoa})$$
   $$\Delta G = -83 \text{ kcal mol}^{-1}$$

   $$2S + 3O_2 \rightarrow 2H_2SO_4 \qquad (\textit{Thiobacillus thiooxidans})$$
   $$\Delta G = -237 \text{ kcal mol}^{-1}$$

(continues)

TABLE 2.6   (*Continued*)

The next reaction involves the degradation of cyanide:

$$2KCN + 4H_2O + O_2 \rightarrow 2KOH + 2NH_3 + 2CO_2 \qquad (Streptomyces)$$
$$\Delta G = -56 \text{ kcal mol}^{-1}$$

All of the preceding reactions illustrate how organisms mediate reactions that can cause or negate pollution. For example, nitrification and sulfur oxidation can result in the production of specific pollutants, i.e., nitrate and sulfuric acid, whereas the destruction of cyanide is obviously beneficial with respect to the mitigation of pollution.

(c) Facultative anaerobic, heterotrophic respiration:
   Pseudomonas denitrificans can achieve this kind of metabolism by utilizing nitrate rather than oxygen as a terminal electron acceptor. Note that these organisms can use oxygen as a terminal electron acceptor if it is available and that aerobic respiration is more efficient than anaerobic respiration.

$$5C_6H_{12}O_6 + 24KNO_3 \rightarrow 30CO_2 + 18H_2O + 24KOH + 12N_2$$
$$\Delta G = -36 \text{ kcal mol}^{-1}$$

(d) Facultative anaerobic autotrophic respiration:

$$S + 2KNO_3 \rightarrow K_2SO_4 + N_2 + O_2 \quad (Thiobacillus \ denitrificans)$$
$$\Delta G = -66 \text{ kcal mol}^{-1}$$

(e) Anaerobic heterotrophic respiration: *Desulfovibrio* is an example of an organism that carries out this type of metabolism.

$$2CH_3CHOH \ COOH + SO_4^{2-} \rightarrow 2CH_3 \ COOH + H_2S + 2HCO_3$$
      lactic acid   acetic acid
$$\Delta G = -40 \text{ kcal mol}^{-1}$$

Organisms can also undergo fermentation, which is also an anaerobic process. But fermentation is not widespread in soil or vadose zone material. Overall, there are many ways in which organisms can and do generate energy. The preceding mechanisms illustrate the diversity of organisms and explain the ability of the environmental community to break down or transform almost any natural substance. In addition, enzyme systems have evolved to metabolize complex molecules, organic or inorganic. These enzymes can also be used to degrade xenobiotics with similar chemical structures. Xenobiotics, which do not degrade easily, are normally chemically different from any known natural substance; hence organisms have not evolved enzyme systems capable of metabolizing such compounds. Many other factors influence the breakdown or degradation of chemical compounds, and these will be discussed in Chapter 16.

Adapted from *Pollution Science*, © 1996, Academic Press, San Diego, CA.

Heterotrophic bacteria derive carbon from preformed organic compounds that are broken down enzymatically. For **chemoheterotrophs,** energy is derived through the oxidation of organic compounds via respiration. There are literally thousands of different types of chemoheterotrophs in the environment, and they are critical to many aspects of environmental microbiology including biogeochemical cycling (see Chapter 14), waste treatment (see Chapter 21), and bioremediation (see Chapter 16). In addition, most pathogenic organisms are chemoheterotrophic. A few microbes, such as the green and purple sulfur bacteria, metabolize in a **photoheterotrophic** mode. These organisms derive energy from light and use organic compounds as a source of reducing power.

Generation of energy through chemical oxidation is referred to as respiration regardless of whether the substrate is inorganic or organic. During the oxidative process, electrons are removed from the substrate and passed via the electron transport chain to a **terminal electron acceptor** (TEA). For **aerobic organisms** the TEA is oxygen. For **anaerobic organisms** the TEA is a combined form of oxygen such as an organic metabolite CO_2, NO_3^- or SO_4^{2-} or an oxidized metal, e.g., Fe^{3+}. Strict anaerobes lack protective enzymes to remove peroxide radicals that originate from O_2 and are lethally affected by oxygen. Other anaerobes do have the protective enzymes but still cannot use O_2 as a TEA. Still other kinds of bacteria, known as **facultative anaerobes,** preferentially use O_2 if it is present but can use other TEAs when O_2 is not available. Generally, facultative anaerobes grow more efficiently in the presence of oxygen. Because soil and vadose zone materials are always discontinuous environments, different levels of oxygen availability (also referred to as redox potential) can occur in close proximity to one another. This results in aerobic and anaerobic microbes being able to function in the same vicinity, which often has important implications in areas of environmental microbiology including biodegradation of xenobiotics (see Chapter 16). Overall, aerobic organisms are ubiquitous in surface soils, whereas oxygen can be limiting in the subsurface, reducing aerobic activity. Collectively, environmental communities of bacteria are often extremely diverse because of the ability to transform organic and inorganic compounds under a variety of redox potential conditions. This diversity, coupled with large numbers, can have beneficial effects as in the case of municipal waste treatment (see Chapter 21) or disastrous results as in the case of sulfur oxidation that results in sulfuric acid production and subsequent metal solubilization following strip-mining operations (see Chapter 15).

2.3 FUNGI

In contrast to bacteria, fungi are eukaryotic organisms. They are ubiquitous in the environment and critically affect human health and welfare. For example, fungi can beneficially affect plants through mycorrhizal associations or adversely affect plants via fungal plant pathogens (see Chapter 18). They are also important in the cycling of organics (see Chapter 14) and bioremediation (see Chapter 16). Fungal pathogens adversely affect human health, and other fungi known as yeasts are utilized in the fermentation of sugars to alcohol in the brewing and wine industries. Overall, fungi are found in fresh water, marine water, or terrestrial habitats including soil and associated dead plant matter.

2.3.1 The Fungal Lifestyle

Fungi are eukaryotic and are fundamentally different from bacteria in that they are more complex. However, soil fungi are similar to bacteria in terms of their large numbers and diversity. Fungi range from microscopic, with a single cell consisting of less than 1 pg of dry weight, to macroscopic, with filaments of a single fungal cell covering several hectares. Overall, fungi are heterotrophic in nature, with different genera metabolizing simple sugars or complex aromatic hydrocarbons. Fungi are particularly important in the degradation of the plant polymers cellulose and lignin, and other complex organic molecules, notably xenobiotics. For the most part they are aerobic, except for the yeasts, which are capable of fermentation.

There are approximately 15,000 species of true soil fungi, a number swollen by the members of the soil basidiomycetes, many of which give rise to macroscopic mushrooms (Bakken, 1997). Fungi are also found in stable symbiotic associations with green algae or cyanobacteria as **lichens,** in which the fungal partner forms a characteristic structure or thallus that encloses and protects the photosynthesizing alga or cyanobacterium. Lichens are important soil colonizers that play critical roles in soil formation and stabilization, particularly in disturbed or denuded sites. Overall, fungi often dominate in terms of soil biomass and can represent a significant portion of the nutrient pool, particularly in oligotrophic soils.

Fungi in general have highly variable growth rates, but as a group they are not as capable of rapid growth as the bacteria. However, some soil genera including *Aspergillus*, *Geotrichum*, and *Candida* have a doubling time of about 1 hour in pure culture (Carlile and Watkinson, 1994). Fungi are particularly important in the degradation of complex organics including lignin and pollutant molecules. The white-rot wood-inhabiting basidiomycete *Phanerochaete chrysosporium* has been particularly well studied in terms of its bioremediation potential (see Chapter 16).

Fungi are also important in environmental microbiology through their interactions with plants, which can be beneficial or detrimental. The mycorrhizal fungi form important mutualistic associations with the roots of plants, supplying them with nutrients and protection from drought and root pathogens. In contrast, root pathogens such as species of *Armillaria*, *Fusarium*, or *Rhizoctonia* cause major losses in agricultural crops such as wheat and cotton (see Chapter 18). Finally, note that fungi are producers of the important pharmaceutical compounds such as penicillin and the cyclosporins.

2.3.2 Taxonomic Diversity of Fungi

The classification of fungi within the biological world has changed over the years. At various times they have been classified as plants or even along with algae and protozoa. There are over 72,000 known species of fungi. A simplified classification of fungi was provided by Thorn (1997). He lists fungi as belonging to four groups: "Motile fungi," "Zygomycota," "Ascomycetes," and "Basidiomycetes."

Motile fungi include organisms related to protozoa and algae. Protozoan fungi include slime molds such as the **myxomycetes.** These have amoeba-like stages that engulf microbes and fungal spores. Water molds related to algae include the **oomycetes,** which can swim through soil water films or aquatic habitats propelled by one or more flagella. Some of these water molds, including *Pythium* and *Phytophthora*, are important plant pathogens.

The **Zygomycota** include three different types of fungi, one of which is the Zygomycetes. An example of a zygomycete is the common bread mold *Rhizopus*, which is characterized as growing very quickly. Other members of the Zygomycota are the Trichomycetes, which are microscopic gut symbionts of insects, and the symbiotic arbuscular-mycorrhizal fungi such as *Glomus aggregatum* (see Chapter 18). Many of the Trichomycetes and mycorrhizal fungi cannot be grown in culture without their plant hosts.

Ascomycetes are traditionally characterized as fungi that reproduce sexually by means of an **ascus** (plural asci). These include some yeasts and molds. The yeasts are ascomycetes that have adapted to unicellar life in a liquid environment, including *Saccharomyces cerevisiae*. The molds include the well-known *Aspergillus* and *Penicillium* species. The species of fungi with no known sexual stages were formally known as the Deuteromycetes, but this nomenclature is generally disfavored by mycologists. Therefore, the fungi formerly known as Deuteromycetes are now classified as asexual forms of Ascomycetes or Basidiomycetes. Lichens formed from Ascomycetes account for approximately 15,000 species of fungi and are important in soil colonization.

Basidiomycetes are the primary agents of lignin degradation (see Chapter 14). They are often characterized by macroscopic fruiting bodies known as mushrooms or puffballs. The basidiomycetes also include most of the ectomycorrhizal fungi and some significant plant pathogens.

2.3.3 Size and Shape of Fungi

Soil fungi range from the microscopic species to widespread filaments that can cover an area of 15 ha. Macroscopic fruiting bodies such as mushrooms or truffles can easily be seen or provide a hearty meal. Fungal numbers can be estimated as reproductive propagules (10^4–10^6 colony-forming units per gram of soil) or as hyphal lengths (100–1000 μm per gram of soil) (Thorn, 1997). **Yeasts** are unicellular fungi that

reproduce asexually through **budding.** During budding, daughter cells separate from mature yeast cells. Molds, in contrast, are characterized by the presence of filamentous structures known as **hyphae** (singular *hypha*). Hyphae are long threadlike structures that entwine together into a mass known as a **mycelium** (Fig. 2.15). A hypha can be 10 to 50 μm in diameter and several millimeters long in the environment. The large diameter of the fungal hyphae distinguishes them from actinomycete hyphae, which are only 1 μm in diameter (characteristic diameter of bacteria). **Nonseptate** hyphae consist of long continuous cells, within which cytoplasm and cell organelles can move freely. Each nonseptate hypha may have several nuclei. In contrast, **septate** hyphae are partitioned into individual compartments by cross-walls or **septa** (singular *septum*). Septa may be solid walls that allow no transfer between compartments, or they may have small pores that allow transfer of organelles and nutrients. **Vegetative hyphae** are responsible for the adsorption of nutrients, and **aerial hyphae** arise vertically from vegetative hyphae. In contrast, in yeasts, filamentous growth occurs by continuous extension of hyphal tips. **Slime molds** have complex life cycles, alternating between a protozoan-like motile stage and a fungus-like spore-forming stage. During their feeding stage, slime molds develop into a multinucleate mass known as a **plasmodium** that is characterized by the lack of a cell wall. These are found in soils and feed by slowly engulfing particles of organic matter as they move in an amoeba-like manner.

2.3.4 Structure of the Fungal Cell

2.3.4.1 Cell Surface

Because they are eukaryotes, fungi differ from bacteria in that they are more complex organisms that contain a nucleus and other specialized cell organelles.

Septate hyphae Non-septate hyphae

FIGURE 2.15 Structural types of hyphae that entwine together, resulting in a mycelium.

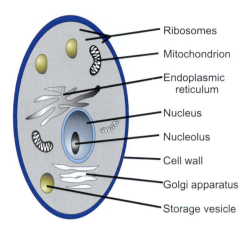

FIGURE 2.16 Structure of a typical fungal cell.

A typical fungal cell is shown in Fig. 2.16. Fungi are characterized by the lack of flagella or other surface appendages. However, fungi can seem to move because as their hyphae grow they extend outward. The outermost region of a fungal cell is the glycocalyx, which is analogous to the slime layer of bacteria. As such, the glycocalyx can be important for protection, surface attachment, and signal reception from the environment. The fungal cell wall is a rigid structure quite distinct from the prokaryotic cell wall (Fig. 2.17). It consists of a thick inner layer of chitin or cellulose and a thinner outer layer of glycoproteins. These differences can be important when community DNA extractions are conducted on soil or vadose zone samples because some lysing agents will not disrupt the fungal cell wall (see Chapter 8.1.3.1). This allows scientists to extract only bacterial DNA from soil. Lining the cell wall is the cytoplasmic membrane, which is similar to prokaryotic membranes except for the addition of sterols. Inside the cell wall and membrane are several

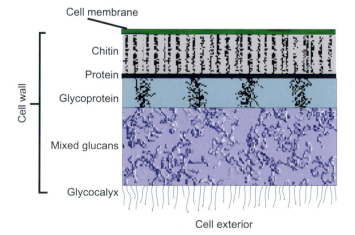

FIGURE 2.17 Structure of a fungal cell wall.

individual membrane-bound organelles, which are characteristic of eukaryotic organisms such as mitochondria, endoplasmic reticulum, Golgi apparatus, and nucleus.

2.3.4.2 Fungal Cell Organelles

The Nucleus

The nucleus is surrounded by a **nuclear envelope,** which consists of two parallel membranes and encloses the **nucleolus.** The nucleolus contains the ribosomal RNA responsible for synthesis of ribosomal subunits. Also contained within the nucleus is a network of fibers known as **chromatin.** Chromatin consists of DNA and protein that together make up the eukaryotic **chromosomes.** When a fungal cell divides through **mitosis,** the chromosomes divide equally into two daughter cells. Fungi possess multiple chromosomes, whereas bacteria have a single chromosome. *Penicillium* spp., for example, contain five.

Endoplasmic Reticulum

The endoplasmic reticulum (ER) is a labyrinth of membrane-bound flat pouches that are differentiated into **rough endoplasmic reticulum (RER)** and **smooth endoplasmic reticulum (SER).** RER originates from the outer membrane of the nuclear envelope and continues as a membranous network throughout the cytoplasm all the way to the cell membrane. Thus, RER allows transportation between the nucleus and the exterior of the cell. It appears rough because of the presence of embedded ribosomes where proteins are synthesized. SER, in contrast, does not contain ribosomes and functions in nutrient processing and synthesis of nonprotein macromolecules such as lipids.

Golgi Apparatus

This organelle consists of a stack of disk-shaped sacs, each of which is membrane bound. The Golgi apparatus results in the formation of proteins that are modified by the addition of polysaccharides and lipids, and ultimately these vesicles are used in packaging materials for export from the cell. A **lysosome** is one type of vesicle that contains digestive enzymes. When the lysosome membrane is intact, these enzymes are inactive. In older cells the membranes leak, releasing hydrolytic enzymes that self-digest and destroy the cell, thereby ensuring the elimination of older or damaged cells.

Mitochondria

Cellular energy is generated in the mitochondria, which contain two sets of membranes. The outer membrane encloses an inner folded membrane, the folds of which are known as **cristae.** The cristae result in a large surface area for the site of cellular respiration, where energy is produced and stored as ATP. The mitochondria also contain circular strands of DNA and 70S ribosomal units that are similar to bacterial ribosomes.

Ribosomes

These are necessary for protein synthesis and consist of 80S units, which in turn consist of 60S and 40S subunits. Ribosomes originate in the nucleolus and are found free in the cytoplasm or associated with the endoplasmic reticulum.

Vacuoles

Vacuoles are membrane-bound vesicles that contain fluids or solid particles that can be stored, digested, or excreted.

2.4 ALGAE

These are a group of photosynthetic organisms that can be macroscopic, as in the case of seaweeds and kelps, or microscopic. The **blue–green algae** are actually classified as bacteria known as the **cyanobacteria.** The cyanobacteria not only are photosynthetic but can also fix atmospheric nitrogen in a free-living state or in a symbiosis with plants known as *Azolla* (Becking, 1978). In terms of their impact in environmental microbiology, the algae are mostly significant in aqueous environments, where they can cause eutrophication. Some algae produce toxins, which can cause death in fish and invertebrates and may cause food poisoning in humans who eat these animals. Large floating communities of microscopic algae are known as **plankton.** Other algal habitats include the surface of soil, rocks, and plants where one finds the **diatoms,** algae that are characterized by the presence of a silicon dioxide cell wall. The capacity of algae to photosynthesize is critical because it forms the basis of aquatic food chains and can also be important in the initial colonization of disturbed terrestrial environments. Overall, the algae are known by common names such as green algae, brown algae, or red algae based on their predominant color. A more technical classification is based on the type of chlorophyll, the type of cell covering, and the nature of their stored foods.

2.4.1 Structure of Algae

A typical algal cell is shown in Fig. 2.18. Algae are aerobic eukaryotic organisms and as such exhibit structural similarities to fungi. Thus, they have a true

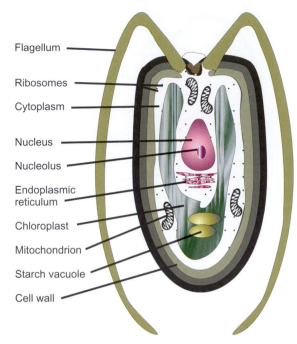

Flagellum
Ribosomes
Cytoplasm
Nucleus
Nucleolus
Endoplasmic reticulum
Chloroplast
Mitochondrion
Starch vacuole
Cell wall

FIGURE 2.18   Structure of a typical algal cell.

nucleus and several membrane-bound cell organelles. The cell walls of algae vary depending on the division. Some algae such as Euglena do not have a cell wall and instead have a thick, flexible membrane known as a **pellicle.** Cell walls of other divisions can consist of cellulose or silicon dioxide in the case of the diatoms.

Most of the cell organelles are structurally and functionally similar to those of the fungi. A notable exception is the **chloroplast.** Chloroplasts are photosynthetic cell organelles that are capable of converting the energy of sunlight into chemical energy through photosynthesis. Chloroplasts are similar to mitochondria in overall structure but are larger and contain specialized pigments. Thus, chloroplasts consist of two membranes, one enclosing the other. The inner membrane is folded into disk-like sacs called **thylakoids** that are arranged in stacks known as **grana.** These structures carry the green pigment chlorophyll and are the site of photosynthesis.

2.5 PROTOZOA

Protozoa are unicellular eukaryotes, meaning that they have characteristic organelles and are large (some are visible with the naked eye). Protozoa characteristically lack cell walls. Although they are single-celled organisms, they are by no means simple in structure, and many diverse forms can be observed among the more than 65,000 named species. Such morphological variability, evolved over hundreds of millions of years, has enabled protozoan adaptation to a wide variety of environments. Throughout this section you will be introduced to the various structures of protozoa, their life cycles and host relationships, as well as the ecological significance of and health consequences associated with their presence in the environment. Because differentiation of protozoa with multiple feeding habits (mixotrophy) is difficult, protozoa are now classified with algae and other unicellular eukaryotes in the same kingdom, Protista. Trophism and structure remain the major identifying factors in Protist classification; however, molecular and immunological methods promise to increase the database, distinction, and knowledge of community structure. Protozoa can be found in nearly all terrestrial and aquatic environments and are thought to play a valuable role in ecological cycles. Many species are able to exist in extreme environments, from polar regions to hot springs and desert soils. In recent years, protozoan pathogens such as *Giardia*, *Cryptosporidium*, and *Microsporidium* have emerged to become the number one concern with regard to safe drinking water.

2.5.1 Protozoan Structure

Protozoa may be free living, capable of growth and reproduction outside any host, or parasitic, meaning that they colonize host cell tissues. Some are opportunists, adapting either a free-living or parasitic existence, as their environment dictates. The size of a protozoan varies from 2 μm, as with the small ciliates or flagellates, to several centimeters, as with amoebae and foraminifera. Therefore, the larger protozoa can be seen with the naked eye, but the aid of a microscope is necessary to observe cellular detail.

In addition, diverse structures have developed to aid in motility and feeding in a plethora of environments. Classifications of protozoa are based mainly on their structural morphology and mechanism of motility (Fig. 2.19). Morphological characterizations are based on colony formation (single existence or in colonies), swimming style (sedentary or motile), external structures (naked, shelled, or scaled), pigmentation (present or absent), and effect on associated organisms (predators or symbionts). Thus, protists are classified into four major groups: (1) flagellates (Mastigophora), (2) amoebae (Sarcodina), (3) sporozoans (Sporozoa), and (4) ciliates (Ciliophora). Advances in molecular analysis have aided in classification based on genetic diversity. Therefore, current morphological classification has resulted in rather great phylogenic diversity between two members of any one group.

A

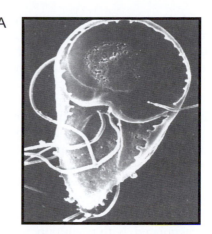

B

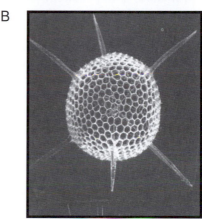

C

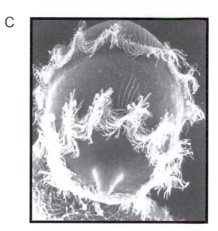

FIGURE 2.19 Basic morphology of protozoa. (A) scanning electron micrograph of a flagellar protozoa, *Giardia*; (B) scanning electron micrograph of testate cilia, *Heliosoma*; (C) electron micrograph of ciliated *Didinium*. Part A reprinted with permission from Cox (1993). Parts B and C reprinted with permission from Sleigh (1989).

Flagellates, as the name implies, use flagella as a locomotive tool. Flagella vary in size but may be as long as 150 μm. Other identifying traits of flagellates include multiplication by binary fission and both heterotrophic and autotrophic feeding mechanisms. Autotrophic flagellates have chloroplasts and are capable of producing their own nutrients via photosynthesis, as in the free-living protozoa *Euglena*. Thus, within the group of flagellates, there are two major divisions: photosynthesizing (Phytomastigophora) and nonphotosynthesizing (Zoomastigophora) protists. *Leishmania* and *Giardia* are two examples of parasitic protozoa with flagellar motility. In contrast, the locomotive tool of amoebae is their pseudopodia. Pseudopodia also aid in ingestion of food materials and provide an extended basis for species classification. The Rhizopoda move by a fluid endoplasm that pushes out pseudopodia. The Actinopoda have a spikelike pseudopodium supported by a microtubule core. These organisms are vacuolated, with pseudopodial structures aiding in movement and feeding, and may be naked or covered in a shell-like outer layer called a **test**. Protozoan shells may be composed of proteinaceous, siliceous, or calcareous substances and have a single chamber or numerous chambers. Active testate amoebae are commonly found in high numbers in soil environments. Some soil protozoa build their shells by excreting substances capable of aggregating soil particles. The family Vahlkampfiidae includes some important free-living soil amoebae, while *Entamoeba* is a well-known parasitic amoeboid.

The Ciliophora, or **ciliates,** are clearly distinct from other protozoan groups. They have hairlike structures, cilia, in an ordered array surrounding the cell and divide by transverse fission. Cilia are similar to flagella, providing locomotion and feeding mechanisms. A well-known ciliate is *Paramecium*, which is generally free living, but some species have adapted to parasitic life cycles. Spore-forming protozoa, Sporozoa, are a wholly parasitic group, living on or in particular hosts, with their own distinctive and complicated life cycles. Symbiotic protozoa may or may not cause death in the host cell; for example, some species have evolved to enable digestion in the gut of domestic livestock. These parasites have no means of locomotion and thus rely on vectors or direct contact, with a susceptible host, to continue their growth and replication. *Toxoplasma, Isospora,* and *Plasmodium* are examples of parasitic, spore-forming protozoa.

2.5.2 Feeding Mechanisms

All protozoa have the equipment necessary to complete essential functions of life, i.e., to obtain energy, metabolize, and reproduce. They are known to exhibit plantlike as well as animal-like nutrition and consume 50–100% of their total volume per hour. The three major feeding categories of protozoa are (1) photoautotrophs, which capture light as an energy source and CO_2 as the principal carbon source; (2) photoheterotrophs, which are phototrophic in energy requirements and must have organic carbon compounds; and (3) chemoheterotrophs, which require chemical energy and organic carbon. Some protozoa are difficult to classify, exemplifying multiple feeding abilities. Most protozoa are heterotrophs, feeding on bacteria, algae, or other protozoa. Cytosomes, or feeding mouths, may be used to draw food into the digestive channels of certain protozoa. Those lacking a mouth tend to engulf their food material and import it into the cell via phagocytosis at the membrane. Some protozoa are also capable of concentrating suspended food particles using their flagella, cilia, or pseudopodia. These motile extensions serve as "filters" to capture food particles from water. Therefore, the motility of the protozoa and the motility of the prey become initiating factors in feeding efficiency. Food particles are also captured, concentrated, and retained directly on sticky protozoan cell wall surfaces until phagocytosis is employed. Processes of diffusion are often utilized, in which particles are taken up from solution osmotically, requiring no motility functions. Pinocytosis is another protozoan feeding mechanism; invaginations remove dissolved nutrients while the vacuole membrane is recycled and transported back to the cytostome in the form of small vesicles. Once in the protozoa, the food is digested by lysosomal enzymes, which requires anywhere from 30 minutes to 16 hours, depending on the protozoan species and the food source.

2.5.3 Cycles of Reproduction

2.5.3.1 Asexual Reproduction

Most protozoa have relatively short life cycles, because the entire cell contributes to reproduction by growth and division into progeny cells. This is considered asexual reproduction and is the typical life cycle of free-living protozoa in the absence of adversity. In asexual reproduction, protozoan cells usually reproduce by binary fission, in which cells increase in size and divide in half to produce daughter cells. Progeny DNA is synthesized and chromosomes are replicated in a series of phases including (1) cell division, (2) cell growth, (3) DNA synthesis, and (4) preparation for the next division (Fig. 2.20). Protozoan binary fission may be simple, as in the naked amoebae which have no definable plane, or quite complex, as in the shell-bearing amoebae requiring replication of their skeletal structures. Binary division may be in a longitudinal or transverse

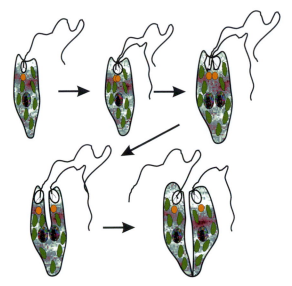

FIGURE 2.20 Schematic representation of asexual reproduction in protozoa: Euglena life-cycle. Simple binary fission in the longitudinal plane.

plane. Some protozoa produce progeny by budding daughter cells. Budding is a form of fission in which the parent cell produces progeny but is only slightly altered in the process. Asexual reproduction is the preferred mechanism of reproduction in free-living protozoa; however, sexual reproduction is often initiated under conditions of adversity, such as aging, starvation, desiccation, or other environmental stress, and is exhibited by all the sporozoa at some stage in their life cycles.

2.5.3.2 Sexual Reproduction

For free-living protozoa, sexual reproduction may be expressed via conjugation, in which genetic material is transferred from one partner to another. Two conjugants of compatible strains come together, exchange pronuclei, and then separate. Eventually, the exconjugants divide by binary fission to produce two identical cells. In other protozoa, male sperm bundles make contact with female eggs and initiate fusion. Possibly as important as increasing numbers, sexual reproduction allows genetic recombinations and reorganization, which can lead to increased adaptability of the mating protozoa.

Parasitic protozoa normally reproduce sexually via complicated and extensive life cycles. Sexual reproduction is a necessary part of their life cycle, often alternating with an asexual phase as in *Cryptosporidium*, which are sporulating, parasitic protozoa with complicated life cycles (Fig. 2.21). The environmental stage of this organism is an oocyst, a metabolically dormant protective phase, often exhibited by parasitic protozoa. The oocyst is approximately 3–5 μm in diameter and stores four sporozoites, each with the capability of infecting a host cell and differentiating into an intracellular structure known as a trophozoite. The trophozoite transforms into a schizont, which produces eight progeny organisms known as merozoites. The merozoites burst from the host cell after this asexual cycle of reproduction and are able to infect neighboring cells. This cycle may continue once more or many times in immunocompromised hosts. Merozoites also differentiate into either microgametocytes (male sex structures) or macrogametocytes (female sex structures) to begin the sexual reproductive stage of the organism. Microgametocytes and macrogametocytes fuse together within the host cell's cytoplasm until fertilization occurs, resulting in a zygote, or oocyst. The oocyst is then shed by the host back into the environment, where it sporulates and repeats the life cycle.

2.5.4 Ecology

Like nearly all aspects of protozoa, their ecological significance in the environment is highly diverse. Free-living protozoa aid in a number of natural processes; contributing to the food chain, microbial population control, and environmental decomposition rates. During feeding, protozoa help to control bacterial biomass in the environment. As vital components of the microbial food web, phototrophic protists (microalgae) constitute a major fraction of the primary productivity in aquatic systems (see Chapter 6); phagotrophic and heterotrophic protozoa act as both predator and prey, aiding in the availability of elements, energy, and nutrients for other members of the microbial community. Researchers have found that in marine environments, ciliates, dinoflagellates, and nanoflagellates often graze 50–100% of daily phytoplankton production. Alternatively phagotrophic protists ingest other heterotrophic protozoa, virus particles, and high-molecular-weight compounds. In addition, parasitic protozoa may cause significant disease or actually help the invaded host, i.e., flagellates, found in the intestines of wood-eating termites, digest cellulose material to an acceptable nutrient for the insect, without which the decomposer would starve to death. This wide range of adaptive traits enables long-term survival and replication of protozoa in a number of different, and often hostile, environments.

2.5.4.1 Survival under Adverse Conditions

A universal requirement of protozoa is water. Water is essential for normal metabolic functioning in all species of protozoa; however, many have adapted to

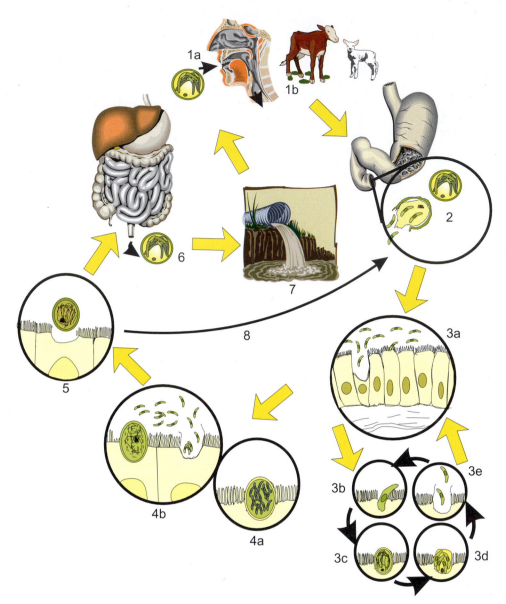

FIGURE 2.21 Diagram of sexual reproduction in protozoa: *Cryptosporidium* life cycle. (1a and 1b) sporulated, infectious oocyst ingested; (2) release of sporozoites due to digestive enzymes; (3a) individual sporozoites infect host cells; (3b) sporozoite embeds itself to the base of the microvilli; (3c) host is induced to engulf parasite; (3d) trophozoite transforms into a schizont, producing eight merozoites; (3e) merozoites burst from host and may 1) reinfect neighboring cells or 2) differentiate into a microgametocyte or a macrogametoocyte; (4a) microgametoocyte produces 12–16 microgametes; (4b) microgametes burst from host cell and fuse with macrogametes until fertilized; (5) the result is a zygote, or oocyst which enters the intestinal tract; (6) oocyst leaves the intestine via feces; (7) it is often then transmitted by water and upon ingestion by a new host repeats the infectious cycle. (8) with *Cryptosporidia* autoinfection also occurs due to *in situ* sporulation.

withstand periods of desiccation by encystment. Encystment is a protective protozoan response to adverse environmental conditions, in which they transform into inactive, nonmotile, environmentally resistant cysts. Although this is not an absolute feature, representatives of each group of protozoa have encystment properties. During encystment, radical changes take place in which cilia, flagella, and feeding vacuoles are lost. Changes also occur in the nuclei and mitochondria. The cell is reduced in volume by as much as 60% and a multilayered cyst wall is formed, after which no respiratory rate can be measured and the protozoa are incapable of immediate growth or reproduction. Although the primary reason for encystment appears to be survival through adverse conditions, another advantage is widespread dispersal to other ecosystems, as cysts can easily be transported via air, water, aquatic birds, or passage through the intestinal tracts of animals. Some soil protozoa, such as *Colpoda*, can remain encysted for years or repeatedly excyst in dewdrops on leaves during the night and encyst during the day when the water evaporates. Encysted protozoa can survive dry heat temperatures up to 120°C, wet heat temperatures up to 50°C, and freezing in liquid nitrogen. Starvation also induces encystment, as do pH changes, changes in the ionic composition of the water, and light. In addition to cyst stages, test structures (shells), as mentioned earlier, have enabled protozoa to survive in harsh ecological niches, such as Antarctic soils, where testate are among the most prevalent protozoan group.

2.5.4.2 Beneficial Roles

With their ability to survive and successfully reproduce in a wide range of common and extreme environments, including salinities exceeding 10%, pH maximums around 10, and even acid mine drainage at pH values less than 3, protozoa have integrated into a number of important ecological roles. One such role is that of recycling nutrients, providing nitrogen, phosphorus and carbon to plants and surrounding microorganisms from metabolic waste products. In this way, protozoa help to sustain bacterial populations and stimulate an increase in decomposition rates. They have been found in eutrophic lakes in concentrations as high as 200 cells/ml constituting up to 60% of the total zooplankton biomass. Because only non–chlorophyll-bearing flagellates occur in soil (photosynthesis is strictly aquatic), grazing, or predation, is the major feeding mechanism of the soil protozoa. Terrestrial protozoa are dominated by the amoebae, heterotrophic flagellates, and ciliates, existing in the desiccation-resistant cyst form or remaining active in water films between soil particles. Total protozoan numbers in soil may range from 10^1 to 10^7 per gram, depending on temperature, moisture, and food source, and the protozoa consume hundreds to thousands of bacterial cells per hour. When predator numbers are very high, the bacterial or fungal community structure may be prevented from increasing, thus decreasing organic decomposition rates. Moderate grazing, however, controls population overgrowth of bacteria and fungi, and although it is detrimental to the individual microorganism, the resultant release of nutrient biomass usually leads to optimal decomposition rates

and increased plant growth. The use of protozoa to treat sewage is another way in which their natural characteristics can be exploited. Protozoa exposed to sewage effluents aid in particulate removal and the reduction of bacteria via flocculation and grazing to provide a higher quality effluent. Sewage treatment plant filters may sustain protozoan populations as high as 5×10^4 cells/ml activated sludge. In this environment, researchers have found 67 species of ciliates alone. A final advantage of the presence of protozoa in the environment is their direct contribution to the food chain as a nutrient source for higher consumers. Common predators of protozoa in soil are nematodes, and aquatic copepods and other crustaceans are known to feed on protozoa in water. In addition, protozoa are known to feed on one another.

Of the more than 65,000 named species of protozoa, approximately 20% are parasitic, requiring a host for completion of their life cycle. Parasitic protozoan relationships may be categorized into three major groupings: (1) commensalism, in which one organism uses another for food, shelter, and transport but causes no ill effects; (2) symbiotic, in which both host and parasite benefit, with the parasite gaining shelter and food while digesting otherwise indigestible food sources, such as cellulose in rumen intestines; and (3) tissue parasitic, in which parasites feed on the tissues of host organisms and may cause mild to serious detrimental illness. Nearly all disease-producing protozoa are tissue parasites.

2.5.5 Disease-Causing Protozoa

Many parasitic protozoa are of obvious public health concern, causing such diseases as malaria, sleeping sickness, Chagas disease, leishmaniasis, giardiasis, and cryptosporidiosis, to name a few (see Chapter 19). They are able to evade host immune responses and have adapted to long-term survival and continual reproduction in their hosts, producing chronic illnesses. A number of flagellated protozoa are intestinal parasites of humans and domestic animals and may be present in the environment in an encysted form. These hearty cysts are known to survive conventional methods of disinfection and thus can be transmitted to their host via a water route. *Entamoeba histolytica* is a parasitic amoeba, with an environmental cyst stage, causing diarrhea and dysentery. *Naegleria* is a free-living amoeba, sometimes present in fresh water, capable of infecting nasal passages of humans and invading brain tissues with potentially fatal results. *Toxoplasma*, a highly nonspecific, invasive protozoan, thought to infect any mammal and all cell types within a given individual, causes blindness and serious illness or death in unborn fetuses. *Cryptosporidium* is responsible for a number of epidemics including the largest U.S. waterborne outbreak to date, in Milwaukee. *Plasmodium* spp. are responsible for the mosquito-borne disease malaria, which affects human populations worldwide. Transmission is most commonly through ingestion or biting arthropods. Many parasitic protozoa enter their host in cyst forms; once they are inside the body, temperature increases to a more desirable level and bile salts act to break down cyst walls, initiating excystation. Parasitic protozoa may colonize the gut, be released into the blood stream, and eventually invade organ tissues or the lymphatic system. Infection in healthy hosts is usually self-limiting and chemotherapeutic agents are effective against some protozoan parasites, but for the immunocompromised, protozoan infections can prove persistent and often fatal. Domestic animals are also at risk of serious illness and death from protozoan infections; e.g., *Histomonas* has a 50–100% kill rate in infected turkeys and *Trichomonas* causes early abortion in 50–100% of infected cows. Control of parasitic protozoa has proved to be problematic, because they tend to be at the smaller end of the protozoan size range, are resistant to disinfection, and have a wide range of transmission routes including water, food, vectors (mosquitoes, ticks, flies, and fleas), and direct person-to-person contact. In addition, a variety of mammals (rodents, dogs, monkeys, cats, apes, humans, cows, and many more) serve as reservoir hosts, aiding in the spread of these parasitic protozoan populations.

QUESTIONS AND PROBLEMS

1. Prior to the advent of antibiotics, bacteriophage were being studied as "antibacterial agents". Debate why this concept may or may not be applicable today.

2. Discuss the advantages and disadvantages associated with microscopic, cultural, and molecular detection of virus populations in the environment.

3. List the mechanisms associated with viruses that aid in their persistence and survivability in the environment.

4. Compare and contrast bacteria and fungi with respect to their outer cell surface and cell wall structure.

5. What is the major implication of the difference between the cell wall structure of gram positive and gram negative bacteria?

6. How do the following organisms generate energy: *Chromatium; Nitrosomonas; Thiobacillus; Bacillus*. Give equations for each energy-producing reaction.

7. Give specific examples of the following kinds of plasmids: resistance plasmids; plant interactive plasmids; degradative plasmids.

8. Compare and contrast the structures of fungi and algae. In what ways are they similar? In what ways are they different?

9. Protozoa are some of the most diverse organisms on the planet. Discuss the specific characteristics of protozoa that have enabled their proliferation under a multitude of adverse conditions.

10. Many of the human protozoan pathogens have a cyst stage. Explain the advantage to the organism and the public health significance of this vital stage in the protozoan life cycle.

References and Recommended Readings

Bakken, L. R. (1997) Culturable and nonculturable bacteria in soil. *In* "Modern Soil Microbiology" (J. D. van Elsas, J. T. Trevors, and E. M. H. Wellington, eds.) Marcel Dekker, New York.

Becking, J. H. (1978) Environmental role of nitrogen fixing blue–green algae and asymbiotic bacteria. *Ecol. Bull.* **26**, 266–281.

Bratbak, G., Heldal, M., Norland, S., and Thingstad, T. F. (1990) Viruses as partners in spring bloom microbial trophodynamics. *Appl. Environ. Microbiol.* **56**, 1400–1405.

Brock, T. D., and Madigan, M. T. (1991) "Biology of Microorganisms." Prentice Hall, Englewood Cliffs.

Carlile, M. J., and Watkinson, S. C. (1994) "The Fungi." Academic Press, San Diego, CA.

Cox, F. E. G. (1993) "Modern Parisitology: A Textbook of Parasitology." Blackwell Scientific Publications, Cambridge, MA.

Davis, B. D., Dulbecco, R., Eisen, H. N., Ginsberg, H. S. (1980) "Microbiology," 3rd ed. Harper & Row, New York.

Despommier, D. D., and Karapelou, J. W. (1987) "Parasite Life Cycles." Springer-Verlag, New York.

Fenner, F., Gibbs, A., and Canberra. (1988) "Portraits of Viruses." S. Karger, Switzerland.

Goyal, S. M., Gerba, C. P., and Bitton, G. (eds.) (1987) "Phage Ecology." John Wiley & Sons, New York.

Hughs, S. S. (1977) "The Virus: A History of the Concept." Neale Watson Academic Publications, New York.

Krawiec, S., and Riley, M. (1990) Organization of the bacterial chromosome. *Microbiol. Rev.* 54, 502–539.

Laybourn-Parry, J. A. (1984) "Functional Biology of Free-Living Protozoa." Croom Helm, Sydney, Australia.

Locke, D. M. (1974) "Viruses: The Smallest Enemy." Crown Publishers, New York.

Nordstrom, K., Molin, S., and Light, J. (1984) "Control of Replication of Bacterial Plasmid: Genetics, Molecular Biology and Physiology of the Plasmid R1 System." *Plasmid* **12**, 71–90.

Neidhardt, F. C., Ingraham, J. L., Schaechter, M.. (1990) "Physiology of the Bacterial Cell: A Molecular Approach." Sinauer Associates, Sunderland, MA.

Paul, E. A., and Clark, F. E. (1989) "Soil Microbiology and Bio-chemistry." Academic Press, San Diego, p. 76.

Paul, J. H., Jiang, S. C., and Rose, J. B. (1991) "Concentration of Viruses and Dissolved DNA from Aquatic Environments by Vortex Flow Filtration. *Appl. Environ. Microbiol.* **57**, 2197–2204.

Pepper, I. L. (1994) Plasmid profiles. *In* "Methods of Soil Analysis. Part 2: Microbiological and Biochemical Properties." Soil Science Society of America, Madison, WI, pp. 635–645.

Radetsky, P. (1991) "The Invisible Invaders: The Story of the Emerging Age of Viruses." Little, Brown, Boston.

Roane, T. M., and Kellogg, S. T. (1996) Characterization of bacterial communities in heavy metal contaminated soils. *Can. J. Microbiol.* **42**, 593–603.

Scott, A. (1987) "Pirates of the Cell." Basil Blackwell, New York.

Silver, S., and Phung, L. T. (1996) Bacterial heavy metal resistance: New surprises. *Annu. Rev. Microbiol.* **50**, 753–789.

Sleigh, M. (1989) "Protozoa and Other Protists." Hodder & Stoughton, New York.

Smit, E., van Elsas, J. D., van Veen, J. A., and de Vos, W. M. (1991) Detection of plasmid transfer from *Pseudomonas fluorescens* to indigenous bacteria in soil by using the bacteriophage OR2F for donor counterselection. *Appl. Environ. Microbiol.* **57**, 3482–3488.

Talaro, K., and Talaro, A. (1993) Procaryotic profiles: The bacteria. *In* "Foundations in Microbiology." Wm. C. Brown Publishers, Dubuque, IA.

Thorn, G. (1997) The fungi in soil. *In* "Modern Soil Microbiology" (J. D. van Elsas, J. T. Trevors, and E. M. H. Wellington, eds.) Marcel Dekker, New York.

White, D. O., and Fenner, F. J. (1994) "Medical Virology," 4th ed. Academic Press, San Diego

CHAPTER

3

Bacterial Growth

RAINA M. MAIER

Bacterial growth is a complex process involving numerous **anabolic** (synthesis of cell constituents and metabolites) and **catabolic** (breakdown of cell constituents and metabolites) reactions. Ultimately, these biosynthetic reactions result in cell division as shown in Fig. 3.1. In a homogeneous rich culture medium, under ideal conditions, a cell can divide in as little as 10 minutes. In contrast, it has been suggested that cell division may occur as slowly as once every 100 years in some subsurface terrestrial environments. Such slow growth is the result of a combination of factors including the fact that most subsurface environments are both nutrient poor and heterogeneous. As a result, cells are likely to be isolated, cannot share nutrients or protection mechanisms, and therefore never achieve a metabolic state that is efficient enough to allow exponential growth.

Most information available concerning the growth of microorganisms is the result of controlled laboratory studies using pure cultures of microorganisms. There are two approaches to the study of growth under such controlled conditions: **batch culture** and **continuous culture.** In a batch culture the growth of a single organism or a group of organisms, called a **consortium,** is evaluated using a defined medium to which a fixed amount of **substrate** (food) is added at the outset. In continuous culture there is a steady influx of growth medium and substrate such that the amount of available substrate always remains the same. Growth under both batch and continuous culture conditions has been well characterized physiologically and also described mathematically. This information has been used to optimize the commercial production of a variety of microbial products includ-

ing antibiotics, vitamins, amino acids, enzymes, yeast, vinegar, and alcoholic beverages. All of these materials are produced in large batches (up to 500,000 liters) also called **large-scale fermentations.**

Although our understanding of growth under controlled conditions is extensive, it is doubtful whether information obtained from laboratory culture studies can be easily extended to an understanding of growth in natural soil or water environments, where the level of complexity is increased tremendously (Fig. 3.2). This complexity arises from a number of factors, including an array of different types of solid surfaces, microenvironments that have altered physical and chemical properties, a limited nutrient status, and consortia of different microorganisms all competing for the same limited nutrient supply (see Chapter 4). So from several different perspectives, the next challenge facing environmental microbiologists is to extend our understanding of microbial growth to natural environments. Such an understanding would facilitate our ability to predict the survival of pathogenic microorganisms in the environment, as well as the movement of environmental contaminants that can be affected by microbial activity (biodegraded or transformed).

3.1 GROWTH IN PURE CULTURE IN A FLASK

Typically, to understand and define the growth of an isolate of a particular microorganism, cells are placed in a flask in which the nutrient supply and environmental conditions are controlled. If the liquid medium supplies all nutrients required for growth and

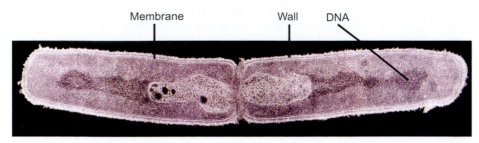

Membrane   Wall   DNA

FIGURE 3.1 Electron micrograph of *Bacillus subtilis,* a gram-positive bacterium, dividing. Magnification, ×31,200. Reprinted with permission from Madigan *et al.* (1997).

environmental parameters are optimal, the increase in numbers or bacterial mass can be measured as a function of time to obtain a growth curve. Several distinct growth phases can be observed within a growth curve (Fig. 3.3). These include the lag phase, the exponential or log phase, the stationary phase, and the death phase. Each of these phases represents a distinct period of growth, and there are associated physiological changes as microorganisms shift from one phase to the next.

3.1.1 The Lag Phase

The first phase observed under batch conditions is the **lag phase**. When an inoculum is placed into fresh medium, growth begins after a period of time called the lag phase. The lag phase is thought to be due to the physiological adaptation of the cell to the culture conditions. This may involve a time requirement for induction of specific messenger RNA (mRNA) and protein synthesis to meet new culture requirements. The lag phase may also be due to low initial densities of organisms that result in dilution of **exoenzymes** (enzymes released from the cell) and of nutrients that leak from growing cells. Normally, such materials are shared by cells in close proximity. But when cell density is low, these materials are diluted and not as eas-

ily taken up. As a result, initial generation times may be slowed until a sufficient cell density, approximately 10^6 cells/ml, is reached.

The lag phase usually lasts from minutes to several hours. The length of the lag phase can be controlled to some extent because it is dependent on the type of medium as well as on the initial inoculum size. For example, if an inoculum is taken from an exponential phase culture in trypticase soy broth (TSB) and is placed into fresh TSB medium at a concentration of 10^6 cells/ml under the same growth conditions (temperature, shaking speed), there will be no noticeable lag phase. However, if the inoculum is taken from a stationary phase culture, there will be a lag phase as the stationary phase cells adjust to the new conditions and shift physiologically from stationary phase cells to exponential phase cells. Similarly, if the inoculum is placed into a medium other than TSB, for example, a mineral salts medium with glucose as the sole carbon source, a lag phase will be observed while the cells reorganize and shift physiologically to synthesize the appropriate enzymes for glucose catabolism. Finally, if the inoculum size is small, for example 10^4 cells/ml, a lag phase will be observed until the population reaches approximately 10^6 cells/ml. This is illustrated in Fig. 3.4, which compares the degradation of phenanthrene in cultures inoculated with 10^7 and with 10^4 colony-forming units (CFU) ml. Although the degradation rate achieved is similar in both cases (compare the slope of each curve), the lag phase was 1.5 days when a low inoculum size was used (10^4 CFU/ml) in contrast to only 0.5 day when the higher inoculum was used (10^7 CFU/ml).

3.1.2 The Exponential Phase

The second phase of growth observed in a batch system is the **exponential phase.** The exponential phase is characterized by a period of cell division when the rate of increase of cells in the culture is proportional to the number of cells present at any particular time. There are several ways in which this concept can be expressed both theoretically and mathematically. One way is to imagine that during exponential

vs.

FIGURE 3.2 Compare the complexity of growth in a flask and growth in a soil environment. Although we understand growth in a flask quite well, we still cannot always predict growth in the environment.

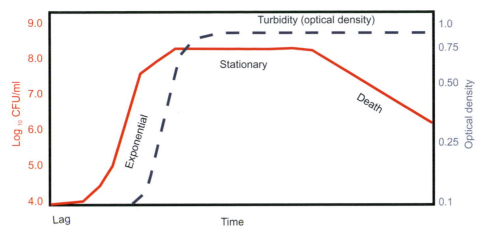

FIGURE 3.3 A typical growth curve for a bacterial population. Compare the difference in the shape of the curves in the death phase (colony-forming units versus optical density).

growth the number of cells increases in the geometric progression 2^0, 2^1, 2^2, 2^4 until, after n divisions, the number of cells is 2^n (Fig. 3.5). This can be expressed in a quantitative manner, for example; if the initial cell number is X_0, the number of cells after n doublings is $2^n X_0$ (see Example Calculation 1). As can be seen from this example, if one starts with a low number of cells exponential growth does not initially produce large numbers of new cells. However, as cells accumulate after several generations, the number of new cells with each division begins to increase explosively.

In the example just given, X_0 was used to represent cell number. However, X_0 can also be used to represent cell mass, which is often more convenient to measure than cell number (see Chapters 10 and 11). Whether one expresses X_0 in terms of cell number or in terms of cell mass, one can mathematically describe cell growth during the exponential phase using the following equation:

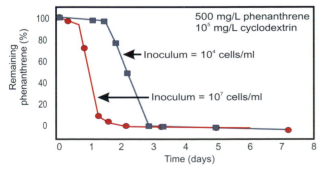

FIGURE 3.4 Effect of inoculum size on the lag phase during degradation of a polyaromatic hydrocarbon, phenanthrene. Because phenanthrene is only slightly soluble in water and is therefore not readily available for cell uptake and degradation, a solubilizing agent called cyclodextrin was added to the system. The microbes in this study were not able to utilize cyclodextrin as a source of carbon or energy. (Courtesy E. M. Marlowe.)

$$\frac{dX}{dt} = \mu X$$

where X is the number or mass of cells (mass/volume), t is time, and μ is the specific growth rate constant (1/time). The time it takes for a cell division to occur is called the **generation time** or the **doubling time**. This equation can be used to calculate the generation time as well as the **specific growth rate** using data generated from a growth curve such as that shown in Fig. 3.3.

The generation time for a microorganism is calculated from the linear portion of a semilog plot of growth versus time. The mathematical expression for this portion of the growth curve is given by Eq. 3.1, which can be rearranged and solved as shown below (Eq. 3.2 to 3.6) to determine the generation time (see Example Calculation 2):

$$\frac{dX}{dt} = \mu X \qquad \text{(Eq. 3.1)}$$

rearrange:

$$\frac{dX}{X} = \mu dt \qquad \text{(Eq. 3.2)}$$

integrate:

$$\int_{X_0}^{X} \frac{dX}{X} = \mu \int_{o}^{t} dt \qquad \text{(Eq. 3.3)}$$

$$\ln X = \mu t + \ln X_0 \quad \text{or} \quad X = X_0 e^{\mu t} \qquad \text{(Eq. 3.4)}$$

for X to be doubled:

$$\frac{X}{X_0} = 2 \qquad \text{(Eq. 3.5)}$$

therefore:

$$2 = e^{\mu t} \qquad \text{(Eq. 3.6)}$$

where t = generation time.

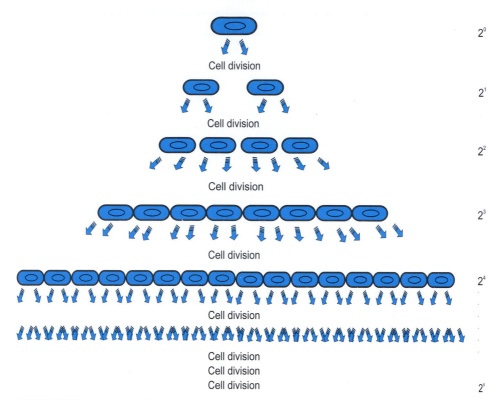

FIGURE 3.5 Exponential cell division. Each cell division results in a doubling of the cell number. At low cell numbers the increase is not very large, however after a few generations, cell numbers increase explosively.

Example Calculation 1

Growth Rate

Problem: **If one starts with 10,000 (10^4) cells in a culture that has a generation time of 2 hours, how many cells will be in the culture after 4 hours, 24 hours, and 48 hours?**

From the equation $X = 2^n X_0$, where X_0 is the initial number of cells, n is the number of generations, and X is the number of cells after n generations:

after 4 hours: n = 4 hours/2 hours per generation
$\qquad\qquad$ = 2 generations:
$\qquad\qquad X = 2^2(10^4) = 4.0 \times 10^4$ cells

after 24 hours: n = 12 generations
$\qquad\qquad X = 2^{12}(10^4) = 4.1 \times 10^7$ cells

after 48 hours: n = 24 generations
$\qquad\qquad X = 2^{24}(10^4) = 1.7 \times 10^{11}$.

This represents an increase of less than one order of magnitude for the 4 hour culture, four orders of magnitude for the 24 hour culture, and seven orders of magnitude for the 48 hour culture!

Example Calculation 2

Specific Growth Rate

Problem: **The following data were collected using a culture of *Pseudomonas* during growth in a minimal medium containing salicylate as a sole source of carbon and energy. Using these data, calculate the specific growth rate for the exponential phase.**

Time (hours)	Culturable cell count (cfu/ml)
0	2.2×10^4
1.2	2.2×10^4
2.2	2.2×10^4
3.2	2.2×10^4
4.2	1.5×10^5
5.2	1.5×10^7
6.2	8.1×10^7
7.2	3.2×10^8
8.2	5.5×10^8
9.2	8.0×10^8
10.2	1.0×10^9
11.2	1.7×10^9
12.2	2.1×10^9
13.2	2.3×10^9
17.2	2.5×10^9
32.2	6.3×10^9
60.2	3.8×10^9
84.2	1.9×10^9

The times to be used to determine the specific growth rate can be chosen by visual examination of a semilog arithmic plot of the data (shown below). Examination of this graph shows that the exponential phase is from approximately 4.2 to 8.2 hours. Using Eq. 3.4, which describes the exponential phase of this graph, one can determine the specific growth rate for this *Pseudomonas*. [Note that Eq. 3.4 describes a line the slope of which is μ, the specific growth rate.] From the data given, the slope of the graph from time 4.2 to 8.2 hours is:

$$\mu = (\ln 5.5 \times 10^8 - \ln 1.5 \times 10^5)/(8.2 - 4.2) = 2.0 \text{ hour}^{-1}$$

It should be noted that the specific growth rate and generation time calculated for growth of the *Pseudomonas* on salicylate are valid only under the experimental conditions used. For example, if the experiment were performed at a higher temperature, one would expect the specific growth rate to increase. At a lower temperature, the specific growth rate would be expected to decrease.

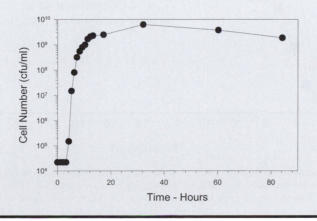

3.1.3 The Stationary Phase

The third phase of growth is the stationary phase. The stationary phase in a batch culture can be defined as a state of no net growth, which can be expressed by the following equation:

$$\frac{dX}{dt} = 0 \qquad \text{(Eq. 3.7)}$$

Although there is no net growth in stationary phase, cells still grow and divide. Growth is simply balanced by an equal number of cells dying. There are several reasons that a batch culture may reach stationary phase. One common reason is that the carbon and energy source or an essential nutrient becomes completely used up. When a carbon source is used up it does not necessarily mean that all growth stops. This is because dying cells can lyse and provide a source of nutrients. Growth on dead cells is called **endogenous metabolism.** Endogenous metabolism occurs throughout the growth cycle, but can be best observed during stationary phase when growth is measured in terms of oxygen uptake or evolution of carbon dioxide. Thus, in many growth curves such as that shown in Fig. 3.6, the stationary phase actually shows a small amount of growth. Again, this growth occurs after the substrate

has been utilized and reflects the use of dead cells as a source of carbon and energy. A second reason that stationary phase may be observed is that waste products build up to a point where they begin to inhibit cell growth or are toxic to cells. This generally occurs only in cultures with high cell density. Regardless of the reason why cells enter stationary phase, growth in the stationary phase is unbalanced because it is easier for the cells to synthesize some components than others. As some components become more and more limiting, cells will still keep growing and dividing as long as possible. As a result of this nutrient stress, stationary phase cells are generally smaller and rounder than cells in the exponential phase (see Chapter 2.2.2).

3.1.4 The Death Phase

The final phase of the growth curve is the **death phase,** which is characterized by a net loss of culturable cells. Even in the death phase there may be individual cells that are metabolizing and dividing, but more viable cells are lost than are gained so there is a net loss of viable cells. The death phase is often exponential, although the rate of cell death is usually slower than the rate of growth during the exponential

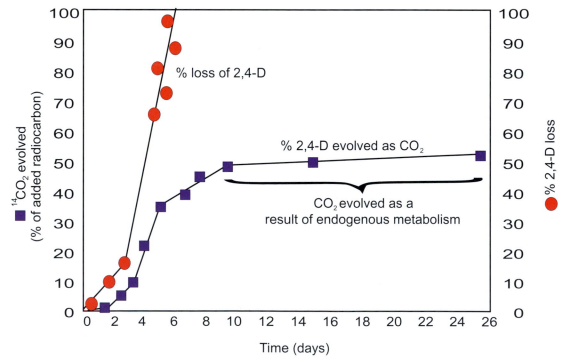

FIGURE 3.6 Mineralization of the broadleaf herbicide 2,4-dichlorophenoxy acetic acid (2,4-D) in a soil slurry under batch conditions. Note that the 2,4-D is completely utilized after 6 days but the CO_2 evolved continues to rise slowly. This is a result of endogenous metabolism. (From Estrella *et al.,* 1993.)

phase. The death phase can be described by the following equation:

$$\frac{dX}{dt} = -k_dX \qquad \text{(Eq. 3.8)}$$

where k_d is the specific death rate.

It should be noted that the way in which cell growth is measured can influence the shape of the growth curve. For example, if growth is measured by optical density instead of by plate counts (compare the two curves in Fig. 3.3), the onset of the death phase is not readily apparent. Similarly, if one examines the growth curve measured in terms of carbon dioxide evolution shown in Fig. 3.6, again it is not possible to discern the death phase. Still, these are commonly used approaches to measurement of growth because normally the growth phases of most interest to environmental microbiologists are the lag phase, the exponential phase, and the time to onset of the stationary phase.

3.1.5 Effect of Substrate Concentration on Growth

So far we have discussed each of the growth phases and have shown that each phase can be described mathematically (see Eq. 3.1, 3.7, and 3.8). One can also write equations to allow description of the entire growth curve. Such equations become increasingly complex. For example, one of the first and simplest descriptions is the **Monod equation**, which was developed by Jacques Monod in the 1940s:

$$\mu = \frac{\mu_{max} S}{K_s + S} \qquad \text{(Eq. 3.9)}$$

where; μ is the specific growth rate (1/time), μ_{max} is the maximum specific growth rate (1/time) for the culture, S is the substrate concentration (mass/volume), and K_s is the half-saturation constant (mass/volume)

Equation 3.9 was developed from a series of experiments performed by Monod. The results of these experiments showed that at low substrate concentrations, growth rate becomes a function of the substrate concentration [note that Eqs. 3.1 to 3.8 are independent of substrate concentration]. Thus, Eq. 3.9 describes the relationship between the specific growth rate, and the substrate concentration. There are two constants in this equation, μ_{max}, the maximum specific growth rate, and K_s, the half-saturation constant, which is defined as the substrate concentration at which growth occurs at one half the value of μ_{max}. Both μ_{max} and K_s reflect intrinsic physiological properties of a particular type of microorganism. They are also dependent on the substrate being utilized and on the temperature of

---

Information Box 1

Both μ_{max} and K_s are constants that reflect:

• The intrinsic properties of the degrading microorganism
• The limiting substrate
• The temperature of growth

The following table gives some representative values of μ_{max} and K_s for growth of different organisms on a variety of substrates.

Organism	Growth temperature	Limiting nutrient	μ_{max} (hr^{-1})	K_s (mg/l)
Escherichia coli	37°C	Glucose	0.8–1.4	2-4
Escherichia coli	37°C	Glycerol	0.87	2
Escherichia coli	37°C	Lactose	0.8	20
Saccharomyces cerevisiae	30°C	Glucose	0.5–0.6	25
Candida tropicalis	30°C	Glucose	0.5	25–75
Candida sp.	0	Oxygen	0.5	0.045–0.45
Candida sp.		Hexadecane	0.5	
Klebsiella aerogenes		Glycerol	0.85	9
Aerobacter aerogenes		Glucose	1.22	1–10

Adapted from Blanch and Clark (1996).

growth (see Information Box 1). Monod assumed in writing Eq. 3.9 that no nutrients other than the substrate are limiting and that no toxic by-products of metabolism build up.

As shown in Eq. 3.10, the Monod equation can be expressed in terms of cell number or cell mass (X) by equating it with Eq. 3.1:

$$\frac{dX}{dt} = \frac{\mu_{max}\, S\, X}{K_s + S} \qquad \text{(Eq. 3.10)}$$

The Monod equation has two limiting cases as shown in Fig. 3.7. The first case is at high substrate concentration where S >> Ks. In this case, as shown in Eq. 3.11, the specific growth rate μ is essentially equal to μ_{max}. This simplifies the equation and the resulting relationship is zero order or independent of substrate concentration:

$$\text{for } S \gg K_s: \quad \frac{dX}{dt} = \mu_{max}\, X \qquad \text{(Eq. 3.11)}$$

Under these conditions, growth will occur at the maximum growth rate. There are relatively few instances in which ideal growth as described by Eq. 3.11 can occur. One such instance is under the initial conditions found in pure culture in a batch flask when substrate and nutrient levels are high. Another is under continuous culture conditions, which are discussed further in Section 3.2. It must be emphasized that this type of growth is unlikely to be found under natural conditions in a soil or water environment, where either substrate or other nutrients are commonly limiting.

The second limiting case occurs at low substrate concentrations where $S \ll K_s$ as shown in Eq. 3.12. In this case there is a first order dependence on substrate concentration (Fig. 3.7):

$$\text{for } S \ll K_s: \quad \frac{dX}{dt} = \frac{\mu_{max}\, SX}{K_s} \qquad \text{(Eq. 3.12)}$$

As shown in Eq. 3.12, when the substrate concentration is low, growth (dX/dt) is dependent on the substrate concentration. Since the substrate concentration is in the numerator, as the substrate concentration decreases, the rate of growth will also decrease. This type of growth is typically found in batch flask systems at the end of the growth curve as the substrate is nearly all consumed. This is also the type of growth that would be more typically expected under conditions in a natural environment where substrate and nutrients are limiting.

The Monod equation can also be expressed as a function of substrate utilization given that growth is

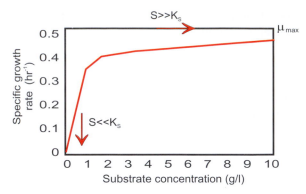

FIGURE 3.7 Dependence of the specific growth rate, μ, on the substrate concentration. The maximal growth rate, $\mu_m = 0.5 \text{ hr}^{-1}$ and $K_s = 0.5 \text{ g/L}$. Note that μ approaches μ_{max} when $S \gg K_s$ and becomes independent of substrate concentration. When $S \ll K_s$, the specific growth rate is very sensitive to the substrate concentration, exhibiting a first-order dependence.

related to substrate utilization by a constant called the **cell yield** (Eq. 3.13):

$$\frac{dS}{dt} = -\frac{1}{Y}\frac{dX}{dt} \qquad \text{(Eq. 3.13)}$$

where Y is the cell yield (mass/mass). The cell yield coefficient is defined as the unit amount of cell mass produced per unit amount of substrate consumed. Thus, the more efficiently a substrate is degraded, the higher the value of the cell yield coefficient (see Section 3.3 for more detail). The cell yield coefficient is dependent on both the structure of the substrate being utilized and the intrinsic physiological properties of the degrading microorganism. As shown in Eq. 3.14, Eqs. 3.10 and 3.13 can be combined to express microbial growth in terms of substrate disappearance:

$$\frac{dS}{dt} = -\frac{1}{Y}\frac{\mu_{max}\, SX}{K_s + S} \qquad \text{(Eq. 3.14)}$$

Figure 3.8 shows a set of growth curves constructed from a fixed set of constants. The growth data used to generate this figure were collected by determining protein as a measure of the increase in cell growth (see Chapter 11). These data were then used to estimate the growth constants μ_{max}, K_s, and Y. Both y and μ_{max} were estimated directly from the data. K_s was estimated using a mathematical model that performs a nonlinear regression analysis of the simultaneous solutions to the Monod equations for cell mass (Eq. 3.10) and substrate (Eq. 3.13). This set of constants was then used to model or simulate growth curves that express growth in terms of CO_2 evolution and substrate disappearance. Such models are useful because they can help to 1) estimate growth constants such as K_s that are difficult to determine experimentally; and 2) quickly

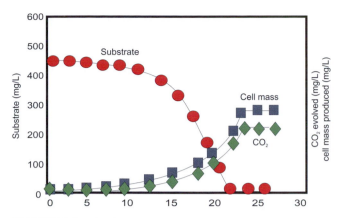

FIGURE 3.8 This figure shows the same growth curve expressed three different ways; in terms of substrate loss, in terms of CO_2 evolution, and in terms of increasing cell mass. The parameters used to generate these three curves were: $\mu_m = 0.29$ hr^{-1}, $K_s = 10$ mg/l, $Y = 0.5$, initial substrate concentration = 500 mg/l, and initial cell mass = 1 mg/l.

understand how changes in any of the experimental parameters affect growth without performing a long and tedious set of experiments.

3.2 CONTINUOUS CULTURE

Thus far, we have focused on theoretical and mathematical descriptions of batch culture growth which is currently of great economic importance in terms of the production of a wide variety of microbial products. In contrast to batch culture, continuous culture is a system that is designed for long-term operation.

Continuous culture can be operated over the long term because it is an open system (Fig. 3.9) with a continuous feed of influent solution that contains nutrients and substrate, as well as a continuous drain of effluent solution that contains cells, metabolites, waste products, and any unused nutrients and substrate. The vessel that is used as a growth container in continuous culture is called a **bioreactor** or a **chemostat.** In a chemostat one can control the flow rate, maintain a constant substrate concentration, as well as provide continuous control of pH, temperature, and oxygen levels. As will be discussed further, this allows control of the rate of growth which can be used to optimize the production of specific microbial products. For example, **primary metabolites** or growth-associated products, such as ethanol, are produced at high flow or dilution rates which stimulate cell growth. In contrast, a **secondary metabolite** or non-growth-associated product such as an antibiotic is produced at low flow or dilution rates which maintains high cell numbers.

Dilution rate and influent substrate concentration are the two parameters controlled in a chemostat to study microbial growth or to optimize metabolite production. The dynamics of these two parameters are shown in Fig. 3.10. By controlling the dilution rate, one can control the growth rate (μ) in the chemostat, represented in this graph as doubling time (recall that during exponential phase the growth rate is proportional to the number of cells present). By controlling the influent substrate concentration, one can control the number of cells produced or the cell yield in the

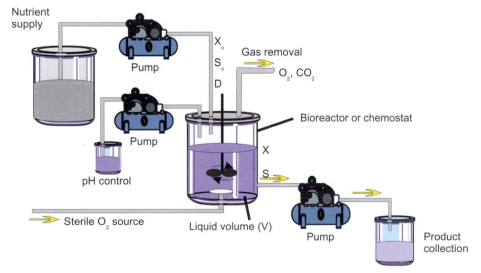

FIGURE 3.9 Schematic representation of a continuously stirred bioreactor. Indicated are some of the variables used in modeling bioreactor systems. X_0 is the dry cell weight, S_0 is the substrate concentration, and D is the flow rate of nutrients into the vessel.

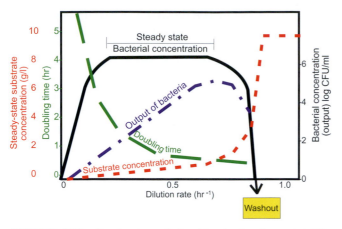

FIGURE 3.10 Steady-state relationships in the chemostat. The dilution rate is determined from the flow rate and the volume of the culture vessel. Thus, with a vessel of 1000 ml and a flow rate through the vessel of 500 ml/hr, the dilution rate would be 0.5 hr^{-1}. Note that at high dilution rates, growth cannot balance dilution and the population washes out. Thus, the substrate concentration rises to that in the medium reservoir (because there are no bacteria to use the inflowing substrate). However, throughout most of the range of dilution rates shown, the population density remains constant and the substrate concentration remains at a very low value (that is, steady state). Note that although the population density remains constant, the growth rate (doubling time) varies over a wide range. Thus, the experimenter can obtain populations with widely varying growth rates without affecting population density. Adapted with permission from Madigan *et al.* (1997).

chemostat (the number of cells produced will be directly proportional to the amount of substrate provided). Because the growth rate and the cell number can be controlled independently, chemostats have been an important tool in studying the physiology of microbial growth and also in the long-term development of cultures and consortia that are acclimated to organic contaminants that are toxic and difficult to degrade. Chemostats can also produce microbial products more efficiently than batch fermentations. This is because a chemostat can essentially hold a culture in the exponential phase of growth for extended periods of time. Despite these advantages, chemostats are not yet widely used to produce commercial products because it is often difficult to maintain sterile conditions over a long period of time.

In a chemostat, the growth medium undergoes constant dilution with respect to cells due to the influx of nutrient solution (Fig. 3.9). The combination of growth and dilution within the chemostat will ultimately determine growth. Thus, in a chemostat, the change in biomass with time is:

$$\frac{dX}{dt} = \mu X - DX \qquad \text{(Eq. 3.15)}$$

where X is the cell mass (mass/volume), μ is the specific growth rate (1/time), and D is the dilution rate (1/time).

Examination of Eq. 3.15 shows that a steady state (no increase or decrease in biomass) will be reached when $\mu = D$. If $\mu > D$, the utilization of substrate will exceed the supply of substrate, causing the growth rate to slow until it is equal to the dilution rate. If $\mu < D$, the amount of substrate added will exceed the amount utilized. Therefore the growth rate will increase until it is equal to the dilution rate. In either case, given time, a steady state will be established where

$$\mu = D \qquad \text{(Eq. 3.16)}$$

Such a steady state can be achieved and maintained as long as the dilution rate does not exceed a critical rate, D_c. The critical dilution rate can be determined by combining Eqs. 3.9 and 3.16:

$$D_c = \mu_{\max}\left(\frac{S}{K_s + S}\right) \qquad \text{(Eq. 3.17)}$$

Looking at Eq. 3.17, it can be seen that the operation efficiency of a chemostat can be optimized under conditions in which $S >> K_s$, and therefore $D_c \approx \mu_{\max}$. But it must be remembered that when a chemostat is operating at D_c, if the dilution rate is increased further, the growth rate will not be able to increase (since it is already at $\mu_{\max}$) to offset the increase in dilution rate. The result will be washing out of cells and a decline in the operating efficiency of the chemostat. Thus, D_c is an important parameter because if the chemostat is run at dilution rates less than D_c, operation efficiency is not optimized, whereas if dilution rates exceed D_c, washout of cells will occur as shown in Fig. 3.10.

3.3 GROWTH UNDER AEROBIC CONDITIONS

Under aerobic conditions, microorganisms utilize substrate by a process known as aerobic respiration (see Table 2.6). The complete oxidation of a substrate under aerobic conditions is represented by the mass balance equation:

$$(C_6H_{12}O_6) + 6(O_2) \rightarrow 6(CO_2) + 6(H_2O) \qquad \text{(Eq. 3.18)}$$
substrate   oxygen   carbon dioxide   water

In Eq. 3.18, the substrate is a carbohydrate such as glucose, which can be represented by the formula $C_6H_{12}O_6$. Oxidation of glucose by microorganisms is more complex than shown in this equation because some of the substrate carbon is utilized to build new cell mass and is therefore not completely oxidized. Thus, aerobic microbial oxidation of glucose can be more completely described by the following, slightly more complex, mass balance equation:

$$a(\text{C}_6\text{H}_{12}\text{O}_6) + b(\text{NH}_3) + c(\text{O}_2) \rightarrow d(\text{C}_5\text{H}_7\text{NO}_2)$$

substrate   nitrogen   oxygen   cell mass
source

$$+ \ e(\text{CO}_2) + f(\text{H}_2\text{O}) \qquad \text{(Eq. 3.19)}$$

carbon dioxide   water

where a, b, c, d, e, and f represent mole numbers.

It should be emphasized that the degradation process is the same whether the substrate is readily utilized (glucose) or only slowly utilized as in the case of a contaminant such as benzene. Equation 3.19 differs from Eq. 3.18 in two ways: it represents the production of new cell mass, estimated by the formula $\text{C}_5\text{H}_7\text{NO}_2$, and in order to balance the equation, it has a nitrogen source on the reactant side, shown here as ammonia (NH_3).

In most cases an increase in cell mass reflects an increase in the number of cells, in which case one can say that the cells are metabolizing substrate under **growth conditions.** However, in some cases, when the concentration of substrate or some other nutrient is limiting, utilization of the substrate occurs without production of new cells. In this case the energy from substrate utilization is used to meet the maintenance requirements of the cell under **nongrowth conditions.** The level of energy required to maintain a cell is called the **maintenance energy** (Niedhardt *et al.*, 1990).

Under either growth or nongrowth conditions, the amount of energy obtained by a microorganism through the oxidation of a substrate is reflected in the amount of cell mass produced, or the cell yield (Y). As discussed in Section 3.2, the cell yield coefficient is defined as the unit amount of cell mass produced per unit amount of substrate consumed. Although the cell yield is a constant, the value of the cell yield is dependent on the substrate being utilized. However, we can often use ballpark estimates of cell yield values. For example, it is generally assumed that approximately 50% of the carbon in a molecule of sugar or organic acid will be used to build new cell mass and 50% will be evolved as CO_2. It is important to remember that this may not be the case for biodegradation of common pollutants because of their unusual chemistry. Examples of cell yield estimates for various substrates that have been reported in the literature are 0.4 for a carbohydrate, 0.05 for pentachlorophenol (very low), and 1.49 for octadecane (very high) (Fig. 3.11). In other words, degradation of some substrates yields more energy for production of cell mass than degradation of others. Why are there such differences in cell yield for the three substrates? As microbes have evolved, standard catabolic pathways have developed for common carbohydrate- and protein-containing substrates. For these types of substrates approximately 50% of the carbon is used to build new cell mass. This translates into a cell yield of approximately 0.4 for a sugar such as glucose (see Example Calculation 3). However, since industrialization began in the late 1800s, many new molecules have been manufactured for which there are no standard catabolic pathways. Pentachlorophenol is an example of such a molecule. This material has been commercially produced since 1936 and is one of the major chemicals used to treat wood and utility poles. To utilize a molecule like pentachlorophenol, which appeared in the environment relatively recently

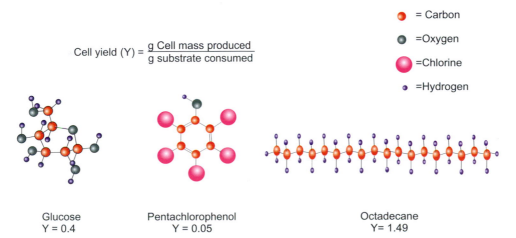

$$\text{Cell yield (Y)} = \frac{\text{g Cell mass produced}}{\text{g substrate consumed}}$$

● = Carbon
● = Oxygen
● = Chlorine
● = Hydrogen

Glucose   Pentachlorophenol   Octadecane
Y = 0.4   Y = 0.05   Y= 1.49

FIGURE 3.11 Cell yield values for various substrates. Note that the cell yield depends on the structure of the substrate.

on an evolutionary scale, a microbe must alter the chemical structure to allow use of standard catabolic pathways. For pentachlorophenol, which has five carbon–chlorine bonds, this means that a microbe must expend a great deal of energy to break the strong carbon–halogen bonds before the substrate can be metabolized to produce energy. Because so much energy is required to remove the chlorines from a highly halogenated compound such as penta-chlorophenol, relatively little energy is left to build new cell mass. This results in a very low cell yield value.

In contrast, why is the cell yield so high for a hydrocarbon such as octadecane? Octadecane is a hydrocarbon typical of those found in petroleum products (see Chapter 16). Because petroleum is an ancient mixture of molecules formed on early earth, standard catabolic pathways exist for most petroleum components, including octadecane. The cell yield value for growth on octadecane is high because octadecane is a saturated molecule (the molecule contains no oxygen, only carbon-hydrogen bonds). Such a highly reduced hydrocarbon stores more energy than a molecule that is partially oxidized such as glucose (glucose contains six oxygen molecules). This energy is released during metabolism, allowing the microbe to obtain more energy from the degradation of octadecane than from the degradation of glucose. This in turn is reflected in a higher cell yield value.

Example Calculation 3

Problem: **A bacterial culture is grown using glucose as the sole source of carbon and energy. The cell yield value is determined by dry weight analysis to be 0.4 (in other words, 0.4 g cell mass was produced per 1 g glucose utilized). What percentage of the substrate (glucose) carbon will be found as cell mass and as CO_2?**

$$a(C_6H_{12}O_6) + b(NH_3) + c(O_2) \rightarrow d(C_5H_7NO_2) + e(CO_2) + f(H_2O) \qquad \text{(Eq. 3.19)}$$

Assume that you start with 1 mole of carbohydrate ($C_6H_{12}O_6$, molecular weight = 180 g/mol), since $g = 1$:

$$\text{(substrate mass) (cell yield)} = \text{cell mass produced}$$
$$(180 \text{ g}) \qquad (0.4) \quad = \qquad 72 \text{ g}$$

From Eq. 3.19, the cell mass can be estimated as $C_5H_7NO_2$, (molecular weight = 113 g/mol), allowing calculation of d:

$$d = \frac{72 \text{ g cell mass}}{113 \text{ g/mol cell mass}} = 0.64 \text{ mol cell mass}$$

In terms of carbon,

for cell mass: (0.64 mol cell mass)(5 mol C/mol cell mass)(12 g/mol C) = 38.4 g carbon

for substrate: (1 mol substrate)(6 mol C/mol sub-strate)(12 g/mol C) = 72 g carbon

The percentage of substrate carbon found in cell mass is

$$\frac{38.4 \text{ g carbon}}{72 \text{ g carbon}} \quad (100) = 53\%$$

and by difference, 47% of the carbon is released as CO_2.

Question

Calculate the carbon found as cell mass and CO_2 for a microorganism that grows on octadecane ($C_{18}H_{36}$), $Y = 1.49$, and on pentachlorophenol (C_6HOCl_5), $Y = 0.05$.

Answer

For octadecane 93% of the substrate carbon is found in cell mass and 7% is evolved as CO_2.
For pentachlorophenol, 10% of the substrate carbon is found in cell mass and 90% is evolved as CO_2.

3.4 GROWTH UNDER ANAEROBIC CONDITIONS

The amount of oxygen in the atmosphere (21%) ensures aerobic degradation for the overwhelming proportion of the organic matter produced annually. In the absence of oxygen, organic substrates can be mineralized to carbon dioxide by fermentation or by anaerobic respiration, although these are less efficient processes than aerobic respiration (see Chapter 2, Table 2.6). In general, anaerobic degradation is restricted to niches such as sediments, isolated water bodies within lakes and oceans, and microenvironments in soils. Anaerobic degradation requires alternative electron acceptors, either an organic compound for fermentation or one of a series of inorganic electron acceptors that can be used for anaerobic respiration (Zehnder and Stumm, 1988). In anaerobic respiration, the terminal electron acceptor used depends on availability, and follows a sequence that corresponds to the electron affinity of the electron acceptors. Examples of alternative electron acceptors in order of decreasing electron affinity are nitrate (nitrate-reducing conditions), manganese (manganese-reducing conditions), iron (iron-reducing conditions), sulfate (sulfate-reducing conditions), and carbonate (methanogenic conditions) (see Table 3.1). In addition to these electron acceptors, certain heavy metals, such as uranium IV, can act as terminal electron acceptors (Lovely, 1993). Currently, the information base for anaerobic biodegradation is much smaller than that for aerobic degradation (Grbić-Galić, 1990).

Often, under anaerobic conditions, organic compounds are degraded by an interactive group or consortium of microorganisms. Individuals within the consortium each carry out different, specialized reactions that together lead to complete mineralization of the compound (Suflita and Sewell, 1991). The final step of anaerobic degradation is methanogenesis, which occurs when other inorganic electron acceptors such as nitrate and sulfate are exhausted. Methanogenesis results in the production of methane and is the most important type of metabolism in anoxic freshwater lake sediments. Methanogenesis is also important in anaerobic treatment of sewage sludge, in which the supply of nitrate or sulfate is very small compared with the input of organic substrate. In this case, even though the concentrations of nitrate and sulfate are low, they are of basic importance for the establishment and maintenance of a sufficiently low electron potential that allows proliferation of the complex methanogenic microbial community.

Mass balance equations very similar to that for aerobic respiration can be written for anaerobic respiration. For example, the following equation can be used to describe the transformation of organic matter into methane (CH_4) and CO_2:

$$C_nH_aO_b + \left(n - \frac{a}{2} - \frac{b}{4}\right) H_2O \rightarrow$$

$$\left(\frac{n}{2} - \frac{a}{8} + \frac{b}{4}\right) CO_2 + \left(\frac{n}{2} + \frac{a}{8} - \frac{b}{4}\right) CH_4$$

$$(Eq. 3.20)$$

where n, a, and b represent mole numbers.

Note that after biodegradation occurs, the substrate carbon is found either in its most oxidized form, CO_2, or in its most reduced form, CH_4. This is called **disproportionation** of organic carbon. The ratio of methane to carbon dioxide found in the gas mixture formed as a result of anaerobic degradation depends on the oxidation state of the substrate used.

TABLE 3.1 Relationship between Respiration, Redox Potential, and Typical Electron Acceptors and Products[a]

Type of respiration	Reduction reaction electron acceptor → product	Reduction potential (V)	Oxidation reaction electron donor → product	Oxidation potential (V)	Difference (V)
Aerobic	$O_2 - H_2O$	+0.81	$CH_2O - CO_2$	−0.47	−1.28
Denitrification	$NO_3^- - N_2$	+0.75	$CH_2O - CO_2$	−0.47	−1.22
Manganese reduction	$Mn^{4+} - Mn^{2+}$	+0.55	$CH_2O - CO_2$	−0.47	−1.02
Nitrate reduction	$NO_3^- - NH_4^+$	+0.36	$CH_2O - CO_2$	−0.47	−0.83
Sulfate reduction	$SO_4^{2-} - HS^-, H_2S$	−0.22	$CH_2O - CO_2$	−0.47	−0.25
Methanogenesis	$CO_2 - CH_4$	−0.25	$CH_2O - CO_2$	−0.47	−0.22

Adapted from Zehnder and Stumm (1988).

[a] Biodegradation reactions can be considered a series of oxidation–reduction reactions. The amount of energy obtained by cells is dependent on the difference in energy between the oxidation and reduction reactions. As shown in this table, using the same electron donor in each case but varying the electron acceptor, oxygen as a terminal electron acceptor provides the most energy for cell growth and methanogenesis provides the least.

Carbohydrates are converted to approximately equal amounts of CH_4 and CO_2. Substrates that are more reduced such as methanol or lipids produce relatively higher amounts of methane, whereas substrates that are more oxidized such as formic acid or oxalic acid produce relatively less methane.

3.5 GROWTH IN THE ENVIRONMENT

How is growth in the natural environment related to growth in a flask or in continuous culture? There have been several attempts to classify bacteria in soil systems on the basis of their growth characteristics and affinity for carbon substrates. The first was by Sergei Winogradsky (1856-1953), the "father of soil microbiology," who introduced the ecological classification system of **autochthonous** versus **zymogenous** organisms. The former metabolize slowly in soil, utilizing slowly released soil organic matter as a substrate. The latter are adapted to intervals of dormant and rapid growth, depending on substrate availability, following the addition of fresh substrate or amendment to the soil. In addition to these two categories, there are the **allochthonous** organisms, which are organisms that are introduced into soil and usually survive for only short periods of time. A second classification system developed to distinguish soil microbes categorizes organisms as either **oligotrophs,** those that grow better at low substrate concentrations, or **copiotrophs,** those that grow better at high substrate concentrations. The most recent theory of classification is founded on the concept of **r** and **K selection.** Organisms that respond to added nutrients with rapid growth rates are designated as r-strategists and correspond to the older definitions of zymogenous or copiotroph. In contrast, K-strategists are characterized by a high affinity for nutrients that are present in low concentration, corresponding to the older definitions of autochthonous or oligotroph.

It is unlikely that autochthonous microbes exhibit the stages of growth observed in batch flask and continuous culture. These microbes metabolize slowly and as a result have long generation times, and they often use energy obtained from metabolism simply for cell maintenance. On the other hand, zymogenous organisms may exhibit high rates of metabolism and perhaps exponential growth for short periods, or may be found in a **dormant** state. Dormant cells are often rounded and small (approximately 0.3 μm) in comparison with healthy laboratory specimens, which range from 1 to 2 m in size (see Chapter 2.2.2). Dormant cells may become **viable but nonculturable** with time because of a lack of nutrients. The existence of viable but nonculturable cells is reflected in the fact that **direct counts** from environmental samples, which include all viable cells, are one to two orders of magnitude higher than **culturable counts,** which include only cells capable of growth on the culture medium used. When nutrients are added, zymogenous microbes quickly take advantage and begin to metabolize actively. In a sense, this is similar to the reaction by microbes in a batch flask when nutrients are added. Thus, these microbes can exhibit the growth stages described in Section 3.2 for batch and continuous culture, but the pattern of the stages is quite different as described in the following sections.

3.5.1 The Lag Phase

The lag phase observed in a natural environment upon addition of a food source, particularly a contaminant, can be much longer than the lag phase normally observed in a batch culture. In some cases, this longer lag phase may be caused by very small initial populations that are capable of utilizing the added contaminant. In this case neither a significant disappearance of the contaminant nor a significant increase in cell numbers will be observed for several generations. Alternatively, degrading populations may be dormant or injured and require time to recover physiologically and resume metabolic activities. Further complicating growth in the environment is the fact that generation times are usually much longer than those measured under ideal laboratory conditions. This is due to a combination of limited nutrient availability and suboptimal environmental conditions such as temperature or moisture that cause stress. Thus, it is not unusual to observe lag periods of months or even years after an initial application of a new pesticide. However, once an environment has been exposed to a particular pesticide and developed a community for its degradation, the disappearance of succeeding pesticide applications will occur with shorter and shorter lag periods. This phenomenon is called **acclimation,** and has been observed with successive applications of many pesticides including the broad-leaf herbicide 2,4-dicholorophenoxyacetic acid (2,4-D).

A second explanation for long lag periods in environmental samples is that the capacity for degradation of an added carbon source may not initially be present within existing populations. This situation may require a mutation or a gene transfer event to introduce appropriate degradative genes into a suitable population. For example, a recent microcosm study documented the transfer of the plasmid pJP4 from an introduced organism to the indigenous soil population. The plasmid transfer resulted in rapid and complete degradation of the herbicide 2,4-D within the microcosm (see Case Study). This is one of the first studies that demonstrated gene transfer to indigenous soil recipients followed by growth and survival of the transconjugants at levels significant enough to affect degradation. There are still few such studies, and the likelihood and frequency of gene transfer in the environment are topics that are currently under debate among environmental microbiologists.

Case Study

Gene Transfer Experiment

Biodegradation of contaminants in soil requires the presence of appropriate degradative genes within the soil population. If degradative genes are not present within the soil population, the duration of the lag phase for degradation of the contaminant may range from months to years. One strategy for stimulating biodegradation is to "introduce" degrading microbes into the soil. Unfortunately, unless selective pressure exists to allow the introduced organism to survive and grow, it will die within a few weeks as a result of abiotic stress and competition from indigenous microbes. DiGiovanni *et al.* (1996) demonstrated that an alternative to "introduced microbes" is "introduced genes." In this study the introduced microbe was *Alcaligenes eutrophus* JMP134. JMP134 carries an 80-kb plasmid, pJP4, that encodes the initial enzymes necessary for the degradation of the herbicide 2,4-D. A series of soil microcosms were set up and contaminated with 2,4-D. In control microcosms there was slow, incomplete degradation of the 2,4-D over a 9-week period (see figure below). In a second set of microcosms, JMP134 was added to give a final inoculum of 10^5 CFU/g dry soil. In these microcosms rapid degradation of 2,4-D occurred after a 1-week lag phase and the 2,4-D was completely degraded after 4 weeks. The scientists examined the microcosm 2,4-D degrading population very carefully during this study. What they found was surprising. They could not recover viable JMP134 microbes after the first week. However, during weeks 2 and 3 they isolated two new organisms that could degrade 2,4-D. Upon closer examination, both of these organisms, *Pseudomonas glathei* and *Burkholderia caryophyllii*, were found to be carrying the pJP4 plasmid! Finally, during week 5 a third 2,4-D degrader was isolated, *Burkholderia cepacia*. This isolate also carried the pJP4 plasmid. Subsequent addition of 2,4-D to the microcosms resulted in rapid degradation of the herbicide, primarily by the third isolate, *B. cepacia*. Although it is clear from this research that the pJP4 plasmid was transferred from the introduced microbe to several indigenous populations, it is not clear how this transfer occurred. There are two possibilities: that there was cell-to-cell contact and transfer of the plasmid via **conjugation** or that the JMP134 cells died and lysed, releasing the pJP4 plasmid. The indigenous populations then took up the released plasmid, a process called **transformation.**

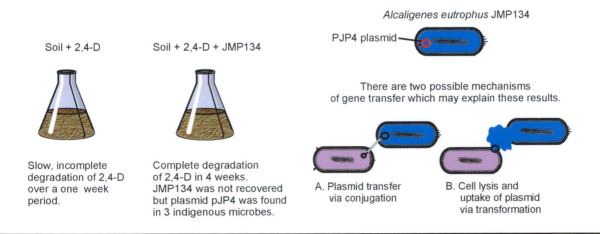

Soil + 2,4-D

Slow, incomplete degradation of 2,4-D over a one week period.

Soil + 2,4-D + JMP134

Complete degradation of 2,4-D in 4 weeks. JMP134 was not recovered but plasmid pJP4 was found in 3 indigenous microbes.

Alcaligenes eutrophus JMP134

PJP4 plasmid

There are two possible mechanisms of gene transfer which may explain these results.

A. Plasmid transfer via conjugation

B. Cell lysis and uptake of plasmid via transformation

3.5.2 The Exponential Phase

In the soil and water environment the second phase of growth, exponential growth, occurs for only very brief periods of time following addition of a substrate. Such substrate might be crop residues, vegetative litter, root residues, or contaminants added to or spilled into the environment. As stated earlier, it is the zymogenous cells, many of which are initially dormant, that respond most quickly to added nutrients. Upon substrate addition, these dormant cells become physiologically active and briefly enter exponential phase until the substrate is utilized or until some limiting factor causes a decline in substrate degradation. As shown in Table 3.2, culturable cell counts increase one to two orders of magnitude in response to the addition of 1% glucose. In this experiment, four different soils were left untreated or were amended with 1% glucose and incubated at room temperature for 1 week.

Because nutrient levels and other factors, e.g., temperature or moisture, are seldom ideal, it is rare for cells in the environment to achieve a growth rate equal to μ_{max}. Thus, rates of degradation in the environment are slower than degradation rates measured under laboratory conditions. This is illustrated in Table 3.3, which compares the degradation rates for wheat and rye straw in a laboratory environment with degradation rates in natural environments. These include: a Nigerian tropical soil that undergoes some dry periods; an English soil that is exposed to a moderate, wet climate; and a soil from Saskatoon, Canada that is subjected to cold winters and dry summers. As shown in Table 3.3, the relative rate of straw degradation under laboratory conditions is twice as fast as in the Nigerian soil, 8 times faster than in the English soil, and 20 times faster than in the Canadian soil. This example illustrates the importance of understanding that there can be a huge difference between degradation rates in the laboratory and in natural environments. This under-

TABLE 3.2 Culturable Counts in Unamended and Glucose-Amended Soils[a]

Soil	Unamended (CFU/g soil)	1% Glucose (CFU/g soil)	Log increase
Pima	5.6×10^5	4.6×10^7	1.9
Brazito	1.1×10^6	1.1×10^8	2.0
Clover Springs	1.4×10^7	1.9×10^8	1.1
Mt. Lemmon	1.4×10^6	8.3×10^7	1.7

[a] Each soil was incubated for approximately 1 week and then culturable counts were determined using R2A agar. (Courtesy E. M. Jutras.)

TABLE 3.3 Effect of Environment on Decomposition Rate of Plant Residues Added to Soil

Residue	Half-life (days)[a]	μ[b] (1/days)	Relative rate[c]
Wheat straw, laboratory	9	0.08	1
Rye straw, Nigeria	17	0.04	0.5
Rye straw, England	75	0.01	0.125
Wheat straw, Saskatoon	160	0.003	0.05

From Paul and Clark (1989).
[a] The half-life is the amount of time required for degradation of half of the straw initially added.
[b] μ is the specific growth rate constant.
[c] The relative rate of degradation of wheat straw under laboratory conditions is assumed to be 1. The degradation rates for straw in each of the soils were then compared with this value.

standing is crucial when attempting to predict degradation rates for contaminants in an environment.

3.5.3 The Stationary and Death Phases

Stationary phase in the laboratory is a period of time where there is active cell growth that is matched by cell death. In batch culture, cell numbers increase rapidly to levels as high as 10^{10} to 10^{11} CFU/ml. At this point, either the substrate is completely utilized or cells have become so dense that further growth is inhibited. In the environment, the stationary phase is most likely of short duration if it exists at all. Recall that most cells never achieve an exponential phase because of nutrient limitations and environmental stress. Rather they are in dormancy or in a maintenance state. Cells that do undergo growth in response to a nutrient amendment will quickly utilize the added food source. However, even with an added food source, cultural counts rarely exceed 10^8 to 10^9 CFU/g soil except perhaps on some root surfaces. At this point, cells will either die or, in order to prolong survival, enter a dormant phase again until new nutrients become available. Thus, stationary phase is likely to be very short if indeed it does occur. In contrast, death phase can certainly be observed, at least in terms of culturable counts. In fact, the death phase is often a mirror reflection of the growth phase. Once added nutrients are consumed, both living and dead cells become prey for protozoa that act as microbial predators. Dead cells are also quickly scavenged by other microbes in the vicinity. Thus, culturable cell numbers increase in response to nutrient addition (see Table 3.2) but will decrease again just as quickly to the background level after the nutrients have been utilized.

QUESTIONS AND PROBLEMS

1. Study example calculations 1, 2, and 3.

2. Draw a growth curve of substrate disappearance as a function of time. Label and define each stage of growth.

3. Calculate the time it will take to increase the cell number from 10^4 CFU/ml to 10^8 CFU/ml assuming a generation time of 1.5 hr.

4. Write the Monod equation and define each of the constants.

5. There are two special cases when the Monod equation can be simplified. Describe these cases and the simplified Monod equation that results.

6. List terminal electron acceptors used in anaerobic respiration in the order of preference (from an energy standpoint).

7. Define disproportionation.

8. Define the term critical dilution rate, D_c, and explain what happens in continuous culture when D is greater than D_c.

9. Compare the characteristics of each of the growth phases (lag, log, stationary, and death) for batch culture and soil systems.

References and Recommended Readings

Blanch, H. W., and Clark, D. S. (1996) "Biochemical Engineering." Marcel Dekker, New York.

DiGiovanni, G. D., Neilson, J. W., Pepper, I. L., and Sinclair, N. A. (1996) Gene transfer of *Alcaligenes eutrophus* JMP134 plasmid pJP4 to indigenous soil recipients. *Appl. Environ. Microbiol.* **62,** 2521–2526.

Estrella, M. R., Brusseau, M. L., Maier, R. S., Pepper, I. L., Wierenga, P. J., and Miller, R. M.. (1993) Biodegradation, sorption, and transport of 2,4-dichlorophenoxyacetic acid in saturated and unsaturated soils. *Appl. Environ. Microbiol.* **59,** 4266–4273.

Gottschal, J. C., and Dijkhuizen, L. (1988) The place of continuous culture in ecological research. *In* "CRC Handbook of Laboratory Model Systems for Microbial Ecosystems," Vol. I (J. W. T. Wimpenny, ed.) CRC Press, Boca Raton, FL.

Grbić-Galić, D. (1990) Anaerobic microbial transformation of nonoxygenated aromatic and alicyclic compounds in soil, subsurface, and freshwater sediments. *In* "Soil Biochemistry," Vol. 6 (J.-M. Bollag and G. Stotzky, eds.) Marcel Dekker, New York, pp. 117–189.

Lovely, D. R. (1993) Dissimilartory metal reduction. *Ann. Rev. Microbiol.* **47,** 263–290.

Madigan, M. T., Martinko, J. M., and Parker, J. (1997) "Brock Biology of Microorganisms," 8th ed. Prentice Hall, Upper Saddle River, NJ.

Neidhardt, F. C., Ingraham, J. L., Schaechter, M.. (1990) "Physiology of the Bacterial Cell: A Molecular Approach." Sinauer Associates, Sunderland, MA.

Paul, E. A. and Clark, F. E. (1989) "Soil Microbiology and Biochemistry." Academic Press, San Diego, CA.

Suflita, J. M., and Sewell, G. W. (1991) Anaerobic biotransformation of contaminants in the subsurface. EPA/600/M-90/024, U.S. Environmental Protection Agency, Ada, OK.

Young, L. Y., and Häggblom, M. M. (1989) The anaerobic microbiology and biodegradation of aromatic compounds. *In* "Biotechnology and Biodegradation" (D. Kamely, A. Chakrabarty, and G. S. Omenn, eds.) Gulf Publishing Company, Houston, TX, pp. 3–19.

Zehnder, A. J. B., and Stumm, W. (1988) Geochemistry and biogeochemistry of anaerobic habitats. *In* "Biology of Anaerobic Microorganisms" (A. J. B. Zehnder, ed.) John Wiley & Sons, New York, pp. 1–38.

4

Terrestrial Environments

RAINA M. MAIER   IAN L. PEPPER

4.1 INTRODUCTION

Terrestrial environments are arguably the richest and most complex of all microbial environments. Although terrestrial environments are extremely varied, from rain forests in Washington State's Olympic National Park to the Sonoran Desert in Arizona, all soils teem with activity and a diversity of microorganisms. Environmental microbiologists have become more interested in this vast diversity of terrestrial microorganisms because of the potential to harness their unique activities, such as pollution abatement, sustainable agriculture, and new applications in biotechnology. For example, it is increasingly apparent that terrestrial microorganisms have the capability to degrade most of the chemical pollutants that humans have added to the environment. The challenge is to stimulate the appropriate activities at a rate that is sufficient to protect human, animal, and ecological health. Discoveries of microorganisms that inhabit extreme environments, such as the deep subsurface or thermal hot springs, have engendered increasing interest in the unusual properties of and enzymes made by these organisms for use in biotechnology. Examples include enzymes for use as detergent additives that work optimally at low temperatures, the continuing search for new antibiotics, and new enzymes to help in food processing. One success story is that of the *Taq* polymerase enzyme isolated from the organism *Thermus aquaticus*, found in a thermal pool in Yellowstone National Park. The *Taq* polymerase, which has an optimal temperature of 72°C, is used in the polymerase chain reaction (PCR) and has annual sales of $80 million.

The key to finding new microorganisms and harnessing their activities is to develop a thorough understanding of the terrestrial environment. To better understand microbial populations and their activity within these environments, one might envision a cross section of the earth (Fig. 4.1) that portrays different regions—surface soils, the unsaturated or vadose zone, and underground aquifers, which are also referred to as the saturated zone. Each of these regions comprises a combination of mineral, solution, and gas phases that is more commonly called a **porous medium.** In this chapter, the characteristics of a porous medium are defined, and the similarities and differences between porous media that make up surface soils and subsurface environments are compared and contrasted. The characteristics of each type of porous medium are then discussed in relation to the numbers, diversity, and activity of microorganisms that populate each medium.

4.2 POROUS MEDIA

All porous media are three-phase systems consisting of: (1) a solid or mineral inorganic phase that is often associated with organic matter, (2) a liquid or solution phase, and (3) a gas phase or atmosphere (Fig. 4.2). All phases interact together when the system is perturbed and then move toward an equilibrium state to create an unique environment. If left undisturbed, this environment remains fairly constant. The unique properties of any porous medium are dependent on the specific composition of each of these phases, which are discussed in the following sections.

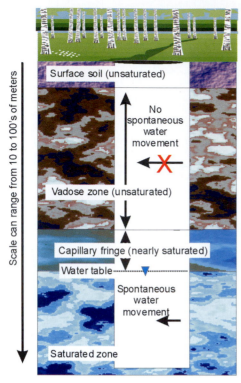

FIGURE 4.1 Cross section of the subsurface showing surface soil, vadose zone, and saturated zone. (Adapted from *Pollution Science*, © 1996, Academic Press, San Diego, CA)

4.2.1 The Solid Phase

Typically, a porous medium contains 45 to 50% solids on a volume basis. Of this solid fraction, 95–99% is the mineral fraction. Silicon (47%) and oxygen (27%) are the two most abundant elements found within the mineral fraction of the earth's crust. These two elements, along with lesser amounts of other elements, combine in a number of ways to form a large variety of minerals. For example, quartz is SiO_2 and mica is $K_2Al_2O_5[Si_2O_5]_3Al_4(OH)_4$. These are primary minerals

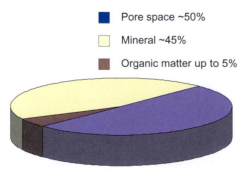

FIGURE 4.2 Three basic components of a porous medium on a volume basis.

that are derived from the weathering of parent rock. Weathering results in mineral particles of different sizes as shown in Table 4.1. The distribution (on a percent by weight basis) of sand, silt, and clay within a porous medium defines its **texture.**

In surface soils, sand, silt, and clay particles are subject to extremely active microbial processes that result in the formation of microbial gums, polysaccharides, and other extracellular microbial metabolites. These materials, along with the microbial cells themselves, act as glues that bind the primary sand, silt, and clay particles together into a secondary structure called **soil aggregates** or **peds.** Aggregate formation is unique for each soil and depends on the parent material, the climate, and the biological activity within the soil. Soils with even modest amounts of clay usually have well-defined aggregates and hence a well-defined **soil structure.** Aggregates usually remain intact as long as the soil is not disturbed, for example, by plowing. In contrast, sandy soils with very low amounts of clay generally have less well defined soil structure. In between the component mineral particles of a porous medium are pore spaces. These pores allow movement of air, water, and microorganisms through the porous medium. Pores that exist between aggregates are called **interaggregate pores,** whereas those within the aggregates are termed **intraaggregate pores** (Fig. 4.3).

Texture and structure are important factors that govern the movement of water, contaminants, and microbial populations in porous media. Of the three size fractions that make up a porous medium, clay particles are particularly dominant in determining the physical and chemical characteristics. For example, clays, which are often composed of aluminum silicates, add both surface area and charge to a soil. As shown in Table 4.1, the surface area of a fine clay particle can be five orders of magnitude larger than the surface area of a 2-mm sand particle. To put this into a microbial perspective, the size of a clay particle is similar to that of a bacterial cell. Clays affect not only the surface area of a porous medium but also the average pore size (Fig. 4.4). Although the average pore size is smaller in a clay soil, there are many more pores than in a sandy soil, and as a result the total amount of pore space is larger in a fine-textured (clay) soil than in a coarse-textured (sandy) soil. However, because small pores do not transmit water as fast as larger pores, a fine-textured medium will slow the movement of any material moving through it, including air, water, and microorganisms (Fig. 4.5). Often, fine-textured regions or layers of materials, e.g., clay lenses, can be found in sites composed primarily of coarser materials, creating very heterogeneous environments. In this case, water will prefer to travel through the coarse material and

TABLE 4.1 Size Fractionation of Soil Constituents

Specific surface area using a cubic model	Soil		
	Mineral constituents	Size	Organic and biologic constituents
0.0003 m²/g	**Sand** Primary minerals: quartz, silicates, carbonates	2 mm	**Organic debris**
0.12 m²/g	**Silt** Primary minerals: quartz, silicates, carbonates	50 µm	**Organic debris, large microorganisms** Fungi Actinomycetes Bacterial colonies

3 m²/g	**Granulometric clay**
	Microcrystals of primary minerals
	Phyllosilicates
	  Inherited; illite, mica
	  Transformed: vermiculite, high-charge smectite
	  Neoformed; kaolinite, smectite
	Oxides and hydroxides
		2 µm	**Amorphous organic matter**
			Humic substances
			Biopolymers
			
			Small microorganisms
			
			Bacteria
			Fungal spores
			Large viruses

30 m²/g	**Fine clay** Swelling clay minerals Interstratified clay minerals Low range order crystalline compounds	0.2 µm	Small viruses

Adapted from Robert and Chenu (1992).

flow around the fine-textured lens. However, clay lenses retain water more tenaciously than sandy materials because of the smaller pore spaces (see Chapter 4.2.3.2). For microorganisms, which are much larger than individual water molecules, a fine-textured hori-zon or lens will inhibit bacterial movement either into or out of the region. Such heterogeneity poses great difficulties when trying to remove contaminants because some finely textured regions are relatively inaccessible to water flow or to microorganisms. Thus, contaminants trapped within very small pores may remain there for long periods of time, acting as a long-term "sink" of contaminant that diffuses out of the pores very slowly with time.

In addition to the importance of clay to the surface area of a porous medium, the surface charge of a porous medium is governed partially by functional groups on clay mineral surfaces. The term **cation-exchange capacity** (CEC) describes the charge associated with both clay and organic particles in a porous medium. For clay particles, CEC is associated with both **isomorphic substitution** and **ionization.** Isomorphic substitution occurs when, for example, a divalent magnesium cation (Mg^{2+}) is substituted for a trivalent aluminum cation (Al^{3+}) within the clay lattice. This results in the loss of one positive charge, which is equivalent to a gain of one negative charge (Fig. 4.6). Other substitutions can also lead to increases in negative charge. Ionization occurs in hydroxyl groups from metal oxides, oxyhydroxides, and hydroxides that are

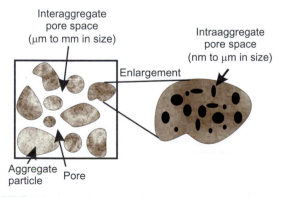

FIGURE 4.3 In surface soils, mineral particles are tightly packed together and even cemented in some cases with microbial polymers forming soil aggregates. The pore spaces between individual aggregates are called interaggregate pores and vary in size from micrometers to millimeters. Aggregates also contain pores that are smaller in size, ranging from nanometers to micrometers. These are called intraaggregate pores.

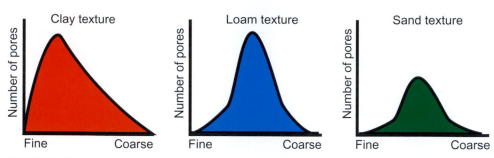

FIGURE 4.4 Typical pore size distributions for clay-, loam-, and sand-textured horizons. Note that the clay-textured material has the smallest average pore size, but the greatest total volume of pore space.

exposed on the clay or lattice surface, also resulting in the formation of a negative charge:

$$Al-OH \rightleftarrows Al-O^- + H^+$$

These are also known as **broken-edge** bonds. Ionizations such as these are pH dependent and increase as the pH increases. Isomorphic substitutions and ionization cause clays to exhibit a net negative charge that can attract positively charged solutes in the porous medium through a process called **cation exchange.** Or-

ganic matter can also contribute to cation exchange through ionization of carboxyl and hydroxide functional groups, a process which is also pH dependent.

The total amount of negative charge in a soil, due to both clay and organic matter content, is usually measured in terms of millimoles of positive charge per kilogram of soil. A CEC of 150 to 200 millimoles positive charge per kilogram soil is an average value for most soils. How does the process of cation exchange work? Common soil cations such as Ca^{2+}, Mg^{2+}, K^+,

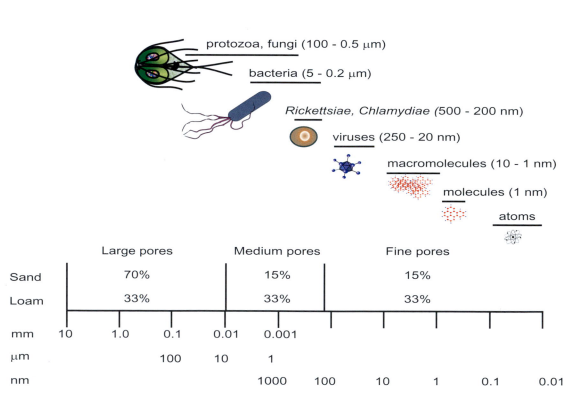

FIGURE 4.5 Comparison of sizes of bacteria, viruses, and molecules with hydraulic equivalent diameters of pore canals. (Adapted from *J. Contam. Hydrol.,* **2,** G. Matthess, A. Pekdeger, and J. Schroeder, Persistence and transport of bacteria and viruses in groundwater—a conceptual evaluation, 171–188, © 1988, with permission from Elsevier Science.)

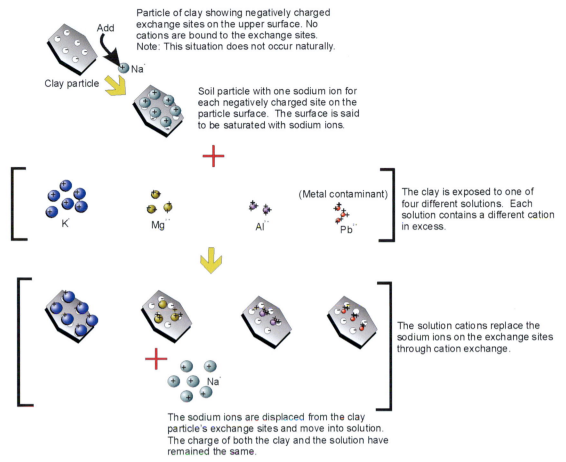

Particle of clay showing negatively charged exchange sites on the upper surface. No cations are bound to the exchange sites. Note: This situation does not occur naturally.

Add

Na

Clay particle

Soil particle with one sodium ion for each negatively charged site on the particle surface. The surface is said to be saturated with sodium ions.

K   Mg   Al   (Metal contaminant) Pb

The clay is exposed to one of four different solutions. Each solution contains a different cation in excess.

Na

The solution cations replace the sodium ions on the exchange sites through cation exchange.

The sodium ions are displaced from the clay particle's exchange sites and move into solution. The charge of both the clay and the solution have remained the same.

FIGURE 4.6 Cation exchange on clay particles. (Adapted from *Pollution Science,* © 1996, Academic Press, San Diego, CA)

Na^+, and H^+ that exist in the soil solution are in equilibrium with cations on exchange sites. If the concentration of a cation in the soil solution is changed, for example, increased, then that cation is likely to occupy more exchange sites, replacing existing cations within the site. Thus, a monovalent cation such as K^+ can replace another monovalent cation such as Na^+, or two K^+ can replace one Mg^{2+}. Note however, that when working with charge equivalents, one milliequivalent of K^+ replaces one milliequivalent of Mg^{2+}. Cation exchange ultimately depends on the concentration of the cation in soil solution and the **adsorption affinity** of the cation for the exchange site. The adsorption affinity of a cation is a function of its charge density, which in turn depends on its total charge and the size of the cation. The adsorption affinities of several common cations are given in the following series in decreasing order:

$$Al^{3+} > Ca^{2+} = Mg^{2+} > K^+ = NH_3^+ > Na^+$$

Highly charged small cations such as Al^{3+} have high adsorption affinities. In contrast, monovalent ions have lower affinities, particularly if they are highly hydrated such as Na^+, which increases the effective size of the cation. The extensive surface area and charge of soil colloids (clays + organic material) are critical to microbial activity since they affect both binding or sorption of solutes and microbial attachment to the colloids.

Sorption is a major process influencing the movement and bioavailability of essential compounds and pollutants in soil. The broadest definition of **sorption** is the association of organic or inorganic molecules with the solid phase of the soil. For inorganic charged molecules, cation exchange is one of the primary mechanisms of sorption (Fig. 4.6). Generally, positively charged ions, for example, calcium (Ca^{2+}) or cadmium (Cd^{2+}), participate in cation exchange. Since sorbed forms of these metals are in equilibrium with the soil solution, they can serve as a long-term source of essential nutrient (Ca^{2+}) or pollutant (Cd^{2+}) that is slowly released back into the soil solution as the soil solution concentration of the cation decreases with time.

Attachment of microorganisms can also be mediated by the numerous functional groups on clays (Fig. 4.7). Although the clay surface and microbial cell surface both have net negative charges, clay surfaces are neutralized by the accumulation of positively charged counterions such as K^+, Na^+, Ca^{2+}, Mg^{2+}, Fe^{3+} and Al^{3+}. Together, these negative and positive surface charges form what is called the **electrical double layer.** Similarly, microbes have an electrical double layer. The thickness of the clay double layer depends on the valence and concentration of the cations in solution. Higher valence and increased cation concentrations will shrink the electrical double layer. Because the double layers of the clay particles and microbial cells repel each other, the thinner these layers are, the less the repulsion between the clay and cell surfaces. As these repulsive forces are minimized, attractive forces such as electrostatic and van der Waals forces allow the attachment of microbial cells to the surface (Gammack *et al.,* 1992). Thus most microbes in terrestrial environments exist attached to soil colloids, rather than existing freely in the soil solution (see Chapter 7.2.3).

4.2.2 Organic Matter

Organic matter in porous media is defined as a combination of (1) live biomass, including animals, microbes, and plant roots; (2) recognizable dead and decaying biological matter; and (3) **humic substances,** which are heterogeneous polymers formed during the process of decay and degradation of plant, animal, and microbial biomass. Because subsurface media contain very little organic matter, this discussion will focus on surface soils. Surface soils have a combination of these types of organic matter, whereas subsurface environments usually contain only small amounts of humic substances as well as a small amount of biomass in the form of microorganisms.

It is the humic fraction of soil organic matter that provides the stable, long-term microbial nutrient base in a porous medium. Humic substances have extremely complex structures that reflect the complexity and diversity of organic materials produced in a typical soil. During degradation of plant, animal, and microbial biomass, a diversity of simple components such as sugars, amino acids, lipids, and phenolic compounds are released (for further detail, see Chapter 14.2.3). Some of these components are degraded rapidly, whereas others resist degradation. Some of the more recalcitrant compounds are spontaneously or enzymatically polymerized, resulting in the formation of the humic substances. Thus humic substances in different soils are never exactly the same and have varying physical and chemical characteristics. Despite these variations, there are some common generalizations that can be made about the structure of humic materials.

Humic substances range in molecular weight from 700 to 300,000 and can be divided into three fractions based on their chemical structure: fulvic acid, humic acid, and humin. These materials are irregular polymers, and their separation is based on differences in molecular weight, oxygen content, and number of acidic functional groups. An example of a humic acid polymer is shown in Fig. 4.8. Overall, humus has a three-dimensional, spongelike structure that contains hydrophobic and hydrophilic areas. As a result, it is generally less polar than water and therefore provides a favorable environment for solutes that are less polar

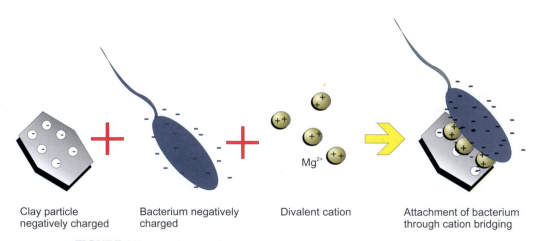

| Clay particle negatively charged | Bacterium negatively charged | Divalent cation | Attachment of bacterium through cation bridging |

FIGURE 4.7   Attachment of a bacterial cell to a clay particle via cation bridging.

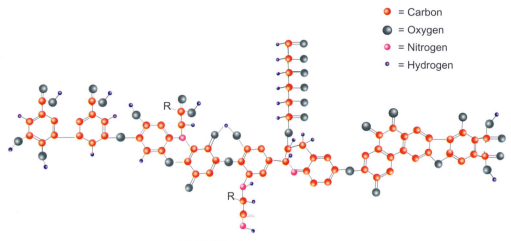

= Carbon
= Oxygen
= Nitrogen
= Hydrogen

FIGURE 4.8   Humus polymer.

than water. This means that humus can sorb nonpolar solutes from the general soil solution through a sorption process called **hydrophobic binding** (Fig. 4.9). Humic substances also contain numerous functional groups, the most important of which are the carboxyl group and the phenolic hydroxyl group, both of which can become negatively charged in the soil solution. As noted earlier, these functional groups, are similar to those found on clays, and can contribute to the pH-dependent CEC of soil and participate in sorption of solutes and attachment of microorganisms by cation exchange, as shown in Figs. 4.6 and 4.7.

Like clay, organic matter and in particular humic substances exert a large influence on microbial activity in a porous medium. Humic substances share two properties with clay: they have a net negative charge and a large surface area. However, because humic materials are carbon based rather than silicon based, they are much more reactive in the sense that they undergo slow but constant change as a result of biological ac-

tivity. Because rates of organic input, primarily from plant production, are similar to rates of organic matter degradation, undisturbed sites establish relatively constant organic matter concentrations. As a result of microbial degradation, the stable fraction of soil organic matter mineralizes at a rate of approximately 2–5% per year, depending mostly on seasonal soil temperatures. Thus, organic matter does more than just provide high surface area and charge in a soil. It also serves as a slow release source of carbon and energy for the autochthonous (indigenous), slow-growing microorganisms in the soil (see Chapter 3.5). Hence the number of microorganisms found in a soil often increases as the soil organic matter content increases.

Numbers of microorganisms are also especially high in the soil directly surrounding plant roots and root hairs. This area, called the **rhizosphere,** is the volume of soil from 1 mm to several millimeters surrounding each root, and contains large concentrations of readily available organic materials. These materials

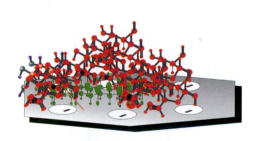

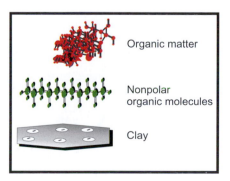

Organic matter

Nonpolar organic molecules

Clay

FIGURE 4.9   Hydrophobic sorption mechanism. Nonpolar organic molecules tend to sorb to organic matter that is associated with solid mineral surfaces by diffusing into the sponge-like interior of organic matter molecules. (Adapted from Schwarzenbach *et al.*, 1993.)

consist of sloughed off root cells and diffused plant metabolites (see Chapter 18.1.3) called root exudates and border cells. Exudates are a mixture of organic acids, sugars, and other soluble plant components that either diffuse out of the root or are released as a result of root damage. Overall, the root and rhizosphere support much of the total microbial and biochemical activity in surface soils. Because most biomass is found at the earth's surface, there is a gradient of organic matter from the soil surface (highest levels) to the lower root zone (decreased levels). Below the root zone, the organic matter content is generally very low. In direct correlation, microbial numbers and activity are highest at the earth's surface and generally decrease with decreasing organic matter content.

4.2.3 The Liquid Phase

4.2.3.1 Soil Solution Chemistry

The soil solution is a constantly changing matrix composed of both organic and inorganic solutes in aqueous solution. The composition of the liquid phase is extremely important for biological activity in a porous medium, for microorganisms are approximately 70% water and most require high levels of water activity (>0.95) for active metabolism. Indeed, all microorganisms, even attached ones, are surrounded by a water film from which they obtain nutrients and into which they excrete wastes. Thus, the amount and composition of the liquid phase ultimately controls both microbial and plant growth.

The soil solution composition reflects the chemistry of the soil as well as the dynamic influx and efflux of solutes in response to water movement. Water movement results from rainfall or irrigation events or from normal groundwater flow, mineral weathering, and organic matter formation and decomposition (Fig. 4.10). This composition is also altered by anthropogenic activities, such as irrigation, fertilizer and pesticide addition, and chemical spills. The chemistry of the soil affects not only the composition of the soil solution but also the form and bioavailability of nutrients. For example, as illustrated in Table 4.2, the form of the common cations found in a soil varies as a function of soil pH. As shown in this table, most cations are found in a more soluble form in acidic environments. In some cases, as for magnesium and calcium, this results in extensive leaching of the soluble form of the cation, leading to decreased concentrations of the nutrient. For many metallic cations, this can lead to increased metal toxicity (see Chapter 17.5). In other cases, as for iron and phosphate, a slightly acidic pH

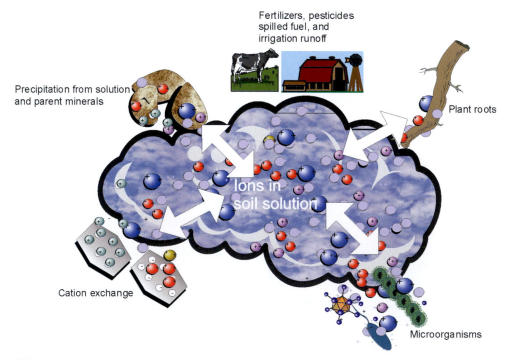

FIGURE 4.10 Paths of dissolution and uptake of minerals in the soil. (Adapted from *Pollution Science,* © 1996, Academic Press, San Diego, CA)

TABLE 4.2   The Form of Common Cations Found in Acid and Alkaline Soils

Cation	Acid soils (low pH)	Alkaline soils (high pH)
Na^+	Na^+	Na^+, $NaHCO_3^0$, NaSO
Mg^{2+}	Mg^{2+}, $MgSO_4^0$, organic complexes	Mg^{2+}, $MgSO_4^0$, $MgCO_3^0$
Al^{3+}	organic complexes, AlF^{2+}, $AlOH^{2+}$	$Al(OH)_4^-$, organic complexes
Si^{4+}	$Si(OH)_4^0$	$Si(OH)_4^0$
K^+	K^+	K^+, KSO_4^-
Ca^{2+}	Ca^{2+}, $CaSO_4^0$, organic complexes	Ca^{2+}, $CaSO_4^0$, $CaHCO_3^+$
Mn^{2+}	Mn^{2+}, $MnSO_4^0$, organic complexes	Mn^{2+}, $MnSO_4^0$, $MnCO_3^0$, $MnHCO_3^+$, $MnB(OH)_4^+$
Fe^{2+}	Fe^{2+}, $FeSO_4^0$, $FeH_2PO_4^+$	$FeCO_3^+$, Fe^{2+}, $FeHCO_3^+$, $FeSO_4^0$
Fe^{3+}	$FeOH^{2+}$, $Fe(OH)_3^0$, organic complexes	$Fe(OH)_3^0$, organic complexes
Cu^{2+}	Organic complexes, Cu^{2+}	$CuCO_3^0$, organic complexes, $CuB(OH)_4^+$, $Cu[B(OH)_4]_4^0$
Zn^{2+}	Zn^{2+}, $ZnSO_4^0$, organic complexes	$ZnHCO_3^+$, $ZnCO_3^0$, organic complexes, Zn^{2+}, $ZnSO_4^0$, $ZnB(OH)_4^+$
Mo^{5+}	$H_2MoO_4^0$, $HMoO_4^-$	$HMoO_4^-$, MoO_4^{2-}

Adapted from Sposito (1989).

provides optimal availability of the element. Overall, the pH range that supports maximum microbial and plant activity is between 6.0 and 6.5.

4.2.3.2 Soil Water Potential

Water is the primary solvent in porous medium systems, and water movement is generally the major mechanism responsible for the transport of chemicals and microorganisms. Water movement in a porous medium depends on the **soil water potential,** which is the work per unit quantity necessary to transfer an infinitesimal amount of water from a specified elevation and pressure to another point somewhere else in the porous medium. The soil water potential is a function of several forces acting on water, including matric and gravitational forces. Soil water potential is usually expressed in units of pressure (pascals, atmospheres or bars). Values of the matric contribution are negative because the reference is generally free water, which is defined to have a soil water potential of zero. Since the matric force decreases the free energy of water in the soil solution, the soil water potential becomes increasingly negative as this force increases.

In the saturated zone, the presence of a regional hydraulic gradient usually results in a general horizontal flow of water. In contrast, flow is generally downward in the unsaturated zone. In an unsaturated zone, where the soil pores are not completely filled with water, there are several incremental forces that affect water movement. These are related to the amount of water present (Fig. 4.11). In very dry soils, there is an increment of adsorbed water that exists as an extremely thin film on the order of angstroms (Å) in

width. This thin film is held very tightly to particle surfaces by surface forces with soil water potentials ranging from -31 to $-10,000$ atm. As a result, this water is essentially immobile. As water is added to this soil, a second increment of water forms as a result of

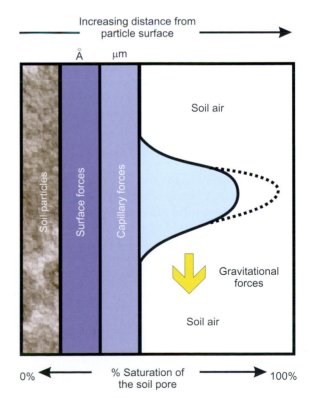

FIGURE 4.11   The continuum of soil water. (Adapted from Dragun, 1988.)

matric or **capillary forces.** This water exists as bridges between particle surfaces in close proximity and can actually fill small soil pores. This results in soil water potentials ranging from -0.1 to -31 atm. This water moves slowly from larger to smaller pores in any direction, and is held against gravitational forces. The next increment of water added, **free water,** can be removed by **gravitational forces.** The soil water potential for free water ranges from 0 to -0.5 atm. Although we have classified these different types of water in categories, in porous media they actually occur as a continuum rather than with sharply defined boundaries.

The amount of water present in the pore space of a medium is another important parameter in understanding the level and type of microbial activity in an environment (see Section 4.2.2 for the importance of organic matter). The optimal environment for active aerobic microbial growth in a porous medium is one in which water is easily available but the medium is not completely saturated. Why is this? As stated earlier, microorganisms obtain nutrients from the water phase surrounding them, so water is absolutely necessary for active microbial growth. Generally, microbial activity in the soil is greatest at -0.1 atm. As the water potential becomes less negative, the soil becomes saturated. In a completely saturated environment, oxygen, which governs aerobic microbial activity, may become limiting because of its limited solubility (9.3 mg/L at 20°C and 1 atm pressure) in water. Because the diffusion of oxygen through water is slow, once the available dissolved oxygen is used up, it is not replenished rapidly. As the water potential becomes more negative than -0.1 atm, water becomes less available because it is held tightly by matric and capillary forces.

4.2.4 Soil Atmosphere

The soil atmosphere has the same basic composition as air: nitrogen, oxygen, and carbon dioxide. As shown in Table 4.3, there is little difference between the atmospheres in a well-aerated surface soil and in the air.

However, plant and microbial activity can greatly affect the relative proportions of oxygen and carbon dioxide in soils that are not well aerated or that are far removed from the soil surface. For example, in a fine clay or a saturated soil, oxygen can be completely removed by the aerobic activity of respiring organisms. During this respiration process, carbon dioxide (CO_2) is evolved, eventually resulting in depressed oxygen (O_2) and elevated CO_2 levels. These changes affect the redox potential of the porous medium, which affects the availability of terminal electron acceptors for aerobic and anaerobic microbes.

Movement of gases in a porous medium occurs by diffusion, where molecules move from high to low concentrations. Diffusion will continue until equilibrium is reached. However, in most porous media, equilibrium conditions do not occur. Instead, as a result of biological activity, differences in amounts of O_2 and CO_2 exist between most porous medium environments and the atmosphere. Thus, for most porous media, there is a net influx of O_2 and a net efflux of CO_2. A second way in which gases can move within a porous medium is by mass flow, which occurs in response to a general change in pressure within a porous medium or in the atmosphere. Pressure changes can result from temperature changes, atmospheric pressure changes, or soil air pressure changes, the latter caused by a change in soil moisture.

4.3 SOIL AND SUBSURFACE ENVIRONMENTS

Surface soils and subsurface environments are both three-phase porous medium systems. Differences between these environments are site dependent, based on a combination of the mineral and biological components of the porous medium. These components depend in turn on the geology, topography, and climate of the site. However, even given such site-dependent differences, there are further general differences in the porous media that make up surface soils and subsur-

TABLE 4.3   Soil Atmosphere

Location	Composition (% volume basis)		
	Nitrogen (N_2)	Oxygen (O_2)	Carbon dioxide (CO_2)
Atmosphere	78.1	20.9	0.03
Well-aerated soil surface	78.1	18–20.5	0.3–3
Fine clay or saturated soil	>79	~0–10	Up to 10

face environments, including both the vadose and saturated zones (see Fig. 4.1). These differences influence the number and activity of microorganisms in these environments. As discussed in Sections 4.2.2 and 4.2.3, organic matter and water are two of the most important parameters controlling the numbers of microorganisms in a porous medium. The following sections discuss specific characteristics of surface soils, the vadose zone, and the saturated zone with respect to water, oxygen, and organic matter content.

4.3.1 Surface Soils

Surface soils are the weathered end product of the **five soil forming factors,** which involve action of climate and living organisms on a specific soil parent material, with a given topography over a given time period. There are three major types of parent material: **igneous rock,** created by the cooling of molten magma; **sedimentary rock,** created by the deposition and cementation of loose mud or sand materials; and **metamorphic rock,** created by intense heat, pressure, or chemical action on igneous or sedimentary material. Parent material near the earth's surface is exposed to a wide variety of temperature variations, rainfall, wind, and biological activity. Such activities over thousands of years result in the breakdown of parent materials into component minerals. These minerals serve as a source of macro- and micronutrients for biological systems as they are established. The breakdown of parent material also results in the formation of physical niches for establishment and growth of microbial, plant, and animal communities. As these communities become established and carry out their life cycles, organic materials are deposited in soils through the death and decay of constituent populations. This organic material is then degraded, resulting in soil organic matter.

During these soil-forming activities, distinct layers or horizons often form. These horizons make up the soil profile, which can be viewed by excavating a section and observing the vertical layering (Fig. 4.12). Generally, soils contain a dark, organic-rich surface layer, designated as the O horizon, then a lighter colored layer, designated as the A horizon, where some humified organic matter accumulates. The layer that underlies the A horizon is called the E horizon because it is characterized by **eluviation,** which is the process of removal of nutrients and inorganics out of the A horizon. Beneath the E horizon is the B horizon, which is characterized by **illuviation,** or the deposition of the substances from the E horizon into the B horizon. Beneath the B horizon is the C horizon, which contains the generally unweathered parent material from which the soil was derived. Below the C horizon is the R horizon, a designation applied to bedrock. Although certain horizons are common to most soils, not all soils contain each of these horizons.

A large proportion of the terrestrial environment contains surface soils that are unsaturated and aerobic except for brief periods of flooding during heavy rains or irrigation. Such surface soils have variable levels of productivity that depend primarily on climate (Fig. 4.13). In general, soils in cooler, wetter climates produce more vegetation than dry warmer climates. This plus the lower degradation rates at lower temperatures results in higher accumulations of organic matter in the cooler climate. Thus levels of organic matter in a mature soil range from <0.1% in a southwestern U.S. desert soil to approximately 5% in the grassland soils found in the prairie states.

A significant portion of the terrestrial environment contains saturated or wetland environments, including swamps, marshes, and bogs. These ecosystems are of increasing interest to environmental microbiologists for their potential to treat polluted waste streams such as sewage effluent and acid mine drainage. Such areas are saturated for most or all of the year because they are at or below the water table. Because wetlands are so productive in terms of vegetative growth and because they also inhibit aerobic microbial activity owing to saturated conditions, they can accumulate levels of organic matter exceeding 20% on a dry weight basis. Wetlands are important ecosystems throughout the world in areas that have a temperate climate. For example, bogs are extensive worldwide, covering 5 to 8% of the terrestrial surface. Bogs are composed of deep layers of waterlogged **peat** and a surface layer of living vegetation (Fig. 4.14). The peat layers are composed of the dead remains of plants that have accumulated over thousands of years. Thus, in these extensive areas, the production of plant material has consistently exceeded the rate of decomposition of plant material. There are several reasons for this. Because these areas are completely submerged, the dissolved oxygen in the water is quickly used up, resulting in extensive anaerobic regions. Under anaerobic conditions, the rate and extent of decomposition of organic material are much lower. A second factor is that many bogs become highly acidic (pH 3.2 to 4.2) as a result of the growth of sphagnum mosses that are an integral part of these areas. The combination of anaerobic and acidic conditions suppresses the growth of most microorganisms that are essential for plant decomposition. Canada has the most extensive bog system in the world, with 129,500,000 hectares or 18.4% of

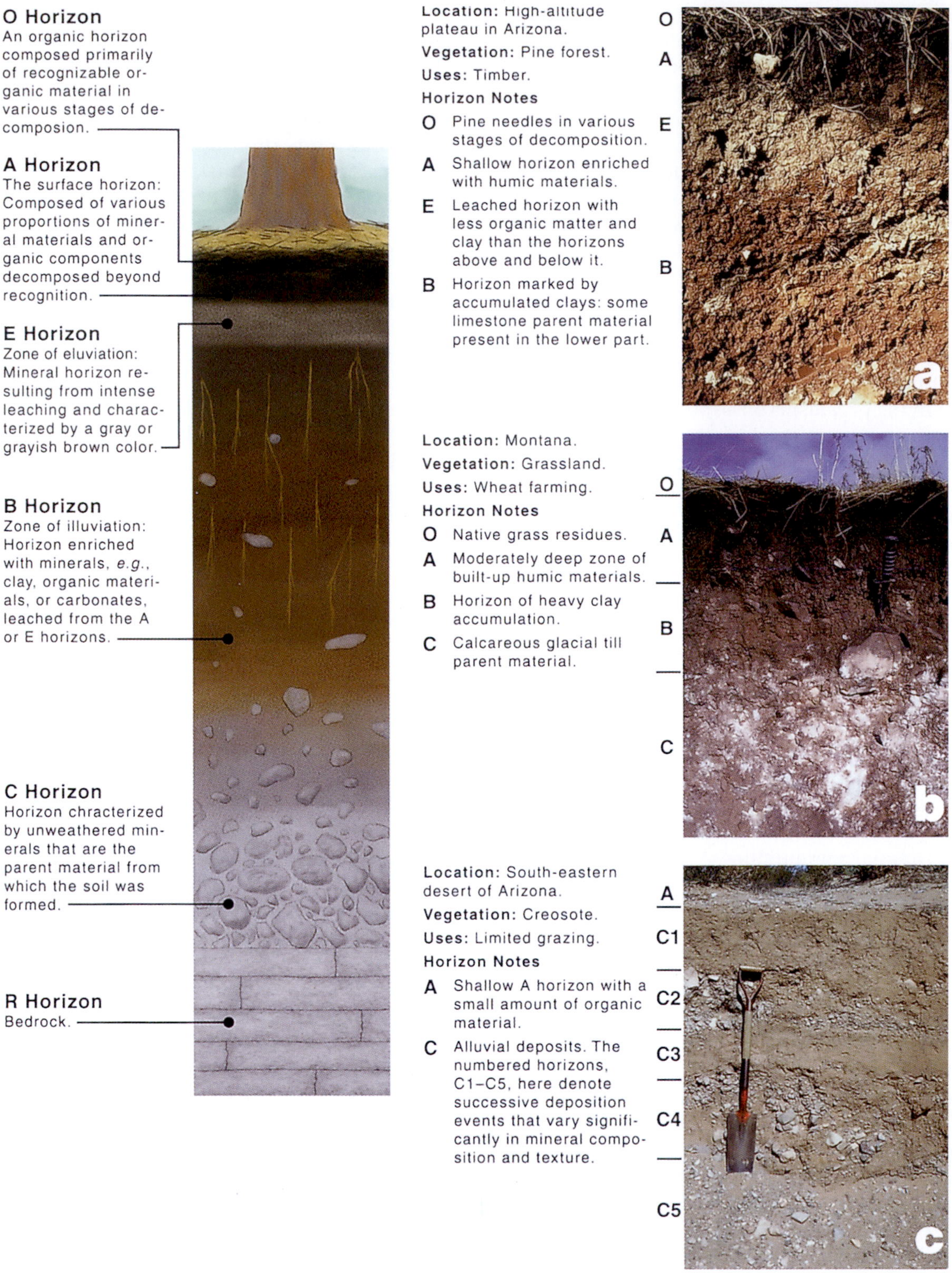

O Horizon
An organic horizon composed primarily of recognizable organic material in various stages of decomposion.

A Horizon
The surface horizon: Composed of various proportions of mineral materials and organic components decomposed beyond recognition.

E Horizon
Zone of eluviation: Mineral horizon resulting from intense leaching and characterized by a gray or grayish brown color.

B Horizon
Zone of illuviation: Horizon enriched with minerals, *e.g.*, clay, organic materials, or carbonates, leached from the A or E horizons.

C Horizon
Horizon chracterized by unweathered minerals that are the parent material from which the soil was formed.

R Horizon
Bedrock.

Location: High-altitude plateau in Arizona.
Vegetation: Pine forest.
Uses: Timber.
Horizon Notes
O   Pine needles in various stages of decomposition.
A   Shallow horizon enriched with humic materials.
E   Leached horizon with less organic matter and clay than the horizons above and below it.
B   Horizon marked by accumulated clays: some limestone parent material present in the lower part.

Location: Montana.
Vegetation: Grassland.
Uses: Wheat farming.
Horizon Notes
O   Native grass residues.
A   Moderately deep zone of built-up humic materials.
B   Horizon of heavy clay accumulation.
C   Calcareous glacial till parent material.

Location: South-eastern desert of Arizona.
Vegetation: Creosote.
Uses: Limited grazing.
Horizon Notes
A   Shallow A horizon with a small amount of organic material.
C   Alluvial deposits. The numbered horizons, C1–C5, here denote successive deposition events that vary significantly in mineral composition and texture.

FIGURE 4.12   Typical soil profiles illustrating different soil horizons. These horizons develop under the influence of the five soil-forming factors and result in unique soils. (From *Pollution Science,* © 1996, Academic Press, San Diego, CA)

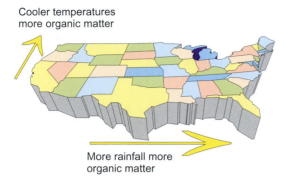

FIGURE 4.13 Effect of climate on organic matter in U.S. soils. The average organic matter content increases to the east and north because of cooler temperatures and higher rainfall. (Adapted from Plaster, 1996.)

its land area composed of bogs. Harvesting of the sphagnum mosses, more commonly known as peat moss, for use as a gardening additive has emerged as an important rural industry in Canada over the past 50 years. Other countries with extensive bog systems include Ireland, which has the most extensive blanket bog system in the world (Fig. 4.14a), and Finland, where 33.5% of the land exists as bogs.

4.3.2 Vadose Zone

The **vadose zone** is defined as the unsaturated oligotrophic environment that lies between the surface soil and the saturated zone. The vadose zone contains mostly unweathered parent materials and has a very low organic carbon content (generally <0.1%). Thus, the availability of carbon and micronutrients is very limited compared with that in surface soils. The thickness of the vadose zone varies considerably. When the saturated zone is shallow or near the surface, the unsaturated zone is narrow or sometimes even nonexistent, as in a wetland area. In contrast, there are many arid or semiarid areas of the world where the unsaturated zone can be up to hundreds of meters thick. These unsaturated regions, especially deep unsaturated regions, may receive little or no moisture recharge from the surface and normally have limited microbial activity because of low nutrient and/or moisture status. However, these regions are receiving more attention from a microbiological perspective, because pollutants that are present from surface contamination must pass through the vadose zone before they can reach groundwater. Thus, the question arises whether these unsaturated zones can be used to alter

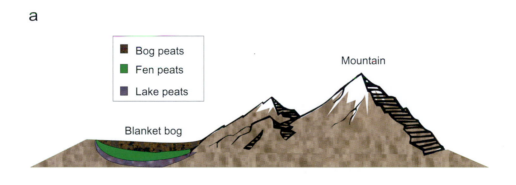

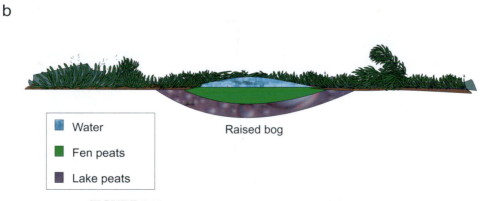

FIGURE 4.14 Blanket bog (a) and raised bog (b) formations.

pollutants moving through them or to restrict their transport to prevent them from reaching groundwater and potential drinking water supplies.

4.3.3 Saturated Zone

The saturated zone lies directly beneath the vadose zone and like the vadose zone is oligotrophic (organic carbon content <0.1%). The boundary between the vadose zone and the saturated zone is not a uniformly distinct one, because the water table can rise or fall depending on rainfall events. The area that makes up this somewhat diffuse boundary is called the **capillary fringe** (see Fig. 4.1). Saturated zones found beneath the vadose zone are commonly called aquifers and are composed of porous parent materials that are saturated with water. Underground aquifers serve as a major source of potable water for much of the world. For example, in the United States, approximately 50% of the potable water supply comes from aquifers.

There are several types of aquifers, including shallow table aquifers and intermediate and deep aquifers that are separated from shallow aquifers by confining layers (Fig. 4.15). Confining layers are composed of

materials such as clay that have very low porosity. Such layers allow little water movement between shallow and deeper aquifers. Of these different types of aquifers, **shallow aquifers** are most closely connected to the earth's surface. They receive water from rainfall events and provide recharge to adjacent streams or rivers. In addition, shallow aquifer systems are very active with rapid groundwater flows (meters per day) and hence usually remain aerobic. Confined aquifers within 300 m of the surface soil are termed **intermediate aquifers.** These have much slower flow rates, on the order of meters per year. It is this aquifer system that supplies a major portion of the drinking and irrigation water in the United States. **Deep aquifers,** those more than 300 m in depth, are characterized by extremely slow flow rates (meters per century). Because so little water flow occurs, these aquifers are usually anaerobic. Deep aquifers are not directly recharged or affected by surface rainfall events.

4.4 GENERAL CHARACTERISTICS OF MICROORGANISMS IN POROUS MEDIA

In surface soils, culturable microorganism concentrations can reach 10^8 per gram of dry soil, although direct counts are generally one to two orders of magnitude larger. These diverse microorganisms have been estimated to represent 10,000 species of bacteria alone (Turco and Sadowsky, 1995). In addition, there are substantial populations of fungi, algae, and protozoa. In general, microbial colonies are found in a nonuniform "patchlike" distribution on soil particle surfaces (Fig. 4.16). Despite the large number of microorganisms found, they make up only a small fraction of the total organic carbon and a very small proportion of the soil volume (0.001%) in most soils.

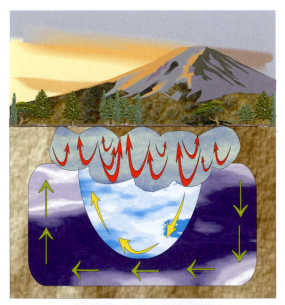

FIGURE 4.15 Shallow, intermediate, and deep aquifer systems. (Adapted from Chapelle, 1993.)

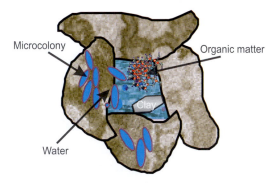

FIGURE 4.16 Niches available for exploitation by subsurface microorganisms. Note the microcolonies inside and outside the micropore formed by these mineral particles. (Adapted from Atlas and Bartha, 1993)

In subsurface environments, the same patchlike distribution of microbes exists that is found in surface soils. Culturable counts range from essentially zero to 10^7 per gram of dry soil depending on the depth and type of porous medium. Direct counts generally range from 10^5 to 10^7 cells per gram of porous medium. Thus, the difference between culturable and direct counts is often much larger in the subsurface than in surface soils. This is most likely due to the presence of **viable but nonculturable microbes.** These microbes exist as a result of the nutrient-poor status of subsurface environments, which is directly reflected in their low organic matter content. Recall that environmental bacteria have diverse, specific, nutritional needs. But they often exist under adverse conditions and as a result may be sublethally injured. Such injured bacteria cannot be cultured by conventional methods and are termed viable but nonculturable. It has been estimated that 99% of all soil organisms may be nonculturable (Roszak and Colwell, 1987). This number may actually be quite conservative, as the status of viable but nonculturable bacteria is difficult to evaluate. Although there is presently a very small database concerning these bacteria, one important point can be made here: any methodology that relies on characterizing environmental organisms via a procedure involving culture may in fact obtain a very small subsection of the total population. Further, this subsection may not be representative of the majority of the community (see Chapter 8.1.3.1).

Environmental microbiologists are currently actively exploring the distribution and activity of microorganisms in surface and subsurface environments. This knowledge is crucial to understanding and predicting the fate of chemical and biological contaminants added to the environment.

4.4.1 Distribution of Microorganisms in Porous Media

The weathered component minerals (which serve as a source of micronutrients) and organic matter (which serves as a carbon and nitrogen source) are two of the primary differences between surface soils and subsurface materials as environments for microorganisms. These differences in nutrient content are reflected in a higher and more uniform distribution of microbial numbers and activity in surface soil environments. Further, microbial density and activity are greater in both surface and subsurface environments that have high recharge from rainfall and water flow.

Microbial distribution is also dependent on soil texture and structure. As soils form, microbes attach to a site that is favorable for replication. As growth and colony formation take place, exopolysaccharides are formed, creating a "pseudoglue" that helps in orienting adjacent clay particles and cementing them together to form a microaggregate (see Fig. 4.16). Although the factors that govern whether a given site is favorable for colonization are not completely understood, several possible factors have been identified that may play a role including nutrient availability and surface properties (see also Chapters 6 and 7). In addition, in surface soils, pore space seems to be an important factor. Pore spaces in microaggregates with neck diameters less than 6 μm have more activity than pore spaces with larger diameters, because the small pore necks protect resident bacteria from protozoal predation. Pore space also controls water content to some extent. Larger pores drain more quickly than smaller pores, and therefore the interior of a small pore is generally wetter and more conducive to microbial activity. It has further been suggested that gram-negative bacteria prefer the interior of microaggregate pore space because of the increased moisture, whereas gram-positive bacteria, which are better adapted to withstand dry conditions, tend to occupy the microaggregate exteriors.

Most microorganisms in porous media are attached. It has been estimated that approximately 80 to 90% of the cells in a porous medium are sorbed to solid surfaces and the remainder are free-living. As stated earlier, attached microbes are found in patches or colonies on particle surfaces. Attachment and growth into colonies confer several advantages for microorganisms, bacteria in particular (Gilbert et al., 1993). Attachment can help protect bacteria from protozoal predation. Attachment and colony formation can also help provide localized concentrations of nutrients that are contained in and recycled among the attached cells within the colony rather than being diluted into the general environment. This is especially important in oligotrophic environments. Another advantage of colony formation is that a microbial colony can alter the immediate microsite environment surrounding the cell, such as pH, to optimize growth conditions. Finally, genetic exchange can occur much more frequently within a colony than between isolated cells in a soil environment.

Although free-living cells are less common, they are an important mechanism of dispersion of microorganisms. As nutrient supplies at a particular surface site are consumed, microorganisms need a mechanism by which cells can disperse to new sites that may have additional food supplies. Fungi spread via spores released from fruiting bodies or via hyphal extension. Bacteria, which undergo only simple cell division, need a different mechanism of dispersal, namely release of free-living daughter cells. In fact, there is evidence that bacterial cells at the surface of a colony undergo changes in their surface properties that cause the release of a newly formed daughter cell after cell division. As these free-living daughter cells grow, their surfaces undergo a chemical change that makes attachment at a new site more favorable.

4.4.2. Metabolic States of Bacteria in Porous Media

The diversity of bacteria that can potentially exist in porous media is enormous because of their different modes of nutrition and different needs with respect to terminal electron acceptors. These differences allow both organic and inorganic substrates to be metabolized under a wide variety of redox conditions. This in turn means that soil is actually a discontinuous environment within which many different kinds of organisms can coexist. Viewed collectively, any given soil or subsurface material will generate a community having large numbers of organisms and great microbial diversity. For each member of the community there must be a **habitat** and a **niche** for that organism to survive. Competition for each habitat and niche ensures the "survival of the fittest" and also that all available substrates will be utilized. Thus, in "normal" soils or sediments, all available nutrients for bacteria are eventually metabolized. Readily degradable materials such as sugars and carbohydrates will exist only transiently, whereas substrates that are more difficult to degrade (e.g., lignin) will be present for longer periods of time. In particular, humic substances (which are the stable component of organic matter) are degraded very slowly at rates of 2–5% per year. It should be noted that in some environments there may a limiting environmental factor, e.g., redox potential, or nutrient that prevents utilization of an abundant nutrient source. In fact, this is true of the vast hydrocarbon accumulations that constitute fossil fuel deposits.

The foregoing discussion illustrates that the metabolism of bacteria within porous media is likely to be limited by lack of available substrate, because by definition any bioavailable nutrient will be metabolized. Even the autochthonous microbes degrading complex organic matter components exhibit only modest levels of activity because of the rate-limiting steps involved in making these components bioavailable. The only time that bacteria within porous media exhibit high rates of metabolism is when fresh substrate is added to the medium or when a specific organism is able to exploit a previously unavailable substrate. An example of the former case occurs when plant litter is incorporated into soil by burrowing animals or insects and is subsequently degraded by zymogenous organisms. An example of the latter case occurs when a genetic mutation or a gene transfer event in a specific microbe results in the expression of enzyme systems that allow degradation of a previously undegradable substrate. In this example, the microbe now has access to plentiful substrate without competition from other organisms and is likely to metabolize at high rates of activity. The resultant growth of the organism is an example of **adaptation.** Similar high rates of metabolism can occur when an organism with specific degradative traits is "introduced" or added to the environment for the sole purpose of achieving biodegradation of an organic contaminant (see Chapter 16.7.5). Overall, then, most bacteria in porous media exist under conditions of limited starvation and stress as a result of competition for available nutrients. This is particularly true for populations in the subsurface, where extremely low amounts of available organic carbon can support only very low levels of microbial activity.

In addition to nutrient-induced stress, bacteria in porous media must combat a variety of extreme abiotic environmental factors. Moisture and temperature within porous media can be highly variable and create adverse conditions for indigenous populations. Soil or subsurface material may be exposed to a variety of organic or inorganic contaminants that can also result in abiotic stress. For example, in the United States, 37% of sites contaminated with organics are also co-contaminated with metals (Riley et al., 1992), and the effect of the double stress often reduces metabolic activity and diversity (Kovalich, 1991).

Because of the harsh physical-chemical environment and the fact that nutrients are usually limiting, most bacteria do not actively metabolize the majority of the time. In fact, they exist under stress to varying extents. Such stress can result in morphological rounding, unbalanced growth, sublethal injury, or even death (Fig. 4.17). Abiotic stress can also result in the formation of true dormant structures, such as endospores in the case of the bacterium *Bacillus*. Depending on the bacterium, such stress can be beneficial or detrimental in terms of its impact on human health and welfare. In the case of bacteria that biodegrade an organic contaminant, such stress is, of course, detri-

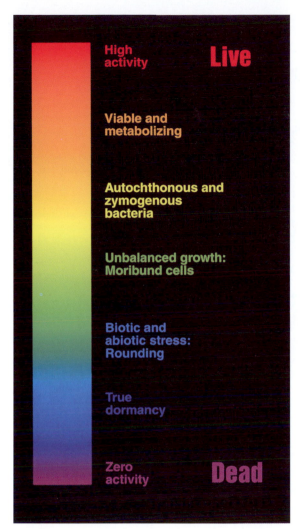

FIGURE 4.17 Physiological states of soil organisms. (From *Pollution Science,* © 1996, Academic Press, San Diego, CA)

4.5 MICROBIAL ACTIVITIES IN POROUS MEDIA

4.5.1 Role of Microorganisms in Surface Soil Formation

Soil formation is a slow process involving physical and chemical weathering and biological processes over thousands or millions of years. Biological processes are especially important in surface soil aggregate formation and soil stability. Fertile, productive soils are characterized by extensive aggregation. There are several ways in which microorganisms participate in the formation and stabilization of soil aggregates and hence soil structure (Robert and Chenu, 1992). As a soil is colonized, soil particles are bound together by filamentous microorganisms that grow over the surfaces of adjacent particles to create an extensive network of hyphae that connect the soil particles involved. In addition, microorganisms can actually cause small clay particles to reorient along the microbial surface. This causes compaction of the clay particles and aids in aggregate formation. A second mechanism by which microbes and plant roots contribute to aggregate formation is through the production of exopolysaccharides. Although it is a complex process, it is a useful exercise to try to imagine how aggregates form because it gives an appreciation for the complexity of the environment and for how microbes exist in soil (Fig. 4.18).

Past agricultural practices have disrupted the soil formation process and have resulted in loss of soil stability and productivity. The effect can be as subtle as reduced organic matter levels in soil due to increased aeration from tillage or as devastating as the impact of human activity in rain forests when deforestation occurs. Rain forests are among the most diverse and productive of the terrestrial environments; however, their soils are nutrient poor owing to extensive leaching by rainwater. As a result, there is little stable soil organic matter in these soils. The organic matter that is present comes from the organic matter within the growing vegetation and the organic matter in freshly fallen and decaying plant litter. Therefore, when a rain forest is cut down to provide fields for agriculture or simply for logging, the small amounts of nutrients remaining in the soil are quickly used up and the soils soon lose structure. Due to subsequent erosion that often occurs following the loss of the plant cover, decreased soil structure, and decreased soil nutrient levels, such soils quickly become unproductive.

mental. However, in the case of pathogens added to soil via municipal sludge amendment, such stress is beneficial in that it limits the survival period of the introduced pathogens (see Chapter 21.2.6) (Pepper *et al.,* 1993).

From our discussion of porous media as environments for microbes, we can infer that indigenous organisms within a particular material will be selected by the specific environment within that material. In addition, the overall metabolic activity within the indigenous community is likely to be low unless the medium is amended with substrate of natural or xenobiotic origin or otherwise manipulated by human activity. Introduced organisms have an even harder time. They are not well adapted and therefore cannot compete with indigenous organisms unless a specific niche is available.

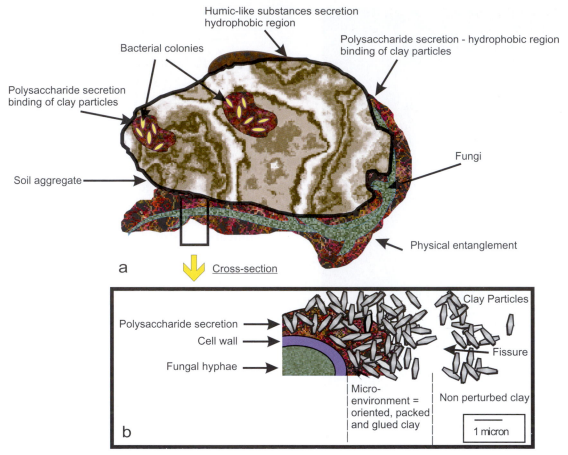

FIGURE 4.18   Microbially mediated aggregation: a) schematic representation of the binding and stabilization of a soil aggregate by microorganisms; b) detail of the microenvironment in the vicinity of a fungus.
(Adapted with permission from Robert and Chenu, 1992.

4.5.2 Role of Microorganisms in Nutrient Cycling

Geologic evidence suggests that the first microorganisms appeared approximately 3.6 billion years
ago. At that time, the earth's atmosphere was hotter
than it is now and contained no oxygen (see Chapter
14.1.2). Photosynthetic microorganisms evolved and
developed the ability to produce oxygen, which accumulated in the atmosphere. Just over 2 billion years
ago this oxygen accumulation resulted in a change
from a reducing to an oxidizing (oxygen-containing)
atmosphere. The evolution of photosynthetic microorganisms (autotrophs) and subsequently the evolution
of plants provided a mechanism for carbon fixation
and a renewable source of carbon-based energy. In addition, oxygen accumulation in the atmosphere allowed the creation of an ozone layer, which reduced
the influx of ultraviolet radiation that was harmful to
life. Consequently, carbon-dependent heterotrophic
microorganisms and higher forms of life developed.

Today, there is a delicate balance between carbon fixation and oxygen production, on the one hand, and
consumption of both fixed carbon and oxygen, on the
other hand. This balance is maintained by the carbon
cycle. Cycling is not unique to carbon. Other elements, including oxygen, nitrogen, sulfur, phosphorus, and iron, are also actively cycled (see Chapter 14).
What is interesting from an environmental microbiology perspective is that these cycles depend in large
part on microbial activity and parts of these cycles are
highly relevant to several economically important areas, including bioremediation, municipal waste disposal, sustainable agriculture, and mining. These topics are discussed in detail in Chapters 15, 16, 17, 18,
and 21.

4.5.3 Role of Microorganisms in Pollution Abatement via Bioremediation

Interest in the use of microorganisms for cleanup
or **bioremediation** of organic-polluted sites has

grown as costs associated with nonbiological remediation technologies have increased. Many of the organic contaminants that are spilled or improperly disposed of in the environment have similarity to naturally occurring carbon-based structures. However, the success of any bioremediation is site specific and depends on the soil type, the indigenous microorganisms, environmental conditions, and the type and quantity of pollutant. In instances in which bioremediation is successful, it can be very cost effective (see Chapter 16.7).

4.5.4 Role of Microorganisms in Municipal Waste Disposal

Landfarming is the practice of disposing of biosolids produced at wastewater treatment plants on agricultural land. This practice takes advantage of indigenous soil microbes to aid in further breakdown of the more recalcitrant organic compounds contained in the sludges left after treatment of municipal waste. These sludges may be composed of either solid or liquid wastes that result from municipal waste treatment. Solid municipal wastes are generated when treated waste is dewatered, dried, or in some cases composted. These solid materials can then be disposed of by transporting them to a field site, where they are disked into the soil. Liquid waste or sludges contain 1 to 8% solids by dry weight. These wastes can also be trucked to disposal field sites and the liquid sludge then injected into the soil. Land disposal of municipal sludges has advantages and disadvantages. Advantages include the application of much needed water and nutrients that come with the sludge material. Disadvantages include the application and potential buildup of pollutants such as lead and mercury that are part of the sludge material. Municipal waste treatment and disposal are discussed in detail in Chapter 21.

4.6 MICROORGANISMS IN SURFACE SOILS

Surface soils are occupied by indigenous populations of bacteria (including actinomycetes), fungi, algae, and protozoa. In general, as the size of these organisms increases from bacteria to protozoa, the number present decreases. It is also known that there may be phage or viruses present that can infect each class of organism, but information on the extent of these infectious agents in surface soils is limited. In addition to these indigenous populations, specific microbes can be introduced into soil by human or animal

activity. Human examples include the deliberate direct introduction of bacteria as biological control agents or as biodegradative agents. Microbes are also introduced indirectly as a result of application of sewage sludge to agricultural fields (see Chapter 21.2.6.1). Animals introduce microbes through bird droppings and animal excrement. Regardless of the source, introduced organisms rarely significantly affect the abundance and distribution of indigenous populations.

The following discussion is an overview of the dominant types of microbes found in surface soils, including their occurrence, distribution, and function.

4.6.1 Bacteria

Bacteria are almost always the most abundant organism found in surface soils in terms of numbers. Culturable numbers vary depending on specific environmental conditions, particularly soil moisture and temperature (Roszak and Colwell, 1987). Culturable bacteria can be as numerous as 10^7 to 10^8 cells per gram of soil, whereas total populations (including viable but nonculturable organisms) can exceed 10^{10} cells per gram. In unsaturated soils, aerobic bacteria usually outnumber anaerobes by two or three orders of magnitude. Anaerobic populations increase with increasing soil depth but rarely predominate unless soils are saturated and/or clogged. Actinomycetes are an important component of bacterial populations, especially under conditions of high pH, high temperature, or water stress. Actinomycete numbers are generally one to two orders of magnitude smaller than the total bacterial population (Table 4.4). One distinguishing feature of this group of bacteria is that they are able to utilize a great variety of substrates found in soil, especially some of the less degradable insect and plant polymers such as chitin, cellulose, and hemicellulose. Indigenous soil bacteria can be classified on the basis of their growth characterisitics and affinity for carbon substrates. As explained in Chapter 3.5, two broad categories of bacteria are found in the environment, those that are K-selected or autochthonous and those that are r-selected or zymogenous. The former metabolize slowly in soil, utilizing slowly released soil organic matter as a substrate. The latter are adapted to intervals of dormancy and rapid growth, depending on substrate availability, following the addition of fresh substrate or amendment to the soil.

Overall estimates have indicated that surface soils can contain up to 10,000 individual species (Turco and Sadowsky, 1995). Not all of these are culturable, and of course information on the nonculturable species is limited. Methodology for the detection of nonculturable-

TABLE 4.4   Characteristics of Bacteria, Actinomycetes, and Fungi

Characteristic	Bacteria	Actinomycetes	Fungi
Population	Most numerous	Intermediate	Least numerous
Biomass	Bacteria and actinomycetes have similar biomass		Largest biomass
Degree of branching	Slight	Filamentous, but some fragment to individual cells	Extensive filamentous forms
Aerial mycelium	Absent	Present	Present
Growth in liquid culture	Yes—turbidity	Yes—pellets	Yes—pellets
Growth rate	Exponential	Cubic	Cubic
Cell wall	Murein, teichoic acid, and lipopolysaccharide	Murein, teichoic acid, and lipopolysaccharide	Chitin or cellulose
Complex fruiting bodies	Absent	Simple	Complex
Competitiveness for simple organics	Most competitive	Least competitive	Intermediate
Fix N	Yes	Yes	No
Aerobic	Aerobic, anaerobic	Mostly aerobic	Aerobic except yeast
Moisture stress	Least tolerent	Intermediate	Most tolerant
Optimum pH	6–8	6–8	6–8
Competitive pH	6–8	>8	<5
Competitiveness in soil	All soils	Dominate dry, high-pH soils	Dominate low-pH soils

From *Pollution Science,* © 1996, Academic Press, San Diego, CA.

able isolates is discussed in Chapter 8.1.3.1 and Chapter 13. Most studies of soil bacteria have focused on culturable organisms, and a massive amount of information is available on key genera.

Tables 4.5, 4.6, and 4.7 identify some of the bacterial genera that are known to dominate typical surface soils and other bacterial genera that are critical to environmental microbiology. Of course, the lists are by no means all inclusive.

4.6.2 Fungi

Fungi other than yeasts are aerobic and are abundant in most surface soils. Numbers of fungi usually range from 10^5 to 10^6 per gram of soil. Despite their lower numbers compared with bacteria, fungi usually contribute a higher proportion of the total soil microbial biomass (Tables 4.4 and 4.8). This is due to their comparatively large size; a fungal hypha can range from 2 to 10 μm in diameter. Because of their large size, fungi are more or less restricted to the interaggregate regions of the soil matrix. Yeasts can metabolize anaerobically (fermentation) and are less numerous than aerobic mycelium-forming fungi. Generally, yeasts are found at populations of up to 10^3 per gram of soil. Because of their reliance on organic sources for substrate, fungal populations are greatest in the surface O and A horizons,

TABLE 4.5   Dominant Culturable Soil Bacteria

Organism	Characteristics	Function
Arthrobacter	Heterotrophic, aerobic, gram variable. Up to 40% of culturable soil bacteria.	Nutrient cycling and biodegradation.
Streptomyces	Gram positive, heterotrophic, aerobic actinomycete. 5–20% of culturable bacteria.	Nutrient cycling and biodegradation. Antibiotic production, e.g., *Streptomyces scabies*.
Pseudomonas	Gram-negative heterotroph. Aerobic or facultatively anaerobic. Possess wide array of enzyme systems. 10–20% of culturable bacteria.	Nutrient cycling and biodegradation, including recalcitrant organics. Biocontrol agent.
Bacillus	Gram-positive aerobic heterotroph. Produce endospores. 2–10% of culturable soil bacteria.	Nutrient cycling and biodegradation. Biocontrol agent, e.g., *Bacillus thuringiensis*.

TABLE 4.6   Examples of Important Autotrophic Soil Bacteria

Organism	Characteristics	Function
Nitrosomonas	Gram negative, aerobe	Converts $NH_4^+ \rightarrow NO_2^-$ (first step of nitrification)
Nitrobacter	Gram negative, aerobe	Converts $NO_2^- \rightarrow NO_3^-$ (second step of nitrification)
Thiobacillus	Gram negative, aerobe	Oxidizes $S \rightarrow SO_4^{2-}$ (sulfur oxidation)
Thiobacillus denitrificans	Gram negative, facultative anaerobe	Oxidizes $S \rightarrow SO_4^{2-}$; functions as a denitrifier
Thiobacillus ferrooxidans	Gram negative, aerobe	Oxidizes $Fe^{2+} \rightarrow Fe^{3+}$

and numbers decrease rapidly with increasing soil depth. As with bacteria, soil fungi are normally found associated with soil particles or within plant rhizospheres.

Fungi are important components of the soil with respect to nutrient cycling and especially decomposition of organic matter, both simple (sugars) and complex (polymers such as cellulose and lignin). The role of fungi in decomposition is increasingly important as the soil pH declines because fungi tend to be more acid tolerant than bacteria (Table 4.4). Some of the common genera of soil fungi involved in nutrient cycling are *Penicillium* and *Aspergillus*. These organisms are also important in the development of soil structure because they physically entrap soil particles with fungal hyphae (see Fig. 4.18). As well as being critical in the degradation of complex plant polymers such as cellulose and lignin, some fungi can also degrade a variety of pollutant molecules. The best-known example of such a fungus is the white rot fungus *Phanerochaete chrysosporium* (see Chapter 16 Case Study). Other fungi, such as *Fusarium* spp., *Pythium* spp., and *Rhizoctonia* spp., are important plant pathogens. Still others cause disease; for example, *Coccidioides immitis* causes

a chronic human pulmonary disease known as "valley fever" in the southwestern deserts of the United States. Finally, note that mycorrhizal fungi are critical for establishing plant–fungal interactions that act as an extension of the root system of almost all higher plants (see Chapter 18.2.5). Without these mycorrhizal associations, plant growth as we know it would be impossible.

4.6.3 Algae

Algae are typically phototrophic and thus would be expected to survive and metabolize in the presence of a light-energy source and CO_2 for carbon. Therefore, one would expect to find algal cells predominantly in areas where sunlight can penetrate, the very surface of the soil. One can actually find algae to a depth of 1 m because some algae, including the green algae and diatoms, can grow heterotrophically as well as photoautotrophically. In general, though, algal populations are highest in the surface 10 cm of soil (Curl and Truelove, 1986). Typical algal populations close to the soil surface range from 5000 to 10,000 per gram of soil (Rouatt *et al.*, 1960). Note that a surface soil where a visible al-

TABLE 4.7   Examples of Important Heterotrophic Soil Bacteria

Organism	Characteristics	Function
Actinomycetes, e.g., *Streptomyces*	Gram positive, aerobic, filamentous	Produce geosmins "earthy odor," and antibiotics
Bacillus	Gram positive, aerobic, spore former	Carbon cycling, production of insecticides and antibiotics
Clostridium	Gram positive, anaerobic, spore former	Carbon cycling (fermentation), toxin production
Methanotrophs, e.g., *Methylosinus*	Aerobic	Methane oxidizers that can cometabolize trichloroethene (TCE) using methane monooxygenase
Alcaligenes eutrophus	Gram negative, aerobic	2,4-D degradation via plasmid pJP4
Rhizobium	Gram negative, aerobic	Fixes nitrogen symbiotically with legumes
Frankia	Gram positive, aerobic	Fixes nitrogen symbiotically with nonlegumes
Agrobacterium	Gram negative, aerobic	Important plant pathogen, causes crown gall disease

TABLE 4.8   Approximate Range of Biomass of Each Major Component of the Biota in a Typical Temperate Grassland Soil

Component of soil biota	Biomass (tons/ha)
Plant roots	Up to 90 but generally about 20
Bacteria	1–2
Actinomycetes	0–2
Fungi	2–5
Protozoa	0–0.5
Nematodes	0–0.2
Earthworms	0–2.5
Other soil animals	0–0.5
Viruses	Negligible

From Killham (1994).

gal bloom has developed can contain millions of algal cells per gram of soil.

Algae are often the first to colonize surfaces in a soil that are devoid of preformed organic matter. Colonization by this group of microbes is important in establishing soil formation processes, especially in barren volcanic areas, desert soils, and rock faces. Algal metabolism is critical to soil formation in two ways: algae provide a carbon input through photosynthesis, and as they metabolize, they produce and release carbonic acid, which aids in weathering the surrounding mineral particles. Further, algae produce large amounts of extracellular polysaccharides, which also aid in soil formation by causing aggregation of soil particles (Killham, 1994).

Populations of soil algae generally exhibit seasonal variations with numbers being highest in the spring and fall. This is because desiccation caused by water stress tends to suppress growth in the summer and cold stress affects growth in the winter. Four major groups of algae are found in soil. The green algae or the Chlorophyta, for example, *Chlamydomonas*, are the most common algae found in acidic soils. Also widely distributed are diatoms such as *Navicula*, which are members of the Chrysophycophyta. Diatoms are found primarily in neutral and alkaline soils. Less numerous are the yellow–green algae such as *Botrydiopsis*, which are also members of the Chrysophycophyta, and the red algae (Rhodophycophyta, e.g., *Prophyridium*). In addition to these algal groups, there are the cyanobacteria (e.g., *Nostoc* and *Anabaena*), which are actually classified as bacteria but have many characteristics in common with algae. The cyanobacteria participate in the soil-forming process discussed in the previous paragraph, and some cyanobacteria also have the capacity to fix nitrogen, a nutrient that is usually limiting in a barren environment. In temperate soils the relative abundance of the major algal groups follows the order green algae > diatoms > cyanobacteria > yellow–green algae. In tropical soils the cyanobacteria predominate.

4.6.4 Protozoa

Protozoa are unicellular, eukaryotic organisms that range up to 5.5 mm in length, although most are much smaller (Table 4.9). Most protozoa are heterotrophic and survive by consuming bacteria, yeast, fungi, and algae. There is evidence that they may also be involved, to some extent, in the decomposition of soil organic matter. Because of their large size and requirement for large numbers of smaller microbes as a food source, protozoa are found mainly in the top 15 to 20 cm of the soil. Protozoa are usually concentrated near root surfaces that have high densities of bacteria or other prey. Soil protozoa are flatter and more flexible than aquatic protozoa, which makes it easier to move around in the thin

TABLE 4.9   Average Length and Volume of Soil Protozoa Compared with Bacteria

Group	Length (μm)	Volume (μm^3)	Shape
Bacteria	<1–5	2.5	Spherical to rod shaped
Flagellates	2–50	50	Spherical, pear shaped, banana shaped
Amoebae			
Naked	2–600	400	Protoplasmic streaming, pseudopodia
Testate	45–200	1000	Build oval tests or shells made of soil
Giant	6000	4×10^9	Enormous naked amoebae
Ciliates	50–1500	3000	Oval, kidney shaped, elongated and flattened

From Ingham (1998).

films of water that surround soil particle surfaces as well as to move into small soil pores.

There are three major categories of protozoa: the flagellates, the amoebae, and the ciliates (see Chapter 2.5). The flagellates are the smallest of the protozoa and move by means of one to several flagella. Some flagellates (e.g., *Euglena*), contain chlorophyll, although most (e.g., *Oicomonas*) do not. The amoebae, also called rhizopods, move by protoplasmic flow, either with extensions called pseudopodia or by whole-body flow. Amoebae are usually the most numerous type of protozoan found in a given soil environment. Ciliates are protozoa that move by beating short cilia that cover the surface of the cell. The protozoan population of a soil is often correlated with the bacterial population, which is the major food source present. Numbers of protozoa reported range from 30,000 per gram of soil from a nonagricultural temperate soil to 350,000 per gram of soil from a maize field to 1.6×10^6 per gram of soil from a subtropical area.

4.7 MICROORGANISMS IN SHALLOW SUBSURFACE ENVIRONMENTS

Although the microorganisms of surface soils have been studied extensively, the study of subsurface microorganisms is relatively new, beginning in earnest in the 1980s. Complicating the study of subsurface life are the facts that sterile sampling is problematic and many subsurface microorganisms are difficult to culture. Some of the initial studies evaluating subsurface populations were invalidated by contamination with surface microbes. As a result, study of subsurface organisms has required the development of new tools and approaches for sterile sampling (see Chapter 8.2.1) and for microbial enumeration and identification. For example, it has been demonstrated that rich media are not suitable for culturing subsurface organisms that are adapted to oligotrophic conditions. To illustrate this point, a study by Balkwill and Ghiorse (1985) compared viable counts from a full-strength PTGY (peptone, trypticase, glucose, yeast extract) agar medium with those from a 1:20 diluted PTGY medium. In all cases the viable counts from the dilute PTGY medium were higher. This difference was approximately one order of magnitude for most samples, with some samples showing a difference of up to three orders of magnitude (1000 times higher).

Because subsurface microbiology is still a developing field, information is limited in comparison with that for surface microorganisms. Yet there is still enough information available to know that subsurface environments, once thought to contain very few if any microorganisms, actually have a significant and diverse population of microorganisms. In particular, shallow subsurface zones, specifically those with a relatively rapid rate of water recharge, have high numbers of microorganisms. The majority of these organisms are bacteria, but protozoa and fungi are also present. As a general rule, total numbers of bacteria, as measured by direct counts, remain fairly constant, ranging between 10^5 and 10^7 cells per gram throughout the profile of a shallow subsurface system. For comparison, numbers in surface soils range from 10^9 to 10^{10} cells per gram. This decrease in numbers is directly correlated with the low amounts of inorganic nutrients and organic matter in subsurface materials. Subsurface eukaryotic counts are also lower than surface counts by several orders of magnitude. Low eukaryotic counts are a result of low organic matter content but, perhaps more importantly, result from removal by physical straining by small soil pores as they move downward (see Chapter 7.2.1). A final point to be made is that both prokaryotic and eukaryotic counts are highest in portions of the subsurface containing sandy sediments. This does not mean that clayey regions are not populated, but the numbers tend to be lower. This may also be due to exclusion and physical straining of microorganisms by small pores in clay-rich media.

Numbers of culturable bacteria in subsurface environments show more variability than direct counts, ranging from zero to nearly equal to direct counts. Thus, in general, the difference between direct and cultural counts in the subsurface is greater than the difference in surface soils (one to two orders of magnitude). Several factors may explain the larger difference between direct and cultural counts in the subsurface. First, because nutrients are much more limiting in the subsurface, a greater proportion of the population may be in a nonculturable state (Fig 4.17). Second, the physiological and nutritional requirements of subsurface organisms are not well understood. Therefore, even though we know that a dilute nutrient medium is better than a rich medium, the type of dilute nutrient medium used may still be inappropriate.

An interesting study of a shallow subsurface environment that included a shallow aquifer illustrates these points. In 1991, Konopka and Turco studied the microbiology of a 26-m bore that was installed at the Purdue University Agronomy Research Center (Figs. 4.19 and 4.20). This site had been in agricultural production for 40 years and contained a silt loam surface soil overlying 17 m of glacial till materials. From 17 to 20 m there was a transition to a sandy layer, which at 20 m was saturated with water. In general, organic

FIGURE 4.19 Slide of borehole at Purdue University (Photo courtesy R. F. Turco, Purdue University.)

tions in shallow subsurface environments, the question becomes what type of microorganisms make up these populations. The dominant population type is aerobic, heterotrophic bacteria, although there are small populations of eukaryotic organisms and small populations of anaerobic and autotrophic organisms. Subsurface microbes are diverse in type, although not

FIGURE 4.20 Stratigraphy of borehole showing sampling intervals and physical makeup. (Adapted from Konopka and Turco, 1991.)

matter and total nitrogen content decreased with depth, the largest decrease occurring at the interface between the A horizon (surface soil) and the E horizon. The researchers compared direct counts, viable counts, and phospholipid content as a function of depth at this site. They found normal fluctuations, approximately 1.5 orders of magnitude, in direct counts ($\sim 5 \times 10^7$ cells per gram) and phospholipid content (~ 400 μg phospholipid per gram) within the depth profile (Fig. 4.21B and C). Such fluctuations may well have been due to changes in the porous medium (see Fig. 4.20). In contrast, viable counts were much lower, averaging $\sim 1 \times 10^4$ cells per gram, a decrease of 3.5 orders of magnitude (Fig. 4.21A). One sample showing higher than normal viable counts (8×10^5 cells per gram) was from a saturated sandy sediment, the type of subsurface conditions that seem most conducive to recovery of viable bacteria.

Given that there are substantial microbial popula-

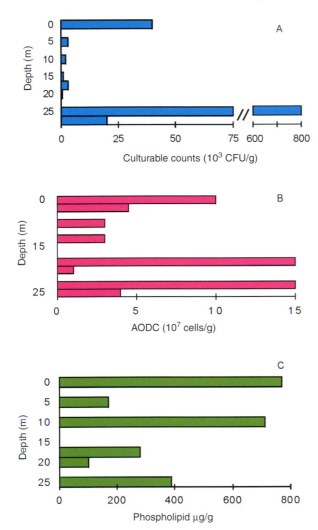

FIGURE 4.21 Distribution of (A) culturable counts on dilute PYGV agar, (B) acridine orange- stained (AODC) direct microscopic counts, and (C) phospholipid profiles as a function of depth in the profile. (Adapted from Konopka and Turco, 1991.)

viable population can quickly adapt to growth on rich nutrient media and that these organisms seem to metabolize a wide variety of organic compounds. These characteristics bode well for the premise that subsurface microorganisms can be harnessed for the *in situ* environmental cleanup of a wide variety of organic chemicals.

4.8 MICROORGANISMS IN DEEP SUBSURFACE ENVIRONMENTS

Until relatively recently, it was thought that the deep subsurface environments contain few if any microorganisms because of the extreme oligotrophic conditions found there. Interest in this area began in the 1920s, when increased consumption of oil led to increased oil exploration and production. Upon examination of water extracted from deep within oil fields, Edward Bastin, a geologist at the University of Chicago, found that significant levels of hydrogen sulfide and bicarbonate were present. The presence of these materials could not be explained on a chemical basis alone, and Bastin suggested that sulfate-reducing bacteria were responsible for the hydrogen sulfide and bicarbonate found in the drilling water. Subsequently, Frank Greer, a microbiologist at the University of Chicago, was able to culture sulfate-reducing bacteria from water extracted from an oil deposit that was hundreds of meters below the earth's surface. Bastin and Greer suggested that these microorganisms were descendants of organisms buried more than 300 million years ago during formation of the oil reservoir. However, their suggestions were largely ignored because the sampling techniques and microbial analysis techniques available at the time could not ensure that the bacteria were not simply contaminants of the drilling process.

Other research hinted at the existence of subsurface microorganisms, most notably the work of Claude Zobell. But not until the 1980s, with the growing concern over groundwater quality, did several new efforts address the questions of whether subsurface microorganisms exist and what range and level of microbial activity occur in the subsurface. Agencies involved in these new studies included the U.S. Department of Energy, U.S. Geological Survey, the U.S. Environmental Protection Agency, the German Federal Ministry of the Interior (Umbeltbundesamt), the Institute for Geological Sciences (Wallingford, England), and the Water Research Center (Medmanham, England). A number of new techniques were developed to facilitate the collection of sterile samples from deep cores in both the saturated and the unsaturated zone, and a great deal of

as diverse as surface organisms. Predominant microorganisms vary from one depth to another within the subsurface. Great diversity in heterotrophic activity has been found; and subsurface microbes have shown the capacity to degrade simple substrates such as glucose as well as more complex substrates such as aromatic compounds, surfactants, and pesticides. The physiology of subsurface microorganisms differs from that of surface microbes in that when subsurface cells are examined, they are rarely dividing and contain few ribosomes or inclusion bodies. This is not surprising considering the nutrient-limited conditions in which subsurface microbes live. What is more unexpected is that a significant portion of the information has been generated concerning the presence and function of microorganisms in deep subsurface environments (see Case Study) (Fredrickson and Onstott, 1996).

4.8.1 Microorganisms in the Deep Vadose Zone

Several studies have looked at deep cores in the unsaturated zone. In one of the first such studies, Frederick Colwell (1989) collected a 70-m core from the eastern Snake River Plain, which is a semiarid, high desert area in southeastern Idaho. Table 4.10 shows a comparison of bacterial numbers in the surface and subsurface samples from this site. Following the pattern described in Section 4.7 for shallow subsurface environments, the direct counts from deep subsurface samples remained high, declining by only one order of magnitude in comparison with the surface samples. In contrast, culturable counts declined by four orders of magnitude to less than 100 colony-forming units per gram of sediment. The majority of the isolates from the subsurface in this study were gram positive and strictly aerobic. In contrast, in surface soils gram-negative bacteria are more numerous. The subsurface atmosphere was found to be similar to ambient surface air in most samples, suggesting that the subsurface was aerobic.

Subsequent studies have largely confirmed these findings and have added some new information. In general, microbial numbers and activities are higher in paleosols (buried sediments) that have had exposure to the earth's surface and plant production. These materials tend to have associated microorganisms and nutrient reserves, even very small ones, that can maintain very slow-growing populations for thousands of years. These later studies also suggest that there are some vadose zone materials, most notably massive basalt samples collected by the Idaho National Engineering Laboratory, that lack viable microorganisms and do not show any detectable metabolic activity. In summary,

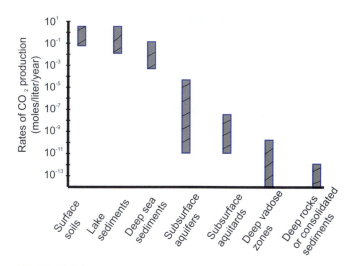

FIGURE 4.22 Ranges of rates of *in situ* CO_2 production for various surface and subsurface environments, as estimated by groundwater chemical analyses and geochemical modeling. (Adapted from Kieft and Phelps, 1997.)

our present understanding of the deep vadose zone is limited. However, it appears that there are areas of the vadose zone that contain microbes that may be stimulated to interact with environmental contaminants, whereas other areas of the vadose zone simply act as a conduit for the downward transport of contaminants.

It must be emphasized that although microbes are present in deep vadose zones, rates of metabolic activity are much lower than rates in surface soils. This is illustrated in Fig. 4.22, which depicts metabolic activity in a range of surface and subsurface environments. Metabolic activity is expressed as the rate of CO_2 production and was estimated by groundwater chemical analysis and geochemical modeling. As can be seen in this figure, the difference between the rates of CO_2 production in a surface soil and in the deep vadose zone is at least nine orders of magnitude.

4.8.2 Microorganisms in the Deep Saturated Zone

Intermediate and deep aquifers are characterized by low rates of recharge and groundwater flow that create a habitat for microorganisms different from that in shallow aquifers. Samples taken from deep cores (see Case Study) show that there is a wide diversity of microorganisms in deep saturated zones. One of the best-studied deep saturated zone sites is the U.S. Department of Energy Savannah River Site in Aiken, South Carolina. A series of boreholes were drilled to collect samples of Cretaceous sediments (70 million to 135 million years old) from depths of almost 500 m. Microbial analysis of core materials showed diverse and numer-

TABLE 4.10 A Comparison of Microbial Counts in Surface and 70-m Unsaturated Subsurface Environments

Sample site	Direct counts (counts/g)	Viable counts (CFU/g)[a]
Surface (10 cm)	2.6×10^6	3.5×10^5
Subsurface basalt–sediment interface (70.1 m)	4.8×10^5	50
Subsurface sediment layer (70.4 m)	1.4×10^5	21

[a] CFU, colony-forming units.

Case Study

Deep Probe—How Low Can Life Go?

U.S. DOE Coring Studies of Deep Cretaceous Sediments of the Atlantic Coastal Plain

In 1987, the U.S. Department of Energy sponsored the drilling of several deep boreholes in South Carolina near the Savannah River nuclear materials processing facility. A team of scientists used a sophisticated sampling device to ensure that the core samples taken from the boreholes were not contaminated with microorganisms from other parts of the borehole or the surface (Fig. 4.23). These samples were then shipped to several laboratories, where microbial analyses were initiated immediately. Results of these analyses have helped confirm the theory that subsurface bacteria are ubiquitous. In addition to the Savannah River Site, microorganisms have been recovered from formations with temperatures as high as 75°C (167°F) and from depths of 2.8 km (1.7 miles). These researchers have found that though microbes are everywhere, their abundance varies considerably depending on the site characteristics.

For example, diverse bacterial communities are commonly found in sedimentary rocks. Sedimentary materials can have a relatively rich supply of nutrients as a result of the formation process. Component sedimentary materials originally at the earth's surface acquire organic matter produced by plants. With time, these loose sands, silts, and clays are buried and consolidated into solid rock. As long as nutrients remain available, microbes living within the pores of the sediments will continue to survive and grow. Sedimentary rocks can also supply oxidized forms of sulfur, iron, and manganese, which can provide energy for growth of lithoautotrophic microorganisms. Over geologic time, these sediments become more deeply buried and compacted. During compaction, pore space is lost and can eventually become cemented with minerals that precipitate from fluids passing through the rock. As a result of this process of sedimentary rock formation, the availability of nutrients declines with depth and the overall rate of metabolism gradually diminishes, except in areas that directly surround rich concentrations of nutrients. This in turn leads to a patchy distribution of microbes. It is therefore not surprising that many early studies of life in the deep subsurface were contradictory. One can imagine how easy it would be to conclude that microbes are not present if a large enough sample is not taken.

One of the questions that has interested deep subsurface researchers is the issue of the origin of microbes in the deep subsurface. Do microbes that are sedimented with materials from the surface survive for thousands and millions of years or are they transported from the surface? It was determined that the groundwater now present at the Savannah River Site had not been in contact with the earth's surface for thousands of years, and in some of the deepest sites it is estimated that residence time is on the order of millions of years. Thus, the answer to this question seems to be that surviving subsurface bacteria have adapted to the nutrient-deprived conditions of the deep subsurface. These bacteria shrink in size and metabolize very slowly, so that they divide only once every century or less. Given this explanation, one can imagine the unique metabolic capabilities may have evolved during adaptation to subsurface life. Environmental microbiologists are examining these organisms with interest to determine whether some of the unique adaptations may be useful in the commercial sector.

ous populations of microbes, with total counts ranging from 10^6 to 10^7 cells per gram of sediment. The types of organisms detected included aerobic and facultatively anaerobic chemoheterotrophs; denitrifiers; methanogens; sulfate-reducers; sulfur-oxidizers; nitrifiers; and nitrogen-fixing bacteria. Low numbers of unicellular cyanobacteria, fungi, and protozoa were also detected in some samples (Balkwill and Boone, 1997).

Despite the high total numbers of bacteria, culturable counts were much lower. For example, Fredrickson *et al.* (1991) found 10^3 to 10^6 colony-forming units per gram of sediment from samples taken from 350 to 413 m in a permeable, saturated Middendorf sediment from this site. Numbers from a low-permeability Cape Fear sediment (450 to 470 m) ranged from nondetectable to 10^4 colony-forming units per gram of sediment. The most abundant culturable forms in these samples were aerobic or facultatively anaerobic

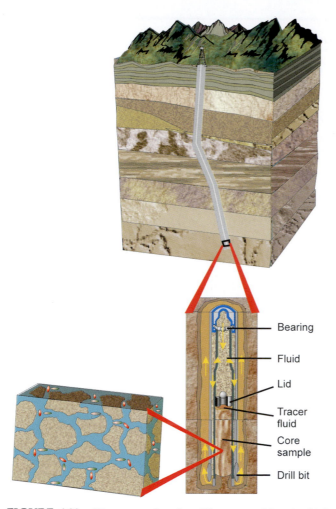

FIGURE 4.23 Slimes, or subsurface lithoautotrophic microbial ecosystems, exist in the pores between interlocking mineral grains of many igneous rocks. Autotrophic microbes (green) derive nutrients and energy from inorganic chemicals in their surroundings and, in turn, many other microbes (red) feed on organics created by autotrophs. (Adapted with permission from Fredrickson and Onstott, 1996.)

TABLE 4.11 Tentative Taxonomic Assignments of Aerobic Chemoheterotrophic Bacteria from the Savannah River Site Based on Phylogenetic Analysis of 16S Ribosomal RNA Gene Sequences

Genus or other taxonomic unit	No. of strains assigned to date[a]
Alpha-Proteobacteria	
Agrobacterium	1
Blastobacter	1
Sphingomonas	1
Uncertain affiliation[a]	7
Beta-Proteobacteria	
Alcaligenes	6
Comamonas	25
Zoogloea	1
Uncertain affiliation[a]	2
Gamma-Proteobacteria	
Acinetobacter	25
Pseudomonas	10
Uncertain affiliation[a]	1
High-G+C gram-positive bacteria	
Arthrobacter	25
Micrococcus	3
Terrebacter	6

From Balkwill and Boone (1997).
[a] Isolate falls within the indicated larger taxonomic unit but cannot be assigned to a specific genus at this time.

large part what is found in surface soils (compare Table 4.11 with Tables 4.5 and 4.7).

Questions and Problems

1. Is soil a favorable environment for microorganisms?

2. Define the terms surface soil, vadose zone, and saturated zone.

3. a) Give at least two examples of aerobic heterotrophic bacteria commonly found in surface soils.
   b) Give at least two examples of anaerobic heterotrophic bacteria commonly found in surface soils.
   c) Give at least two examples of aerobic autotrophic bacteria commonly found in surface soils.
   d) Give at least two examples of anerobic autotrophic bacteria commonly found in surface soils.
   e) Describe the contribution of each of the microbes you have listed to the ecology of the soil environment.

chemoheterotrophs. The bacteria from these samples were able to metabolize simple sugars, organic acids, complex polymers such as the storage product β-hydroxybutyric acid, and surfactants such as Tween 40 and Tween 80. Thus, subsurface microbes exhibit diverse metabolic capabilities.

Some specific types of organisms have been identified through the analysis of 16S ribosomal RNA gene sequences (see Chapter 13.3.2.2), although much work remains to be done to completely understand the types and distribution of subsurface microbes. Table 4.11 gives a list of tentative taxonomic assignments for aerobic isolates from the Savannah River Site (Balkwill and Boone, 1997). What is interesting about these data is that they suggest that subsurface microbes reflect in

4. What components of a soil make it reactive?

5. Compare and contrast microbial numbers (both cultural and direct counts) found in surface soils, the vadose zone, and the saturated zone.

6. Compare and contrast levels of microbial activity found in surface soils, the vadose zone, and the saturated zone.

7. What are viable but nonculturable microbes?

8. What are the major activities of microbes in a soil?

References and Recommended Readings

Atlas and Bartha. (1993) "Microbial Ecology." Benjamin Cummings, Redwood City, CA.

Balkwill, D. L., and Boone, D. R. (1997) Identity and diversity of microorganisms cultured from subsurface environments. *In* "The Microbiology of the Terrestrial Deep Subsurface" (P. S. Amy and D. L. Haldeman, eds.) CRC Press, Boca Raton, FL, pp. 105–117.

Bone, T. L., and Balkwill, D. L. (1988) Morphological and cultural comparison of microorganisms in surface soil and subsurface sediments at a pristine study site in Oklahoma. *Microbiol. Ecol.* 16, 49–64.

Chapelle, F. H. (1993) "Ground-Water Microbiology and Geochemistry." John Wiley & Sons, New York.

Colwell, F. S. (1989) Microbiological comparison of surface soil and unsaturated subsurface soil from a semiarid high desert. *Appl. Environ. Microbiol.* 55, 2420–2423.

Curl, E. R., and Truelove, B. (1986) "The Rhizosphere." Springer-Verlag, New York, p. 105.

Dragun, J. (1988) "The Soil Chemistry of Hazardous Materials." Hazardous Materials Control Research Institute, Silverspring, MD.

Fredrickson, J. K., and Onstott, T. C. (1996) Microbes deep inside the earth. *Sci. Am.* October, 68–73.

Fredrickson, J. K., Balkwill, D. L., Zachara, J. M., Li, S.-M. W., Brockman, F. J., and Simmons, M. A. (1991) Physiological diversity and distributions of heterotrophic bacteria in deep Cretaceous sediments of the Atlantic Coastal Plain. *Appl. Environ. Microbiol.* 57, 420–411.

Gammack, S. M., Paterson, E., Kemp, J. S., Cresser, M. S., and Killham, K. (1992) Factors affecting the movement of microorganisms in soils. *In* "Soil Biochemistry" (G. Stotzky and J.-M. Bollag, eds.), Vol. 7. Marcel Dekker, New York, pp. 263–305.

Ghiorse, W. C., and Wilson, J. T. (1988) Microbial ecology of the terrestrial subsurface. *Adv. Appl. Microbiol.* 33, 107–172.

Gilbert, P., Evans, D. J., and Brown, M. R.W. (1993) Formation and dispersal of bacterial biofilms *in vivo* and *in situ*. *J. Appl. Bacteriol Symp. Suppl.* 74, 67S–78S.

Ingham, E. R. (1998) Protozoa and nemadees. *In* "Principles and Applications of Soil Microbiology," (D. M. Sylvia, J. J. Fuhrmann, P. G. Hartel, and D. A. Zuberer, eds.) Prentice-Hall, Upper Saddle, NJ, pp. 114–131.

Kieft, T. L., and Phelps, T. J. (1997) Life in the slow lane: Activities of microorganisms in the subsurface. *In* "The Microbiology of the Terrestrial Deep Subsurface" (P. S. Amy and D. L. Haldeman, eds.) CRC Lewis, Boca Raton, FL, pp. 137–163.

Killham, K. (1994) "Soil Ecology." Cambridge University Press, Cambridge.

Konopka, A., and Turco, R. (1991) Biodegradation of organic compounds in vadose zone and aquifer sediments. *Appl. Environ. Microbiol.* 57, 2260–2268.

Kovalich, W. (1991) Perspectives on risks of soil pollution and experience with innovative remediation technologies. Fourth World Congress of Chemical Engineering, Karlsruhe, Germany, June 16–21, 1991, pp. 281–295.

Matthess, G., Pekdeger, A., and Schroeder, J. (1988). Persistence and transport of bacteria and viruses in groundwater—a conceptual evaluation. *J. Contam. Hydrol.* 2, 171–188.

Pepper, I. L., Josephson, K. L., Bailey, R. L., Burr, M. D., and Gerba, C. P. (1993) Survival of indicator organisms in Sonoran Desert soil amended with sewage sludge. *J. Environ. Sci. Health.* A28, 1287–1302.

Plaster, E. J. (1996) "Soil Science and Management." Delmar Publishers, New York, p. 24.

Reeves, R. H., Reeves, J. Y., and Balkwill, D. L. (1995) Strategies for phylogenetic characterization of subsurface bacteria. *J. Microbiol. Methods* 21, 235–251.

Riley, R. G., Zachara, J. M., and Wobber, F. J. (1992) Chemical contaminants on DOE lands and selection of contaminant mixtures for subsurface science research. Report DOE/ER–0547T, U.S. Department of Energy, Washington, DC.

Robert, M., and Chenu, C. (1992) Interactions between soil minerals and microorganisms. *In* "Soil Biochemistry" (G. Stotzky and J.-M. Bollag, eds.), Vol. 7. Marcel Dekker, New York, pp. 307–418.

Roszak, D. B., and Colwell, R. R. (1987) Survival strategies of bacteria in the natural environment. *Microbiol. Rev.* 51, 365–379.

Rouatt, J. W., Katznelson, H., and Payne, T.M.B. (1960) Statistical evaluation of the rhizosphere effect. *Soil Sci. Soc. Am. Proc.* 24, 271–273.

Schwarzenbach, R. P., Gschwend, P. M., and Imboden, D. M. (1993) "Environmental Organic Chemistry." JohnWiley & Sons, New York.

Sposito, G. (1989) "The Chemistry of Soils." Oxford University Press, New York.

Turco, R. F., and Sadowsky, M. (1995). The microflora of bioremediation. *In* "Bioremediation: Science and Applications" (H. D. Skipper and R. F. Turco, eds.). Special Publ. No. 43, Soil Science Society of America, Madison, WI, pp. 87–102.

5

Aeromicrobiology

SCOT E. DOWD   RAINA M. MAIER

5.1 INTRODUCTION

In the 1860s a new area of biology was conceived based on the work of Louis Pasteur. Pasteur was able to show that invisible, airborne particles were responsible for the mysterious fermentative reactions in his experiments, which were, up until that point, considered by scientists of the time to be spontaneous generation of life (Ariatti and Comtois, 1993). In the 1930s, F. C. Meier coined the term **aerobiology** to describe a project that involved the study of life in the air (Boehm and Leuschner, 1986). Since then, aerobiology has been defined by many as the study of the aerosolization, aerial transmission, and deposition of biological materials. Others have defined it more specifically as the study of diseases that may be transmitted via the respiratory route (Dimmic and Akers, 1969). Despite the variations in definition, this relatively new science is becoming increasingly important in many aspects of such diverse scientific fields as public health, environmental science, industrial engineering, agricultural engineering, biological warfare, and space exploration.

The first part of this chapter will introduce the basics of aerobiology, including the nature of aerosols, the fundamentals of the **aeromicrobiological (AMB) pathway,** common bioaerosol sampling methods, and aerobiological transport modeling. The remainder of the chapter will then focus on a subset of the science that we shall term aeromicrobiology. **Aeromicrobiology,** as defined for the purpose of this text, involves various aspects of **intramural** (indoor) and **extramural** (outdoor) **aerobiology** as they relate to the airborne transmission of environmentally relevant microorgan-

isms, including viruses, bacteria, fungi, yeasts, and protozoans.

5.2 IMPORTANT AIRBORNE PATHOGENS

Generally, as indicated by the previous definitions of aerobiology, one usually associates airborne microorganisms with disease occurrence in humans, animals, or plants. Numerous plant pathogens are spread by the aeromicrobiological pathway (Table 5.1). Up to 70% of all plant diseases are caused by fungi such as wheat rusts that can be spread by airborne transmission. Aerial transmission is capable of transporting these phytopathogens many thousands of kilometers. The impact of airborne plant pathogens, especially fungi, on the economy of the agricultural industry is in the billions of dollars each year.

There are also numerous airborne pathogenic microorganisms that infect animals (Table 5.2). Infection of pets and livestock by airborne microorganisms also costs the public and livestock owners billions of dollars each year. For example, foot-and-mouth disease virus is known to be transmitted by the aeromicrobiological pathway. In 1967, an outbreak of foot-and-mouth disease in England that lasted only 4 months affected over 2300 farms and resulted in the loss of almost 450,000 animals. Finally, the airborne transmission of many pathogens (Table 5.3) such as *Legionella pneumophila, Mycobacterium tuberculosis,* and newly recognized pathogens such as the Sin Nombre virus (hantavirus) is associated with human infection and disease.

TABLE 5.1   Important Airborne Plant Pathogens

Plant Diseases	Pathogens
Fungal Diseases	
Dutch Elm disease	*Ceratocystis ulmi*
Apple rust	*Gymnosporangium* spp.
Potato late blight	*Phytophthora infestans*
Banana leaf spot	*Mycosphaerella musicola*
Blossom infection	*Sclerotinia laxa*
Cedar rust	*Gymnosporangium* spp.
Leaf rust	*Puccinia recondita*
Crown rust of oats	*Puccinia coronata*
Fusiform rust of southern pines	*Cronartium fusiforme*
Loose smut of wheat	*Ustilago tritici*
Beef downy mildew	*Perospora* spp.
Downy mildew	*Pseudoperonospora humuli*
Maize rust	*Puccinia sorghi*
Annosus root rot	*Fomes annosus*
Powdery mildew of barley	*Erysiphe graminis*
Southern corn leaf blight	*Helminthosporium maydis*
Stem rust of wheat and rye	*Puccinia graminis*
Tonbacco blue mold	*Peronospora tabacina*
Sigatoka disease of bananas	*Mycosphaerella musicola*
White pine blister rust	*Cronartium ribicola*

TABLE 5.2   Important Airborne Animal Pathogens

Diseases of Animals	Pathogens
Bacterial Diseases	
Tuberculosis	*Mycobacterium bovis*
Glanders	*Actinobacillus mallei*
Brucellosis	*Brucella* spp.
Salmonellosis	*Salmonella* spp.
Fungal Diseases	
Aspergillosis	*Aspergillus* spp.
Cryptococcosis	*Cryptococcus* spp.
Coccidioidomycosis	*Coccidioides immitis*
Viral Diseases	
Canine herpes	Herpesviridae
Eastern equine Encephalomyelitis	Alphavirus
Hog cholera	Pestivirus
Influenza	Influenza virus
Feline distemper	Morbillivirus
Fowl plaque	
Rabies	Rhabdoviridae
Canine distemper	Morbillivirus
Newcastle disease	
Infectious bronchitis	Influenza, others
Foot and mouth disease	Aphthovirus
Rhinderpeste	Morbillivirus
Ephemeral fever	
Infectious laryngotracheitis	

TABLE 5.3   Important Airborne Human Pathogens

Human Diseases	Pathogens
Bacterial Diseases	
Brucellosis	*Brucella melitensis*
Pulmonary tuberculosis	*Mycobacterium tuberculosis*
Glanders	*Actinobacillus mallei*
Pneumonia	*Clamydia psittaci*
Pneumonia	*Klebsiella pneumoniae*
Pulmonary anthrax	*Bacillus anthracis*
Staph, respiratory infection	*Staphylococcus aureus*
Strep. respiratory infection	*Streptococcus pyogenes*
Legionellosis	*Legionella* spp.
Meningococcal infection	*Neisseria meningitidis*
Pneumonic plague	*Yersinia pestis*
Typhoid fever	*Salmonella typhi*
Whooping cough	*Bordetella pertussis*
Tuleremia	*Francisella tularensis*
Diptheria	*Corynebacterium diptheriae*
Fungal Diseases	
Aspergillosis	*Aspergillus fumigatus*
Blastomycosis	*Blastomyces dermatiridi*
Coccidioidomycosis	*Coccidioides immitis*
Cryptococcosis	*Cryptococcus neoformans*
Histoplasmosis	*Histoplasma capsulatum*
Nocardiosis	*Nocardia asteriodes*
Sporotrichosis	*Sporotrchum schenckii*
Viral Diseases	
Influenza	Influenza virus
Hemorrhagic fever	Bunyavirus
Hantavirus pulmonary syndrome	Hantavirus
Hepatitis	Hepatitis virus
Chicken pox	Herpes virus
Common cold	Picornavir
Yellow fever	Flavivirus
Dengue fever	Flavivirus
Lyssa fever	Lyssavirus
Pleurodynia	Coxsackievirus, Echovirus
Rift Valley fever	Phlebovirus
Rubella	Rubivirus
Measles	Morbillivirus
Protozoal Diseases	
Pneumocystosis	*Pneumocystis carinii*

Many of these pathogens such as the rhinovirus, which causes the common cold, affect almost everyone during their lifetime. The purchase of medications and treatments for symptoms of the common cold keeps numerous companies, such as those that make tissue paper and cold medicines, economically viable. This indicates how important the AMB pathway is to public health, not to mention the economy.

5.3 IMPORTANT AIRBORNE TOXINS

Also to be considered by aeromicrobiologists are microbial toxins. One example is a toxin from *Clostridium botulinum* (botulinal A toxin) that is a potential biological warfare agent (SIPRI, 1979). Botulinal toxin is a neurotoxin that is normally associated with ingestion of contaminated food. However, the lethal dose is so small that aerosolization can also be a means of dissemination. The lethal dose for botulinal toxin by inhalation is 0.3 μg of toxin, with death expected 12 hours after exposure. Symptoms are caused by inhibition of the production of acetylcholine at nerve endings. Death is a result of asphyxiation caused by the paralysis of respiratory muscles. Another toxin produced by bacteria is staphylococcal enterotoxin. This toxic protein is highly resistant to inactivation in association with the AMB pathway. It is rarely associated with long-lasting effects but can incapacitate those exposed to it. On occasion this toxin can be fatal. The lethal dose is estimated to be 25 μg by inhalation. The symptoms include cramping, vomiting, and diarrhea, which occur within 1 hour of exposure by aerosolization.

Other examples of airborne toxins are lipopolysaccharides (LPS) (Hurst *et al.*, 1997). LPS is derived initially from the cells of gram-negative bacteria and, when released into the air, can cause various respiratory distress syndromes. LPS molecules, also referred to as endotoxins, are highly antigenic biological agents that, when associated with airborne particles such as dust, are often associated with acute respiratory symptoms such as chest tightness, coughing, shortness of breath, fever, and wheezing. There are many sources associated with the production of high levels of LPS, such as cotton mills, haystacks, sewage treatment plants, solid waste handling facilities, swine confinement buildings, poultry houses, and even homes and office buildings. LPS is liberated when gram-negative bacteria in these environments are lysed but can also be released when they are actively growing.

Lipopolysaccharides are one of the major constitutive elements of the outer membranes of gram-negative bacteria. Because gram-negative bacteria are ubiquitous in the environment, LPS is also ubiquitous and is considered by some to be the most important aerobiological allergen. LPS (Fig. 5.1) has three major components: a lipid A moiety, which is a disaccharide of phosphorylated glucosamines with associated fatty acids; a core polysaccharide; and an O-side chain. The lipid A moiety and the core polysaccharide are similar among gram-negative bacteria, but the O-side chain

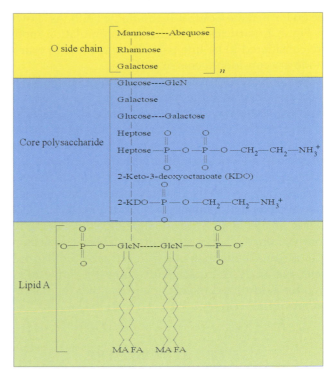

FIGURE 5.1 Schematic structural representation of the lipopolysaccharide from *Salmonella typhimurium*. The number of repeating units (n) in the side chain varies from 10 to 40. The sugars found in the side chain vary among bacterial species, whereas the composition of the core polysaccharide is usually the same. There is a molecule of beta-hydroxymyristic acid (MA), which is a 14-carbon fatty acid attached to each N-Acetylglucosamine (GlcN) residue. Other fatty acids (FA) are attached to these residues as well.

varies among species and even strains. It is the O-side chain that is responsible for the hyperallergenic reaction. The dose of LPS required to initiate the toxic effect by inhalation is less than 10 ng.

5.4 NATURE OF BIOAEROSOLS

To better understand bioaerosols and the AMB pathway, one must understand the nature of aerosols in general. Bioaerosols vary considerably in size, and composition depends on a variety of factors including the type of microorganism or toxin, the types of particles they are associated with such as mist or dust, and the gases in which the bioaerosol is suspended. Bioaerosols in general range from 0.02 to 100 μm in diameter and are classified on the basis of their size. The smaller particles (<0.1 μm in diameter) are considered to be in the **nuclei mode,** those ranging from 0.1 to 2 μm are in the **accumulation mode,** and larger particles are considered to be in the **coarse mode** (Committee on Particulate Control Technology, 1980). As shown in Fig. 5.2, particles in nuclei or accumulation mode are considered to be fine particles and those in coarse mode are considered coarse particles.

The composition of bioaerosols can be liquid or solid or a mixture of the two and should be thought of as microorganisms associated with airborne particles or as airborne particles containing microorganisms. This is because it is rare to have microorganisms (or toxins) that are not associated with other airborne particles such as dust or mist. This information is derived from particle size analysis experiments, which indicate that the average diameter of airborne bacterial particles is greater than 5 μm (Fengxiang *et al.*, 1992). By comparison, the average size of a soil borne bacterium, 0.3 to 1 μm, is less than one fifth this size. Similar particle size analysis experiments show the same to be true for aerosolized microorganisms other than bacteria, including viruses.

5.5 THE ATMOSPHERE

The AMB pathway by its nature involves the atmosphere. The layer of most interest and significance in aeromicrobiology is the **boundary layer,** which is the name given to the earth's atmosphere extending to a height of about 0.1 km from the earth's surface. It should be noted, however, that airborne transport of microorganisms is by no means limited to this layer

and it is not uncommon to have microorganisms associated with layers of the **troposphere** above the turbulent boundary layer. However, it is the surface boundary layer that is largely responsible for the transport of particles over both short and long distances. The boundary layer consists of three parts: the laminar boundary layer, the turbulent boundary layer, and the local eddy layer (Fig. 5.3).

The **laminar boundary layer** is a layer of still air associated with the earth and all projecting surfaces, whether solid or liquid. This layer can be anywhere from 1 μm to several meters thick, depending on weather conditions. Still conditions cause the thickness of this layer to increase, and windy conditions minimize it to a very close association with surfaces. The **turbulent boundary layer** is the layer that is considered to be always in motion and responsible for horizontal transport phenomena (wind dispersion), which occur whenever microorganism-associated particles are launched either indoors or outdoors. In the lower levels of the turbulent layer, the linear flow of air is interrupted by surface projections and their associated laminar boundary layers. This interaction results in the formation of friction against the airflow. This friction, which is apparent in the form of local areas of "swirling" turbulence, gives the final layer its name. The **local eddy layer** is the actual zone of interaction between the still laminar boundary layer of surface projections and the turbulent boundary layer.

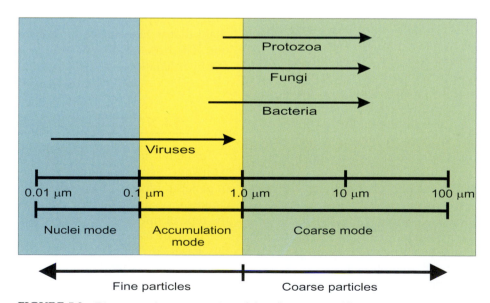

FIGURE 5.2 Diagrammatic representation of the relative sizes of bioaerosols. The depictions of the various kinds of organisms are indicative of their potential sizes when associated with airborne particles (rafts). The terminologies used to describe the various sizes of the bioaerosols are also indicated.

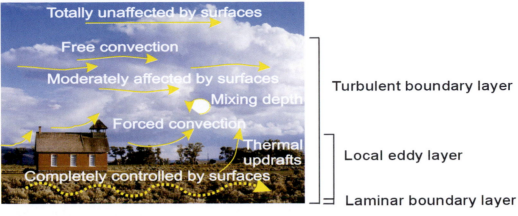

FIGURE 5.3 Atmospheric layers and airflow patterns. The figure shows the airflow patterns associated with the boundary layer. Upper air layers are relatively unaffected by objects such as buildings and airflow is unhindered. The middle layers are only moderately affected, and lower layers known as the local eddy layers are increasingly controlled by the presence of surface objects. The laminar boundary layer is a thin layer of still air associated with all surfaces. This laminar layer is still even under extremely windy conditions though the thickness decreases with increasing turbulence.

5.6 AEROMICROBIOLOGICAL PATHWAY

The aeromicrobiological pathway describes the **launching** of bioaerosols into the air, the subsequent **transport** via **diffusion** and **dispersion** of these particles, and finally their **deposition**. An example of this pathway is that of liquid aerosols containing the influenza virus launched into the air through a cough, sneeze, or even through talking. These virus-associated aerosols are dispersed by a cough or sneeze, transported through the air, inhaled, and deposited in the lungs of a nearby person, where they may begin a new infection (Fig. 5.4). Traditionally, the deposition of viable microorganisms and the resultant infection are given the most attention, but all three processes

(launching, transport, and deposition) are of equal importance in understanding the aerobiological pathway.

5.6.1 Launching

The process whereby particles become suspended within the earth's atmosphere is termed launching. Because bioaerosols must be launched into the atmosphere to be transported, it is important to understand this process. The launching of bioaerosols is mainly from terrestrial and aquatic sources, with greater airborne concentrations or atmospheric loading being associated with terrestrial sources than with aquatic sources. Some researchers speculate that there may even be atmospheric sources of bioaerosols in addition to

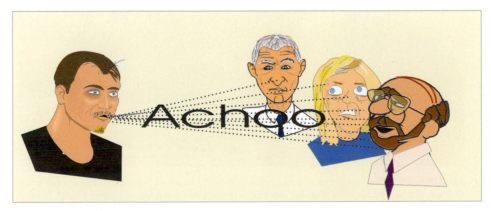

FIGURE 5.4 As shown in this figure, a cough or sneeze launches infectious microbes into the air. Anyone in the vicinity may inhale the microbes resulting in a potential infection.

terrestrial and aquatic ones. This phenomenon is related to the limited potential for microorganisms to reproduce while airborne. This, however, is an area of aeromicrobiology for which there is little available information.

Launching into the surface boundary layers can include, but is certainly not limited to, diverse mechanisms such as: air turbulence created by the movement of humans, animals, and machines; the generation, storage, treatment, and disposal of waste material; natural mechanical processes such as the action of water and wind on contaminated solid or liquid surfaces; and the release of fungal spores as a result of natural fungal life cycles.

Airborne particles can be launched from either point, linear, or area sources. A **point source** is an isolated and well-defined site of launching such as a pile of biosolid material (Fig. 5.5), before it is applied over the field. Point sources tend to display a general conical-type dispersion (Fig. 5.6). Point sources can be further defined on the basis of the type of launching phenomenon: 1) **instantaneous point sources,** for example, a single event such as a sneeze, or 2) **continuous point sources,** from which launching occurs over extended periods of time, such as the biosolid pile.

In contrast to point sources, **linear sources,** and **area sources,** involve larger, less well defined areas. When considered on the same size scale, linear and area sources display more particulate wave dispersion as opposed to the conical type of dispersion displayed by point sources. Linear and area sources can also be divided into instantaneous and continuous launching points of origin. For example, an instantaneous linear source might be a passing aircraft releasing a biological warfare agent (Fig. 5.7). A continuous area source might be exemplified by release of bioaerosols from a large field that has received an application of biosolids.

5.6.2 Transport

Transport or dispersion is the process by which kinetic energy provided by the movement of air is transferred to airborne particles, with resultant movement from one point to another. This "energy of motion" gained by airborne particles is considerable and can result in dissemination of airborne microorganisms over long distances. Transport of bioaerosols can be defined in terms of time and distance. **Submicroscale transport** involves short periods of time, under 10 minutes, as well as relatively short distances, under 100 m. This type of transport is common within buildings or other confined spaces. **Microscale transport** ranges from 10 minutes to 1 hour and from 100 m to 1 km and is the most common type of transport phenomenon. **Mesoscale transport** refers to transport in

FIGURE 5.5 This is a photo of a large pile of biosolid material, which is being loaded onto a mechanical spreader. This biosolid material will then be placed upon a nearby field. These areas, because of the physical disturbance caused by the machinery, are associated with large bioaerosol loading rates. The biosolid pile is an example of a continuous point source because bioaerosols can be generated over a long period of time and the source is relatively small and well defined. (Photo courtesy of S. D. Pillai.)

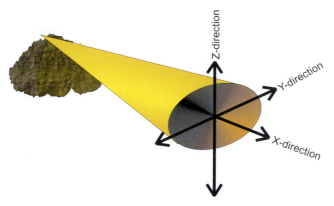

FIGURE 5.6 Schematic representation of the type of bioaerosol distribution expected from a point source. This figure shows the three plains of diffusion: 1) x-direction is the mean direction in which the wind is blowing; 2) y-direction is the lateral diffusion; and 3) z-direction is the vertical diffusion.

the submicroscale and microscale. It should be noted, however, that some viruses, spores, and spore-forming bacteria have have been shown to enter into mesoscale and even macroscale transport. The hoof-and-mouth disease outbreak in England described in Section 5.2 involved transport by wind of over 60 km (mesoscale) (Hugh-Jones and Wright, 1970). Other studies have shown that coliforms aerosolized from sewage treatment plants have been transported over 1.2 km (macroscale). Another interesting example is influenza **pandemics.** Pandemics are epidemics that occur over a wide geographic area, and influenza has been shown to spread from east to west around the world. These pandemics have been positively correlated with the prevailing trade winds (Fig. 5.8). This can be considered global transport and illustrates how important the aeromicrobiological pathway can be in microbiological dissemination, especially in relation to disease.

terms of days and distances up to 100 km, and in **macroscale transport,** the time and distance are extended even further. Because most microorganisms have limited ability to survive when suspended in the atmosphere, the most common scales considered are

As bioaerosols travel through time and space, different forces act upon them such as diffusion, inactivation, and ultimately deposition. **Diffusion** is the scattering and/or dissipation of bioaerosols in response to a concentration gradient as well as gravity, and is gen-

FIGURE 5.7 This photo shows a linear bioaerosol source using the example of the release of biological warfare agents. This is an illustration of an instantaneous linear bioaerosol release. (Corel Gallery www.corel.com)

FIGURE 5.8 This figure shows the estimated path of the 1957 Asian flu as it migrated around the globe. This indicates the ability of microorganisms (in this case virus) to spread worldwide. The Asian flu of 1957 was believed to have started in central China as indicated by the black circle. In early 1998, a new influenza virus strain known as the Avian flu made a cross-species jump from chickens to humans. There was considerable concern that this new human pathogen would follow a similar path as the Asian flu causing another pandemic.

erally aided by airflow and atmospheric **turbulence.** The amount of turbulence associated with airflow and, thus, the relative amount of diffusion that may occur in association with particulates such as bioaerosols can be estimated using the method of Osbert Reynolds. Reynolds found that factors associated with mean **wind velocity,** the **kinetic viscosity** of the air, and the relative dimension of the interfering structures could provide an indication of the amount of turbulence associated with linear airflow. Without turbulence, airborne particles from a point source would travel in a concentrated stream directly downwind. The **Reynolds equation** is written as follows:

$$\text{Reynolds number} = \frac{\text{velocity} \times \text{dimension}}{\text{viscosity}} \quad \text{(Eq. 5.1)}$$

Consider, for instance, a situation in which there are relatively high winds (500 cm/sec) that are passing over a small bush (24 cm). Because the occurrence of frictional turbulence associated with an object depends on the wind velocity being high enough and the object it is flowing over being large enough, we find that at normal air viscosity (0.14 cm^2/sec) the Reynolds number **(Re)** becomes

$$\text{Re} = \frac{500\,\text{cm/sec} \times 24\,\text{cm}}{0.14\,\text{cm}^2/\text{sec}} = 85{,}700 \quad \text{(Eq. 5.2)}$$

The limiting value for the Reynolds equation is usually considered to be 2000, with values above this number indicating turbulent conditions. The higher the value, the higher the relative turbulence of the airflow, and the higher the turbulence of the airflow, and the greater the microorganism-associated particle diffusion that occurs per unit time. In the preceding ex-

ample, one would expect a great deal of turbulence around the bush, which would increase the diffusion rates of passing bioaerosols.

When dealing with particulate transport over time and distance, Tayler (1915) indicated that diffusion during horizontal transport could be viewed as an increase in the standard spatial deviation of particles from the source over time. What does this mean? For an instantaneous point source under the influence of a mean wind direction, spread would be a standard spatial deviation from a linear axis (x) extending from the source (origin) in the mean direction of wind flow, with diffusion caused by turbulence occurring in the lateral (y) and vertical (z) axes (Fig. 5.6). The standard deviation of particulate diffusion cannot be considered constant over a particular spatial orientation but is instead dependent on the time taken to reach the particular distance. Mathematical models that attempt to estimate the transport of airborne particles use this basic premise as a foundation for predictions. To picture this concept, imagine standing at the door of a room, where someone is holding a smoking candle. If there is no air current in the room the smoke will still eventually reach you at the door but it will be very diffuse as it is spreading in every other direction as well. However, if there is a fan behind the person holding the smoking candle and this fan is pointed at the door, then the smoke from the candle will be carried by this air current. It will travel the same distance as it did before, but it will travel faster, undergo less diffusion, and as a result be more concentrated when it reaches you. This is the principle of time-dependent diffusion as indicated by Tayler's theory.

5.6.3 Deposition

The last step in the AMB pathway is deposition. An airborne bioaerosol will eventually leave the turbulence of the suspending gas and will ultimately be deposited on a surface by one or a combination of interrelated mechanisms. These mechanisms are discussed in the following sections and include: gravitational settling, downward molecular diffusion, surface impaction, rain deposition, and electrostatic deposition, to name a few of the most important. These processes are linked in many ways, and even though viewed separately, they all combine to create a constant, if not steady, deposition of particles.

5.6.3.1 Gravitational Settling

The main mechanism associated with deposition is the action of gravity on particles. The force of gravity acts upon all particles heavier than air, pulling them down and essentially providing spatial and temporal limitations to the spread of airborne particles. Steady-state gravitational deposition (Fig. 5.9) in the absence of air movement can be described in very simplistic terms by **Stokes law,** which takes into account gravitational pull, particle density, particle diameter, and air viscosity. The Stokes equation is solved in terms of terminal velocity (v) and can be used to predict the behavior of particles having diameters of biological significance (1 to 100 μm). Stokes law is written as follows:

$$v = \frac{\rho d^2 g}{18\eta}, \text{cm/sec} \qquad \text{(Eq. 5.3)}$$

where ρ is the particle density (g/cm^3); d the particle diameter, cm; g the acceleration due to gravity (cm/sec^2); and η the viscosity of air, (g/cm-sec).

Consider an airborne clostridial spore with a radius of 0.0001 cm (~2 μm diameter) with a particle density close to 1.3 g/cm^3. Under normal gravitational acceleration, which is 981 cm/sec^2, and under conditions of normal air viscosity (18°C), which is 1.8×10^{-4} g/cm-sec, the terminal velocity can be calculated:

$$v = \frac{1.3 \text{ g/cm}^3 \times (0.0002 \text{cm})^2 \times 981 \text{cm/sec}^2}{18(1.8 \times 10^{-4} \text{ g/cm-sec})} \qquad \text{(Eq. 5.4)}$$

$$= 0.016 \text{ cm/sec}$$

Thus, the clostridial spore will settle at 0.016 cm/sec under normal conditions. Larger particles will obtain higher terminal velocities and thus settle out of the

AMB pathway faster. It should be noted, however, that for particles of microbiological relevance that are exposed to winds above 8×10^3 m/hr, gravitational deposition may be negligible unless the particles cross out of the laminar flow via processes such as downward molecular diffusion or increase in density because of condensation reactions such as rain deposition.

5.6.3.2 Downward Molecular Diffusion

Downward molecular diffusion, as indicated by the name, can be described as a randomly occurring process caused by natural air currents and eddies that promote and enhance the downward movement of airborne particulates (Fig. 5.10). These random movements exist even in relatively still air and tend to be in the downward direction because of gravitational effects. As a result, measured rates of gravitational deposition tend to be greater than those predicted by the Stokes equation. The increase in the rate of deposition is due to the added effects of downward molecular diffusion. Molecular diffusion is also influenced by the force of the wind. Molecular diffusion-enhanced deposition rates tend to increase with increasing wind speed and turbulence.

5.6.3.3 Surface Impaction

Surface impaction is the process by which particles make contact with surfaces, such as leaves, trees, walls, and computers. With impaction there is an associated loss of kinetic energy. In nature, it is rare to find flat, smooth surfaces on which wind currents are unobstructed. Thus, surface impaction is a very critical

FIGURE 5.9 Schematic representation of gravitational settling, which is a function of the earth's gravitational pull, particle density, particle diameter, and the viscosity of air. This figure does not take into account random air movement. Stokes equation was developed to give an estimate of the terminal velocity achieved by particles as a function of gravitational settling.

FIGURE 5.10 Schematic representation of downward molecular diffusion, a naturally occurring process caused by the air currents and eddies that promote and enhance gravitational settling of airborne particles. Although molecular diffusion can occur in any direction, due to the effects of gravity, the overall trend of the process results in net downward movement and deposition.

factor influencing transport and deposition, especially for bioaerosols.

Impaction potential is the relative likelihood that an airborne object will collide with another object in its path. Impaction does not necessarily result in permanent deposition, however. Once a particle collides with an object, it has the potential to bounce. Bouncing off a surface causes the particle to reenter the air current at a lower rate, which can have one of two effects: (1) it can allow subsequent downward molecular diffusion and gravitational settling to occur, resulting in deposition on another nearby surface, or (2) it can allow the particle to escape the surface and once again reenter the air current. Studies have shown that impaction is influenced by the velocity and size of the particle as well as the size and shape of the surface it is approaching. To visualize the two extremes of this effect, consider a small airborne particle traveling slowly toward another small round object (Fig. 5.11). This small particle has much less chance of impaction with the small object than a larger particle traveling at a greater velocity toward a much larger flat object. Thus, the larger particle has a higher impaction potential.

5.6.3.4 Rain and Electrostatic Deposition

Rainfall and electrostatic charge also can affect deposition. Rainfall deposition occurs as a condensation reaction between two particles (raindrop and bioaerosol), which combine and create a bioaerosol with a greater mass, which settles faster. This can be described mathematically using the Stokes equation (Eq. 5.1). In the example presented, a clostridial spore alone has a calculated terminal velocity of 0.016 cm/sec. The same spore (bioaerosol), if it condensed with

another particle such as a water droplet, has a greater mass and thus a greater terminal velocity. For instance, if the clostridial spore were to condense with a water droplet that doubled the bioaerosol density from 1.3 to 2.6 g/cm^3, the terminal velocity would be increased from 0.016 to 0.032 cm/sec. The overall efficiency of rain deposition also depends on the spread area of the particle plume. Larger, more diffuse plumes undergo stronger impaction than smaller, more concentrated plumes. Rain deposition is also affected by the intensity of the rainfall. The heavier the rainfall, the greater the overall rates and numbers of the condensation reactions and the greater the subsequent increase in rain deposition.

Electrostatic deposition also condenses bioaerosols, but is based on electrovalent particle attraction. All particles tend to have some type of associated charge. Microorganisms typically have an overall negative charge associated with their surfaces at neutral pH. These negatively charged particles can associate with other positively charged airborne particles, resulting in electrostatic condensation. The major phenomenon occurring may be a coagulation effect between particles (much like the condensation of the clostridial spore with the water droplet), which would increase the bioaerosol mass and enhance deposition. It might also be assumed that as an electromagnetically charged bioaerosol comes into close proximity with an electromagnetically charged surface, electroattractive or electrorepulsive influences may be present.

5.7 MATHEMATICAL MODELING

When microorganism-associated particles are aerosolized or launched into the atmosphere, the major question becomes, how far will they travel and in what concentration? Although the answer to this question is complex, mathematical equations have been developed to attempt to simulate this process. Initial models such as Tayler's (1915) time dependent diffusion model were simplistic in reasoning but provided a solid foundation which subsequent modelers have built upon to modify and develop more sophisticated models. Modeling the aerobiological pathway is intended for use in prediction of airborne bioaerosol concentrations in the vicinity of a contaminated source. Models are useful in that they can help determine effective sampling locations and in some cases provide valuable information when sampling and analysis are not possible (Gerba and Goyal, 1982).

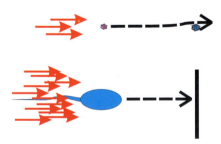

FIGURE 5.11 Schematic representation of forces that influence impaction potential. In the top figure a small particle represented by a icosahedral virus is traveling slowly toward another small particle (target) also represented by an icosahedral virus. The small size and slow velocity of the red virus and the small size and round shape of the target result in a low impaction potential. On the other hand a large object with greater mass, represented as a bacterium, traveling at a high velocity toward a large flat surface has a very high impaction potential.

5.7.1 Point Source Modeling

The classical model of plume spread from a point source was developed in 1961 by Pasquill, who described the transport of airborne particles in general. The model requires the input of four variables: (1) the mean wind speed; (2) the atmospheric stability class (which influences diffusion); (3) the downwind distance from the origin of the point of interest (sampler); and (4) the source height. Pasquill's equation for the transport of inert particles is given as follows:

$$\chi(x,y,z) = \frac{Q}{2\pi\sigma_y\sigma_z\bar{u}} \exp\left[-\frac{1}{2}\left(\frac{y}{\sigma_y}\right)^2\right]$$

$$\left\{\exp\left[-\frac{1}{2}\left(\frac{z-H}{\sigma_y}\right)^2\right]\right. \qquad \text{(Eq. 5.5)}$$

$$\left.+ \exp\left[-\frac{1}{2}\left(\frac{z+H}{\sigma_x}\right)^2\right]\right\}$$

where χ is the particle concentration at a particular location (x, y, z) downwind (Fig. 5.6) from the source (particles/m^3), H is the height of the source (m), Q is the rate of release from the source (particles/sec), $\bar{u}$ is the mean wind velocity (m/sec), and σ_x and σ_y refer to the diffusion coefficient in the y and z directions measured in meters (m).

The assumptions of the model are that: (1) the particles display a Gaussian distribution in both vertical and horizontal planes, (2) the particles reflect from the ground, (3) there is a continuous point source, (4) wind velocity and direction are constant (or averaged) for the modeled time and distance, (5) the modeled surface is flat, (6) gravitational settling is negligible, and (7) the particle and wind velocities are essentially the same.

To illustrate how this equation can be used, consider the pile of biosolid material shown in Fig. 5.5. Assume that the biosolid pile is releasing *Salmonella* cells at a rate of 2×10^3 cells/sec (Q) and there is an air sampling device located 10 m downwind (x) that is situated at ground level ($H = 0$). We will assume the wind has a mean velocity (μ) of 5 m/sec and the diffusion of the spores is equal (2 m) in the y and z directions at the time of sampling. The equation can now be solved for particle concentration in one direction (x, downwind) which allows both y and z to be set to zero thus simplifying the equation to:

$$\chi(x, 0, 0) = \frac{2 \times 10^3 \text{ cells/sec}}{2\pi\,(2\text{ m})_y\,(2\text{ m})_z\,(5\text{ m/sec})} \qquad \text{(Eq. 5.6)}$$

$$= 16 \text{ cells/m}^3$$

Lighthart and Frisch (1976) modified Pasquill's equation in order to better model the transport of microorganisms. They did this by incorporating a factor that accounts for biological decay (BD), caused by environmental stress.

$$\chi(x, y, z: H)_{BD} = \chi(x, y, z: H)\exp(-\lambda_1 t) \qquad \text{(Eq. 5.7)}$$

where λ_1 is the BD rate constant (per second), and t is the time. Consider, for example, that the time taken to reach the sampler in the previous example is 2 seconds (based on the average wind velocity) and the BD rate constant (Section 5.9) for the *Salmonella* cells is equal to 1.92×10^{-4}. Using these values to calculate the effect of BD (Eq. 5.8) one can see that a 2 second exposure of *Salmonella* has little effect on its viability.

$$16 \text{ cells/m}^3\,[\exp\,(-1.92 \times 10^{-4})(2\text{ sec})] = 16 \text{ cells/m}^3 \qquad \text{(Eq. 5.8)}$$

If the bioaerosol sampler were positioned 100 m (instead of 10 m) downwind from the source, the *Salmonella* cell concentration would decrease. This is a result of diffusion causing a lateral and vertical expansion of the plume. The effect of diffusion on the measured concentration of *Salmonella* is shown in Fig. 5.12. A second factor influencing the *Salmonella* concentration is biological decay. At 100 m downwind source, the *Salmonella* will have been airborne for 20 seconds. If both diffusion and biological decay are considered, Eq. 5.6 and Eq. 5.7 predict there will be 0.16 cells/m^3 100 m downwind from the source. Note that again, BD has an insignificant effect on the predicted concentration.

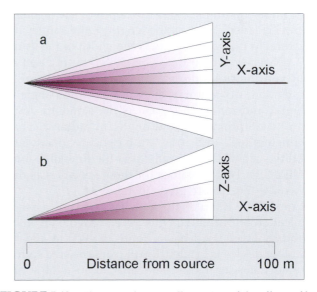

FIGURE 5.12 This is a schematic illustration of the effects of lateral (a) and vertical (b) diffusion on the downwind concentration of a microbe. As shown in this figure, as the distance from the source increases, the microbe concentration on the x-axis decreases due to lateral and vertical spreading. Note that darker shades of color represent higher concentrations of microbes.

5.7.2 Area Source Modeling

The previous model is used to describe plume spread from a point source. Another model, by Parker *et al.* (1977), describes continual launching from line or area sources such as a land application site immediately after biosolids have been applied. This model is as follows:

$$\chi(x > x_0, y) = \frac{Q}{\sqrt{2\pi \bar{u} \sigma_z\{x\} y_0}}$$

$$\times \left\{ 1 + 2 \sum_{n=1}^{\infty} \left\{ \exp\left[-\frac{1}{2}\left(\frac{2nH_m}{\sigma_z\{x\}}\right)^2 \right] \right\} \right\}$$

$$\times \left\{ \text{erf}\left[\frac{y_0/2 + y}{\sqrt{2}\,(\sigma_y\{x\})} \right] \right.$$

$$\left. + \text{erf}\left[\frac{y_0/2 - y}{\sqrt{2}\,(\sigma_y\{x\})} \right] \right\} \times \exp(-K\bar{t}) \quad \text{(Eq. 5.9)}$$

where

$$\sigma_z\{x\} \left\{ \begin{array}{ll} \dfrac{\sigma E' x_0}{\ln\left[\dfrac{\sigma E'(x + x_0) + \sigma_{z0}}{\sigma E'(x) + \sigma_{z0}} \right]} & \text{when; } x < 3x_0 \\ \\ or \ \sigma E'(x + x_0/2) + \sigma_{z0} & \text{when; } x \geq 3x_0 \end{array} \right\}$$

(Eq. 5.10)

and

$$\sigma_y\{x\} = \sigma A'(x + x_0/2) \quad \text{(Eq. 5.11)}$$

In these equations, H_m is the depth of the turbulent boundary layer (m), x_o is the length of the source area collateral to the wind (m), y_o is the width of the source area perpendicular to the wind direction (m), and σ_{z0} is the vertical source dimension (m). For Eq. 5.10, $\sigma E'$ is the standard deviation of the wind elevation angle (in radians), and for Eq. 5.11, $\sigma A'$ is the standard deviation of the horizontal wind angle (in radians). These parameters ($\sigma E'$ and $\sigma A'$) are based on atmospheric stability such as that described by the Reynolds number (Eq. 5.1). Typical values for these variables as they are used in Eq. 5.9 are given in Table 5.4.

Consider a 100m² field that has received a recent application of biosolids. On a given day the turbulent boundary layer (H_m) has a depth of 1000 m and the sampling device is located 100 m from the area source. We shall consider σ_{z0} to be 0 (zero) because the source is at ground level. Otherwise, conditions are the same as those described for Eq. (5.5). Solving Eq. 5.10 and considering that $x < 3x_o$, we obtain the following:

$$\sigma_z\{x\} = \frac{0.3491(100m)}{\ln \dfrac{0.3491(100m + 100m) + 0}{0.3491(100m) + 0}} = 50.4 \text{ m}$$

(Eq. 5.12)

$$\sigma_y\{x\} = 0.1745(100m + 100m/2) = 26.2 \text{ m} \quad \text{(Eq. 5.13)}$$

Inserting these variables into the original equation, we obtain the following:

$$\chi(x > x_0, y) = \frac{2 \times 10^3 \text{ cells/sec}}{\sqrt{2\pi(5m/sec)(50.4)(100m)}}$$

$$\times \left\{ 1 + 2 \sum_{n=1}^{\infty} \left\{ \exp\left[-\frac{1}{2}\left(\frac{2n(1000m)}{50.4m}\right)^2 \right] \right\} \right\}$$

$$\times \left\{ \text{erf}\left[\frac{100m/2 + 0}{\sqrt{2}(26.2)} \right] \right. \quad \text{(Eq. 5.14)}$$

$$\left. + \text{erf}\left[\frac{100m/2 - 0}{\sqrt{2}(26.2)} \right] \right\}$$

$$\times \exp(-1.92 \times 10^{-4})(20 \text{ sec}) = 0.06 \text{ cells/m}^3$$

5.7.3 Indoor Air Modeling

The point source and area source models and variations of these models can be used for a wide variety of outdoor situations. Indoor modeling of bioaerosols, however, is very different and more complex in many ways. This is because buildings are isolated environments made up of many individual compartments. In each of these compartments, distinctly different air mixing patterns can occur, which increases the complexity of the modeling efforts. Central heating and cooling systems can introduce additional sources of air turbulence and means for the dissemination of organisms. Heating and cooling systems can also be the actual origins of microbial contamination, as in the case of legionellosis (see later).

A general model for indoor aerosols was introduced by Nazaroff and Cass (1986). The model describes a set of interconnected chambers. Within each chamber (i), the rate of change in the bioaerosol concentration (C_{ijk}) over time, for each organism (k), and for each section of the chamber (j), is given by:

$$\frac{d(C_{ijk})}{dt} = S_{ijk} - (L_{ijk})(C_{ijk}) \quad \text{(Eq. 5.15)}$$

where S_{ijk} is the source term indicating the origin of bioaerosols, which includes direct emission, advective

TABLE 5.4   Variables for Area Source Modeling Relating to Atmospheric Stability Conditions

Atmospheric stability	Parameters		
	$\sigma A'$	$\sigma E'$	H_m
Stable	0.0524	0.0524	30
Transitional	0.08727	0.1745	100
Unstable	0.1745	0.3491	1,000

transport from other chambers, advective transport from the extramural environment, and coagulation of mass from smaller particles into the section (*j*); and L_{ijk} is the sum of all sinks, including loss to the surfaces, removal by ventilation and filtration, and various forms of deposition. If the input parameters, S_{ijk} and L_{ijk}, and the initial conditions, $C_{ijk}(t_o)$, are provided, the equation provides a solution for the change in S_{ijk} over time (Zannetti, 1990). Because of the inherent complexity of modeling the indoor environment, the description of the derivatives of this equation (other than the basic mathematics involved after the final values for L_{ijk}, S_{ijk}, and C_{ijk} have been inserted) is beyond the scope of this discussion.

5.8 SAMPLING DEVICES FOR THE COLLECTION OF BIOAEROSOLS

Many devices have been designed for the collection of bioaerosols. Choosing an appropriate sampling device is based on many factors, such as availability, cost, volume of air to be sampled, mobility, sampling efficiency (for the particular type of bioaerosol), and the environmental conditions under which sampling will be conducted. Another factor that must be taken into account, especially when sampling for microorganisms, is the overall biological sampling efficiency of the device. This factor is related to the maintenance of microbial viability during and after sampling. In this section, several types of commonly used samplers are described on the basis of their sampling methods: impingement, impaction, centrifugation, filtration, and deposition. **Impingement** is the trapping of airborne particles in a liquid matrix; **impaction** is the forced deposition of airborne particles on a solid surface; **centrifugation** is the mechanically forced deposition of airborne particles using inertial forces of gravity; **filtration** is the trapping of airborne particles by size exclusion; and **deposition** is the collection of airborne particles using only naturally occurring deposition forces. The two most commonly used devices for microbial air sampling are the all glass **AGI-30 impinger** (Ace Glass, Vineland, NJ) and the **Anderson six-stage impaction sampler (6-STG,** Andersen Instruments Incorporated, Atlanta, GA.).

5.8.1 Impingement

The AGI-30 (Fig. 5.13) operates by drawing air through an inlet that is similar in shape to the human nasal passage. The air is transmitted through a liquid medium where the air particles become associated with the fluid and are subsequently trapped. The AGI-

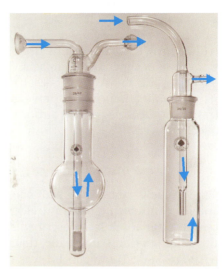

FIGURE 5.13   This is a schematic representation of two all glass impingers (AGI). The impinger on the right is the classic AGI-30 impinger. Arrows indicate the direction of air flow. The air enters the impinger drawn by suction. As bioaerosols impinge into the liquid collection medium contained in the bottom of the impinger, the airborne particles are trapped within the liquid matrix. (Photo courtesy Ace Glass Inc., Vineland, NJ.)

30 impinger is usually run at a flow rate of 12.5 l/min at a height of 1.5 m, which is the average breathing height for humans. The AGI-30 is easy to use, inexpensive, portable, reliable, easily sterilized, and has high biological sampling efficiency in comparison with many other sampling devices. The AGI-30 tends to be very efficient for particles in the range of 0.8 to 15 μm. The usual volume of collection medium is 20 ml and the typical sampling duration is approximately 20 minutes, which prevents evaporation during the sampling of warm climates or freezing of the liquid medium when sampling at lower temperatures. Another feature of the impingement process is that the liquid and suspended microorganisms can be concentrated or diluted, depending on the requirements for analysis. Liquid impingement media can also be divided into subsamples in order to test for a variety of microorganisms by standard cultural and molecular methods such as those described in Chapters 10 and 13. The impingement medium can also be optimized to increase the relative biological recovery efficiency. This is important, because during sampling the airborne microorganisms, which are already in a stressed state because of various environmental pressures such as ultraviolet (UV) radiation and desiccation, can be further stressed if a suitable medium is not used for recovery. Sampling media range from simple to complex. A simple medium is 0.85% NaCl, which is an osmotically balanced, sampling medium used to prevent osmotic shock of recovered organisms. A more complex

medium is peptone (1%) which is used as a resuscitation medium for stressed organisms. Finally, enrichment or defined growth media can be used to sample selectively for certain types of organisms. The major drawback when using the AGI-30 is that there is no particle size discrimination, which prevents accurate characterization of the sizes of the airborne particles that are collected.

5.8.2 Impaction

Unlike the AGI-30, the Andersen six-stage impaction sampler (Andersen 6-STG) provides accurate particle size discrimination. It is described as a multi-level, multiorifice, cascade impactor (Jensen *et al.*, 1992). The Andersen 6-STG (Fig. 5.14) was developed by Anderson in 1958 and operates at a input flow rate of 28.3 l/min. The general operating principle is that

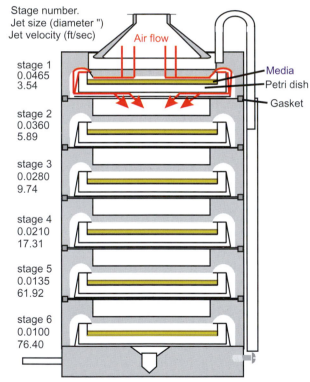

FIGURE 5.14 This is a schematic representation of the Andersen six stage impaction air sampler. Air enters through the top of the sampler and larger particles are impacted upon the surface of the petri dish on stage 1. Smaller particles, which lack sufficient impaction potential follow the air stream to the subsequent levels. As the air stream passes through each stage the air velocity increases thus increasing the impaction potential so that particles are trapped on each level based upon their size. Therefore larger particles are trapped efficiently on stage 1 and slightly smaller particles on stage 2 and so on until even very small particles are trapped on stage 6. The Andersen six stage thus seperates particles based upon their size.

air is sucked through the sampling port and strikes agar plates. Larger particles are collected on the first layer, and each successive stage collects smaller and smaller particles by increasing the flow velocity and consequently the impaction potential. The shape of the Andersen sampler does not conform to the shape of the human respiratory tract, but the particle size distribution can be directly related to the particle size distribution that occurs naturally in the lungs of animals. The lower stages correspond to the alveoli and the upper stages to the upper respiratory tract. The Andersen sampler is constructed of stainless steel with glass petri dishes, allowing sterilization, ease of transport, and reliability. It is useful over the same particle size range as the AGI-30 (0.8 to over 10 μm), corresponding to the respirable range of particles. It is more expensive than the AGI-30, and the biological sampling efficiency is somewhat lower because of the method of collection, which is impaction on an agar surface. Analysis of viruses collected by impaction is also somewhat difficult, because after impaction, the viruses must be washed off the surface of the impaction medium and collected before assay. In contrast, bacteria or other microorganisms can be grown directly on the agar surface. Alternatively, these microbes can be washed off the surface and assayed using other standard methodologies as described in Chapter 9. The biggest single advantage of the Andersen 6-STG sampler is that particle size determinations can be obtained. Thus, the two reference samplers (AGI-30 and Andersen 6-STG) complement each other's deficiencies.

5.8.3 Centrifugation

Centrifugal samplers use circular flow patterns to increase the gravitational pull within the sampling device in order to deposit particles. The Cyclone, a tangential inlet and return flow sampling device, is the most common type (Fig. 5.15). These samplers are able to sample a wide range of air volumes (1–400 l/min), depending on the size of the unit. The unit operates by applying suction to the outlet tube, which causes air to enter the upper chamber of the unit at an angle. The flow of air falls into a characteristic tangential flow pattern, which effectively circulates air around and down along the inner surface of the conical glass housing. As a result of the increased centrifugal forces imposed on particles in the airstream, the particles are sedimented out. The conical shape of the upper chamber opens into a larger bottom chamber, where most of this particle deposition occurs. Although these units are able to capture some respirable-size particles, in order to

Return
air flow

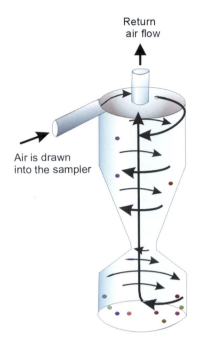

Air is drawn
into the sampler

FIGURE 5.15 This is a schematic representation of a tangential inlet and returned flow centrifugal air sampler. Air is drawn into the sampler at an angle (tangential) to the walls of the device so that it circulates around and down the walls. As it circulates the decrease in the diameter of the sampling body causes a dramatic increase in the velocity of the air and subsequently on particle's terminal velocity. This increase in gravitational settling potential causes the particles to be trapped in the lower collection chamber because their "centrifugally increased" mass prevents them from exiting with the return air flow.

trap microorganisms efficiently, the device must be combined with some type of metered fluid flow that acts as a trapping medium. This unit, when used by someone proficient, can be effective for microbiological air sampling. It is relatively inexpensive, easily sterilized, and portable, but it lacks high biological sampling efficiency and particle sizing capabilities. Analysis is performed by rinsing the sampler with an eluent medium, collection of the eluent, and subsequent assay by standard methodologies.

5.8.4 Filtration and Deposition

Filtration and deposition methods are both widely used for microbial sampling for cost and portability reasons. Filter sampling requires a vacuum source and involves passage of air through a filter, where the particles are trapped. Membrane filters can have variable pore sizes tend to restrict flow rates. After collection, the filter is washed to remove the organisms before analysis. Filtration sampling for microorganisms is not highly recommended because it has a low over-

all sampling efficiency and it is not portable. However, in many cases the low cost makes it an attractive method.

One case in which filtration is routinely used is in sampling for airborne LPS. The sampling and analysis procedure for airborne LPS levels is slightly different from methods used for analysis of airborne microorganisms. The most efficient means of sampling is usually filter collection using polyvinyl chloride or glass fiber membrane filters. Quantification analysis is usually done using a chromogenic *Limulus* amoebocyte lysate assay (Hurst *et al.*, 1997). This system uses a *Limulus* amoebocyte lysate obtained from blood cells of horseshoe crabs. The lysate contains a enzyme-linked coagulation system, which is activated by the presence of LPS. With the addition of a substrate, the system is able to quantitate, by luminescence, the amount of environmental LPS by comparison with a standard curve.

Deposition sampling is by far the easiest and most cost-effective method of sampling. Deposition sampling can be accomplished merely by opening an agar plate and exposing it to the wind, which results in direct impaction, gravity settling, and other depositional forces. The problems with this method of sampling are: low overall sampling efficiency because it relies on natural deposition, no defined sampling rates or particle sizing, and an intrinsic difficulty in testing for multiple microorganisms with varied growth conditions. Analysis of microorganisms collected by depositional sampling is similar to impaction sample analysis.

5.9 MICROBIAL SURVIVAL IN THE AIR

The atmosphere is an inhospitable climate for microorganisms mainly because of desiccation stress. This results in a limited time frame in which microbes can remain biologically active. Many microorganisms, however, have specific mechanisms that allow them to be somewhat resistant to the various environmental factors that promote loss of biological activity. Spore-forming bacteria, molds, fungi, and cyst-forming protozoa all have specific mechanisms that protect them from harsh gaseous environments, increasing their ability to survive aerosolization. For organisms that have no such specific mechanisms the survival in aerosols can often be measured in seconds. In contrast, organisms with these mechanisms can survive indefinitely.

As a result, viability is highly dependent on the environment, the amount of time the organism spends in the environment, and the type of microorganism. In addition, microbes may be viable but nonculturable (see Chapters 3.5, 4.4.2, and 10.1.2), but for simplicity in this chapter we will use the term viable rather than the term culturable. Many environmental factors have been shown to influence the ability of microorganisms to survive. The most important of these are relative humidity and temperature. Oxygen content, specific ions, UV radiation, various pollutants, and AOFs (air-associated factors) are also factors in the loss of biological activity. Each of these factors is discussed in the following sections.

The loss of biological activity can be termed **inactivation** and can generally be described using the following equation:

$$X_t = X_o^{-kt}$$ (Eq. 5.16)

where X_t represents the viable organisms at time t; X_o is the starting concentration; and k is the inactivation constant which is dependent on the particular species of microorganisms as well as a variety of environmental conditions.

5.9.1 Relative Humidity

The **relative humidity** or the relative water content of the air has been shown to be of major importance in the survival of airborne microorganisms. Wells and Riley (1937) were among the first to show this phenomenon, indicating that as the relative humidity approaches 100%, the death rate of *Escherichia coli* increases. In general, it has been reported that most gram-negative bacteria associated with aerosols tend to survive for longer periods at low relative humidities. The opposite tends to be true for gram-positive bacteria, which tend to remain viable longer in association with high relative humidities (Theunissen *et al.*, 1993). Thus, the ability of a microorganism to remain viable in a bioaerosol is related to the organism's surface biochemistry. One mechanism that explains loss of viability in association with very low relative humidity is a structural change in the lipid bilayers of the cell membrane. As water is lost from the cell, the cell membrane bilayer changes from the typical crystalline structure to a gel phase. This structural phase transition affects cell surface protein configurations and ultimately results in inactivation of the cell (Hurst *et al.*, 1997).

Early studies by Loosli *et al.* (1943) showed that the influenza virus was also adversely affected by an increase in relative humidity. More recent work suggests that viruses possessing enveloped nucleocapsids (such as the influenza virus) have longer airborne survival when the relative humidity is below 50%, whereas viruses with naked nucleocapsids (such as the enteric viruses) are more stable at a relative humidity above 50% (Hers and Winkler, 1973). It should be noted that viruses with enveloped nucleocapsids tend to have better survival in aerosols than those without. Some viruses are also stable in the AMB pathway over large ranges of relative humidity, which makes them very successful airborne pathogens.

5.9.2 Temperature

Temperature is a major factor in the inactivation of microorganisms. In general, high temperatures promote inactivation, mainly associated with desiccation and protein denaturation, and lower temperatures promote longer survival times. When temperatures approach freezing, however, some organisms lose viability because of the formation of ice crystals on their surfaces. The effects of temperature are closely linked with many other environmental factors, including relative humidity.

5.9.3 Radiation

The main sources of radiation damage to microorganisms including bacteria, viruses, fungi, and protozoa are the shorter UV wavelengths, and ionizing radiation such as x rays. The main target of UV irradiation damage is the nucleotides that make up DNA. Ionizing radiation or x rays cause several types of DNA damage, including single strand breaks, double strand breaks, and alterations in the structure of nucleic acid bases. UV radiation causes damage mainly in the form of intrastrand dimerization, with the DNA helix becoming distorted as thymidines are pulled toward one another (Freifelder, 1987). This in turn causes inhibition of biological activity such as replication of the genome, transcription, and translation.

Several mechanisms have been shown to protect organisms from radiation damage. These include association of microbes with larger airborne particles, possession of pigments or carotenoids, high relative humidity, and cloud cover, all of which tend to absorb or shield bioaerosols from radiation. Many types of organisms also have mechanisms for repair of the DNA damage caused by UV radiation. An example of an organism that has a radiation resistance mechanism is *Dienococcus radiodurans*. *D. radiodurans* is a soil bacterium that is considered the most highly radiation-resistant organism that has been yet been isolated. An important component of its radiation resistance is the ability to enzymatically repair damage to chromosomal DNA. The repair mechanism used by these bacteria is so highly efficient that much of the metabolic energy of the cell is dedicated exclusively to this function.

5.9.4 Oxygen, OAF, and Ions

Oxygen, open air factors (OAFs), and ions are environmental components of the atmosphere that are difficult to study at best. In general, it has been shown that these three factors combine to inactivate many species of airborne microbes. **Oxygen toxicity** is not related to the dimolecular form of oxygen (O_2), but is instead important in the inactivation of microorganisms when O_2 is converted to more reactive forms (Schwartz, 1971; Cox and Heckley, 1973). These include superoxide radicals, hydrogen peroxide, and hydroxide radicals. These radicals arise naturally in the environment from the action of lightning, UV radiation, pollution, etc. Such reactive forms of oxygen cause damage to DNA by producing mutations, which can accumulate over time. The repair mechanisms described in the previous section are responsible for control of the damaging effects of reactive forms of oxygen.

Similarly, the **open air factor** (OAF) is a term coined to describe an environmental effect that cannot be replicated in laboratory experimental settings. It is closely linked to oxygen toxicity and has come to be defined as a mixture of factors produced when ozone and hydrocarbons (generally related to ethylene) react. For example, high levels of hydrocarbons and ozone causing increased inactivation rates for many organisms, probably because of damaging effects on enzymes and nucleic acids (Donaldson and Ferris, 1975; May *et al.*, 1969). Therefore, OAFs have been strongly linked to microbial survival in the air.

The formation of other ions, such as those containing chlorine, nitrogen, or sulfur occurs naturally as the result of many processes. These include the action of lightning, shearing of water, and the action of various forms of radiation that displace electrons from gas molecules, creating a wide variety of anions and cations not related to the oxygen radicals. These ions have a wide range of biological activity. Krueger *et al.* (1957) showed that positive ions cause only physical decay of microorganisms, e.g., inactivation of cell surface proteins, whereas negative ions exhibit both physical and biological effects such as internal damage to DNA.

5.10 EXTRAMURAL AEROMICROBIOLOGY

Extramural aeromicrobiology is the study of microorganisms associated with outdoor environments. In the extramural environment, the expanse of space and the presence of air turbulence are two controlling factors in the movement of bioaerosols. Environmental factors such as UV radiation, temperature, and relative humidity modify the effects of bioaerosols by limiting the amount of time aerosolized microorganisms will remain viable. This section is an overview of extramural aeromicrobiology that includes several topics: the spread of agricultural pathogens; the spread of airborne pathogens associated with waste environments; and germ warfare. This section is not intended as a review of all aspects of extramural aeromicrobiology but instead attempts to show the wide diversity of the science.

5.10.1 Agriculture

Contamination of crops and animals via bioaerosols has a huge economic impact worldwide. The list of agricultural pathogens in Table 5.1 shows the diversity of airborne microorganisms that infect plants and animals. As the earth's population increases, the need for a larger, more stable supply of food becomes increasingly important. Rice and wheat are the two major staple crops that are paramount to world food security. Major pathogens of such crops are the **wheat rust fungi.** These spore-forming fungi cause some of the most devastating of all diseases of wheat and other grains. In 1993, one type of wheat rust (leaf rust) was responsible for the loss of over 40 million bushels of wheat in Kansas and Nebraska alone. Even with selective breeding for resistance in wheat plants, leaf rust continues to have major economic impacts. The high concentration of wheat in areas ranging from northern Texas to Minnesota and up into the Dakotas makes this whole region highly susceptible to rust epidemics.

Spores of wheat rust (Fig. 5.16) are capable of spreading hundreds if not thousands of kilometers through the atmosphere (Ingold, 1971). The airborne spread of rust disease has been shown to follow a predictable trend, which starts during the fall with the planting of winter wheat in the southern plains. Any rust-infected plant produces thousands of spores, which are released into the air (Fig. 5.17) by either natural atmospheric disturbance or mechanical disturbance during the harvesting process. Once airborne, these spores are capable of long-distance dispersal, which can cause downwind deposition onto other susceptible wheat plants. The generation time of new spores is measured in weeks, after which, new spores are again released from vegetative fungi into the AMB pathway. For example, during the harvest of winter wheat in Texas, the prevailing wind currents are from south to north, which can allow rust epidemics to spread into the maturing crops farther north in Kansas and up into the young crops in the Dakotas (Fig. 5.18). This epidemic spread of wheat rust and the resulting

FIGURE 5.16 This photo shows seven slices of leaves of wheat in various stages of infection by wheat rust. The upper slices are in the first stages of infection, while the lower slices show increasingly infected leaves. Wheat rust is a fungal pathogen, which is spread by the aeromicrobiological pathway. Infection with the phytopathogens causes losses in the millions of dollars every year to the agricultural industry. (From McIntosh *et al.*, 1995, with kind permission from Kluwer Academic Publishers.)

economic destruction produced are indicative of the impact that airborne microbial pathogens can have on agriculture. In addition, this example indicates the extent and rate at which the AMB pathway can spread such contagion.

A factor that complicates the control of such diseases is that chemical treatment for the control of pathogens is viewed as undesirable. This is because many pesticides have extremely long half-lives and their residence in an ecosystem can be extremely harmful. Therefore, instead of using wheat rust fungicides, attempts are being made to breed strains of wheat that are more resistant to the fungi. Another method used for controlling phytopathogenic (plant pathogenic) fungi is spore monitoring as a disease con-

FIGURE 5.17 This photo shows a field of wheat, which is highly infected by phytopathogenic wheat rust. The field is being harvested by a hay machine, which is releasing a cloud of rust spores into the aeromicrobiological pathway. These spores can spread thousands of miles and infect other crops downwind causing catastrophic losses to wheat crops. (From McIntosh *et al.*, 1995, with kind permission from Kluwer Academic Publishers.)

FIGURE 5.18 In this map of the United States, the arrow indicates the northern path of wheat rust infections as spread by the aeromicrobiological pathway. As depicted here, the wheat rust infection begins in the winter harvest in Texas and spreads northward with the prevailing wind currents. The epidemic spread of these phytopathogens infects maturing crops in Kansas and then moves up into the young crops in the Dakotas.

trol strategy. In this approach, the life cycle of the fungi especially the release of spores is monitored, and fungicide application is timed to coincide with spore release. This approach minimizes use of harmful chemicals. Thus, efficient AMB pathway sampling, monitoring, detection, and modeling have the ability to aid in the control of airborne pathogens.

The airborne spread of pathogenic microorganisms is also highly important in the animal husbandry industry. The occurrence of foot-and-mouth disease is an example of the importance of bioaerosols in the spread of airborne disease. It has long been thought that bioaerosol spread is linked primarily to respiratory pathogens, but there is growing evidence that gastrointestinal pathogens are also important in airborne transmission of disease among animals. One example of bioaerosol spread of a gastrointestinal pathogen is transmission of *Salmonella typhimurium* among calves that are housed individually in small pens (Hinton *et al.*, 1983). The potential for bioaerosol spread of this pathogen was recognized because the initial symptoms resembled those of pneumonia and appeared randomly within these animals, two factors that are not characteristic of oral transmission. Oral transmission generally occurs sequentially from one pen to the next, whereas aerial transmission can carry organisms past nearby pens, infecting calves randomly. Furthermore, Wathes *et al.* (1988), showed that *S. typhimurium* could survive for long periods in an airborne state, and calves and mice exposed to aerosolized *S. typhimurium* developed symptoms, proving that gastrointestinal pathogens could be spread by aerosolization. Finally, Baskerville *et al.* (1992) showed that aerozolized *Salmonella enteritidis* could infect laying hens. These hens showed clinical symptoms and were shedding the test strain of salmonellae in their feces within a few days. Thus, the AMB pathway can be important even in the spread of diseases for which pathogens not normally considered airborne.

5.10.2 Waste Disposal

Waste disposal is a multibillion dollar industry in the United States. However, there are many hazards inherent in the treatment and disposal of wastewater (Fig. 5.19) and **biosolid** material (Fig. 5.20). Major hazards associated with waste effluents are pathogenic microorganisms including bacteria, viruses, protozoa, and helminths. Wastewater treatment plants utilize activated sludge and trickling filter systems and all of these treatment processes potentially create relatively large amounts of aerosols, which have been shown to include pathogenic microorganisms. Other aspects of

FIGURE 5.19 This is a photo depicting the application of secondary treated wastewater onto agricultural lands. This method is highly efficient at conserving water and has been shown to improve the fertility of soils. Due to the presence of pathogens in wastewater, and the nature of these land application systems, there are high concentrations of bioaerosols generated. Currently, however, there is little epidemiological and microbial risk assessment information available to determine if there may be health concerns for populations living in the vicinity of such operations, though there is a growing base of information on the concentrations and types of pathogens found in these bioaerosols. (Reprinted with permission from Corel Gallery www.corel.com)

the treatment process such as composting and land disposal are also associated with the generation of aerosols containing pathogenic microorganisms.

One of the primary methods for the disposal of biosolids is agricultural land application. This type of disposal is steadily becoming one of the more widely used alternatives because of restrictions on ocean and landfill dumping. Dowd *et al.* (1997b), showed the presence of a wide variety of airborne pathogens and pathogen indicators that were collected with an AGI-30 sampler, including *Salmonella* sp., *Clostridium* sp., *E. coli*, and F+ coliphages. The major concern associated with the aerosolization process in relation to waste disposal operations is the exposure of waste disposal workers to pathogenic microorganisms, although nearby population centers are also potential exposure risks. At present, epidemiological studies are being conducted to determine the health risks associated with wastewater and waste disposal operation, but no clear-cut risk has yet been identified.

5.10.3 Germ Warfare

Biological warfare has been a combat weapon for centuries. As early as A.D. 1346, Tartars besieging the

FIGURE 5.20   These photos show an agricultural field in Tucson, Arizona, that has just received an application of biosolid material. The biosolid material shown just after placement in the upper photo is approximately 15% solids and can contain considerable populations of microbial pathogens. The lower photo shows the same field being tilled illustrating the potential for the generation of aerosols containing microorganisms. The practice of biosolid placement on agricultural land is highly beneficial. The organic and inorganic content of the biosolid material is a valuable source of fertilizer for non-food crops and the disposal of the biosolid material in this manner is highly economical considering the increasing costs of disposing of such biosolid material in landfills. Thus, the public and the U.S. Environmental Protection Agency are rapidly accepting the benefits of using this biosolid material on agricultural lands. (Photo courtesy C. Etsitty.)

walled city of Kaffa used catapults to launch plague-infested bodies into the city. During the French and Indian War, the English offered blankets to Native Americans who had taken control of a military outpost. The English had purposely exposed the blankets to the smallpox virus knowing the lack of immunity of the native people.

The United States began field-scale experiments in the 1950s using inert substances (fluorescent dyes) to simulate biological warfare agents. These aerosols were released into the air circulation system of a subway system and into the air off the coast of San Francisco. The studies indicated that 100% of the associated populations inhaled what was considered lethal doses of the inert indicators. In Russia, an accident at a biological warfare research institute caused the widespread exposure of nearby populations to a genetically modified strain of *Bacillus anthracis*. In 1995, Tokyo police found large quantities of *Clostridium botulinum* toxin during one of several raids on a terrorist-controlled facility. In the late 1990s, Iraq was investigated of having production facilities for biological warfare agents. Thus, we see that biological warfare is a reality in today's society.

Detection of biological warfare agents is an area that requires sophisticated equipment and training. The war between the United States and Iraq known as Desert Storm is a recent example of these facts. The United States needed the ability to detect biological warfare agents in order to give early warning to troops in case biological warfare agents were released. Unfortunately, the technology available, including the level of sensitivity, and the specificity required for rapid detection of biological and chemical warfare agents was and still is inadequate. With this in mind, the United States is presently developing a more advanced anti-biological warfare defense program.

5.11 INTRAMURAL AEROMICROBIOLOGY

The home and workplace are environments in which airborne microorganisms create major public health concerns. In comparison with the extramural environment, intramural environments have limited circulation of external air and much less UV radiation exposure. Indoor environments also have controlled temperature and relative humidity, which are generally in the ranges that allow extended microbial survival. Thus, these conditions are suitable for the accumulation and survival of microorganisms within many enclosed environments, including office buildings, hospitals, laboratories, and even spacecraft. In this section, we will consider these three diverse areas as examples of current topics related to intramural aeromicrobiology. Again, it should be noted that this section does not cover all aspects of intramural aeromicrobiology but instead attempts to show the wide diversity of the science.

5.11.1 Buildings

Many factors can influence bioaerosols and therefore how "healthy" or how "sick" a building is. These include: the presence and/or efficiency of air filtering devices, the design and operation of the air circulation systems, the health and hygiene of the occupants, the amount of clean outdoor air circulated through the building, the type of lighting used, the ambient temperature in the building, and the relative humidity.

Some pathogens are uniquely adapted for survival and transmission in the intramural environment. One good example of such an organism is *Legionella pneumophilia*, the causative agent of both Legionnaires disease and Pontiac fever. **Legionnaires disease** or legionellosis is a pneumonia that causes disease in up to 5% of those exposed. Of those who contract the disease, up to 39% die from the infection. Pontiac fever is associated with flu-like symptoms and affects up to 100% of those exposed, although it is generally not associated with mortality. The causative agent of both diseases is a poorly staining, gram-negative *bacillus* called *L. pneumophila*. This organism is named in association with the first highly characterized outbreak of the disease, which occurred in 1976 at an American Legion convention in Philadelphia.

Legionella spp. are ubiquitous in the environment. They are found in association with lakes, ponds, and streams and have even been found in deep terrestrial subsurface environments. In addition to natural reservoirs, there are many human-made systems within which legionellae can find a niche. These include cooling towers, evaporative condensers, plumbing systems, whirlpools, shower heads, and hot-water faucets (Bollin *et al.*, 1985). In the case of the American Legion convention, the reservoir for the organism that caused the outbreak was a poorly maintained cooling tower, which provided optimal conditions for *Legionella* proliferation. Because of the poor design of the air circulation system at the convention, this proliferation led to the subsequent aerosolization and spread of the organisms throughout the building.

What conditions promote the proliferation of *Legionella* spp.? Stagnant water and temperatures in the range of 35-46°C are factors that can lead to the rapid multiplication of background levels of *Legionella* spp. Another interesting aspect of the ecology of *Legionella* is that they can grow intracellularly within other microbes including amoebae, cyanobacteria, and protozoa. How can growth and spread of *Legionella* spp. be avoided? Several strategies can be used. In the maintenance of hot-water plumbing systems, operating temperatures should be greater than 50°C. All potential places where water can stagnate in water pipes should be avoided. For cooling towers, the recommendations involve the installation of ozonization units, dry convective heat exchange designs, and the avoidance of any design that could potentially mix the wet system with the supply air. Biocidal agents such as chlorine or copper can also be effective when used regularly at low levels.

5.11.2 Spaceflight

Microorganisms are associated with almost every aspect of life on earth and have been found thriving in the harshest and most extreme environments. Spacecraft are no exception to this rule. However, remember that although some microorganisms are pathogenic, others are necessary and beneficial. For example, microorganisms can be harnessed for many valuable purposes including food production, oxygen production, waste treatment, and air purification.

The use of microorganisms as components of air purification systems is a good example of a beneficial use for microorganisms in association with the AMB pathway. **Biological air filtration (BAF)** (Fig. 5.21) is a method of air purification that is currently under investigation for use during spaceflight (Binot and Paul, 1989). This system essentially uses microorganisms to remove organic contaminants from the airstream. The BAF is a closed two-phase system. The airborne organic contaminants are passed into the closed system, where they pass through a membrane filter into a liquid phase. The liquid phase is made up of microorganisms and nutrients that develop into an associated biofilm on the membrane surface. Different bacterial strains or consortia of bacteria have been isolated that can completely mineralize most of the contaminants of concern to the space industry. This includes a greater than 99% reduction of toluene, chlorobenzene, and dichloromethane in the airstream. In a real-time study in a spaceflight simulation facility, the BAF system removed methane, acetone, toluene, and isopropanol, among many other compounds. The BAF system appears to be flexible and able to adapt to different levels and types of contaminants. Even when a new compound is introduced that is toxic, the BAF system has demonstrated relatively rapid systemic recovery. Thus, the system has adaptability and high efficiency, which are the two most critical factors needed in such an air purification system.

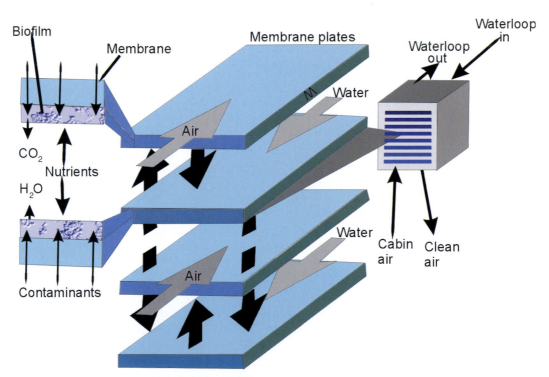

FIGURE 5.21   This is a schematic representation of a typical biological air filtration (BAF) device. The figure shows that it is a dual-phase air purification system that operates by diffusion of air pollutants out of the vapor (or air) phase through the gas permeable membrane and into the liquid phase, where the biofilm containing specialized microbial communities is able to degrade them. Within each of the membrane plates is a nutrient solution, which keeps the biofilm healthy. The far left of the diagram depicts this system of membrane, biofilms, and nutrient matrices. The middle of the figure shows how the membrane plates and the vapor phase are oriented in relation to each other and the far right of the figure indicates the housing that would contain this system. These type of air filters are being tested by NASA for their ability to purify air within closed environmental systems such as would be represented by a space station. (Adapted from Binot and Paul, 1989.)

5.11.3 Public Health

Pathogens in the AMB pathway can be a potential source of deadly diseases, but AMB pathway also has potential for use in immunization against disease. For example, influenza (flu) strikes millions every year and kills about 20,000 of the most vulnerable patients. Injectable flu vaccines are currently being used that decrease the severity of the disease, but they are not widely used because they require a painful injection in the arm. Currently in development is a flu vaccine that is delivered by nasal spray. Aerosol delivery may actually be a better method for vaccination against respiration-associated diseases because the vaccine is delivered onto the mucous membranes, which are the first line of defense against respiratory infection. Thus nasal sprays may increase the levels and specificity of the immune response, especially in these vulnerable areas. Researchers have already tested this type of flu vaccine nose spray on 320 children and found it to be successful. It proved to be not only safe and effective, but also was well accepted by the children as an alternative to an injection. Experiments are also under way to use nasal sprays in delivery of vaccines against para-influenza and respiratory syncytial viruses, both major causes of pneumonia in children (Recer, 1997).

5.11.4 Hospitals and Laboratories

Hospitals and microbiology laboratories are the two indoor environments with perhaps the greatest potential for the aerosolization of pathogenic microorganisms. Hospitals, because they are centers for the treatment of patients with diseases, have a high percentage of individuals, including patients and staff, who are active carriers of infectious, airborne pathogens. Of particular concern are neonatal wards, surgical transplant wards, and surgical theaters, all critical areas where the control of nosocomial infection is imperative. Illustrating this point is a study by Portner et al. (1965) that evaluated airborne microbial concentrations in surgical theaters, industrial clean rooms, typical industrial manufacturing areas, and a horizontal laminar flow clean room designed for the space industry. The surgical theater had by far the highest counts of pathogenic airborne microbial contaminants, followed by the industrial manufacturing area, the industrial clean room, and finally the laminar flow room, which had the lowest counts of airborne microbes.

The microbiological laboratory is also a breeding ground for pathogenic agents. Because of this, handling procedures have been developed and refined to protect laboratory workers. However, even under the strictest of conditions, aerosolization events may occur. In 1988, for instance, eight employees in a clinical microbiological laboratory developed acute brucellosis (Staszkiewicz et al., 1991). *Brucella melitensis* was the causative agent of a variety of symptoms including fever, headache, weakness, profuse sweating, chills, arthralgia, depression, and hepatitis. A survey of the laboratory and the personnel showed that a cryogenically stored clinical isolate of *Brucella* sp. had been thawed and subcultured without the use of a biosafety cabinet. Other than this, the laboratory worker claimed to have used good technique. This example demonstrates the ease with which a bioaerosol can spread within areas where pathogens are handled for research and clinical purposes, and indicates the importance of bioaerosol control methodologies. The following sections describe how bioaerosol formation and spread is actually controlled in the laboratory.

5.12 BIOAEROSOL CONTROL

The control of airborne microorganisms can be handled in a variety of ways. Launching, transport, and deposition are all points at which the airborne spread of pathogens can be controlled. The mechanisms used to control bioaerosols include ventilation, filtration, UV treatment, biocidal agents, and physical isolation. These are discussed in the following sections.

5.12.1 Ventilation

Ventilation is the method most commonly used to prevent the accumulation of airborne particles. This mechanism involves creating a flow of air through areas where airborne contamination occurs. This can mean simply opening a window and allowing outside air to circulate inward, or use of air-conditioning and heating units that pump outside air into a room. Ventilation is considered one of the least effective methods for controlling airborne pathogens, but is still very important. Ventilation relies on mixing of intramural air with extramural air to reduce the concentration of airborne particles. However, in some cases the addition of extramural air can actually increase airborne particles. For example, one study showed that hospitals in Delhi, India that relied on ventilation alone contained airborne fungal loads that were higher inside the hospital than those outside. This indicates that ventilation alone may not be sufficient to significantly reduce circulating bioaerosols. Thus, for most public buildings, especially hospitals, other forms of bioaerosol control need to be implemented.

5.12.2 Filtration

Unidirectional airflow filtration is a relatively simple and yet effective method for control of airborne contamination. Some filters, for example, high-efficiency particulate air (**HEPA**) filters (Fig. 5.22), are reported to remove virtually all infectious particles. These types of filters are commonly used in biological safety hoods. However, because of their high cost, they are not often used in building filtration systems. Instead, other filtration systems that rely on baghouse filtration (a baghouse works on the same principle as a vacuum cleaner bag) are used. Typically, air filters (baghouse, HEPA, etc.) are rated using the **dust-spot percentage,** which is an index of the size of the particles efficiently removed by the filter, with higher percentages representing greater filtration efficiencies. The typical rating for the filters used in most buildings is 30 to 50%. Studies have shown that a 97% dust-spot rating is required to effectively remove virus particles from the air. Other factors that influence filtration efficiency are related to the type of circulation system and how well it mobilizes air within the building, the type of baghouse system used, and the filter material chosen (nylon wound, spun fiberglass, etc.) as well as the filter's nominal porosity (1 μm, 5 μm, etc.). All these factors combine to influence the efficiency of the air filtration and removal of particles including bioaerosols. In spite of the high level of efficiency that can be achieved with filtration, many systems still cannot stop the circulation of airborne microorganisms, especially viruses, and added treatments may be required to ensure that air is safe to breathe.

5.12.3 Biocidal Control

Biocidal control represents an added treatment that can be used to eradicate all airborne microorganisms, ensuring they are no longer viable and capable of causing infection. Many eradication methods are available, for example, superheating, superdehydration, ozonation, and UV irradiation. The most commonly used of these methods is **UVGI** or **ultraviolet germicidal radiation.** UVGI has been shown to be able to control many types of pathogens, although some microbes show various levels of resistance. The control of contagion using UV irradiation was tested in a tuberculosis (TB) ward of a hospital. Contaminated air was removed from the TB ward through a split ventilation duct and channeled into two animal holding pens that contained guinea pigs. One pen received air that had been treated with UV irradiation; the other received untreated air. The guinea pigs in the untreated-air compartment developed TB, but none of the animals in the UV-treated compartment became infected. The American Hospital Association (1974) indicated that, properly utilized, UV radiation can kill nearly all infectious agents, although the effect is highly dependent on the UV intensity and exposure time. Thus, major factors that affect survival (temperature, relative humidity, UV radiation, ozone, etc.) in the extramural environment can be used to control the spread of contagion in the intramural environment.

5.12.4 Isolation

Isolation is the enclosure of an environment through the use of positive or negative pressurized air gradients and airtight seals. Negative pressure exists when cumulative airflow travels into the isolated region. Examples of this, as previously mentioned, are the isolation chambers of the tuberculosis wards in hospitals used to protect others outside the TB wards from the infectious agent generated within these negative-pressure areas. This type of system is designed to protect other people in the hospital from the pathogens (*Mycobacterium tuberculosis*) present inside the isolation area. Air from these rooms is exhausted into the atmosphere after passing through a HEPA filter and biocidal control chamber.

Positive-pressure isolation chambers work on the opposite principle by forcing air out of the room, thus protecting the occupants of the room from outside

FIGURE 5.22 These are different types of high efficiency particulate air filters (HEPA filters). These filters can remove most infectious particles and allergens from the air. HEPA filters, despite their cost, are being used in an increasing number of ways. They are found in hospitals, office building, and laboratories and now even in homes. Some of the newer home vacuum cleaners may contain HEPA filters. (Photo courtesy Hepa Corporation www.hepa.com)

contamination. One can reason that the TB ward is a negative-pressure isolation room while the rest of the hospital, or at least the nearby anterooms, are under positive-pressure isolation. Other examples are the hospitals critical care wards for immunosuppressed patients such as organ transplant, human immunodeficiency virus (HIV)-infected, and chemotherapy patients. These areas are protected from exposure to any type of pathogen or opportunistic pathogens. The air circulating into these critical care wards is filtered using HEPA filters, generating purified air essentially free of infectious agents.

5.13 BIOSAFETY IN THE LABORATORY

Microbiological laboratories are rooms or buildings equipped for scientific experimentation or research on microorganisms. Many microbiological laboratories work specifically with pathogenic microorganisms, some of which are highly dangerous, especially in association with the AMB pathway. Also, many types of equipment, such as centrifuges and vortexes, that are commonly used in microbiological laboratories can promote the aerosolization of microorganisms. Thus, laboratories and specialized equipment used in these laboratories (*e.g.*, biosafety cabinets) are designed to control the spread of airborne microorganisms. There are essentially four levels of control designed into laboratories, depending on the type of research being conducted. These levels of control are termed **biosafety levels 1-4**, with 1 being the lowest level of control and 4 the highest level of control. Within these laboratories, biosafety cabinets are essentially isolation chambers that provide safe environments for the manipulation of pathogenic microorganisms. In this section we will discuss biosafety cabinets and biosafety suits, followed by a short discussion of the actual biosafety levels imposed to achieve specific levels of control.

5.13.1 Biological Safety Cabinets

Biological safety cabinets (BSCs) are among the most effective and commonly used biological containment devices in laboratories that work with infectious agents (U.S. Department of Health and Human Services, 1993). There are two basic types of biosafety cabinets currently available (Class II, and Class III), each of which has specific characteristics and applications that dictate the type of microorganisms it is equipped to contain. Properly maintained biosafety cabinets provide safe environments for working with microorganisms. Class II biosafety cabinets are characterized by having considerable negative-pressure airflow that provides protection from

infectious bioaerosols generated within the cabinet, and Class III biosafety cabinets are characterized by total containment. Class I cabinets are also in existence, but they are no longer produced and are being replaced by Class II cabinets for all applications.

Class II biosafety cabinets (Fig. 5.23), of which there are several types, are suitable for most work with moderate-risk pathogens (Table 5.5). Class II biosafety cabinets operate by drawing airflow past the worker and down through the front grill. This air is then passed upward through conduits and downward to the work area after passing through a HEPA filter. Room air is also drawn into the cabinet through the top of the unit, where it joins the circulating air and passes through the HEPA filter and into the work area. About 70% of the air circulating in the work area is

FIGURE 5.23 This is a photo of a Class II biosafety cabinet designed for the containment of many types of biohazardous agents. This cabinet allows access to the work area through the opening in the front window. HEPA filters purify air circulating in and exhausted from the cabinets. These cabinets protect the laboratory worker and the environment from aerosolized infectious particles generated within the cabinet. (Photo courtesy S. E. Dowd.)

TABLE 5.5   Examples of Classification of Biological Agents According to Risk

Class	Type of Agent	Agent
Class I	Bacterial Fungal Protozoal	All those which have been assessed for risk and do not belong in higher classes
	Viral	Influenza virus reference strains Newcastle virus Parainfluenza virus 3, SF4 strain
Class II	Bacterial	*Actinobacillus* spp. *Bordetella* spp. *Borrelia* spp. *Campylobacter* spp. *Clostridium* spp. *E. coli* spp. *Klebsiella* spp. *Listeria* spp. *Mycobacteria* spp. *Shigella* spp. *Vibrio* spp. *Salmonella* spp.
	Fungal	*Cladosporium* spp. *Blastomyces* spp. *Penicillium* spp. *Cryptococcus* spp. *Microsporum* spp.
	Protozoal	*Cryptosporidium* spp. *Giardia* spp. *Encephalitozoon* spp. *Enterocytozoon* spp. *Babesia* spp. *Echinococcus* spp. *Entamoeba* spp. *Fasciola* spp. *Leishmania* spp. *Plasmodium* spp. *Schistosoma* spp. *Trypanosoma* spp.
	Viral	Adenoviruses Corona viruses Cowpox virus Coxsackie A and B viruses Echoviruses Hepatitis viruses A, B, C, D,& E Epstein–Barr virus Influenza viruses Vaccinia virus Rhinoviruses
Class III	Bacterial	*Bartonella* spp. *Brucella* spp. *Codiella burnetti* *Mycobacterium bovis* *Mycobacterium tuberculosis* *Pseudomonas mallei* *Rickettsia* spp. *Yersinia pestis*
	Fungal	*Coccidioides immitis* *Histoplasma capsulatum*
	Protozoal	None

(Continues)

TABLE 5.5   *(Continued)*

Class	Type of Agent	Agent
	Viral	Dengue virus Lymphocytic choriomeningitis virus Monkey pox virus Yellow fever virus
Class IV	Bacterial Fungal Protozoal	None None None
	Viral	Absettarov Hemmorrhagic fever agents Ebola fever virus Guanarito Hanzalova Lassa virus Marburg virus Tick borne encephalitis viruses Herpesvirus simiae

Adapted from Univ. of Pennsylvania Biological Safety Manual.

then removed by passing it through the rear grill of the cabinet, where it is discharged into the exhaust system. The remaining 30% is passed through the front grill, essentially recirculating in the cabinet (Fig. 5.24). Laboratory personnel require special training in order to properly use Class II cabinets and to ensure proper containment of bioaerosols. One of the major hazards associated with Class II cabinets is the potential for the disruption of the negative airflow. Many mechanical actions can disrupt the protective airflow, such as repeated insertion and withdrawal of arms, opening or closing of doors in the laboratory, or even someone walking past the cabinet while it is in use. Any of these actions can potentially allow the escape of bioaerosols from the cabinet. Practices that minimize the release of bioaerosols include the following:

1. Delaying manipulation of materials for 1 minute after inserting arms or hands inside the cabinet;
2. Keeping the arms raised slightly off the front grill of the cabinet to prevent room air from flowing directly into the work area;
3. Performing all operations at least 4 inches in front of the front grill;
4. Running the cabinet 24 hours a day to help to maintain the room-cabinet air balance (at a minimum, cabinets should be turned on at least 5 minutes before use);
5. Wiping work surface, interior walls, the surfaces of all introduced objects, and interior surface of the window with 70% ethanol or other appropriate disinfectant prior to work with pathogens;

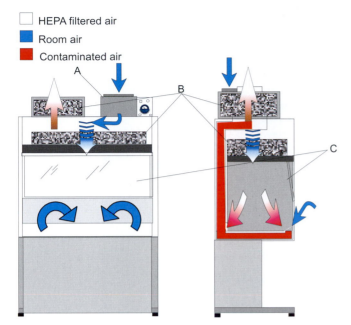

☐ HEPA filtered air
■ Room air
■ Contaminated air

FIGURE 5.24 This is a schematic representation of the airflow paths within a typical class II biosafety cabinet. Room air is drawn in from the top and from the front as indicated by the blue arrows. The non-purified atmospheric air (blue) entering from the top of the cabinet is drawn in by an air pump (A), and then is purified by a HEPA filter (B) as it enters the workspace. Non-purified air from the room (blue) entering from the front of the cabinet passes into the front grill and passes up through the top of the cabinet where it also passes through a HEPA filter before entering the workspace (C). This prevents the experiments in the workspace from being contaminated by airborne room contaminants. As the purified room air is exposed to the work environment and becomes contaminated (red) it is passed through yet another HEPA filter before being exhausted to the atmosphere. This pattern of airflow and purification insures that the worker and the atmosphere are not exposed to the biohazards contained within the biosafety cabinet.

6. Placing plastic-backed absorbent toweling on the work surface to reduce splatter and aerosol formation;

7. Placing all materials toward the rear of the cabinet, with bulky items placed to one side and aerosol-generating objects such as centrifuges and vortexes as far to the back of the cabinet as possible (Fig. 5.25 shows a typical cabinet workspace layout);

8. Open flames should be avoided in biosafety cabinets, because they cause turbulence, which disrupts the pattern of air supplied to the work surface. Open flames are unnecessary in the nearly microbe-free environment created within the biosafety cabinet work area. Touch-plate burners equipped with pilot lights or small electric furnaces can be used to decontaminate bacteriological loops when necessary;

9. Finally, for decontamination of the hood, the

surfaces of all materials should be wiped with 70% ethanol or other appropriate disinfectant before removal, biohazardous waste materials should be contained within appropriate biohazard bags or containers and their surfaces wiped down with disinfectant before removal from the cabinet, all interior surfaces should then be wiped down including the interior of the window, and finally gloves and gowns should be removed and the investigator's hands washed.

This list is not meant as an exhaustive account of the procedures required in the use of biosafety cabinets; rather, it serves as an indication of the extent of the containment procedures required to safely handle pathogens.

The Class III biosafety cabinet (Fig. 5.26) is a completely enclosed environment that offers the highest degree of personnel and environmental protection from bioaerosols. Class III cabinets are used for high-risk pathogens (Table 5.5). All operations in the work area of the cabinet are performed through attached rubber gloves. Class III cabinets use complete isolation to protect workers. All air entering the cabinet is filtered using a HEPA filter, and the air leaving the cabinet is filtered by two HEPA filters in series. The exhaust may also include biocidal treatment such as incineration following the HEPA filtration to further ensure complete biological inactivation. In addition to these safeguards, Class III cabinets are connected with airtight seals to all other laboratory equipment (such as incubators, refrigerators, and centrifuges) that is needed for working with the pathogens while using the cabinet. The Class III cabinet must also be connected to autoclaves and chemical dunk tanks used to sterilize or disinfect all materials entering or exiting the cabinet.

Another type of containment that typically provides the same level of protection as a Class III biosafety hood is the biological safety suit. The biological suit, unlike biosafety cabinets, operates under positive pressure created by an external air supply, thus protecting the wearer. Like the biosafety cabinets, the biosafety suit isolates the laboratory worker wearing it from bioaerosols. Biosafety suits are typically used in airtight complete biocontainment areas, and are decontaminated by means of chemical showers upon exiting the biohazard area. Some biosafety suits are portable (Fig. 5.27) and can be used in environments outside the laboratory such as "hot zones" (epidemiological areas that are currently under the influence of epidemic cases of diseases caused by high-risk pathogens) so that microbiologists and physicians working in these areas can minimize their risk of exposure to pathogens. As in biosafety cabinets, the air

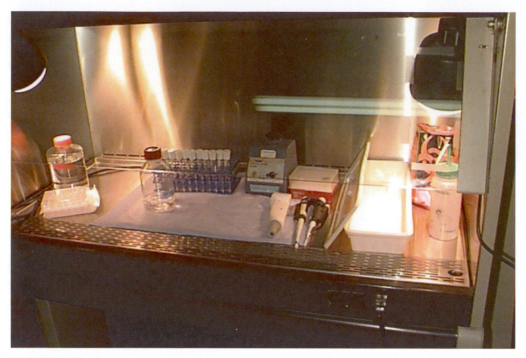

FIGURE 5.25 This is a photo showing the typical experimental layout in a biosafety hood. This type of set up minimizes the aerosolization and cross contamination within the biosafety hood workspace. The biological disposal bag and tray are on the far right of the cabinet indicating the area of highest contamination. The middle is the active workspace with bulky materials and equipment such as centrifuges and vortexes places as far toward the back of the workspace as possible without obstructing the vents. The far right of the workspace is designated for sterile media. Thus, the flow from contaminated to sterile is from right to the left. (Photo courtesy of S. E Dowd).

entering and leaving the biosafety suit passes through two HEPA filters.

5.13.2 Biosafety Laboratories

Biosafety laboratories are carefully designed environments where infectious or potentially infectious agents are handled and/or contained for research or educational purposes. The purpose of a biosafety laboratory is to prevent the exposure of workers and the surrounding environment to biohazards. There are four levels of biohazard control, which are designated as biosafety levels 1 through 4.

Biosafety level 1, as defined by the Centers for Disease Control (U.S. Department of Health and Human Services: CDC-NIH, 1993), indicates laboratories where well-characterized agents that are not associated with disease in healthy adult humans are handled. In general, no safety equipment is used other than sinks for hand washing and only general restrictions are placed on public access to these laboratories. Work with the microorganisms can be done on bench tops using standard microbiological techniques. A good example of a biosafety 1 laboratory is a teaching laboratory used for undergraduate microbiology classes.

Biosafety 2 indicates an area where work is performed using agents that are of moderate hazard to humans and the environment (Fig. 5.28). These laboratories differ from biosafety 1 laboratories in that the personnel have specialized training in the handling of pathogens and access to the work areas is limited. Many procedures that may cause aerosolization of pathogenic microorganisms are conducted in biological safety level II cabinets or other physical containment equipment to protect the laboratory workers.

Biosafety 3 indicates laboratories where agents that can cause serious or fatal disease as a result of AMB exposure are handled. As with biosafety 2, all personnel are specifically trained to handle pathogenic microorganisms. All procedures involving these infectious agents are conducted in biological safety level II cabinets or other physical containment devices. These facilities also have permanent locks to control access, negative airflow, and filtered ventilation in order to protect the public and the surrounding environments. With certain pathogens used in biosafety 3 laboratories, Class III safety hoods may also be used and clothes must be changed before leaving the premises.

Biosafety 4 is the highest level of control and is indicated for organisms that have high potential for life-

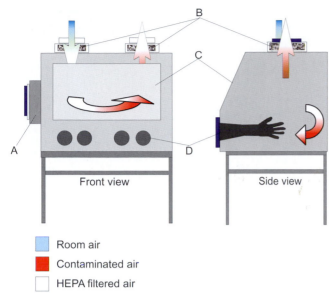

- ■ Room air *(blue)*
- ■ Contaminated air *(red)*
- □ HEPA filtered air

FIGURE 5.26 This is a schematic representation of a Class III biological safety cabinet. This cabinet is completely sealed from the environment. Any materials entering or leaving the cabinet are passed through a chemical dunk tank or autoclave (A) in order to sterilize them and prevent environmental contamination. Air Entering or leaving these cabinets is passed through HEPA filters (B). Access to the workspace is by means of rubber gloves (D) and the workspace is visualized through a sealed window (C). These biosafety cabinets are utilized when working with highly pathogenic microorganisms to protect workers and the environment. Class III cabinets can be used to work with all biohazardous agents except those specifically designated for level 4 containment.

threatening disease in association with aerosolization. To work in these facilities, personnel must have specialized training beyond that required for biosafety levels 2 and 3. Biosafety level 4 laboratories (Fig. 5.29) are 100% isolated from other areas of a building and may even be separated from other buildings altogether. Work in these areas is confined exclusively to Class III biological safety cabinets unless one-piece positive-pressure ventilation suits are worn, in which case Class II biosafety cabinets may be used. These laboratories are also specially designed to prevent microorganisms from being disseminated into the environment. The laboratories have complete containment and require personnel to wear specialized clothing, which is removed and sterilized before leaving the containment areas. Personnel are also required to shower before leaving the facility. In general, all air into and out of these laboratories is sterilized by filtration and germicidal treatment. These facilities represent the ultimate in our ability to control the AMB pathway.

5.13.3 Biological Agent Classification

For any microorganism, defined degrees of risk associated with its use indicate the type of containment

FIGURE 5.27 This is a photo of a portable biosafety suit designed to protect the wearer from biohazardous materials in the environment. This type of suit is equipped with a portable air supply. This type of suit is designed to protect workers performing field work in "Hot Zones" where active aeromicrobiological spread of highly contagious biological agents is occurring. (Photo courtesy Center for Disease Control and Prevention, Atlanta, GA.)

needed to ensure the safety of laboratory workers, the public, and the environment. There are five classes of organisms. Class I microorganisms are those that pose little or no hazard under ordinary conditions of handling and can be safely handled without special apparatus or equipment. In contrast, Class II are agents of low potential hazard that may cause disease if accidentally inocu-

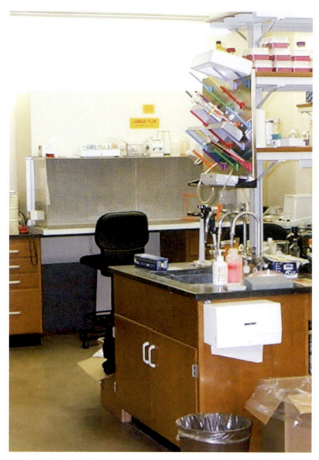

FIGURE 5.28 This is a photo of a typical biosafety level 2 laboratory. There are sinks for washing hands and access in controlled to prevent public access. The difference between biosafety level 1 and level 2 laboratories is that moderately hazardous biological agents are handled and personnel receive specialized training. In level 1 laboratories only non-pathogenic, well characterized organisms are handled and no specialized training is needed. Level 1 laboratories are often used as microbiology classrooms where young scientists are first introduced to safe handling practices. (Photo courtesy of S. E. Dowd)

FIGURE 5.29 This is a photo showing a technician in a biosafety level 4 laboratory. The technician is wearing a biosafety suit with an external air supply and is working in a Class II biosafety cabinet with double HEPA filtered exhaust combined with biocidal incineration. These conditions are used with the most dangerous (Class IV) pathogens. The aeromicrobiological conditions are completely controlled and not even the smallest viable organism can escape from these laboratories. (Photo courtesy Center for Disease Control and Prevention, Atlanta, GA.).

lated or injected but that can be contained by ordinary laboratory techniques. Class III agents are those that require special containment; they are associated with aerosol disease transmission and special permits are required to import them from outside the country. Class IV agents are those that require extreme containment and are extremely hazardous to laboratory personnel or may cause serious epidemic disease. Finally, Class V agents are restricted foreign pathogens whose importation, possession, or use is prohibited by law.

5.14 OTHER AREAS OF INTEREST

This chapter has covered several areas that can be affected by the AMB pathway. Although it is beyond the scope of the chapter to cover all areas, there are several environmental problems associated with the AMB

pathway. For example, it has been suggested that airborne bacteria and fungi are causative agents in the formation of calcium oxalate films on stone monuments (Pinna, 1993). The formation of this film is caused by metabolic transformations of proteins and lipids associated with the stone. These calcium oxalate films are extensive, yellow-brown in color, and resist disinfection, and they diminish the beauty of the monuments (see Fig. 15.1). In subterranean environments, cultural relics such as tombs, prehistoric caves, and underground churches are subject to biodeterioration caused by contamination with airborne microorganisms (Monte and Ferrari, 1993). The conservation of underground funeral accouterments, wall paintings, etc. is critical for historic purposes. One tomb located in Veio, Italy that was discovered in 1843 and was prized for its beautiful artwork, frescoes, and wall paintings has been completely destroyed by the airborne contamination that occurred after it was opened. Deposition of airborne microorganisms are also responsible for degradation of papers, parchment, leather and adhesives in libraries. This becomes an especially important cause of degradation especially if the relative humidity in the library rises above 65%. Museums also experience the same problems associated with the decay of paintings and other works of art (Nugari *et al.*, 1993).

QUESTIONS AND PROBLEMS

1. List some beneficial microorganisms not described in the text that are spread via the aeromicrobiological pathway?

2. Describe the AMB pathway for hantavirus including launching mechanisms, transportation mechanisms, and deposition mechanisms.

3. Give an example of a continuous linear source and an example of an instantaneous area source of bioaerosols.

4. Considering a windspeed of 1.5 m/s, an object that is 12 cm tall, and normal air viscosity, determine whether conditions around the object would be considered turbulent.

5. Consider an airborne virus and an airborne protozoan with a radius of 30 nm and 1 μm and particle densities of 2.0 and 1.1 g/cm^3, respectively. Under normal gravitational acceleration calculate the terminal velocity for each.

6. Assume that a pile of feces on the ground is releasing virus at a rate of 2×10^1 particles/sec. An air sampling device is located 100 m downwind at ground level. The wind is blowing with a mean velocity of 1 m/sec. Diffusion of the spores is equal (1 m) in the y and z directions in a plume at the time of sampling. The inactivation rate constant is equal to 0.016/sec. Determine the downwind virus concentration in particles/m^3.

7. There is a great need for rapid detection systems able to monitor air for the presence of biological warfare agents. Design a rapid detection system capable of detection of bacterial pathogens in air.

8. Describe the difference between level III and level IV biosafety laboratories relating to the control of the AMB pathway.

9. Consider ways to control the AMB pathway onboard a space station to prevent the spread of pathogenic microorganisms.

10. Can HEPA filters be relied upon to trap all microorganisms? Why or why not?

11. Draw a schematic of two types of hospital wards (a cancer ward protecting immunocompromised patients and an isolation ward containing people infected with Ebola virus) in relation to the flow of air and placement of AMB pathway control devices.

References and Recommended Readings

American Hospital Association. (1974) "Infection Control in the Hospital," 3rd ed. American Hospital Association, Chicago, pp 69–117.

Andersen, A. A. (1958) New sampler for the collection, sizing, and enumeration of viable airborne particles. *J. Bacteriol.* **76,** 471–484.

Ariatti, A., and Comtois, P. (1993) Louis Pasteur: The first experimental aerobiologist. *Aerobiologia* **9,** 5–14.

Baskerville, A., Humphrey, T. J., Fitzgeorge, R. B., Cook, R. W., Chart, H., Rowe, B., and Whitehead, A. (1992) Airborne infection of laying hens with *Salmonella enteritidis* phage type 4. *Vet. Rec.* **130,** 395-398.

Binot, R., and Paul, P. (1989) BAF—An advanced ecological concept for air quality control. 19th Intersociety Conference on Environmental Systems, SAE Technical Paper 891535.

Boehm, F., and Leuschner, R. M. (1986) "Advances in Aerobiology." Proceedings of the 3rd International Conference on Aerobiology, Birkhäuser Verlag, Boston.

Bollin, G. E., Plouffe, J. F., Para, M. F., and Hackman, B. (1985) Aerosols containing *Legionella pneumophila* generated by shower heads and hot-water faucets. *Appl Environ. Microbiol.* **50,** 1128–1131.

Brenner, K. P., Scappino, P. V., and Clark, C. S. (1988) Animal viruses, coliphages and bacteria in aerosols and wastewater at a spray irrigation site. *Appl. Environ. Microbiol.* **54,** 409-415.

Burge, H. A. (1995) "Bioaerosols." Lewis Publishers, Boca Raton, LA.

Committee on Particulate Control Technology. (1980) "Controlling Airborne Particles." Environmental Studies Board, National Research Council, National Academy of Sciences, Washington, DC.

Cox, C. S. (1987) The aerobiological pathway of microorganisms. Wiley Interscience. New York.

Cox, C. S., and Heckley, R. J. (1973) *Can. J. Microbiol.* **19,** 189–194.

Dimmic, R. L., and Akers, A.B. (1969) "An Introduction to Experimental Aerobiology" Wiley Intescience. New York.

Donaldson, A. I., and Ferris, N. P. (1975) *J. Hyg.* **74,** 409–416.

Dowd, S. E., Pepper, I. L., Gerba, C. P., and Pillai, S. D. (1999) Mathematical modeling and risk assessment of airborne microbial pathogens in association with the land application of biosolids. *J. Environ. Quality.* In review.

Dowd, S. E., Widmer, K. W., and Pillai, S. D. (1997) Thermotolerant clostridia as an airborne pathogen indicator during land application of biosolids. *J. Environ. Qual.* **26,** 194–199.

Eckart, P. (1996) "Spaceflight Life Support and Biospherics." Microcosm Press, Torrence, CA.

Edmonds, R. L. (1979) "Aerobiology: The Ecological Systems Approach." Dowden, Hutchinson & Ross, Stroudsburg, PA.

Fengxiang, C., Qingxuan, H., Lingying, M., and Junbao, L. (1992) Particle diameter of the airborne microorganisms over Beijing and Tianjin area. *Aerobiologia.* **8,** 297–300.

Freifelder, D. M. (1987) "Microbial Genetics" Jones and Bartlett, Portolla Valley, CA.

Gerba, C. P., and Goyal, S. M. (1982) "Methods in Environmental Virology." Marcel Dekker, New York.

Gregory, P. H. (1973) "The Microbiology of the Atmosphere." Leonard Hill Books, Aylesbury, Bucks.

Hers, J. F. P., and Winkler, K. C. (1973) "Airborne Transmission and Airborne Infection." Oosthoek Publishing Co., Utrecht, The Netherlands.

Hinton, M., Ali, E. A., Allen, V., and Linton, A. H. (1983) *J. Hyg.* **91,** 33.

Hugh-Jones, M. E., and Wright, P. B. (1970) Studies on the 1967-8 foot and mouth disease epidemic: The relation of weather to the spread of disease. *J. Hyg.* **68,** 253–271.

Hurst, C. J., Knudsen, G. T., McInerney, M. J., Stetzenbach, L. D., and Walter, M. V. (1997) "Manual of Environmental Microbiology." ASM Press, Washington, DC, pp. 661–663.

Ingold, C. T. (1971) "Fungal Spores—Their Liberation and Dispersal." Clarendon Press, Oxford.

Jensen, P. A., Todd, W. F., Davis, G. N., and Scarpino, P. V. (1992) Evaluation of eight bioaerosol samplers challenged with aerosols of free bacteria. *J. Ind. Hyg. Assoc.* **53,** 660-667.

Krueger, A. P, Smith, R. F., and Go, I. G. (1957) *J. Gen. Physiol.* **41,** 359–381.

Kundsin, R. B. (1980) Airborne contagion. *Ann. N. Y. Acad. Sci.* 353 1–2.

Lighthart, B., and Frisch, A. S. (1976) Estimation of viable airborne microbes downwind from a point source. *Appl. Environ. Microbiol.* **31,** 700–704.

Loosli, C. G., Lemon, H. M., Robertson, O. H., and Appel, E. (1943) Experimental airborne infection. I. Influence of humidity on survival of virus in air. *Proc. Soc. Exp. Biol. Med.* **53,** 205–206.

May, K. R., Druett, H. A., and Packman, L. P. (1969) *Nature* **221,** 1146–1147.

McIntosh, R. A., Wellings, C. R., and Park, R. F. (1995) "Wheat Rusts: An Atlas of Resistance Genes." Kluwer Academic Pub., East Melborne, Australia.

Monte, M., and Ferrari, R. (1993) Biodeterioration in subterranean environments. *Aerobiologia* **9,** 141–148.

Muilenberg, M., and Burge, H. (1996) "Aerobiology." Lewis Publishers, New York.

Nazaroff, W. M., and Cass, G. R. (1986) *Environ. Sci. Technol.* **20,** 924–934.

Nazaroff, W. M., and Cass, G. R. (1989) Mathematical modeling of indoor aerosol dynamics. *Environ. Sci. Technol.* **23,** 157–166.

Nilsson, S., and Raj, F. (1983) "Nordic Aerobiology." Almqvist & Wiksell, Stockholm.

Nugari, M. P., Realini, M., and Roccardi, A. (1993) Contamination of mural paintings by indoor airborne fungal spores. *Aerobiologia* **9,** 131–139.

Parker, D. T., Spendlove, J. C., Bondurant, J. S., and Smith, J. H. (1977) Microbial aerosols from food processing waste spray fields. *J. Water Pollut. Control Fed.* **49,** 2359–2365.

Pasquill, F. (1961) The estimation of the dispersion of windborne material. *Meteorol. Mag.* **90,** 33–49.

Pillai, S. D., Widmer, K. W., Dowd, S. E., and Ricke, S. C. (1996) Occurrence of airborne bacteria and pathogen indicators during land application of sewage sludge. *Appl. Environ. Microbiol.* **62,** 296–299.

Pinna, D. (1993) Fungal physiology and the formation of calcium oxalate films on stone monuments. *Aerobiologia* **9,** 157–167.

Portner, S. M., Hoffman, R. K., and Phillips, C. R. (1965) Microbial control in assembly areas needed for spacecraft. *Air Eng.* **7,** 46–49.

Recer, P. (1997) Flu vaccine nose spray shows promise, researchers say. Associated Press, San Diego Daily Transcripts communication.

Riley, R. L., and O'Grady, F. (1961) "Airborne Infection, Transmission and Control." Macmillan, New York.

Schwartz, J. M. (1971) *Cryobiology* **8,** 255–264.

SIPRI. (1979) "The Problem of Chemical and Biological Warfare," Vol. II. Humanities Press, New York.

Staszkiewicz, J., Lewis, C. M., Colville, J., Zeros, M., and Band, J. (1991) Outbreak of *Brucella melitensis* among microbiology laboratory workers in a community hospital. *J. Clin. Microbiol.* **29,** 287-290.

Tayler, G. I. (1915) Eddy motion in the atmosphere. *Philos. Trans. R. Soc. Ser. A* **215,** 1–26.

Theunissen, H. J. J., Lemmens-Den Toom, N. A., Burggraaf, A., Stolz, E., and Michel, M. F. (1993) Influence of temperature and relative humidity on the survival of *Chlamydia pneumoniae* in aerosols. *Appl. Environ. Microbiol.* **59,** 2589–2593.

U.S. Department of Health and Human Services: CDC-NIH. (1993) "Biosafety in Microbiological and Biomedical Laboratories," 3rd ed., HHS Publication No. (CDC) 93-8395. U.S. Government Printing Office, Washington, DC.

Voyles, B. A. (1993) "The Biology of Viruses." Mosby—Year Book, St. Louis.

Wathes, C. M., Zaidan, W. A. R., Pearson, G. R., Hinton, M., and Todd, N. (1988) Aerosol infection of calves and mice with *Salmonella typhimurium.* Vet. Rec. **123,** 590–594.

Wells, W. F., and Riley, E. C. (1937) An investigation of the bacterial contamination of the air of textile mills with special reference to the influence of artificial humidification. *J. Ind. Hyg. Toxicol.* **19,** 513–561.

Zannetti, P. (1990) "Air Pollution Modeling: Theories, Computational Methods and Available Software." Computational Mechanics Publishing, New York.

6

Aquatic and Extreme Environments

SCOT E . DOWD   DAVID C. HERMAN   RAINA M. MAIER

6.1 INTRODUCTION

Aquatic microbiology, in broad terms, is the study of microorganisms and microbial communities in water environments. Aquatic environments occupy more than 70% of the earth's surface. Although most of this area is occupied by oceans, there is also a broad spectrum of other aquatic environments including estuaries, harbors, major river systems, lakes, wetlands, streams, springs, and aquifers. Water is essential for life and may arguably be our most important natural resource. Aquatic environments, in addition to providing water for drinking, provide necessary resources for agriculture, mining, power generation, semiconductor manufacturing, and virtually every other industry. Thus, protection and preservation of these environments are vital for the continuation of life as we know it. Microbiota are the primary producers in the aquatic environment (and are responsible for approximately one half of all primary production on the earth). They are also primary consumers. The microbiota that inhabit aquatic environments include bacteria, viruses, fungi, algae, and other microfauna. Identifying the microbiological composition of the aquatic environment and determining the physiological activity of each component are the first steps in understanding the ecosystem as a whole (Christian and Capone, 1997). The approach taken in the first part of this chapter is to define the organization, composition, and functioning of the community structural habitats that are characteristic of aquatic environments. These habitats are the planktonic, sediment, microbial mat, and biofilm communities. The second part of this chapter contains descriptions and general microbial characteristics of the three main aquatic environments: (1) inland surface waters (lakes, rivers, streams), (2) seas (oceans, harbors etc.), and (3) groundwaters. Finally, examples, of extreme aquatic environments are given along with features that enable microorganisms to survive in these locations.

6.2 MICROBIAL HABITATS IN THE AQUATIC ENVIRONMENT

6.2.1 Planktonic Environment

Plankton refers to the microbial communities suspended in the water column. Photoautotrophic organisms within this community including both eukaryotes (algae) and prokaryotes (cyanobacteria) are collectively referred to as **phytoplankton**. Suspended heterotrophic bacterial populations are referred to as **bacterioplankton**, and protozoan populations make up the **zooplankton**. Together these three groups of organisms make up the microbial planktonic community. Figure 6.1 shows the relationship and interdependence of the various microbial components within a general planktonic food web. Phytoplankton are the **primary producers** in the food web, using photosynthesis to fix CO_2 into organic matter. In the planktonic microbial community, this primary production is the major source of organic carbon and energy, which is transferred to other trophic levels within the web. The organic compounds produced by phytoplankton can be divided into two classes, particulate or dissolved, depending on their size. **Particulate organic matter (POM)** compounds are large macromolecules such as

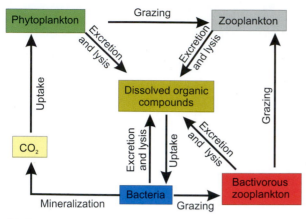

FIGURE 6.1 The microbial loop in the planktonic food web. The microbial loop represents a pathway in which the dissolved organic products are efficiently utilized. The role of bacterioplankton is to mineralize important nutrients contained within organic compounds and to convert a portion of the dissolved carbon into biomass. Grazing by bactivorous protozoans provides a link to higher trophic levels. (Modifed from Fuhrman, 1992.)

polymers, which make up the structural components of the cells, including cell walls and membranes. **Dissolved organic matter (DOM)** is composed of smaller compounds, such as amino acids, carbohydrates, organic acids, and nucleic acids, which are rapidly taken up by microbes and recycled.

6.2.1.1 Primary Production

The amount of primary production within a given water column depends on a variety of environmental factors. These factors include the availability of essential inorganic nutrients, particularly nitrogen and phosphorus; water temperature; and the turbidity of the water, which affects the amount of light transmitted through the water column. Open oceans have relatively low primary productivity because of low levels of the essential nutrients nitrogen and phosphorus. The exceptions are areas where currents cause upwelling of water from the bottom of the ocean, bringing with it nutrients from the deep sea. Coastal areas are productive because of the introduction of dissolved and particulate organic material from river outflows and surface runoff from the terrestrial environment. Higher nutrient loading also results from the decomposition of aquatic plants rooted in shallow waters. Freshwater lakes, like the open seas, are often low-productivity environments, particularly those with large, deep, nutrient-poor (**oligotrophic**) bodies of water. In contrast, smaller and shallower freshwater bodies tend to be nutrient rich (**eutrophic**). Nutrient loading, which causes eutrophication of lakes, can

be a natural evolutionary process or the result of human activity. Sources of natural nutrient loading include terrestrial runoff, rivers that feed into the lake, and plant debris (such as leaves). Nutrient loading resulting from human activities includes the disposal of municipal wastewater and runoff of fertilizers from agricultural fields. Both of these nutrient sources contain high levels of nitrogen and phosphorus, the nutrients that are most often limiting in the aquatic environment.

6.2.1.2 Secondary Production

In a typical food web, phytoplankton (primary producers) are consumed by microfauna (zooplankton), which in turn are consumed by progressively larger organisms, such as fish or other filter feeders. This is called the **grazing food chain.** However, the actual transfer of carbon and energy between trophic levels is much more complex than what is implied by the grazing food chain. A substantial portion of the carbon fixed by photosynthesis (>50%) is released into the water column in the form of dissolved organic matter (see Fig. 6.1). This DOM is rapidly utilized by heterotrophic bacteria (bacterioplankton) present within the planktonic community. The utilization of dissolved organic compounds released from phytoplankton or from other sources (see later) is a pathway in the aquatic food web referred to as the **microbial loop.** In this loop, bacterioplankton mineralize a portion of the organic carbon into CO_2 and assimilate the remainder to produce new biomass. This production of bacterial biomass is referred to as **secondary production.** Secondary production is a major pathway for the utilization of photosynthates, as well as a pathway for the transfer of carbon and energy to higher trophic levels in the aquatic environment. Thus, the microbial loop serves in the efficient utilization of DOM released into the water column by various mechanisms.

The source of the DOM pool in the planktonic environment is primarily phytoplankton, but zooplankton and bacterioplankton also contribute. Aquatic fauna contribute to the dissolved organic carbon pool through excretion and the lysis of dead cells. Among the phytoplankton, it is known that both "healthy" cells and "stressed" cells (those under some form of environmental stress) can release DOM into the water column. Another suggested mechanism is that "sloppy" feeding habits of aquatic animals that prey on phytoplankton may allow a portion of the DOM to be released into the water column. Finally, dissolved organic compounds are also released during the lysis of phytoplankton and bacterioplankton by viruses.

6.2.2 Benthic Habitat

The benthos is a transition zone between the water column and the mineral subsurface. This interface collects the organic material that settles from the water column or that is deposited from the terrestrial environment. The interface is a diffuse and noncompacted mixture of organic matter, mineral particulate material, and water. This zone is characterized by a dramatic increase in the concentration of microorganisms (as much as five orders of magnitude) compared with the planktonic environment. The actual microbial concentration, however, depends on the availability of organic material and the availability of oxygen. Below the interface zone, microbial numbers may decline because of oxygen depletion. Anaerobic microbial processes, that utilize nitrate, sulfate, or iron as terminal electron acceptors (see Chapter 3.4), characterize deeper sediment layers, and in even deeper layers there is a methanogenic consortium.

The benthic habitat is an important feature of the aquatic environment. The cycling of essential nutrients, such as such as C, N, and S, in this area is dependent on a combination of aerobic and anaerobic microbial transformations (Fig. 6.2). The benthic environment can support and often favors the formation of conjoined aerobic and anaerobic microenvironments. Oxygen depletion by microbial activity at the organic-

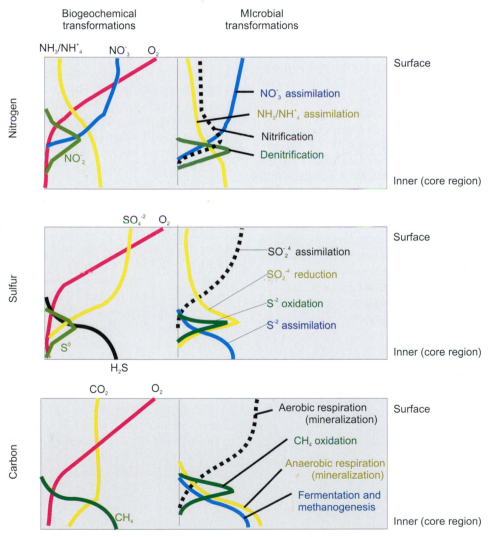

FIGURE 6.2 Biogeochemical profiles and major carbon, nitrogen, and sulfur transformations that can be predicted for environments in which oxygen levels are highest at the "surface" layer and are depleted by microbial activity to create anoxic conditions in the "inner" region. (Adapted with permission from Pearl and Pinckney, 1996.)

rich interface creates anaerobic microenvironments that support the activity of facultative and strictly anaerobic microbes. Specific physiological groups of microorganisms are strategically positioned relative to the oxic–anoxic interface to take advantage of such microenvironmental niches. In this way, the sediment zone acts to support a physiologically diverse aquatic microbial community. For example, fermentative bacteria metabolize DOM into organic acids, such as acetic acid, and CO_2. Organic acids can act as electron donors for a group of strictly anaerobic bacteria, which utilize CO_2 as the final electron acceptor in anaerobic respiration, thus generating methane (CH_4). The methanogenic activity in turn supports the activity of the methane-oxidizing bacteria, which, under aerobic conditions, can utilize methane and other one-carbon compounds as an energy source, regenerating CO_2. Methanotrophic activity is localized at the sediment–water interface zone in order to use CH_4 released from the anaerobic zone and the oxygen available in the water column. This is an example of biogeochemical cycling on the scale of a small habitat (see Chapter 14).

The decomposition of organic material in the sediment layer also generates ammonia from organic debris. The fate of ammonia is controlled by (1) the assimilation of ammonia as a source of essential nitrogen by planktonic and sediment microorganisms and (2) nitrification of ammonia as an energy source by chemoautrophic microorganisms (*Nitrosomonas* and *Nitrobacter* species) (see also Chapter 14.3). These organisms sequentially oxidize ammonia to nitrate (NO_3^-), a process known as nitrification. Nitrification is often localized at the sediment–water interface because it relies on the presence of oxygen and the release of ammonia by the decomposers. The control of ammonia compounds can be important, especially in alkaline environments, where the undissociated NH_4OH form can be toxic to aquatic animals. The activity of the ammonia-oxidizing or nitrifying bacteria, *Nitrosomonas* and *Nitrobacter*, is highly sensitive to the presence of certain DOM, including naturally occurring and industrial chemicals. Therefore, the inhibition of nitrification (ammonia oxidation), which can be detected by an accumulation of ammonia or NO_2, provides a sensitive indicator of the environmental impact of certain toxic pollutants.

6.2.3 Microbial Mats

The previous section described the organization of the sediment–planktonic interface into microenvironments in which the combined aerobic and anaerobic activity is able to support a diversity of microbial populations. Microbial mats are an extreme example of an interfacial aquatic habitat in which many microbial groups are laterally compressed into a thin mat of biological activity. These groups interact with each other in close spatial and temporal physiological couplings. The thickness of mats ranges from several millimeters to a centimeter in depth. They are also vertically stratified into distinct layers. Microbial mats have been found associated with environments such as the surface–planktonic interface of hot springs, deep-sea vents, hypersaline lakes, and marine estuaries (Fig. 6.3). By supporting most of the major biogeochemical cycles, the mats are considered largely self-sufficient (Stal, 1995; Pearl and Pinckney, 1996).

Stal (1995) describes an example of a laminated form of a microbial mat (Fig. 6.4). Cyanobacteria (photosynthetic prokaryotes) occupy the upper zone of the mat, where they have access to sunlight. This primary production by cyanobacteria is the most important source of organic compounds within the microbial mat community. The photosynthetic activity of the cyanobacteria creates an oxygenic environment in the upper layer of the mat. Photosynthetic activity creates oxygen-supersaturated conditions during the day, but at night, in the absence of sunlight, microbial respiration rapidly depletes all the available oxygen. An important process in the decomposition of DOM released by the cyanobacteria involves the activity of sulfate-reducing bacteria. Anaerobic respiration by sulfate-reducing bacteria has been considered a strictly anaerobic process; however, in microbial mats, sulfate-reducing bacteria are active in the oxygenic upper layer. Considering that anoxic conditions develop rapidly at night, there may be a temporal separation between oxygenic photosynthesis and anaerobic activity. It is also possible that a high demand for oxygen by heterotrophic activity can create microenvironments in which anaerobic activity is supported. This type of temporal and spatial separation of aerobic and anaerobic activity is also important in nitrogen cycling. A major source of nitrogen for the mat community is the fixation of atmosphere nitrogen by bacteria that function only under strictly anaerobic conditions. The reduced nitrogen and sulfide products of anaerobic activity provide energy sources for the nitrifying and sulfur-oxidizing bacteria, which are present in close association with the cyanobacteria and are dependent on oxygen to complete the sulfur and nitrogen cycles.

A layer of sediment rich in oxidized iron may form directly underneath the cyanobacterial layer. The origin of the oxidized iron layer is not known, but it appears to form a barrier between aerobic and anaerobic phototrophic activities. Purple sulfur bacteria can form a distinctive layer just below the aerobic–anaerobic interface. The purple sulfur bacteria are also photosyn-

FIGURE 6.3 This photo shows a variety of microbial mats in a hot spring. In the foreground there is a microbial mat that is purple in color. (From Corel Gallery www.corel.com)

thetic, but compared with the cyanobacteria, these bacteria have a less developed photosynthetic process and they are not capable of the photolysis of water. The purple sulfur bacteria utilize a reduced sulfur

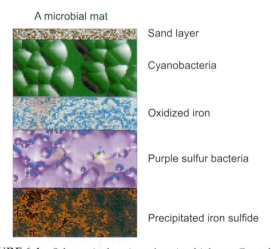

A microbial mat

Sand layer

Cyanobacteria

Oxidized iron

Purple sulfur bacteria

Precipitated iron sulfide

FIGURE 6.4 Schematic drawing of a microbial mat. Cyanobacteria form the surface layer of the microbial mat but may be covered by a layer of sediment, organic debris, or even cyanobacterial sheaths containing a pigment that acts as a sunscreen, blocking excessive ultraviolet radiation. Often a layer of oxidized iron appears below the cyanobacteria, followed by a layer of purple sulfur bacteria that thrive under anaerobic conditions. An extensive zone of sediment enriched in iron sulfide is often present.

compound (sulfide) as an electron donor. The sulfide is provided by the activity of the sulfate-reducing bacteria. Purple sulfur bacteria are readily visible in microbial mats, or other anaerobic environments, because of their conspicuous purple pigmentation. Below the purple sulfur bacteria is an extensive black layer enriched by the precipitation of iron sulfide (FeS).

The microbial mats are unique communities because the interdependent microbial components form clearly stratified and distinctively colored zones. Mats are often found in extreme environments or in environments where conditions fluctuate rapidly. The cyanobacteria are known to be tolerant of extreme conditions, such as high temperatures or highly saline waters, and thrive in locations where competition from other microbial groups and predation by grazing organisms are limited by the inhospitable environment. The microbial mats are also of evolutionary significance. Fossilized microbial mats, known as **stromatolites,** dating back 3.5 billion years were among the first indications of life on earth (see Chapter 14.1.2). At that time, the earth's atmosphere lacked oxygen, and the stromatolites from that era were probably formed with anoxygenic phototrophic bacteria (purple and green sulfur bacteria).

6.2.4 Biofilms

A **biofilm** is a layer of organic matter and microorganisms formed by the attachment and proliferation of bacteria on the surface of an object. In most cases this surface is submersed in nonsterile water or surrounded by a moist environment. Solid surfaces suitable for bacterial colonization in the natural environment include inert surfaces, such as rocks and the hulls of ships and even living surfaces, such as the submerged portions of aquatic plants. Biofilms are characterized by the presence of bacterial extracellular polymers, which can create a visible "slimy" layer on a solid surface (Marshall, 1992). The secretion of extracellular polysaccharides (glycocalyx) provides a matrix for the attachment of bacterial cells and forms the internal architecture of the biofilm community. The exopolymer matrix is also an integral component influencing the functioning and survival of biofilms in hostile environments. Biofilms have been extensively studied for their role in nutrient cycling and pollution control within the aquatic environment, as well as for their beneficial or detrimental effects on human health.

Biofilm development is a complicated issue and is usually initiated by the attachment of bacteria to a solid surface (Fig. 6.5). Permanent attachment of bacteria requires two stages: (1) **reversible attachment,** which is a transitory physicochemical attraction, and (2) **irreversible attachment,** which is a biologically mediated stabilization reaction (see Chapter 7.2.3). Reversible attachment is a function of the initial attraction of a bacterium to a surface and is controlled by several competing forces, including hydrophobic, electrostatic, and van der Waals forces (Marshall, 1985). The van der Waals forces attract an object to a solid surface; while hydrophobic and electrostatic forces can be attractive or repulsive, depending on the surface properties of the bacterium and the solid surface. The properties of a solid surface may be modified by the presence of an adherent conditioning film of organic compounds. Bacterial attachment is promoted when DOM of limited water solubility coats the surface of solid substrates. The source of these hydrophobic organic compounds can be the decomposition of organic material, excretion by living organisms, or lytic products from dead organisms. The organic substrates attached to solid surfaces are a source of nutrients for bacteria, especially in oligotrophic environments where nutrients are limited. In specific cases, it is believed that the initial contact between the cell and the organic conditioning film is made using microbial appendages, such as pili or flagella (Korber *et al.*, 1995).

The initial reversible attachment of organisms to surfaces can, with time, become permanent. Irreversible attachment is initiated by the excretion of extracellular polymers by the reversibly attached bacteria. The extracellular polymers create a matrix that surrounds the cell and forms a strong chemical bridge

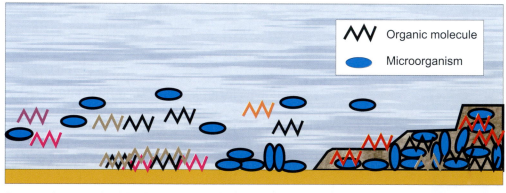

| Clean surface | Conditioning film of dissolved organic matter | Phase I primary microbial colonization | Phase II irreversible microbial attachment | Mature biofilm |

FIGURE 6.5   Representation of biofilm formation. Dissolved organic molecules of a hydrophobic nature accumulate at the solid surface–water interface and form a conditioning film. Bacteria approach the solid surface because of water flow and/or active motility. The initial adhesion (phase I) is controlled by various attractive or repulsive physicochemical forces leading to passive, reversible attachment to the surface. An irreversible attachment is a biological, time-dependent process related to the proliferation of bacterial exopolymers forming a chemical bridge to the solid surface (phase II). By a combination of colonization and bacterial growth, the mature biofilm is formed. It is characterized by cell clusters surrounded by water-filled voids. (Adapted from Marshall, 1992, and Marshall, 1997.)

to the solid surface. The fibrous, anionic extracellular matrix can have several functions for the adherent cells. The matrix can act as an ion-exchange resin to filter and collect essential nutrients. It can protect the attached community from environmentally stressful conditions such as desiccation or changes in pH and temperature. It can also provide some protection of the cells from predation by protozoans. Attachment has been reported to increase, inhibit, or have no effect on the growth rate of attached cells when compared with planktonic cells (van Loosdrecht *et al.*, 1990). The positive effect of attachment reported for some bacteria may be the result of indirect factors such as an increase in the local concentration of nutrients or the protective effects of the biofilm exopolymeric matrix. It has also been suggested that attachment may activate the expression of specific genes that are beneficial to the microorganism in the altered physicochemical conditions of the biofilm environment (Goodman and Marshall, 1995).

The transition from initial microbial attachment to the development of a mature biofilm is due not only to the proliferation of the attached cells but also to the continued deposition of cells from the bulk fluid. Examination of mature biofilms using microscopic techniques has revealed a complex organization. The biofilm cells have been shown to be organized into column-shaped clusters embedded within the extracellular polymer matrix and surrounded by large void spaces (Wolfaardt *et al.*, 1994; Korber *et al.*, 1995). The void spaces form channels, which can carry limiting nutrients, such as oxygen, into the exopolymeric matrix. The presence of void spaces increases the biofilm surface area and the efficiency with which nutrients and gases are transferred between the biofilm and the surrounding water. The exact nature of the biofilm architecture depends on numerous factors, including the type of solid surface, the microbial composition of the biofilm, and environmental conditions.

Attachment and biofilm formation as a survival strategy are well illustrated by the distribution of bacteria in the nutrient-poor environment of high alpine streams (Geesey *et al.*, 1978). Bacteria are predominantly found attached to rock surfaces, where they can take full advantage of the continuous renewal of nutrients provided by water flow, and thus act as a biological filter to remove DOM from flowing waters. The filtration of DOM from the water by these attached communities represents a water purification system in natural environments. This system has long been exploited for use in purifying water from municipal (sewage) or industrial sources. In systems known as trickling filters and fluidized bed reactors, the biomass of the biofilm is maximized by providing a porous network of solid supports for bacterial attachment (see Chapter 21.2.2.1). The porous network is meant to provide an expansive area for biofilm development and increase the area of contact between the biofilm and DOM in the wastewater stream. Biofilter systems have also been used for the degradation of a wide range of pollutants, including hydrocarbons, pesticides, and industrial solvents, that may be present in wastewater streams.

Microbial colonization at a solid–liquid interface can be beneficial, as in nutrient cycling and water purification processes. However, in some instances, biofilm development can be detrimental, and there is a need to control the growth of biofilm cells using antibacterial substances. For example, biofilm control is required in industries that make extensive use of pipelines. Many industries use water pipelines in cooling towers or heat exchange structures, and the presence of a thick biofilm inside these pipes can lower the flow capacity of the pipes and decrease heat-exchange efficiency. Compared with planktonic cells, biofilm formations are more resistant to antibacterial substances, such as antibiotics and disinfectants. The reduced efficacy of these agents may be due to their inability to penetrate the extracellular matrix material or to an altered physiological state of the attached bacterial cells. The resistance of bacteria in biofilms can present problems in the use of antibiotics to treat diseases and in the use of chemical disinfectants to sterilize medical devices, such as surgical tools or medical implants. Control of biofilm development in drinking water distribution systems is a well-documented problem. Attached bacteria grow by utilizing low concentrations of DOM present in water distribution systems. These biofilms can harbor opportunistic pathogens and require high doses of disinfectant for their control. The high concentration of disinfectant, in turn, can cause public and environmental health problems (see Chapter 23). The development of innovative technologies has focused on ways to enhance the effectiveness of disinfectants so that lower doses can be used or to remove the disinfectant from the water supply after the biofilm has been destroyed (Mittelman, 1995). Alternative approaches include the impregnation of materials, such as plastics, with biocides or antibiotics so that colonization of the surface is inhibited.

The support of mixed microbial groups, which interact as a consortium within the architecture of the biofilm, provides a distinct advantage for the survival of diverse physiological types within the biofilm environment. The advantage of adherence of microbial consortia is evident with the biodegradation of cellulose fibers. Bacteria that possess cellulolytic capabilities are able to attach directly to the cellulose fibers. These cells are then surrounded by other cells capable of utilizing the glucose monomers released by the depolymerization of cellulose. Another good example is the positioning of different physiological types in separate aerobic or anaerobic microenvironments that are engaged in complementary metabolic activities. The products of one form of metabolism are exchanged across the aerobic–anaerobic interface, providing the essential requirements for the growth of a different physiological type. The contribution of biofilm structure to the corrosion of metal surfaces is an example of this level of organization and is discussed in Section 15.2.

6.3 AQUATIC ENVIRONMENTS

6.3.1 Freshwater Environments

Freshwater environments are inland bodies of water (such as springs, rivers and streams, and lakes) that are not directly influenced by marine waters. The science that focuses on the study of freshwater habitats is called **limnology** and the study of freshwater microorganisms is **microlimnology.** There are two types of freshwater environments. The first is standing water, or **lentic habitats,** such as lakes, ponds, and bogs. The second is running water, or **lotic habitats,** including springs, streams, and rivers. These freshwater environments have very different physical and chemical makeups and correspondingly characteristic populations and communities of microorganisms. In addition, no single description of these microbial communities can accurately describe each and every case. For instance, the microbial community in a lake in Egypt is not the same as the microbial community in one of the Great Lakes in the northeastern United States. In this section we define various freshwater environments and provide ranges of numbers and the types of populations of the microbes that inhabit them.

6.3.1.1 Springs

Springs form wherever subterranean water reaches the earth's surface (Fig. 6.6). There are many types of springs. For instance, melted snow and ice in mountainous regions feed **cold springs** such as scree springs. In contrast, groundwater springs are fed by surface-linked aquifers. Warm or hot springs originating from volcanic areas or great depths are known as **thermal springs.** A good example are the geysers found in Yellowstone National Park (Fig. 6.7). Springs are also characterized with respect to their chemical properties, especially if they have a distinctive mineral or chemical composition. Examples include **sulfur** and

FIGURE 6.6 Example of a freshwater spring. This photograph shows subterranean water bubbling forth and the photosynthetic populations associated with this type of environment. Primary producers predominate in these environments, where there is typically little organic carbon originating from the subsurface source. (From Corel Gallery www.corel.com)

magnesium springs, acid springs, and **radioactive springs.** Microorganisms, especially bacteria and algae, are often the only inhabitants of springs.

In general, photosynthetic populations dominate spring environments with photosynthetic bacteria and algal communities ranging from 10^2 to 10^8 organisms/ml. These primary producers are present in highest concentrations (10^6 to 10^9 organisms/ml) along the shallower edges of the spring and in association with rock surfaces, where light is available and inorganic nutrients are in highest concentrations (Kaplan and Newbold, 1993; Rheinheimer, 1985). Although heterotrophs are also present, because the nutrient level, especially DOM, is low, numbers are usually quite low (10^1 to 10^6 organisms/ml) compared with other surface waters. As they mature and die, photosynthetic populations provide the initial source

FIGURE 6.7 Example of an erupting geyser in Yellowstone National Park. These hot springs are commonly caused by collection of rainwater under pressure in hot rocks below ground. The water turns partly to steam and causes the water above it to overflow, thereby reducing the pressure and leading to an eruption. (From Corel Gallery www.corel.com)

6.3.1.2 Rivers and Streams

Springs, as they flow away from their subsurface source, merge with other water sources to form streams and rivers that eventually flow into other bodies of water such as lakes or seas (Fig. 6.8). These water sources can be other springs, terrestrial runoff, and even hyporheos. A **hyporheos** is defined as an interface where the subsurface water table meets and interacts with the streambed. It can provide an active exchange interface between the surface water and subsurface aquatic environments.

As a stream progresses and becomes larger, it tends to accumulate organic matter and heterotrophic populations. These heterotrophic populations are inoculated from the surrounding terrestrial environment, and the profile of microorganisms in river water often resembles associated terrestrial microbial communities.

Most physical characteristics of rivers and streams, such as temperature, volume, velocity, and chemical makeup, are determined by the geographic and climatic conditions of the area through which they flow. For instance, mountain streams, because of the steep terrain through which they flow, have fast currents and low temperature, whereas rivers that flow through plains may flow very slowly and have higher temperatures. Rivers are usually not very deep (only a few meters), but larger rivers can have pools over 50 m deep. The volume of rivers and streams is highly dependent on seasonal fluctuations. All of these variables influence the microbial populations found in rivers and streams. For instance, current affects the development of planktonic communities. In lotic habitats these communities develop only if the local current is extremely slow or nonexistent. In most cases, stable and localized planktonic communities in rivers are found only in pools.

Streams, like most surface water environments, contain primary producer communities, especially when light can penetrate to the bottom of the stream. Photosynthetic populations range from 10^0 to 10^8 organisms/ml and tend to be present as attached communities associated with biofilms because of the flowing nature of the water column. Phytoplankton (free-living) communities also exist in streams, but because of the constant water movement, they are not spatially stable populations (Rheinheimer, 1985).

As streams evolve into rivers, they tend to acquire more DOM. The increase in DOM limits the penetration of light and consequently begins to limit photoautotrophic populations. However, heterotrophic populations begin to increase in response to increased DOM. Because of their flow patterns, stream and river

of organic matter for downstream heterotrophic populations. However, the largest portion of DOM found in surface fresh water originates from surrounding terrestrial sources. This organic input, which originates from sources such as plant exudates, dead plants, animals, and microbial biomass, is transported into lotic habitats by mechanisms such as terrestrial runoff, seepage, and wind deposition. Thus, we have the image of spring water starting with very low concentrations of DOM and heterotrophs. The DOM and the heterotrophic populations steadily increase as the spring moves away from the source as inputs of terrestrial organic matter and microbial biomass continues to accumulate (Kaplan and Newbold, 1993).

FIGURE 6.8 Example of a river. As the river flows, organic matter is accumulated. Much of this organic matter comes from surrounding terrestrial environments. It can be seen from the photograph that there is considerable organic matter just in the form of plant life. This level of organic matter leads to increases in the ratio of heterotrophic populations, which more often than not predominate in these environments. (From Corel Gallery www.corel.com)

waters are for the most part well aerated. Therefore, heterotrophic populations are predominantly aerobic or facultatively anaerobic. In general, the concentration of heterotrophs in streams and rivers ranges from 10^4 to 10^9 organisms/ml, with microbial numbers increasing as DOM increases. Sewage outflows are areas where this is especially evident. Downstream of these outflows, heterotrophic populations often increase two to three orders of magnitude. Although isolated pools that form in rivers act as DOM and POM sinks and support fairly stable heterotrophic planktonic communities, the only truly stable populations in the lotic habitats of streams and rivers are the biofilm and sediment (benthic) communities (Rheinheimer, 1985).

6.3.1.3 Lakes

Physical and Chemical Characteristics

Lakes vary in depth from a few meters to more than 1000 m (Fig. 6.9). Lakes can also vary considerably in surface area, from small ponds of only a few square meters to huge lakes that cover areas of up to 100,000 km². Although often regarded as **lentic** or nonflowing environments, lakes have inflows and outflows, wind-generated turbulence, and temperature-generated mixing, all of which create a dynamic environment. Salt lakes, such as the Great Salt Lake in Utah, are distin-

guished by a high salt content and are examples of extreme environments (see Section 6.4.3). Other lakes are also characterized by chemical composition, such as bitter lakes that are rich in $MgSO_4$, borax lakes that are high in $Na_2B_4O_7$, and soda lakes that are high in

FIGURE 6.9 Example of a typical oligotrophic lake. Lakes are very dynamic aquatic environments with extensive phototrophic and heterotrophic populations with equally dynamic interactions. Whether the phototrophic or heterotrophic populations predominate depends on environmental factors. (From Corel Gallery www.corel.com)

NaHCO$_3$. Lakes are by far the most complex of the freshwater environments. As a result, the microbial communities and their interactions are equally complex and diverse.

Lakes are divided into subsections based upon morphometric (depth, dimension, geology of shores, currents, etc.) and physicochemical (temperature, pH, oxygen content) parameters (Fig. 6.10). The edge of the lake, where sunlight can penetrate to the bottom, is known as the **littoral zone.** The air–water interface including the upper few millimeters of the water column is known as the **neuston layer.** The general structure of the neuston is shown in Fig. 6.11. The neuston is known to accumulate nutrients. The very top layer of the neuston is a thin lipid layer (10 nm deep) that is created because it is energetically favorable for nonpolar organic molecules to align at the air–water interface. Adjacent to this is a slightly thicker layer (100 nm) containing proteins and polysaccharides. Together, these layers form a thin gel-like matrix at the air–water interface. Bacteria attach to this organic layer in a firm but reversible manner (Maki, 1993). Thus, we also have the image of the neuston layer as a biofilm, where organic molecules "condition" the air–water interface, allowing bacteria to attach. The **limnetic zone** refers to the surface layer of open water away from the littoral zone where light readily penetrates. The area below the limnetic zone, where light intensity is less than 1% of sunlight (the light compensation point), is known as the **profundal zone.** Finally, the **benthic zone** consists of the lake bottom and the associated sediments.

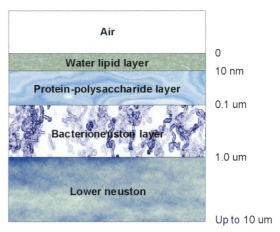

FIGURE 6.11 Schematic representation of the neuston. This is the upper layer of aquatic environments and can range from 1 to 10 μm in depth. Most scientists consider the neuston an extreme environment (see Section 6.4.1) because of many factors, including intense solar radiation, large temperature fluctuations, and the natural accumulation of toxic substances including chemicals, organic matter, and heavy metals. The upper layer that interacts with the atmosphere consists of a water–lipid mixture that has increased surface tension. Below this is a layer of organic matter that accumulates from organic matter rising up the water column.

Temperature is very important, especially in lakes, giving rise to another classification scheme. The three regions in this scheme are the upper zone or **epilimnion,** the lower zone or **hypolimnion,** and the **thermocline,** which is a middle zone characterized by a rapid change in temperature (Fig. 6.12). Because water is most dense at 4°C, temperature-induced density stratification occurs at the thermocline in the summer and the winter. In the summer, the epilimnion, which is heated by sunlight, is typically warm and oxygen rich. This zone is usually characterized by intensive

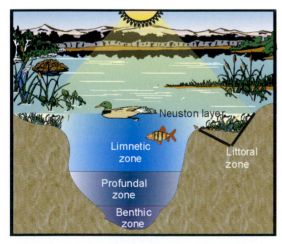

FIGURE 6.10 Schematic representation of a typical lake showing common designations based on sunlight. Other designations for zones are based on features such as temperature, oxygen concentration, and pH. However, the most common are those shown here primarily because of the controlling influence sunlight has on these environments.

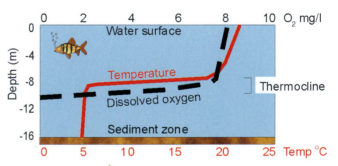

FIGURE 6.12 Idealized profiles of temperature and oxygen in a temperate region, eutrophic lake. Stratification is due to thermal warming of the upper layers in the summer months. Cooling of the upper layer in the fall and early winter breaks the mixing barrier and allows the sediment zone to be reoxygenated. (Adapted from Wetzel, 1983.)

primary productivity that can deplete the epilimnion of mineral nutrients, resulting in nutrient-limiting conditions. The characteristics of the epilimnion are reversed in the hypolimnion, which has low temperature and oxygen levels, lack of light penetration, and a high mineral nutrient content. This stratification would tend to make lakes very static, but as fall and winter approach, the warm waters of the epilimnion cool until they reach the temperature, and consequently the density, of the hypolimnion. When this happens the thermocline breaks down and allows mixing of the epilimnion and the hypolimnion. In the winter a layer of ice forms at the top of the lake and the epilimnion is formed in the region of 0°C (ice layer) to 4°C. The hypolimnion remains at 4°C or warmer, and again a thermocline is formed and no mixing occurs. In the spring, as the lake thaws and the two zones reach a similar temperature, mixing occurs once again. In essence, the turnover and mixing of these two layers allow reoxygenation of the hypolimnion and replenishment of mineral nutrients in the epilimnion.

Microbial Characteristics

Lakes are the most extensively studied aquatic environments and in many cases are the most complex microbially. Lakes contain extensive primary and secondary productive populations that interact dynamically. In the littoral zone the planktonic community is composed predominantly of algae and secondarily of cyanobacteria. Even the attached communities are dominated by the presence of filamentous and epiphytic algae. Thus, the primary productivity in the littoral zone is high. The limnetic habitat is also dominated by phytoplankton, which form distinct community gradients based upon the wavelength and the amount of light that penetrates to a given depth. Figure 6.13 shows characteristic photosynthetic organisms (phytoplankton) and their light absorption spectra. As an example of how these microbial community gradients become established, consider *Chlorobium*, a green sulfur bacterium. *Chlorobium* can utilize longer wavelengths of light than many other phototrophs. They are also anaerobic organisms, requiring H_2S rather than H_2O for photosynthesis (see Section 14.4.3.2). Thus, they have a competitive advantage in establishing a niche at depths lower in the water column or even in the surfaces of sediments, where only small amounts of light penetrate, little or no oxygen is present, but hydrogen sulfide is available.

In addition to their phototrophic populations, lakes have extensive heterotrophic communities. The amount of secondary production is directly related to primary production. In general, secondary production in the photic zone is 20 to 30% of primary production.

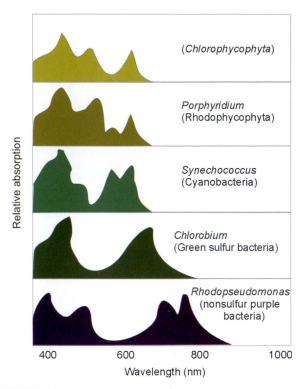

FIGURE 6.13 Graph showing the light absorbance spectrum of common phytoplanktonic algae and photosynthetic bacteria. It can be seen that each of these groups has a different profile. This enables groups to take advantage of their niche. In general organisms that are capable of utilizing longer wavelengths are found deeper in the water column. Thus, they do not have to compete with organisms higher in the water column that absorb the shorter wavelengths. (Adapted from Atlas and Bartha, 1993.)

Heterotrophic concentrations vary with depth, but there are three areas that generally have elevated numbers of heterotrophs. The first is the neuston layer, where accumulations of proteins and fatty acids create localized eutrophic conditions of which heterotrophic organisms take advantage. The second area where heterotrophic populations are markedly higher is the thermocline, which is just below the zone of highest primary productivity. This is because organic debris tends to settle and accumulate at the thermocline as it does in the neuston layer. The third area of a typical lake that is characterized by higher concentrations of heterotrophs is the upper layer of the benthos. The populations here are primarily anaerobic.

In comparing oligotrophic and eutrophic lakes, there are some striking differences. Figure 6.14a and b compare the major bacterial populations found in typical oligotrophic and eutrophic lakes along with their relative concentrations (Konopka, 1993). In terms of primary productivity, oligotrophic lakes have higher rates (20 to 120 mg carbon/m^3/day) than eutrophic lakes (1 to 30 mg carbon/m^3/day). This is because eu-

A   B

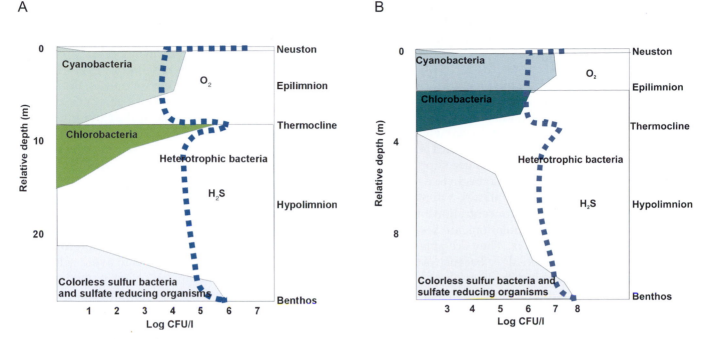

FIGURE 6.14 (A) Schematic representation of bacterial distribution in a typical oligotrophic lake. Notice especially the distribution and concentrations of the photosynthetic populations. Also note the lower concentration of heterotrophs in the upper zone, where cyanobacteria predominate. The large increase in the heterotrophic population between the epilimnion and the hypolimnion is related to the presence of a zone where organic matter accumulates. This area is known as a thermocline and is a zone where the sunlight-warmed surface water (less dense) and the deeper colder water (more dense) meet, forming a density gradient where organic matter accumulates. (B) Schematic representation of a typical eutrophic lake. The figure shows the same groups of organisms as in (A) indicating the localization and relative concentrations throughout the water column. Notice that both the photosynthetic and the heterotrophic populations are considerably higher in a eutrophic lake. (Adapted from Rheinheimer, 1985.)

trophic lakes have much higher levels of organic matter, which causes turbidity and interferes with light penetration. As might be expected, in terms of secondary productivity, eutrophic lakes have much higher rates (190 to 220 mg carbon/m^3/day) than oligotrophic lakes (1 to 80 mg carbon/m^3/day) (Atlas and Bartha, 1993).

Apart from their bacterial and algal populations, streams, rivers, and lakes also contain fungal, protozoan, and viral populations, which interact and contribute to the functioning of the food web. Fungi are rarely planktonic but they serve as parasites of planktonic algae, preventing overpopulation and allowing light to penetrate farther into the water column. Some fungi have simple lifestyles, colonizing surfaces and often forming fungal lawns, whereas other fungi such as *Zoophagus insidians* have a more complex lifestyle. *Z. insidians* live on filamentous green algae in rivers and lakes. The fungi have long hyphae, which branch off and extend into the water column. The hyphae actually act like fishing lines because when they are touched by the cilia of rotifers they quickly secrete a

sticky substance, essentially hooking or attaching to the animal. At this point, and with great rapidity, the hyphae grow into the mouth of the rotifer, where a fungal mycelium forms, absorbing the contents of the animal's body.

Viruses in freshwater environments can be very abundant and can utilize bacteria, cyanobacteria, and microalgae as their hosts. Viruses affect both population dynamics and community composition in the planktonic environment. The impact of viral infection is a complex issue and an area of current study. Much attention has been focused on the interaction between bacterioplankton and viruses. The total viral population can exceed that of the bacterioplankton by two orders of magnitude, and viral population densities tend to fluctuate with the population densities of their hosts. Viruses that infect bacterioplankton (bacteriophage) also vary in host specificity. In other words, some bacteriophage can be infective to a variety of bacterial species and genera, whereas others may be infective to only one particular species. Virus-induced lysis of bacterioplankton can account for approximately 20 to 50% of bacterial mortality (Suttle, 1994; Hennes and Simon, 1995). The DOM released by the lysis of microbial cells is an important nutrient source for noninfected bacteria, as discussed in Section 6.2.1.2 (Middledoe *et al.*, 1996).

Protozoa are also important predators of aquatic microorganisms, bacteria and algae in particular. As bacterial and algal concentrations increase, they provide an abundance of food for protozoa, whose populations begin to increase. As the food source is used up, the numbers of protozoa are no longer supported by bacterial and algal biomass and so begin to decline. As protozoal numbers fade, this allows a renewed increase in bacterial and algae concentrations. It should be noted that protozoan populations are several logs lower than bacterial numbers. They are able to affect the numbers of bacteria and algae because each protozoan is able to consume hundreds of bacteria and algae per day. It should also be noted that viral concentrations tend to mimic protozoan ones. Thus, it can be assumed that together the protozoan and viral populations help to control the concentrations and biomass of the bacterial and algal communities, providing population balance in freshwater ecosystems.

6.3.2 Brackish Water

Brackish water is a broad term used to describe water that is more saline than fresh water but less saline than true marine environments. Often these are transitional areas between fresh and marine waters. An **estuary,** which is the part of a river that meets the sea, is the best known example of brackish water. Estuaries are highly variable environments because the salinity can change drastically over a relatively short distance. Dramatic change can also occur at a given point in the estuary as a function of the time of day or season of the year. For example, at high tide, the salt content at a given spot in the estuary will increase as ocean water moves into the area. In contrast, seasonal increases in fresh water due to rainfall or snowmelt will decrease the salinity at a given point in the estuary. The varia-

tion in salinity can range from 10‰ to 32‰, with the average salinity of fresh water being 0.5‰ (see Information Box). In order to survive in these environments, microbes and plants in an estuary must be adapted to the fluctuations in salinity. Despite this, estuaries are very productive environments. Specific examples of highly productive brackish water environments are mangrove swamps such as those found in the everglades of Florida. The salinity in these swamps is usually very close to that of seawater. Mangrove swamps are named for the characteristic trees that grow in the saturated soils of the swamp. These trees have specially adapted roots that grow up from the water to allow gas exchange above the water so that the trees can continue to obtain oxygen and "breathe." A second important adaptation is that mangrove trees inhibit salt transport into the roots to avoid salt stress. Mangrove swamps are an important transition community because they help filter contaminants and nutrients from the water, they stabilize sediments, and protect the shoreline from erosion. They also provide an active habitat for more than 250 animal and 180 bird species.

In general, estuarine primary production (10 to 45 mg carbon/m^3/day) is not always enough to support the secondary populations. Estuaries tend to be turbid because of the large amount of organic matter brought in by rivers and the mixing action of tides (Ducklow and Shiah, 1993). As a result, light penetration is poor. Numbers of primary producers are variable, ranging from 10^0 to 10^7 organisms/ml, and these populations also vary considerably in relation to depth and proximity to existing littoral zones. Despite low primary productivity, because availability of substrate is not limited, heterotrophic activity is high, ranging from 150 to 230 mg carbon/m^3/day. Local runoff and organic carbon are brought in abundantly by the rivers that flow into the estuaries. In fact, the supply of nutrients can be so great that in many cases estuaries can actually become anoxic for whole seasons during the year (Ducklow and Shiah, 1993). As a result of the steady and abundant carbon supply, numbers of sec-

---

What is Salinity?

The average salt concentration in the ocean is approximately 3.5%. This is more precisely expressed in terms of salinity. Salinity ‰, is defined as the mass in grams of dissolved inorganic matter in 1 kg of seawater after all Br$^-$ and I$^-$ have been replaced by the equivalent quantity of Cl$^-$, and all HCO$_3^-$ and CO$_3^{2-}$ have been converted to oxide. In terms of salinity, marine waters range from 33 to 37‰, with an average of 35‰.

ondary producers fall into a much narrower range from 10^6 to 10^8 organisms/ml.

6.3.3 Marine Water

6.3.3.1 Physical and Chemical Characteristics

Marine water environments, like those of lakes, are highly diverse. Marine water is characterized by salinity between 33 and 37‰ and can range in depth up to 11,000 m in the deepest of ocean trenches. Oceans are not static in size or shape; considerable mixing (especially of the surface layers) and movement are caused by the action of tides, currents, temperature upwelling, and winds. Because the oceans are so expansive and their surface area is so great, the effect of sunlight is important. The ocean is divided into two zones, the **photic zone**, through which light can penetrate, and the lower **aphotic zone**. Light is able to penetrate to a depth of 200 m, depending on the turbidity of the water. In coastal areas, where the amount of suspended particulate matter in the water is high, light may penetrate less than 1 m.

Other designations for classifying zones in marine environments are based on habitats. Four major habitats are important from a microbiological standpoint. At the surface of the sea (air–water interface) is the habitat referred to as the **neuston** (Fig. 6.11). The **pelagic zone** is a broad term used to describe the water column or planktonic habitat. The pelagic habitat is subdivided on the basis of the precise depth in the water column. The habitat in the upper 100 m of the water column is known as the **epipelagic zone** (i.e., the photic zone). A large proportion of the organisms in the epipelagic zone are photosynthetic. Further depths are designated as mesopelagic, bathypelagic and abyssopelagic habitats. Finally, the **benthopelagic zone** (benthos) is the sea–sediment interface. Apart from the pelagic zone and the neuston layer, the third major habitat is the **epibiotic** habitat, which refers to surfaces on which attached communities occur. The fourth is the **endobiotic** habitat, which pertains to organisms found within the tissues of other larger organisms such as fish. One interesting endobiotic bacterium, *Epulopiscium fishelsoni,* is discussed in Section 6.5.1.

6.3.3.2 Microbial Characteristics

The ocean contains diverse microbial habitats. Also, depending on location (compare the Mediterranean with the Antarctic ocean), markedly different microbial populations predominate. As a rule, especially in deep waters, microbial concentrations are highest within the neuston and drop markedly below this region. Immediately below the neuston, the numbers are on average close to 10^7 organisms/ml and decrease by more than a log at a depth of 100 m. As a result of organic input from the terrestrial environment, total bacterial numbers are on average one order of magnitude higher in coastal water than in the open ocean. This is especially true for harbors and near populated areas (Rheinheimer, 1985). If we consider the vertical distribution of the heterotrophic populations we find that, as in lake environments, numbers increase at the thermocline that exists below the zone of primary production (50 m). As mentioned in Section 6.3.1.3, lake thermoclines are also regions of high heterotrophic activity because of the accumulation of organic matter. At greater depths, the numbers of heterotrophs quickly diminish until, at a depth of 200 m, concentrations are very low. Heterotroph numbers increase again immediately above the ocean floor.

Thermocline-induced stratification is not as dramatic in coastal waters because of the mixing of water by winds, currents, and temperature. For this reason, bacterial numbers are uniform at all depths except when the weather is very calm for long periods of time. In addition, seasonal fluctuations in bacterial numbers, which are not observed in the open ocean, are common near the coast. In general, there are two times of the year when there is an increase in bacterial populations in coastal waters, late spring–early summer and late summer–early fall. These are also times when the phytoplankton are most active. Thus, primary production is intimately tied to the overall productivity of the ocean. Figure 6.15 shows the relationship between chlorophyll *a*, a photosynthetic pigment, and heterotrophic numbers. It can be seen that an increase in phototrophs is followed by an increase in heterotrophs. This tends to be true in many aquatic environments where the concentration of primary producers ranges from 10^0 organisms/ml in some benthic environments to 10^8 organisms/ml in surface zones. Heterotrophic populations generally vary considerably with depth as well, although not as much as the photosynthetic populations. The general range is from 10^1 organisms/ml in zones characterized as oligotrophic to 10^8 organisms/ml in zones where organic matter is present. Oceans have profiles similar to those presented in Fig. 6.14a and b, depending on whether the marine environment is oligotrophic like the open ocean or eutrophic like coastal waters, especially coastal waters where sewage outpours may be present.

Phytoplankton abundance and community compositions, as mentioned in Section 6.3.1.3, vary depending on the season. Interesting phenomena related to phytoplankton in the ocean are **algal blooms.** These

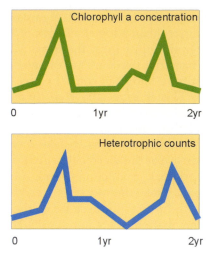

FIGURE 6.15 Diagram of the interrelationship between the concentration of chlorophyll *a*, a photosynthetic pigment, and heterotroph concentration. The concentration of chlorophyll in water is related to the amount of primary production. This in turn influences the amount of secondary production by heterotrophic populations. In this figure, it can be seen that as the chlorophyll *a* concentration increases, it is closely followed by an increase in heterotrophic populations. Thus, secondary production is intimately tied to primary production. (Adapted from Rheinheimer, 1985.)

occur when a eutrophic water body and the appropriate environmental conditions (usually warm, sunny, and calm) occur. In these nutrient-rich and agreeable environmental conditions, it is common for certain algae or cyanobacteria to proliferate rapidly, resulting in blooms. Such blooms are often a natural part of the yearly cycling of lake and ocean ecosystems. On the other hand, when uncharacteristically high eutrophication events occur, algal blooms can adversely affect the water quality. Such water tends to be scummy and smelly, making it unpleasant for recreation and even dangerous for fishing, boating, or swimming. In the worst cases, algal blooms may be composed of algae that produce potent toxins. The **"red tide"** is an example of such an event. This is an immense algal bloom that is composed of red-pigmented dinoflagellates. These algae produce powerful toxins that are taken up by fish, causing extensive fish kills. These toxins can also affect marine birds, marine mammals, and even humans who consume fish caught in these waters. Types of human illness caused in humans by algal toxins include skin rashes, eye irritation, vomiting, diarrhea, fever, and joint pain.

Fungi, protozoa, and viruses in their roles as predators and parasites are also important components of marine waters. Bacteriophage in particular are highly prevalent, often occurring in concentrations one to two logs higher than the bacterial populations at any given location in the water column. Viruses that infect phy-toplankton are very important from an organic cycling standpoint because the organic matter released by phytoplankton after they are infected and lysed promotes secondary productivity. Secondary producers and even algae and larger marine animals such as fish and crabs are also targets of marine viruses. Such viruses can be responsible for considerable economic losses for the fishing industry.

Protozoa also act as important bacterial predators in the marine environment. In addition, fungi have recently been found to be widely distributed in the marine habitat. Marine fungi feed on both plants and animals in addition to bacteria and algae. A relatively small number of genera are found in the marine environments, but these can often be present at surprisingly high concentrations. Certain species of fungi can be isolated with concentrations up to 10^3 to 10^4 organisms/ml, especially in carbon-rich areas of the water column and in organic-rich benthic habitats. Another surprising fact is that fungi have been found at depths well below 1000 m.

6.3.3.3 Deep-Sea Hydrothermal Vents

In 1977 geologists first described deep-sea hydrothermal vents (Fig. 6.16). These are areas on the ocean floor where, driven by magma-derived hydrothermal convection, hot water laced with minerals flows up through cracks and fissures. The cracks, which are known as **hydrothermal vents,** often have a buildup of chemicals resembling chimneys surrounding them. Water, reaching temperatures up to 400°C, is emitted from these vents at rates of 1 to 2 m/sec. This mineral-rich, hydrothermal water forms a dark cloud of mineral precipitates as it mixes with seawater. Thus, these hydrothermal vent chimneys are known as "black smokers." It was surprising to find that in this environment, which has no light and extremely high temperature and pressure, whole self-contained ecosystems consisting of microscopic and macroscopic life have developed.

There is currently vigorous debate about the actual source of primary organic carbon and the process for its introduction into these ecosystems. There are three main schools of thought, each of which is based as much on theory as on actual scientific data. The hypothesis that currently has the most credibility is the **chemoautotrophic theory.** Chemoautotrophs utilize CO_2 as a major and usually exclusive source of carbon (Karl, 1995). In this hypothesis, primary organic carbon production in hydrothermal vent communities is based on bacterial use of H_2S, H_2 and CH_4 as electron donors, and CO_2 as a carbon source. Thus, according to this hypothesis, the entire food web in a hydrother-

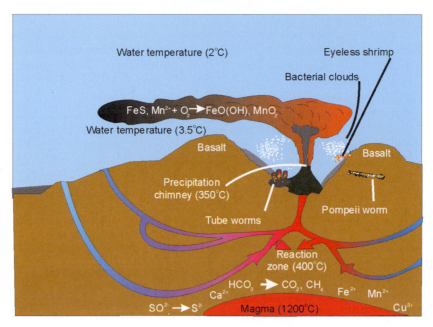

FIGURE 6.16 Schematic representation of a hydrothermal vent community. This figure depicts a black smoker rising from the ocean floor and a plume of chemical-rich superheated water rising from it. Also depicted are characteristic organisms, including bacterial clouds that primarily consist of heterotrophic bacteria. Other creatures that are often found in hydrothermal vent communities are shown, including tube worms, eyeless shrimp, and Pompeii worms. Each of these is reliant on microbial primary or secondary productivity. The tube worms, as described in the text, rely on sulfur bacteria for primary production of carbon which is shared symbiotically with the worm. The eyeless shrimp consume sulfur bacteria from microbial mats around vents, and Pompeii worms rely on enzymes produced by thermophilic bacteria for their own thermal protection.

mal vent community is based on chemoautotrophy, not photoautotrophy as in surface environments (see Section 14.2.2).

Another hypothesis for the formation of primary organic carbon is based on an abiotic process called **organic thermogenesis.** This hypothesis followed from the observation that organic-free rocks initially collected near hydrothermal vents, were able to synthesize sugars and amino acids from paraformaldehyde and urea at high temperatures in the presence of carbonates (Degens, 1974). This observation led Ingmanson and Dowler (1980) to think that the production of carbon in hydrothermal vent communities is the result of the combination of physicochemical factors found in the vent environment. Specifically, these factors include: a concentrated source of energy in the form of heat; a considerable amount of ionizing radiation is present; there is a ready supply of all the required precursors (e.g., CO_2, N_2, NH_3, H_2, CH_4); and the advective flow around vents provides rapid cooling and the subsequent quenching of thermogenic reactions necessary for organic synthesis reactions. Once CO_2 is converted to simple sugars in this thermogenic process,

chemoheterotrophs can utilize the carbon to build biomass. Thus, thermogenesis results in primary production by heterotrophic bacteria.

The third major theory, called the **advective plume hypothesis,** is that primary production in hydrothermal vents is based on photoautotrophy. This theory is strange on the surface because such deep-sea vents are essentially isolated from solar radiation. The theory, however, is based on the assumption that settling of organic carbon from near the ocean surface occurs. This carbon, which settles from near-surface phototrophic layers, is concentrated around hydrothermal vents by advection (uprising in the water column due to heat), which essentially draws water and DOM in from relatively great distances (Lonsdale, 1977). In this case, primary production would also result from heterotrophic activity. In summary, it may very well be that all three of these hypotheses are valid and each mechanism contributes to some extent to primary production in the hydrothermal vent community.

The vent communities support macrofauna that rely on the bacterial populations as a source of organic carbon. There are at least three major mechanisms for transfer of this bacterial carbon and energy to the next trophic level. The first is an endosymbiotic relationship between vent bacteria and an invertebrate that has been dubbed "tube worm". Tube worms are large tube-shaped creatures that grow from the seafloor. These worms have no mouth, gut, or any other digestive system and depend completely on bacteria for their nutrition. Instead of consuming the bacteria, the worms have interior surfaces that are colonized by massive quantities (3×10^{11} bacteria per ounce of tissue) of sulfur-oxidizing chemoautotrophs (Karl, 1995) (see Ch. 14.4.3.1). The worm's body is filled with blood containing large amounts of hemoglobin that binds H_2S. The blood transports the H_2S to the bacteria, which oxidize it and fix CO_2 into organic compounds that nourish the worm. The second method is termed microbial gardening. In this case, bacterial cultures are maintained by mussels and other invertebrates on specialized appendages such as tentacles and gills. These invertebrates periodically harvest and consume the bacteria, retaining small inocula to initiate the next crop. The third mechanism for carbon transfer to higher trophic levels is direct consumption of free-living bacterial cells, filaments, or mats. Crabs, amphipods, predatory fish, and even other microorganisms, including bacteria, have been observed to feed directly on the chemoautotrophic or chemoheterotrophic primary producers (Karl, 1995).

Free-living bacteria around hydrothermal vents have been seen in such great numbers that the area was described as a bacterial snowstorm. These concentrations range from 10^4 to 10^8 bacteria/ml, depending on the proximity to the mouth of the vent, with the highest concentrations occurring within 1 m of the vent mouth (Baross and Deming, 1995). In addition to the clouds of bacteria, bacterial mats several inches thick are found surrounding the vents. Thus, extensive bacterial communities are found in this extreme environment. The cellular mechanisms that help these microorganisms survive in this high-pressure, high-temperature environment are described further in Section 6.4. Primary production around hydrothermal vents when all forms of chemoautotrophy are considered is estimated to be on the order of 2 mg carbon per liter, although this may be somewhat conservative (Baross and Deming, 1995).

6.3.4 Subterranean Water

The groundwater environment is found inland in the subsurface zone and includes shallow and deep aquifers. The characteristics and microbial communities of the groundwater environment have been dis-

cussed in Sections 4.3.3, 4.7, and 4.8.2. Briefly, microorganisms are the sole inhabitants of these environments and bacteria are the dominant type of microbe present. In groundwater, unlike other aquatic environments that have substantial planktonic populations, most of the bacterial populations are attached or only transiently suspended. In general, levels of microbial activity are low, especially in intermediate and deep aquifers. As shown in Fig. 4.22, activity is orders of magnitude lower in these aquifers than in other aquatic habitats (Fig. 4.22). This is due to low nutrient levels. Many subsurface environments may even be considered extreme from a nutrient perspective (see Section 6.4.6).

6.4 ENVIRONMENTAL DETERMINANTS THAT GOVERN EXTREME ENVIRONMENTS

There are two definitions of extreme environments. The first characterizes an environment as extreme if the environmental conditions are at one of two extremes (high or low). These environmental conditions can include pH, temperature, salinity, pressure, and nutrients. The second definition refers to environments in which conditions select for extremely low microbial diversity. In general, growth and multiplication are the primary factors that influence the survival of any organism and its progeny. Extremophile is a term applied to organisms that have successfully adapted to environments where it is difficult or impossible for other organisms to survive. Thus, extremophiles have been selected over time for characteristics that allow them to grow and multiply in a variety of extreme environments. This section will briefly describe some of these environments and the physiological adaptations used by extremophiles to compete or survive in their particular niche.

6.4.1 Air–Water Interface

The air–water interface is a unique habitat that is often considered an extreme environment for many reasons, including high levels of solar radiation; accumulation of toxic substances (e.g., heavy metals, pesticides); large temperature, pH, and salinity fluctuations; and competition. As a rule, the air–water interface, also referred to as the neuston, actually contains higher concentrations of organisms than other layers of the water column. There is some debate, however, about the percentage of viable bacteria in the neuston. Indeed, most reports suggest that the numbers may be higher but the ratio of metabolic activity to total counts is lower in the neuston than in the planktonic habitat. The neuston accumulates nutrients and especially attracts nonpolar organic and inorganic molecules, which form a film at the air–water interface. Bacteria attach to this layer because it concentrates nutrients. However, in addition to the nutrient accumulation, the neuston tends to accumulate toxins. Among these toxins are nonpolar organic molecules, including pesticides such as DDT and petroleum hydrocarbons, as well as metals such as Cd, Cu, Mn, Hg, Pb, Se, and Cr. Thus, the microorganisms that inhabit the neuston have developed unique metabolic, genetic, and functional strategies that allow them to survive the extremes found in this environment. These strategies include the use of pathways that catabolize toxic compounds and provide resistance to metals that accumulate at the interface. Some microbes have developed efficient DNA repair mechanisms to combat DNA damage caused by exposure to ultraviolet radiation. Finally, neuston inhabitants also need to be able to respond and quickly adapt to variations in environmental conditions (Maki, 1993).

6.4.2 High Temperature

There are many examples of environments with extreme temperatures. Environments with high temperatures (>70°C) include terrestrial and submarine hot springs, some of which can reach temperatures of 100°C, and hydrothermal vents, which can reach temperatures in excess of 300°C. Such high temperatures are inhospitable for most forms of life except for certain bacteria and archaebacteria. Most notable of these organisms are the genera *Thermus*, *Methanobacterium*, *Sulfolobus*, *Pyrodictium*, and *Pyrococcus*. The last two of these genera are especially adapted to high temperatures and can live at temperatures in excess of 100°C. Another species of **thermotolerant** bacteria, namely *Thermus aquaticus*, is especially renowned because of its thermotolerant DNA polymerase. This enzyme has been patented and is used around the world in the polymerase chain reaction (PCR) (see Section 13.3.2).

Many mechanisms allow microorganisms to survive at temperatures that would normally denature proteins, cell membranes, and even genetic material. In terms of proteins, thermophiles have an increased number of salt bridges (cations that bridge charges between amino acid residues). These bridges help the protein to remain folded even at high temperatures. In terms of cell membranes, thermophilic eubacteria have increased amounts of saturated fatty acids in their membranes that allow the membranes to remain stable at high temperatures. Extreme thermophiles, almost

all of which are archaebacteria, have an entirely different cell membrane composed of repeating units of the five-carbon compound isoprene (Fig. 6.17) that are ether linked to glycerol phosphate. Finally, in terms of DNA, thermophiles contain special DNA binding proteins that arrange the DNA into globular particles that are more resistant to melting. Another factor that is common to all hyperthermophiles tested so far is a unique DNA gyrase. This gyrase acts to induce positive supercoils in DNA, theoretically providing considerable heat stability (Kikuchi and Asai, 1984; Bouthier del la Tour *et al.*, 1990)

There are numerous biotechnological applications for enzymes isolated from thermotolerant microorganisms and the number of applications is growing rapidly, especially in commercial industry. One example already mentioned is the thermostable DNA polymerase used in the PCR. Other examples include proteases, lipases, amylases, and xylanases that are used in the agricultural, paper, pharmaceutical, water purification, bioremediation, mining, and petroleum recovery industries.

6.4.3 High Solute

Halotolerant, or high-salt-tolerant, organisms require salt concentrations for growth that are substantially higher than that found in seawater (Table 6.1). One of the best known examples of this type of environment is Utah's Great Salt Lake; another is the Dead Sea, which lies between Israel and Jordan. *Halobacterium* and *Haloanaerobium* are two examples of halotolerant bacteria. In addition to bacteria, some algae and fungi are known to be halotolerant. In general, halotolerance is not a requirement for high solute concentration but is specific for Na^+. The main mechanism of salt tolerance displayed by bacteria is internal sequestration of high concentrations of a balancing solute to equal the salt concentration found

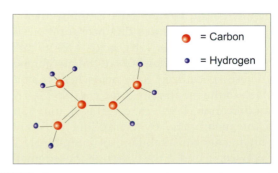

FIGURE 6.17 Molecular structure of isoprene, the monomer subunit from which archaebacterial cell membrane components are assembled.

TABLE 6.1   Ranges of Microbial Salt Tolerance

Category	Salt range (M)	Optimum salt concentration (M)
Nonhalophile	0–1.0	<0.2
Slight halophile	0.2–2.0	0.35
Halophile	1.4–4.0	2.5
Extreme halophile	2.0–5.2	>3.0

external to the cell (Grant *et al.*, 1998). Examples of these balancing solutes include K^+, which is important in halotolerant bacteria, and glycerol, which is important in halotolerant eukaryotes (Galinski, 1993). A second mechanism of salt tolerance of halophiles involves proteins that are acidic and typically have low proportions of nonpolar amino acids. Thus, for these proteins to be active, high salt concentrations are needed to balance their charge and acidity (Grant *et al.*, 1998). Because of these macromolecular modifications, halotolerant bacteria are usually unable to survive in environments lacking high salt concentrations. Thus, many are considered obligate halophiles.

6.4.4 Low pH

Acidic environments, such as acid hot springs, the gastrointestinal tract, mining waste streams, acid mine wastewater, and various mineral oxidizing environments, are populated by bacteria such as *Thiobacillus*, which are discussed extensively in relation to acid mine drainage (see Section 15.3.1). Other examples of acidophiles are *Clostridium acetobutylicum* and *Sarcina ventriculi*, which are obligate anaerobes that ferment sugars. In addition to bacteria, some fungi, algae, and protozoa are also known to be acid tolerant. Strategies used by microorganisms to deal with high or low pH values usually involve modifications of the cell membrane. The first of these modifications is to the structure of membrane components to allow them to be acid tolerant. This includes the incorporation of very long chain dicarboxylic fatty acids (32–36 carbons), which make up more than 50% of the membrane fatty acids. These specialized fatty acids help inhibit acid hydrolysis of the membrane (Jung *et al.*, 1993). The second modification involves control of ion transport across the membrane. By controlling ion transport, these organisms can maintain an internal pH in the range between 5 and 7, even though the external environment can have a pH less than 2 (Norris and Johnson, 1998).

6.4.5 High Pressure

Deep-sea environments are characterized by high pressure and cold temperatures. Microbes that live in this environment are called **barophiles.** Barophiles have developed unique mechanisms that allow them to tolerate the high pressures of more than 1000 bars found in deep-sea trenches (normal atmospheric pressure is 1 bar). Bacteria retrieved from depths greater than 2000 m actually grow better under high pressure than at normal atmospheric pressure. In addition to being pressure tolerant, barophiles are for the most part also **psychrophilic,** meaning that they grow better at low temperature. Finally, these microbes are adapted to darkness, as they live at depths that are essentially isolated from solar radiation.

Organisms that live under these deep ocean conditions, like other extreme microbes, have developed unique mechanisms for survival. Many of these mechanisms involve changes in macromolecular structure and function. For example, long-chain polyunsaturated fatty acids are found in high concentration in the membranes of barophiles. These modified fatty acids maintain the membrane in a fluid state under a pressure and temperature that would otherwise tend to gel or crystallize them. Interestingly, many of the mechanisms used by these organisms are the opposite of the mechanisms utilized by thermotolerant organisms. Although little is actually known experimentally about the mechanisms involved in replication, transcription, and translation in barophiles, it appears that they involve control of salt concentrations (Yayanos, 1998).

6.4.6 No Nutrient

Another extreme environment, which is of great importance in today's world, is ultrapure or nutrient-free water. Ultrapure water is used in the semiconductor, medical, and many other industries. Contamination of ultrapure water by microorganisms can be devastating to these industries. For instance, microbial contamination of ultrapure water can cause flaws in the crystal design of computer chips, lowering the efficiency of the chips. Because there are virtually no nutrients in ultrapure water, it is considered an extreme environment. Indeed, very few organisms are able to survive, let alone proliferate, in ultrapure water. One of these is the bacterium *Caulobacter* and another is *Pseudomonas fluorescens*. In distilled water, for instance, even the limited exchange of nutrients (CO_2) from the atmosphere provides enough nutrients to allow limited growth. Springwater bottling plants are also of concern because the low-nutrient spring water is able to provide an even more favorable environment for proliferation and colonization of many of these organisms, even opportunistic pathogens such as *Pseudomonas aeruginosa*.

6.5 AQUATIC MICROBES IN THE NEWS

6.5.1 Giant Marine Bacteria

A number of aquatic microbes are unique, or interesting for one reason or another. One of these is a bacterium known as *Epulopiscium fishelsoni*. *E. fishelsoni* is a symbiotic bacterium isolated from the gut of a Red Sea surgeonfish. One of the seminal laws in microbiology, and life in general, is that for every rule there is an exception. *E. fishelsoni* provides exceptions for two of the main rules governing bacteria. *E. fishelsoni* is 0.6 mm long and 0.08 mm in diameter. So, unlike every other known bacterium, this organism does not require a microscope to be visualized; it can be seen with the naked eye. In addition to their large size, these bacteria appear to have viviparous progeny, as opposed to undergoing typical bacterial division. Two small "baby bacteria" develop inside the cytoplasm of the mother cell and are released through a slit in one end of the cell (Clements *et al.*, 1989; Fishelson *et al.*, 1985; Montgomery and Pollak, 1988). This phenomenon is thought to have evolved from *Epulopiscium's* ancestors as a modification of sporulation. Thus, instead of forming inert endospores, as their relatives do, these prokaryotes form internal vegetative progeny, which are then released from the cell. Another organism, *Metabacterium polyspora*, which has been isolated from the gastrointestinal tract of guinea pigs, has an intermediate stage of propagation between that of endospore-forming *Clostridium* sp. and that of the internal viviparous reproduction of *E. fishelsoni*. This organism is able to produce multiple endospores by asymmetric division at both poles of the cell and by symmetric division of the endospores at an early stage of their development. *Epulopiscium* is still unique, however, because of its ability to give birth to live offspring in a process that mimics to that in higher life forms.

The initial work on *E. fishelsoni* was done by Fishelson (Fishelson *et al.*, 1985), who initially thought this organism was a protozoan. Later, using molecular methodologies, such as PCR and *in situ* hybridization, the genes encoding the ribosomes of this organism were characterized and it was found that the genetic sequence was that of a prokaryote. Phylogenetic analysis has further classified this organism as a relative of the clostridia. *E. fishelsoni* offers unique opportunities for molecular and cellular microbiology as well as for environmental microbiology. Because of its large size, it is feasible that intracellular probes can be inserted within the bacterial cell without disrupting its cellular machinery to allow the study of cellular processes such as membrane transport, cellular responses to environmental stimuli, and even genetic regulation processes. Such experiments are not as easy to perform with typical bacteria, which have only one millionth the cellular volume of *E. fishelsoni*.

6.5.2 Aquatic Microbes: Food for the Future

Spirulina is a cyanobacterium (phytoplankton) that is rapidly becoming a popular dietary product. *Spirulina* is a traditional food for several cultures and can be found growing naturally in warm-water alkaline volcanic lakes. Scientific studies suggest that *Spirulina* protein may be one of the highest quality types of protein available, because these cyanobacteria are up to 70% protein by weight and their protein contains all of the essential amino acids. In addition to being a high quality protein source, *Spirulina* is considered to be a vitamin and mineral "gold mine." *Spirulina* has more beta-carotene than carrots, 28 times as much iron as beef liver, the highest whole-food concentration of vitamin B_{12}, and even high levels of glycogen (a natural source of energy).

NASA has considered use of *Spirulina* as a food of choice for space flight. The United Nations (UN) considers *Spirulina* as a possible solution to global protein shortages. To this end, *Spirulina* farms are being constructed in underdeveloped countries. The per-acre protein yield of *Spirulina* is 10 times that of soybeans and 200 times that of beef (Table 6.2). The human body is also able to assimilate over 95% of the protein derived from *Spirulina*, whereas we can assimilate only about 20% of the available protein found in beef. Studies also suggest the possibility that consumption of these cyanobacteria can strengthen the immune system, enhancing the ability to fight infection. These other health benefits may be related to the presence of phytonutrients. Phytonutrients are special compounds produced naturally by plants and other organisms that are able to enhance and strengthen the body's natural resistance to disease, support cardiovascular health, lower cholesterol, aid in digestion, and even act as natural antioxidants. As a result, *Spirulina* is used as a supplement in many dietary supplements, including protein drinks.

TABLE 6.2   Comparison of *Spirulina* Aquatic Farming with Other Important Agricultural Products

Protein source[a]	Land area needed (m²)	Water needed (l)	Total energy input required (10^7 kilojoules)	Total energy output generated (10^7 kilojoules)
Spirulina	0.6	2,100	3.5	23
Soybeans	16	9,000	11.7	13.8
Corn	22	12,500	55.7	16.5
Grain-fed (feedlot) beef	190	105,000	456.7	16

[a] Values are related to the production of 1 kg of protein.

To meet the increasing demand for *Spirulina*, the number of *Spirulina* farms are proliferating. One example is California's Earthrise Farms. At Earthrise Farms, *Spirulina* is grown in numerous specially designed pond systems, each of which is larger than a football field. These growing ponds function like immense chemostats, maintaining a constantly growing pure culture of the cyanobacterium (Fig. 6.18). One of the primary concerns in growing cyanobacteria commercially is the need to maintain a pure culture. This is done using specially designed pond systems and carefully balancing the pond's environmental conditions

FIGURE 6.18 This photo shows five football-field-size *Spirulina* ponds on a *Spirulina* farm. During the growing season (April to October) ponds are harvested daily. The paddlewheels shown on the right are used to mix the water and nutrients in each pond. (Photo courtesy of Earthrise Farms, Calipatria, CA. www.earthrise.com.)

to favor the growth of *Spirulina* over other, potentially toxic organisms. Thus, these ponds are supplied with purified mineral- and nutrient-rich water and are kept aerated by mechanical paddles. To enhance growth rates further, carbonated water is bubbled through the ponds to increase the amount of carbon dioxide available for photosynthesis. Unlike other crops that are harvested once or twice a year, *Spirulina* can be harvested 24 hours a day during peak growing seasons. After harvesting, Spirulina cells are dried by a flash evaporation system and packaged for sale. If processed and packaged appropriately, *Spirulina* can be stored at room temperature for 5 years or more without degradation of the nutrients in the product.

Finally, *Spirulina* production has environmental advantages over normal crop production. Because Spirulina is an aquatic microorganism, it does not require fertile land and hence does not cause soil erosion or groundwater contamination. Table 6.2 shows the amount of land required by other major food sources to produce 1 kg of protein compared with that required to produce 1 kg of protein from *Spirulina*. As a further example of the efficiency of farming this cyanobacterium, consider that production of 1 kg of corn results in loss of 22 kg of fertile topsoil. In addition, *Spirulina,* in spite of being an aquatic microorganism, uses less water per kilogram of protein produced than other crops (Table 6.2). Thus, *Spirulina* requires less room, soil, and water than traditional agricultural production of protein.

QUESTIONS AND PROBLEMS

1. Define the four types of microbial habitats found in aquatic environments.

2. What is meant by the term microbial loop?

3. Describe the process of biofilm formation.

4. Describe the types and numbers of microbes that predominate in the different freshwater environments described (e.g., springs, rivers, lakes).

5. What is a thermocline?

6. Describe how marine environments differ from freshwater environments physically, chemically, and microbially.

7. Define what is meant by an extreme environment.

8. Describe microbial adaptations to life at high temperature.

9. Why is microbial growth in water that contains little or no nutrients of concern?

10. Describe microbial adaptations to life at high pressure.

References and Recommended Readings

Atlas, R. M., and R. Bartha. (1993) "Microbial Ecology." Benjamin Cummings, Redwood City, CA.

Baross, H. A., and Deming, J. W. (1995) Growth and high temperatures: Isolation and taxonomy, physiology and ecology. *In* "The Microbiology of Deep-Sea Hydrothermal Vents" (D. M. Karl, ed.) CRC Press, Boca Raton, FL, pp. 169–217.

Bouthier de la Tour, C., Portemer, C., Nadal, M., Stetter, K. O., Forterre, P., and Duguet, M. (1990) Reverse gyrase, a hallmark of the hyperthermophilic archaebacteria. *J. Bacteriol.* **172**, 6803–6808.

Christian, R. R., and Capone, D. G. (1997) Overview of issues in aquatic microbial ecology. *In* "Manual of Environmental Microbiology" (C. J. Hurst, G. R. Knudsen, M. J. McInerney, L. D. Stetzenbach, and M. V. Walter, eds.) American Society for Microbiology Press, Washington, DC, pp. 358–365.

Chróst, R. J. (1990) Microbial exoenzymes in aquatic environments. *In* "Aquatic Microbial Ecology: Biochemical and Molecular Approaches" (J. Overbeck, and R. J. Chróst, eds.) Springer-Verlag, New York, pp. 47–78.

Clements, K. D., Sutton, D. C., and Choat, J. H. (1989) Occurrence and characteristics of unusual protistan symbionts from surgeonfishes (Acanthuridae) of the Great Barrier Reef. *Aust. Mar. Biol.* **102**, 812–826

Degens, E. T. (1974) Synthesis of organic matter in the presence of silicate and lime. *Chem. Geol.* **13**, 1–10.

Ducklow, H. W., and Shiah, F. (1993) Bacterial production in estuaries. *In* "Aquatic Microbiology: An Ecological Approach" (T. E. Ford, ed.) Blackwell Scientific Publications, Cambridge, MA, pp. 261–287.

Fishelson, L., Montgomery, W. L., and Myrberg, J. A. A. (1985) A unique symbiosis in the gut of tropical herbivorous surgeonfish (Acanthurdiae: Teleostei) from the Red Sea. *Science* **229**, 49–51.

Fuhrman, J. (1992) Bacterioplankton roles in cycling of organic matter: The microbial food web. *In* "Primary Productivity and Biogeochemical Cycles in the Sea" (P. G. Falkowski, and A. D. Woodhead, eds.) Plenum, New York, pp. 361–383.

Galinski, E. A. (1993) Compatible solutes of halophilic eubacteria: Molecular principles, water–solute interactions, stress protection. *Experientia* **49**, 487–496.

Geesey, G. G., Mutch, R., Costerton, J. W., and Green, R.B. (1978) Sessile bacteria: An important component of the microbial population in small mountain streams. *Limnol. Oceanogr,* 1214–1223.

Goodman, A. E., and Marshall, K. C. (1995) Genetic responses of bacteria at surfaces. *In* "Microbial Biofilms" (H. M. Lappin-Scott, and J.W. Costerton, eds.) Cambridge University Press, Cambridge, pp. 80–98.

Grant, W. D., Gemmel, R. T., and McGenity, T. J. (1998) Halophiles. *In* "Extremophiles: Microbiological Life in Extreme Environments" (K. Horikoshi and W. D. Grant, eds.) Wiley-Liss, New York, pp. 47–92.

Hennes, K. P., and Simon, M. (1995) Significance of bacteriophages for controlling bacterioplankton growth in a mesotrophic lake. *Appl. Environ. Microbiol.* **61**, 333–340.

Ingmanson, D. E., and Dowler, M. J. (1980) Unique amino acid composition of Red Sea brine. *Nature* **286**, 52–52.

Jung, H. S., Lowe, S. E., Hollingsworth, R., and Zeikus, J. G. (1993) *Sarcina ventriculi* synthesizes very long chain dicarboxylic acids in response to different forms of environmental stress. *J. Biol. Chem.* **268**, 2828–2835.

Kaplan, L. A., and Newbold, J. D. (1993) Biogeochemistry of dissolved organic carbon entering streams. *In* "Aquatic Microbiology: An Ecological Approach" (T. E. Ford, ed.) Blackwell Scientific Publications, Cambridge, MA, pp. 139–165.

Karl, D. M., ed. (1995) "The Microbiology of Deep-Sea Hydrothermal Vents." CRC Press. Boca Raton, FL, pp. 35–124.

Kikuchi, A., and Asai, K. (1984) Reverse gyrase—a topoisomerase which introduces positive superhelical turns into DNA. *Nature* **309**, 677–681.

Konopka, A. E. (1993) Distribution and activity of microorganisms in lakes: Effects of physical processes. *In* "Aquatic Microbiology: An Ecological Approach" (T. E. Ford, ed.) Blackwell Scientific Publications, Cambridge, MA, pp. 47–68.

Korber, D. R., Lawrence, J. R., Lappin-Scott, H. M., and Costerton, J. W. (1995) Growth of microorganisms on surfaces. *In* "Microbial Biofilms" (H. M. Lappin-Scott, and J.W. Costerton, eds.) Cambridge University Press, Cambridge, MA, pp. 15–45.

Lonsdale, P. (1977) Clustering of suspension-feeding macrobenthos near abyssal hydrothermal vents at oceanic spreading centers. *Deep-Sea Res.* **24**, 857–863.

Madsen, E. L., and Ghiorse, W. C. (1993) Groundwater microbiology: Subsurface ecosystem processes. *In* "Aquatic Microbiology: An Ecological Approach" (T. E. Ford. ed.) Blackwell Scientific Publications, Cambridge, MA, pp. 167–213.

Maki, J. S. (1993) The air–water interface as an extreme environment. *In* "Aquatic Microbiology: An Ecological Approach" (T. E. Ford, ed.) Blackwell Scientific Publications, Cambridge, MA, pp. 409–440.

Marshall, K. C. (1985) Mechanisms of bacterial adhesion at solid-liquid interfaces. *In* "Bacterial Adhesion: Mechanisms and Physiological Significance" (D. C. Savage, and M. Fletcher, eds.) Plenum, New York, pp.131–161.

Marshall, K. C. (1992) Biofilms: An overview of bacterial adhesion, activity, and control at surfaces. *Am. Soc. Microbiol. (ASM) News* **58**, 202–207.

Marshall, K. C. (1997) Colonization, adhesion, and biofilms. *In* "Manual of Environmental Microbiology" (C. J. Hurst, G. R. Knudsen, M. J. McInerney, L. D. Stetzenbach, and M. V. Walter, eds.) American Society for Microbiology Press, Washington, DC, pp. 358–365.

Middledoe, M., Jorgensen, N.O.G., and Kroer, N. (1996) Effects of viruses on nutrient turnover and growth efficiency of noninfected marine bacterioplankton. *Appl. Environ. Microbiol.* **62**, 1991–1997.

Mittelman, M. W. (1995) Biofilm development in purified water systems. *In* "Microbial Biofilms" (H. M. Lappin-Scott, and J. W. Costerton, eds.). Cambridge University Press, Cambridge, pp. 133–147.

Montgomery, W. L., and Pollak, P. E. (1988) *Epulopiscium fishelsoni*, n. g., n. s., a protist of uncertain taxonomic affinities from the gut of an herbivorous reef fish. *J. Protozool.* **35**, 565–569.

Norris, P. R. and Johnson, D. B. (1998) Acidophilic microorganisms. *In* "Extremophiles: Microbiological Life in Extreme Environments" (K. Horikoshi and W. D. Grant, eds.) Wiley-Liss, New York, pp. 47–92.

Pearl, H. W., and Pinckney, J. L. (1996) A mini-review of microbial consortia: Their roles in aquatic production and biogeochemical cycling. *Microb. Ecol.* **31**, 225–247.

Rheinheimer, G. (1985) "Aquatic Microbiology," 3rd ed. John Wiley & Sons. New York, pp. 1–247.

Stal, L. R. (1995) Physiological ecology of cyanobacteria in microbial mats and other communities. *New Phytol.* **131**, 1–32.

Stal, L. J., and Caumette, P., eds. (1994). "Microbial Mats: Structure, Development, and Environmental Significance." NATO ASI Series G: Ecological Sciences, Vol. 35. Springer-Verlag, New York.

Suttle, C. A. (1994) The significance of viruses to mortality in aquatic microbial communities. *Microb. Ecol.* **28**, 237–243.

van Loosdrecht, M., Lyklema, J., Norde, W., and Zehnder, A. J. B. (1990). Influence of interfaces on microbial activity. *Microbiol. Rev.* **54**, 75–87.

Wetzel, R. G. (1983) "Limnology," 2nd ed. Saunders College Publishing, New York.

Wolfaardt, G. M., Lawrence, J. R., Robarts, R. D., Caldwell, S. J., and Caldwell, D. E. (1994) Multicellular organization in a degradative biofilm community. *Appl. Environ. Microbiol.* **60**, 434–446.

Yayanos, A. A. (1998) Empirical and theoretical aspects of life at high pressure in the deep sea. *In* "Extremophiles: Microbiological Life in Extreme Environments" (K. Horikoshi and W. D. Grant, eds.) Wiley-Liss, New York, pp. 47–92.

7

Microbial Transport

DEBORAH T. NEWBY   IAN L. PEPPER   RAINA M. MAIER

7.1 INTRODUCTION

Transport of microbes or of their genetic information through soil and vadose zones is a complex issue of growing concern. The introduction of microorganisms into an environmental system is a potentially powerful tool for the manipulation of a variety of processes. These include enhanced biodegradation of organic contaminants, remediation of metal-contaminated sites, improvement of soil structure, increased crop production through symbiotic relations, and even biological control of plant-pathogenic organisms. Microbial inoculation with the intent of deriving such benefits is referred to as **bioaugmentation.** However, microbes are not always introduced intentionally and their introduction does not always produce desirable effects. For example, the introduction of pathogens via sludge or animal wastes can have adverse effects, especially if the pathogens pass through the terrestrial matrix and enter the groundwater. A secondary concern with the transport of indigenous or introduced microbes is that they may either decrease or increase the transport potential of chemical pollutants, making prediction of pollutant fate more difficult.

Soils have intentionally been inoculated with microbes for years, but the results have often been disappointing. There are two primary reasons for this. The first is that microbes inherently like to adhere to surfaces, which reduces their transport. The second is that when non-native microbes are introduced, they often do not replicate and eventually die (Bitton and Gerba, 1984). A common observation is that after introduction, a rapid decline in numbers of inoculant cells occurs. This is often accompanied by a decline in the av-

erage activity per cell of the introduced microbes. An exception to this observed behavior occurs when there is an ecological niche for the introduced microbes, in which case the population establishes itself and replication occurs. For example, if a contaminant-degrading microbe is introduced into a site that contains no indigenous contaminant degrader, there exists a potential niche for the introduced microbe. Whether or not there is growth, under optimal transport conditions microbes may move sufficient distances and survive long enough to prove useful or to be a source of concern, as the case may be.

Any potential for microbial transport through the subsurface calls into question the quality of groundwater, which has historically been considered a reliable and safe source of water, protected naturally from surface contamination. Increased industrial, agricultural, and mining activities; the practice of land disposal of sewage effluents, sludges, and solid wastes; septic tank effluents; and urban runoff have all led to microbial contamination of groundwater. Specific examples include outbreaks of hepatitis A, viral gastroenteritis, cholera, typhoid fever, and giardiasis that have all been traced to groundwater contamination (Bitton and Gerba, 1984). As exemplified by this list, microbes of concern include viruses, bacteria, and protozoa. Although soil does serve as a natural filter that limits microbial transport to a certain extent, its retention capabilities are obviously limited as evidenced by these outbreaks.

Just as it is important to assess the fate of chemical pollutants in soil and subsurface regions, it is essential to investigate the factors controlling dispersion of microorganisms and their genes. What are the controlling factors regarding microbial transport? And what are the implications of microbial transport with respect to remediation approaches, agricultural practices, and the safety of groundwater? In this chapter, the key factors that determine the transport of microorganisms and nucleic acids through the soil profile and vadose zones are examined. These include factors that control microbial adhesion to and detachment from solid surfaces. Among others these factors are the physical-chemical properties of both the porous medium surfaces and the surrounding solution; the surface properties of the microbe; and the percent water saturation and resulting water movement. Microbial survival and activity levels are also key parameters in the assessment of transport potential. Approaches to facilitating microbial transport are addressed, as well as methodologies for studying microbial transport. Finally, models that describe microbial transport are introduced and their usefulness in the prediction of microbial transport is discussed.

7.2 FACTORS AFFECTING MICROBIAL TRANSPORT

Transport of microorganisms through composite systems such as soils, vadose zone materials, and sediments is governed by a variety of abiotic and biotic factors. These factors include but are not limited to the following: adhesion processes, filtration effects, physiological state of the cells, porous medium characteristics, water flow rates, predation, and intrinsic mobility of the cells. A simplistic approach to assessing transport would be to group the factors into two main categories—hydrogeological and microbial. The first category pertains to abiotic factors such as soil characteristics and water flow; the second focuses primarily on characteristics influencing microbial survival and/or potential for activity. However, ultimately it is the extensive interplay between hydrogeological and microbial factors that determines the extent of microbial transport.

Understanding how these various factors influence transport is the first step toward either assessing the feasibility or risk of microbial transport, or actually delivering microbes to a target site.

7.2.1 Microbial Filtration

Transport of microbes and other contaminants occurs within the pore spaces of a soil or subsurface material. One mechanism by which microbial transport is limited is physical straining or **filtration** of cells by small pores. Filtration of bacterial cells has been shown

to be statistically correlated with bacterial size (Gannon *et al.*, 1991a). Filtration becomes an important mechanism when the limiting dimension of the microbe is greater than 5% of the mean diameter of the soil particles (Herzig, *et al.*, 1970). Thus, for a sandy soil with particle diameters of 0.05 to 2.0 mm, filtration will have a relatively small impact on the retention of bacteria of diameter approximately 0.3 to 2 μm. However, in a soil containing a significant portion of silt or clay particles (particle diameters range from 0.2 to 50 μm), filtration will be a major mechanism of bacterial cell removal. In contrast, studies of factors affecting the movement of particles less than 50 nm in diameter, such as viral particles, have shown that filtration has little effect on movement (Gerba *et al.*, 1991). An example of typical pore sizes found in a sandy loam is shown in Fig. 4.5.

Cell shape, defined as the ratio of cell width to cell length, has also been shown to influence bacterial transport through a porous medium. Weiss *et al.* (1995) examined the transport of 14 strains of bacteria suspended in artificial groundwater through columns packed with quartz sand. A comparison of the distributions of size and shape of cells in the effluent with those in the influent suspensions revealed that cells in the effluent were smaller and rounder.

Another consequence of microbial filtration is **micropore exclusion**. This term merely says that bacteria may be excluded from the microporous domain of structured (e.g., aggregated, macroporous) porous media (see Fig. 4.3). Most bacteria range from 0.3 to 2 μm in diameter, and micropores or pore throats located in the microporous domain of structured media can be much smaller in size. As a result, bacterial cells are physically excluded from the micropores (Fig. 7.1). Thus, the location and rate of microbial activities can vary over a relatively small scale within a porous medium. In other words, microbial activity within the micropores that exclude microbes can be expected to be nonexistent, whereas an immediately adjacent site that is colonized may have an extremely high rate of activity. There can be both beneficial and detrimental consequences of micropore exclusion in a soil.

Ferguson and Pepper (1987) demonstrated that this phenomenon was beneficial in terms of increased ammonium fertilization efficiencies. They amended soils with ammonium (NH_4^+) and clinoptilolite zeolite, which contains intraaggregate pores of 1 nm (10^{-9} m) in size within its rigid lattice. This zeolite has the potential to protect NH_4^+ physically sorbed to its internal exchange sites because the internal pores are too small for microbes to enter. Theoretically, microbial exclusion from the internal pores should reduce NH_4^+ loss by limiting two important microbially mediated reactions. These include nitrification, an aerobic reac-

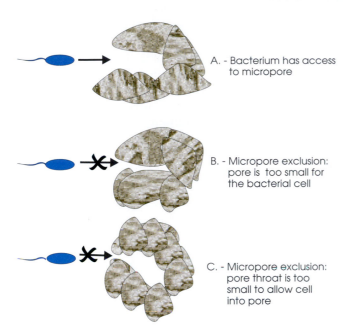

A. - Bacterium has access to micropore

B. - Micropore exclusion: pore is too small for the bacterial cell

C. - Micropore exclusion: pore throat is too small to allow cell into pore

FIGURE 7.1 Exclusion of a bacterial cell from microporous domains in structured porous media.

tion, whose products (NO_2^- and NO_3^-) are very mobile and volatilization to gaseous forms of nitrogen under anaerobic conditions (see Chapter 14.3). Results from this work showed that amendment of soils with clinoptilolite reduces NH_4^+ loss and thus increases the efficiency of ammonium fertilization.

Micropore exclusion of bacteria can also have negative impacts. For example, when contaminants diffuse into micropores that exclude bacteria, they become unavailable for biodegradation. Because diffusion occurs slowly, this is a problem that normally worsens as the contact time between the contaminant and the porous medium increases. This process, known as **contaminant aging,** results in slower rates of contaminant degradation. Contaminant aging is exemplified by the occurrence of "residues" of pesticides in agricultural fields long after the pesticides were applied. For example, residues of EDB (1,2-dibromoethane) were found to persist in fields for as long as 19 years after application, even though EDB is volatile and readily degraded (Steinberg *et al.*, 1987). It was suggested that the residual EDB was trapped in regions of the soil that rendered it unavailable for biological transformation.

7.2.2 Physiological State

A variety of factors influence the effective diameter or limiting dimension of a microbe and thus its transport potential. The physiological state of microbial cells is one factor that plays a significant role in deter-

mining their size. When nutrients are not limiting, most cells produce exopolymers that coat the outer surface of the cell. Exopolymers increase the effective diameter of a cell; may adhere to pore surfaces, in essence decreasing pore size; and may modify solid surfaces to promote attachment. All of these, in addition to cell proliferation, may lead to **pore clogging.** Pore clogging can severely limit bioaugmentation efforts, in which selected organisms are inoculated into a site. Once a pore is clogged, further penetration of bacteria is restricted, as is water flow, which in turn may lead to an altered path of water flow (Vandevivere and Baveye, 1992). This is especially true when microbes are directly injected into subsurface zones rather than applied over a large surface area. There are numerous other examples of problems generated by pore clogging. These include poor septic tank performance related to clogging of the drain field, clogging of nutrient injection wellheads used for *in situ* bioremediation, and reduced rates of groundwater infiltration in recharge basins.

Under starvation conditions, bacteria typically decrease in size to 0.3 μm or even smaller and shed their glycocalyx or capsule layer. This may increase their transport potential, because both cell size and surface properties are changed. This may be an important consideration in movement of soil organisms since nutrients are often limiting in porous media, and thus a large proportion of the bacteria present are in starved or semistarved states (see Chapter 4.4). In 1981, Torrella and Morita coined the term **ultramicrobacteria** to describe bacteria that were less than 0.3 μm in diameter. These small bacteria demonstrated slow growth and did not significantly increase in size when inoculated onto a nutrient-rich agar medium. Although this term was initially used to describe bacteria isolated from seawater, today it is also used to describe bacteria isolated from soil systems. Ultramicrobacteria not only occur in natural environments but also can be created in the laboratory by placing normal cells under starvation conditions. It is thought that a large proportion of environmental isolates can be starved into ultramicrobacteria and then resuscitated successfully. Herman and Costerton (1993) subjected a *p*-nitrophenol degrader isolated from a waste lagoon to starvation by placing it in phosphate-buffered saline for 10 weeks. The difference in cell size and morphology before and after starvation is clearly shown in Fig. 7.2A and B. The starved cells were then resuscitated by adding *p*-nitrophenol to the medium (Fig. 7.2C and D).

To demonstrate the difference in transport of ultramicrobacteria and normal cells, MacLeod *et al.* (1988) examined the movement of starved and normal *Klebsiella pneumoniae* cells through glass bead columns.

The starved cells not only were smaller, but also demonstrated a reduction in glycocalyx production as compared with the vegetative cells. As expected, the starved cells penetrated further into the column than did the vegetative cells. Figure 7.3 depicts the observed reduction in column permeability as *K. pneumoniae* cells (10^8 ml^{-1}) in different metabolic states were injected into the column. The ability of the cells to cause a reduction in permeability within the column was shown to be dependent on the length of starvation prior to inoculation and on the volume of cells at a given concentration injected through the core. Cell distribution within the columns also differed depending on nutrient status. Starved cells were evenly distributed throughout the column, whereas the vegetative cells were found in much higher num-

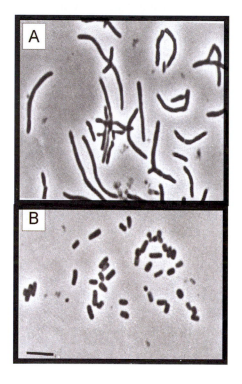

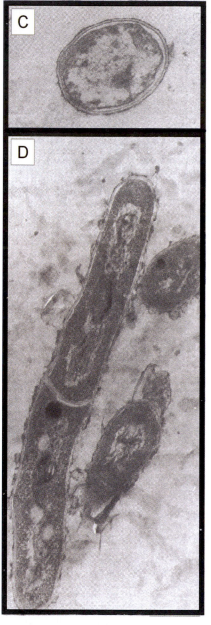

FIGURE 7.2 (A) Phase-contrast micrograph of an isolate grown on *p*-nitrophenol as the sole carbon source. (B) Phase-contrast micrograph of the *p*-nitrophenol degrader after 10 weeks of starvation in phosphate-buffered saline. (C) Electron micrograph of the starved *p*-nitrophenol-degrading isolate. (D) Electron micrograph of the starved *p*-nitrophenol-degrading isolate after resuscitation on *p*-nitrophenol. (Modified with permission from Herman and Costerton, 1993.)

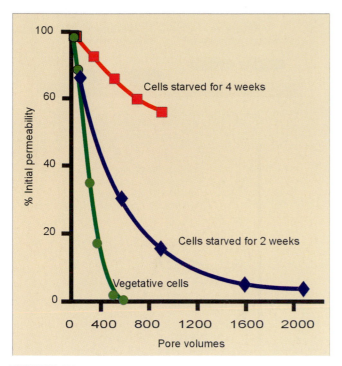

FIGURE 7.3 The different permeability reduction profiles of fused glass bead cores injected with *K. pneumoniae* cells either in a vegetative state or starved for a period of time in phosphate-buffered saline. (Modified with permission from MacLeod *et al.,* 1988.)

bers near the inlet end of the column (Fig. 7.4). Upon nutrient stimulation, the starved cells were found to enter a state of growth accompanied by increased polymer production. This work demonstrates that inoculation with starved cells followed by nutrient stimulation has potential for increasing bioaugmentation efforts. For example, a starved cell that migrates farther through the terrestrial profile has an increased likelihood of reaching a targeted contaminated site. Once at the site, the contaminant can serve as a nutrient source, inducing the microbe to enter a metabolically active state.

The ability to create ultramicrobacteria and then resuscitate the cells has also been studied with respect to creation of biobarriers, for example, to improve oil recovery. In this case, after oil is initially flushed from a geologic formation, removal of further oil residuals becomes more difficult because flow paths have been established. At this point, ultramicrobacteria could be injected into the formation. These small, unencapsulated organisms move relatively easily through established flow paths. They would then be resuscitated by nutrient injection, grow and divide, thereby plugging pores and forcing flow through other regions of the geologic formation.

7.2.3 Microbial Adhesion—The Influence of Cell Surface Properties

To understand microbial adhesion and its influence on transport, a variety of chemical interactions between the microbial cell surface and porous medium surfaces must be considered. The key cell surface factors influencing adhesion are charge and hydrophobicity. In addition, some studies have indicated that the composition of the lipopolysaccharide layer and the presence of specific proteins in cell surfaces, appendages, or extracellular polymers can also play a role in determining transport potential. There is wide variation in cell surface properties among genera and even species. Furthermore, microbial cell surfaces are dynamic in nature and adapt to shifts in environmental parameters through phenotypic and genotypic modifications.

The adhesion of microbes to soil particles and vadose zone materials requires an initial interaction between the cell and a particle surface (van Loosdrecht *et al.,* 1990). Once a cell is in the vicinity of a surface, this initial interaction can occur in one of three ways: diffusion, convective transport, or active movement of

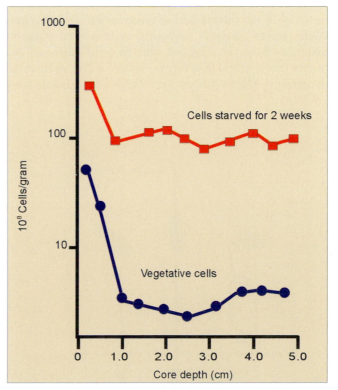

FIGURE 7.4 Differences in the DNA-derived cell distribution in cores injected with either a vegetative cell culture of *K. pneumoniae* or a cell suspension that was starved in phosphate-buffered saline for 2 weeks. (Modified with permission from MacLeod *et al.,* 1988.)

the cell. **Diffusion** is a result of Brownian motion and allows random interactions of cells with surfaces. The effective rate of diffusion is on the order of 40 μm/hr. **Convective transport** is due primarily to water movement and can be several orders of magnitude faster than diffusive transport. Finally, motile cells close to a surface may come in contact with the surface in response to a chemotactic chemical gradient or in some cases by chance through **active movement** (Fig. 7.5).

Once contact between the cell and a particle surface has been made, adhesion can take place. Adhesion is a physicochemical process, and depending on the mechanisms involved, can be reversible or irreversible. **Reversible adhesion,** often thought of as initial adhesion, is controlled primarily by the balance of the following interactions: **electrostatic interactions, hydrophobic interactions,** and **van der Waals forces.** These are explained in detail in the following sections.

In general, electrostatic interactions are repulsive because both cell and particle surfaces are negatively charged. In contrast, hydrophobic interactions and van der Waals forces tend to be attractive. Initial adhesion occurs when attractive forces overcome repulsive forces. Porous medium properties in conjunction with cell surface properties determine the relative importance of each of these interactions. Figure 7.6 illustrates how the interaction of electrostatic and van der Waals forces governs reversible adhesion at various distances between the cell and particle surfaces. As can be seen from this figure, when a cell surface is in actual contact with, or very close to, a particle sur-

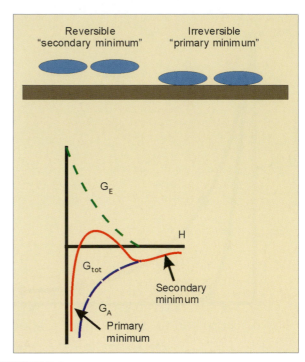

FIGURE 7.6 Gibbs energy of interaction between a sphere (in this case a bacterial cell) and a flat surface having the same charge, according to DLVO theory. G_E, electrostatic interaction; G_A, van der Waals interaction; G_{TOT}, total interaction; H, shortest separation distance between the two surfaces. (Modified with permission from van Loosdrecht *et al.*, 1990.)

face, the attractive forces are very strong, creating a primary minimum. The forces governing the primary minimum are short-range forces such as hydrogen bonding and ion pair formation. As the two surfaces are separated slightly, e.g., by several nanometers, repulsive forces grow quickly and prohibit adhesion. At slightly longer distances, another more shallow minimum exists called the secondary minimum. It is the secondary minimum that is responsible for the initial reversible adhesion of microbes. As can be seen from Fig. 7.6, the cell and particle surfaces are not in actual contact at the secondary minimum. As a result, cells can be removed from the surface easily, for example, by increasing the water flow velocity or by changing the chemistry of the porous medium solution, e.g., ionic strength.

After initial adhesion, cells can become irreversibly attached to a particle surface. **Irreversible attachment** is a time-dependent process that occurs as a result of the interaction of cell surface structures such as fimbriae with the solid surface or as a result of the production of exopolymers that cement the microbial cell to the surface (Fig. 7.7). The role of reversible and irreversible attachment in biofilm formation is discussed further in Section 6.2.4.

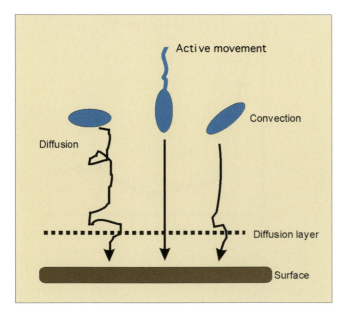

FIGURE 7.5 Different ways in which a cell can approach a solid surface. (Modified with permission from van Loosdrecht *et al.*, 1990.)

FIGURE 7.7 Irreversible attachment is mediated by physical attachment of cells to a surface, which can occur via production of exopolymers or special cell surface structures such as fibrils. (Modified with permission from van Loosdrecht *et al.,* 1990.)

7.2.3.1 Electrostatic Interactions

Electrostatic interactions occur between charged particles. The force (F) between a pair of charges, q_1 and q_2, separated in an environmental system by a distance r, is given by **Coulomb's law:**

$$F = k \, \frac{q_1 \, q_2}{\epsilon \, r^2}$$

where k is a constant whose value depends on the units used and ϵ is the dielectric constant. The value of the latter constant is dependent on the medium between the two charged particles. The dielectric constant accounts for the fact that the charged microbe is separated from the soil by the soil solution, which contains water and other ions. This increases the distance between the charged particles of interest and in essence screens the charges from one another, decreasing the electrostatic interaction. If the charges on the two interacting species are opposite, one positive and the other negative, then the force is negative, indicating attraction between the two particles. However, if both particles carry the same charge, the force is positive, signifying repulsion. Cations in the matrix solution can also serve as bridges linking two negatively charged species that would otherwise be repelled (see Chapter 4.2.1).

In terms of microbial transport, repulsion is the dominant electrostatic interaction because both the porous medium and microbial cell surfaces are generally negatively charged. For example, clay particles typically display a net negative surface charge. This charge can be attributed in part to **isomorphic substitutions** in the crystal lattice of the mineral. Isomorphic substitution is the replacement of particular ions in the clay structure with other ions of similar size. In clays, common substitutions are Al^{3+} for Si^{4+} and Mg^{2+} for Al^{3+}. The fraction of the net charge generated by these substitutions is negative and is independent of the soil solution conditions. Negative charge can also result from an imbalance of complexed proton and hydroxyl charges on the surface, in particular at the exposed edges of the mineral. Charge generated in this manner varies with the soil pH. As the pH of the soil solution increases, more protons dissociate from these sites, making the net charge increasingly negative. Another contribution to negative charge comes from associated organic matter, which contains numerous carboxyl functional groups that dissociate with increases in pH. As a result of these combined contributions, the vast majority of soils have a net negative surface charge, although some highly weathered soils can actually have a net positive charge at sufficiently low pH.

As already mentioned, microbial cell surfaces are generally negatively charged. The negative charge comes primarily from lipoteichoic acids on the surface of gram-positive bacteria and lipopolysaccharides on the surface of gram-negative bacteria. Viral protein coats are also predominantly negatively charged. The overall charge on a microbe can be measured by electrostatic interaction chromatography or by electrophoretic mobility. Surface charge varies between types and species of microbes and can be affected by the pH of the matrix solution.

7.2.3.2 van der Waals Forces

Interactions between neutral molecules generally result from **van der Waals forces.** Van der Waals forces occur because while neutral molecules have no net charge or permanent dipole moment, they do have a dynamic distribution of charge. As two molecules approach, this charge distribution can become favorable for interaction between the two molecules (Fig. 7.8A). What actually occurs as two molecules approach is that the van der Waals attractive forces increase to a maximum, then decrease and become repulsive (Fig. 7.8B). The van der Waals radius is defined as one half the distance between two equivalent atoms at the point of the energy minimum (where attractive forces are at a maximum). Van der Waals radii range from one to several angstroms in length, so these forces are effective only over short distances. Although individual van der Waals interactions are weak, the total attraction between two particles is equal to the sum of all

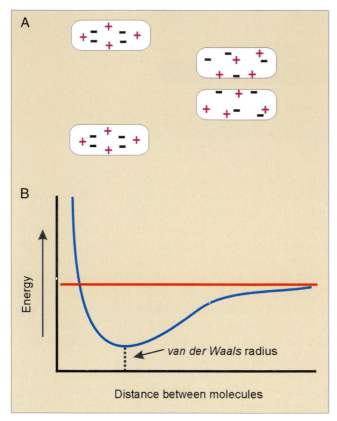

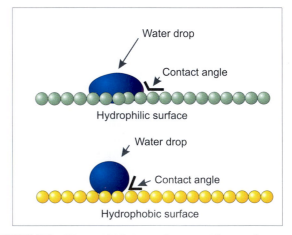

FIGURE 7.8 (A) For a neutral molecule the charge distribution in a molecule can vary to produce a net electrostatic attraction, allowing the molecules to approach very closely. This is a very weak attraction called the van der Waals force. Van der Waals forces can become strong if they are numerous enough. (B) As two molecules approach each other, the van der Waals attractive force increases to a maximum, then decreases and becomes repulsive.

attractive forces between every atom of one particle and every atom of the other particle. Thus, total van der Waals interactions can be quite strong.

7.2.3.3 Hydrophobic Interactions

Hydrophobic interactions refer to the tendency of nonpolar groups to associate in an aqueous environment. Nonpolar entities associate with one another in water not because they have a high affinity for each other, but because polar water molecules associate strongly with themselves and exclude nonpolar molecules. The influence of hydrophobic interactions on microbial transport varies depending on the hydrophobicity of the cell surface as well as on characteristics of the matrix surface and pore solution. Because both particle and cell surfaces are partly hydrophobic in nature, hydrophobic interactions are considered attractive in nature. The hydrophobicity of particle surfaces is governed to a large extent by the amount of organic matter present. Increasing amounts

of organic matter increase the degree of hydrophobicity of the porous medium. The hydrophobicity of bacterial cell surfaces can vary significantly between species and is influenced by the type and extent of the glycocalyx and by the presence of specific molecules that exist on the cell surface, for example, lipopolysaccharides. In addition, it is important to realize that the hydrophobicity of a particular bacterial cell is not a fixed property but can change depending on growth rate and nutritional status (van Loosdrecht *et al.*, 1987). Hydrophobicity can be measured in a variety of ways, including contact angle determination, which is done by examining the shape of a drop of water that is placed on a layer of bacterial cells (Fig. 7.9). Other methods commonly used to assess hydrophobicity include phase partitioning (BATH test) and hydrophobic interaction chromatography. The **BATH** or bacterial adherence to hydrocarbon test is a relatively simple test that measures the partitioning of microbial cells between a water phase and an organic phase.

As a result of hydrophobic forces, cells tend to partition from the aqueous phase and accumulate at the solid–water interface, resulting in decreased transport potential. Van Loosdrecht *et al.* (1990) examined adhesion of a variety of bacteria with different cell surface properties to two surfaces, one hydrophobic (polystyrene) and one hydrophilic (glass). Both cell surface charge and cell surface hydrophobicity were considered in this series of experiments. As shown in Fig. 7.10A, cell surface hydrophobicity was the dominant force in determining adhesion to the hydrophobic polystyrene surface. For cells with high cell surface hydrophobicity, adhesion was strong regardless of the charge on the cells. For cells with low cell surface hy-

FIGURE 7.9 Water, which is a polar material, spreads out on a hydrophilic or polar surface but forms a round bead on a hydrophobic or nonpolar surface. The angle that describes the interaction of a water droplet with a surface is called the contact angle.

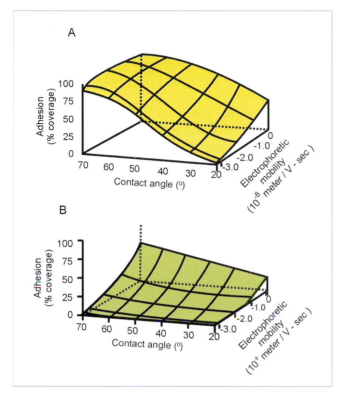

FIGURE 7.10 Relationship between bacterial adhesion to (A) sulfated polystyrene (hydrophobic) and (B) glass (hydrophilic) and bacterial surface characteristics as determined by contact angle measurement and electrophoretic mobility. (Modified with permission from van Loosdrecht *et al.*, 1990.)

drophobicity, cell surface charge played a greater role in adhesion. In this case, increasing cell surface charge interfered with cell adhesion. Adhesion to the hydrophilic surface, glass, was quite different. In this case, cells with a high surface charge showed very little adhesion and cell surface hydrophobicity had little impact. For cells that carried relatively little charge, hydrophobicity had more impact. In this case, as hydrophobicity increased, adhesion increased. In summary, two trends in cell adhesion can be inferred from this study: (1) adhesion typically decreases with decreasing hydrophobicity of either the solid surface or the cell surface and (2) adhesion generally increases with decreasing cell surface charge. Knowledge of the combined effects of hydrophobicity and cell surface charge can be used to predict initial microbial adhesion of a particular microbe.

Similarly, hydrophobic effects and electrostatic repulsion govern the sorption of viruses depending on viral and subsurface characteristics. A study conducted by Bales *et al.* (1993) on PRD-1 and MS-2 bacteriophage transport through silica bead columns demonstrates the variability of viral transport. These

two phage differ in their surface characteristics. PRD-1 is an icosahedral lipid-containing hydrophobic phage of diameter 62 nm. MS-2 is an icosahedral phage of diameter 26 nm and has both hydrophobic and hydrophilic moieties on its coat. In testing the transport of these two viruses, it was found that at pH 5.5 PRD-1 adhered very strongly, apparently controlled by hydrophobic interactions, whereas adhesion of the more hydrophilic MS-2 was much less significant. It should be noted that phage adhesion is very dependent on pH, as discussed in Section 7.2.4. However, for a phage such as PRD-1, which is very hydrophobic, pH has less effect. For example, although increasing pH from 6.5 to 7.6 increased PRD-1 removal, a large amount of this lipid-containing phage remained attached. This relative insensitivity to pH suggests that under the experimental conditions, hydrophobic effects were dominant over electrostatic interactions for PRD-1. In general, adhesion of viruses can be quite significant and recovery of viruses that have been added to a porous medium is observed over long periods of time even after virus input has ceased. This suggests that slow detachment of viruses can result in long-term release of the viruses. Accordingly, virus adhesion in porous media may pose a threat to the safety of groundwater for extended periods of time.

In natural systems, dissolved organic matter, most often present in the form of polymers, can also influence bacterial adhesion by adsorbing to bacteria and/or solid surfaces (Dexter, 1979). The polymeric coating may affect adhesion by changing the electrostatic, the van der Waals, and/or the hydrophobic interactions between the bacterium and the solid surface. When polymers adsorb and coat both bacteria and the solid surface completely, adhesion is reduced because of an extra repulsive interaction due to steric hindrance (Fletcher, 1976). However, if only one of the surfaces is covered with polymers, or if both surfaces are partly covered with polymers, one polymer molecule may attach to both surfaces, thus forming a "bridge" between the two surfaces. This reduces the Gibbs free energy of adhesion (Fig. 7.6) and results in a strong bond (Dexter, 1979).

7.2.4 Impact of pH on Microbial Transport

The pH of the matrix solution within a porous medium does not seem to have a large effect on bacterial transport. However, viral transport can vary greatly depending on the pH of the porous medium solution. The difference in impact of pH on the transport of bacteria and viruses can be attributed to a variety of factors. Remember that the primary interaction limiting bacterial transport is filtration, not adsorption as it is for viruses. In addition, in contrast to viruses, bacteria have very chemically diverse surfaces, and thus a change in pH would not be expected to alter the net surface charge to the same extent as for the more homogeneous viral surface. Finally, the overall charges on the surfaces of bacteria and viruses differ and will be affected differently by pH changes. This can be expressed in terms of the **isoelectric point (pI).** The pI is the pH at which the net charge on a particle of interest is zero. For bacteria, the pI usually ranges from 2.5 to 3.5, so the majority of cells are negatively charged at neutral pH. At pH values more acidic than the isoelectric point, a microbe becomes positively charged. This will reduce its transport potential because of increased sorption. This will not happen often with bacterial cells in environmental matrices, given their low pI values; the pH would have to decrease to 2.5 or lower to alter cell surface charge significantly. However, viruses display a wider range of isoelectric points (see Table 7.1) making their net surface charge much more dependent on changes in pH.

The sorption of a wide variety of bacteriophage and viruses to nine different soil types was examined by Goyal and Gerba (1979). Results indicate that adsorption to a given soil is very dependent not only on virus type but also on the particular strain. Furthermore, this study shows that although several soil characteristics apparently affect adsorption, soil pH is the single most important factor influencing viral adsorption to soil. In general, based on this study, soils with a pH < 5 tend to favor virus adsorption.

7.2.5 Impact of Ionic Strength on Transport

It has already been established that the net charge on both particle and cell surfaces is negative, which causes electrostatic repulsion between these two surfaces. If the composition of the porous medium solution were pure water, the repulsive forces between bacteria and viruses and particle surfaces would be much greater than they actually are. This is because repulsive forces are reduced by the presence of cations in the soil solution. The concentration of anions and cations in solution is referred to as the **ionic strength** of the medium. Soil solution ionic strength influences transport primarily through two mechanisms—by altering the size of the diffuse double layer and by influencing soil structure (see Chapter 4.2).

The negative charges present on mineral particles electrostatically attract cations in the solution. Thus, in the immediate vicinity of the negatively charged surface, there is an excess of cations and a deficit of anions. Further from the particle surface, the cation concentration decreases until it reaches that of the bulk solution.

TABLE 7.1 Isoelectric Points of Selected Viruses

Virus	pI
Reovirus 3 (Dearing)	3.9
Rhinovirus 2	6.4
Polio 1 (Bruenders)	7.4, 3.8
Polio 1 (Mahoney)	8.2
Polio 1 (Chat)	7.5, 4.5
Polio 1 (Brunhilde)	4.5, 7.0
Polio 1 (Lsc)	6.6
Polio 2 (Sabin T2)	6.5, 4.5
Echo 1 (V239)	5.3
Echo 1 (V248)	5.0
Echo 1 (V212)	6.4
Echo 1 (R115)	6.2
Echo 1 (4CH-1)	5.5
Echo 1 (Farouk)	5.1
Coxsackie A21	6.1, 4.8
Parvovirus AA4. X 14	2.6
Vaccinia (Lister)	5.1
Vaccinia (Lister)	3.8
Smallpox (Harvey)	5.9
Smallpox (Harvey)	3.3
Influenza A (PRS)	5.3
Shope papilloma	5.0
T2 bacteriophage	4.2
T4 bacteriophage	4–5
MS-2 bacteriophage	3.9
PRD-1	3–4
QB[a]	5.3
ϕX174[a]	6.6
PM2[a]	7.3

Data compiled from Gerba, 1984, and Ackermann and Michael, 1987.

[a] Ackermann and Michael, 1987.

Thus, the porous medium solution is often thought of as a double layer as depicted in Fig. 7.11. The impact of ion concentration on this diffuse double layer will ultimately play a critical role in the transport of microbes. As the ionic strength of the bulk soil solution increases, the difference in cation concentration between the cation-rich layer and the bulk layer is reduced, and thus there is a tendency for cations to diffuse away from particle or cell surfaces. This causes a general compression of the diffuse double layer because the interacting cations and anions neutralize one another. The result is a decreased electric potential, which increases the likelihood of attachment of cells to the surfaces.

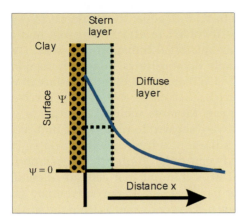

FIGURE 7.11   Illustration of the diffuse double layer. The diffuse double layer is a combination of the charge layer on the surface and the charge in solution. The first monolayer of ions in contact with the surface, the Stern layer, is held tightly to the surface. The second layer, the diffuse layer, responds to the remaining charge on the surface but is held more loosely than the Stern layer ions. This figure shows the energy (ψ) required to bring an ion from the bulk solution to the surface as a function of distance from the surface. Immediately next to the surface the decrease in potential is linearly related to increasing distance from the surface (the Stern layer). As the distance from the surface is increased further, the potential decreases exponentially. (Modified with permission from Tan, 1993, p. 198 by courtesy of Marcel Dekker, Inc.)

In addition to the overall ionic strength, the type of ion contributing to ionic strength is important. This is because the radii of hydration of a cation in the soil solution affects the extent of the diffuse double layer and thus the soil structure. The radius of hydration of a particular cation is a function of surface charge density and refers to the radius of the cation and its complexed water molecules (Fig. 7.12). In general, monovalent cations have lower surface charge densities and thus larger radii of hydration than divalent cations. Thus, in the presence of high concentrations of monovalent cations such as Na^+, clays tend to be dispersed. Dispersed clays create puddled soils, which are sticky when wet and hard when dry. As a dispersed soil dries, compaction may occur, which reduces pore spaces, inhibiting soil aeration and reducing the capacity for water flow. This adversely affects the transport potential of microbes. On the other hand, the presence of divalent cations such as Ca^{2+} and Mg^{2+}, with smaller radii of hydration, leads to flocculated soils, which have increased pore space and thus favor transport.

The impact of ionic strength on microbial transport is demonstrated by the following examples. Bai *et al.* (1997) found that fewer cells were recovered in a column study when 2 mM NaCl was used as the percolating solution as compared with the use of artificial groundwater with a lower ionic strength. This observation can be explained in terms of cation concentra-

tion and associated electrostatic interactions. In addition, experiments indicate that viruses either do not readily sorb or are released from soil particle surfaces suspended in low-ionic-strength solutions (Landry *et al.*, 1979; Goyal and Gerba, 1979).

How can ionic strength vary in a porous medium? One example is a rainfall event. When it rains, the added water will generally lower the ionic strength of the soil solution. In addition, a rainfall event results in increased rates of water flow. Both decreased ionic strength and increased flow rate will promote microbial transport.

7.2.6 Cellular Appendages

Bacteria may have a variety of appendages such as pili, flagella, or fimbriae. Flagella are responsible for bacterial motility while fimbriae and pili are involved in attachment. These appendages may all play a role in microbial transport through the terrestrial profile. The influence of bacterial motility on overall transport is generally minimal because extensive continuous

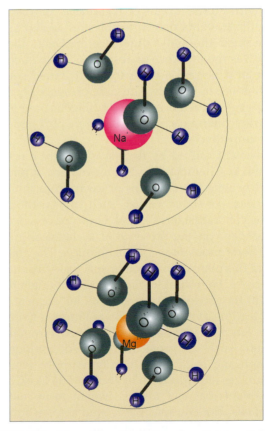

FIGURE 7.12   The radius of hydration of a cation in soil depends on the charge density of the atom. In the example shown, magnesium has a higher charge density than sodium and thus attracts water molecules more strongly resulting in a larger radius of hydration.

water films would be needed to support microbial movement. Although continuous films exist, they are present only in soils with high soil moisture contents. In flowing systems, the primary mechanism of transport is water flow, commonly referred to as advective transport. **Advective transport,** will be many orders of magnitude greater than transport due to motility which typically occurs on a micrometer scale.

In nonflowing systems where no advective transport occurs, one would intuitively expect motility to increase transport potential, although this increased transport may be only over a very small scale. Increased transport by motile versus nonmotile strains has been observed. For example, Jenneman *et al.* (1985) found that a motile strain of *Enterobacter aerogenes* penetrated nutrient-saturated sandstone cores up to eight times faster than did a nonmotile strain of *Klebsiella pneumoniae*. However, there are metabolic and physiological differences between these two strains, thus, motility alone may not account for the differences in transport rates. In a similar study, penetration through nutrient-saturated sand-packed cores under static conditions was evaluated (Reynolds *et al.*, 1989). In this study, transport through the column was four times faster with motile strains of *Escherichia coli* than with nonmotile mutants defective only in flagellar synthesis. Because the only difference between the two strains was the presence or absence of flagella, the difference in penetration rate can be attributed solely to the difference in motility. Thus, the presence of cellular appendages involved in motility (flagella) can lead to measurable increases in microbial transport under certain circumstances.

Movement caused by flagella is usually a result of chemotaxis. **Chemotaxis** is the movement of microbes toward beneficial substances or away from inhibitory substances. As stated earlier, such movement occurs only over short distances in soil. This type of movement is dependent on the presence of a chemical gradient within continuous films of soil solution. The ability to move in this manner may confer survival advantages to the microbe. For example, chemotaxis is thought to play a role in the movement toward and subsequent infection of legume roots by *Rhizobium*, a nitrogen-fixing bacterium. *Rhizobia* derive a variety of benefits from this association, such as protection from desiccation and predation and a generous nutrient supply, while the plant utilizes nitrogen fixed by the bacteria (see Chapter 18). A plant may not always benefit from microbial chemotaxis. For example, plant exudates can stimulate activation of pathogenic fungal spores to form vegetative propagules which through chemotaxis are attracted to the root system, promoting fungal infection and ultimately damaging if not killing the plant.

In contrast to flagella, the presence of cellular appendages involved in attachment (pili and fimbriae) can reduce microbial transport potential. The following example demonstrates this phenomenon. Rosenberg *et al.* (1982) observed that the surface of wild-type *Acinetobacter calcoaceticus* RAG-1 cells is covered with fine fimbriae. Mutants without fimbriae or wild-type cells that were sheared to remove the fimbriae showed reduced adhesion to hydrophobic surfaces compared with wild-type cells, implicating the fimbriae in attachment. It is thought that cellular appendages can penetrate the electrostatic barrier thereby facilitating attachment at greater distances from the surface. Functional groups (hydrophobic groups or positive charge sites) on the appendages may facilitate interaction with surfaces leading to increased sorption. Thus, the presence of appendages may actually decrease microbial transport in some cases. These examples demonstrate the variable effects of appendages on transport.

7.2.7 Sedimentation

Another mechanism that may influence microbial transport through the subsurface is **sedimentation,** the gravitational deposition of particles on matrix surfaces. Sedimentation occurs when the density of the microbe is greater than that of the liquid medium and flow velocities are sufficiently low. With high flow rates and short time scales, such as found in many column experiments, bacteria are assumed to demonstrate neutral or near-neutral buoyancy (Hornberger *et al.*, 1992). Thus, advective processes are generally thought to overwhelm any significant effect of sedimentation. Accordingly, although sedimentation may play a role in bacterial transport in very slow-moving water over extended periods of time, the mechanism is frequently ignored by researchers.

7.2.8 Hydrogeological Factors

A basic understanding of the chemical and physical properties of the different layers of the terrestrial profile is necessary to assess microbial transport potential in the environment. This profile consists of soil, the vadose zone (unsaturated zone), and the saturated zone or groundwater. Soil texture and structure, porosity, water content and potential, and water movement through the profile are key hydrogeological factors influencing microbial transport. These factors are all described in detail in Chapter 4. It is the combination of these factors that makes each site unique. The specific soil and vadose zone layers within a site serve as protective or attenuating zones with regard to contamination of groundwater by microbes (or chemical pollutants) via a variety of mechanisms, including filtration and adhesion. In addition to the site-specific makeup of the porous medium, the distance between the soil surface and the vadose–groundwater interface is often a critical factor for determining pollution potential: the greater the distance, the less likely it is that groundwater contamination will occur.

Terms used to describe flow and transport of dissolved and particulate substances are commonly applied to describe the transport of microbes (Fig. 7.13). **Advection,** the movement of the bulk pore fluid and its dissolved and suspended constituents, is primarily responsible for microbial transport. **Dispersion** is the combined result of mechanical mixing and molecular diffusion. Mechanical mixing results from the **path tortuosity** and velocity differences within the pore that depend on pore size and location of the microbe as depicted in Fig. 7.14. Spreading due to molecular diffusion, the random movement of very small particles suspended in a fluid, results from the pres-

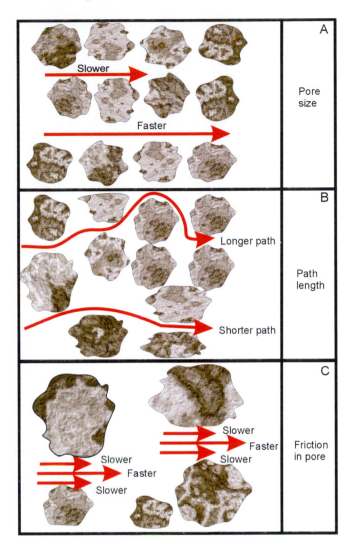

FIGURE 7.14 Factors causing mechanical dispersion at the scale of individual pores. A) microbes are transported through small pores more slowly than through large pores; B) depending on pore sizes and shapes, path lengths can vary considerably; C) flow rates are slower near the edges of the pore than in the middle. (Modified with permission from Fetter, 1993, © MacMillan Magazines Limited.)

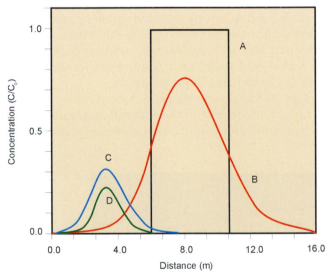

FIGURE 7.13 Effects of various processes on contaminant transport. This figure shows the theoretical distribution of a short pulse of microbes added to a saturated soil column 16 m long. The ordinate represents the relative concentration where C = the microbial concentration in the solution phase at a given point in the column, and C_o = the influent concentration of microbes. The abcissa represents distance along the column from 0 to 16 m. Pulse A represents microbes that have moved through the column influenced only by advection. Pulse B represents the combined influence of advection and dispersion on microbial distribution. Note that no microbes are lost from the solution phase in either pulse A or B. Pulse C represents addition of adsorption to advective and dispersive processes. In this case microbes are lost to the solid phase and the resulting pulse is smaller and retarded. Finally, pulse D represents the addition of decay to the other three processes which further removes microbes from the solution phase. (Modified with permission from Yates and Yates, 1991.)

ence of a concentration gradient. It is generally considered negligible with regard to bacterial transport, but can be significant in the transport of smaller particles (<1 μm) such as viruses. Finally, **adsorption** represents the removal of microbes from the bulk solution by reversible and irreversible adhesion.

Because microbes are transported along with the soil solution primarily through advection, the flow rate and degree of saturation of the soil can play significant roles in determining transport potential. In general, higher water content and greater flow velocities result in increased transport. For example, virus penetration was studied through columns packed with loamy sand soil. Virus applied under unsaturated flow reached a maximum depth of 40 cm, compared with a penetration depth of 160 cm during saturated flow through the same soil columns (Lance and Gerba, 1984). It seems plausible that during unsaturated flow, viruses move through the soil only in thin films of water at the matrix surfaces. The increased proximity of the viruses to soil particles would increase the potential for virus adsorption and thus decrease the virus penetration depth. In addition, lowering the infiltration rate into soils while maintaining saturated conditions can restrict virus movement (Lance *et al.*, 1982). It should be noted that in unsaturated soil or the vadose zone, water movement is primarily vertical because of the force of gravity. However, water movement in the saturated zone or groundwater is horizontal because of differences in pressure or elevation.

The flow rate of water through a saturated column can be calculated using Darcy's law:

$$Q = K \, \frac{\Delta H \, A \, t}{z} \qquad \text{(Eq. 7.1)}$$

where
Q is the volume of water moving through the column (m^3),
A is the cross-sectional area of the column (m^2),
t is time (days),
ΔH is the hydraulic head difference between inlet and outlet (m),
K is the hydraulic conductivity constant (m/day), and
z is the length of the column (m).

A typical saturated column experiment is depicted in Fig. 7.15. **Hydraulic conductivity** can be defined as the ease with which water moves through soil. A hydraulic conductivity greater than 10 m/day is considered large, and a value less than 0.10 m/day indicates limited water flow (Wierenga, 1996). Darcy's law may also be applied to unsaturated soils; however in this case, the hydraulic conductivity in the preceding equation is no longer constant. This is because the unsaturated hydraulic conductivity of a soil, $K(h)$, is a nonlinear function of the matric potential, which in turn is related to the water content. Figure 7.16 shows typical $K(h)$ values for a coarse-textured soil (sand) and a fine-textured soil (clay). At saturation, the pores are filled with water. Thus, the coarse-textured soil has a higher conductivity because it contains more large pores, where the water is held less tightly. When water is no longer added to the system, these large pores drain first, resulting in a pronounced decrease in hydraulic conductivity. The hydraulic conductivity values eventually cross at a matric potential of approximately -5×10^3. From this point on (at higher matric potentials) the clay soil has a higher hydraulic conductivity because predominantly only the smaller pores remain filled with water. Because considerably more water is present in clay soils under these conditions, there is an increased probability of a continuous water film remaining, which facilitates transport.

Darcy's law was developed for steady flow, where Q is constant. However, in the subsurface, conditions are dynamic and thus Q is constant only over short periods

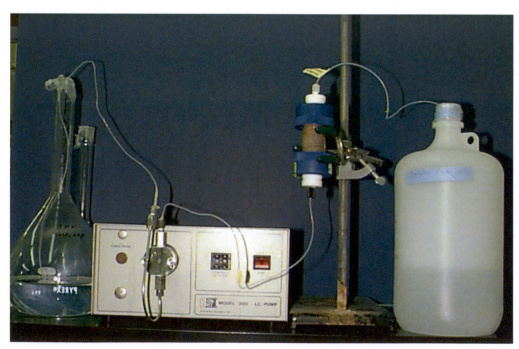

FIGURE 7.15 A typical column used for saturated flow experiments. (Photo courtesy R. M. Maier.)

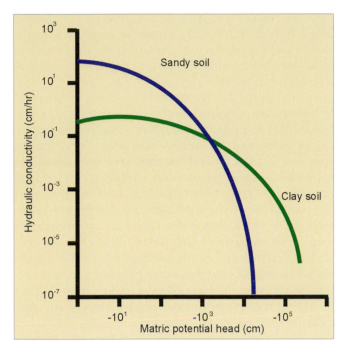

FIGURE 7.16 The hydraulic conductivity of a soil is dependent on the texture and the moisture content of the soil. This figure compares the hydraulic conductivity of a sand and a clay soil as a function of moisture content. (Modified with permission from "Soil Physics," Jury *et al.*, © 1991, John Wiley & Sons, Inc.)

of time. To account for changing flow, the flow equation is written in differential form to yield the Darcy flux:

$$q = -K \frac{\partial H}{\partial z} \qquad \text{(Eq. 7.2)}$$

where $q = Q/At$ (m/day) and $\partial H / \partial z$ is the hydraulic gradient (m/m). By definition, q is the volume of water moving through a 1-m^2 face area per unit time. However, because water moves only through pore space and not through solids, the actual velocity of water moving through soil is considerably higher than q, the Darcy velocity. The pore water velocity is proportional to pore size; however, the average pore water velocity is generally defined as

$$v = \frac{q}{\theta_w} \qquad \text{(Eq. 7.3)}$$

where
v is the pore water velocity
q is the flow rate per unit area determined for Darcy's law
θ_w is the water-filled porosity

In saturated soils, θ_w is equal to the total porosity, so the pore water velocity is approximated well by the Darcy velocity. However, in unsaturated soils there is a marked increase in pore water velocity over Darcy velocity.

Another factor to consider is hydrologic heterogeneity arising as a function of soil structure. Variations in structure such as cracks, fissures, and channels can greatly affect flow rates by creating preferred flow paths, with increased flow velocities. This phenomenon is termed **preferential flow.** Such structural inconsistencies can greatly increase transport.

7.2.9 Persistence and Activity of Introduced Microbes

The introduction of microbes into a soil often has a dramatic effect on their survival and activity. Typically, there is a rapid decline in numbers of inoculant cells, and there is often a decrease in the average activity per cell of the surviving introduced microbes. This phenomenon has been attributed to the scarcity of available nutrients and the hostility of the soil environment to incoming microbes. Temperature, pH, soil texture, moisture content, and the presence of indigenous organisms including plants, earthworms, and microbes can all affect the survival and metabolic potential of introduced microbes during transport and upon deposition. Evidence suggests that microbial adhesion to particle surfaces tends to provide some degree of protection from adverse factors. For example, many essential microbial nutrients are positively charged ions (i.e., NH_4^+, Na^+, K^+, Mg^{2+}). These cations are electrostatically attracted to the negatively charged particle surfaces and thus tend to accumulate at the solid–liquid interface. Similarly, molecules slowly released from organic matter associated with matrix particles can serve as microbial nutrients. Access to and utilization of these solid phase–associated nutrients may account, in part, for the increased survival of sorbed microorganisms. Accordingly, the effective density and activity levels that introduced organisms achieve depend on the prevailing environmental conditions and on the niche the organisms can establish for themselves.

Viral inactivation, for example, has been shown to be directly correlated with increases in temperature (Yates *et al.*, 1985), with sorbed viruses demonstrating decreased temperature sensitivity compared with free viruses. Temperature can also affect microbial activity by altering the moisture content of the terrestrial system. Although microbes require a certain amount of moisture to avoid desiccation, increases in moisture content above a certain optimal level have actually been shown to lead to a decrease in microbial numbers in a natural soil (Postma and van Veen, 1990). This may be due to oxygen depletion as more pores become water filled. In addition, microbial predators such as protozoa tend to be more active at higher soil moisture contents, perhaps because increased water tends to provide a mechanism for movement between pores. Microbial sorption to particles and within small pores is thought to provide protection from protozoa, which are typically larger and thus may be excluded from certain pores or bacterial sorption sites. Indeed, in the presence of protozoa, higher percentages of particle-associated bacteria have been observed (Postma *et al.*, 1990). The population dynamics between soil bacteria and protozoa resulting from an "ideal" predator–prey relationship is depicted in Fig. 7.17. The addition of a nutrient source, such as plant residue, to the soil can stimulate growth of the bacteria, leading to a peak in their population density. The bacteria in turn can serve as a food source for the protozoa. Thus, a lagged pulse in protozoal numbers is generally observed.

Immobilization of bacterial cells in a carrier material such as polyurethane or alginate has been investigated as an improved inoculation technique leading to increased survival and degradation capabilities of the inoculum (Hu, *et al.*, 1994; van Elsas, *et al.*, 1992). For introduction into the soil environment, the carrier material presumably confers some protection against harmful physicochemical and biological factors to the inoculum. Regardless of the carrier medium, it is critical that the encapsulated cells remain viable during storage and following introduction into the soil. In addition, in some cases it is desirable for the cells to be released from the matrix upon delivery to the target site.

Alginate, although not as strong as polyurethane, has the advantage of being biodegradable and thus potentially serving as a nutrient source or supplement for the entrapped microorganisms as well as promoting slow release of the organisms into soil (Bashan, 1986). Increased survival of cells entrapped in an alginate matrix versus cells added directly to the soil has been demonstrated (van Elsas *et al.*, 1992). The use of alginate beads containing additional protective reagents (e.g., skim milk, bentonite clay) or reagents that provide a selective environment for the inoculant cells (e.g., antibiotics, easily utilized C sources) has also been suggested.

Although the soil environment is often detrimental to introduced organisms and thus their transport, certain biotic components of the terrestrial profile can increase movement of added microbes. For instance, channels formed by earthworms have been shown to increase transport by creating regions of preferential flow (Thorpe *et al.*, 1996). Similarly, bacterial transport has been shown to be stimulated by root growth (Hekman *et al.*, 1995). Water movement through channels formed by root growth and/or in films along the root surfaces contributes to the increased bacterial dispersion. The region of soil in close proximity to roots often differs from the bulk soil by a variety of properties including increased nutrient levels. Thus, roots provide not only preferred paths for water movement but also a favorable niche for bacterial growth. Accordingly, this region, termed the **rhizosphere,** typically demonstrates increased bacterial populations (see Chapter 18). An increased population of surviving inoculum increases transport potential. However, a percentage of the cells in this region are strongly attached to soil particles, probably as a result of polysaccharide production by both the root and the associated microorganisms. Even though earthworms and plant roots can increase microbial transport, it is important to put their impact on microbial transport in perspective. As shown by Madsen and Alexander (1982), advective water movement can cause up to 100-fold greater dispersal of microorganisms than plants or earthworms.

7.3 FACTORS AFFECTING TRANSPORT OF DNA

The survival and transport potential of introduced microbes are both issues of concern. However, it is important to realize that a dead or inactivated microbe usually breaks open, releasing its genetic material to the environment. Upon lysis, there is potential for the genetic material to be transported or sorbed to colloids where it can remain protected from degradation (Ogram *et al.*, 1988). Free or desorbed nucleic acids may be reincorporated into other microbes via transformation. This can result in the expression of genes encoded by these nucleic acids or in the potential transport within the intact recipient cell.

Sorption of free DNA depends on several factors, including the mineralogy of the matrix material, ionic

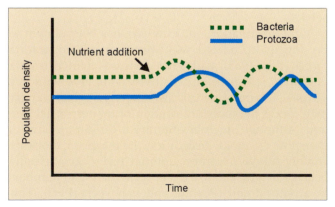

FIGURE 7.17 An "ideal" prey–predator relationship depicting soil bacterial population density as a function of soil protozoan density.

strength and pH of the soil solution, and length of the DNA polymer. DNA has a pK$_a$ of approximately 5. At pH values equal to the pK$_a$ the DNA is neutral, and at lower pH values it is positively charged. In either of these states the DNA is subject to adsorption to colloids and to intercalation into certain minerals, such as montmorillonite. This is enhanced by the fact that the pH of the microenvironment surrounding a soil particle may be as much as two or three units below the pH of the bulk solution. However, at higher pH values the DNA is negatively charged and is repelled from the negatively charged surfaces. Ogram *et al.* (1988) determined that the surface pH of some natural soils and sediments may be near the pK$_a$ of DNA, and thus significant amounts of DNA could remain nonsorbed and be present in the aqueous phase. The same group also found that higher molecular weight DNA was sorbed more rapidly and to a greater extent than lower molecular weight DNA. Depending on the specific conditions and soil sample, DNA sorption can be highly variable (Ogram *et al.*, 1988).

7.4 NOVEL APPROACHES TO FACILITATE MICROBIAL TRANSPORT

As detailed in previous sections, the transport of microbes through the terrestrial profile is limited by a multitude of physical, chemical, and biological factors in nature. However, for a number of applications including oil recovery and bioremediation, delivery of viable microbes may be critical to the success of the application. As a result, strategies have been developed to attempt to optimize the "natural conditions" that favor transport. Several novel approaches designed to facilitate microbial transport through the terrestrial profile are being investigated. Formation of ultramicrobacteria, biosurfactants, and gene transfer are among those that show potential.

7.4.1 Ultramicrobacteria

It has been known for the past 20 years or so that marine bacteria react to starvation by dividing and shrinking to one third their normal size. Such bacteria are referred to as ultramicrobacteria (UMBs). Furthermore, it has been found that 65% of all isolates obtained from soil can be placed in a nutrient-deprived medium such as phosphate-buffered saline and form UMBs. After several weeks of starvation, a distinct morphological change takes place in these cells. As shown in Fig. 7.2, the cells shrink to approximately 0.3 μm in size and become rounder. They also lose their glycocalyx layer, thereby becoming less

sticky. These bacteria can then be resuscitated by providing a carbon source. They recover both morphologically and physiologically. The interesting feature of UMBs is that they exhibit several characteristics that enhance their transport. First, they are smaller than normal metabolically active cells and hence are less subject to removal by filtration. Second, the UMB cell surface is less sticky because it has no glycocalyx layer. Such UMBs have been shown to penetrate farther into sandstone cores than their vegetative counterparts.

Interest in UMBs first centered on their potential for use in oil recovery. It has been proposed that local bacteria could be used to form UMBs and be injected into porous oil-bearing strata, followed by nutrient injection (Lappin-Scott *et al.*, 1988). After nutrient addition, the cells should grow and recover in size. Such growth can create plugging within a geological formation, which would allow subsurface oil flow to be rechanneled, improving oil recovery. Another potential use proposed for UMBs is in closure of mine tailing heaps. One of the problems in capping a mine tailing heap is that initial revegetation efforts are often difficult under the low-pH conditions of the mine tailings. The low pH may be intrinsic to the mining process or driven by the activity of the sulfur oxidizers, obligate aerobes that create acid mine drainage via the oxidation of sulfur compounds within the tailings (see Chapter 15.3.1). UMBs may have the potential to form a living biobarrier to prevent oxygen diffusion into the mine tailing heap. A preliminary experiment to evaluate whether this approach is feasible showed that UMBs were able to penetrate farther into the mine tailing material than normal bacteria (76 cm versus 10 cm). In addition, mine tailing effluent characteristics before UMB treatment were pH 3.2 and E_h (redox potential) 150 mV. After UMB treatment the pH was increased to 6.4 and the E_h was decreased to 15 mV. These results suggest that UMBs were able to colonize and form a biobarrier to prevent further acid mine drainage formation (Fig. 7.18). This is a critical step in the restoration of mine tailing disposal sites.

7.4.2 Surfactants

Another approach involves the use of a chemical additive, specifically a surfactant, to increase the transport potential of microbes. Bai *et al.* (1997) investigated the influence of an anionic monorhamnolipid **biosurfactant** on the transport of three *Pseudomonas* strains with various hydrophobicities through soil under saturated conditions. The surfactant for this study was extracted from *Pseudomonas aeruginosa* American Type Culture Collection (ATCC) 9027 culture supernatants and purified before use. Columns packed with sterile

A

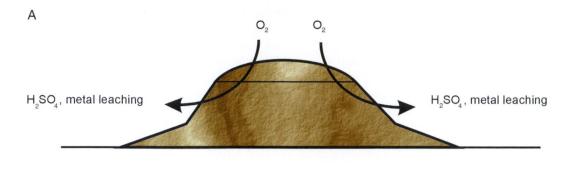

B

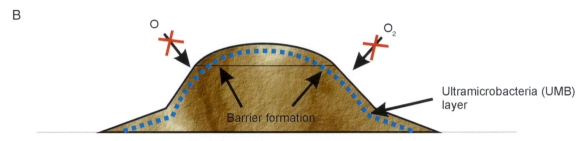

FIGURE 7.18   The use of ultramicrobacteria in capping a mine tailing disposal site. When oxygen and rainwater can freely penetrate the mine tailing surface, acid mine drainage forms, causing effluents from the waste heap to be characterized by low pH and leached metal (A). After injection and resuscitation of ultramicrobacteria, oxygen is quickly utilized, maintaining anaerobic conditions that suppress the formation of acid mine drainage (B).

sand were saturated with sterile artificial groundwater, and then three pore volumes of ^{3}H-labeled bacterial suspensions with various rhamnolipid (RL) concentrations were pumped through the column. Four additional pore volumes of rhamnolipid solution were then applied. Rhamnolipid enhanced the transport of all cell types tested but to varying degrees. Recovery of the most hydrophilic strain increased from 22.5 to 56.3%, recovery of the intermediate strain increased from 36.8 to 49.4%, and recovery of the most hydrophobic strain increased from 17.7 to 40.5%. Figure 7.19 shows the breakthrough curves for the most hydrophilic strain at different rhamnolipid concentrations.

In this experiment it was found that the surface charge density of the bacteria did not change in the presence of the rhamnolipid, but the negative surface charge density of the porous medium increased. Thus, reduced bacterial sorption may be due to one of several factors including an increase in surface charge density caused by rhamnolipid adsorption; solubilization of extracellular polymeric glue; or reduced availability of sorption sites on porous surfaces. The advection–dispersion transport model used to interpret these results suggests that the predominant effect of rhamnolipid was to prevent irreversible adsorption of cells.

Similarly, Jackson *et al.* (1994) found that the ad-

dition of an anionic surfactant, sodium dodecyl benzenesulfonate (DDBS), facilitated transport of an isolate of *Pseudomonas pseudoalcaligenes* through a processed soil matrix under pressurized flow conditions. They found in their column studies that below a depth of 6 inches, bacterial density was from one to five orders of magnitude greater when the bacteria were suspended in surfactant solution instead of water. As suggested by batch experiments, the contribution of DDBS to enhanced bacterial transport may be related to its ability to prevent adsorption and reduce straining. The reduction in adsorption may be due to binding of the anionic (negatively-charged) surfactant to the cell wall, causing the net negative charge of the cell surface to increase. This would increase the electrostatic repulsion between the cell and the soil surface. The surfactant may also be able to prevent flocculation of the bacterial cells by dissolving natural polymers that are involved in forming bridges between bacterial cells. Similarly, the surfactant may solubilize polymers responsible for permanent attachment to the soil surface.

7.4.3 Gene Transfer

Bacteria are intentionally introduced into soil systems in order to manipulate components and/or processes that occur within the soil profile. Typically,

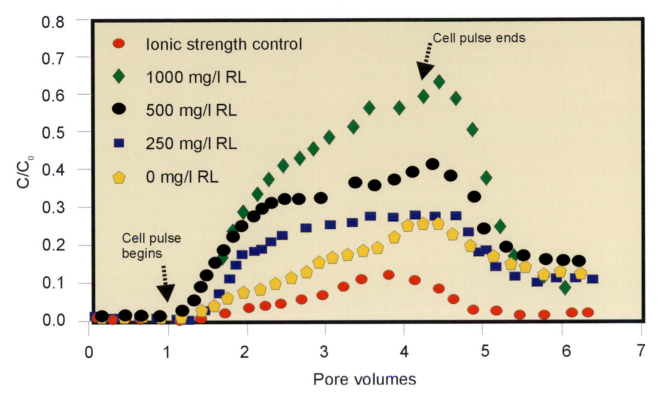

FIGURE 7.19 The effect of a rhamnolipid (RL) biosurfactant on transport of a *Pseudomonas* sp. where C_o is the CFU/ml in the influent solution, and C is the CFU/ml in the effluent solution. A pore volume is the amount of liquid it takes to fill all of the soil pores in the column. (Modified with permission from Bai *et al.*, 1997.)

enhancement of microbial activities, e.g., organic degradation or metal resistance/immobilization, is the driving force behind their introduction. It may be possible to circumvent some of the factors that limit microbial transport and thus the success of bioremediation in soil via genetic exchange. Gene transfer between organisms may occur through **conjugation, transduction,** or **transformation.** Transfer events such as these may make it possible to distribute genetic information more readily through the soil. Gene transfer events between an introduced organism and indigenous soil recipients has been shown to occur and in some cases this results in increased degradation of a contaminant (DiGiovanni *et al.*, 1996) (see Chapter 4 Case Study).

Studies have addressed the transport potential of transconjugants, which arise when indigenous bacteria receive a plasmid from an introduced donor through conjugation. In a column study involving a donor inoculum at the column surface, Daane *et al.* (1996) found that transconjugants were limited to the top 5 cm of the column. However, when earthworms were also introduced into the column, not only did the depth of transport of donor and transconjugants increase, depending on the burrowing behavior of the earthworm species, but also the number of transconjugants found increased by approximately two orders of magnitude.

In a separate study, Lovins *et al.* (1993) examined the transport of a genetically engineered *Pseudomonas aeruginosa* strain that contained plasmid pR68.45 and the indigenous recipients of this plasmid in nonsterile, undisturbed soil columns. The surface of the column was inoculated and unsaturated flow conditions were maintained. Transconjugants survived longer in the columns and were found to have moved farther down the column than the donor. The greater survival rate of transconjugants would be expected because these organisms have previously adapted to the particular conditions of the soil. The increased transport could be the result of plasmid transfer to smaller, more mobile bacteria.

In addition, consecutive gene transfer events between indigenous microbes have been suggested as a mode of transfer (DiGiovanni *et al.*, 1996). This would be especially feasible when microbes are present in high densities, such as stationary microbes growing within a biofilm on soil surfaces or in the rhizosphere.

7.4.4 Other Ideas

In addition to the approaches already suggested, a variety of other ideas for facilitating microbial transport through the subsurface are being investigated. The use of adhesion-deficient mutants (DeFlaun and Ensley, 1993) and *in situ* electrophoresis (Condee and DeFlaun, 1994) are just two examples. The latter idea capitalizes on the negative charge of the microbes. A direct-current (DC) electric field is set up *in situ* with the negatively charged cathode near the inoculation point. As the electric current passes through the soil, the microbes migrate toward the positively charged anode. Although this approach may be useful over short distances, its application to field sites would be problematic.

Another potential approach to increasing transport of inoculated microbes is to add a substrate pulse to the soil. Recent research in our laboratory has shown that addition of a benzoate or salicylate (20 mg/l) pulse to inoculated soil columns results in enhanced transport of these microbes.

7.5 METHODOLOGY FOR STUDYING TRANSPORT

In situ transport experiments involving microorganisms are difficult to conduct. There are many obstacles associated with sampling and manipulating a complex environmental system that make the determination of critical transport factors difficult. In addition, the release of microbes into the environment may disrupt the natural dynamic soil equilibrium. The release of **genetically engineered microbes (GEMs)** is strictly regulated because of uncertainty regarding biological safety. Thus, most transport studies are conducted using either columns or lysimeters. These systems are contained, may be designed to facilitate sampling at a variety of depths, and can be manipulated more readily to determine the influence of specific factors on transport.

7.5.1 Columns

The degree to which column studies resemble natural soil systems varies greatly. Typical columns used are chromatography columns that are adapted for transport studies or commercially availabe soil columns (Soil Measurement Systems, Tucson, AZ). The design of the column and the matrix material it contains are dependent on the goals of the researcher. Additional factors that should be considered are column length and diameter. A typical column apparatus

used for unsaturated flow studies is shown in Fig. 7.20, while Fig. 7.15 shows one designed for saturated flow.

The column itself is generally made of plastic, brass, stainless steel, glass, or a combination of these materials. Screens placed at the ends of the column allow regulation of flow conditions and also retain the porous medium packed into the column. Screens can be made of plastic, stainless steel, or nylon mesh. The column is packed with the porous medium of interest, which is typically glass or silica beads (of constant or mixed dimensions), natural soil, or vadose zone materials. For example, laboratory columns packed with 0.5-mm silica beads have been used to approximate sandy soils for microbial retention studies (Kinoshita *et al.*, 1993). The column must be packed carefully and uniformly in order to minimize formation of macropores that result in preferential flow and to obtain a specific bulk density. Sandy soils are often used for microbial transport studies to allow reasonably high flow rates and to prevent column plugging via filtration. When performing site-specific research, soil from the site of interest should be used as the packing material in order to obtain the most relevant information. In order to mimic the natural soil system, the soil should be packed to the bulk density of that in the natural environment, which is generally from 1.3 to 1.5 g/cm^3. Another approach is to use intact soil cores, which can be obtained by driving an empty column into the soil and then carefully extracting the soil-filled column and capping the end appropriately (Lovins *et al.*, 1993). However, this methodology is usually limited to coarse-textured soils due to compaction problems with fine-textured soils. In addition, it is very difficult to recover an intact core that does not exhibit preferential flow.

A pump is generally used to deliver fluid into the column. For saturated flow systems, the fluid is usually delivered from the bottom to the top of the column to help displace any gas bubbles within the column that might interfere with water flow. After a column is packed, it is generally saturated with several pore volumes of 0.01 N $CaCl_2$, which mimics the ionic strength of a normal soil solution. $CaCl_2$ is used rather than NaCl to prevent dispersion of clays within the soil, which can cause column plugging. For unsaturated columns, a vacuum chamber located at the outflow end of the column can be used to provide a near-constant degree of saturation throughout the column. Flow for unsaturated columns is generally from top to bottom because for unsaturated flow gas is present by definition.

For **vertical transport studies,** inoculation of a column for a microbial transport study can be accom-

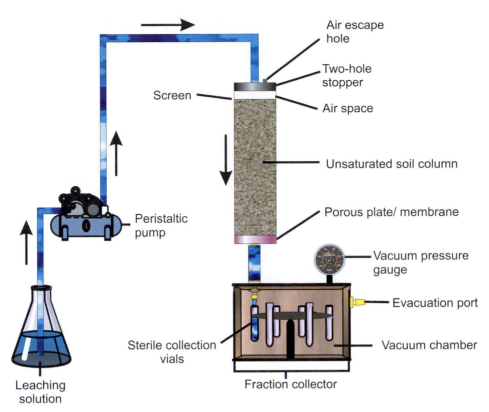

Air escape hole

Two-hole stopper

Screen

Air space

Unsaturated soil column

Porous plate/ membrane

Vacuum pressure gauge

Peristaltic pump

Evacuation port

Vacuum chamber

Sterile collection vials

Leaching solution

Fraction collector

FIGURE 7.20 The setup used to run an unsaturated soil column. (Modified with permission from Wierenga, 1996.)

plished in a variety of ways. Gannon *et al.* (1991b) inoculated their columns by placing a circular mound of dry sterile soil on top of a wetted column. An aliquot of inoculum was placed in the center of that mound, followed by the addition of a layer of sterile soil to make the top of the column surface approximately level. Trevors *et al.* (1990) used another approach. This group inoculated soil portions, mixed them thoroughly, and allowed them to air dry to the moisture content of the soil already packed in the column. The inoculated soil was then packed onto the top of the soil columns to the bulk density of the remainder of the columns and to depths ranging from 1 to 2.6 cm. A layer of gravel was then placed at the surface of the column to prevent channeling of water during percolation. Alternatively, the inoculum can be pumped into the column under either saturated or unsaturated flow conditions as described by Vandevivere and Baveye (1992) or Bai *et al.* (1997).

The column is then maintained under flow conditions for a specified time. Sampling of the column can be achieved in a variety of ways. **Destructive sampling** of the column involves disassembling the column to obtain samples. Destructive sampling disrupts the matrix to such an extent that the experiment must then be terminated. In some cases, the soil is carefully

removed by pushing it out of the column. Care must be taken because this approach can cause compaction of the soil. Trevors *et al.* (1990) used a small ethanol-sterilized saw to dissect small microcosm columns into three sections: 0 to 1 cm, 1 to 5.5 cm, and 5.5 to 10 cm. The soil within each section was then homogenized prior to analysis. Alternatively, Gannon *et al.* (1991b) took five vertical cores from each soil column using a 5-ml syringe and then divided each core into sections taken from depths of 0.0 to 1.7, 1.7 to 3.4, and 3.4 to 5.0 cm to assess microbial transport. Finally, some columns are constructed in sections that can be attached to each other (i.e., screwed together) and then separated for easier analysis of the soil.

Nondestructive sampling methods include monitoring the effluent from the column or using sampling ports to assess the presence of the microbe, nucleic acid, or contaminant of interest from a site within the column. For example, during the course of their experiment, Trevors *et al.* (1990) monitored the percolation water exiting their soil columns for bacterial counts before destructively sampling columns at designated time intervals. Bai *et al.* (1997) also monitored the column effluent to assess transport in the presence of a chemical additive. Tensiometers can be inserted through sampling ports to monitor the soil-water pressure, indicating the degree of saturation at each port. These ports, located at a variety of depths along the column, can be used for the introduction of a syringe to sample the soil solution or as sites for the removal of small soil cores. However, especially in the latter case, this type of sampling may create altered flow paths or flow rates, which may influence the extent of transport. In all types of sampling, whenever possible, analysis of samples should be limited to portions of the sample not in immediate contact with the column surface, because transport at this interface may be different from that through the matrix itself.

Columns can also be used for **horizontal transport studies** in a similar manner. For this type of experiment, the inoculum can be injected through a well that runs down the center of the column or in a mound of soil at the center of the column. The column is then sampled laterally at variable distances from the center of the column. Some studies have shown that little horizontal transport occurs in columns inoculated at the center, indicating that movement of cells along the column wall does not appear to contribute to measurements of the transport of bacteria (Gannon *et al.*, 1991b; Trevors *et al.*, 1990).

It is essential that aseptic techniques are used when sampling columns. Introduction of exogenous microbes during destructive sampling will result in contamination of the samples and may skew the transport data obtained. During nondestructive sampling, contamination can also occur and poses the same threat to the validity of the transport data. Introduction of contaminant microbes can alter the stability of the nucleic acid component of the soil or the persistence of the microbe of interest.

The transport of organisms through columns will vary depending on the organism studied as well as the type of column used. Smith *et al.* (1985) observed that under saturated flow conditions disturbed cores (packed columns) retained at least 93% of the bacterial cells applied, whereas intact cores retained only 21 to 78%, depending on the matrix texture and structure. Increased heterogeneity in soil structure and flow velocity are probably key factors accounting for the increased transport within the intact soil cores.

The following examples, conducted under saturated flow conditions, demonstrate the limited bacterial movement that generally occurs through packed columns. Gannon *et al.* (1991b) examined the mobility of 19 strains of bacteria through columns containing Kendaia loam soil and found that only 0.01 to 15% of added cells passed through a 5-cm long column after four pore volumes of water had been applied. Trevors *et al.* (1990) examined the transport of a genetically en-

gineered *Pseudomonas fluorescens* strain through soil columns packed with either loamy sand or loam soil using groundwater for column percolation. The distribution of the introduced cells throughout the columns was largely established after the first groundwater flush and did not change substantially with subsequent percolations. This suggests that after one flush, the bulk of the introduced population has either adhered to soil particles or entered relatively protected soil pores. In either case, only a small portion of introduced cells would remain in more open spaces and be available for elution with later percolations. For both soils, more than 90% of surviving introduced cells were retained within a 9-cm pathway, indicating that each soil had high retention of bacteria.

7.5.2 Lysimeters

Lysimeters provide an additional approach to monitoring transport by bridging the gap in scale between laboratory column studies and field studies. Lysimeters are large tanks that are packed in a fashion similar to columns. The tanks are placed into excavated holes or pits where, normally, the upper surface of the tank is at the same level as the ground surface. For a pictorial representation of a lysimeter, see the lysimeter Case Study included in this chapter. The placement of lysimeters at the study site allows approximation of *in situ* conditions while still separating the experimental system from the environment. The surface of the lysimeter may be covered or left exposed, allowing studies under natural precipitation conditions or under irrigated water flow conditions. **Weighing lysimeters** are simply lysimeters that rest on scales. They are generally used to measure changes in mass that can be attributed directly to plant water use and/or evaporation (i.e., evapotranspiration), and to study water movement through soil profiles. **Nonweighing lysimeters,** those that are not placed on scales, are also often used to assess water movement, but they are less accurate because they rely on measured differences between inputs of water (influent) and effluent that is collected from the bottom of the lysimeter. Nonweighing lysimeters often have drain outlets connected to suction pumps to allow effluent collection. Regardless of the type of lysimeter, their uses can be extended to study transport of chemicals and/or microbes and also *in situ* remediation processes. Ports for the insertion of tensiometers and solution collection probes can be placed at a variety of locations around the periphery of the lysimeter. The lysimeter should ideally be housed in such a way that the ports can be readily accessed.

CASE STUDY

Lysimeter Case Study

The University of Arizona has two large weighing lysimeters (Fig. 7.21) filled with Vinton fine sand at its Karsten Center for Turfgrass Research. Each lysimeter is 4.0 m in depth, 2.5 m in diameter, and contains 96 sampling ports. These lysimeters are much deeper than most lysimeters, providing a unique opportunity to study water movement and solute transport through deep soil profiles. Each lysimeter is housed in such a way that the many monitoring devices installed during packing of the lysimeter tanks can be accessed from any angle, even from below. The monitoring devices include time domain reflectometer (TDR) probes for the determination of the volumetric water content of the soil, stainless steel solution samplers, ceramic solution samplers, wick samplers, and compaction gauges to monitor potential settling of soil solids. In addition, a stainless steel neutron probe access tube was installed in the center of each tank to allow neutron probe monitoring of water content profiles. The size and design of these lysimeters give them great potential for conducting a variety of studies. Currently, the transport of the contaminant trichloroethylene (TCE) and its remediation are being evaluated. The lysimeters allow characterization of the microbial population of the soil within the lysimeter and investigation of changes in the microbial population that occur during biodegradation of the TCE.

FIGURE 7.21   Lysimeters constructed at the University of Arizona to study solute movement. In (a) and (b) a lysimeter is lowered into the study site with the upper surface level with the ground. (c) The lysimeter in position in the ground. Tensiometers and solution collection probes dot the side of the lysimeter now filled with soil. The entire lysimeter rests on a scale, enabling accurate measurements of water content changes in the soil caused in particular by water lost through surface evaporation and root water uptake. Photographs: M. Young (a) and (b), M. R. Stoklos (c). (Modified with permission from Wierenga, 1996.)

7.5.3 Field Studies

In order to realistically assess microbial transport potential in the environment, it is necessary to study transport *in situ*. Although laboratory column studies allow factors that affect transport to be varied individually, field studies, in which multiple factors may change simultaneously, generally provide more relevant information. However, because of difficulties involved in manipulating individual parameters *in situ*, field studies are generally conducted only after extensive laboratory studies. Examination of intact soil cores, obtained either from the surface or through a borehole, and pore water sampling are techniques often utilized for field studies.

A sterilized auger can be used to obtain intact soil cores (see Chapter 8.1.1). As with column studies, contamination during sampling is a constant concern and aseptic techniques should be practiced at all times. Intact soil cores can be taken at different depths and at time intervals designed to assess conditions prior to, during, and/or following a specified treatment. Soil cores can be used for a wide variety of studies, an example of which is presented next.

Land application of anaerobically digested sewage sludge is used to dispose of all treated municipal sewage in Tucson, Arizona. A major benefit of the sludge is that it serves as a nutrient source for the growth of field crops such as cotton. However, the survival and vertical transport potential of bacterial pathogens introduced into the environment by this practice are of concern because they pose a potential public health risk. In order to assess the risk of groundwater contamination, Pepper *et al.* (1993) conducted field studies to determine the survival and vertical transport of **indicator organisms** through a nonirrigated agricultural soil amended with sludge. Soil cores were taken from surface (0–50 cm) or subsurface (50-cm increments through 250-cm depth) soils immediately following injection with anaerobically digested liquid sludge (98% water). Cores were also taken prior to sludge application and at designated time intervals after application. Figure 7.22 depicts the results obtained for one site. Fecal coliforms migrated only a slight extent through the soil profile, indicating that the soil acted as an efficient bacterial filter. Note that

regrowth of introduced organisms within the soil profile was observed at this site following rainfall events. This has potential to result in increased transport and thus be of concern, depending on the particular site characteristics.

In a similar study, Straub *et al.* (1995) used a **semi-nested polymerase chain reaction (PCR)** (see Chapter 13.3.2.5) to detect the presence of enteroviruses in soil core samples obtained from cotton fields amended by surface application of anaerobically digested sludge. The use of this method enabled the detection of viruses, following irrigation and rainfall events, that had been transported to depths of 100 to 150 cm beneath the surface. Additional significance is placed on these results because PCR detection occurred 3 months after sludge amendment, implying that at least some of the viral nucleic acids remained intact over this time period. However, when cell culture assays were conducted on the samples, none contained infectious enterovirus. The implications of PCR detection of viruses in the field are still being debated nationally.

Obviously, there are limitations on the depth that can be sampled using an auger to obtain intact soil cores. However, this problem can be largely circumvented by taking soil cores from the inside of deep boreholes. This sampling approach has been used successfully in South Carolina near the Savannah River nuclear material processing facility to examine the microbiology of the deep subsurface (see Chapter 4.8 and Case Study—deep probe—"how low can life go?").

Examination of soil pore water collected from field sites can also provide useful information. Soil pore water is typically obtained by insertion of a solution sampler. Sampling pore water from unsaturated zones is more difficult than from saturated zones because it requires application of an external force to remove the water. However, soil pore water can easily be obtained following irrigation or rainfall events. The most common method involves applying suction to the soil solution through a porous membrane made of ceramic, stainless steel, Teflon, porous glass, or plastic material. In a single-chamber suction sampler, this membrane is located at the base of a collection tube to which a line connected to a pressure–vacuum pump and a separate discharge line going to a sample bottle are also attached (see Fig. 7.23). The discharge line is clamped while suction is applied to the soil solution through the pressure–vacuum line, drawing pore water into the body of the sample tube. The discharge line is then opened and pressure rather than suction is applied. This forces the soil solution that accumulated in sample tube into the discharge line and into the sample bottle. More complex double-chamber samplers can also be used, but the basic principles remain the same.

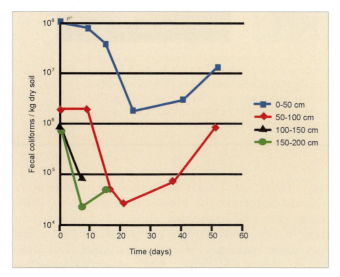

FIGURE 7.22 Survival of fecal coliforms at four soil depths. (Modified with permission from Pepper *et al.*, 1993.)

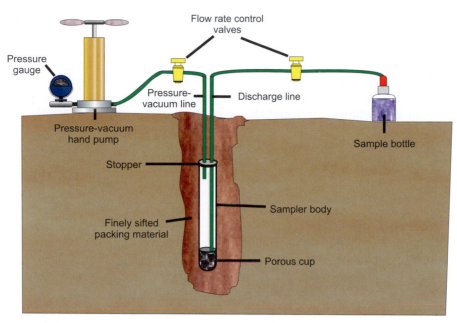

FIGURE 7.23 Example of an experimental setup that allows *in situ* sampling of soil pore water, which can subsequently be examined for microbial or chemical contaminants. (Modified with permission from Soilmoisture Equipment Corp., Santa Barbara, CA.)

Once installed, a solution sampler can be used repetitively to sample the same site at the same location. However, several limitations to this method exist. For example, it is difficult to extract enough solution from a dry soil (i.e., >0.5 bar). In addition, the soil solution may interact with the porous wall of the sampler. Such an interaction may result in loss of the microbe of interest. In one study, the use of porous ceramic samplers resulted in 74% loss of MS-2 virus, whereas with sintered stainless steel samplers with 2 μm pore size there was no virus loss (Powelson, *et al.,* 1990). So it may be possible to sample viruses. However, larger microbes will probably be strained out by the solution sampler. Thus, pore water sampling is generally restricted to the assessment of chemical and viral components of the soil solution.

7.5.4 Tracers

Tracers, chemical or particulate in nature, are often used to estimate microbial transport potential. Tracers are chosen so that their transport will closely mimic that of the microbe of interest. Their use is especially informative with regard to abiotic processes that influence movement of bacteria and viruses through subsurface media. Tracers are advantageous, particularly for field studies, for a variety of reasons. They can be added to a system in high numbers, their transport can be monitored without introducing a risk of infection, and they are typically easy to detect. A number of dif-

ferent tracers have been used in microbial transport studies, including microspheres, halides, proteins, and dyes. The use of microbe-sized microspheres has the advantage over the use of dissolved tracers since the microspheres should follow the same flow paths as bacteria, even in highly heterogeneous subsurfaces. However, their surfaces may interact with subsurface particles very differently from bacteria. Studies indicate that the tracer choice is often critical in determining relevant estimates of transport but that all tracers have limitations in terms of mimicking microbial transport through the terrestrial profile.

Hinsby *et al.* (1996) compared the transport of two bacteriophage with that of fluorescent microspheres through fractured clay. The observed flow patterns were very similar, suggesting that microsphere transport can be indicative of viral transport. Powelson *et al.* (1993) also compared the transport of a tracer with that of two phage. This group used potassium bromide, a conservative chemical tracer, and the bacteriophage MS-2 and PRD-1 which were selected because of their low adsorption to soils and their long survival time in the environment. Both the viruses and the conservative chemical tracer arrived at sampling depths in irregular patterns, indicating preferential flow. Virus breakthroughs were later than bromide except when viruses were added after pore clogging had reduced infiltration of the surface-applied sewage effluent. This study demonstrates the variability of relative transport rates that can exist between microbes and a tracer.

The ratio of the time it takes for the maximum concentration of a **conservative tracer** (a tracer that is nonsorbing) to be detected in the column effluent to the corresponding time for the coinjected microorganism defines the **retardation factor** for the microbe. This factor can be used for comparisons involving the same microbe and different soils or different microbes and the same soil. For example, Bai *et al.* (1997) found that in the absence of rhamnolipid (a surfactant) the retardation factors for three *Pseudomonas* strains through sandy soil ranged from 3.13 to 2.12. This is in comparison to a value of 1 which indicates no retardation. In addition, the retardation factor can be used to assess the occurrence of preferential flow. Preferred flow paths are suggested when the retardation factor is less than 1.0, indicating enhanced microbe transport relative to mean flow velocity (tracer transport). Gerba *et al.* (1991) suggested a worst-case value of 0.5 for the retardation factor when using models to predict microbial transport. One possible explanation for increased microbial transport is pore-size exclusion. Microbes may be excluded from smaller pores, where on average water travels more slowly (see Fig. 7.14A). Thus, they are forced to travel in the larger pores, with velocities that are higher than that of the soil solution as a whole. Thus, transport of microbes can be faster than that of a conservative tracer through the same porous medium.

7.6 MODELS FOR
MICROBIAL TRANSPORT

Accidental release of organisms into the environment, as well as their deliberate introduction, has inspired the development of models to describe microbial transport through the subsurface. Models can be used to predict not only the extent of movement of microbes but also the time required for the microorganisms to arrive at a specific location. Mathematical models designed to predict microbial transport through the terrestrial profile should take into consideration a variety of factors that influence both survival and sorption. In addition, they should account for changing conditions that the microbe encounters as it migrates through the terrestrial profile. Laboratory and field studies should be conducted in an attempt to verify the accuracy of conceptual transport models. Such studies indicate that although models can be useful tools, they can also generate predictions that are off by orders of magnitude. Accordingly, model predictions should be treated with caution.

In general, the more parameters considered in a model, the greater potential the model will have for producing relevant transport predictions. For example, Corapcioglu and Haridas (1986) introduced a highly comprehensive equation designed to predict microbial movement in soils. Many factors that influence microbial movement, including survival, chemical perturbations, clogging and declogging of the soil, soil moisture, pH, soil composition, and microbial growth potential, were taken into consideration. However, increasing complexity means increased time and expense, because more experiments must be conducted to generate the input data for the numerous parameters. In cases in which various parameters are not known and/or cannot be experimentally determined, estimates must be made, increasing the potential for error. Accordingly, it makes sense to decide on the transport model that will be applied to an environmental system before collecting data so that appropriate parameters can be measured whenever possible.

The presence of different microbes within a population can further complicate the application of models. In this situation, one microbe can be selected as "representative" of the population. However, the choice of a single microbe potentially leads to significant error, since it is unlikely that it is truly representative of all microbes present. Typically, a microbe that exhibits transport characteristics at one end of the spectrum or the other is chosen. The choice of an organism that does not travel extensively may be useful in assessing problems that may be encountered with bioaugmentation efforts. On the other hand, a microbe known to have high transport potential may be chosen when attempting to determine the minimum distance between a sewage release point and a well for drinking water. Another approach involves modeling the transport of each microbe individually and then combining the results to determine an overall concentration at a specified time and location. This method may be more accurate but is also much more involved.

A term is generally incorporated in equations to reflect the survival characteristics of the microorganism. This term represents the loss of the microbe, via death or inactivation, resulting from adverse chemical, physical, or biological processes. Microbial growth counteracts a portion of this decay, and thus a net rate decay term (i.e., net decay rate = growth rate − death or inactivation rate) is often used. Microbial decay rates have been found to vary by several orders of magnitude, making it necessary to evaluate survival characteristics for each particular microbe. Contaminant transport models, specifically advection–dispersion models and filtration models, are often modified for application to microbial transport. Factors that influence microbial transport can be incorporated into equations governing either of these models. However, the fundamental basis of each model differs, and thus under certain conditions use of a particular model may be preferable.

7.6.1 Advection–Dispersion Models

Unmodified advection–dispersion models assume that the contaminant is in solution and thus has the same average velocity as the matrix solution. Values for average velocity and dispersion of a contaminant are generally obtained from conservative tracer tests. These values may not be appropriate for microbial transport because microbes are not dissolved but instead suspended in the liquid medium. In order to obtain accurate adsorption data for input into the advection–dispersion equation, it may be necessary to conduct site- and microbe-specific adsorption studies. Furthermore, both irreversible sorption and reversible sorption should be considered. The advection–dispersion model often incorporates a decay term along with terms that account for transport with the bulk flow (advection), transport resulting from diffusion and mechanical mixing (dispersion), and adsorption. Advection–dispersion equations can be expanded in order to take into account hydrogeological heterogeneities in addition to the variety of factors that determine microbial survival. Figure 7.13 illustrates the influence of advection, dispersion, adsorption, and decay on the transport of a contaminant. A relatively simple advection–dispersion equation is

$$R_f \, \partial C / \partial t = -V \, \partial C / \partial x + D \, \partial^2 C / \partial x^2 \pm R_x$$

where
C is the concentration of microbe (mass/volume),
x is the distance traveled through the porous medium (length),
V is the average linear velocity constant (length/time),
R_x is the microbial net decay term (mass/time-vol),
t is time, and
R_f is the retardation factor, accounting for reversible interaction with the porous medium.

Bai *et al.* (1997) used a one-dimensional advection–disperison model to assess bacterial transport through a sandy soil in the presence of rhamnolipid (Fig. 7.19). They found that three parameters were especially important: R, the retardation factor, which represents the effect of reversible adsorption on cell transport, and two irreversible sticking rate constants, one for instantaneous sorption and the other for rate-limited sorption. They found that all three constants decreased with increasing rhamnolipid concentration; however, the rate-limited sorption sites were affected the most.

7.6.2 Filtration Models

Filtration models, on the other hand, assume that the contaminant is particulate in nature and that its removal is dependent on physical straining and sorption processes. These processes are often combined into a filtration coefficient for the system. Such models take into account mechanisms by which colloids (i.e., microbes) come in contact with particle surfaces and the relative size of the microbe compared with the pores in the medium. The premise of these models is that as the microbial suspension passes through the terrestrial profile, microbes will be removed. Yao *et al.* (1971) demonstrated that filtration models may be applicable in terms of predicting bacterial immobilization during transport through the subsurface. A general filtration equation is

$$\partial C / \partial x = \lambda C$$

where
C is the concentration of colloid (mass/volume),
x is the distance traveled through the porous medium (length), and
λ is the filter coefficient (1/length).

According to this equation, microbial removal would be exponential with depth, as is often observed. As with the advection–dispersion equation, terms can be incorporated into the equation to account for net microbial decay.

QUESTIONS AND PROBLEMS

1. Compare and contrast the major factors influencing bacterial versus viral transport through the terrestrial profile. Which type of microbe would you expect to find deeper in the profile following surface application?

2. Choose either a bacterium or virus and design a column experiment to assess its transport potential. Your discussion should include items such as column design, type of matrix material, flow conditions, percolating solution, inoculation and sampling approaches, etc. Support your choices.

3. What are UMBs? How can they be used to facilitate bioremediation of contaminated sites?

4. Since both soil particles and microbes generally have a net negative surface charge, why is sorption to matrix material often a factor limiting microbial transport?

5. Why do microbes introduced to a site often die within a few days to weeks? What impact does

this have on the transport potential of the introduced microbes?

6. Discuss the advantages and disadvantages of using a lysimeter to assess microbial transport potential as compared to the use of a column.

References and Recommended Readings

Ackermann, H. W., and Michael, S. D. (1987) "Viruses of Prokaryotes" CRC Press, Boca Raton, FL, pp. 173–201.

Bai, G., Brusseau, M. L., and Miller, R. M. (1997) Influence of a rhamnolipid biosurfactant on the transport of bacteria through a sandy soil. *Appl. Environ. Microbiol.* **63**, 1866–1873.

Bales, R. C., Li, S., Maguire, K. M., Yahya, M. T., and Gerba, C. P. (1993) MS-2 and poliovirus transport in porous media: Hydrophobic effects and chemical perturbations. *Water Resour. Res.* **29**, 957–963.

Bashan, Y. (1986) Alginate beads as synthetic inoculant carriers for slow release of bacteria that affect plant growth. *Appl. Environ. Microbiol.* **51**, 1089–1098.

Bellin, C. A., and Rao, P. S. C. (1993) Impact of bacterial biomass on contaminant sorption and transport in a subsurface soil. *Appl. Environ. Microbiol.* **59**, 1813–1820.

Bitton, B., and Gerba, C. P. (1984) "Groundwater Pollution Microbiology." John Wiley & Sons, New York.

Condee, C. W., and DeFlaun, M. F. (1994) Directed transport of bacteria through aquifer solids using an electric field. Envirogen. Inc., Lawrenceville, NJ. 94th ASM General Meeting.

Corapcioglu, M. Y., and Haridas, A. (1986) Microbial transport in soils and groundwater: A numerical model. *Adv. Water Resour.* **8**, 188–200.

Daane, L. L., Molina, J. A. E., Berry, E. C., and Sadowsky, M. J. (1996) Influence of earthworm activity on gene transfer from *Pseudomonas* fluorescens to indigenous soil bacteria. *Appl. Environ. Microbiol.* **62**, 515–521.

DeFlaun, M. F., and Ensley, B. D. (1993) Development of adhesion-deficient trichloroethylene (TCE) degrading bacteria for *in situ* applications. Envirogen, Inc., Lawrenceville, NJ. 93rd ASM General Meeting.

Dexter, S. C. (1979) Influence of substratum critical surface tension on bacterial adhesion—*in situ* studies. *J. Colloid Interface Sci.* **70**, 346–354.

DiGiovanni, G. D., Neilson, J. W., Pepper, I. L., and Sinclair, N. A. (1996) Gene transfer of *Alcaligenes eutrophus* JMP 134 plasmid pJ4 to indigenous soil recipients. *Appl. Environ. Microbiol.* **62**, 2521–2526.

Ferguson, G. A., and Pepper, I. L. (1987) Ammonium retention in sand amended with clinoptilolite. *Soil. Sci. Soc. Am. J.* **51**, 231–234.

Fetter, C. W. (1993) Mass transport in saturated media. *In* "Contaminant Hydrogeology," Macmillan, New York, pp. 43–114.

Fletcher, M. (1976) The effects of proteins on bacterial attachment to polystyrene. *J. Gen. Microbiol.* **94**, 400–404.

Fletcher, M. (1991) The physiological activity of bacteria attached to solid surfaces. *In* "Advances in Microbial Physiology," Vol. 2 (A. H. Rose and D. W. Tempest, eds.) Academic Press, New York, pp. 54–85.

Gannon, J. T., Manilal, V. B., and Alexander, M. (1991a) Relationship between cell surface properties and transport of bacteria through soil. *Appl. Environ. Microbiol.* **57**, 190–193.

Gannon, J. T., Mingelgrin, U., Alexander, M., and Wagenet, R. J. (1991b) Bacterial transport through homogeneous soil. *Soil Biol. Biochem.* **23**, 1155–1160.

Gerba, C. P. (1984) Applied and theoretical aspects of virus adsorption to surfaces. *Adv. Appl. Microbiol.* **30**, 133–168.

Gerba, C. P., and Bitton, G. (1984) Microbial pollutants: Their survival and transport pattern to groundwater. *In* "Groundwater Pollution Microbiology" (G. Bitton and C. P. Gerba, eds) John Wiley & Sons, New York, pp. 65–88.

Gerba, C. P., Yates, M. V., and Yates, S. R. (1991) Quantitation of factors controlling viral and bacterial transport in the subsurface. *In* "Modeling the Environmental Fate of Microorganisms," (C. J. Hurst, ed.) ASM Press, Washington, DC, pp. 77–88.

Goyal, S. M., and Gerba, C. P. (1979) Comparative adsorption of human enteroviruses, simian rotavirus, and selected bacteriophages to soils. *Appl. Environ. Microbiol.* **38**, 241–247.

Hekman, W. E., Heijnen, C. E., Burgers, S. L. G. E., van Veen, J. A., and van Elsas, J. D. (1995) Transport of bacterial inoculants through intact cores of two different soils as affected by water percolation and the presence of wheat plants. *FEMS Microbiol. Ecol.* **16**, 143–158.

Herman, D. C., and Costerton, J. W. (1993) Starvation–survival of a *p*-nitrophenol–degrading bacterium. *Appl. Environ. Microbiol.* **59**, 340–343.

Herzig, J. P., Leclerc, D. M., and LeGolf, P. (1970) Flow of suspensions through porous media—application to deep filtration. *Ind. Eng. Chem.* **62**, 8–35.

Hinsby, K., McKay, L. D., Jorgensen, P., Lenczewski, M., and Gerba, C. P. (1996) Fracture aperture measurements and migration of solutes, viruses, and immiscible creosote in a column of clay-rich till. *Ground Water* **34**, 1065–1089.

Ho, W. C., and Ko, W. H. (1985) Soil microbiostasis; effects of environmental and edaphic factors. *Soil Biol. Biochem.* **17**, 167–170.

Hornberger, G. M., Mills, A. L., and Herman, J. S. (1992) Bacterial transport in porous media: Evaluation of a model using laboratory observations. *Water Resour. Res.* **28**, 915–923.

Hu, Z., Korus, R. A., Levinson, W. E., and Crawford, R. L. (1994) Adsorption and biodegradation of pentachlorophenol by polyurethane-immobilized *Flavobacterium. Environ. Sci. Technol.* **28**, 491–496.

Jackson, A., Roy, D., and Breitenbeck, G. (1994) Transport of a bacterial suspension through a soil matrix using water and an ionic surfactant. *Water Res.* **28**, 943–949.

Jenneman, G. E., McInerney, M. J., and Knapp, R. M. (1985) Microbial penetration through nutrient-saturated Berea sandstone. *Appl. Environ. Microbiol.* **50**, 383–391.

Jury, W. A., Gardner, W. R., and Gardner, W. H. (1991) Water movement in soil. *In* "Soil Physics," 5th ed. John Wiley & Sons, New York, pp. 73–121.

Kinoshita, T., Bales, R. C., Yahya, M. T., and Gerba, C. P. (1993) Bacteria transport in a porous medium: Retention of *Bacillus* and *Pseudomonas* on silica surfaces. *Water Res.* **27**, 1295–1301.

Lance, J. C., and Gerba, C. P. (1984) Virus movement in soil during saturated and unsaturated flow. *Appl. Environ. Microbiol.* **47**, 335–337.

Lance, J. C., Gerba, C. P., and Wang, D. C. (1982) Comparative movement of different enteroviruses in soil columns. *J. Environ. Qual.* **11**, 347–351.

Landry, E. F., Vaughn, J. M., Thomas, M. Z., and Beckwith, C. A. (1979) Adsorption of enteroviruses to soil cores and their subsequent elution by artificial rainwater. *Appl. Environ. Microbiol.* **38**, 680–687.

Lappin-Scott, H. M., Cusack, F., and Costerton, J. W. (1988) Nutrient resuscitation and growth of starved cells in sandstone cores: A novel approach to enhanced oil recovery. *Appl. Environ. Microbiol.* **54**, 1373–1382.

Lovins, K. W., Angle, J. S., Wiebers, J. L., and Hill, R. L. (1993) Leaching of *Pseudomonas aeruginosa* and transconjugants containing pR68.45 through unsaturated, intact soil columns. *FEMS Microbiol. Ecol.* **13,** 105–112.

Lynch, J. M. (1981) Promotion and inhibition of soils aggregate stabilization by microorganisms. *J. Gen. Microbiol.* **126,** 371–375.

MacLeod, F. A., Lappin-Scott, H. M., and Costerton, J. W. (1988) Plugging of a model rock system by using starved bacteria. *Appl. Environ. Microbiol.* **54,** 1365–1372.

Madsen, E. L., and Alexander, M. (1982) Transport of *Rhizobium* and *Pseudomonas* through soil. *Soil Sci. Soc. Am. J.* **46,** 557–560.

Ogram, A., Sayler, G. S., Gustin, D., and Lewis, R. J. (1988) DNA adsorption to soils and sediments. *Envrion. Sci. Technol.* **22,** 982–984.

Pepper, I. L., Josephson, K. L., Bailey, R. L., Burr, M. D., and Gerba, C. P. (1993) Survival of indicator organisms in Sonoran Desert soil amended with sewage sludge. *J. Environ. Sci. Health* **28,** 1287–1302.

Postma, J., and van Veen, J. A. (1990) Habitable pore space and survival of *Rhizobium leguminosarum* biovar *trifolii* introduced into soil. *Microb. Ecol.* **19,** 149–161.

Postma, J., Hok-A-Hin, C. H., and van Veen, J. A. (1990) Role of microniches in protecting introduced *Rhizobium leguminosarum* biovar *trifolii* against competition and predation in soil. *Appl. Environ. Microbiol.* **56,** 495–502.

Powelson, D. K. , Simpson, J. R., and Gerba, C. P. (1990) Virus transport and survival in saturated and unsaturated flow through soil columns. *J. Environ. Qual.* **19,** 396–401.

Powelson, D. K., Gerba, C. P., and Yahya, M. T. (1993) Virus transport and removal in water during aquifer recharge. *Water Res.* **27,** 583–590.

Reynolds, P. J., Sharma, P., Jenneman, G. E., and McInerney, M. J. (1989) Mechanisms of microbial movement in subsurface materials. *Appl. Environ. Microbiol.* **55,** 2280–2286.

Rosenberg, M., Bayler, E. A., Dellarea, J., and Rosenberg, E. (1982) Role of thin fimbriae in adherence and growth of *Acinetobacter calcoaceticus* RAG-1 on hexadecane. *Appl. Environ. Microbiol.* **44,** 929–937.

Smith, M. S., Thomas, G. W., White, R. E., and Ritonga, D. (1985) Transport of *Escherichia coli* through intact and disturbed soil columns. *J. Environ. Qual.* **14,** 87–91.

Steinberg, S. M., Pignatello, J. J., and Sawhney, B. L. (1987) Persistence of 1,2-dibromomethane in soils: Entrapment in intraparticle micropores. *Environ. Sci. Technol.* **21,** 1201–1208.

Straub, T. M., Pepper, I. L., and Gerba, C. P. (1995) Comparison of PCR and cell culture for detection of enteroviruses in sludge-amended field soils and determination of their transport. *Appl. Environ. Microbiol.* **61,** 2066–2068.

Tan, K. H. (1993) Colloidal chemistry of inorganic soil constituents. *In* "Principles of Soil Chemistry," 2nd ed. Marcel Dekker, Inc., New York, pp. 129–205.

Thorpe, I. S., Prosser, J. I., Glover, L. A., and Killham, K. (1996) The role of the earthworm *Lumbricus terrestris* in the transport of bacterial inocula through soil. *Biol. Fertil. Soils* **23,** 132–139.

Torrella, F., and Morita, R. Y. (1981) Microcultural study of bacterial size changes and microcolony and ultramicrocolony formation by heterotrophic bacteria in sea water. *Appl. Environ. Microbiol.* **41,** 518–527.

Trevors, J. T., Van Elsas, J. D., Van Overbeek, L. S., and Starodub, M. (1990) Transport of a genetically engineered *Pseudomonas fluorescens* strain through a soil microcosm. *Appl. Environ. Microbiol.* **56,** 401–408.

Vandevivere, P., and Baveye, P. (1992) Saturated hydraulic conductivity reduction caused by aerobic bacteria in sand columns. *Soil Sci. Soc. Am. J.* **56,** 1–13.

van Elsas, J. D., Trevors, J. T., Jain, D., Wolters, A. C., Heijnen, C. E., and van Overbeek, L. S. (1992) Survival of, and root colonization by, alginate-encapsulated *Pseudomonas* fluorescens cells following introduction into soil. *Biol. Fertil. Soils* **14,** 14–22.

van Loosdrecht, M. C. M., Lyklema, J., Norde, W., Schraa, G., and Zehnder, A. J. B. (1987) The role of bacterial cell wall hydrophobicity in adhesion. *Appl. Environ. Microbiol.* **53,** 1893–897.

van Loosdrecht, M. C. M., Norde, W., Lyklema, J., and Zehnder, A. J. B. (1990) Hydrophobic and electrostatic parameters in bacterial adhesion. *Aquat. Sci.* **52,** 103–114.

Weiss, T. H., Mills, A. L., Hornberger, G. M., and Herman. J. S. (1995) Effect of bacterial cell shape on transport of bacteria in porous media. *Environ. Sci. Technol.* **29,** 1737–1740.

Wierenga, P. J. (1996) *In* "Pollution Science" (I. L. Pepper, C. P. Gerba, and M. L. Brusseau, eds.) Academic Press, San Diego, pp. 45–62.

Yates, M. V., Gerba, C. P., and Kelley, L. M. (1985) Virus persistence in groundwater. *Appl. Environ. Microbiol.* **49,** 778–781.

Yates, M. V., and Yates, S. R. (1991) Modeling microbial transport in the subsurface: A mathematical discussion. *In* "Modeling the Environmental Fate of Microorganisms," (C. J. Hurst, ed.) ASM Press, Washington, DC, pp. 48–76.

Yao, K. M., Habibian, M. T., and O'Melia, C. R. (1971) Water and waste water filtration: Concepts and applications. *Environ. Sci. Technol.* **11,** 1105–1112.

8

Environmental Sample Collection and Processing

IAN L. PEPPER   CHARLES P. GERBA   RAINA M. MAIER

8.1 SOILS AND SEDIMENTS

Soils are discontinuous, heterogeneous environments that contain large numbers of diverse organisms. As described in Chapter 4, soil microbial populations vary with depth and soil type, with surface soil horizons generally having more organisms than subsurface horizons. Populations not only vary with depth and soil type, but also vary from site to site and even within sites because of natural microsite variations that can allow very different microorganisms to coexist side by side. Because of the great variability in populations, it is often necessary to take more than one sample to obtain a representative microbial analysis of a site. Therefore the overall sampling strategy will depend on many factors, including the goal of the analyses, the resources available, and the site characteristics. The most accurate approach is to take many samples within a given site and perform a separate analysis of each sample. However, in many instances time and effort can be conserved by combining the samples taken to form a **composite** sample that is used to limit the number of analyses that must be performed. Another approach often used is to sample a site sequentially over time from a small defined location to determine temporal effects on microbes. Because so many choices are available, it is important to delineate a sampling strategy to ensure that quality assurance is addressed. This can be done by developing a **quality assurance project plan (QAPP).** As shown in Table 8.1, a QAPP involves delineating the details of the sampling strategy, the sampling methods, and the subsequent storage of all samples.

8.1.1 Sampling Strategies and Methods for Surface Soils

Bulk soil samples are easily obtained with a shovel or, better yet, a soil auger (Fig. 8.1). Soil augers are more precise than simple shovels because they ensure that samples are taken to exactly the same depth on each occasion. This is important, as several soil factors can vary considerably with depth, such as oxygen, moisture and organic carbon content and soil temperature. A simple hand auger is useful for taking shallow soil samples from areas that are unsaturated. Given the right conditions, a hand-auger can be used to depths of 6 feet, although some soils are simply too compacted or contain too many rocks to allow sampling to this depth. In taking samples for microbial analysis, consideration should be given to contamination that can occur as the auger is pushed into the soil. In this case, microbes that stick to the sides of the auger as it is inserted into the soil and pushed downward may contaminate the bottom part of the core. To minimize such contamination, one can use a sterile spatula to scrape away the outer layer of the core and use the inner part of the core for analysis. Contamination can also occur between samples, but this can be avoided by cleaning the auger after each sample is taken. The cleaning procedure involves washing the auger with water, then rinsing it with 75% ethanol or 10% bleach, and giving it a final rinse with sterile water.

Because soils are heterogeneous and an auger has a limited diameter, the sample collected may not be truly representative of the site. Therefore, several sam-

TABLE 8.1   Collection and Storage Specifications for a Quality Assurance Project Plan (QAPP)[a]

Sampling strategies: Number and type of samples, locations, depths, times, intervals

Sampling methods: Specific techniques and equipment to be used

Sample storage: Types of containers, preservation methods, maximum holding times

[a] The QAPP normally also includes details of the proposed microbial analysis to be conducted on the soil samples.

ples may be collected and formed into a composite sample to reduce the total number of samples and associated costs of the analyses that are performed. Composite samples are obtained by collecting equal amounts of soil from samples taken over a wide area and placing them in a bucket or plastic bag. The whole soil mass is then mixed and becomes the composite sample. To reduce the volume of samples to be stored, a portion of the composite sample can be removed, and this becomes the sample for analysis. In all cases, samples should be stored on ice until processed and analyzed .

In some instances, a series of experimental plots or fields need to be sampled to test the effect of a soil amendment, such as fertilizer, pesticide, or sewage sludge, on microbial populations. In this case, a soil sample must be taken from each of several plots or fields to compare control nontreated plots with plots that have received an amendment. For example, a researcher might be interested in the influence of inorganic nitrogen fertilizers on soil nitrifying populations. The investigator would then sample an unamended plot (the control) for comparison with a plot that had been treated with inorganic fertilizer. Another example would be the case in which soil amended with sewage sludge is sampled for subsequent viral pathogen analysis. In either example, multiple samples or replicates always give a more refined

estimate of the parameters of interest. However, field-work can be costly and the number of samples taken must be weighed against the cost of analysis and the funds available. In the examples given, two-dimensional sampling plans can be used to determine the number and location of samples taken. In two-dimensional sampling, each plot is assigned spatial coordinates and set sampling points are chosen according to an established plan. Some typical two-dimensional sampling patterns, including random, transect, two-stage, and grid sampling, are illustrated in Fig. 8.2.

Random sampling involves choosing random points within the plot of interest, which are then sampled to a defined depth. **Transect sampling** involves collection of samples in a single direction. For example, transect sampling might be useful in a riparian area, where transects could be chosen adjacent to a streambed and at right angles to the streambed. In this way the influence of the stream on microbial populations could be evaluated. In **two-stage sampling** an area is broken into regular subunits called primary units. Within each primary unit, subsamples can be taken randomly or systematically. This approach might be useful, for example, when a site consists of a

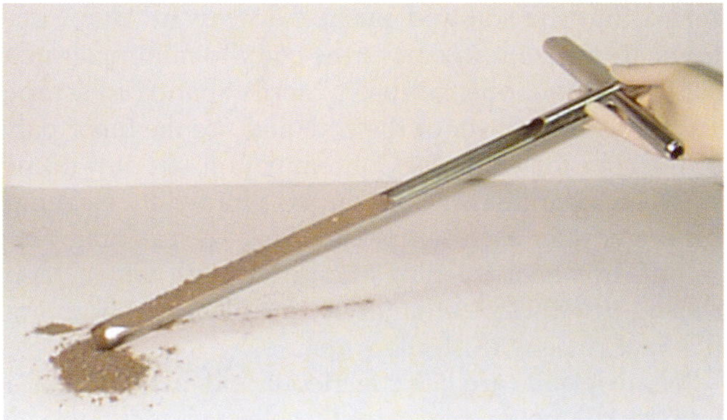

FIGURE 8.1   Hand auger. (Photo courtesy R. M. Maier.)

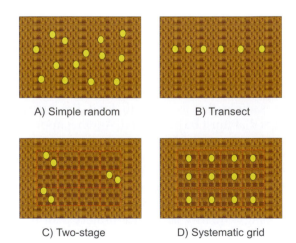

A) Simple random   B) Transect

C) Two-stage   D) Systematic grid

FIGURE 8.2   Alternative spatial sampling patterns. (From Pollution Science, © (1996) Academic Press, San Diego, CA.)

hillside slope and a level plain and there is likely to be variability between the primary units. The final example of a sampling pattern is **grid sampling,** in which samples are taken systematically at regular intervals at a fixed spacing. This type of sampling is useful for mapping an area when little is known about the variability within the soil.

Two-dimensional sampling does not give any information about changes in microbial populations with depth. Therefore, three-dimensional sampling is used when information concerning depth is required. Such depth information is critical when evaluating sites that have been contaminated by improper disposal or spills of contaminants. Three-dimensional sampling can be as simple as taking samples at 50-cm depth increments to a depth of 200 cm, or can involve drilling several hundred meters into the subsurface vadose zone. For subsurface sampling, specialized equipment is needed and it is essential to ensure that subsurface samples are not contaminated by surface soil.

Finally, note that there is a specialized zone of soil that is under the influence of plant roots. This is known as the **rhizosphere** and is of special interest to soil microbiologists and plant pathologists because of enhanced microbial activity and specific plant–microbe interactions (see Chapter 18.1.2). Rhizosphere soil exists as a continuum from the root surface (the **rhizoplane**) to a point where the root has no influence on microbial properties (generally 2–10 mm). Thus, rhizosphere soil volumes are variable and are difficult to sample. Normally, roots are carefully excavated and shaken gently to remove bulk or nonrhizosphere soil. Soil adhering to the plant roots is then considered to be rhizosphere soil. Although this is a crude sampling mechanism, it remains intact to this day. As a result, the sampling of rhizosphere remains a major experimental limitation, regardless of the sophistication of the microbial analyses that are subsequently performed.

8.1.2 Sampling Strategies and Methods for the Subsurface

Mechanical approaches using drill rigs are necessary for sampling the subsurface environment. This significantly increases the cost of sampling, especially for the deep subsurface. As a result, few cores have been taken in the deep subsurface and these coring efforts have involved large teams of researchers (see Chapter 4, Case Study). The approach used for sampling either deep or shallow subsurface environments depends on whether the subsurface is saturated or unsaturated. For unsaturated systems, air rotary drilling

can be used to obtain samples from depths up to several hundred meters (Chapelle, 1992). In air rotary drilling, a large compressor is used to force air down a drill pipe, out the drill bit, and up outside the borehole (Fig. 8.3). As the core barrel cuts downward, the air serves to blow the borehole cuttings out of the hole and also to cool the core barrel. This is important, because if the core barrel overheats, microbes within the sample may be effectively sterilized, posing difficulty for subsequent microbial analysis. In normal air drilling, small amounts of water containing a surfactant are injected into the airstream to control dust and help cool the drill bit. However, this increases the possibility of contamination, so cores such as the one drilled in Idaho's Snake River Plain, have been drilled with air alone (Colwell, 1989). To help keep the core barrel cool, the coring was simply done very slowly to avoid overheating. To help maintain sterile conditions and prevent contamination from surface air, all air used in the coring process was prefiltered through a 0.3-μm high-efficiency particulate air (HEPA) filter (see Chapter 5.12.2). Immediately after the core was collected the surface layer was scraped away with a sterile spatula, and then a subcore was taken using a 60-ml sterile plastic syringe with the end removed. The samples were immediately frozen and shipped to

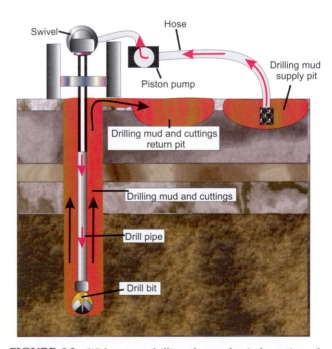

FIGURE 8.3 With rotary drilling the mechanical rotation of a drilling tool is used to create a borehole. Either air (air rotary drilling) or a fluid often called a drilling mud (mud rotary drilling) is forced down the drill stem to displace the borehole cuttings to the outside of the drill and upward to the surface. This figure illustrates mud rotary drilling.

a laboratory, where microbial analyses were initiated within 18 hours of collection.

Saturated subsurface environments are sampled somewhat differently because the sediments are much less cohesive than those found in unsaturated regions. Therefore, the borehole must be held open so that an intact core can be taken and removed at each desired depth. For sampling of depths down to 100 feet, hollow-stem auger drilling with push-tube sampling is widely used (Fig. 8.4). The auger consists of a hollow tube with a rotating bit at the tip that drills the hole. The outside of the hollow auger casing is reverse threaded so that the cuttings are pushed upward and out of the hole as drilling proceeds. As the borehole is drilled, the casing of the auger is left in place to keep the borehole open. Thus, the casing acts as a sleeve into which a second tube, the core barrel, is inserted to collect the sample when the desired depth has been reached. The core barrel is basically a sterile tube that is placed at the tip of the hollow-stem auger, driven down to collect the sediment sample, and then re-trieved. Drilling can then continue to the next desired depth and the coring process repeated. Each core collected is capped, frozen, and sent to a laboratory for study. To avoid contamination of samples, the outside of the core is scraped away or the core may be sub-cored.

For cores that are deeper than 100 feet, mud rotary coring is used (Chapelle, 1992). In this case, the hole is again bored using a rotating bit. However, drilling fluids are used to remove the borehole cuttings and to apply pressure to the walls of the borehole to keep it from collapsing. Mud rotary drilling has been used to obtain sediment samples to 3000 feet beneath the soil surface. An example of such a core is one taken from the deep subsurface sediments of the Southeast Coastal Plain in South Carolina (see Chapter 4.8.2). During this coring, samples were retrieved from depths ranging from 1198 to 1532 feet. In order to ensure the integrity of the cores obtained, the drilling fluids were spiked with two tracers, potassium bromide and rhodamine dye. The use of these two tracers al-

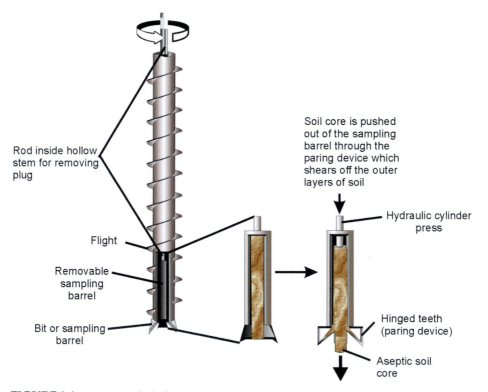

Rod inside hollow stem for removing plug

Soil core is pushed out of the sampling barrel through the paring device which shears off the outer layers of soil

Hydraulic cylinder press

Flight

Removable sampling barrel

Bit or sampling barrel

Hinged teeth (paring device)

Aseptic soil core

FIGURE 8.4 Diagram of a hollow-stem auger. Note the reverse threading on the outside of the auger. This is used to displace the borehole cuttings upward to the surface. This type of auger was used at Purdue University to collect core samples to a depth of 26 m for microbial and soil analysis as described by Konopka and Turco (1991) (see Fig. 4.19). A subcore of each core collected is taken using a split spoon sampler or a push tube. In either case, the outside of the core must be regarded as contaminated. Therefore, the outside of the core is shaved off with a sterile spatula or a subcore can be taken using a sterile plastic syringe. Alternatively, as shown in this figure, intact cores are automatically pared to remove the outer contaminated material, leaving an inner sterile core.

lowed researchers to evaluate how far the drilling fluids had penetrated into the cores. Any areas of the cores that are contaminated with tracer must be discarded. The cores were retrieved in plastic liners, frozen, and sent for immediate analysis.

It is important to emphasize that coring either saturated or unsaturated environments is a difficult process for several reasons. First, it may take years to plan and obtain funding to proceed with cores such as those described here for the Snake River Plain and the Southeast Coastal Plain. The actual drilling and recovery of samples is an engineering problem whose sophistication has only been touched upon in this section. Also keep in mind that the cores obtained are not always truly representative of the sediments from which they are taken. For example, a 4-foot core may be compressed considerably in the coring process so that it is difficult to identify exactly the depth from which it was taken. A second difficulty in obtaining representative samples is due to horizontal heterogeneities in the subsurface material. Such heterogeneities can mean that two samples taken a few feet apart may have very different physical, chemical, and microbiological characteristics. Finally, for microbial analysis it is not enough merely to retrieve the sample; the sample must be uncontaminated and the logistics of sample storage and analysis must be considered as well.

8.1.3 Sample Processing and Storage

Microbial analyses should be performed as soon as possible after collection of a soil to minimize the effects of storage on microbial populations. Once removed from the field, microbial populations within a sample can and will change regardless of the method of storage. Reductions in microbial numbers and microbial activity have been reported even when soil samples were stored in a field moist condition at 4°C for only 3 months (Stotzky et al., 1962). Interestingly in this study, although bacterial populations changed, actinomycete populations remained unchanged.

The first step in microbial analysis of a surface soil sample usually involves sieving through a 2-mm mesh to remove large stones and debris. However, to do this samples must often be air dried to facilitate the sieving. This is acceptable as long as the soil moisture content does not become too low, because this can also reduce microbial populations (Sparkling and Cheshire, 1979). Following sieving, short-term storage should be at 4°C prior to analysis. If samples are stored, care should be taken to ensure that samples do not dry out and that anaerobic conditions do not develop, because this too can alter microbial populations. Storage up to

21 days appears to leave most soil microbial properties unchanged (Wollum, 1994), but again time is of the essence with respect to microbial analysis. Note that routine sampling of surface soils does not require sterile procedure. These soils are continually exposed to the atmosphere, so it is assumed that such exposure during sampling and processing will not affect the results significantly.

More care is taken with processing subsurface samples for three reasons. First, they have lower cultural counts, which means that an outside microbial contaminant may significantly affect the numbers counted. Second, subsurface sediments are not routinely exposed to the atmosphere and microbial contaminants in the atmosphere might substantially contribute to microbial types found. Third, it is more expensive to obtain subsurface samples and often there is no second chance at collection. Subsurface samples obtained by coring are either immediately frozen and sent back to the laboratory as an intact core or processed at the coring site. In either case, the outside of the core is normally scraped off using a sterile spatula or a subcore is taken using a smaller diameter plastic syringe. The sample is then placed in a sterile plastic bag and analyzed immediately or frozen for future analysis.

8.1.3.1 Processing Soil and Sediment Samples for Bacteria

Traditional methods of analyzing microbial populations have usually involved either cultural assays utilizing dilution and plating methodology on selective and differential media or direct count assays (see Chapter 10). Direct counts offer information about the total number of bacteria present but give no information about the number or diversity of populations present. Plate counts allow enumeration of total cultural or selected cultural populations and hence provide information on the different populations present. However, data indicate that often less than 1% of soil bacteria are culturable (Amann et al., 1995), so cultural information offers only a piece of the picture. In addition, the type of medium chosen for cultural counts will select for the populations that grow best on that particular medium. Thus, the choice of medium is crucial in determining the results obtained. This is illustrated by the data in Table 8.2, which show that whereas direct counts from a series of sediment samples spanning a 4.9-m depth were similar, the culturable counts varied depending on the type of medium used. A nutritionally rich medium, PTYG, made from peptone, trypticase, yeast extract, and glucose, consistently gave counts that were one to three orders of

TABLE 8.2   Total and Viable Cell Counts of Bacteria in Sediment Samples

Depth (m)	Saturation status	AODC[a] cells/g dry weight	Culturable counts (CFU/g dry weight)		
			PTYG[b]	Dilute PTYG[c]	SSA[d]
1.2	Unsaturated	$6.8 \pm 4.9 \times 10^6$	$3.4 \pm 0.9 \times 10^4$	$1.9 \pm 0.4 \times 10^5$	$1.3 \pm 0.2 \times 10^5$
3.1	Interface[e]	$3.4 \pm 2.6 \times 10^6$	$2.0 \pm 0.5 \times 10^4$	$2.6 \pm 0.2 \times 10^6$	$2.9 \pm 0.6 \times 10^6$
4.9	Saturated	$6.8 \pm 4.3 \times 10^6$	$2.6 \pm 0.7 \times 10^3$	$3.5 \pm 0.1 \times 10^6$	$4.1 \pm 0.2 \times 10^6$

Adapted from Balkwill and Ghiorse (1985).

[a] AODC, acridine orange direct counts.

[b] PTYG, a nutritionally rich medium composed of peptone, trypticase, yeast extract, and glucose.

[c] Dilute PTYG, a 1:20 dilution of PTYG medium.

[d] SSA, soil extract agar. This medium was made by autoclaving a 1:2 suspension of surface soil in distilled water and then centrifuging and filtering the extract to clarify it.

[e] This sample was taken at the interface between the unsaturated and the saturated zone.

magnitude lower than counts from two different low-nutrient media that were tested. These were a 1:20 dilution of PTYG and a soil extract agar made from a 1:2 suspension of surface soil. These data reflect the fact that most soil microbes exist under nutrient-limited conditions.

A powerful technique that has been developed in recent years involves obtaining and studying total DNA extracted from bacteria within a soil sample (see Chapter 13). This approach allows information to be obtained about the types of populations present and their genetic potential. As with any technique, there are limitations to the data that can be obtained with DNA extraction. Therefore, many researchers now use DNA extraction in conjunction with direct and cultural counts to maximize the data obtained from an environmental sample. There are two approaches to the isolation of bacterial DNA from soil samples (Fig. 8.5). The first is based on fractionation of bacteria from soil followed by cell lysis and DNA extraction. The second method involves *in situ* lysis of bacteria within the soil matrix with subsequent extraction of the DNA released from cells. Both approaches are discussed further in the following sections.

Extraction of Cells from Soil and Subsequent DNA Recovery

In this approach, whole cells are removed from the sample and then lysed to recover DNA. An in-depth protocol for extraction of bacteria from soil is provided by Holben (1994). The first step is homogenization of the soil or sediment sample to break up aggregates and dislodge both fungal and bacterial cells. To homogenize a sample, normally 10 g of soil is placed in a buffer containing polyvinylpolypyrrolidone (PVPP). PVPP complexes with humic acids and aids in their removal, an important step because humic materials of-

ten interfere with DNA analysis. The soil or sediment particles as well as fungal cells are removed by a low-speed centrifugation step, followed by high-speed centrifugation to recover the bacterial cells. This process is referred to as **differential centrifugation.** Often multiple rounds of homogenization and differential centrifugation are necessary to ensure that maximum numbers of bacteria are recovered.

After cells are recovered, they are lysed by addition of the enzyme lysozyme. The cellular debris is removed by centrifugation and the free DNA extracted from the supernatant and purified. The yield of DNA obtained varies with any particular sample but is often in the range of 1–5 μg per gram of surface soil. To put this in perspective, the protocol will recover about 33% of the total bacterial population existing in a sandy loam soil (Holben *et al.*, 1988). Soils high in clay will not yield such high recoveries because tightly sorbed bacterial cells are removed with soil particles during the low-speed centrifugation step. Sorption of cells also affects the specific populations of bacteria that are recovered. Rapidly metabolizing bacteria that produce new cells via binary fission are less tightly sorbed and preferentially recovered. In contrast, older cells that are not actively growing tend to be sorbed more tightly and are lost (Holben, 1994). Thus, this process selects for bacteria on the basis of their stickiness or degree of sorption to clay colloids, and not all cell types are extracted with the same efficiency. Finally, note that this process is highly labor intensive (6–8 hours is required for six samples) and does not lend itself to studies in which many samples need to be analyzed.

The procedure does, however, have some advantages over the *in situ* lysis method (Table 8.3). First, only bacterial DNA is ultimately recovered because the bacteria are separated from fungal cells prior to cell

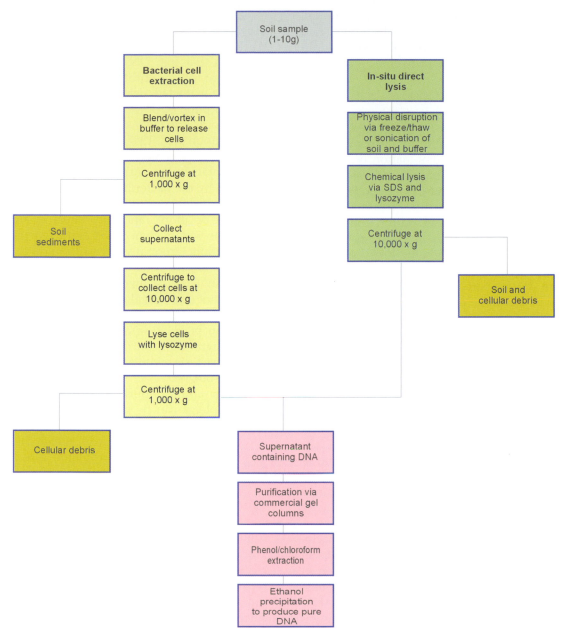

FIGURE 8.5   Methods for obtaining community DNA from soil.

TABLE 8.3   **Comparison of Bacterial Fractionation and *In Situ* Lysis Methodologies for the Recovery of DNA from Soil**

Issue	Bacterial fractionation	*In situ* lysis
Yield of DNA	1–5 μg/g	1–20 μg/g
Representative of community	Less representative because of cell sorption	More representative, unaffected by cell sorption
Source of DNA recovered	Only bacteria	Mostly bacteria
Degree of DNA shearing	Less shearing	More shearing
Average size of DNA fragments	50 kb	25 kb
Degree of humic contamination	Less contaminated	More contaminated
Ease of methodology	Slow, laborious	Faster, less labor intensive

lysis. The method also yields viable cells from the soil bacterial community that could be subsampled and analyzed in a variety of ways. Note, however, that intact nonviable cells will also be recovered.

In Situ Lysis and DNA Recovery

In this approach, the microbes of interest must be lysed within the soil and their DNA released prior to extraction of DNA from the sample. Lysis methodology has usually involved a combination of physical and chemical treatments. For bacteria, physical treatments have involved freeze–thaw cycles and/or sonication or bead beating, and chemical treatments have often utilized a detergent such as sodium dodecyl sulfate (SDS) or an enzyme such as lysozyme or proteinase (Moŕe *et al.*, 1994). Following lysis, cell debris and soil particles are removed by centrifugation, and the DNA in the supernatant is precipitated with ethanol. The DNA can be purified by passage through homemade or commercial columns packed with ion-exchange resins or gels. These steps are crucial for removal of humic materials, which would inhibit DNA analysis. Further purification can be achieved with phenol–chloroform/isoamyl alcohol extractions, followed once more by ethanol precipitations (Xia *et al.*, 1995). Pure samples of DNA are necessary to allow subsequent molecular analyses such as with the polymerase chain reaction (PCR) (see Chapter 13.3.2). However, regardless of what purification methodology is employed, each step in the purification process causes loss of DNA. Thus, purified DNA is obtained only at the expense of DNA yield.

A major advantage of the direct lysis approach is that it is less labor intensive and faster than the bacterial fractionation method. It also generally results in higher yields of DNA (1–20 μg of DNA per gram of soil). Finally, this approach seems to represent the bacterial community better than the fractionation method because it is not affected by sorption of cells to clay colloids. As a result, direct lysis is the more commonly used approach in DNA recovery. Although direct lysis has many advantages, it also has some problems. One problem associated with DNA extraction is distinguishing free from cellular DNA. Free DNA released from microbes that lysed naturally some time before

the DNA extraction can sometimes be protected from degradation by sorption to soil particles (Lorentz and Wackernagel, 1987). However, the available evidence indicates that free DNA (extracellular) is either degraded rapidly or sorbed to clay colloids and is not normally extracted using these procedures (Holben, 1994). Sorption of DNA from lysed cells by clay or humic colloids can also reduce the yield of extracted DNA (Ogram *et al.*, 1987). In fact, a major disadvantage of the direct lysis method is that the DNA is generally more contaminated with humic material than is DNA obtained by the bacterial fractionation method. A further disadvantage is that DNA isolated by direct lysis tends to be more randomly sheared and of lower molecular weight than DNA extracted following cell fractionation. Originally, it was thought that DNA obtained by *in situ* lysis might also contain fungal and protozoan DNA. However, the chemicals normally used do not lyse fungal cells, and protozoa are present in soils at much lower densities than bacteria. Thus a protozoan population of 10^4 per gram is unlikely to make a significant contribution to DNA obtained from 10^8–10^9 bacteria, even allowing for the larger genome size of protozoa.

Once a sample of purified DNA is obtained from the soil sample, it can be quantified by ultraviolet (UV) spectroscopy. Normally, readings are made at wavelengths of 260 and 280 nm from which the purity and quantity of DNA can be estimated (see Information Box).

Once the amount of DNA per mass of soil is known, estimates can be made of the microbial population. For bacteria such as *Escherichia coli*, a typical chromosome contains 4–5 million base pairs, equivalent to about 9 fg (9×10^{-15} g) of DNA. However, the amount of DNA per cell varies and other estimates are lower, approximately 4 fg per cell. The amount of DNA per cell can also vary because of chromosome replication occurring faster than cell division, resulting in two or three chromosomes per cell (Krawiec and Riley, 1990). These theoretical DNA estimates can be used to relate total extracted DNA to the number of microbes within a sample. Table 8.4 shows the total DNA extracted from four soils amended with glucose. The amounts of DNA obtained increased with the amount of soil or-

---

Information Box

The amount of DNA is estimated from the 260 nm reading. An absorbance reading of 1.0 is equivalent to 50 μg of DNA per ml of solution.

The purity of DNA is estimated from the ratio of the reading at 260 nm to that at 280 nm. A value >1.7 indicates relatively pure DNA. The maximum theoretical value is 2.0.

From E. M. Jutras, University of Arizona, unpublished data.
[a] Amendments were added at time zero and DNA was obtained via direct lysis.

TABLE 8.4 Total Community DNA Extracted from Four Soils Amended with 1% Glucose and 0.1% Potassium Nitrate[a]

Soil	\multicolumn{5}{c}{Extracted DNA (μg/g soil) with time (days)}
	0	2	4	6	8
Clay loam	0.12	0.04	0.21	1.3	0.52
Silt loam	17.80	17.6	16.6	18.4	19.9
Sandy loam	0.63	0.60	1.90	5.50	1.90
Loam	1.30	0.90	1.30	4.20	7.70

ganic matter (silt loam and loam), presumably because of higher sustainable bacterial populations. Extracted DNA also decreased in soils high in clay (clay loam), most likely because of sorption of DNA by soil colloids (Ogram *et al.*, 1987). Overall, the influence of the amendments can be seen over time as microbial populations increase through growth, resulting in more extractable DNA. The theoretical number of bacterial cells that the extracted DNA represents can be calculated.

For example, at time zero for the clay loam soil:

Extracted DNA = 0.12 μg/g soil
Therefore, if each cell has 4 fg of DNA,
Number of cells = $\dfrac{0.12 \times 10^{-6} \text{ g DNA/g soil}}{4 \times 10^{-15} \text{ g DNA/cell}}$
= 3.0×10^7 cells/g soil

Similar values at time zero for the other soils are 1.6×10^8 cells/g soil (sandy loam), 3.3×10^8 cells/g soil (loam), and 4.5×10^9 cells/g soil (silt loam).

Overall, the processing of soil to obtain bacterial DNA represents an alternative method for subsequent analysis of bacterial communities. Comparable approaches to obtain fungal community DNA have not been achieved. This may be due in part to the difficulty of lysing fungal hyphae without also lysing bacteria.

8.1.3.2 Processing Soil Samples for Fungal Hyphae and Spores

As with bacteria, it is also impossible to culture all species of viable fungi from a soil or sediment sample. Cultural methods for fungi are described in Chapter 10.3. However, several approaches have been developed for direct isolation of fungal hyphae or spores from soil. The first is a soil washing methodology. This involves saturating small volumes of soil with sterile water. Aggregates of soil are gently teased open with a fine jet of water, allowing heavier soil particles to sediment and finer particles to be decanted off. The procedure is repeated several times until only the heavier particles remain. These are then spread in a film of sterile water and examined under a dissecting microscope. Sterile needles or very fine forceps can then be used to obtain any observed fungal hyphae.

For spores, a different approach can be used. Soil samples are placed in separate sterile boxes each of which contains a number of sieves of graded size. The soil samples are washed vigorously in each box, and soil of defined size is retained on each sieve. Spores are determined empirically by plating successive washings (2 minutes per wash) from each sieve. Because hyphae are retained by the sieves, any fungal colonies that arise must be due to the presence of spores. Information on fungi present as hyphae can be obtained by plating the washed particles retained by the sieves. However, these washing methods are labor intensive and rely on trial and error in terms of which size fractions are most relevant with respect to individual spore size.

8.1.3.3 Processing Sludge, Soil, and Sediment for Viruses

To assess fully and understand the risks from pathogens in the environment, it is necessary to determine their occurrence in sewage sludges (biosolids), soils to which wastewater or sludge is applied, or marine sediments that may be affected by sewage outfalls or sludge disposal. In fact, the U.S. Environmental Protection Agency currently requires monitoring of sludge for enteroviruses for certain types of land application. To detect viruses on solids it is first necessary to extract them by processes that will cause their desorption from the solid. As with microporous filters (see Chapter 8.2.2.1), viruses are believed to be bound to these solids by a combination of electrostatic and hydrophobic forces (see Chapter 7.2.3). To recover viruses from solids, substances are added that will break down these attractive forces, allowing the virus to be recovered in the eluting fluid (Berg, 1987; Gerba and Goyal, 1982).

The most common procedure for sludges involves collecting 500–1000 ml of sludge and adding $AlCl_3$ and HCl to adjust the pH to 3.5. Under these conditions, the viruses bind to the sludge solids, which are removed by centrifugation and then resuspended in a beef extract solution at neutral pH to elute the virus. The eluate is then reconcentrated by flocculation of the proteins in the beef extract at pH 3.5, resuspended in 20–50 ml, and neutralized. A major problem with sewage sludge concentrates prepared in this manner is that they often contain substances toxic to cell culture. A diagram of the details of this procedure is shown in Fig. 8.6. Similar extraction techniques are used for the

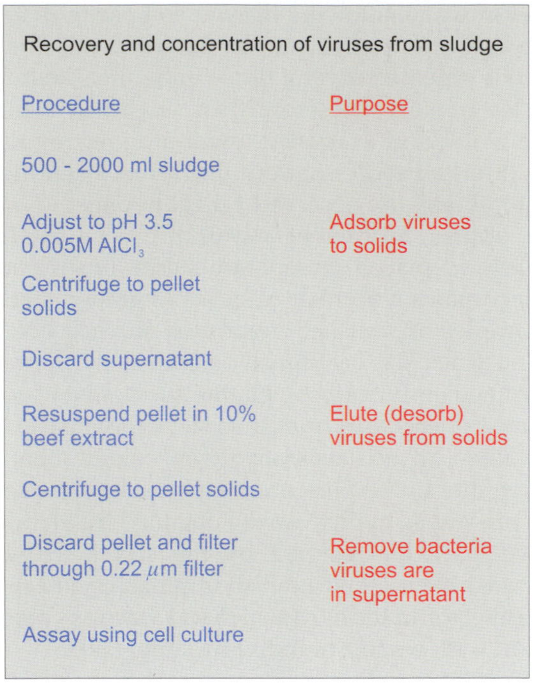

Recovery and concentration of viruses from sludge

Procedure	Purpose
500 - 2000 ml sludge	
Adjust to pH 3.5 0.005M AlCl₃	Adsorb viruses to solids
Centrifuge to pellet solids	
Discard supernatant	
Resuspend pellet in 10% beef extract	Elute (desorb) viruses from solids
Centrifuge to pellet solids	
Discard pellet and filter through 0.22 μm filter	Remove bacteria viruses are in supernatant
Assay using cell culture	

FIGURE 8.6 Procedure for recovery and concentration of viruses from sludge.

recovery of viruses from soils and aquatic sediments (Hurst *et al.,* 1997).

8.2 WATER

8.2.1 Sampling Strategies and Methods for Water

Sampling environmental waters for subsequent microbial analysis is somewhat easier than sampling soils for a variety of reasons. First, because water tends to be more homogeneous than soils, there is less site-to-site variability between two samples collected within the same vicinity. Second, it is often physically easier to collect water samples because it can be done with pumps and hose lines. Thus, known volumes of water can be collected from known depths with relative ease. Amounts of water collected depend on the environmental sample being evaluated but can vary from 1 ml to 1000 liters. Sampling strategy is also less complicated for water samples. In many cases, because water is mobile, a set number of bulk samples are simply collected from the same point over various time intervals. Such a strategy would be useful, for example, in sampling a river or a drinking water treatment plant. For marine waters, samples are often collected sequentially in time within the defined area of interest.

Although the collection of the water sample is relatively easy, processing the sample prior to microbial analysis can be more difficult. The volume of the water sample required for detection of microbes can sometimes become unwieldy because numbers of microbes tend to be lower in water samples than in soil samples (see Chapter 6). Therefore, strategies have been developed to allow concentration of the microbes within a water sample. For larger microbes, including bacteria and protozoan parasites, samples are often filtered to trap and concentrate the organisms. For bacteria this often involves filtration using a 0.45-μm membrane filter (Chapter 10.2.1.3). For protozoan parasites coarse woven fibrous filters are used. For viruses water samples are also filtered, but because viral particles are often too small to be physically trapped, collection of the viral particles depends on a combination of electrostatic and hydrophobic interactions of the virus with the filter. The different requirements for processing of water samples for analysis of viruses, bacteria, and protozoa are outlined in the next two sections.

8.2.2 Processing of Water Samples for Virus Analysis

The detection and analysis of viruses in water samples is often difficult because of the low numbers encountered and the different types that may be present. There are four basic steps in virus analysis: sample collection, elution, reconcentration, and virus detection. For **sample collection,** it is often necessary to pass large volumes of water (100 to 1000 liters) through a filter because of the low numbers of viruses present. The viruses are concentrated from the water by adsorption onto the filter. Recovery of virus from the filter involves **elution** of the virus from the collection filter as well as a **reconcentration** step to reduce the sample volume before assay. Viruses can be detected using cell culture or molecular methods such as PCR. However, both methods can be inhibited by the presence of toxic substances in the water that are concentrated along with the viral particles. Many strategies have been developed to overcome the difficulties associated with analysis of viruses, but they are often time consuming, labor intensive, and costly. For example, the cost of enterovirus detection ranges from $600 to $1000 per sample for drinking water. Another problem with analysis of viruses is that the precision and accuracy of the methods used suffer from the large number of steps involved. In particular, the efficiency of viral recovery associated with each step is dependent on the type of virus that is being analyzed. For example, hepatitis A virus may not be concentrated as efficiently as rotavirus by the same process. Variability also results from the extreme sensitivity of these assays. Methods for the detection of viruses in water have been developed that can detect as little as one plaque-forming unit in 1000 liters of water. On a weight-to-weight basis with water, this is a sensitivity of detection of one part in 10^{18}. For comparison, the limit of sensitivity of most analytical methods available for organic compounds is about 1 μg/liter. This corresponds to one part in 10^9.

8.2.2.1 Sample Collection

Virus analysis is performed on a wide variety of water types. Types of water tested include potable water, ground and fresh surface waters, marine waters, and sewage. These waters vary greatly in their physical–chemical composition and contain substances either dissolved or suspended in solution, which may interfere with our ability to employ various concentration methods. The suitability of a virus concentration method depends on the probable virus density, the volume limitations of the concentration method for the type of water, and the presence of interfering substances. A sample volume of less than 1 liter may suffice for recovery of viruses from raw and primary sewage. For drinking water and relatively nonpolluted waters, the virus levels are likely to be so low that hundreds or perhaps thousands of liters must be sampled to increase the probability of virus detection. Various methods employed for virus concentration from water are shown in Table 8.5.

Most methods used for virus concentration depend on adsorption of the virus to a surface, such as a filter or mineral precipitate, although hydroextraction and ultrafiltration have been employed (Gerba, 1987). Field systems for virus concentration usually consist of a pump for passing the water through the filter (at rates of 5 to 10 gallons per minute), a filter housing, and a flowmeter (Fig. 8.7A). The entire system can usually be contained in a 20-liter capacity ice chest.

The class of filters most commonly used for virus collection from large volumes of water is adsorption-elution microporous filters, more commonly known as **VIRADEL** (for **vir**us **ad**sorption–**el**ution) (APHA, 1995). VIRADEL involves passing the water through a filter to which the viruses adsorb. The pore size of the filters is much larger than the viruses, and adsorption takes place by a combination of electrostatic and hydrophobic interactions (Gerba, 1984). Two general types of filters are available: **electronegative** (negative surface charge) and **electropositive** (positive surface charge). Electronegative filters are composed of either cellulose esters or fiberglass with organic resin

binders. Because the filters are negatively charged, cationic salts ($MgCl_2$ or $AlCl_3$) must be added in addition to lowering the pH to 3.5. This reduces the net negative charge usually associated with viruses allowing adsorption to be maximized (see Chapter 2.1.4.1). Such pH adjustment can be cumbersome, as it requires modifying the water prior to filtering and use of additional materials and equipment such as pH meters. The most commonly used electronegative filter is the Filterite. Generally it is used as a 10-inch (25.4-cm) pleated cartridge with either a 0.22- or 0.45-μm nominal pore size rating.

Electropositive filters are ideal when concentrating viruses from seawater and waters with high amounts of organic matter and turbidity (Gerba *et al.*, 1978). Electropositive filters may be composed of fiberglass or cellulose containing a positively charged organic polymeric resin, which creates a net positive surface charge to enhance adsorption of the negatively charged virus. These filters adsorb viruses efficiently over a wide pH range without a need for polyvalent salts. However, they cannot be used with seawater or water with a pH exceeding 8.0–8.5 (Sobsey and Glass, 1980). The electropositive 1MDS Virozorb is especially manufactured for virus concentration from water. VIRADEL filter methods suffer from a number of limitations. Suspended matter in water tends to clog the filters, thereby limiting the volume that can be processed and interfering with the elution process. Dissolved and colloidal organic matter in some waters can interfere with virus adsorption to filters, presumably by competing with viruses for adsorption sites. Finally, the concentration efficiency varies depending on the type of virus, presumably because of differences in the isoelectric point of the virus, which influences the net charge of the virus at any pH.

8.2.2.2 Sample Elution and Reconcentration

Adsorbed viruses are usually eluted from the filter surfaces by pressure filtering a small volume (1–2 liters) of an eluting solution through the filter. The eluent is usually a slightly alkaline proteinaceous fluid such as 1.5% beef extract adjusted to pH 9.5 (Fig. 8.7B). The elevated pH increases the negative charge on both the virus and filter surfaces, which results in desorption of the virus from the filter. The organic matter in the beef extract also competes with the virus for adsorption on the filter, further aiding desorption. The 1- to 2-liter volume of the elutant is still too large to allow sensitive virus analysis, and therefore a second concentration step (reconcentration) is used to reduce the volume to 20–30 ml before assay. The elution-reconcentration process is shown in detail in Fig. 8.8.

TABLE 8.5   Methods Used for Concentrating Viruses from Water

Method	Initial volume of water	Applications	Remarks
Filter adsorption–elution			
Negatively charged filters	Large	All but the most turbid waters	Only system shown useful for concentrating viruses from large volumes of tap water, sewage, seawater, and other natural waters; cationic salt concentration and pH must be adjusted before processing
Positive charged filters	Large	Tap water, sewage, seawater	No preconditioning of water necessary at neutral or acidic pH levels
Adsorption to metal salt precipitate, aluminum hydroxide, ferric hydroxide	Small	Tap water, sewage	Have been useful in reconcentration
Charged filter aid	Small	Tap water, sewage	40-liter volumes tested, low cost; used as a sandwich between prefilters
Polyelectrolyte PE60	Large	Tap water, lake water, sewage	Because of its unstable nature and lot-to-lot variation in efficiency for concentrating viruses, method has not been used in recent years
Bentonite	Small	Tap water, sewage	
Iron oxide	Small	Tap water, sewage	
Glass powder	Large	Tap water, sewage	Columns containing glass powder have been made that are capable of processing 400-liter volumes
Positively charged glass wool	Small to large	Tap water	Positively charged glass wool is inexpensive; used in pipes or columns.
Protamine sulfate	Small	Sewage	Very efficient method for concentrating reoviruses and adenoviruses from small volumes of sewage
Hydroextraction	Small	Sewage	Often used as a method for reconcentrating viruses from primary eluates
Ultrafiltration			
Soluble filters	Small	Clean waters	Clogs rapidly even with low turbidity
Flat membranes	Small	Clean waters	Clogs rapidly even with low turbidity
Hollow fiber or capillary	Large	Tap water, lake water, seawater	Up to 100 to 1000 liters may be processed, but water must often be prefiltered
Reverse osmosis	Small	Clean waters	Also concentrates cytotoxic compounds that adversely affect assay methods

Overall, these methods can recover enteroviruses with an efficiency of 30–50% from 400- to 1000- liter volumes of water (Gerba *et al.*, 1978; Sobsey and Glass, 1980).

8.2.2.3 Virus Detection

Several options are available for virus detection that are described in detail in Chapter 10.5. Briefly, virus can be detected by inoculation of a sample into an animal cell culture followed by observation of the cells for cytopathogenic effects (CPE) or by enumeration of clear zones or plaque-forming units (PFU) in cell monolayers stained with vital dyes (i.e., dyes that stain only living non–virus-infected cells). The PFU method allows more adequate quantitation of viruses because they can be enumerated. PCR can also be used to detect viruses directly in either the sample concentrates or the animal cell culture. The overall procedure for sampling and detecting viruses in water is shown in Fig. 8.8.

8.2.3 Processing Water Samples for Detection of Bacteria

Processing water samples for bacteria is much simpler than the processing required for viruses. Typ-

A

B

FIGURE 8.7 (A) Field Viradel system for concentrating viruses from water. (B) Elution of virus from filter with beef extract. (Photo courtesy C. P. Gerba.)

ically, bacteria are collected and enumerated by one of two different procedures: the membrane filtration and most probable number (MPN) methodologies. Membrane filtration, as the name implies, relies on collection and concentration of bacteria via filtration. In the MPN method, samples are generally not processed prior to the analysis. In both procedures, bacteria are detected via cultural methods that are described in Chapter 10, Sections 1.3 (MPN) and 2.1.3 (membrane filtration technique).

8.2.4 Processing Water Samples for Detection of Protozoan Parasites

As with enteric viruses, it requires ingestion of only a few protozoan parasites to cause infection in humans. As a result, sensitive methods for analysis of protozoa are required. Current methods have been developed primarily for the concentration and detection of *Giardia* cysts and *Cryptosporidium* oocysts (Hurst *et al.*, 1997). The first step usually involves collection and filtration of large volumes of water (often hundreds of liters). During filtration, the cysts or oocysts are entrapped in a 1-μm nominal porosity spun fiber filter (25 cm in length) (Fig. 8.9) made of Orlon or polypropylene. The precise volume of water collected depends on the investigation. For surface waters it is common to collect 100–400 liters, and volumes in excess of 1000 liters have been collected for drinking water analysis. Usually, a pump running at a flow rate of 4–12 liters per minute is used to collect a sample. The filter cartridge is placed in a plastic bag, sealed, stored on ice, and sent to the laboratory to be processed within 72 hours.

In the laboratory the cysts and oocysts are extracted by cutting the filter in half, unwinding the yarn, and placing the fibers in an eluting solution of Tween 80, sodium dodecyl sulfate, phosphate-buffered saline. The filters are either hand washed in large beakers or washed in a mechanical stomacher. The 3- to 4-liter resulting volume is then centrifuged to concentrate the oocysts, which are then resuspended in 10% formalin or 2.5% buffered potassium dichromate (for *Cryptosporidium*). Cysts and oocysts are stable for many weeks when stored in formalin or potassium dichromate. A great deal of particulate matter is often concentrated along with the cysts and oocysts and the pellet is further purified by density gradient centrifugation with Percoll–surcose, sucrose, or potassium citrate solutions. Aliquots of the concentrated samples are then filtered through a membrane cellulose acetate filter, which collects the cysts and oocysts in its surface. The cysts and oocysts are stained with fluorescent monoclonal antibodies and the filter placed under an epifluorescence microscope (Fig. 8.10). Fluorescent bodies of the correct size and shapes are identified and examined by differential interference contrast microscopy for the presence of internal bodies (i.e., trophozites or sporozites). The entire procedure is shown in detail in Fig. 8.11.

Although this method is time consuming and cumbersome, it has been successful in isolation of *Giardia* and *Cryptosporidium* from most of the surface waters in the United States (Rose *et al.*, 1997). This method also appears capable of concentrating the protozoan parasite

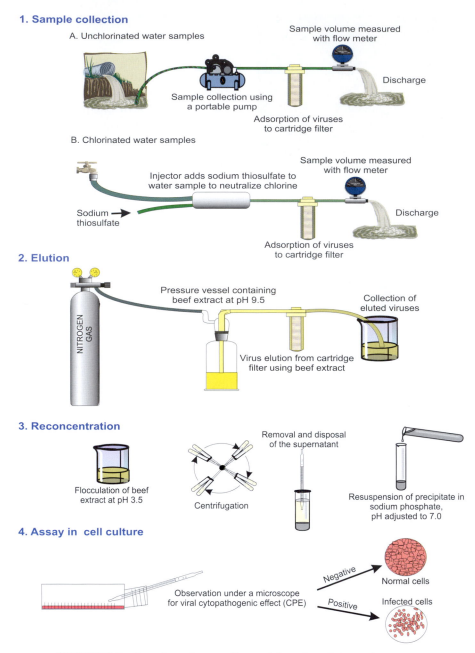

FIGURE 8.8 Procedures for sampling and detection of viruses from water.

microsporidium from water. There are several disadvantages associated with this method of analysis. These include frequent low recovery efficiencies (5–25%); a 1- to 2-day processing time; interference caused by fluorescent algae; inability to determine viability; and finally a limitation in the sample volume that can be collected from some sites because of high turbidity and filter clogging. For these reasons methods to improve concentration efficiency are under development including calcium carbonate flocculation, membrane filtration, ultrafiltration, and immunomagnetic separation.

8.3 DETECTION OF MICROORGANISMS ON FOMITES

Fomites are inanimate objects that may be contaminated with infectious organisms and serve in their transmission. Clothing, dishes, toys, tabletops, and hypodermic needles are examples of common fomites. Fomites can range in size from as small as a particle of household dust to as large as an entire floor surface and in complexity from a flat tabletop to a delicate medical instrument. The involvement of

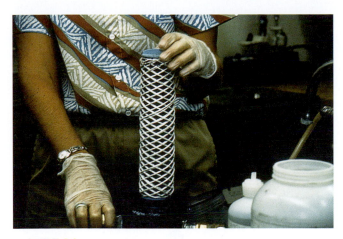

FIGURE 8.9 Spun fiber filter for concentrating *Giardia* cysts and *Cryptosporidium* oocysts. (Photo courtesy of C. P. Gerba, University of Arizona.)

fomites in disease transmission was recognized long before the identification of some pathogenic microorganisms. At the turn of the century, spread of smallpox among laundry workers was not uncommon. In 1908 outbreaks of smallpox were traced to imported raw cotton contaminated with variola virus in crusts or scabs (England, 1984). Fomites are also believed to be important in the transmission of respiratory viruses, such as rhinovirus. An outbreak of hepatitis B virus, typically a blood-borne virus associated with blood transfusion, was associated with computer cards as the probable agents of transfer. These cards, when handled, inflict small wounds on the fingertips, allowing transmission and entry of the pathogen into a new host (Pattison *et al.*, 1974). Growth of enteric pathogenic bacteria in household sponges and on utensils or surfaces used for food preparation has long been recognized as an important route for transfer of these organisms to other foods or surfaces or self-inoculation when the fingers that have handled a sponge or utensil are brought to the mouth.

Fomites may become contaminated with pathogenic microorganisms by direct contact with infectious body secretions or fluids, soiled hands, contaminated foods, or settling from the air. For fomites to serve as vehicles of microbial disease, the organisms must be able to survive in association with the fomites and be successfully transferred to the host. Survival of organisms on a surface is influenced by temperature, humidity, evaporation, desiccation, light, ultraviolet radiation, the physical and chemical properties of the surface, and the substance in which the organism is suspended. Enteric and respiratory pathogens may survive from minutes to weeks on fomites, depending

on the type of organism and the previously listed factors (Sattar and Springthorpe, 1997).

Sampling of fomites is essential in the food manufacturing industry to assess sanitation practices and is in common use in the food service and health care industries to evaluate cleaning and disinfection efficacy. It is also useful in epidemiological investigation and evaluation of hard surface disinfectants. The approaches most commonly used for detection of bacteria on fomites involve **Rodac agar** plates and the swab–rinse technique. Rodac dishes are petri dishes in which the agar fills the entire dish to produce a convex surface, which is then pressed against the surface to be sampled. Selective media can be used for isolation of specific groups of organisms (e.g., m-Fc media for fecal coliforms). After incubation the colonies are counted and reported as colony-forming units (CFU) per cm^2. The swab–rinse method was developed in 1917 for studying bacterial contamination of eating utensils (England, 1984). The method is also suitable for sampling of viruses. A sterile cotton swab is moistened with a buffer or other solution and rubbed over the surface to be sampled. The tip of the swab is then placed aseptically in a container with a sterile collection solution, the container is shaken, and the rinse fluid is assayed on an appropriate culture medium or by a molecular technique such as the PCR method.

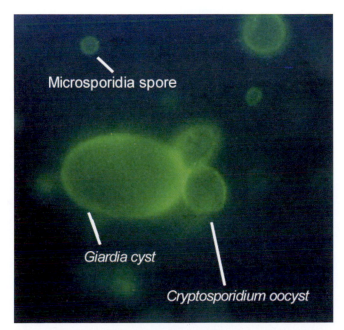

FIGURE 8.10 Immunofluorescence of *Giardia* cysts, *Cryptosporidium* oocysts, and *Microsporidia* spores an emerging waterborne pathogen. (Photo courtesy S. E. Dowd).

a

1. Sample collection

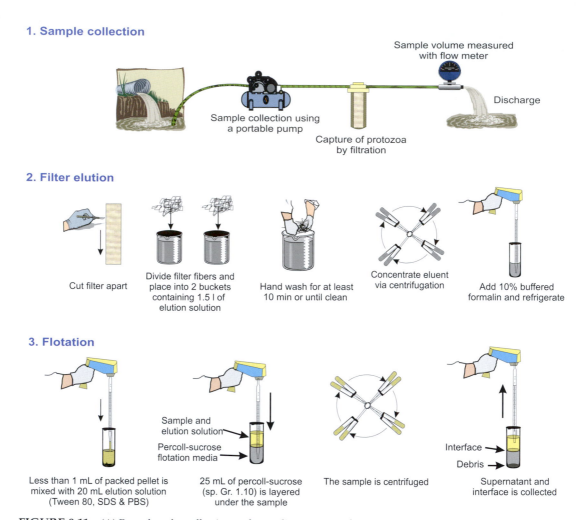

2. Filter elution

3. Flotation

FIGURE 8.11 (A) Procedure for collection and sampling protozoa from water. (B) Procedure for antibody staining and examination of *Giardia* and *Cryptosporidium* (see facing page).

b **4. Indirect antibody staining**

Place sample onto 0.45 mm cellulose acetate membrane

Stain with primary antibody for 30 min

Rinse and stain with secondary antibody for 30 min.

Rinse with ethanol series to clear membrane

5. Examination

After antibody labeling, the slide is examined using microscopic methodologies

Examination is performed to determine characteristic size and shape of fluorescing microbes using UV epifluorescent microscopy

Cryptosporidium sp.
4-6 um

Giardia sp.
8-12 um

Examination is performed to determine characteristic internal structure using Differential Interference Contrast (DIC) microscopy

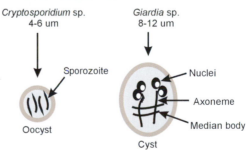

Sporozoite

Oocyst

Nuclei

Axoneme

Median body

Cyst

Up to 4 sporozoites can be observed within a *Cryptosporidium* oocyst

Nuclei, axonemes, and median bodies are characteristically observed in *Giardia* cysts

FIGURE 8.11 (*Continued*)

QUESTIONS AND PROBLEMS

1. How many samples and what sampling strategy would you use to characterize a portion of land of area 500 km? Assume that the land is square in shape with a river running through the middle of it which is contaminating the land with nitrate.

2. Discuss the reasons that deep subsurface sampling is more difficult than surface sampling.

3. A soil is extracted for its community DNA and is found to contain 0.89 μg DNA per g soil. How many bacterial cells does this theoretically involve?

4. When would one utilize electropositive filters for concentrating viruses from environmental samples?

5. When would one utilize electronegative filters for concentrating viruses from environmental samples?

6. In what ways is it easier to sample water for subsequent microbial analysis than it is to sample soil?

7. If you collected a surface soil sample from a desert area in the summertime, when daytime temperatures were in excess of 40°C, how would you store the soil and how would you get the soil ready for microbial experiments to be conducted one month later? Discuss the pros and cons of various strategies.

8. What sampling strategy would you use to give the most complete picture of all bacteria found in a soil sample?

References and Recommended Readings

Amann, R. I., Ludwig, W., and Schleifer, K. H. (1995) Phylogenetic identification and in situ detection of individual microbial cells without cultivation. *Microbiol. Rev.* **59**, 143–169.

APHA. (1995) "Standard Methods for the Examination of Water and Wastewater," 19th ed. American Public Health Association, Washington, DC.

Balkwill, D. L., and Ghiorse, W. C. (1985) Characterization of subsurface bacteria associated with two shallow aquifers in Oklahoma. *Appl. Environ. Microbiol.* **50**, 580–588.

Berg, G. (1987) "Methods for Recovering Viruses from the Environment." CRC Press, Boca Raton, FL.

Chapelle, F. H. (1992) "Ground-Water Microbiology and Geochemistry." John Wiley & Sons, New York.

Colwell, F. S. (1989) Microbiological comparison of surface soil and unsaturated subsurface soil from a semiarid high desert. *Appl. Environ. Microbiol.* **55**, 2420–2423.

England, B. L. (1984) Detection of viruses on fomites. In "Methods in Environmental Virology" (C. P. Gerba and S. M. Goyal, eds.) Marcel Dekker, New York, pp. 179–220.

Gerba, C. P. (1984) Applied and theoretical aspects of virus adsorption to surfaces. *Adv. Appl. Microbiol.* **30**, 33–168.

Gerba, C. P. (1987) Recovering viruses from sewage, effluents, and water. In "Methods for Recovering Viruses from the Environment" (G. Berg, ed.) CRC Press, Boca Raton, FL, pp. 1–23.

Gerba, C. P., and Goyal, S. M. (1982) "Methods in Environmental Virology." Marcel Dekker, New York.

Gerba, C. P., Farah, S. R., Goyal, S. M. Wallis, C., and Melnick, J. L. (1978). Concentration of enteroviruses from large volumes of tap water, treated sewage, and seawater. *Appl. Environ. Microbiol.* **35**, 540–548.

Holben, W. E. (1994) Isolation and purification of bacterial DNA from soil. In "Methods of Soil Analysis," Part 2, "Microbiological and Biochemical Properties." SSSA Book Series No. 5. Soil Science Society of America, Madison, WI, pp. 727–750.

Holben, W. E., Jannson, J. K., Chelm, B. K., and Tiedje, J. M. (1988) DNA probe method for the detection of specific microorganisms in the soil bacterial community. *Appl. Environ. Microbiol.* **54**, 703–711.

Hurst, C. J., Knudsen, G. R., McInerney, M. J., Stetzenbach, L. D., and Walter, M. V. (1997) "Manual of Environmental Microbiology." ASM Press, Washington, DC.

Katznelson, E., Fattal, B., and Hostovesky, T. (1976) Organic flocculation: An efficient second-step concentration method for the detection of viruses in tapwater. *Appl. Environ. Microbiol.* **32**, 638–639.

Krawiec, S., and Riley, M. (1990) Organization of the bacterial chromosome. *Microbiol. Rev.* **54**, 502–539.

Lorentz, M. G., and Wackernagel, W. (1987) Adsorption of DNA to sand, and variable degradation rate of adsorbed DNA. *Appl. Environ. Microbiol.* **53**, 2948–2952.

Moré, M. I., Herrick, J. B., Silva, M. C., Ghiorse, W. C., and Madsen, E. L. (1994) Quantitative cell lysis of indigenous microorganisms and rapid extraction of microbial DNA from sediment. *Appl. Environ. Microbiol.* **60**, 1572–1580.

Ogram, A., Sayler, G. S., and Barkay, T. (1987) The extraction and purification of microbial DNA from sediments. *J. Microbiol. Methods* **7**, 57–66.

Pattison, C. P., Boyer, K. M., Maynard, J. E., and Kelley, P. C. (1974) Epidemic hepatitis in a clinical laboratory: Possible association with computer card handling. *J. Am. Med. Assoc.* **230**, 854–857.

Pepper, I. L., Gerba, C. P., and Brusseau, M. L., eds. (1996) Pollution Science, Academic Press, San Diego, CA.

Rose, J. B., Lisle, J. T., and LeChevallier, M. (1997) Waterborne cryptosporidiosis: Incidence, outbreaks, and treatment strategies. In "*Cryptosporidium* and Cryptosporidiosis" (R. Fayer, ed.) CRC Press, Boca Raton, FL, pp. 93–109.

Sattar, S. A., and Springthorpe, V. S. (1997) Transmission of viral infections through animate and inanimate surfaces and infection control through chemical disinfection. In "Modeling Disease Transmission and Its Prevention by Disinfection" (C. J. Hurst, ed.) Cambridge University Press, Cambridge, pp. 224–257.

Sobsey, M. D., and Glass, J. S. (1980) Poliovirus concentration from tap water with electropositive adsorbent filters. *Appl. Environ. Microbiol.* **40**, 201–210.

Sparkling, G. P., and Cheshire, M. V. (1979) Effects of soil drying and storage on subsequent microbial growth. *Soil Biol. Biochem.* **11**, 317–319.

Stotzky, G., Goos, R. D., and Timonin, M. I. (1962) Microbial changes occurring in soil as a result of storage. *Plant Soil* **16**, 1–18.

Wollum, A. G. (1994) Soil sampling for microbiological analysis. In "Methods of Soil Analysis," Part 2, "Microbiological and Biochemical Properties." SSSA Book Series No. 5. Soil Science Society of America, Madison, WI, pp. 2–13.

Xia, X., Bollinger, J., and Ogram, A. (1995) Molecular genetic analysis of the response of three soil microbial communities to the application of 2,4-D. *Mol. Ecol.* **4**, 17–28.

9

Microscopic Techniques

TIMBERLEY M. ROANE   IAN L. PEPPER

9.1 HISTORY OF MICROSCOPY

Microscopy had its origins in the 17th century when the Dutch discovered the ability to magnify objects by combining convex and concave glass lenses. Who actually invented the first Dutch microscope is not clear; however, in 1611, Johannes Kepler, a German mathematician and astronomer, found that magnification could be achieved with the use of a convex ocular with a convex objective lens and created the Kepler ocular. Giovanni Faber coined the word "microscope" itself in 1625 in reference to the ability to see small things. The first use of the microscope to see microorganisms was by the Dutch merchant Antonie van Leeuwenhoek. At the age of 40, van Leeuwenhoek began experimenting with glass lenses. He eventually made over 400 different microscopes. With these microscopes, van Leeuwenhoek was the first to see "animalcules," today known as microorganisms.

The Dutch physicist and astronomer Christiaan Huygens made a crucial discovery in the 17th century. The Huygens eyepiece consists of two convex lenses, each with the convex side facing the objective. The lower lens provides a brighter, smaller image from the objective lens and the upper Huygens lens then focuses the image. The Huygens design is still used today in eyepieces with magnifications of 10× or less.

With the development of the microscope and with the increasing quality of lenses and understanding of resolution and magnification, different types of microscopies were developed. A type of dark-field microscopy was first used by van Leeuwenhoek. Erasmus Bartholin in 1669 used polarization microscopy to view calcite crystals. Many others used polarization microscopy to examine soil structure. From the first light microscope in the 1600s to the development of the first electron microscope built in the 1920s, microbiology has become an exciting, rapidly advancing field. The use of van Leeuwenhoek's microscopes to see animalcules marks the true beginning of microbiology. In spite of current microscopes that use electrons, x rays, and lasers to examine microorganisms, the basic light microscope is still a necessity in every microbiology laboratory.

9.2 THEORY OF MICROSCOPY

Regardless of the microscope, microscopy relies heavily on user interpretation. While this makes it a highly subjective tool, microscopy can provide extremely useful information about microorganisms. The human eye alone can resolve about 150 μm between two points. The objective of the microscope is to increase the resolution of the human eye. **Resolution** is the smallest distance between two points visible to the eye, aided or unaided by a microscope. Similarly, **resolving power,** a function of the wavelength of light and the aperture of the objective lens used in viewing a specimen (see Information Box), is the ability to distinguish two points as separate. The resolving power of most light microscopes is 0.2 μm.

An **aberration** in a microscope refers to the inability to image a point in an object as a point. The light microscope has five kinds of aberrations: spherical, coma, astigmatism, curvature of field, and distortion. Aberrations are functions of the lenses in an optical system and severe aberrations result in decreased res-

---

Information Box
Theoretical Resolving Power

The resolving power (RP) is the minimum distance that an optical system can distinguish.

$$RP = \frac{\text{wavelength of light}}{2 \times \text{numerical aperture of objective lens}}$$

For example, for the light microscope

$$RP = \frac{500\ nm}{2 \times 1.25} = 200\ nm = 0.2\ \mu m$$

---

olution. However, in the light microscope, corrective lenses eliminate aberrations so that the theoretical resolving power can be achieved. For example, spherical aberration is the most common aberration. **Spherical aberration** results when light rays pass through a lens at different points on the lens, resulting in light rays of different focal lengths (Fig. 9.1a). Recall that the wavelength of light determines resolution in the light microscope. Light of varying focal planes or wavelengths results in poor resolution of two points in an object. The use of a light diaphragm corrects spherical aberrations by focusing light rays to a single focal plane (Fig. 9.1b). The ability to stack correcting lenses in the light

microscope has eliminated aberrations, allowing the theoretical resolution to be achieved. The electromagnetic lenses in electron microscopes have the same aberrations as the glass lenses of light microscopes. However, there are no glass lenses in the electron microscope and so these aberrations are not as easy to correct. Consequently, although the theoretical resolution of the electron microscope is 0.0002 nm, the actual working resolution is only 0.2 nm.

Perhaps the most important aspect of microscopy is **illumination** of the sample. Without illumination, the specimen cannot be visualized. In light microscopy, transmitted light or reflected light may be used. The source of illumination can be white light, ultraviolet light. Realize, however, that all microscopic techniques rely on the manipulation of light (or electrons in electron microscopy) to influence the resolution of a specimen. In Köhler illumination, for example, a series of condenser lenses and diaphragms are used to focus light rays onto the specimen, increasing not only illumination but also resolution (Fig. 9.2).

Magnification is the ability to enlarge the apparent size of an image, and useful magnification is a function of the resolving power of the microscope and the eye, which can be stated as

$$\frac{\text{Limit of resolution by eye}}{\text{Limit of resolution of microscope}} = \frac{0.15\ mm}{0.0002\ mm} = 750\times$$

Although the ability to magnify an image is almost infinite, most of this magnification is blurred because resolution is limited by the wavelength of the light, and so it becomes empty magnification.

Contrast refers to the ability to distinguish an object from the surrounding medium, and without it both resolution and magnification become unimportant. More specifically, two points in an object that are resolved separately are not seen separately unless their images are contrasted against their surrounding medium. Colorless contrast is based on the intensity or amplitude of light waves and gives contrast in terms

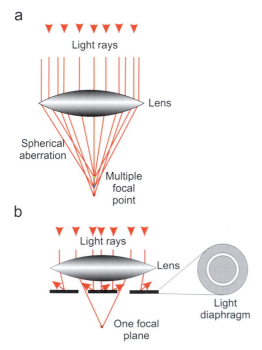

FIGURE 9.1   (a) Spherical aberrations are inherent in any lens, resulting in multiple focal planes of light. (b) The addition of a light diaphragm eliminates spherical aberrations in light microscopes by focusing the light onto one focal plane.

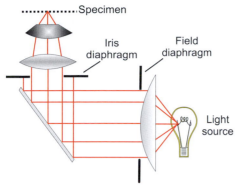

FIGURE 9.2 Schematic of Köhler illumination. Multiple diaphragms inserted between the light source and the condenser focus light directly on the specimen. (Adapted from Zieler, 1972, McCrone Research Institute, Microscope Publications Division.)

of black, white, and grays. Many of the advances in microscopy have been for the sole purpose of increasing contrast, e.g., in phase-contrast microscopy. Dyes and stains are also used to increase contrast with all types of microscopes.

9.3 BASIC MICROSCOPIC TECHNIQUES

One of the most common pieces of laboratory equipment is the microscope. In microbiology, the microscope is critical for the determination of cellular morphology, motility and classification. In the struggle to see more and more detail, microscopy has become a scientific field unto itself.

9.3.1 Types of Microscopes

Optical microscopes, also known as light microscopes, have multiple lenses, including ocular, objective, and condenser lenses (Fig. 9.3). By varying these lenses and light sources, five types of microscopes can be defined: bright-field, dark-field, phase-contrast, differential interference, and fluorescence microscopes. Characteristics of each of these microscopes are given in Table 9.1.

The **bright-field** microscope is the most common of all light microscopes. Images seen with this microscope are the result of light being transmitted through a specimen. The specimen absorbs some of the light,

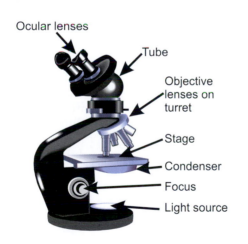

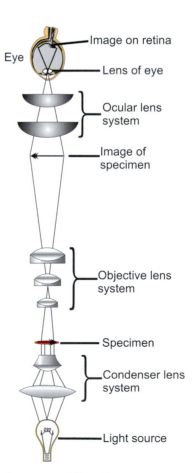

FIGURE 9.3 A typical compound light microscope and its optics.

TABLE 9.1   Comparison of Types of Microscopy

Microscope	Maximum practical magnification	Resolution	Important features
Visible light as source of illumination			
Bright-field	2,000×	0.2 μm (200 nm)	Common multipurpose microscope for live and preserved stained specimens; specimen is dark, field is white; provides fair cellular detail
Dark-field	2,000×	0.2 μm	Best for observing live, unstained specimens; specimen is bright, field is black; provides outline of specimen with reduced internal cellular detail
Phase-contrast	2,000×	0.2 μm	Used for live specimens; specimen is contrasted against gray background; excellent for internal cellular detail
Differential interference	2,000×	0.2 μm	Provides brightly colored, highly contrasting, three-dimensional images of live specimens
Ultraviolent rays as source of illumination			
Fluorescent	2,000×	0.2 μm	Specimens stained with fluorescent dyes or combined with fluorescent antibodies emit visible light; specificity makes this microscope an excellent diagnostic tool
Electron beam forms image of specimen			
Transmission electron microscope (TEM)	1,000,000×	0.5 nm	Sections of specimen are viewed under very high magnification; finest detailed internal structure of cells and viruses is shown; used only on preserved material
Scanning electron microscope (SEM)	100,000×	10 nm	Whole specimens are viewed under high magnification; external structures and cellular arrangement are shown; generally used on preserved material

and the rest of the light is transmitted up through the ocular lens. The specimen will appear darker than the surrounding brightly illuminated field. Bright-field microscopy is most commonly used to examine morphology, and perform Gram stains. **Dark-field** microscopy can be used to increase the contrast of a transparent specimen. By inserting a central stop before the condenser, some but not all of the light from the condenser is prevented from reaching the objective (Fig. 9.4). Only light scattered from the edges of the specimen is viewed. Thus, the specimen appears as a bright image against a dark background. Dark-field microscopy is often used to visualize live specimens that have not been fixed or stained. For example, dark-field microscopy has been used to quantify the motility of bacteria and protozoa and to monitor the growth of bacterial microcolonies (Korber *et al.,* 1990). Although gross morphology can be delineated, internal details are not revealed. Murray and Robinow (1994) describe the nature of dark-field microscopy and its applications.

To examine fine internal detail, a **phase-contrast microscope** can be used. This microscope takes advantage of the fact that although many internal cell components are transparent, they have different den-

sities. Different densities interact differently with light, thereby creating contrast between internal cellular components and the surrounding medium (Fig. 9.5). Phase-contrast microscopy uses a series of diaphragms

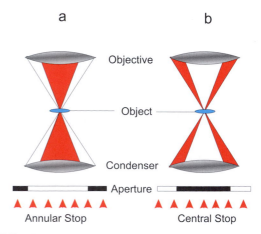

FIGURE 9.4   Differing from bright-field microscopy which uses an annular stop (a), phase contrast microscopy uses a central stop (b), allowing some but not all of the light from the condenser to reach the objective. (Adapted from Rochow and Tucker, 1994.)

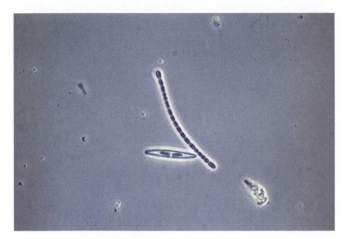

FIGURE 9.5   Phase-contrast image of a free-living nitrogen-fixing cyanobacterium (40 μm length) and the algal cell known as a diatom (12 μm length). (Photo courtesy P. Rusin.)

for separating and recombining direct versus diffracted light rays (Fig. 9.6). Köhler illumination is used to focus the light source on one focal plane. Light rays through the Köhler diaphragm are focused as a hollow cone onto the specimen. In the back focal plane of the objective, there is an annular diaphragm or a diffraction plate. The phase of light rays entering the diffrac-

tion plate, also called a phase plate, is altered. The degree of retardation of light through the plate results in either lightening or darkening of the specimen. Phase-contrast microscopy has been used to estimate bacterial growth rates in biofilms (Caldwell and Lawrence, 1986).

Other types of light microscopy include differential interference contrast (DIC) microscopy and fluorescence microscopy. In **DIC microscopy,** the illuminating beam is split into two separate beams. One beam passes through the specimen, creating a phase difference between the sample beam and the second or reference beam. The two beams are then combined so that they interfere with each other. DIC can allow the detection of small changes in depth or elevation in the sample, thus giving the perception of a three-dimensional image (Fig. 9.7). **Fluorescence microscopes** utilize ultraviolet (UV) light sources. This type of light is used to illuminate fluorescent dyes, such as acridine orange or fluorescein, which in turn emit visible light. Specimens stained with a fluorescent dye appear as brightly colored images against a dark background. Fluorescence microscopes are often used in conjunction with immunologic procedures, as discussed in Chapter 9.5.

9.3.2 Preparation of Samples for Microscopic Observation

Sample preparation for microscopy can be as simple as mounting on a glass slide or as complex as the thin sectioning and mounting on a copper grid as with transmission electron microscopy. In general, sample preparation varies depending on the goals of the observer and on the type of microscope available.

Viable microorganisms are generally viewed via **wet mounts.** Here cells are suspended in water, saline,

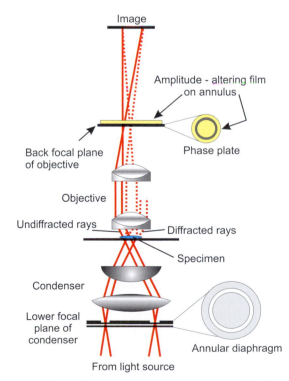

FIGURE 9.6   A phase-contrast microscope equipped with an amplitude-altering film on the phase plate to increase specimen contrast. (*Handbook of Chemical Microscopy, Vol. 1,* 4th ed. C. W. Mason. Copyright © 1983. Adapted by permission of John Wiley & Sons, Inc.)

FIGURE 9.7   A differential interference contrast (DIC) image of *Cryptosporidium* with associated sporozoites. (Photo courtesy P. Rusin.)

or some other liquid medium. The liquid maintains viability and allows locomotion. Wet mounts can be done on a simple glass slide with a coverslip or on specially constructed slides. For example, in the latter case, a drop containing the specimen can be placed on a glass coverslip and then a slide with a concave depression is placed on top of the coverslip. Upon inversion of both coverslip and slide, the drop hangs from the coverslip. The drop is not affected by the glass slide, due to the concave depression, thereby creating a **hanging drop.** Hanging drop slides are useful in monitoring bacterial motility. Wet mounts are often viewed with phase-contrast or DIC microscopes to maximize specimen contrast.

Although morphology can be determined in a wet mount, this is often difficult because of the lack of contrast and detail between the specimen and the surrounding medium and also because the microbes are moving. Another way to view morphology and other internal structures is to fix and stain specimens on glass slides. This allows observation of only nonviable organisms, because the fixing and staining process kills them. Fixation of cells involves spreading a thin film of a liquid suspension of cells onto a slide and air drying it, producing a **smear.** The smear is then fixed on the slide by gently heating it over a flame for a few seconds. The smear is normally stained by addition of one of a variety of dyes that enable cellular detail to be seen. Many different categories of stains are available, but in general they can be classified as **basic dyes,** which have a positive charge, or **acidic dyes,** with a negative charge. Cell components that are negatively charged such as nucleic acids attract basic dyes, and those that are positively charged, for example some cell membrane components, attract acidic dyes.

Stains can also be categorized as **positive** or **negative.** Positive stains are those that attach to the specimen, giving it color, while the background remains unstained. The most important types of positive stains are **simple stains** that involve a single dye such as methylene blue and result in all cells and all cell components appearing the same color, in this case blue, against an unstained background. Simple stains are useful for size and morphological assessments and for cell enumeration. Negative stains, in contrast, stain the background, thus highlighting the specimen as a silhouette. Generally, negative stains, as the name implies, are negatively charged and are repelled by the negatively charged surface of most cells. Examples include the acidic dye India ink, but overall these types of stains are not as important as positive stains, which allow internal detail to be seen.

Differential stains utilize two different dyes designated as the primary dye and the counterstain. The **Gram stain,** developed by Hans Christian Gram, is the most important differential stain. The Gram stain utilizes crystal violet and safranin dyes to classify bacteria into one of two major categories. **Gram-positive** bacteria stain purple, whereas **gram-negative** bacteria stain red (see Fig. 2.13). In both cases, the differential staining is due to differences in cell wall components (see Chapter 2.2.4.2). For many environmental isolates, the Gram stain may be inconclusive, and such isolates are designated as gram variable. In this case, both red and purple cells may be seen, as is true of *Arthrobacter* spp., which are common soil organisms.

Finally, there are a number of **special stains.** These special stains are used to identify specific cell components, such as bacterial capsules and spores. One such stain is the **acid-fast stain** which was developed to identify difficult-to-stain bacteria. These organisms do not stain with commonly used dyes, such as those used in the Gram stain. Acid-fast bacteria are those that when stained with carbolfuchsin cannot be destained, even with acid. This property is typical of *Mycobacterium* spp., which have mycolic acids on their cell surface. Mycobacteria are of particular interest because they are causative agents of several serious human diseases, including tuberculosis and leprosy. They are also common soil isolates that are slow growing but seem to have the ability to degrade many organic contaminants.

9.4 *IN SITU* SOIL MICROSCOPY

Historically, direct examination of microorganisms *in situ*, or within their environment, has been an important tool for microbiologists. Both light and electron microscopies allow direct examination of the form and arrangement of microorganisms in their environments. However, quantitation of actual microbial numbers has been difficult because interfering colloids and soil particles potentially mask large numbers of organisms. An example of direct examination is the **buried slide technique.**

Rossi *et al.* first introduced the buried slide in 1936. In this technique, a glass microscope slide is embedded in a soil or sediment sample. After a period of incubation, the slide is carefully removed with minimal disturbance, and soil particles with attached microbes can be viewed directly under the light microscope. Although this method is more than 60 years old, it is still useful in illustrating the abundance of microorganisms in soil and their relationship to each other and to soil particles. Details of this technique can be found in Pepper *et al.* (1995). A variation of the buried slide technique is the pedoscope technique. Here, optically flat capillary tubes (the tubes are square so that all light passing through the tube has the same distance to travel) are buried in soil. Because soil microorganisms grow in pores or within soil aggregates, the relationships seen on the surface of a typical flat glass slide may not be truly representative of the natural state. The pedoscope capillary tubes overcome this by resembling soil pore spaces.

Thin sectioning of soil samples (this procedure is described in Section 9.8.2.2) is time consuming and challenging; however, no better technique for viewing microorganisms in their natural state exists. The sections, depending on their thickness, can be viewed with either a light microscope or an electron microscope.

9.5 *IN SITU* DETECTION WITH MOLECULAR PROBES AND EPIFLUORESENCE MICROSCOPY

One of the latest innovations in terms of microscopic methods used to detect environmental microorganisms is the use of molecular probes. The basis of this methodology is that a fluorescently labeled biomolecule can be used to bind to a specific target. Examples of biomolecules used include proteins, nucleic acids, polysaccharides, and lipids. These labeled biomolecules are chosen to bind selectively a particular antigen, carbohydrate, or nucleic acid sequence, thus providing a means of detecting these biological targets.

The detection reagents used can be classified as primary or secondary. **Primary detection reagents** bind directly to a specific target (Fig. 9.8a) and can be detected by intrinsic fluorescence. Examples include fluorescent antibodies and fluorescent lectins. Although many biomolecules bind selectively to a biological target, they must first be conjugated to a fluorescent dye, a step that is time consuming and in some cases difficult to do. Instead, a **secondary detection reagent,** defined as a molecule that can be indirectly linked to the molecule of interest, can be employed (Fig. 9.8b). Many secondary detection reagents rely on immunological approaches (see Chapter 12). In all cases, the labeled secondary detection reagent binds to a target-specific molecule that contains specific binding sites or **haptens.** The target-specific molecule, as the name implies, also binds specifically to the target, thereby linking the fluorescently labeled secondary detection reagent to the target. In some cases, the target-specific molecule can have multiple binding sites for the detection reagent, which can increase the sensitivity of detection. Although there are several techniques for labeling reagents, in environmental microbiology fluorescent labels are of paramount importance. These include fluorescently labeled antibodies (see Chapter 12) and fluorochrome-labeled nucleic acid probes. Figure

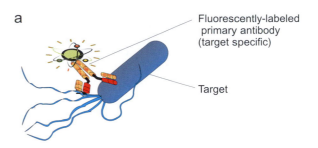

a

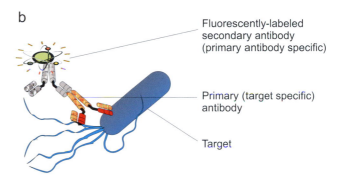

Fluorescently-labeled
primary antibody
(target specific)

Target

b

Fluorescently-labeled
secondary antibody
(primary antibody specific)

Primary (target specific)
antibody

Target

FIGURE 9.8 Examples of reagents used in fluorescence microscopy. These can be nonspecific, such as the fluorescent dye acridine orange, which binds nucleic acids, or specific, such as fluorescently labeled antibodies. (a) Binding of a fluorescently labeled antibody (primary reagent). (b) Binding of a fluorescently labeled antibody (secondary reagent) to the primary antibody.

9.9 shows an example of the use of a fluorescent antibody to detect rhizobia.

Fluorescent *in situ* hybridization (FISH) involves the use of fluorescently labeled nucleic acid probes to target ribosomal RNA (rRNA) within morphologically intact cells. Single bacterial cells with fluorochrome-labeled probes hybridized to their rRNA can be identi-

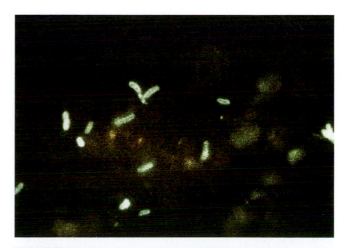

FIGURE 9.9 Use of fluorescent antibodies coupled to fluorescein isothiocyanate to detect antigens. Here rhizobia fluorescence in response to UV irradiation is shown. (Photo courtesy I. L. Pepper.)

fied within mixed populations with the use of fluorescence microscopy (Amann *et al.*, 1990). This is done through selective targeting of regions of rRNA, which consist of evolutionarily conserved and variable nucleotide regions. Thus, by choosing the appropriate rRNA probe sequence, FISH can be used to detect all bacterial cells (a universal probe) or a single population of cells (a strain specific probe). However, sensitivity can be problematic in soils or sediments and is restricted if cells are not actively growing. This results because the number of target rRNA copies within a cell is dependent on metabolic activity. This technique can be useful in detecting airborne bacteria, such as *Pseudomonas aeruginosa* and *Escherichia coli* (Lange *et al.*, 1997), and in the quantitation and identification of organisms.

9.6 DIRECT BACTERIAL COUNTS

Microbiologists are often interested in determining numbers of microorganisms associated with a given environment or process. In pure culture, this can easily be done by cultural procedures (see Chapter 10). For environmental samples, enumeration cannot be addressed as easily because of the mixed populations of organisms and, in some cases, the complexity of the environmental matrix itself. There are two main methods for determining microbial numbers. The first involves the indirect or viable count based on culturable assays as discussed in Chapter 10. The second method, known as the **direct count,** involves direct microscopic observations.

Direct count procedures usually provide numbers that are higher than culturable counts because direct counts include viable, dead, and viable but nonculturable organisms. With water samples, organisms that are relatively large can be observed in counting chambers that hold specific volumes, such as a hemocytometer or a Petroff–Hauser chamber. Direct microscopy of soil microorganisms involves separating the organisms from soil particles. Soils are first treated with a dispersing agent, such as Tween 80, and vortexed to remove organisms from soil particles and to disrupt soil aggregates. A known volume of the resulting soil suspension is then placed on a glass microscope slide or mixed with agar in a counting chamber to be examined with the microscope (Jones and Mollison, 1948).

A number of stains enhance the efficacy of direct microscopy. In direct microscopy, fluorescent stains are used rather than simple stains, such as Rose Bengal. For example, fluorescein diacetate is used to estimate fungal biomass in soil (Soderstrom, 1977). Perhaps the most widely used stain for direct microscopy of bacte-

ria is **acridine orange** (AO), used in obtaining acridine orange direct counts (AODCs). Acridine orange intercalates with nucleic acids, and bacteria stained with AO appear either green (high amounts of RNA) or orange (high amounts of DNA) (Fig. 9.10a). Originally, it was thought that the green or orange color correlated with the viability of the organism; however, this has not been established. Two other important stains used in direct counting are 4,6-diamidino-2-phenylindole (DAPI) and fluorescein isothiocyanate (FITC).

In soil suspensions, regardless of the fluorescent stain, there are problems associated with the presence of clay colloids, which can either autofluoresce or nonspecifically bind the fluorescent stain. Colloids also mask the presence of soil microorganisms. At high dilutions, colloidal interference decreases; however, the numbers of organisms are also diluted, which may result in low counts that are not statistically valid. At low dilutions, the numbers of organisms are higher, but so is **colloidal interference**. In some cases, colloidal interference in water samples can be minimized with filtration through a Nuclepore filter. This type of filtration is designed to remove particulates larger than 2 μm, but allow passage of bacteria that are smaller.

As previously mentioned, the use of epifluorescence for direct microscopy usually results in numbers that may be two orders or more of magnitude greater than culturable counts. The difference between direct and culturable counts is attributed to **viable but nonculturable organisms** (Roszak and Colwell, 1987a,b). Microorganisms are thought to be rendered viable but nonculturable when exposed to extreme stress or when a nutritional requirement is not met under either environmental or laboratory conditions. Numbers of viable but nonculturable organisms can be estimated in pure culture by doing two assays, a total direct count with acridine orange, for example, and a **direct viability count** (DVC). In a DVC, the numbers of viable or actively growing cells are determined by incubation with nalidixic acid (Kogure *et al.*, 1979). Nalidixic acid inhibits DNA synthesis, resulting in elongated growing cells that cannot divide. Consequently, upon fluorescent staining and microscopic observation, the cells that appear elongated are viable (Fig. 9.10b). The difference between an AODC and a DVC is a rough estimate of the viable but nonculturable bacteria. There may be some dead cells in the AODC count, but it is thought that dead cells degrade rapidly and so are few in number. The overall concept of viable but nonculturable organisms is important in environmental microbiology for two reasons. First, viable but nonculturable pathogens may still be able to infect and cause disease. Second, because environmen-

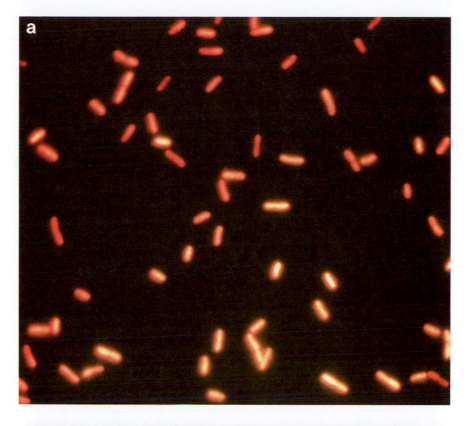

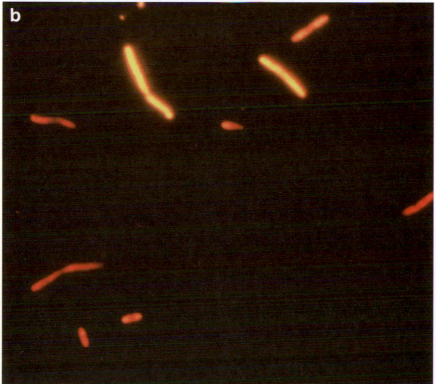

FIGURE 9.10 Acridine orange direct counts of (a) normal bacteria and (b) bacteria treated with nalidixic acid. The treated bacteria appear as elongated cells because of inability to replicate in the presence of nalidixic acid. (Photo courtesy K. L. Josephson.)

tal microbiologists often only study cultured isolates, many important environmental species that are not culturable are ignored.

9.7 ESTIMATION OF BIOMASS USING MICROSCOPIC COUNTS

Scientists are often interested in estimates of living microbial biomass in certain environments. Estimates can be calculated in terms of bacterial or fungal biomass as carbon as shown in Table 9.2. To calculate bacterial biomass, some assumptions must be made. Approximate bacterial volumes must be determined using average cell lengths and diameters. Approximate bacterial numbers are determined using direct count microscopy. For fungi, an estimate of fungal hyphal lengths per gram of sample must be known. In addition, estimates of the solids content for each organism have to be made. It should be noted

TABLE 9.2 Equations for Calculating Biomass

Calculation of bacterial numbers in soil:

$$N_g = N_f \frac{A}{A_m} \frac{V_{sm}}{V_{sa}} D \frac{W_W}{W_d}$$

N_g = number of bacteria per gram dry soil
N_f = bacteria per field
A = area (mm^2) of smear (or filter)
A_m = area (mm^2) of microscope field
V_{sm} = volume (ml) of smear of filter
V_{sa} = volume (ml) of sample
D = dilution
W_w = wet weight soil
W_d = dry weight soil

Bacterial biomass as carbon:

$$C_b = N_g V_b e S_c \frac{\%C}{100} \times 10^{-6}$$

C_b = bacterial biomass carbon (μg/g-soil)
N_g = number of bacteria per gram soil
V_b = average volume (μm^3) of bacteria (r^2L; r = bacterial radius, L = length)
e = density (1.1×10^{-3} in liquid culture)
S_c = solids content (0.2 in liquid culture, 0.3 in soil)
$\%C$ = carbon content (45% dry weight)

Calculation of fungal biomass carbon:

$$C_r = \pi r^2 L S_c \, \%C \times 10^{10}$$

C_r = fungal carbon (μg carbon/g-soil)
r = hyphal radius (often 1.13 μm)
L = hyphal length (cm/g-soil)
e = density (1.1 in liquid culture, 1.3 in soil)
S_c = solids content (0.2 in liquid culture, 0.25–0.35 in soil)

Adapted from Paul and Clark (1989).

that biomass estimates can also be made using chemical fumigation methods (see Chapter 11.4.2.3) or DNA content (see Chapter 13). The obvious limitations associated with estimating biomass are in estimating organism numbers. This problem is exacerbated in soils with colloidal interference.

9.8 ADVANCED MICROSCOPIC TECHNIQUES

Microscopy has come a long way since the rudimentary microscopes of the 17th century. Increasingly powerful microscopes allow scientists to see DNA, molecular structures, and even structures at the atomic level. Microscopy has developed into an immense scientific field and only a few of the recent advances in microscopy are presented here.

9.8.1 Polarization Microscopy

Anisotropic light originates from specimens that have asymmetry in their crystal lattice properties. Anisotropy is observable in liquid and solid crystals; strained glasses; stressed plastic materials; crystallized resins and polymers; refracting surfaces; synthetic filaments; and biological fibers, cells, and tissues. **Polarized light** is light in one plane that can be used to examine anisotropy in sample materials. Polarization microscopy is traditionally used in determining the optical properties of soil particles in soil identification (Fig. 9.11). The optical anisotropy of individual crystals reflects the bonding patterns of units, e.g., molecules or elements, and usually involves differences in at least two crystallographic directions (at least two directions of polarized light). Multiple anisotropic crystals have optical characteristics above and beyond those of individual crystals. Anisotropy observed in a sample can provide more information about the sample than ordinary unpolarized light.

For example, a result of light polarization is **molecular birefringence.** Molecular birefringence is manifested by long or flat molecules, especially polymeric macromolecules, and is particularly applicable in the examination of microbially produced extracellular polymers. In molecular birefringence, when polarized light encounters a series of atomic dipoles arranged in chains, as in long molecules, the strength of the dipoles causes the light to vibrate lengthwise along the chain, resulting in greater polar anisotropy at the poles. However, side chains on the molecules tend to reduce the strength of birefringence in the main chain of the molecule, resulting in less polar anisotropy. The patterns and strengths of anisotropy evident in a sample

FIGURE 9.11 Polarization microscopy used to view sand grains in a sandy loam soil. The various colors are the result of light interference, which can be used to identify individual minerals. Magnification 400×. (Photo courtesy T. M. Roane.)

can give indications of the purity and elemental structure of the sample.

The resolution of a specimen's effects on polarized light depends on producing plane polarized light with a polarizer and examining the effects with an analyzer. Two polarizers are used in polarization microscopy: a polarizer and an analyzer. For transmitted light, the polarizer is placed between the light source and the substage condenser lens. The analyzer is placed between the objective and ocular lenses. Most polarizers are produced from polarizing film, e.g., Polaroid, between two protective pieces of thin glass. When polarized light from the first polarizer is vibrating in a direction that allows it to pass through the analyzer, the field of view in the microscope will be black as the polarizers are crossed with respect to their directions of vibration of light (Fig. 9.12). Thus, in a polarizing microscope, contrast is the result of various interference phenomena throughout the sample. Light interference or retardation at each point in a crystal results in contrast and color on a dark background. In accordance with the Michel–Levy interference spectra based on light retardation through varying sample thicknesses, light interference gives first-

order gray, high order white, and color to the sample image.

9.8.2 Electron Microscopy

An electron microscope is an optical device for producing high resolution detail in a specimen. Instead of light, electron microscopes utilize electrons to form images. The extremely short wavelength and focusability of electron beams are responsible for the theoretically high resolving power of electron microscopes. The increased resolution allows a functional magnification of up to 1,000,000× for the observation of fine structure and detail. Although electron microscopes are conceptually similar to light microscopes, there are some fundamental differences between using light versus electronic illumination (Table 9.3). In the electron microscope, an electron gun aims a beam of electrons at a specimen placed in a vacuum sample chamber. A series of coiled electromagnets are used to focus the beam. As in light microscopy, poor contrast is a problem in electron microscopes, so samples are often stained to increase contrast. Images produced in the electron microscope are in shades of gray, although

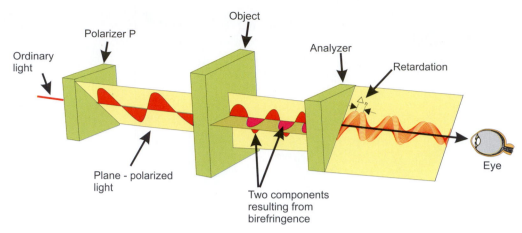

FIGURE 9.12 Schematic showing the function of the polarizer and the analyzer in polarization microscopy. (Adapted from Mason, 1983.)

computerized color may be added in newer scopes. The two most common types of electron microscopy are scanning electron microscopy and transmission electron microscopy.

9.8.2.1 Scanning Electron Microscopy

In the scanning electron microscope (SEM), an image is formed as an electron probe scans the surface of the specimen (Fig. 9.13), producing secondary electrons, backscattered electrons, x rays, Auger electrons, and photons of various energies. The SEM uses these signals to produce three-dimensional surface characteristics of specimens (Fig. 9.14). There are several advantages of the SEM. These include a large depth of field; the ability to examine bulk samples with low

magnification; lifelike images; and in addition, because there are no lenses after the sample, there are fewer aberrations, so the sample itself determines the resolution.

In a typical SEM, an electron gun and multiple condenser lenses produce an electron beam whose rays are aligned through electromagnetic scan coils. Electron-accelerating voltages in the gun range from 60 to 100 kV (kilovolts). A tungsten filament, heated to approximately 2700°K, is the illumination source within the gun. Heating the filament causes electrons to be released from the tip of the filament. An image of the surface topography of the specimen is generated by electrons that are reflected (backscattered) or given off (secondary electrons). Contrast in the SEM is enhanced by coating the sample with a thin layer of a conductive metal, e.g., gold or palladium, or even carbon. Image formation itself is the result of **rastering** the electron beam (from 2 to 200 Å in diameter) back and forth along the specimen surface (Fig. 9.15). A visual image corresponding to the signal produced by the interaction between the beam spot and the specimen at each point along each scan line is simultaneously built up on the face of a cathode ray tube in the same way a television picture is generated. As with all microscopy, interpretation of SEM images is subjective, especially because SEM images include high resolution, contrast, and varying depths of focus resulting in topography. Sample preparation for the SEM is relatively straightforward. The general sequence involves: (1) sample fixation with an aldehyde solution, (2) dehydration of the sample (because the sample must be under vacuum in the SEM), (3) mounting of the specimen on a metal stub, and (4) coating of the specimen with a thin layer of electrically conductive material.

TABLE 9.3   Comparison of Optical Microscopes and Electron Microscopes

Characteristic	Optical	Electron
Illuminating beam	Light beam	Electron beam
Wavelength	7500 Å (visible) 2000 Å (ultraviolet)	0.086 Å (20 kV) 0.037 Å (100 kV)
Medium	Atmosphere	Vacuum
Lens	Glass lens	Electrostatic lens
Resolving power	2000 Å	3 Å
Magnification	110 × 2000×	90 × 1,000,000×
Focusing	Mechanical	Electrical
Viable specimen	Yes	No
Specimen requires staining or treatment	Yes/no	Always
Colored image produced	Yes	No

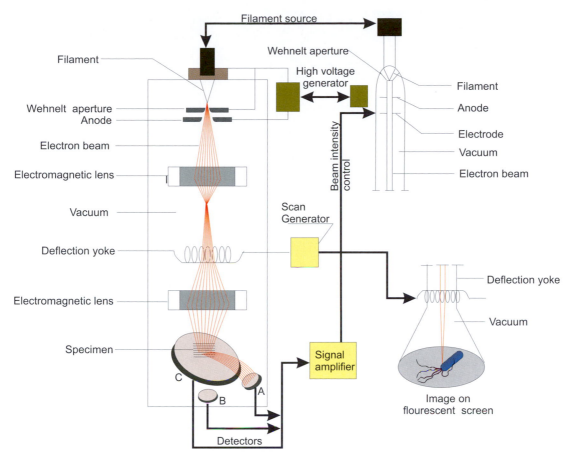

FIGURE 9.13 A typical scanning electron microscope (SEM) and its similarity to a television. They are based on the same principles. (Courtesy FEI Company.)

Other scanning microscopes include the scanning tunneling microscope, developed in the 1970s, and the atomic force microscope, invented in 1986. Both use probes to examine specimen surfaces. Such microscopes have the ability to view single atoms with a magnification of 1,000,000×. **Scanning tunneling microscopes** have a tungsten probe that hovers near the surface of the specimen and follows its topography while simultaneously giving off an electric signal of its pathway, creating a specimen image. These types of microscopes are commonly used to detect defects on silicon computer chips, but they have also been used to view DNA. The **atomic force microscope** has a diamond–metal probe that is forced down onto the surface of the sample, like a needle on a record. Electrical signals again generate an image. Atomic force microscopes are increasingly being used with biological samples to examine, for example, the molecular structure of cell membranes.

9.8.2.2 Transmission Electron Microscopy

In the SEM, electrons interacting with the surface of the specimen form the image. In transmission electron microscopy (TEM) (Fig. 9.16), the image is formed by electrons passing through the specimen. Consequently, the specimens must be thin sectioned to allow the passage of electrons. TEM is often used to view detail of internal structures (Fig. 9.17), the result of selective absorption of electrons by different parts of the specimen.

Sample preparation for the TEM is much more extensive than for the SEM. A sample initially undergoes **fixation** in glutaraldehyde or formaldehyde to preserve structure. Fixation also protects the sample from damage that may occur during the rest of the preparation. Following fixation, the sample is **dehydrated** by replacing water with, most commonly, ethanol. The ethanol acts as a solvent between the aqueous environment of the cell and the hydrophobic embedding medium. **Embedding** involves resin infiltration, where the ethanol is replaced by a highly miscible plastic embedding agent and is later cured at a high temperature (~70°C). Curing causes the embedding medium to polymerize and become solid. A **microtome** equipped with either a glass or diamond knife is

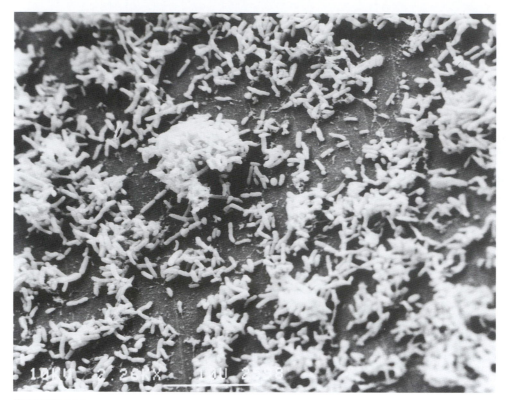

FIGURE 9.14   Scanning electron microscope image of a biofilm of a *Rhizobium* bacterium. Each bacterium is approximately 2 μm. (Photo courtesy I. L. Pepper.)

then used to make thin sections approximately 90 nm thick for viewing under the TEM.

As previously mentioned, all lenses are subject to aberrations. The electromagnetic lenses used in electron microscopy are no exception. Unlike those in light microscopy, however, aberrations in the electron microscope are difficult to resolve because of the inability to add corrective lenses to the optics. Consequently, whereas in the light microscope the theoretical and achievable resolving powers are similar, in the electron microscope the theoretical resolution is not reached.

9.8.2.3 Elemental Analysis

One of the advantages of electron microscopy is the ability to perform microanalysis x-ray spectrometry with an **energy-dispersive spectrometer** (EDS). In EDS, when an electron beam of sufficient energy en-

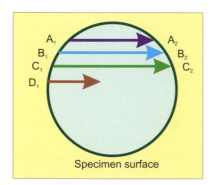

FIGURE 9.15   Direction of scanning of the electron beam on a sample surface in a scanning electron microscope. (Diagram courtesy D. Bentley.)

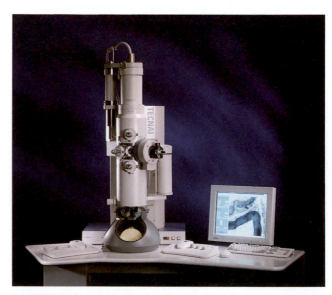

FIGURE 9.16   A typical transmission electron microscope (TEM). (Photo courtesy FEI Company.)

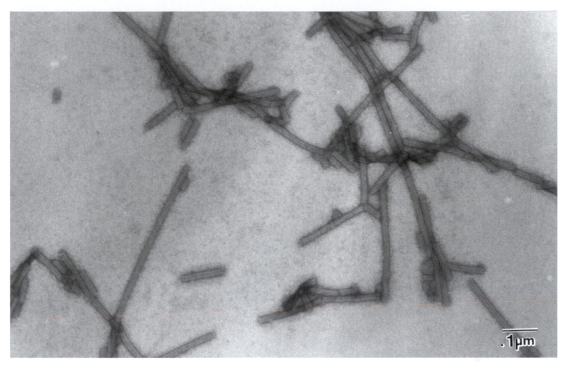

FIGURE 9.17 Transmission electron micrograph of the tobacco mosaic virus. (Photo courtesy C. E. Stauber.)

counters a surface, x-ray photons of characteristic energies may be emitted via inner-shell ionization (Fig. 9.18a and b). The result is a fingerprint of x-ray energies specific for a particular element. By comparison with fingerprints of known elements, an unknown element in a sample can be identified and quantitated. Energy-dispersive spectrometry can be performed with both transmission and scanning electron microscopes. No additional sample preparation is needed beyond that necessary for the electron microscope. EDS applications in environmental microbiology are broad and are discussed in more detail in Chapter 17.

9.8.3 Confocal Scanning Microscopy

A recent development in microscopy is the **confocal scanning microscope** (CSM). The confocal scanning microscope gives higher resolution, increased contrast, and thinner planar views than other forms of light microscopy. Because three-dimensional views can be generated, the CSM readily lends itself to digital processing, by which images of thin optical sections can be reassembled into a composite, three-dimensional image. These images may be viewed as a whole or as individual sections for greater detail. Confocal scanning microscopy is commonly used in bright-field, dark-field, and fluorescence microscopies and is being developed for transmission and polarization applications.

In confocal (having the same foci) microscopy, two lenses are used to focus the specimen. The confocal scanning microscope has the ability to take optical sections at successive focal planes (known as a Z series). Apertures are used so that only a small area of the specimen is focused at any given time. Light from the plane of focus enters the eyepiece, eliminating any scattered light, which has the tendency to blur images. The focused light beam moves across the specimen, scanning it. Scanning is required because only a small volume is illuminated at any given time and a number of these small volumes must be collected for a complete specimen image.

9.8.4 Photography

Photography is an important aspect of all microscopy because it is used to document perceived detail and information about specimens. It is often accomplished by traditional means of exposing and developing photographic film, but the use of **digital imaging** is becoming increasingly popular. With the development of more sophisticated computers and imaging systems, images can be digitized; this process has higher resolution than traditional photographic methods and creates a more accurate reproduction of a microscopic image. The use of computers in the digital processing of microscopic images has allowed auto-

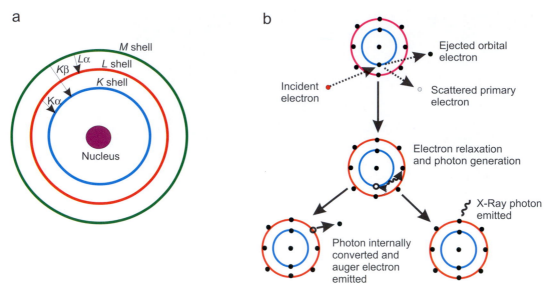

FIGURE 9.18 (a) Electronic transitions in an atom and (b) electron interactions as an electron from the electron beam encounters an atom. X-ray generation is used in electron dispersive spectrometry. (Adapted from Goldstein and Yakowitz, 1975.)

mated image processing. Automated image processing allows the three-dimensional reconstruction of images when successive specimen transects are taken.

Whether created by traditional or digital means, a **micrograph** is a reproduction of an image formed by a particular kind of microscope. A photomicrograph is an image taken by a light microscope. An electron micrograph is an image taken by an electron microscope. An x-ray micrograph results from x-ray microscopy. Micrography provides a means of permanently recording an image for both artistic and scientific purposes.

9.8.5 Other Imaging Systems

In **flow cytometry,** microscopic detection of cells or other particles is required as the cells pass through a laser detector. Flow cytometry was first discovered in the 1950s and its uses include the detection of a variety of microorganisms, including bacteria and parasites. As a cell passes through the detector's laser beam, the amount of light scattered in the forward direction and a direction at a 90° angle is measured. These measurements respectively correlate with the size and internal complexity of the particle. The instrument can also measure the fluorescent light emitted by each particle. Data in the flow cytometer are collected as light energy, converted to electrical energy, and then plotted on user-defined histograms.

In flow cytometry, particles are separated and flow singly through the detector. Flow cytometer cell sorters have the ability to detect desired cells or particles among unwanted ones. The cell sorter vibrates the sample stream, causing it to break into droplets. Information about the particles of interest, such as light scattering and fluorescence criteria, is programmed into the cytometer computer so that when the particle is encountered, the instrument electrically charges the droplet carrying that particle. Oppositely charged deflection plates pull the particles of interest out of the uncharged sample stream toward the charged plate and ultimately deflect them onto a glass microscope slide or into a collection tube. The droplets containing unwanted particles flow into a waste collection tank.

Flow cytometry is commonly used in environmental microbiology. The FITC method relies on the binding of a fluorescein (FITC)-conjugated antibody to antigens present in the sample. For example, this technique has been used to quantitate *Cryptosporidium* and *Giardia* present in environmental samples (see Fig. 8.10 and Chapter 19.3.1). The FITC-stained sample suspension is aspirated by the flow cytometer and each particle in the sample is examined in the instrument's laser beam. The fluorescein molecule, when excited by the 488-nm laser light, in return emits light at 525 nm. The light energy is detected in the flow cytometer and quantitated. The cysts and oocysts of these organisms are identified by their 90° light scattering and additional FITC fluorescence properties. Collection by the flow cytometer on a glass slide allows additional microscopic analysis for identification.

Another imaging system commonly used in environmental microbiology is the **gel imager.** Gel imagers, although they do not contain a microscopic optical system, rely on cameras to increase imaging capabilities for nucleic acids and proteins via polymer gels and blots. Gel imagers are used in the analysis of polymerase chain reaction products and Southern, Western, and Northern blots from gene probing (see Chapter 13).

QUESTIONS AND PROBLEMS

1. List 4 possible applications in environmental microbiology for each of the following microscopic techniques: bright field, fluorescence, electron, and *in situ* microscopy.

2. Direct microscopic counts using acridine orange (AODC) are often 2–3 orders of magnitude greater than viable counts. Why?

3. If you wanted to determine whether or not a specific membrane protein was being produced by a microorganism, which microscopic techniques might you use and why?

4. List at least 4 things about a microorganism you can learn from viewing it with bright field microscopy.

5. Which is most important in microscopy-resolution, contrast or magnification? Why?

6. List two prokaryotic structures that would stain in response to an acidic dye? A basic dye?

References and Recommended Readings

Amann, R. I., Binder, B. J., Olson, R. J., Chisholm, S. W., Devereux, R., and Stahl, D. A. (1990) Combination of 16S rRNA–targeted oligonucleotide probes with flow cytometry for analysing mixed microbial populations. *Appl. Environ. Microbiol.* **56,** 1919–1925.

Caldwell, D. E., and Lawrence, J. R. (1986) Growth kinetics of *Pseudomonas fluorescens* microcolonies within the hydrodynamic boundary layers of surface microenvironments. *Microb. Ecol.* **12,** 299–312.

England, B. M. (1991) Scanning electron microscopy. *Mineral. Rec.* **22,** 123–132.

Goldstein, J. I., and Yakowitz, H. (1975) "Practical Scanning Electron Microscopy." Plenum, New York.

Jones, P. C. T., and Mollison, J. E. (1948) The technique for the quantitative estimation of soil microorganisms. *J. Gen. Microbiol.* **2,** 54–64.

Kogure, K., Simidu, U., and Taga, N. (1979) A tentative direct microscopic method for counting living marine bacteria. *Can. J. Microbiol.* **25,** 415–420.

Korber, D. R., Lawrence, J. R., Zhang, L., and Caldwell, D. E. (1990) Effect of gravity on bacterial deposition and orientation in laminar flow environments. *Biofouling* **2,** 335–350.

Lange, J. L., Thorne, P. W., and Lynch, N. (1997) Application of flow cytometry and fluorescent *in situ* hybridization for assessment of exposures to airborne bacteria. *Appl. Environ. Microbiol.* **63,** 1557–1563.

Mason, C. W. (1983) "Handbook of Chemical Microscopy," Vol. 1, 4th ed. John Wiley & Sons, New York.

Murray, R. G. E., and Robinow, C. F. (1994) Light microscopy. *In* "Methods for General and Molecular Bacteriology" (P. Gerhardt, R. G. E. Murray, W. A. Wood, and N. R. Krieg, eds.) American Society for Microbiology, Washington, DC, pp. 8–20.

Paul, E. A., and Clark, F. E. (1989) "Soil Microbiology and Biochemistry." Academic Press, San Diego.

Pepper, I. L., Gerba, C. P., and Brendecke, J. W. (1995) "Environmental Microbiology Laboratory." Academic Press, San Diego.

Rochow, T. G., and Tucker, P. A. (1994) "Introduction to Microscopy by Means of Light, Electrons, X Rays or Acoustics." Plenum, New York.

Rossi, G., Riccardo, S., Gesue, G., Stanganelli, M., and Wang, T. K. (1936) Direct microscopic and bacteriological investigations of the soil. *Soil Sci.* **41,** 53–66.

Roszak, D. B., and Colwell, R. R. (1987a) Metabolic activity of bacterial cells enumerated by direct viable count. *Appl. Environ. Microbiol.* **53,** 2889–2983.

Roszak, D. B., and Colwell, R. R. (1987b) Survival strategies of bacteria in the natural environment. *Microbiol. Rev.* **51,** 365–379.

Soderstrom, D. E. (1977) Vital staining of fungi in pure culture and in soil with fluorescein diacetate. *Soil Biol. Blochem.* **9,** 59–63.

Zieler, H. W. (1972) "The Optical Performance of the Light Microscope," Part 1. Microscope Publications, Chicago.

CHAPTER

10

Cultural Methods

KAREN L. JOSEPHSON   CHARLES P. GERBA   IAN L. PEPPER

10.1. CULTURAL METHODS FOR ENUMERATION OF BACTERIA

The microbial ecologist is often interested in isolating or enumerating viable microorganisms present in an environmental sample. This information may help establish the level of soil fertility or degree of contamination of soil or water by metals, toxic organics, or pathogens. In addition, cultural methods can be used to evaluate the diversity of microbial communities or quantitate a specific organism of interest. However, unlike the assay of pure culture samples, culture techniques involving the diverse microbial communities found in the environment are complex, and the numbers obtained need to be qualified by the culture technique that is used.

To determine the appropriate culture method, it is necessary to define which specific microorganism or group of microorganisms is to be enumerated. The methods for enumerating bacteria, fungi, algae, and viruses are very different. In addition, within each group special enumeration techniques may be needed, as in the case of anaerobic or autotrophic bacteria. Precise, standard methods are necessary for assaying indicator organisms such as fecal coliforms or pathogens such as *Salmonella* spp.

The type of environmental sample under analysis must also be considered when enumerating microorganisms. Different techniques may be necessary for extraction of the organisms from the sample, and, depending on the number of organisms of interest present in the particular sample, dilution or concentration of the sample may be necessary (see Chapter 8). Selective or differential growth media (see Chapter

10.2.1.2) may also be necessary to observe a small population of a specific organism when an abundance of other organisms are also present in the sample. For example, it is usually necessary to enumerate *Salmonella* in sewage samples in which a large number of coliforms are present.

In this section we will discuss the two basic approaches to viable count procedures for bacteria: the standard plate count and the most probable number (MPN) technique. We will also discuss the advantages and disadvantages of the two methods and present selected protocols for culturing specific bacteria from environmental samples.

10.1.1 Enumeration and Isolation Techniques

10.1.1.1 Extraction of Cells from Soil

For accurate viable plate counts from soil it is necessary to have efficient extraction of the microorganisms that are attached to soil particles or are present in the pores of soil aggregates. Extraction of cells from soil is discussed in detail in Chapter 8. Methods often employed include hand or mechanical shaking with or without glass beads, mechanical blending, and sonication. Different extracting solutions may be used depending on the pH and texture of the soil. A surfactant such as Tween 80 (Difco) may be used, often with the use of a dispersing agent such as sodium pyrophosphate.

10.1.1.2 Dilution Step

The extraction step is followed by serial dilution of the sample to separate the microorganisms into indi-

213

vidual reproductive units. Sterile water, physiological saline, and buffered peptone or phosphate solutions are a few of the solutions commonly used for this step. Although water is convenient and commonly used, it is not preferred because it does not prevent osmotic shock during the dilution process.

10.1.1.3 Concentration Step for Water Samples

Some environmental samples, such as marine water or drinking water, contain low bacterial numbers and require concentration rather than dilution before enumeration (see Chapter 8.2.1). In this method, a specified volume of water is filtered through a membrane using a vacuum. The bacteria are trapped on the membrane, which is placed on the agar medium or a cellulose pad soaked in medium to allow growth of individual colonies. This is the basis of the **membrane filtration technique** (see Section 10.2.1.3 and Chapter 20.2.2). Different volumes of water may need to be filtered to obtain the correct concentration of bacteria on the membrane for isolation and counting purposes. In this situation, it is critical to select a type and size of membrane appropriate to the bacteria to be collected. Often a nitrocellulose filter of pore size 0.45 μm is used. Care must be taken during processing of the sample to cause minimal stress to the organism with respect to such factors as processing time, vacuum pressure, and desiccation.

10.1.2 Plating Methods

After dilution or concentration, the sample is added to petri dishes containing a growth medium consisting of agar mixed with selected nutrients. Two different methods are used for application of the diluted sample to the growth medium. In the **spread plate** method, a 0.1-ml aliquot of selected dilutions of the sample is uniformly spread on top of the solid agar with the aid of a sterile glass rod (Fig. 10.1). Alternatively, in the **pour plate** method, 1-ml aliquots of appropriate sample dilutions are mixed with molten agar (45°C) in a petri dish and allowed to solidify (Fig. 10.2). The spread plate technique is advantageous in that it allows colonies to develop on the surface of the agar, making it easier to distinguish different microorganisms on the basis of morphology. It also facilitates further isolation of the colonies. The spread plate method generally gives bacterial counts that are higher than the pour plate method (for the same size of inoculant), perhaps because of improved aeration and desegregation of clumps of bacteria with this method.

After plating, the samples are incubated under specified conditions, allowing the bacteria to multiply into macroscopic, isolated colonies known as colony-

forming units (CFUs). Because it is assumed that each colony-forming unit originates from a single bacterial cell, it is critical for the organisms to be separated into discrete reproductive units in the dilution step prior to plating. In reality, however, colonies may arise from chains or clusters of bacteria, resulting in underestimation of the true bacterial number. The total number of bacteria of interest is calculated from the number of colonies found on a specific dilution. The range of colonies acceptable for counting is 30–300 on a standard 150-mm-diameter agar plate. Below 30, accuracy is reduced; above 300, accuracy increases only slightly and, in fact, numbers may be reduced by overcrowding and competition between organisms growing on the plate. An example of a typical dilution and plating calculation is shown in Information Box 1.

The process of isolating microorganisms from environmental samples often necessitates further isolation of individual colonies arising from the spread or pour plate method to allow further characterization or confirmation. In this technique, a sterile inoculating loop is used to pick colonies from the original agar plate and the loop of bacteria is streaked to dilution on a new agar plate. This process is shown in Fig. 10.3.

The standard plate count technique for the enumeration of microorganisms is one of the oldest and most widely used techniques in microbiology. Despite its popularity, the dilution and plating technique has been subject to much scrutiny and criticism almost since its inception. One of the main criticisms is that only a small fraction of the total population, as observed microscopically, can be cultured on laboratory media. It is well documented that only 1–10% of the number of cells observed with direct microscopic counts can be recovered as viable bacteria using plate count techniques (see Chapter 9.6). This large discrepancy between techniques may be due to several factors. First of all, environmental samples with low substrate levels may be populated with nutrient-starved or oligotrophic organisms, which are very slow growing or nonculturable and so do not appear as colonies in a reasonable time frame. Second, viable organisms may become nonculturable because of stress imposed on them during the extraction and dilution process. Another reason is that competition that occurs among bacteria to form colonies may suppress the growth of some microbes. In addition, all microbes have different optimal and minimal growth conditions with respect to nutrient and electron acceptor requirements. Thus, some microbes grow very slowly or not at all on the plating medium. Because most natural environments are oligotrophic, the nutrient status of the plating medium is an important point to consider when performing viable counts. Many soil isolates grow better on a nutrient-poor

Step 1. Make a 10-fold dilution series.

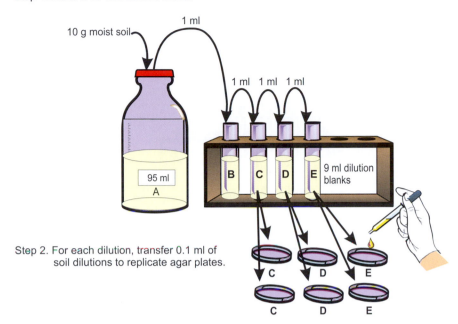

Step 2. For each dilution, transfer 0.1 ml of
       soil dilutions to replicate agar plates.

Step 3a. A glass spreading rod is flame sterilized.

Step 3b. Sample is spread on the surface of
        the agar. This is done by moving the
        spreader in an arc on the surface of the
        agar while rotating the plate.

Step 4. Incubate plates under specified conditions.

Step 5. Count dilutions yielding 30-300 colonies
       per plate. Express counts as CFUs per
       g dry soil.

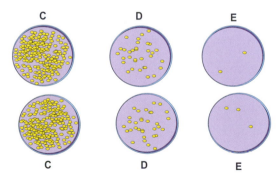

FIGURE 10.1   Dilution and spread plating technique. Here, soil that initially contains billions of microbes is diluted prior to being spread plated to enable discrete colonies to be seen on each plate. Numbers of colonies on each plate can be related to the original soil microbial population. (Adapted from Pepper *et al.*, 1995).

Step 1. Make a 10-fold dilution series.

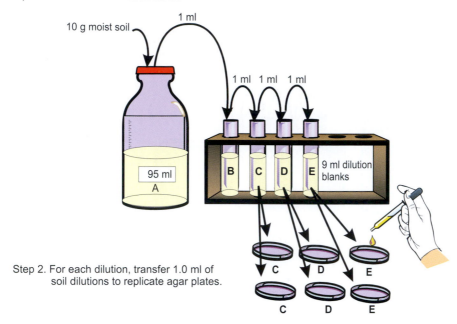

Step 2. For each dilution, transfer 1.0 ml of
       soil dilutions to replicate agar plates.

Step 3a. Add molten agar cooled to 45°C
        to the dish containing the soil
        suspension.

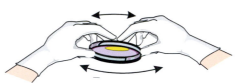

Step 3b. After pouring each plate, replace
        the lid on the dish and gently swirl
        the agar to mix in the inoculum and
        completely cover the bottom of the
        plate.

Step 4. Incubate plates under specified conditions.

Step 5. Count dilutions yielding 30-300 colonies
       per plate. Express counts as CFUs per
       g dry soil.

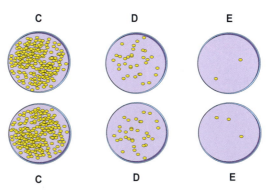

FIGURE 10.2   Dilution and pour plating technique. Here, the diluted soil suspension is incorporated directly in the agar medium rather than being surface applied as in the case of spread plating. (Adapted from Pepper *et al.,* 1995).

medium than on a rich medium. Finally, variation in recovery of viable bacteria can be due to unsatisfactory release of the bacteria from soil particles during the extraction step or to poor separation of the bacteria during the dilution step. This allows clumps of bacteria to give rise to single colonies, resulting in underestimation of the colonies present. It is impossible to determine which of these factors is responsible for the observed 1- to 2-log difference in direct and viable counts. As a rule, the cells that cause this difference in counts

---

Information Box 1
Dilution and Plating Calculations

A 10-gram sample of soil with a moisture content of 20% on a dry weight basis is analyzed for viable culturable bacteria via dilution and plating techniques. The dilutions were made as follows:

Step			Dilution
10 g soil	$\rightarrow$	95 ml saline (solution A)	10^{-1} (weight/volume)
1 ml solution A	$\rightarrow$	9 ml saline (solution B)	10^{-2} (volume/volume)
1 ml solution B	$\rightarrow$	9 ml saline (solution C)	10^{-3} (volume/volume)
1 ml solution C	$\rightarrow$	9 ml saline (solution D)	10^{-4} (volume/volume)
1 ml solution D	$\rightarrow$	9 ml saline (solution E)	10^{-5} (volume/volume)

1 ml of solution E is pour plated onto an appropriate medium and results in 200 bacterial colonies.

$$\textbf{Number of CFU} = \frac{1}{\text{dilution factor}} \times \text{number of colonies}$$

$$= \frac{1}{10^{-5}} \times 200 \text{ CFU/g moist soil}$$

$$= 2.00 \times 10^7 \text{ CFU/g moist soil}$$

But, for 10 g of moist soil,

$$\text{Moisture content} = \frac{\text{moist weight} - \text{dry weight (D)}}{\text{dry weight (D)}}$$

Therefore,

$$0.20 = \frac{10 - D}{D} \quad \text{and}$$

$$D = 8.33 \text{ g}$$

$$\textbf{Number of CFU per g dry soil} = 2.00 \times 10^7 \times \frac{1}{8.33} = 2.4 \times 10^7$$

---

are referred to as viable but nonculturable, as it is assumed that dead cells are rapidly degraded and utilized (Roszak and Colwell, 1987).

10.1.3 Most Probable Number Technique

The **most probable number (MPN)** technique is sometimes used in place of the standard plate count method to estimate microbial counts in the environment. In this method the sample to be assayed is dispersed in an extracting solution and successively diluted, as in the plate count. This method relies on the dilution of the population to extinction followed by inoculation of 5—10 replicate tubes containing a specific liquid medium with each dilution. After incubation, the tubes are scored as $+/-$ for growth on the basis of such factors as turbidity, gas production, and appearance or disappearance of a substrate (Fig. 10.4). Scoring a tube positive for growth means that at least one culturable organism was present in the dilution used for its inoculation. The number of positive and negative tubes at each dilution is used to calculate the number present in the original sample through the use of published statistical MPN tables or computer programs designed to simplify the analysis. MPN tables can be found in American Public Health Association (1992).

MPN methodology is useful because it allows estimation of a population of bacteria based on a process-related attribute such as nitrification by *Nitrosomonas* and *Nitrobacter* spp. or nitrogen fixation by *Rhizobium* spp. It is mandatory to use MPN analysis when enumerating a microorganism requiring broth enrichment prior to cul-

Step 1 Sterilize inoculating loop

A   B

Step 2 Obtain culture from an agar plate (A) or from broth (B).

Step 3 Make successive streaks on an agar plate
        to isolate single colonies

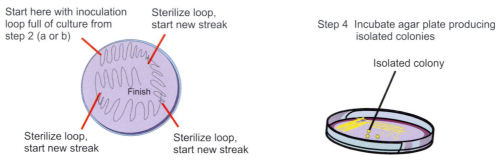

Start here with inoculation   Sterilize loop,
loop full of culture from   start new streak
step 2 (a or b)

Step 4 Incubate agar plate producing
        isolated colonies

Isolated colony

Finish

Sterilize loop,   Sterilize loop,
start new streak   start new streak

FIGURE 10.3   Isolation of a bacterial colony using the streak plate technique.

turing. It is also essential when assaying many foods, sediments, and soils. However, the MPN technique is very labor intensive and results are usually less precise than those obtained with direct plating methods.

10.2 CULTURAL MEDIA FOR BACTERIA

10.2.1 General Media Used for Culturing Bacteria

Media for the cultivation of microorganisms must contain substances that will support their growth, and the media available are as diverse as the microorganisms themselves. Many media are available from commercial manufacturers such as Difco (Detroit, MI) or BBL (Cockeysville, MD), and variations and new formulations are published by researchers and available through journals and volumes such as the "Handbook of Microbiological Media" (Atlas, 1993). The major components of microbiological media include (1) a source of carbon for incorporation in biomass, such as glucose for heterotrophic bacteria or CO_2 for autotrophic bacteria; (2) nitrogen, which is needed for growth and commonly supplied as ammonia, proteins, amino acids, peptones, or extracts from plants or meat; (3) buffers to maintain a suitable pH; and (4) growth factors such as defined trace minerals or

metals, or undefined factors such as those found in extracts made from the environmental samples themselves. Agar is the most common solidifying agent used in media. It is a polysaccharide of an extract from marine algae. Agar exists as a liquid at high temperatures but solidifies on cooling to 38°C. Although agar ideally does not supply nutritional value to the medium, variations in growth may be observed with different types of agar. Bacteria may also be cultured in a broth medium lacking this solidifying agent. Many media also contain selective components that favor the growth of specific organisms while inhibiting the growth of nontarget organisms (see Chapter 10.2.1.2).

While cultural media generally are used to enumerate or isolate specific microorganisms, they also can be used in metabolic fingerprinting analysis to allow identification of microbial isolates. Such systems include API strips (Analytab Products) and Enterotube (Roche) which are routinely used for clinical identification of gram-negative microorganisms. For environmental microbiology, the most commonly used of the metabolic fingerprinting systems is **Biolog.** Biolog is used for identification of single isolates as well as for analysis of community composition. The basis for the Biolog system is a 96-well microtiter plate where 92 of the wells contain different carbon substrates. Each well is inoculated with the same isolate or community sample. If the substrate is utilized, the well turns purple. The plates

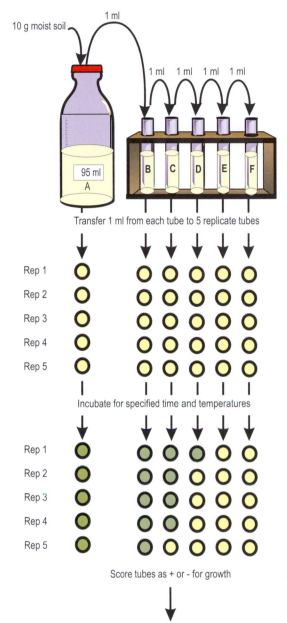

FIGURE 10.4   Most probable number technique. Here, an environmental sample is diluted to extinction. Diluted samples are used to inoculate replicate tubes at each dilution. The presence or absence of the microorganism of interest at a given dilution can be analysed statistically to estimate the original population in the environmental sample. (Adapted by permission Pepper *et al.*, 1995).

can be read either manually or automatically using a plate reader. The results are then compared (in the case of isolates) with a database provided by Biolog. One of the drawbacks of this system for environmental isolates is that the database is not yet large enough to provide identification for most isolates. In the case of community samples, Biolog is used to provide a comparison of the

community before and after a perturbation or simply to monitor the community for a period of time.

10.2.1.1 Heterotrophic Plate Counts

Heterotrophic bacteria obtain energy and carbon from organic substances, and it is often desirable to obtain a count of the total heterotrophic population from soil or water. Heterotrophic plate counts give an indication of the general "health" of the soil as well as an indication of the availability of organic nutrients within the soil. Two basic types of media can be used for this analysis: **nutrient-rich** and **nutrient-poor media.** Examples of nutrient-rich media are nutrient agar (Difco), peptone–yeast agar (Atlas, 1993), and soil extract agar amended with glucose and peptone (Atlas, 1993). These media contain high concentrations of peptone, yeast, and/or extracts from beef or soil. Nutrient-poor media are often called **minimal media** and contain as much as 75% less of these ingredients, often with substitutions such as casein, glucose, glycerol, or gelatin. Examples of minimal media are R2A agar (BBL, Difco), m-HPC agar (Difco), and soil extract agar with no amendments. In many cases, higher colony counts are obtained with a minimal medium because a large number of oligotrophic organisms in the environment cannot be cultured on a rich medium (Table 8.2). This is true for most water and subsurface porous medium samples, for which the nutrient-poor medium, R2A, is often used (Fig. 10.5). Growth on a nutrient-poor medium may take longer (5–7 days) and

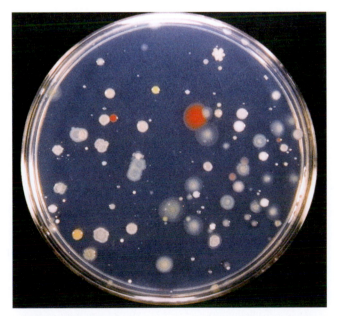

FIGURE 10.5   Heterotrophic colonies on an R2A agar plate. A number of discrete colonies with diverse morphology arise after dilution and plating from soil. (From *Pollution Science*, © 1996 Academic Press, San Diego, CA.)

the colony sizes are often smaller than for a rich medium. In addition, the community of isolates from a single sample may be entirely different when plated on the two media types.

10.2.1.2 Culturing Specific Microbial Populations

It is often necessary to detect or enumerate a specific population of microorganisms, or even a very specific bacterial isolate, from the total population of bacteria found in an environmental sample. This may necessitate culturing with one or more specialized media and often requires that a specified sequence of steps be performed to ensure maximum culturability of the target environmental organism (See Case Study). We will now define the media involved in these steps and explain why they are important.

Pre-enrichment or resuscitation medium is a liquid medium that allows the microorganisms in the environmental sample to begin to actively metabolize and increase in number. There are two purposes for a pre-enrichment medium. First, damaged cells are given time and the necessary nutrients to repair and grow. This step is important because direct inoculation of a sample into a selective enrichment medium (discussed next) may result in the death of some bacteria. This is especially critical for sublethally injured organisms, which may not be recoverable under selective conditions. The second purpose of pre-enrichment is to increase the number of cells. This step is crucial when trying to enumerate low numbers of a target organism in environmental samples.

Enrichment medium is a liquid medium that promotes the growth of a particular physiological type of

microorganism present in a mixture of organisms while suppressing the growth of competitive background flora. It aids in the detection of the desired environmental isolate when the specimen contains a high population of normal flora. Enrichment media may be elective or selective. **Elective enrichment medium** allows growth of a single or limited type of bacteria based on a unique combination of nutritional or physiological attributes. **Selective enrichment medium** involves the use of inhibitory substances or conditions to suppress or inhibit the growth of most organisms while allowing the growth of the desired organism. A detailed description of selective agents is given next.

A **selective plating medium** is a modification of an agar medium to suppress or prevent the growth of one group of organisms while fostering the growth of the desired organism. Antibiotics are among the most widely used and effective selection agents. Cycloheximide is often used in heterotrophic media to inhibit the growth of fungi from soil. Table 10.1 lists some of the more common antibiotics, with their spectrum and mode of action. Other examples of selective agents often added to growth media are metals such as mercury or lead, which are usually used in the presence of a minimal salts medium. Selection may also be accomplished by the use of toxic chemicals such as high salt concentrations or dyes. For example, the dye crystal violet inhibits most gram-positive bacteria while allowing the gram-negative bacteria to grow. Adjustments in pH or osmotic conditions are also used for selection. Finally, selectivity may be based on incubation conditions of the inoculated growth medium. These may include temperature or oxygen levels.

TABLE 10.1 Common Antibiotics Often Used in Selective Media

Name	Spectrum	Mode of action
Chloramphenicol	Broad spectrum	Inhibits protein synthesis by binding to 50S ribosomal subunit
Erythromycin	Mostly gram-positive	Inhibits protein synthesis by binding to 50S ribosomal subunit
Tetracycline	Broad spectrum	Inhibits protein synthesis by binding to 30S ribosomal subunit
Streptomycin	Broad spectrum	Inhibits protein synthesis by binding to 30S ribosomal subunit
Polymyxin	Gram-negative bacteria, especially *Pseudomonas*	Disrupts cell membrane
Nalidixic acid	Gram-negative bacteria	Inhibits DNA synthesis
Novobiocin	Gram-negative bacteria	Inhibits DNA synthesis
Trimethoprim	Broad spectrum	Inhibits purine synthesis
Rifampicin	Gram-positive bacteria	Inhibits RNA synthesis
Penicillin	Mostly gram-positive bacteria	Inhibits cell wall peptidoglycan synthesis

A **specialized isolation medium** contains formulations that meet the nutritional needs of specific groups of organisms, such as *Staphylococcus* or *Corynebacterium*, thereby allowing differentiation and identification.

A **differential medium** contains ingredients to allow distinction of different microbes growing on the same medium. It includes an indicator, usually for pH, to distinguish certain groups of organisms on the basis of variations in nutritional requirements and the production of acid or alkali from various carbon sources. This results in a distinguishing morphological characteristic of the colony, usually color. In addition, it may support the growth of a selected group of bacteria while inhibiting the growth of others. Thus, a medium can be selective as well as differential.

Although culturing procedures for pure laboratory strains or clinical samples are fairly straightforward, the same techniques may not be successful for culturing the same organisms found in soil or water. Often it is necessary to follow a series of steps or modifications in procedures along with a combination of media to obtain optimal culturing of the target environmental organism. Furthermore, once colonies are isolated by cultural methods, they may be characterized by physiological (see Chapter 11), immunological (see Chapter 12), or molecular (see Chapter 13) techniques. In the next sections we present five examples of the application of these methods to culture specific microbes from soil and water samples. These examples were chosen to illustrate a wide range of bacteria and problems encountered in the environment. They are (1) the isolation of fecal coliforms from water as indicators of wastewater contamination; (2) the isolation of a pathogen, *Salmonella*, from sewage sludge; (3) the isolation of a fluorescent *Pseudomonas*, a common bacterium found in soil; (4) the isolation from soil of nitrifying bacteria, a population that is among the first affected by stress conditions; and (5) the isolation of 2,4-dichlorophenoxyacetic acid (2,4-D)–degrading bacteria to represent the isolation of an organism with specific enzymatic activity. This is far from an exhaustive list, but the selection represents a variety of cultural methodologies and their application to environmental samples.

10.2.1.3 Identification of Fecal Coliforms from Water Samples Using the Membrane Filtration Technique

Coliform bacteria are nonpathogenic bacteria that occur in the feces of warm-blooded animals. Their presence in a water sample indicates that harmful pathogenic bacteria may also be present. Coliforms are found in numbers corresponding to the degree of pollution, are relatively easy to detect, and are overall hardier than pathogenic bacteria. For these reasons, coliforms are important in that researchers may look for the presence of these "indicator organisms" in an environmental sample to assess water quality prior to or in place of culturing other organisms.

The membrane filtration technique is most commonly used for the detection of fecal coliforms in water samples (see Chapter 20, Fig. 20.3). The volume of water required to enumerate fecal coliforms varies for different samples, but a volume of 100 ml is normally required for drinking water, lakes, and reservoirs; 10 ml for marine water; and 0.1 ml for sewage. After sample filtration, the membrane containing the trapped bacteria is subjected to cultural methods. For fecal coliform analysis, the membrane is applied to a petri dish containing m-FC agar (Difco) and incubated at 45°C for 24 hours. Growth from fecal coliforms results in blue colonies, and nonfecal coliform colonies are gray to cream colored. Colonies are viewed and counted on a microscope under low magnification. The fecal coliform–membrane filtration procedure uses an enriched lactose medium with 1% rosalic acid to inhibit the growth of noncoliform colonies, and an elevated incubation temperature is a critical component in the selection for fecal coliforms.

This is a published standard method ("Standard Methods for the Examination of Water and Wastewater," 9222D), but there are also variations to help recover injured bacteria. These include an enrichment–temperature acclimation in which the membrane is incubated on nonselective agar at 35°C for 2 hours prior to incubation on the m-FC medium, and deletion of the rosalic acid suppressive agent from the m-FC medium. Individual colonies must also be confirmed to be fecal coliforms.

10.2.1.4 Enumeration of Salmonella from Sewage Samples Using the MPN Technique

Salmonella is an enteric bacterium that is pathogenic to humans and causes a wide range of symptoms, primarily gastroenteritis. It can be transmitted by drinking improperly disinfected water or contaminated recreational water, but infection in the United States is primarily due to foodborne transmission because *Salmonella* infects both beef and poultry. Municipal sewage sludge contains many microorganisms, including *Salmonella*. Some cities now apply treated sewage to agricultural soil to improve nutrient quality. Thus, it has become necessary to monitor the sewage for indicator organisms and pathogens such as *Salmonella* to assess health risks to the general population. As you will see from the following example, culture methods for *Salmonella* are varied, complex, and time consuming and require final confirmatory tests (Fig. 10.6).

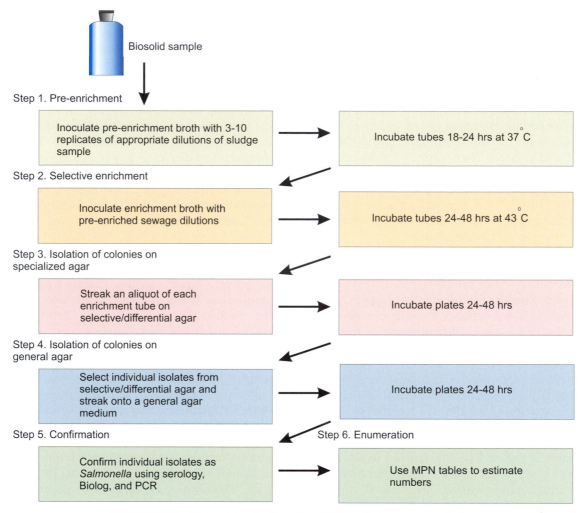

FIGURE 10.6   Protocol for the detection of *Salmonella* from sewage sludge.

The traditional method for detecting *Salmonella* spp. relies on an enrichment and plating technique with subsequent estimation using MPN tables. Selected volumes of sewage, generally ranging from 10 to 0.01 ml, are added to a pre-enrichment broth and incubated for 18–24 hours at 37°C. Laboratory studies have found peptone and lactose broth to be most effective for the resuscitation of most *Salmonella* spp. Aliquots from the pre-enrichment phase are added to a selective enrichment broth. Numerous media are available for this step, all with different selection techniques. Included are the use of the inhibitory agent sodium selenite in broth and brilliant green and tetrathionate, which are contained in Mueller–Hinton tetrathionate broth (Difco). Rappaport- Vassiliadis (RV) (Difco) broth is also widely used; selection is based on resistance of *Salmonella* to malachite green, $MgCl_2$, and ability to grow at a pH of 5.0. Recently, RV broth was modified to include the antibiotic sodium novobiocin with incu-

bation at an elevated temperature of 43°C. This medium, known as NR10 broth, has shown much promise for isolation of *Salmonella* from marine water and possible deletion of the pre-enrichment phase of analysis (Alonso *et al.*, 1992).

Growth from the enrichment phase is plated onto a selective or differential medium such as Hektoen enteric agar (Difco), brilliant green bile agar (Difco), or xylose–lysine–deoxycholate agar (Difco). Characteristics used for differential purposes are production of H_2S and the inability of *Salmonella* to ferment lactose. Selective agents include triphenylmethane dyes, antibiotics such as sulfadiazine, or sodium deoxycholate salts.

Individual colonies arising from this medium are streaked onto conventional growth media to allow confirmation using biochemical tests such as BIOLOG, serological identification (see Chapter 12), or confirmation using nucleic acid methods such as the polymerase chain reaction (PCR) (see Chapter 13).

10.2.1.5 Isolation of Fluorescent Pseudomonads from Soil

Pseudomonas are gram-negative aerobic chemo-heterotrophs and are commonly found in the environment. They exhibit diverse enzymatic systems and are capable of conducting many biochemical transformations. Fluorescent pseudomonads are characterized by the production of siderophores. Siderophores are compounds that are produced under low levels of iron. They have the ability to chelate the iron and transport it into the microbial cell. These microbes also have a yellow–green pigment that diffuses through agar media during growth and fluoresces under ultraviolet light. Pigment production is enhanced by iron deprivation.

For enumeration of fluorescent pseudomonads, soil is added to an extracting solution and a dilution series is performed as outlined earlier (see Section 10.1.1). The spread plate technique is used to isolate individual colonies on selective media (see Section 10.1.2). One medium appropriate for isolation is called S1 medium (Gould *et. al.*, 1985). Components of the media include sucrose, glucose, Casamino acids, sodium lauryol sarcosine (SLS), trimethoprim, various salts, and agar. The SLS prevents the growth of gram-positive organisms and the trimethoprim is an effective inhibitor of facultative gram-negative organisms. Sucrose and glycerol provide an osmotic stress that selects for the fluorescent pseudomonads. Fluorescent colonies are identified on the medium with the use of ultraviolet light.

10.2.1.6 Isolation of Nitrifying Organisms from Soil

Nitrosomonas and *Nitrobacter* are chemoautotrophic organisms found in soil and water and are responsible for the oxidation of ammonium to nitrite (*Nitrosomonas*) and nitrite to nitrate (*Nitrobacter*). This process, known as nitrification, is important because it can affect plant growth beneficially, but nitrate also contributes to potable water contamination (see Chapter 15.6). Direct plating techniques are difficult, because even with use of strictly inorganic substrates in the medium, slight amounts of organic material introduced by inoculation with the environmental sample allow growth of faster growing heterotrophic organisms. One approach used to overcome this problem is a lengthy serial enrichment technique that includes soil enrichment, an initial culture enrichment step, a final enrichment step, isolation on agar, and rigorous purity checks (Fig. 10.7). To begin the isolation of *Nitrobacter* from soil, fresh field soil is leached with sodium nitrite on a daily basis for 2–3 weeks to enrich

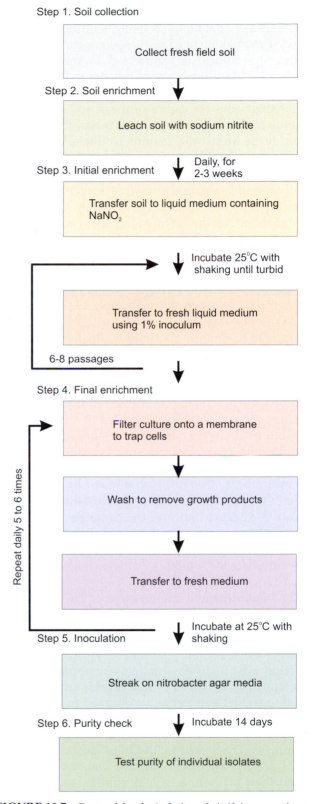

FIGURE 10.7 Protocol for the isolation of nitrifying organisms.

for *Nitrobacter*. The enriched soil is transferred to a liquid medium containing $NaNO_2$. Cultures are incubated with shaking at 25°C. As soon as turbidity is detected, the culture is transferred to fresh medium using a 1% inoculum. Prompt transfer reduces the development of heterotrophs in the sample. The initial enrichment goes through a series of six to eight passages to fresh medium using a 1% inoculum each time. After these steps, the culture is filtered through a membrane to trap the cells and washed to remove growth products. The membrane is transferred to fresh medium and again undergoes five or six passages into fresh medium after filtration of cells onto a membrane. The final enrichment culture is streaked on *Nitrobacter* agar medium (Atlas, 1993) and tiny colonies appearing after 14 days are indicative of *Nitrobacter*. Purity of the test culture must be checked by using media with various carbon and nitrogen sources and osmotic strengths for selection (Schmidt *et al.*, 1973). Note that in this method selection occurs for the isolates most adapted to the culture conditions, which may not necessarily represent the major composition of nitrifiers in the original sample.

Numbers of *Nitrobacter* spp. in a soil sample can also be estimated using an MPN technique. After extraction, the soil dilutions are added to replicate tubes of broth medium containing potassium buffers, magnesium salts, trace elements, iron, $NaNO_2$ as the inorganic substrate for oxidation, and Bromthymol Blue as the indicator. A drop in pH monitored by a change in color of the tubes from blue–green to yellow over a period of 3–6 weeks indicates the oxidation of NH_4^+ to NO_2^-. Dilutions are scored as $+/-$ for growth, indicated by the presence of NO_2, and the population is estimated using MPN tables.

10.2.1.7 Isolation and Enumeration of 2,4–D Degrading Bacteria in Soil

Researchers are often interested in determining whether bacteria that perform a known metabolic function exist in an environmental sample. The most common approach used to address this question is the dilution and plating technique outlined in Fig. 10.1. As an example, the research community is very interested in the fate of pesticides that are introduced into the environment. The herbicide 2,4-D (2,4-dichlorophenoxyacetic acid) is one such pesticide that has been commonly used in fate and ecology studies (see Chapter 3, Case Study). A cultural medium used for enumeration and isolation of viable 2,4-D degraders is a selective, differential enrichment medium called eosin–methylene blue (EMB)–2,4-D agar (Neilson *et al.*, 1994). This agar contains minimal salts and 2,4-D as the carbon

source. Indicators eosin B and methylene blue allow selection for gram-negative bacteria and differentiation of 2,4-D–degrading colonies, which turn black. In essence, this is using 2,4-D as an elective carbon source and the indicators as both selective agents and differential agents that distinguish on the basis of the black colonies that arise during 2,4-D degradation.

10.3 CULTURAL METHODS FOR FUNGI

Fungi are ubiquitous in nature and can be found in samples taken from soil, sediments, and aquatic environments such as lakes, ponds, rivers, marine water, wastewater, and well water. They are heterotrophic organisms, mostly aerobic or microaerophilic in nature. Fungi exist in a variety of morphological and physical states, which makes them difficult to quantitate and identify by cultural techniques. Cultural methods for fungi are similar to those for bacteria but must be modified to restrict bacterial growth. This is normally done by the addition of antibiotics or dyes, such as Rose Bengal, or by lowering the pH of the medium. Cultural methods for fungi are normally used to obtain pure fungal isolates. Cultural methods for fungi are not, however, appropriate for quantitative dilution and plating analyses, because counts can be highly biased by the presence of spore-forming fungi. In this case, colonies can arise from spores, and the number of culturable colonies obtained will not be a true reflection of the number of colonies in the environmental sample. As we have discussed previously, for cultural assays of bacteria it is assumed that each colony-forming unit originates from a single bacterial cell. However, in the same enumeration technique for fungi, fungal colonies may develop from a single spore, an aggregate of spores, or a mycelial fragment containing more than one viable cell. Keeping in mind this limitation, plating techniques are still a valuable tool for the assessment of fungi in the environment. Techniques for culturing fungi include the pour plate technique (see Section 10.1.2), the spread plate technique (see Section 10.1.2), and the membrane filtration technique (see Section 10.1.1.3). These protocols as they relate specifically to the cultivation of fungi can be found in "Standard Methods for the Examination of Water and Wastewater" (Methods 9610B, C, and D, respectively). Modifications of the dilution and plating technique include directly picking fungal hyphae from samples with subsequent plating on nutrient agar or washing the soil (or plant roots) and placing a small amount of the washed sample on the agar medium.

Fungi are often isolated on a nonselective agar medium which allows the isolation of the maximum number of fungal taxa from the sample under study. As with bacteria, a wide selection of nutrient media are available. The medium of choice for cultivating a wide variety of fungal species is Neopeptone–Glucose–Rose Bengal Aureomycin agar (Atlas, 1993) or Rose Bengal streptomycin agar (Martin, 1950) (Fig. 10.8). Czapek–Dox medium (Difco Manual, 1984) is often used for the cultivation of *Aspergillus*, *Penicillium*, and related fungi.

General cultural practices include the use of nitrate as a source of nitrogen and adjustment of the medium to an acidic pH. As already mentioned, it is often necessary to suppress the growth of bacteria to aid in fungal enumeration by the addition of growth inhibitors such as crystal violet or Rose Bengal or a broad-spectrum antibiotic such as streptomycin. A lowered incubation temperature of 4–10°C also helps fungi outcompete bacteria on the growth medium. The optimal number of colonies to count per plate is usually about 100.

Often, the scientist is interested in the isolation of a specific physiological group of fungi or a specific taxon such as Basidiomycetes and mycorrhizal fungi. A wide range of selective media have been developed for this purpose (Seifert, 1990). The MPN technique (see Section 10.1.3) is often used for the quantification of mycorrhizal fungi, which do not grow well on plating media. In this technique, the soil dilutions are used to inoculate a host plant rather than a nutrient medium. After a specified incubation, plant roots are observed for colonization by the fungi, and calculations are performed using the MPN tables to estimate the numbers present in the original sample. After the

initial isolation of fungi from the environmental sample, individual colonies must be selected for further purification and identification.

10.4 CULTURAL METHODS FOR ALGAE AND CYANOBACTERIA

Algae are unicellular or multicellular phototrophic eukaryotic microorganisms that occur in fresh and marine water and moist soil (see Chapter 2.4). In contrast, the so-called blue–green algae are not true algae—rather they belong to a group of bacteria known as cyanobacteria. The majority of soil algae are obligate photoautotrophs, using light to manufacture organic compounds from inorganic nutrients (phototrophic eukaryotes). Thus, their nutritional requirements include water, light, oxygen, carbon dioxide, and inorganic nutrients. A small population of algae are photoheterotrophic, requiring organic compounds for growth. Algae and cyanobacteria can be enumerated by dilution and plating techniques (see Section 10.1.2) as well as MPN assay (see Section 10.1.3) with some suggested modifications. Because of the filamentous nature of some algal species, it is sometimes necessary to use more forceful methods for extraction than with bacteria, prior to dilution and culturing techniques. These include grinding soil samples with a mortar and pestle prior to extraction, as well as a longer and more vigorous homogenizing step using a blender or glass beads. When enumerating microalgal colony-forming units on solid media, the addition of antibiotics such as penicillin and streptomycin to the algal medium is recommended to ensure their growth in the presence of the faster growing heterotrophic bacteria and fungi. Alternatively, low doses of UV radiation or exposure to nonspecific bactericides such as formaldehyde or sodium lauryl sulfate may be used to suppress growth of bacteria. An incubation period of 1–3 weeks or even longer may be necessary, requiring special care that desiccation of the medium does not occur. Recommended air temperature is 20 to 25°C, with a photoperiod of either 12 hours light–12 hours dark, or 16 hours light–8 hours dark.

When isolating algae from environmental samples, the approach may be to select for a specific alga or to obtain a quantitative index of the entire community. In the latter case, an MPN technique using a growth medium consisting of a soil–water mix is often used. Many media are available for specific enrichment and plating, including Bold's Basa medium for green and yellow–green microalgae (and cyanobacteria) (Fig. 10.9).

There have been only a few attempts to isolate and culture the cyanobacteria, but culturable methods do

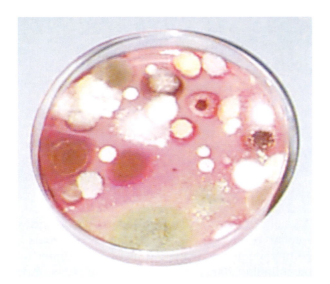

FIGURE 10.8 Fungal colonies cultured on Rose Bengal streptomycin agar. (Photo courtesy K. L. Josephson.)

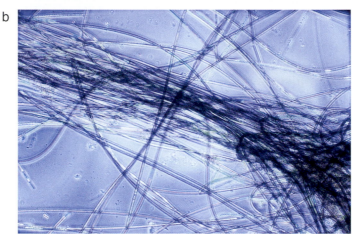

FIGURE 10.9   (a) Cyanobacteria (blue-green algae) grown in full fluorescent light in BG11 broth. (b) An example of a cyanobacterium, *Lynabya* seen using phase contrast microscopy. (Photo courtesy P. Rusin.)

exist. A good review of such methods was presented by Castenholz (1988), including several media suitable for freshwater cyanobacteria. A typical medium (BG-11) is as follows:

NaNO$_3$	1.5 g
K$_2$HPO$_4$	40.0 mg
MgSO$_4$ − 7H$_2$O	75.0 mg
CaCl$_2$ − H$_2$O	36 mg
Citric acid	6.0 mg
Ferric ammonium citrate	6.0 mg
EDTA (disodium salt)	1.0 mg
Na$_2$CO$_3$	20.0 mg
Trace metals	1.0 ml
Purified agar	10.0 g
Distilled water	1.0 L
Cycloheximide	20 mg

After sterilization, the pH should be 7.1.

Trace metals:	
H$_3$BO$_3$	2.86 g
MnCl$_2$ − 6H$_2$O	1.81 g
ZnSO$_4$ − 7H$_2$O	222 mg
Na$_2$MoO$_4$ − 2H$_2$O	39 mg
CuSO$_4$ − 5H$_2$O	79 mg
Co(NO$_3$)$_2$ − 6H$_2$O	49.4 mg
Distilled water	1.0 L

Cycloheximide is usually added to preclude growth of eukaryotic algae, diatoms, and protozoa. The medium should be incubated at 24–26°C under fluorescent light of 2000–3000 lux (cool white light or daylight fluorescent light). For agar plates, samples are usually diluted and applied as a spread plate.

10.5 CELL CULTURE–BASED DETECTION METHODS FOR VIRUSES

Living cells are necessary for virus replication, and originally human volunteers, laboratory animals, or embryonated hens' eggs were the only techniques available to study and isolate human viruses. However, the advent of animal cell culture techniques after World War II revolutionized the study of animal viruses; for the first time, viruses could be grown and isolated without the need for animals. Two main types of cell culture are used in virology, the first of which is **primary cell culture.** In this procedure, cells are removed directly from an animal and can be used to subculture viral cells for only a limited number of times (20–50 passages). Thus, a source of animals must be available on a continuous basis to supply the needed cells. Primary monkey kidney cell cultures (commonly from rhesus or green monkeys) are permissive or allow the replication of many common human viruses and were once routinely used for the detection of enteric viruses in the environment. A second type of cell culture is known as **continuous cell culture.** Continuous cell lines are also derived from animals or humans, but may be subcultured indefinitely. Such cell lines may be derived from normal or cancerous tissue.

Cell cultures are initiated by dissociating small pieces of tissue into single cells by treatment with a proteolytic enzyme and a chelation agent (generally trypsin and EDTA). The dispersed cells are then suspended in cell culture medium composed of a balanced salt solution containing glucose, vitamins, amino acids, a buffer system, a pH indicator, serum (usually fetal calf), and antibiotics. The suspended cells may be placed in flasks, tubes, or petri dishes, depending on the needs of the laboratory. Plastic flasks and multiwell plates are especially prepared for cell culture use and are currently the most common type of container used (Fig. 10.10). The cells attach to the surface of the vessel and begin replicating. Replication ceases when a single layer or **monolayer** of cells occupies all the available surface of the vessel.

Unfortunately, not all animal viruses will grow in the same cell line, and others, including Norwalk virus, have never been grown in cell culture. Viruses require specific receptors for attachment and replication with the host cell, and if these are not present, no replication takes place. For example, rotavirus grows well in the MA-104 cell line but not the BGM cell line. The most commonly used cell line for enterovirus detection in water is the BGM cell line. In recent years, the cell line, CaCO2, originating from a human colon carcinoma, has been found to grow more types of enteric viruses (hepatitis A, astroviruses, adenoviruses, enteroviruses, rotaviruses) than any other cell line and is seeing increased use in environmental virology. Cell lines commonly used to grow human viruses are shown in Table 10.2.

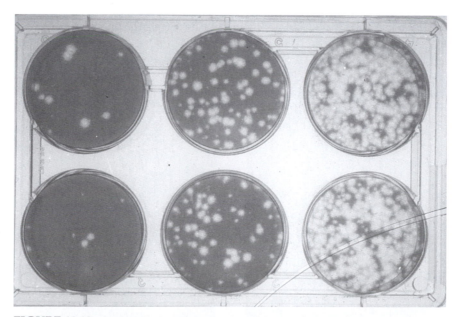

FIGURE 10.10 Multiwell plates for cell culture for virus detection. These wells show increasing dilutions (right to left) of virus on a monolayer. Each clear zone, or plaque, theoretically arises from a single infectious virus particle, i.e. a plaque-forming unit (PFU). (Photo courtesy C. P. Gerba.)

TABLE 10.2 Commonly Used Continuous Cell Cultures for Isolation and Detection of Enteric Viruses

Cell culture line	Adeno	Astro	Coxsackie A	Coxsackie B	Echo	Polio	Rota	Hepatitis A
BGM	−[a]	−	−	+	+	+	−	−
BSC-1	−	*	−	*	*	+	*	*
CaCO2	+	+	*	+	+	+	+	+
Hep-2 (HeLa)	+	*		+	*	*	*	*
RD	*	*	+	−	+	*	*	
RfhK	*	*	*	*	*	*	*	+

[a] +, growth and/or production of cytopathogenic effect (CPE); −, no growth and/or production of CPE; *, no data.

The two most common methods for detecting and quantifying viruses in cell culture are the **cytopathogenic effect (CPE)** method and the **plaque-forming unit (PFU)** method. Cytopathic effects are observable changes that take place in the host cells as a result of virus replication. Such changes may be observed as changes in morphology, including rounding or formation of giant cells, or formation of a hole in the monoloayer due to localized lysis of virus-infected cells (Fig. 10.11 and 10.12). Different viruses may produce very individual and distinctive CPE. For example, adenovirus causes the formation of grapelike clusters, and enteroviruses cause rounding of the cells. The production of CPE may take as little as 1 day to 2–3 weeks, depending on the original concentration and type of virus. Not all viruses may produce CPE during replication in a given cell line. In addition, the virus may grow to high numbers, but no visible CPE is observed. In these instances, other techniques for virus detection must be used. Alternatives include the use of an immunoassay that detects viral antigens or the use of PCR to detect viral nucleic acid.

Generally, laboratory strains of viruses have been selected for their ability to grow in cell culture easily and rapidly with the production of CPE. In contrast, viruses isolated from environmental samples often come directly from infected animals or humans and do not grow as rapidly in cell culture. Often, **"blind passages"** of cell culture showing no CPE are passed on to fresh monolayers of cells before a CPE is observed. Two procedures can be used to quantify viruses utilizing CPE. The first is the **serial dilution endpoint** or **TCID$_{50}$ method,** which involves adding serial dilutions of virus suspension to host cells and subsequently observing the production of CPE over time. The titer or endpoint is the highest viral dilution capa-

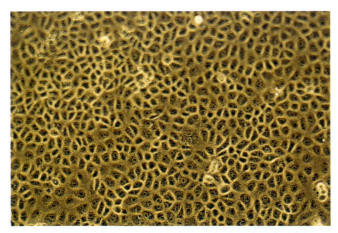

FIGURE 10.11 This figure shows a normal uninfected cell culture monolayer. Compare with Fig. 10.12. (Photo courtesy M. Abbaszadegan.)

FIGURE 10.12 This figure shows an infected cell culture monolayer exhibiting CPE. Compare with Fig. 10.11. (Photo courtesy M. Abbaszadegan.)

CASE STUDY
_____

Evaluation of Pathogens in Household Kitchens Using Cultural Methodologies*

Background

Previous laboratory studies by different research groups have drawn attention to the high prevalence of bacterial contamination in the home. The results of these studies, along with an increase in the incidence of foodborne disease, have generated increased interest in bacterial contamination and dissemination, specifically in the kitchen. A recent 2-year study by researchers at the University of Arizona evaluated 10 "normal" kitchens in the United States for the presence of five different bacteria. The bacteria of interest were total heterotrophs, *Staphylococcus, Pseudomonas* spp, total coliforms, and fecal coliforms. The surfaces in the kitchen assayed were the sink basin, faucet handle, table, countertop, refrigerator door, oven control, cutting board, and sponge.

Kitchen Cleaning Regimen

Phase I: The primary objective of the first phase of this study was to establish baseline populations of common bacterial pathogens and indicator organisms on a variety of surfaces located within the kitchen. Ten selected households that did not currently use any disinfectant cleaners were monitored over a 9-month period. Participants were advised to maintain normal cleaning practices throughout this phase of the study.

Phase II: In this phase, bacterial pathogens and indicator organism populations in household kitchens were characterized as a function of casual disinfectant cleaner use. All homes were provided with an antimicrobial kitchen cleaner. Families were encouraged to use the product, but use of the product was not mandated. Households chose when and how to use the cleaner (i.e., "casual use").

Phase III: This phase evaluated the efficacy of the antimicrobial cleaner under controlled use of the kitchen cleaning product. All homes continued to use the kitchen cleaner as they wished. However, immediately before swabbing, there was controlled cleaning using the antimicrobial product. All surfaces except the sponge and cutting board were sprayed with cleaner and allowed to stand for 5–10 minutes before the surface was wiped clean with a paper towel. The sink surface was swabbed twice, once before and once after cleaning. All other surfaces were swabbed after product use, except the sponge and the cutting board, which, as previously stated, did not receive controlled cleaning with the product.

Collection and Processing of Swab Samples

Dacron swabs were moistened with filter-sterilized neutralizing solution prior to swabbing each surface. Samples were vortexed to remove the bacteria from the swab and eluents were serially diluted in tris–peptone buffer. To culture the various bacteria, appropriate dilutions were grown on the appropriate media.

Isolation of Bacteria

Total heterotrophic bacteria were enumerated on R2A agar and incubated 7 days at 28°C. For detection of *Pseudomonas aeruginosa*, samples were plated on m-PAC agar and incubated at 41.5°C for 72 hours. Selected colonies were confirmed by growth on milk agar at 35°C for 24 hours. *Staphlococci* were cultured on VJ agar at 35°C for 48 hours and then enumerated. Total coliforms were grown on m-T7 agar at 35°C for 24 hours. Fecal coliforms were grown on m-FC agar at 44.5°C for 24 hours.

Fecal coliforms and *Escherichia coli* were also inoculated into Colilert tubes. This test relies on the substrates O-nitrophenyl-β-d-galactopyranoside (ONPG) and 4-methylumbelliferyl-B-D-glucuronide (MUG) to

*Reprinted from Josephson *et al.*, 1997, with permission from Blackwell Science, Ltd.

(continued)

Evaluation of Pathogens in Household Kitchens Using Cultural Methodologies (*continued*)

detect total coliforms and fecal coliforms, respectively. This system recovers injured or stressed bacteria, detects a single viable organism, and minimizes false positives due to *Klebsiella*. Samples positive for *E. coli* were confirmed by using *lamB* PCR primers, which are specific for *E. coli* (Josephson *et al.*, 1997) (see Chapter 13.2.1).

Salmonella and *Campylobacter* were detected using enrichment and plating techniques.

Results and Discussion

The data collected during phase I of this study reflect the background populations of a variety of bacteria isolated from common kitchen surfaces. Almost all locations at all households exhibited contamination. The fact that all kitchen surfaces exhibited some degree of bacterial contamination illustrates the potential for cross-contamination of surfaces through handling of foods and subsequent normal usage of the kitchen. Two items that were consistently contaminated were the sink and the sponge, perhaps due to the constant available moisture within these regimes. Because many households routinely use sponges for washing dishes and wiping off other surfaces, there is a high potential for the transfer of bacterial contaminants from the sponge to other kitchen surfaces. Mean counts and the number of samples found to be positive for fecal coliforms and *Staphylococcus* are shown in Figs. 10.13 and 10.14, respectively.

Apart from *E. coli* (which is a normal gut flora, but can be pathogenic), the incidence of potential pathogens in households was low. *Salmonella* was detected and confirmed on one occasion and *Campylobacter* on two occasions. In contrast, *E. coli* was found routinely in households with the Colilert detection system. These data show that hygiene is relevant and important in the kitchen even though the overall incidence of pathogens was low.

Casual use of an antimicrobial disinfectant cleaner (phase II) did not result in any consistent reduction in the incidence of bacterial contamination (Fig. 10.15). These results can be explained by the haphazard manner in which most disinfectants are casually used. Specifically, the intermittent use of the cleaner meant that if surfaces were swabbed after disinfectant use but before subsequent food preparation, the degree of contamination was likely to be low. If swabbing occurred after food preparation without a subsequent cleaning event, bacterial contamination was likely to be high.

This hypothesis is supported by data from phase III, in which product use was mandated. Here, in almost every case, the incidence of contamination was decreased dramatically, particularly in the percentage of sam-

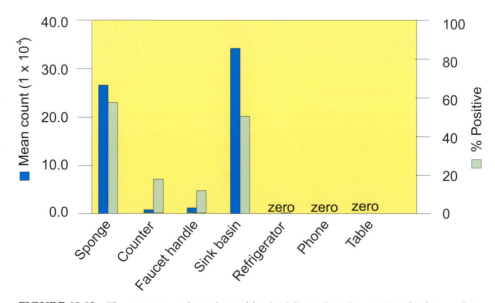

FIGURE 10.13 The presence and numbers of fecal coliforms found on various kitchen surfaces in homes not using disinfectant cleaner.

(continued)

Evaluation of Pathogens in Household Kitchens Using Cultural Methodologies (*continued*)

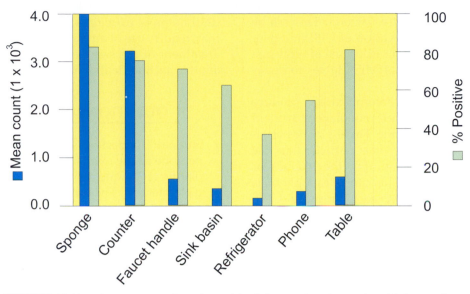

FIGURE 10.14   The presence and numbers of *Staphylococcus* found on various kitchen surfaces in homes not using disinfectant cleaner.

ples that showed contamination. It is important to note that numbers of bacteria reported in phase III reflect mean values of samples in which contamination occurred. In many cases there was zero contamination, but these cases were not averaged into CFU counts.

In summary, this study illustrated several important points with respect to kitchen hygiene. Without disinfectant use, kitchen surfaces are likely to be contaminated with a variety of bacteria including fecal coliforms, and *E. coli* and to a lesser extent with specific pathogens such as *Campylobacter* or *Salmonella*. Casual use of antimicrobial agents is unlikely to reduce the risk of these infectious agents. The success of the "mandated" use of such products suggests that cleaning with a disinfectant after each incidence of food preparation is more likely to reduce potential bacterial contamination. This would appear to be true particularly after high-risk foods including meat and seafood have been handled.

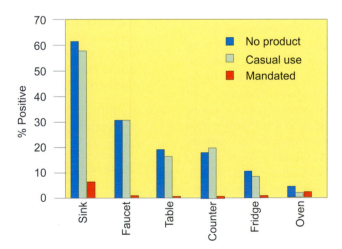

FIGURE 10.15   The percentage of samples showing contamination with fecal coliforms with no, casual, or mandated use of a disinfectant.

ble of producing a CPE in 50% of the tissue culture vessels or wells and is referred to as the medium tissue culture infective dose or $TCID_{50}$. The second method is the most probable number or **MPN** method, which is similar to that used to quantify bacteria. This can be used with viral cells by observation of CPE in monolayers inoculated with different dilutions of viral suspension and subsequent use of MPN tables to determine viral numbers.

Some viruses, such as enteroviruses, produce plaques or zones of lysis in cell monolayers overlaid with solidified nutrient medium. These plaques originate from a single infectious virus particle; thus the titer can be quantified by counting the plaques or plaque-forming units (PFU). This can be thought of as analogous to enumeration of bacteria using CFU. There are variations of the basic plaque-forming unit assay, but usually a vital dye, which only stains living cells, is used. As the virus infection in the monolayer spreads and cells lyse, a clear plaque appears in a stained monolayer of living cells (Fig. 10.10). A system commonly used for enteroviruses utilizes an agar overlay containing media components, serum, and the dye neutral red. A virus kills and lyses infected cells, leaving a clear plaque in a red-stained cell monolayer, because the dye does not stain dead cells.

Cell culture is not 100% effective in detecting all of the viral particles that can be observed through use of an electron microscope. This is because not all the particles are infectious. A noninfectious viral particle may contain no nucleic acid in the capsid or the nucleic acid may be incomplete. Other reasons for the inefficiency of cell culture may be that there is a lack of cell receptors or the cell culture incubation period is insufficient for all of the viruses to attach to viral receptors on the host cells. Further, the plaquing efficiency or production of a CPE is influenced by the salts, additives, or enzymes added to the medium. For example, good plaque formation for rotaviruses requires the addition of trypsin to the overlay medium. The susceptibility of cell lines can be enhanced by the addition of 5-iododeoxyuridine for enteric viruses (Benton and Ward, 1982). Generally, for laboratory-grown viruses the ratio of PFU or infectious virus to total viral particles is 1:100, but it may be much greater for viruses from the host or environment (1:10,000) (Ward *et al.*, 1984). For this reason, techniques such as PCR, which can detect 1–10 copies of a viral genome, are more sensitive than conventional cell culture methods.

Use of cell culture for detection of viruses in environmental samples is also often made difficult by the presence of bacteria, molds, and toxic substances. To overcome problems with other microorganisms, antibiotics are added or the samples are filtered. Unfor-

tunately, this sometimes results in loss of viruses that may be associated with solids. Problems with substances toxic to the cell culture are also difficult to overcome. The easiest solution is to dilute the sample until no more toxicity is observed. Unfortunately, this may require use of an excessive amount of cell culture. Treatment of the sample with Freon or chloroform is another technique that has often been employed.

QUESTIONS AND PROBLEMS

1. Student "A" performs a soil extraction with peptone as the extracting solution. A dilution and plating technique is performed using the spread plate method. The following results are obtained:

   a. 150 colonies are counted on an R2A plate at the 10^5 dilution

   b. 30 colonies are counted on a nutrient agar plate at 10^4 dilution

   Calculate the number of viable bacteria per gram of soil for R2A and nutrient agar. Why are the counts for nutrient agar lower than for the R2A media?

2. Student "B" performs a soil extraction with peptone as the extracting solution. A dilution and plating technique is performed using the pour plate method. The number of bacteria per gram of soil is calculated to be 8×10^8. What dilution plate contained 80 colonies?

3. Student "C" used direct microscopic methods to count bacteria in the same soil as student "A" above. How would student "C's" numbers compare to student "A's" numbers and why?

4. The last student also extracted the same soil in peptone but let the extraction sit overnight at room temperature before diluting and plating. How would these numbers compare to student "A's" numbers and why?

5. Identify the major errors associated with dilution and plating techniques.

References and Recommended Readings

Alonso, J. L., Batella, M. S., Amoros, I., and Rambach, A. (1992) *Salmonella* detection in marine waters using a short standard method. *Water Res.* **26**, 973–978.

American Pubic Health Association. (1995) "Standard Methods for the Examination of Water and Wastewater," 19th ed. APHA, Washington, DC.

Atlas, R. M. (1993) "Handbook of Microbiological Media." CRC Press, Boca Raton, FL.

Benton, W. H., and Ward, R. L. (1982) Induction of cytopathogenic-ity in mammalian cell lines challenged with culturable enteric viruses and its enhancement by 5-iododeoxyuridine. *Appl. Environ. Microbiol.* **43,** 861–868.

Castenholz, R. W. (1988) Culturing methods for cyanobacteria. *In* "Methods in Enzymology" (J. N. Abelson and M. I. Simon, eds.), Vol. 167 Academic Press, San Diego, pp. 68–93.

Difco Manual. (1984) "Dehydrated Culture Media and Reagents for Microbiology." Difco Laboratories, Detroit, MI.

Fricker, C. R. (1987) A review. The isolation of salmonellas and campylobacters. *J. Appl. Bacteriol.* **63,** 99–116.

Gerba, C. P., and Goyal, S. M. (1982) "Methods in Environmental Virology." Marcel Dekker, New York.

Gould, W. D., Hagedorn, C., Bardinelli, T. R., and Zablotowica, R. M. (1985) New selective medium for enumeration and recovery of fluorescent pseudomonads from various habitats. *Appl. Environ. Microbiol.* **49,** 28–32.

Josephson, K. L., Rubino, J. R., and Pepper, I. L. (1997) Characterization and quantification of bacterial pathogens and indicator organisms in household kitchens with and without the use of a disinfectant cleaner. *J. Appl. Bacteriol.* **83,** 737–750.

Martin, J. P. (1950) Use of acid rose bengal and streptomycin in the plate method for estimating soil fungi. *Soil Sci.* **69,** 215–232.

Metting, J. R., and Blaine, F. (1994) Algae and cyanobacteria. *In* "Methods of Soil Analysis," Part 2, "Microbiological and Biochemical Properties." Soil Science Society of America, Madison, WI, pp. 427–458.

Neilson, J. W., Josephson, K. L., Pepper, I. L., Arnold, R. G., DiGiovanni, G. D., and Sinclair, N. A. (1994) Frequency of horizontal gene transfer of a large catabolic plasmid (pJP4) in soil. *Appl. Environ. Microbiol.* **60,** 4053–4058.

Parkinson, D. (1994) Filamentous fungi. *In* "Methods of Soil Analysis," Part 2, "Microbiological and Biochemical Properties." Soil Science Society of America, Madison, WI, pp. 329–347.

Pepper, I. L., Gerba, C. P. and Brendecke, J. W. (1995) "Environmental Microbiology: A Laboratory Manual" Academic Press, San Diego, CA.

Roszak, D. B., and Colwell, R. R. (1987) Survival strategies of bacteria in the natural environment. *Microbiol. Rev.* **51,** 365–379.

Schmidt, E. L., Molina, J. A. E., and Chiang, C. (1973) Isolation of chemoautrophic nitrifiers from Moroccan soils. *Bull. Ecol. Res. Comm.* (Stock.) **17,** 166–167.

Seifert, K. A. (1990) Isolation of filamentous fungi. *In* "Isolation of Biotechnological Organisms from Nature" (D. P. Labeda, ed) McGraw-Hill, New York, pp. 21–51.

Ward, R. L., Knowlton, D. R., and Pierce, M. J. (1984) Efficiency of human rotavirus propagation in cell culture. *J. Clin. Microbiol.* **19,** 748–753.

11

Physiological Methods

DAVID C. HERMAN   RAINA M. MAIER

11.1 INTRODUCTION

The focus of this chapter is on measurement of microbial activity via physiological methodologies both in pure culture and within environmental samples. Why do scientists want to make such measurements? For pure culture work, a measure of microbial activity allows one to determine the maximum level of activity that can be achieved given the conditions imposed on the pure culture. For environmental samples, measurement of microbial activity provides a relative measure of the actual or potential level of activity within that environment. For an undisturbed environment, the activity measured will reflect a basal level of activity. This basal activity level is dependent on a variety of biological, chemical, and physical parameters, as well as the nutrient status of the environment. In fact, measurement of microbial activity can be used to evaluate the activity dynamics of the basal microbial community in response to a variety of stimuli, including naturally variable parameters such as moisture level and anthropogenically imposed variables such as addition of fertilizers or contaminant spills. In an ecological sense, measurement of microbial activity in an undisturbed environment allows one to determine the microbial contribution to nutrient cycling (see Chapter 14) within an ecosystem. For example, in the case of carbon cycling, the focus of microbial activity measurements is to determine accurately the production of new biomass by each microbial component and estimate the amount of energy that is being stored in a particular group of micro- or macroorganisms. This information allows determination of the transfer of energy between trophic levels in the food chain within a particular ecosystem.

For disturbed environments, measurement of microbial activity can provide an indication of the general health of the environment and can be used to evaluate the impact of a disturbance on the microbial community. For example, microbial activity has been used to evaluate the impact on soil quality of certain land use practices such as farming, logging, and mining. Microbial activity is also an important indicator in evaluating the process of restoration of disturbed sites. One example of this is **intrinsic bioremediation** or **natural attenuation,** the process by which indigenous microbial populations degrade spilled pollutants within a natural environment (see Chapter 16.7). Other areas where microbial activity measurements are often used include engineered systems, such as municipal wastewater treatment systems, compost systems, and biological reactor systems, all of which apply microbial processes for a specific purpose.

The goal of this chapter is to examine different types of microbial activity measurements in both pure culture and environmental samples. Although activity measurement in pure culture is relatively straightforward, environmental communities contain a diversity of microorganisms and physiological types including aerobic and anaerobic heterotrophs and autotrophs (see Chapter 2.2.6). Thus, any measurement of microbial activity in an environmental sample can be more or less inclusive of different classes and physiological types of microorganisms. As a result, it is important to understand the different activity measurements and their relative specificity. In this chapter, activity measurements in pure culture will be discussed first, including measurement of: substrate disappearance, terminal electron acceptor (TEA) utilization, cell mass increase, and carbon dioxide evolution. Following this, activity measurements in environmental samples will be addressed. Environmental activity measurements have been divided into four broad categories: (1) carbon respiration, (2) incorporation of radiolabeled tracers in cellular macromolecules, (3) adenylate energy charge, and (4) enzyme activity assays. Which of these measurements is chosen is a function of the practical or research question being asked.

11.2 MEASURING MICROBIAL ACTIVITY IN PURE CULTURE

As discussed in Chapter 3, growth in pure culture can be described by the generalized equation representing respiration.

$$\text{Substrate} + \text{nitrogen} + \text{TEA} \rightarrow \text{cell mass}$$
$$+ \text{carbon dioxide} + \text{water source} \quad \text{(Eq. 11.1)}$$

In this equation, the TEA (terminal electron acceptor) is oxygen for aerobic conditions or one of several alternate TEAs for anaerobic conditions (see Table 3.1). Examination of this equation shows that there are several ways in which microbial activity can be measured. These include substrate disappearance, TEA utilization, biomass production, and carbon dioxide evolution. Each of these approaches is discussed briefly in the following sections.

11.2.1 Substrate Disappearance

11.2.1.1 Heterotrophic Substrates

For heterotrophic activity substrate is carbon-based. Such substrate can be measured in many different ways, depending on the chemistry of the substrate molecule and whether a single component or multiple components are present. For example, in a single-solute system, aromatic compounds such as benzene and all benzene derivatives can be measured easily and sensitively by **UV spectrophotometry** (Fig. 11.1). A UV spectrophotometer measures how much light is absorbed while passing through a sample. The spectrophotometer can be set to a particular wavelength of light that is chosen on the basis of the absorbance spectrum of the compound. For instance, the absorbance spectrum of benzoate, shown in Fig. 11.2, indicates that the optimal detection wavelength is 224 nm (note that 224 nm is chosen because there is a great deal of interference by salts such as NO_3^- and SO_4^{2-} below 200 nm). Analogously, a **fluorimeter** can be used to measure fluorescent molecules such as polyaromatic hydrocarbons (PAHs) within single-component systems (Fig. 11.3). A fluorimeter offers increased sensitivity over UV spectroscopy. Often, samples from pure cultures and especially environmental samples contain compounds

FIGURE 11.1   UV Spectrophotometer. (Photo courtesy R. M. Maier.)

that interfere with the quantitation of the target compound. This can be considered a multiple-component system. For example, humic compounds present in soil and water samples contain many aromatic molecules that interfere strongly with measurement of aromatic substrates. Alternatively, there may be multiple components of interest in a sample containing a complex mixture of compounds such as might be found in gasoline or other fuel oils. In this case, a separation or **chromatography** step is required prior to detection of the target compound. There are different types of chromatography, including liquid chromatography and gas chromatography. One common instrument based on liquid chromatography is the **high-performance liquid chromatograph (HPLC)** (Fig. 11.4). The HPLC

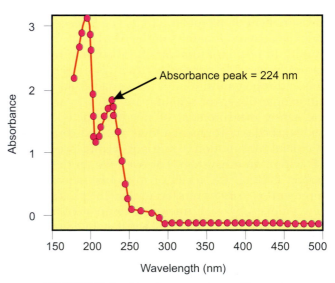

FIGURE 11.2   Absorbance spectrum of benzoate.

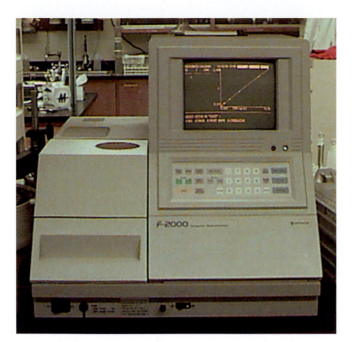

FIGURE 11.3 Fluorimeter. (Photo courtesy R. M. Maier.)

forces liquid solvent through a packed column under high pressure to separate the components within a mixture. The separation is dictated by the type of column packing and is generally based on charge or hydrophobicity. The resulting chromatogram can be analyzed to provide quantitative information about the amount of each component present in the sample (Fig. 11.5). For the example chromatogram shown, a UV detector was used for analysis; however, several other types of detectors can be used. These include the **photodiode array detector,** which is based on measuring light wavelength absorption but uses a detector array to rapidly measure the entire wavelength spectrum instead of just a single wavelength. Other detectors used in HPLC include **refractometers,** which

are primarily for sugar analysis; **evaporative light scattering detectors,** used for surfactants and polymers; **conductivity meters,** used for ionic or charged compounds; and **fluorescence detectors,** used for analysis of polyaromatic hydrocarbons and fluorescent dyes.

A second instrument based on chromatographic separation of solutes is the **gas chromatograph (GC)** (Fig. 11.6). In gas chromatography, the mobile phase is a mixture of gases instead of liquids. Because the mobile phase is gaseous, compounds amenable to GC analysis must be volatile so that they can move through the GC column with the mobile phase. Therefore, compounds that can be analyzed by GC are characterized by lower boiling points and higher volatilities. The GC can have two types of detectors, the **flame ionization detector (FID)** and the **electron capture detector (ECD).** The FID is suitable for analysis of hydrocarbons such as **BTEX** (benzene, toluene, ethylbenzene, xylene), polyaromatic hydrocarbons, or aliphatic hydrocarbons. The ECD detector is used for analysis of halogenated materials, primarily chlorinated compounds such as trichloroethylene (TCE). It should be noted that for some of these hydrocarbons both HPLC and GC can be used for analysis but for others, such as the aliphatic hydrocarbons, which do not absorb in the UV range, only GC can be used. In other instances, such as for organic acids that have low volatility, HPLC is the method of choice.

Finally, it should be mentioned that both the GC and the HPLC can be used in conjunction with a **mass spectrometer** to allow identification of an unknown compound (Fig. 11.7). Mass spectrometers bombard a compound with electrons, causing it to break up and fragment into smaller pieces. The fragmentation pattern can be interpreted by an expert to determine what the original molecule looked like. For routine analyses, most mass spectrometers contain libraries of fragmentation patterns that can be used to provide a best fit match or prediction of the target compound being analyzed.

11.2.1.2 Chemoautotrophic Substrates

In contrast to chemoheterotrophic activity, for which energy is provided by oxidation of organic or carbon-based substrate for chemoautotrophic activity, the energy providing substrate is an inorganic mineral. Important microbial activities that contribute to the cycling of minerals in soil and water environments include chemoautotrophic oxidation of ammonia (nitrification) and sulfur (sulfur oxidation). The following sections discuss the approaches used to measure these two chemoautotrophic substrates.

FIGURE 11.4 High performance liquid chromatograph. (Photo courtesy R. M. Maier.)

a

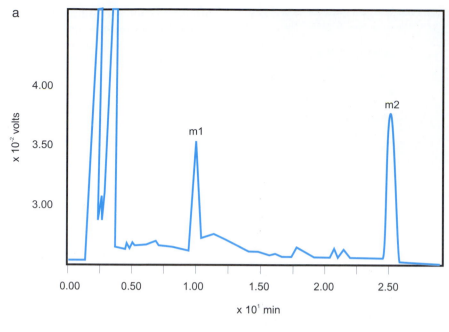

b

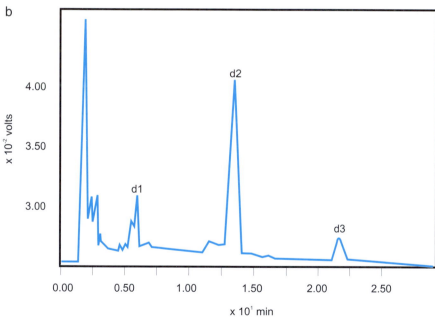

FIGURE 11.5 HPLC chromatogram. These are typical chromatograms of a microbial surfactant, rhamnolipid (see Fig. 17.6 for the chemical structure), produced by *Pseudomonas aeruginosa*. (a) is monorhamnolipid (one rhamnose sugar per molecule) and (b) is dirhamnolipid (two rhamnose sugars per molecule). Note that each chromatogram has more than one peak. This is because microbes make a mixture of surfactants that vary in the length of the fatty acid tails. The longer the tail, the more hydrophobic the molecule is and the longer it is retained on the column. Thus, for monorhamnolipid, the peak labeled m2 has longer lipid tails than the monorhamnolipid in the peak labeled m1. A reverse phase C18 column was used for sample separation. Elution was isocratic (meaning the mobile phase composition remained constant during analysis) using a mobile phase of acetonitrile-water (40 : 60 v/v) at a flow rate of 1 ml/min. A UV detector set at 214 nm was used for detection of the surfactant. (Chromatograms courtesy Y. Zhang.)

FIGURE 11.6   Gas chromatograph. (Photo courtesy R. M. Maier.)

Nitrification

Nitrification, the oxidation of ammonia to nitrite and nitrate (see Chapter 14.3.4), can be measured in several ways. A direct assay involves addition of ^{15}N-labeled ammonia and measurement of the formation of ^{15}N-nitrite and ^{15}N-nitrate as the ^{15}N-ammonia is oxidized (Paerl, 1998). These ^{15}N-labeled products are measured by mass spectroscopy. However, the sensitivity of this measurement is low and thus large quantities of ammonia that far exceed those found naturally may be required in an experiment. A second approach to the measurement of nitrification is to incubate samples with $^{14}CO_2$ to measure CO_2 fixation. Some of the samples are further treated with a chemical reagent such as N-serve. N-serve inhibits nitrifying microbes which are very sensitive to chemical inhibitors (see Table 15.2). Carbon dioxide fixation is then compared

FIGURE 11.7   Mass spectrometer. (Photo courtesy R. M. Maier.)

in the presence and absence of the nitrification inhibitor to determine nitrification activity.

Sulfur Oxidation

Sulfur oxidation (see Chapter 14.4.3.1) can be measured by addition of elemental sulfur (S^0) to a sample and measuring the amount of sulfate (SO_4^{2-}) produced during incubation. However, this is a complicated process. One such assay measures sulfate as a precipitate with barium chloride in acid solution (Kelly and Wood, 1998). Following precipitation, the barium in solution is measured by **atomic absorption spectroscopy** (AA), an instrument that is used to measure metals. While the AA has specific lamps that measure specific metals, it cannot distinguish between the different species of a given metal in solution. In this assay, as the amount of sulfate increases as a result of sulfur oxidation, the amount of barium in solution decreases through precipitation with the sulfate. A second approach to measurement of sulfur oxidation is to determine CO_2 uptake in the presence of sulfur compounds. As described above for nitrification, samples can be incubated with $^{14}CO_2$ to measure CO_2 fixation as an estimate of sulfur oxidation.

11.2.2 Terminal Electron Acceptors

A variety of TEAs are available for microbial use and levels of microbial activity are reflected in the disappearance of the TEA and the appearance of the reduced TEA. Measurement of several common TEAs is discussed in the following sections.

11.2.2.1 Oxygen Uptake

The majority of environmental activities measured are under aerobic conditions where oxygen is the required TEA. Oxygen, unlike many other nutrients, is relatively insoluble in water and can become limiting to growth unless it is constantly replenished. Thus, aeration is an important consideration in both pure culture and environmental systems. Normal dissolved oxygen levels in an aqueous sample that is equilibrated with air range up to 10 mg/L depending on sample temperature and atmospheric pressure. For example, the concentration of oxygen in water that is in equilibrium with ambient air at one atmosphere and 25°C is 8.1 mg/l. In general, aerobic activity is maintained until the dissolved oxygen concentration in the sample falls below 1 to 2 mg/l. Because oxygen is a limiting nutrient, it is a good measure of growth and activity. Measurement of oxygen in a pure culture is relatively rapid, routine, and cheap to perform. This technique is perhaps used most extensively in the wastewater treatment industry to determine the biological oxygen demand (BOD) of wastewater samples (see Section 11.4.2.4). As discussed in the following sections, oxygen levels in an aqueous sample can be measured using an oxygen probe, by colorimetric methods, or by manometry.

Oxygen Probes

An oxygen probe is basically an electrode covered by a gas-permeable membrane combined with a meter that converts the electrical signal into an analytical measurement. Oxygen probes do not measure the absolute amount of oxygen in a sample; rather, they measure oxygen amounts relative to an oxygen-saturated solution. A variety of commercially available oxygen probes can be used to measure dissolved oxygen. These include probes that are used for routine BOD measurement (Fig. 11.8a) and microprobes that can be used for small sample amounts or can be inserted into a soil column through a port (Fig. 11.8b). Advances in technology have produced microprobes for the detection of O_2, as well as nitrogen compounds (N_2O, NO_3, NH_4), sulfide, hydrogen, and pH (Revsbech and Jørgensen, 1986). The oxygen microelectrodes can be as small as 20 μm diameter, allowing the determination of gas flux across critical interfaces.

Colorimetric Assays

Alternatively, a variety of colorimetric assays are available. In this case a sample is mixed with a series of reagents to produce an oxygen-dependent color. A titration method known as the Winkler method can be used to determined dissolved oxygen in water samples. Winkler titrations are based on the precipitation of dissolved oxygen using manganous sulfate and a potassium hydroxide-potassium iodide mixture (Wetzel and Likens, 1991). The oxygen precipitate, manganic basic oxide, reacts with sulfuric acid to form a manganic sulfate, which in turn reacts with potassium iodide to liberate iodine. The number of moles of iodine liberated is equivalent to the number of moles of oxygen present in the water sample. The liberated iodine can be determined spectrophotometrically in clear water samples or can be determined by titration with sodium thiosulfate. The basic Winkler method may be affected by the presence of oxidizing or reducing materials that can occur in natural waters and especially in polluted waters. Modifications of the basic method are described in the APHA, AWWA, WEF (1995) manual. There are also several kits available for oxygen measurement that are based on a colorimetric reaction. An example of such an assay is shown in Fig. 11.9.

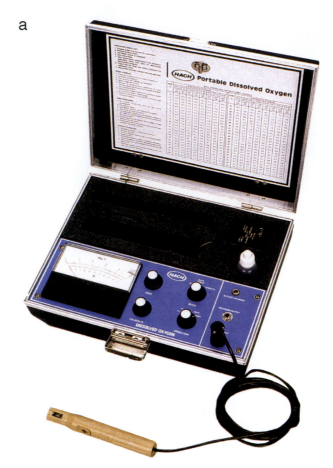

a

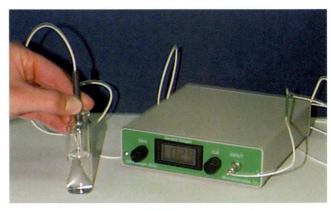

FIGURE 11.8 (a) Oxygen probe for BOD measurement. (b) Oxygen probe (microprobe). (Photos courtesy R. M. Maier.)

Manometry

Manometric techniques quantify microbial respiration by measurement of O_2 depletion from the atmosphere within a sealed flask. The technique is based on the ideal gas law:

$$PV = nRT$$

where P is the pressure in the system, V is the volume of the system, R is the gas constant, T is the tempera-

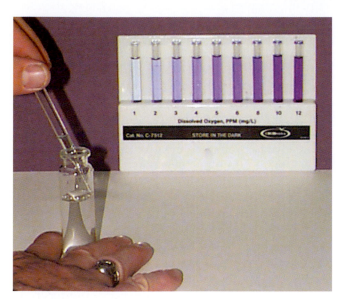

FIGURE 11.9   Oxygen measurement using a Chemets Kit. (Photo courtesy R. M. Maier.)

ture, which is held constant, and n is the number of moles of gas. A bacterial culture is placed in a sealed environment where the amount of oxygen consumed can be measured by monitoring the change in gas pressure (P) or gas volume (V), given that one variable is held constant and the other allowed to fluctuate.

The **Warburg method** is a manometric technique in which the gas volume is held constant and the measured change in pressure reflects the amount of oxygen consumed by the bacteria as they respire. A diagram of a constant-volume Warburg type of respirometer is shown in Fig. 11.10. An alternative manometric technique is to hold the gas pressure constant and allow the gas volume to fluctuate. An example of a constant-pressure manometer is the **Gilson manometer,** with which CO_2 production can be trapped by an alkali solution and volume changes due to O_2 consumption are recorded by the movement of a plastic rod required to maintain the system at a constant pressure. The original Warburg- or Gilson-type manometers were designed for laboratory experiments on small volumes of defined media and are not well suited for use with environmental samples. However, the system can in some cases be modified to handle soil samples, and one example is described in detail by Anderson (1982).

Alternate Terminal Electron Acceptors

Although not nearly as commonly measured as oxygen, perhaps because the methodologies are much more complex and time-consuming, there are methods available to quantitate alternate TEAs. For example, in denitrification, nitrate (NO_3^-) serves as the terminal electron acceptor and is reduced sequentially to nitrite

(NO_2^-), nitric oxide (NO), nitrous oxide (N_2O) and nitrogen (N_2) (see Chapter 14.3.5.2). One common assay for denitrification involves the use of acetylene (C_2H_2) which blocks the reduction of nitrous oxide to nitrogen (Paerl, 1998). Thus, nitrous oxide accumulates in the system and can be measured using a gas chromatograph with an electron capture detector. A disadvantage of using this method in environmental samples is that acetylene also inhibits nitrification. Since nitrification and denitrification are closely coupled (e.g., the nitrate produced by nitrification is used in denitrification), this inhibition may reduce or eliminate denitrification in the sample. Therefore, in some instances, denitrification is estimated by simply measuring nitrification (see Section 11.2.1.2). A second problem with this technique is that it is not effective at low nitrate concentrations (less than 0.63 mg/l).

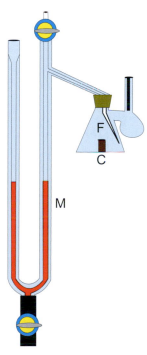

FIGURE 11.10   Warburg constant volume respirometer. In this setup, a flask (F) of known volume is connected to a manometer (M) containing a liquid of known density. The flask holds a suspension of bacterial cells, and has a center well (C) which contains an alkali trap for CO_2 so that CO_2 evolved does not interfere with O_2 measurement. Oxygen consumption will result in a drop in pressure within the flask, as recorded by a drop in the fluid level within the manometer. Critical factors are to 1) hold the temperature in the flask constant by incubating the flask in a water bath between manometric readings, and 2) shake the flask during incubation to ensure that oxygen exchange between the atmosphere and bacterial suspension is not a limiting factor during microbial respiration. CO_2 production can also be measured in a Warburg constant pressure-type manometer. This is done by comparing pressure changes in two identical flasks, but only one flask contains an alkali trap for CO_2. The difference in pressure decrease between flasks in which CO_2 is present or absent can be used to calculate CO_2 production.

Similarly, there are methodologies available for measuring sulfate (SO_4^{2-}) reduction (Tabatabai, 1994a), iron and manganese reduction (Ghiorse, 1994), and methanogenic activity (Zinder, 1998).

11.2.3 Cell Mass

An increase in cell mass in a pure culture is often quantified by performing culturable plate counts or direct microscopic counts. This approach is also often used to estimate the cell mass in environmental samples. Methods for determining an increase in cell number using culturable plate counts or direct microscopic counts are discussed in detail in Chapter 10. Another common approach to estimating cell mass is to measure the turbidity or the protein content in the culture being studied. Although these measurements are not suitable for environmental samples, they are easy, rapid, and reproducible when used for measuring cell mass increases in pure culture.

11.2.3.1 Turbidity

A rapid estimate of cell mass can be obtained by measuring the turbidity of a bacterial suspension. Turbidity measurements are generally made using a colorimeter or spectrophotometer, both of which work by directing a light beam through the sample. The bacteria in the sample scatter the light beam and lower the intensity of the light coming through the suspension. At low bacterial densities, there is a direct linear relationship between the number of bacteria and the amount of light scattered; thus, as the number of bacteria increases, the turbidity of the suspension increases. At high bacterial densities, this relationship becomes nonlinear. Therefore, a standard curve must be constructed to determine the linear range for turbidity measurement for each organism being measured. In theory, a wide range of wavelengths can be used to measure turbidity, although the practical range is between 490 and 550 nm. The sensitivity of the measurement increases at lower wavelengths, but there are often yellow or brown pigments produced during growth that can interfere with turbidity measurements at lower wavelengths.

11.2.3.2 Protein

A variety of methods are available for measurement of protein (Daniels et al., 1994). Currently, the most commonly used assays are based on the Folin reaction (Lowry assay; Lowry et al., 1951); the Coomassie Blue reaction (Bradford assay; Bradford, 1976); and the bichinchoninic acid-copper reaction (Smith et al., 1985). Each of these assays quantifies the amount of protein present by measuring color produced by a reaction of the protein present with the assay reagents. For example, the Lowry assay is based on the development of a blue color resulting from the reaction of the Folin reagent with aromatic amino acids contained in the protein sample. The sample is measured in a spectrophotometer at 550 nm. Similarly, the Bradford assay measures the binding of Coomassie Brilliant Blue G dye to protein amino groups, which causes the development of a blue color that can be measured at 595 nm. The bichinchoninic assay is based on the fact that proteins react with Cu^{2+} to produce Cu^{+}, which then reacts with bichinchoninic acid to form a purple product that is measured at 562 nm.

The method chosen for a particular application depends on the types of samples and potential for interference by sample components. The basic procedure for protein analysis is first to lyse cells in order to release the protein. This can be done by placing an aliquot of cells into 1 N NaOH and heating the suspension at 90°C for 10 minutes. It is a good practice to wash the cells prior to lysis to minimize the presence of materials that may interfere with the protein assay. Once the cells are lysed, the protein assay is performed and the resulting sample measured in a spectrophotometer to determine protein quantity.

11.2.4 Carbon Dioxide Evolution

In general, the CO_2 trapping methodologies described here have been used to measure aerobic mineralization. However, the methods could be adapted to measurement of CO_2 evolved under anaerobic conditions as well.

11.2.4.1 CO_2 Trapping

An alkaline trap, usually composed of sodium hydroxide or another strongly basic solution, can be used to trap CO_2 produced during mineralization (Fig. 11.11a). The alkaline solution traps carbon dioxide by transforming it into the bicarbonate (HCO_3^-) form:

$$CO_2 + OH^- \rightarrow HCO_3^- \qquad \text{(Eq. 11.2)}$$

The trapped carbonate can be detected by using titrimetric, gravimetric, or conductimetric measurements. Perhaps the most common approach to quantifying trapped CO_2 is titration of the alkaline solution that was used to trap CO_2 with a standardized acid solution. The HCO_3^- that was formed in the trapping solution is first precipitated with barium chloride. Then the solution is titrated with hydrochloric acid in the presence of a pH-sensitive dye such as phenolphthalein to indicate the end point of the titration. The amount of CO_2 trapped is then calculated from the reduction of the original hydroxyl ion concentration in the trapping solution. Alternatively, a simple gravimetric method is to precipitate the carbon dioxide as barium carbonate by the addition of an excess of $BaCl_2$. The precipitate can be collected, dried, and weighed. A second gravimetric approach is to trap carbon dioxide in granular soda lime and monitor the gain in mass. Conductimetric quantification of CO_2 trapping is based on the reduction of the number of hydroxyl ions (OH^-), which is directly related to the amount of CO_2 trapped and can be quantified by a decrease in conductivity of the trapping solution.

11.2.4.2 $^{14}CO_2$ Trapping

The use of ^{14}C-radiolabeled substrates allows very sensitive and specific measurements of $^{14}CO_2$ evolution. In this case the CO_2 evolved is specific for the substrate added. This technique is often used to evaluate the feasibility and rates of biodegradation of organic contaminants. The generation of $^{14}CO_2$ offers confirmation that it is actually the organic contaminant in question that is being degraded. In this case, a biometer flask such as shown in Fig. 11.11a can be used, and the alkali within the sidearm trap is assayed for radioactivity using liquid scintillation counting (Fig. 11.12). Alternatively, a simple alkaline trap can be made by inserting a test tube containing alkali into a flask (Fig. 11.11b). A slightly more complex system is the stripping chain shown in Fig. 11.11c. In this case, the first two traps are filled with a general scintillation cocktail to remove any ^{14}C-labeled volatile organic compounds that may have been produced during biodegradation. The final two traps are filled with a phenethylamine-based cocktail that selectively traps CO_2. Although this system is more complex than those shown in Fig. 11.11a and b, it offers confirmation that the radioactivity assayed is actually $^{14}CO_2$.

11.3 CHOOSING THE APPROPRIATE ACTIVITY MEASUREMENT FOR ENVIRONMENTAL SAMPLES

Measuring the activity of a diverse population of microbes within an environmental sample is a very different prospect from measuring the microbial activity of a single isolate in the laboratory setting. As discussed in Section 11.2, the microbial activity of a pure culture in a defined medium can easily be determined by measuring cell number (culturable plate counts or direct counts) or biomass (turbidity or protein); however, these measurements are not always practical or even realistic for environmental samples. For example, protein analysis of an environmental sample reflects sources of biomass other than the microbial population, including plant debris or microscopic animals.

Although both culturable and direct counts are commonly used to enumerate bacteria at a particular location, they are less helpful for measuring microbial activity. Culturable plate counts underestimate bacterial numbers in an environmental sample because this process enumerates only the cells compatible with the growth medium and culture conditions. Further, although many bacteria are viable in their environmental setting, they are nonculturable in the artificial environment of the laboratory. It should be noted that despite the drawbacks of culturable counts for measurement of total activity within a sample, this technique is still useful for estimation of specific physiological groups, such as sulfate-reducing organisms or hydrocarbon-degrading organisms. Direct counts as a measure of total microbial activity are problematic because they include dormant and moribund cells. Also, it is often a small portion of the total population that is responsible for a given activity, and it is difficult to estimate the activity of this part of the population with direct counts. Finally, both viable and direct counts fail to enumerate cells that adhere tightly to soil particles or other surfaces. As a result of these problems, the measurement of microbial activity in environmental samples has focused more on the quantification of metabolic activity, such as respiration or synthesis of cellular macromolecules, which provides a more direct reflection of the level of metabolic activity within a sample (Table 11.1). These types of measurements are explained in detail in the following sections of the chapter.

11.4 CARBON RESPIRATION

A primary activity of the microbial community in undisturbed environments is the utilization of naturally occurring organic materials. **Heterotrophic activity** is a measure of the extent to which microbial populations utilize these organic materials as a carbon and energy source. This process is called **respiration** and involves the consumption of oxygen (under aerobic conditions) or an alternate terminal electron acceptor (under anaerobic conditions) and production of carbon dioxide as shown in Eq. 11.1. As discussed in Chapter 3.3, the carbon from a substrate will be partially evolved as CO_2 and partially utilized to build new cell material. For heterotrophic activity under aerobic conditions, one can use the general assumption

FIGURE 11.11 (a) Biometer flask for trapping CO_2. This is a widely adopted system for measurement of CO_2. At intervals over the incubation period the CO_2 trapping solution is withdrawn from the sidearm and then replaced with fresh alkaline solution. As the trapping solution is withdrawn, air is drawn into the flask to replenish the oxygen supply within the sealed flask. The replacement air is drawn through a CO_2 absorbing filter (ascarite). To determine respiration, the amount of CO_2 generated is determined by titration of the trapping solution, and then the cumulative CO_2 production over the incubation period can be calculated. (b) Flask with a test tube trap containing alkali. (c) Stripping chain for $^{14}CO_2$ collection. (Photos courtesy R. M. Maier.)

that 50% of the carbon goes to CO_2 and 50% goes to cell mass. Under anaerobic conditions a smaller portion of the carbon is used to build cell mass and a larger portion goes to CO_2 and methane (see Chapter 3.4). Thus microbial activity within environmental samples can be monitored by measuring TEA consumption or CO_2 evolution. However, for both CO_2 and TEA, the relationship between the measured parameter and the amount of new cell material formed is specific to the organism, the particular substrate, and environmental conditions that influence growth rate. Therefore, a conversion efficiency must be estimated to quantify bacterial production absolutely. Even if an absolute value for microbial activity cannot be deter-

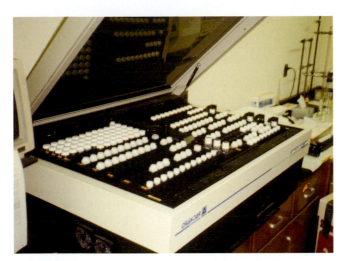

FIGURE 11.12 Liquid scintillation counter. (Photo courtesy of R. M. Maier.)

mined, measurement of the flux of respiratory gases can provide a relative level of microbial activity.

11.4.1 Measurement of Respiration Gases, CO_2 and O_2, in Laboratory and Field Studies

As described in Sections 11.2.2 and 11.2.4, there are several approaches to measuring CO_2 and TEA, in particular O_2, in pure culture. Most of these approaches can also be adapted for use with environmental samples. The presence of respiration gases can also be determined by gas chromatographic analysis, typically using a thermal conductivity detector. Chromatography requires that an air sample is collected in a gastight syringe and then injected into a stream of inert gas, which carries the sample through a column that has been packed with selective material. The column packing material acts to separate the individual gaseous components of the sample and allows simultaneous monitoring of O_2 and CO_2 within the same sample. Carbon dioxide in atmospheric samples can also be monitored using an infrared gas analyzer, which detects CO_2 by the absorption of a specific electromagnetic wavelength. Some specific approaches for measurement of respiration gases in terrestrial and aquatic environments are detailed in the following sections.

11.4.1.1 Terrestrial Environments

Microcosm Studies

Microbial activity within soil or subsurface materials can be measured under controlled conditions in a laboratory or can be measured *in situ* (in the field). In a laboratory study, a sample of the porous medium is typically incubated in a sealed, airtight enclosure, usually referred to as a **microcosm**. Some examples of microcosms are shown in Fig. 11.13. This approach allows one to design complex experiments that can be performed and replicated under relatively controlled conditions. For example, microcosm studies performed in the laboratory allow the standardization of environmental parameters, such as soil moisture content and temperature for all samples. In the establishment of laboratory microcosms, a common practice is to sieve the soil through a 2-mm sieve to homogenize the sample and to remove large stones or plant debris. The soil can then be dried and adjusted to a set moisture content. However, the disadvantage of preparing the soil in this manner is that soil structural features are altered, such as soil aggregation and pore size distribution, and this may have an effect on microbial activity.

The use of sealed microcosms allows the measurement of microbial activity within the sample as determined by the flux of CO_2 and/or O_2 within the headspace atmosphere. Headspace gas samples can be withdrawn using a gastight syringe, and the CO_2 or O_2 concentrations can be determined using gas chromatography. An alternative method is to trap the CO_2 produced in a basic solution using a trap such as that shown in Fig. 11.11. One problem that can arise during long-term incubation periods is depletion of oxygen within the headspace of sealed, airtight flasks. To address this problem, flow-through incubation systems have been devised to allow headspace gases to be replenished while still allowing quantification of microbial respiration. In this case, CO_2-free air is used as the flow-through gas, and any CO_2 in the air exiting the flask is a direct result of microbial activity. The CO_2 in the effluent air can be trapped in alkali and quantified or fed directly into a CO_2 detection device, such as an infrared detector (Brooks and Paul, 1987). A second problem in some alkaline soils is that CO_2 evolution may be underestimated, because the equilibrium shown in Eq. (11.2) is dependent on pH. In soils with a pH above 6.5, a significant portion of the CO_2 produced by mineralization will be retained in the soil in the bicarbonate (HCO_3^-) form. Flow-through systems, which provide continuous replenishment of the atmosphere within the incubation flasks, reduce the retention of CO_2 in the soil by maintaining a low concentration of CO_2 in the atmosphere of the microcosm.

Respiration measurements to determine microbial activity in soil should be corrected for possible nonbiological sources of gas exchange. The nonbiological, spontaneous generation of CO_2 can occur in certain soils, especially when moisture is added to a dry soil,

TABLE 11.1   Several Popular Methods Used for the Measurement of Microbial Activity in Environmental Samples[a]

Test	Basis of test	Application	Advantages	Disadvantages
Measurement of respiration gases	Measurement of oxygen utilization or CO_2 production in an environmental sample. The flux of respiration gases provides an indication of overall metabolic activity, which can reflect the level of microbial activity.	Basal respiration measurements reflect microbial metabolism of organic substrates present in the environment. However, respiration by other components, such as plant roots in soil or algae in water, will also be included, depending on the environment.	Field chambers can be built and installed *in situ*, to monitor flux of respiration gases in relatively undisturbed samples. The addition of an organic substrate to the sample can be used to indicate the level of potential microbial activity.	Incubation of samples in a closed chambers (microcosms) in order to monitor the flux of respiration gases can create an artificial environment. CO_2 production can be underestimated due to pH-dependent retention of inorganic carbon as bicarbonate.
Respiration of radiolabeled substrates	Metabolism of a radiolabeled substrate is monitored by measuring the evolution of labeled-CO_2.	This method is used to determine the potential for metabolism of a foreign substrate, such as an organic pollutant. Also, overall heterotrophic potential can be estimated by determining the turnover of organic substrates that occur naturally in the environment.	The use of radiolabeling imparts a high sensitivity to the measurement, thus short incubation periods can be used. The use of specific radiolabeled compounds shows the potential for degradation of that specific substrate.	The concentration of substrate added is often greater than the concentration present in the environment, thus the rate of metabolism may be overestimated unless corrective procedures are used.
Microelectrodes	Probes with tips <20 μm in diameter can be inserted into environmental samples to provide a continuous monitoring of activity.	Microelectrodes have been designed to measure specific respiratory activities, including oxygen utilization and nitrate respiration.	Can monitor real-time activity at critical interfaces in biological systems. Especially useful to monitor the interdependence of aerobic and anaerobic processes in the environment.	The instrumentation is delicate and relatively expensive.
Incorporation of radiolabeled thymidine into cellular DNA	Microorganisms will scavenge DNA precursors, such as thymidine, from their environment. By radiolabeling thymidine, the rate of incorporation into DNA can be measured.	The rate of DNA synthesis provides a reasonable estimate of the rate of cell division, thus providing an estimate of microbial biomass production.	When using a short incubation time, thymidine incorporation is thought to measure bacterial DNA production because the rate of bacterial incorporation of thymidine is thought to be much faster than for other organisms which may be present in environmental samples.	Not all bacteria will incorporate exogenously supplied thymidine into DNA. Also, estimating microbial activity requires the development of a conversion factor relating thymidine incorporation to biomass production.
Adenylate energy charge (AEC)	AEC is a weighted ratio of ATP to total adenylates. ATP is quantified using a luciferin-luciferase substrate-enzyme system.	AEC values reflect a continuum between an active microbial community (AEC >0.8) and a community with a high proportion of dead or moribund cells (AEC <0.4).	AEC can establish the presence of a metabolically active community without the need to incubate the sample or add a surrogate substrate.	AEC is not necessarily a direct measure of microbial activity because adenylates are present in all living organisms and can also be released by decaying cells.

(Continues)

TABLE 11.1 (*Continued*)

Test	Basis of test	Application	Advantages	Disadvantages
Dehydrogenase assay	This assay measures the rate of oxidation-reduction reactions (electron transport chain activity) by monitoring the reduction of a tetrazolium salt by actively respiring microorganisms.	The rate of reduction of tetrazolium salts reflects the overall activity by all respiring microorganisms.	Actively respiring microorganisms can be visualized microscopically by the deposition of pigmented tetrazolium salts reduction products within the cell. The assay has a high sensitivity, and can be used to measure activity in low productivity environments.	The amount of tetrazolium salts reduced depends on many factors including sample incubation conditions. Direct comparison of activity between samples can be made only when the assay is performed under identical conditions.
Hydrolysis of fluorescein diacetate	Hydrolysis of fluorescein diacetate is performed by a variety of enzymes including esterases, proteases, and lipases.	The assay measures enzymatic activity, thus providing an estimate of total microbial activity in an environmental sample.	Fluorescein diacetate hydrolysis produces a highly fluorescent product, fluorescein, which is easily detected.	Enzymes involved in hydrolysis can be intracellular or extracellular.

[a] This table is meant to highlight advantages and disadvantages of these techniques, and should not be considered as comprehensive.

or from soils containing free calcium carbonate. This is a particular problem with arid soils that are high in calcium carbonate ($CaCO_3$). Nonbiological oxygen uti-

FIGURE 11.13 Examples of microcosms used in studies under aerobic conditions. As shown, a microcosm can be any shape or size depending on the size of the sample and the headspace required. In some cases much larger microcosms may be used so that several samples can be removed and analyzed during the experiment. In other cases, volatile substrates are added to the microcosm. When volatile substrates are used it is important that the cap seal be airtight and also that the cap seal does not absorb the volatile material. To address these problems a teflon liner can be placed into the cap. There are also microcosms that can be used under anaerobic conditions. These are usually serum vials that can be sealed with a crimp-top cap. To impose anaerobic conditions, a needle is inserted into a septum in the crimp top and the flask is flushed exhaustively with an inert gas (e.g., N_2) to drive out the oxygen.

lization can occur in the spontaneous oxidation of certain chemical elements, such as iron or copper. The separation of chemical reactions from biological production requires the use of sterile controls. The most common approaches to sterilization of soil are to autoclave the soil three times on three consecutive days or to treat the soil with a chemical agent such as mercuric chloride ($HgCl_2$) to inhibit microbial activity (Rozycki and Bartha, 1981).

Field Studies

Alternatively, field studies can be performed by placing a field chamber over a plot of surface soil and using this setup to make *in situ* measurements (Fig. 11.14). In this way the soil structure is not altered by the sampling procedure, and field respiration rates of the indigenous population are more reliably determined. Samples of atmosphere from within the field chamber can be withdrawn using a gastight syringe and then injected into a gas chromatograph for CO_2 and O_2 determination. In some cases it may be desirable to measure respiration gases below the soil surface. In this case a gas sampling probe can be inserted beneath the soil surface and used to withdraw gas samples at discrete depths in the soil profile. However, care must be taken to avoid the creation of preferential flow paths, which could result in the introduction of aboveground atmosphere into the subsurface.

There are several difficulties inherent in field studies. Because environmental parameters, such as soil moisture and temperature, cannot be readily con-

FIGURE 11.14 A field chamber that allows measurement of respiration gases. (From *Pollution Science,* 1996, Academic Press, San Diego, CA.)

Another factor that must be considered in aquatic systems is the close association between photosynthetic and heterotrophic microorganisms. In order to prevent photosynthetic generation of oxygen, which would interfere with the measurement of heterotrophic activity using oxygen depletion, incubations should be performed in the dark. However, in this case the oxygen demand measured will include both heterotrophic activity and oxygen utilization in the dark respiration cycle of photosynthetic organisms. Because of these inherent difficulties in the measurement of respiration gases in aquatic systems, a better choice of activity measurement is usually one of the tracer methods described in the following sections. These techniques have the advantage of requiring much shorter incubation periods and are more specific for heterotrophic bacteria.

trolled, field respiration measurements have more variability than laboratory microcosms measurements. It is also difficult to perform control studies in the field because sterilization of a soil plot is difficult. Finally, in many environments the respiration of plant roots may contribute significantly to the flux of respiration gases. Because the contributions of heterotrophic bacteria and plant roots cannot be separated in field measurements, CO_2 production has been viewed as a measure of the gross soil metabolic activity.

11.4.1.2 Aquatic Environments

Oxygen depletion is a common means of determining microbial activity in aquatic environments, in part because of the ease with which dissolved oxygen can be determined. However, many aquatic systems are oligotrophic, especially marine environments, and thus the number and activity of heterotrophic bacteria are limited. Therefore, water samples must be incubated in sealed microcosms for extended periods, 12 hours or often longer, in order to detect significant oxygen depletion. Incubation of water samples for the extended periods required to record oxygen demand may create an artificial environment because of alteration of several time-dependent variables in the system. These include a change in the quantity and the form of organic carbon substrate, changes in grazing pressure resulting from the presence of bactivorous zooplankton, and the exclusion of larger forms of grazing animals from the microcosm (Pomeroy *et al.,* 1994). With time, these effects can create deviations from the initial respiratory rate. To avoid this type of error, incubation times should be kept as short as possible.

11.4.2 The Application of Respiration Measurements in Environmental Microbiology

11.4.2.1 Basal Rate of Microbial Activity in Soil Samples

Concern about the transport of organic pollutants into subsurface environments and recognition of the potential role of bacteria in the remediation of pollutants in the subsurface have led to an exploration of the microbial life in subsurface environments. In one study described by Kieft *et al.* (1995), the existence of microorganisms was investigated in sediments ranging in depth from 173.3 to 196.8 m below the surface at a site in south–central Washington State. To determine the existence and distribution of microorganisms a battery of tests were performed on each subsurface sample, including acridine orange direct counts (AODCs), the mineralization of glucose, and basal respiration. Basal respiration was measured as the biological CO_2 production in samples that had received no nutrient amendment. In a parallel experiment, the soil was amended with glucose to determine if nutrient addition was required to detect microbial activity. Basal respiration was determined by transferring 5.0 g of subsurface sediment into 70-ml vials or microcosms, which were then sealed with rubber septa. After 3, 7, and 14 days of incubation (22°C), a sample of headspace gas was withdrawn and analyzed for CO_2 content using gas chromatography. The possibility of nonbiological CO_2 production was investigated using poisoned controls in which $HgCl_2$ (250 μg/ml) was added to stop microbial activity. Basal respiration in the deep subsurface samples ranged from <0.001 to

0.664 μg CO_2/g dry soil/hr. The highest rates of basal respiration were detected in subsurface samples with the highest total organic carbon content, which were fine-grained sediments originating from lake (lacustrine) deposits. Subsurface deposits originating from buried soil (paleosol) or river (fluvial) sediments were found to have much lower basal respiration. There was strong agreement in the pattern of microbial distribution as revealed by basal respiration, cell number determined by microscopic counts (AODCs), and glucose mineralization. It should be noted that basal respiration is not an *in situ* measurement, and it should be expected that the results obtained will differ from *in situ* respiration rates, which are affected by the physical and environmental parameters of the particular site.

Basal respiration has also been used as an indicator of soil "health" or condition. Dulohery *et al.* (1996) used large (0.5 by 1.0 m) field chambers (Fig. 11.15) to determine CO_2 production in forest soils that had been affected by logging activity. Tree harvesting involves the use of heavy equipment, which can alter forest soils in many ways, including removal of topsoil and compaction of the soil profile. A comparison of soil respiration was made between pristine sites and sites where logging activities had taken place. Basal levels of CO_2 production ranged from 52 to 257 mg/m^2/hr

in undisturbed forest soil sites, depending on the season. CO_2 production was significantly reduced in sites affected by heavy equipment. For example, in areas that had been used as skid trails to haul logs from the site, soil respiration had declined an average of 34%. Attempts to improve soil conditions after logging using fertilizer application and soil tillage were evaluated.

11.4.2.2 Substrate-Induced Microbial Activity

Microbial respiration has been used to evaluate the response of microbial populations to the introduction of organic pollutants, such as petroleum hydrocarbons. In this case, the desired response is that the hydrocarbons be used as a carbon source by microorganisms. This response can be measured as an increase *in situ* respiration above the basal respiration level of the indigenous microbial community. Hinchee and Ong (1992) used subsurface probes to monitor microbial respiration in a number of sites where the subsurface was contaminated with a variety of different hydrocarbon mixtures, including jet fuel and crude oil. The soil probes (Fig. 11.16) were installed to monitor atmospheric gases between 1 and 5 m below the ground surface, depending on the site. Probes were installed in areas already contaminated with hydrocarbons and, in order to estimate the basal respiration rate, in areas

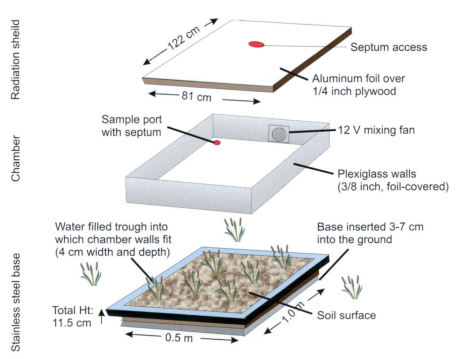

FIGURE 11.15 A static chamber used for sampling forest-floor gas fluxes. (Adapted with permission from Dulohery *et al.*, 1996.)

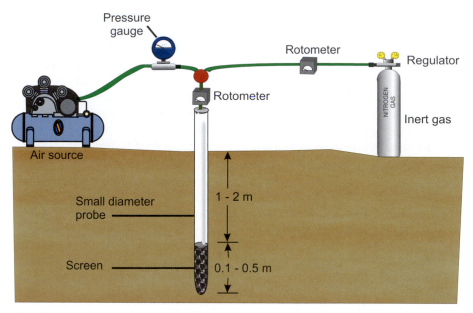

FIGURE 11.16   A subsurface probe used to remove gas samples from various depths within the soil profile. This setup was used to measure in situ respiration in a hydrocarbon-contaminated site. (Adapted with permission from Hinchee and Ong, 1992.)

where there was no contamination. Prior to measurement, air was injected through the probe into the subsurface for 24 hours to ensure that the subsurface contained an adequate amount of oxygen. The air injection was then turned off, and the microbial utilization of O_2 and production of CO_2 were monitored by withdrawing gas samples at intervals of 2 to 8 hours. Respiration rates were calculated on the basis of O_2 utilization (% O_2/hr) and revealed that more than 50% of the available oxygen was utilized within 20 to 80 hours in the hydrocarbon-contaminated sites. This was a significant increase over the approximate 10% reduction in the available oxygen in a comparable but uncontaminated site.

As discussed in Section 11.4, one needs to have a conversion factor to determine the rate of substrate utilization from oxygen uptake. In this study, an estimate of the rate of hydrocarbon biodegradation based on O_2 utilization was made using a reference hydrocarbon, hexane. The stoichiometric relationship for the oxidation of hexane indicated that 9.5 moles of O_2 were required for the complete mineralization of 1 mole of hexane. Based on this relationship, the rate of hexane degradation ranged from 0.4 to 13.0 mg hexane/kg soil/day for sites contaminated with jet fuel and 3.6 to 19.0 mg hexane/kg soil/day for sites contaminated with crude oil. The results provide an approximation of the rate at which petroleum hydrocarbons are biodegraded in this particular subsurface site if adequate aeration is provided. Such an estimate is

important in determining the rate at which the site can be remediated biologically and whether this rate is high enough to contain the contaminant spill and prevent further migration of the hydrocarbons.

Hydrocarbon biodegradation rates based on CO_2 production were also determined but were less consistent than rates based on O_2 utilization. CO_2 production (% CO_2/hr) was unexpectedly low in contaminated sites where the soil pH was above 7.0, indicating the possibility of CO_2 retention in the bicarbonate form by the soil.

11.4.2.3 Microbial Biomass Determination

A measurement commonly performed to give an indication of soil condition and the potential for metabolic activity is that of the amount of microbial biomass in a soil. This measurement can be made using two different approaches, both based on CO_2 evolution. One approach is a **chloroform fumigation-incubation method** (see Information Box). A soil sample is first fumigated with chloroform to kill the indigenous microorganisms. Upon reinoculation of the soil, the dead microbial biomass becomes available for microbial consumption. The amount of CO_2 produced by the consumption of this organic material is monitored and used to calculate the initial quantity of microbial biomass. In this procedure, a sample of fumigated soil and a sample of nonfumigated soil, which serves as a control to determine the basal respiration level, are incubated for 10 days in a sealed microcosm. The microorganisms responsible for the respiration will be the very small proportion of microbes that survived the fumigation treatment, or the fumigated soil will be inoculated with a starter culture from the non-fumigated soil. The amount of CO_2 released by mineralization is quantified using an alkaline trap or by GC analysis of headspace gases. The production of CO_2 in the nonfumigated soil represents the basal rate of mineralization, and CO_2 production in the fumigated soil represents primarily the mineralization of microbial biomass. Therefore, the amount of respiration, corrected for the basal rate, is a measure of the amount of microbial biomass present prior to the fumigation step. The amount of carbon held in microbial biomass is calculated as

$$\text{Biomass C} = \frac{\text{Fc} - \text{Ufc}}{\text{Kc}} \qquad \text{(Eq. 11.3)}$$

where

biomass C = the amount of carbon trapped in microbial biomass

Fc = CO_2 produced by the fumigated soil sample

Ufc = CO_2 produced by the nonfumigated soil sample

Kc = fraction of biomass C mineralized to CO_2

The value of Kc is of considerable importance because it expresses the proportion of total metabolized organic carbon that is mineralized. Literature values for Kc range from 0.41 to 0.45. These estimates are usually based on tracer experiments in which the mineralization of radiolabeled substrates, such as glucose or bacterial cells, by microorganisms isolated from the soil is determined in a liquid culture. The proportion of radiolabeled substrate that is mineralized to $^{14}CO_2$ or assimilated into microbial biomass can then be determined.

An alternative method for the estimation of microbial biomass is the **substrate-induced respiration method.** This method estimates the amount of carbon held in living, nonresting heterotrophic cells by determining the initial respiration response when glucose, a readily metabolized organic carbon source, is added to the soil. The incubation time is limited to several hours, with samples of the mineralized CO_2 taken at regular intervals. The rate of glucose respiration in the soil is expected to increase with time, reflecting both the initial metabolism of the substrate and the fueling of cell division, which results in an increase in the number of metabolizing cells. Thus, the lowest and usually initial rate of glucose respiration is used as an indicator of the microbial biomass present in the soil sample. Calculation of microbial biomass is based on a

---

Information Box
Measurement of microbial biomass in soil—the chloroform fumigation method

Step 1. Collect a representative soil sample (see Chapter 8).
Step 2. Perform a fumigation experiment as outlined below.

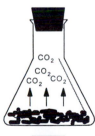

1. Fumigate a portion of the soil sample with chloroform.
2. Inoculate.
3. Incubate for 10 days.
4. Determine CO_2 as an estimate of biomass mineralization = Fc.

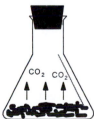

1. Incubate a second portion of the soil sample for 10 days.
2. Determine CO_2 as an estimate of basal mineralization levels = Ufc.

Step 3. Estimate Kc by performing a mineralization experiment using ^{14}C-labeled glucose and a soil inoculum (see Section 11.2.4).

correlation between the substrate-induced respiration method and the chloroform fumigation-incubation method, described earlier, or other methods for determining microbial biomass in soil. A commonly used relationship was provided by Anderson and Domsch (1978):

$$y = 40.04x + 0.37$$

where y is biomass C (mg/100 g dry weight soil) and x is the respiration rate (ml CO_2/100 g dry weight sediment/hr).

The main advantage of the substrate induction method, compared with the chloroform fumigation method, is that microbial biomass can be estimated in a much shorter time. To implement this method, preliminary studies are required to determine the appropriate glucose concentration to add to the soil. Increasing glucose concentrations saturate the initial uptake capacity of the microbial population, resulting in a plateau in glucose mineralization. The lowest glucose concentration that yields the maximum respiratory response must be independently determined for each soil type and then applied to that soil in order to standardize the substrate induction method between different soil types. Examples of estimates of microbial biomass range from 12.8 to 203 mg biomass-C in 100 g dry soil, depending on soil type and also on the method used to determine microbial biomass (Martens, 1987).

The substrate-induced respiration method has been modified to determine the individual contribution of bacteria and fungi to total heterotrophic activity within soil by the use of selective antibiotics. Glucose-induced respiration is determined in the presence of either streptomycin, which inhibits prokaryotes, or cycloheximide, which inhibits eukaryotes. Dominance of fungal glucose-induced respiration was evident in a soil sample from a semiarid region (Johnson *et al.*, 1996) and in the early stages of mineralization of plant litter (Beare *et al.*, 1990). In contrast, bacteria are thought to dominate in the rhizosphere and in subsurface sediments. However, it must be recognized that the glucose-responsive, antibiotic-sensitive microbial population may not represent the entire microbial population within an environment, and the use of antibiotics can only yield estimates of the relative contribution of fungi and bacteria to glucose-induced respiration (Johnson *et al.*, 1996).

11.4.2.4 Biological Oxygen Demand

The 5-day BOD test was developed as a means of monitoring wastewater quality. The BOD test quantifies the oxygen required to metabolize dissolved or-ganic carbon present after wastewater treatment. Thus, the BOD test can provide an indication of the impact that wastewater treatment plant effluents may have on the receiving waters to which they are discharged. This test has become an industry standard in terms of evaluating water quality and has an established protocol (APHA, AWWA, WEF, 1995). Simply stated, water samples, usually 250 to 300 ml, are incubated in sealed bottles for 5 days at a constant temperature, $20 \pm 1°C$. The BOD is calculated from the difference in dissolved oxygen concentration measured at the beginning and at the completion of the incubation period. The emphasis of the BOD test is on the determination of the oxygen demand created by the presence of dissolved organic material. To that end, it may be necessary to add inorganic nutrients or even a "seed" solution of heterotrophic bacteria to ensure that dissolved organic material is, in fact, degraded.

The oxygen demand within the sediment of freshwater and estuarine environments can have a major impact on oxygen levels in surface waters, especially in shallow lakes and in rivers receiving organic-based waste material. The oxygen demand is created by microbial activity, as well as invertebrate respiration and nonbiological oxidation. Dissolved oxygen standards for the protection of fish and the aquatic ecosystem as a whole have been instituted for lakes and rivers affected by organic waste generated by human activity, and the implementation of water quality standards necessitates quantification of sediment oxygen demand. Bowman and Delfino (1980) described an experimental apparatus for the determination of sediment oxygen demand in which water circulates over a layer of sediment in a sealed container (Fig. 11.17). The closed loop of circulating water passes by a dissolved-oxygen probe, which continuously monitors dissolved oxygen levels.

11.4.2.5 Depth-Dependent Processes in Biofilms

Respiration measurements have also been used to better understand how biofilms work. Mixed community microbial mats, or biofilms, are formed on many different surfaces as bacteria attach to and proliferate into a densely packed film (see Chapter 6.2.4). Biofilms can be found on almost any surface imaginable, from rock surfaces in streambeds to filter systems used in wastewater treatment to the surfaces of our teeth. The flux of respiratory gases within and around biofilms has revealed their importance in terms of organic carbon and nutrient cycling.

The study of biofilms has been made easier by the development of microelectrodes that can measure different respiration gases. For example, depth-

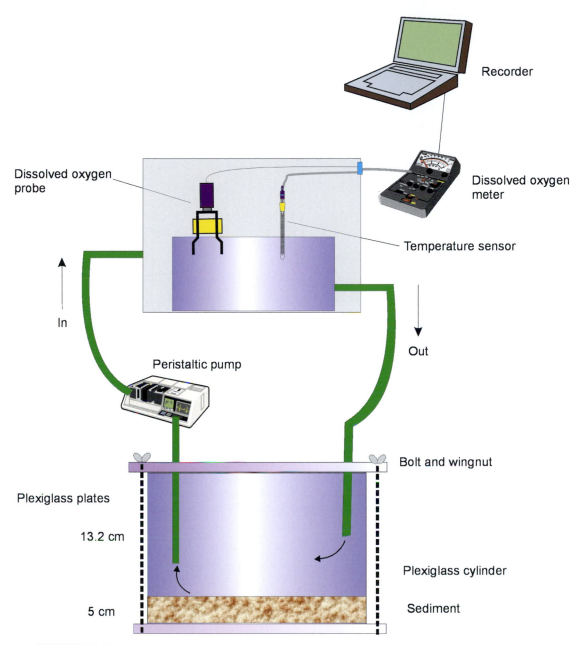

FIGURE 11.17 Apparatus used to determine sediment oxygen demand. (Adapted with permission from Bowman and Delfino, 1980.)

dependent activity in a microbial mat was studied by Nielsen *et al.* (1990) using a combined O_2 and N_2O microelectrode with a sensor tip that was 20 μm in diameter (Fig. 11.18). In contrast, the biofilms being studied were several centimeters thick. The microelectrode was lowered into the biofilm at intervals as small as 50 μm to record specific respiration activity. Results revealed that oxygen was depleted by microbial activity in the surface layers of the mat and that beneath the surface, conditions were anoxic. This allowed the development of denitrifying bacterial populations in the in-

terior of the biofilm. These bacteria can respire under anaerobic conditions using NO_3^- as a terminal electron acceptor in place of O_2. Anaerobic respiration of NO_3^-, referred to as denitrification, was estimated using a N_2O microelectrode and actelyene gas. The addition of acteylene gas blocks the reduction of NO_3^- during denitrification so that the intermediate N_2O is formed (see Section 11.2.2.2). Thus, the rate of N_2O production provides an estimate of the rate of anaerobic respiration of NO_3^- (i.e., μmol NO_3^- utilized/cm^3/hr).

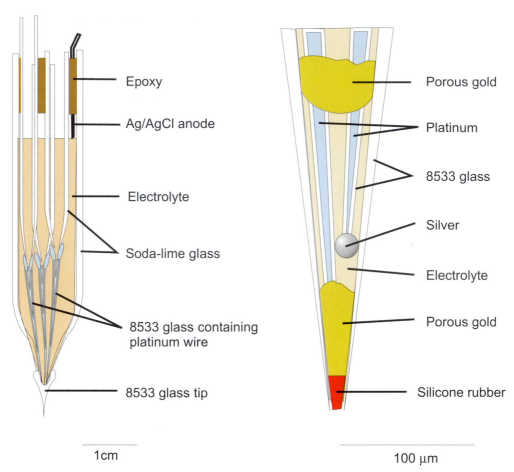

FIGURE 11.18   Microelectrodes used to study biofilms. On the left is the microsensor used for analysis of O_2 and N_2O. There are three cathodes within the outer casing; one is shown behind the plane of the two others. On the right, the tip of the microsensor is enlarged 250 times. (Adapted with permission from Revsbech *et al.*, 1988.)

Schramm *et al.* (1996) also revealed the anaerobic nature of microbial activity in the interior of a microbial biofilm using a N_2O microelectrode. In this case, the N_2O microelectrode was coated with immobilized bacteria, which were selected for their ability to reduce NO_3^- metabolically, but only as far as N_2O. Insertion of this microelectrode into the interior of a biofilm recorded the respiration of NO_3^-, a process which could occur only in the absence of oxygen. Using these tools, it has been possible to better understand how the structure of a biofilm affects its function. Specifically, the rapid utilization of oxygen by the abundant bacteria in the upper layers of the biofilm creates an anaerobic environment in the deeper layers that supports the anaerobic respiration of NO_3^-.

11.4.3 Tracer Studies to Determine Heterotrophic Potential

The use of ^{14}C-labeled carbon substrates was discussed in Section 11.2.4.2 as a sensitive way to measure biodegradation of a specific organic compound. This approach can also be used to determine heterotrophic potential in environmental samples. Many radiolabeled substrates are commercially available, including sugars such as glucose, organic acids such as lactic acid, amino acids, and even many representative organic pollutants. Mineralization of these substrates can be quantified by measurement of the evolution of $^{14}CO_2$.

The addition of a ^{14}C-labeled substrate to an environmental sample reveals the presence of a degrading population in that sample and also indicates the rate at which the substrate can be mineralized. However, the rate of substrate mineralization is related to the amount of added substrate up to a saturation limit (see Fig. 3.7). Thus, depending on the level of substrate added, this measure may represent basal metabolism or may represent the potential for a microbial response to higher levels of added substrate. So the dilemma associated with the use of radiolabeled substrates is that although they can provide a very sensitive measurement of substrate utilization, the rate of utilization is usually proportional to the amount of labeled substrate added to the environmental sample. This is especially true for compounds such as glucose, which are very rapidly metabolized by microbial populations. If the fundamental concern of the environmental microbiologist is to determine the rate at which naturally occurring organic substrates are utilized, he or she must be very careful not to overestimate basal levels of activity by adding levels of a carbon source that exceed those already present in the system.

To address this problem, a kinetics-based analysis of labeled substrate mineralization has been developed to calculate the indigenous rate of glucose mineralization in environmental samples. A saturation kinetics model was adapted so that the stimulated rates of substrate mineralization can be extrapolated back to the rate at which a substrate is mineralized in an undisturbed sample. This analysis is referred to as the determination of heterotrophic potential (Wright and Burnison, 1979; Ladd et al., 1979). When a radiolabeled substrate, such as ^{14}C-labeled glucose, is added to a soil or water sample, the rate at which $^{14}CO_2$ is evolved is monitored using an alkaline trapping technique. In the case of a water sample, the amount of radiolabeled substrate assimilated into biomass can be monitored by filtering the bacterial cells from the water sample and then measuring the radioactivity incorporated into the biomass. Combining the mineralization with the assimilation data provides a measure of the total amount of substrate utilized by the cell. If a single concentration of ^{14}C-glucose is considered, then the rate of uptake by a microbial community can be expressed as:

$$v = \frac{f(S_n + A)}{t} \qquad \text{(Eq. 11.4)}$$

where
v is the rate of uptake (or respiration) by the microorganisms (mass/volume/time)
t is the incubation time
f is the fraction of labeled substrate taken up (or respired) in time t
A is the amount of added substrate (mass/volume)
S_n is the naturally occurring concentration of the substrate (mass/volume)

If S_n is known and A is added such that $A << S_n$, then an assumption can be made that the natural rate of uptake is not significantly altered by the presence of A. Using a short incubation time, the fraction of labeled substrate taken up (or respired) can be determined, and can be expressed as

$$T_n = \frac{S_n}{v_n} \cdot \frac{A}{} = \frac{t}{f} \qquad \text{(Eq. 11.5)}$$

where v_n is the natural rate of substrate uptake (or respiration), and T_n is the substrate turnover time due to uptake by the natural population at the natural substrate concentration.

However, in most cases S_n is not known, and A will be added at a concentration much greater than S_n. A situation in which $A >> S_n$ would result in the stimulation of substrate utilization rates above what would be considered the natural level of microbial activity by the indigenous population. Therefore, a single concentration tracer study would not be appropriate for determining T_n. However, T_n can be calculated by measuring the substrate utilization rate with added substrate, and then extrapolating back to the natural level of microbial activity. The basis of this extrapolation technique is Michaelis-Menten kinetics, which state that as the concentration of a substrate increases, the rate of activity of an enzyme will also increase until a plateau is reached. An analogy can be drawn between the transformation of a substrate by an enzyme and the uptake of an organic substrate by bacteria. Therefore, the rate of uptake of an organic substrate by bacteria increases as the concentration of the added substrate is increased until the mechanism of substrate uptake has been saturated. Michaelis-Menten kinetics can be adapted for heterotrophic potential studies using the modified Lineweaver-Burk transformation, which states that

$$\frac{S_n + A}{v_n} = \frac{1}{V_{max}} A + \frac{K + S_n}{V_{max}} \qquad \text{(Eq. 11.6)}$$

where V_{max} is the theoretical maximal rate of substrate uptake by the microbial population, and K is a transport constant defined as the substrate concentration at which $v = \frac{1}{2} V_{max}$.

Combining equations 11.5 and 11.6 will give

$$\frac{t}{f} = \frac{1}{V_{max}} A + \frac{K + S_n}{V_{max}} \qquad \text{(Eq. 11.7)}$$

The value of (t / f) can be plotted over several levels of added substrate A, and fitted with a regression line (Fig. 11.19). The line is extrapolated back to the x-axis. The slope of the line is $(1/V_{max})$, the y-intercept is $[(K + S_n) / V_{max}]$, and the x intercept is $(K + S_n)$.

This kinetic approach generates parameters that describe the in situ rate of heterotrophic activity. The natural turnover time, T_n, can be calculated when $A = 0$ using equations 11.4 and 11.6

$$\frac{t}{f} = \frac{K + S_n}{V_{max}} = \frac{S_n}{v} \qquad \text{(Eq. 11.8)}$$

Other useful parameters include, V_{max}, the theoretical maximum rate of substrate uptake, which reflects the abundance and activity of the microbial population.

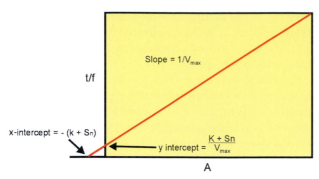

$$\text{Slope} = 1/V_{max}$$

t/f

$\text{x-intercept} = -(k + S_n)$

$\text{y intercept} = \dfrac{K + Sn}{V_{max}}$

A

FIGURE 11.19   A Lineweaver-Burke plot.

The quantity $(K + S_n)$ may be used as an estimate of the natural substrate concentration, S_n, assuming that K is very small $(K \ll S_n)$, which is not necessarily true in all cases.

These parameters can be used to compare the heterotrophic activity of different microbial populations in different environments. However, certain assumptions and limitations must be considered. First, the kinetic approach assumes no significant change in microbial number and no significant depletion of the added substrate during the incubation period. To meet these assumptions, short incubation periods are required. In highly productive environments, incubations of 1 hour may be adequate, but the incubation period must be increased in oligotrophic environments. The temperature of incubation will also affect the measured parameters. All samples should be incubated at the same temperature; preferably at *in situ* temperatures so that the results are environmentally relevant.

It should also be noted that the heterotrophic potential approach measures only a single substrate and analyzes a diverse microbial population by calculating kinetic parameters as if they were uniform in the population. An assumption is made that the uptake of a substrate by a natural community follows saturation kinetics as if it were a single enzymatic reaction that is not influenced by the presence of other substrates in the environment (Van Es and Meyer-Reil, 1982).

Measurement of heterotrophic activity has been particularly important in understanding how carbon is cycled through the aquatic environment. Heterotrophic bacterial populations have been shown to dominate the utilization of dissolved organic carbon in aquatic environments. Determination of dissolved organic carbon utilization rates revealed that the production of new biomass through heterotrophic processes can equal the amount of biomass formed through primary production by photosynthetic mi-

croorganisms. Heterotrophic activity measurements have increased our understanding of the distribution of microbial communities in different ecological zones within the aquatic environment. It has been shown that bacterial abundance and the rate of carbon cycling are greatest in freshwater sediments, as compared with the planktonic, or free-floating, environments. From this, we now understand the ecological importance of sediment-associated bacterial communities in carbon and energy cycling.

11.4.4 Anaerobic Respiration as an Indicator of Microbial Activity

As discussed earlier, soil respiration under aerobic conditions can be monitored using CO_2 production or O_2 utilization as an indicator of carbon utilization. However, many environments are completely anaerobic or contain anaerobic niches, and there are situations in which monitoring the activity of anaerobic respiration is critical. For example, whereas anaerobic environments can support the biotransformation of organic pollutants, in some cases this results in the formation of more toxic metabolites. Anaerobic respiratory pathways include the use of NO_3^-, Fe^{3+} (and other metals), SO_4^-, and CO_2 as terminal electron acceptors. For some of these terminal electron acceptors, a gaseous intermediate or end product is formed that can be used as a measure of anaerobic respiration. For example, nitrous oxide (N_2O) is an intermediate in the bacterial reduction of NO_3^- to N_2 (dentrification). Acetylene gas is used to inhibit the complete denitrification of NO_3^-, resulting in the accumulation of N_2O gas (see Section 11.2.2.2). Production of N_2O in samples treated with acetylene gas is proportional to the rate of denitrification. In terms of the soil condition, the presence of N_2O gas suggests a reducing environment in which the indigenous bacteria can utilize NO_3^- as an alternative electron acceptor. The production of N_2O also indicates that nitrification processes at some point resulted in an adequate amount of NO_3^-.

The reduction of CO_2 during anaerobic respiration also produces methane (CH_4) (see Chapter 3.4). The utilization of CO_2 as a terminal electron acceptor is limited to a group of bacteria called the methanogens and this process requires a strongly reduced environment. The carbon substrates compatible with methanogenesis are simple one- or two-carbon compounds that are produced as a result of the activities of several different groups of bacteria, including fermentative and acidogenic microorganisms. Despite the complexity of the microbial interactions leading to methane production, methanogenic environments are common in soils and are also well characterized for use in the treatment of large volumes of organic waste. Methane production can be a problem in some instances and in fact commonly occurs within municipal land-fill sites. As landfills age, they often develop highly productive methanogenic populations, and the release of copious amount of CH_4 can create a fire hazard. The most sensitive means of determining N_2O and CH_4 is by gas chromatography. Microelectrodes specific for N_2O are less sensitive than gas chromatography, but have been used to investigate the location of denitrifing activity in inhibited samples (Nielsen *et al.*, 1990).

11.5 INCORPORATION OF RADIOLABELED TRACERS INTO CELLULAR MACROMOLECULES

Quantification of cellular constituents, such as protein or nucleic acids, can be used to monitor the increase in biomass of a bacterial population. This approach is often used to monitor the growth of a pure culture on defined media in the laboratory. In the natural environment, numerous sources of these constituents, including plant debris and soil animals, can contribute to the total protein or nucleic acids, and, therefore, these assays would not be specific to bacteria. A way to make these assays more specific to bacteria is to measure the incorporation of radiolabeled tracer molecules into cellular macromolecules. This technique is based on the fact that many species of heterotrophic bacteria can scavenge preformed molecules such as nucleotides or amino acids from their environment, incorporating them directly in cellular constituents. Monitoring the rate at which radiolabeled nucleotides or amino acids are incorporated into essential cellular macromolecules can be used as a means of determining microbial activity. Examples of specific molecules used as tracers are the nucleoside thymidine, labeled with tritium (^3H) which is incorporated into DNA, and the amino acid leucine, labeled with either tritium or carbon-14 (^{14}C), which is incorporated into protein.

There are several advantages of measuring microbial activity using radiolabeled tracer molecules. This technique requires a short incubation period, which is more convenient for field studies and which reduces artifacts created by extended incubation of environmental samples. Further, the use of radiolabeled tracers increases the sensitivity of the measurements, allowing the quantification of very low levels of microbial activity such as those found in extreme environments such as Antarctica (Tibbles and Harris, 1996). There are also several potential sources of error

in the tracer method including (1) the fact that not all bacteria are capable of assimilating tracers into macromolecules, (2) the possible nonspecific incorporation of the label into cellular macromolecules other than the intended target, and (3) the extent of isotope dilution, which may vary both spatially and temporally within a sample site.

11.5.1 Incorporation of Thymidine into DNA

The rate at which a tracer is incorporated into a cellular macromolecule such as DNA provides an indication of the rate of formation of that macromolecule. The measured rate of formation will provide an accurate measure of bacterial growth only if the formation rate is directly related to cell division. In a state of growth known as **balanced growth,** all cell constituents increase at the same rate. Therefore, the doubling time of any cellular macromolecule would equal the doubling time of the whole cell. Although balanced growth can be achieved under controlled laboratory conditions, sustained balanced growth is unlikely under environmental conditions. However, the synthesis of DNA and protein is strongly enough related to cell division that, even when growth is unbalanced, their synthesis is expected to provide a reasonable reflection of cell division.

The scavenging of the thymidine nucleoside and its incorporation into DNA by heterotrophic bacteria have been reviewed by Azam and Fuhrman (1984), Moriarty (1986), and Robarts and Zohary (1993). Once transported across the cell membrane, the thymidine is converted into thymine monophosphate by the action of the enzyme thymidine kinase. Further phosphorylation results in incorporation into DNA as the thymine base. In very general terms, the procedure involves adding [^3H]thymidine to a water sample containing planktonic bacteria or to a slurry prepared by mixing soil or sediment with water. The incubation time is usually limited to several hours, and the incorporation of labeled thymidine by the cells is terminated by the addition of a chemical inhibitor or by placing the sample in an ice bath. DNA is then extracted from the cells and the radioactivity quantified to determine the amount of label that has been incorporated. One problem with this technique is that during the labeling and uptake procedure, nonspecific labeling of other macromolecules, such as proteins and RNA, can occur. In this case, purification of the DNA may be required to improve the accuracy of the activity measurement.

One of the aspects of this assay that makes it useful

for measurement of bacterial heterotrophic activity is that the rate of thymidine incorporation into bacterial DNA is much higher than the rate for other organisms tested, such as algae, fungi, or protozoa. Therefore, if the uptake study is limited to a short incubation periods (several hours), it is generally believed that the incorporation of these precursors into cellular macromolecules will reflect the activity of growing heterotrophic bacteria. However, not all growing bacteria will incorporate exogenously supplied thymidine into DNA. Exceptions have been noted, particularly within the genus *Pseudomonas,* which may limit the usefulness of this technique in certain environments.

A further complication in the interpretation of the results of this assay is that there are both external and internal pools of thymidine or thymidine metabolites that can compete with the added [^{3}H]-thymidine for incorporation into DNA. The internal pool is created by cells in a *de novo* process in which the nucleotides that form DNA are synthesized from cellular components. If this internal nucleotide pool is large, it may compete with scavenged labeled thymidine for incorporation into DNA. There can also be an external pool of thymidine available for uptake. This external pool is composed of extracellular thymidine probably released from dying organisms. This is especially true for sediments, where thymidine becomes sorbed to sediment particles. The existence of the external and internal pools of thymidine metabolites will result in dilution of the added [^{3}H]thymidine and cause underestimation of the rate of DNA synthesis. Isotope dilution, which refers to the size of the pool into which [^{3}H]thymidine is diluted, can be estimated (Moriarty, 1986; Robarts and Zohary, 1993) and used to correct estimates of DNA synthesis. Further, the addition of sufficient quantities of exogenous labeled thymidine may cause competitive inhibition of the *de novo* synthesis route and thereby limit the size of the internal thymidine pool.

The thymidine incorporation rate can be converted into a measure of microbial activity using a conversion factor that relates the number of new cells formed to a given amount of thymidine incorporated (i.e., cells formed per mole thymidine incorporated). This conversion factor is derived on the basis of measurements, or best possible estimates, of the amount of DNA per cell, the thymine content of a cell, and the extent of dilution of the labeled thymidine. An alternative way to determine the conversion factor is to relate the [^{3}H]thymidine incorporation rate directly to cell division based on a separate and independent measure of the increase in bacterial numbers. Common estimates of this conversion factor range from 1.3 to 2.0×10^{18} cells/mole [^{3}H]thymidine incorporated.

Based on thymidine incorporation, bacterial growth can range from less than 1×10^6 cells/g soil/day in an oligotrophic aquifer environment to greater than 1×10^9 cells/g soil/day in a marine sediment (Thorn and Ventullo, 1988). The rate of thymidine incorporation can also be used without conversion to estimate cell division. The thymidine incorporation rate provides a relative measure of microbial activity that can be used in controlled experiments to evaluate the impact of a specific factor, such as the toxicity of heavy metals to microbial activity in environmental samples (Díaz-Raviña *et al.,* 1994).

11.5.2 Incorporation of Leucine into Protein

The rate of leucine incorporation into cellular protein has also been used as a measure of microbial activity (Kirchman *et al.,* 1985; Chin-Leo and Kirchman, 1988). Studies have indicated that a majority of bacteria scavenge leucine from their environment and that most of the assimilated radioactive label will be incorporated into proteins. Further, the *de novo* synthesis of leucine is inhibited by sufficient quantities of exogenous leucine.

11.6 ADENYLATE ENERGY CHARGE

Adenosine triphosphate (ATP) is a compound synthesized by actively growing cells as a means of short-term energy storage and transfer. ATP captures metabolic energy in the form of high-energy phosphate bonds and is transported to sites within the cell where energy is required to drive a biochemical reaction. Adenosine diphosphate (ADP) and adenosine monophosphate (AMP) are precursors of ATP, and together the three forms represent the cellular adenylates. ADP and AMP cycle between the sites where high-energy phosphate bonds are added to form ATP and sites where the phosphate bonds are broken to transfer energy to a metabolic process. The ATP content of the cell varies depending on its level of activity, with rapidly growing cells having a higher ATP content than stressed cells. A fairly constant relationship between ATP and cell biomass (10–12 mol ATP/g biomass C) has been measured for soils incubated under specific conditions (Jenkinson, 1988). However, a strong correlation between ATP content and microbial biomass may not hold for microbial populations under conditions of environmental stress, such as in dry or excessively wet soils (Inubushi *et al.,* 1989; Rosacker and Kieft, 1990). Thus, the detection of total adenylates (ATP, ADP, and AMP) may provide a better indicator of microbial biomass.

In terms of determining microbial activity, the relative abundance of ATP compared with its precursors, ADP and AMP, indicates how rapidly the highest energy state (ATP) is formed. The ATP/ADP/AMP ratio provides a biochemical basis for assessing the physiological and nutritional status of organisms. A measure of the **adenylate energy charge** (AEC) ratio is the weighted ratio of cellular adenylates:

$$AEC = \frac{ATP + \frac{1}{2}ADP}{ATP + ADP + AMP}$$

High AEC values (>0.8) reflect an active community, intermediate values (0.4 to 0.8) reflect cells in a resting state, and low values (<0.4) reflect a high proportion of dead or moribund cells (Kieft and Rosacker, 1991).

The quantification of AEC from environmental samples can be used as an indicator of bacterial biomass and has some advantages and disadvantages in comparison with approaches already discussed. A major benefit of the use of AEC is that it provides a measure of microbial activity in environmental samples without the requirement to introduce a substrate and/or the need to incubate the sample for a period of time, as required in respiration or radiolabeled tracer activity measurements. It should be noted, however, that cellular adenylates are present in all living organisms, and its measurement is not selective for microbial populations. Rapid degradation of cellular adenylates from organic debris will minimize nonmicrobial sources. A preincubation of the soil under aerobic conditions may be necessary to ensure that cellular adenylates from sources such as plant root fragments are given an opportunity to be degraded (Sparling et al., 1985).

The quantification of ATP is based on the transfer of energy to a luciferin–luciferase substrate–enzyme system. This system was originally isolated from the abdomen of the common firefly. With the energy supplied by ATP, the enzyme luciferase acts on the substrate luciferin to produce radiant energy. The light emitted can be detected and quantified and is directly proportional to the amount of available ATP. Quantification of ATP from environmental samples requires an extraction procedure, followed by the concentration of cellular components into a buffer solution (Nannipieri et al., 1990). The amount of ATP present is then determined directly using the luciferin-luciferase assay. ADP and AMP in the sample are then converted to ATP by enzymatic reactions and quantified.

AEC has been used to characterize microbial activity in both soil and subsurface environments. AEC analysis in subsurface samples from various depths revealed a wide range of values from relatively high values (0.76), reflecting an active population, to much

lower values (0.23), reflecting dead or moribund microbial populations (Kieft and Rosacker, 1991). In surface soils, AEC analysis has indicated a range of microbial activity from inactive (0.5) (Rosacker and Kieft, 1990) to active (0.8) (Brookes et al., 1983; Ciardi et al., 1991). Drying or saturating the soil can cause a decrease in AEC, to below 0.4 in some soils (Inubushi et al., 1989; Ciardi et al., 1991). However, Rosacker and Keift (1990) reported that air drying caused only a temporary decrease in AEC due to a transient increase in cellular AMP concentration. Increased AMP may have been the result of cellular catabolism of RNA, indicating that stressed cells may utilize endogenous metabolism of cellular macromolecules in order to survive. Such transient increases in AMP by stressed cells may affect the interpretation of AEC results.

The value of AEC measurement is that it can establish the presence of metabolically active microbial populations in environmental samples. For example, AEC has been used to characterize microbial populations in environments contaminated with pollutants, such as aviation fuel (Webster et al., 1992). AEC measurements in polluted environments can reveal the presence of microorganisms that are resistant to the adverse affects of the pollutants. These microorganisms may also be actively metabolizing the pollutants, thus removing them from the environment.

11.7 ENZYME ASSAYS

Enzymes are specialized proteins that combine with a specific substrate and act to catalyze a biochemical reaction. In the soil and sediment environments, enzymatic activity is essential for energy transformation and nutrient cycling reactions. For example, enzymes catalyze the hydrolysis of certain nitrogen-, phosphorus- or sulfur-containing organic compounds, releasing the ammonia, phosphate, or sulfate constituents, which are then available for assimilation by other organisms. Also of note are enzymes that catalyze the hydrolysis of plant constituents, such as cellulose, starch, and other polysaccharides, releasing the monomeric sugar units, such as glucose, which provide important energy sources for microorganisms. Several reviews have examined enzymatic reactions in the soil environment and discussed their importance in terms of soil fertility (Burns, 1978; Tabatabai, 1994b; Morra, 1997).

One means of monitoring enzymatic reactions is to measure the conversion of a specific substrate into a product. An example is monitoring nitrate reductase activity through the disappearance of nitrate and the formation of elemental nitrogen. More commonly, enzymatic reactions are detected using a bioassay procedure that is specific for a particular class of enzymes. A bioassay utilizes a surrogate substrate that is transformed by a specific class of enzymes, producing a product that has specific properties for detection. An example is the conversion of p-nitrophenol phosphate by phosphatase enzymes, producing p-nitrophenol and phosphate. The indicator product, p-nitrophenol, can be quantified spectrophotometrically.

The enzymatic activity within an environmental sample can originate from a variety of sources. Enzymatic activity can be directly associated with actively growing microorganisms, including bacteria, fungi, and actinomycetes. These enzymes may be contained within the cell and are often located within the cell membrane. Enzymes may also be released from the cell, in which case they are called **extracellular enzymes.** Extracellular enzymes are released from actively growing cells to hydrolyze large polymers, such as the plant polymers, cellulose, hemicellulose, and lignin, in order to facilitate their uptake for further metabolism. Alternatively, enzymes may be found outside cells as a result of the decay and disintegration of bacterial, animal, or plant cells (Nannipieri *et al.*, 1990). Enzymes may be associated for a short time with moribund cells. These enzymes may also become stabilized on clay or humic particles within the soil

structure and can remain viable for a period of time (Tabatabai, 1994b). Therefore, any bioassay of enzymatic activity in environmental samples will measure activity from all sources and provide an indication of total enzymatic potential within that sample.

Characterizing and quantifying enzyme activity in environmental samples can reflect the health of the environment in terms of nutrient cycling and can also reflect soil fertility parameters, such as crop yield. However, there are problems associated with using enzyme assays to directly quantify microbial activity (Nannipieri *et al.*, 1990). First of all, enzymatic activity associated with actively growing microorganisms can not be easily separated from the activity of extracellular enzymes stabilized in the soil environment or enzymes associated with decaying cells. Another problem is that enzyme assays often require the addition of a surrogate substrate and, as a consequence, the assay determines the potential enzymatic activity and not the actual level of activity in the sample. Enzyme assays are also specific for a particular substrate-enzyme combination and may not reflect the overall activity of all types of microorganisms. The simultaneous determination of a large number of enzyme assays may be more representative of overall microbial activity, but this approach is more labor intensive.

Many different enzyme assays have been developed to detect either specific or general microbial ac-

TABLE 11.2 General and Specific Enzyme Assays That Can Be Used to Measure Microbial Activity

Enzyme	Substrate	Description of assay
Dehydrogenase	Triphenyltetrazolium	Dehydrogenases convert triphenyltetrazolium chloride to triphenylformazan; the triphenylformazan is extracted with methanol and quantitated spectrophotometrically.
Phosphatase	p-Nitrophenol phosphate	Phosphatases convert the p-nitrophenol phosphate to p-nitrophenol, which is extracted in aqueous solution and quantitated spectrophotometrically.
Protease	Gelatin	Gelatin hydrolysis, as an example of proteolytic activity, can be measured by the determination of residual protein.
Amylase	Starch	The amount of residual starch is quantitated spectrophotometrically by the intensity of the blue color resulting from its reaction with iodine.
Chitinase	Chitin	Production of reducing sugars is measured using anthrone reagent.
Cellulase	Cellulose Carboxymethylcellulose	Production of reducing sugars is measured using anthrone reagent. Cellulases alter the viscosity of carboxymethylcellulose, a quantity that can be measured.
Nitrogenase	Acetylene	Nitrogenase, besides reducing dinitrogen gas (N_2) to ammonia (NH_3), is also capable of reducing acetylene (C_2H_2) to ethylene (C_2H_4); the rate of formation of ethylene can be monitored using a gas chromatograph, and the rate of nitrogen fixation can be calculated using an appropriate conversion factor.
Nitrate reductase	Nitrate	Dissimilatory nitrate reductase can be assayed by the disappearance of nitrate or by measuring with a gas chromatograph the evolution of denitrification products, such as nitrogen gas and nitrous oxide, from samples; denitrification can be blocked at the nitrous oxide level by the addition of acetylene, permitting a simpler assay procedure.

From Atlas and Bartha (1993).

tivity in environmental samples. Some examples are given in Table 11.2. The first several assays in this table measure general microbial activity and the latter ones are assays for specific activity. One commonly used assay for general microbial activity is the dehydrogenase assay.

11.7.1 Dehydrogenase Assay

Dehydrogenases are intracellular enzymes that catalyze oxidation-reduction reactions required for the respiration of organic compounds. Because dehydrogenases are inactive when outside the cell, this assay is considered a measure of microbial activity. Dehydrogenase reactions can be detected using a water soluble, almost colorless, tetrazolium salt, which when reduced forms a reddish colored formazan product that can be detected in a variety of ways. Tetrazolium salts compete with other electron acceptors for the reducing power of the electron transport chain. Thus, measurements of the reduction of tetrazolium salts will reflect electron transport chain activity. Hence, this measurement is an index of the general level of activity of a large part of the microbial community, but it is not a direct measure of microbial growth in terms of the production of new biomass. It should also be recognized that all respiring organisms have an electron transport chain, including both aerobic and anaerobic microorganisms. Also included in the measurement are eukaryotic populations, such as algae and fungi. Therefore, the measurement of tetrazolium reduction provides an overall indication of electron transport chain activity.

A commonly utilized tetrazolium salt is 2-(p-iodophenyl)-3-(p-nitrophenyl)-5-phenyltetrazolium chloride (INT), which is transformed into an intensely colored, water-insoluble formazan (INT-formazan). INT has been used to measure microbial activity in surface waters (Posch *et al.*, 1997), soil and sediment samples (Trevors, *et al.*, 1982; Songster-Alpin and Klotz, 1995), subsurface sediments (Beloin *et al.*, 1988), and biofilms (Blenkinsopp and Lock, 1990). Environmental samples are suspended in a solution containing INT and incubated for a matter of hours. Incubations between 1 and 12 hours are usually sufficient and, as always, the shorter the incubation time the less chance that the microbial community will undergo significant changes in activity. The production of INT-formazan can be detected by microscopic examination of the red INT-formazan deposits that form within the cells or by quantifying total INT-formazan production. Microscopic examination gives an indication of the physiological status of the microbial population by determining the percentage of total cells that are actively

respiring (Zimmerman, *et al.*, 1978). Total cell number can then be determined using a counterstain, such as acridine orange, which stains all cells. The difference between the INT-formazan–containing cells and the total cell count is the proportion of the population that is metabolically active. This method is very sensitive and can be used to detect electron transport chain activity even when low numbers of microorganisms are present or when samples are incubated at low temperatures (Trevors, 1984). For example, Posch *et al.* (1997) studied planktonic bacteria in a high mountain lake that was covered by a layer of ice at the time of sampling. The lake water was exposed to INT and incubated at the *in situ* temperature, 2°C, in order to determine the physiological status of the planktonic community. They obtained the surprising result that as much as 25% of planktonic bacteria in this oligotrophic lake were actively respiring. However, when using this method, it should be recognized that some of the cells considered inactive may in fact be respiring very slowly or not utilizing INT as an electron acceptor (Posch *et al.*, 1997).

An alternative to microscopic examination of cells is quantification of the amount of INT-formazan produced. An environmental sample can be exposed to INT for a period of time, and then a solvent, such as methanol, is used to extract the INT-formazan from the cells. INT-formazan can be detected spectrophotometrically and the total production calculated from a standard curve. In terms of quantifying total INT-formazan production, this measurement provides a relative index by which electron transport chain activities in different samples can be compared. However, INT has a low efficiency as an electron acceptor, and formazan production can be affected by numerous factors including the concentration of INT used, incubation time, incubation temperature, pH of the sample, and whether the sample was incubated under aerobic or anaerobic conditions. Therefore, a direct comparison of activities of different samples can be made only if identical methods and experimental conditions are used (Trevors, 1984).

Several other forms of tetrazolium salts are available and provide certain advantages. For example, tetrazolium salt sodium 3'-{1-[(phenylamino)-carbonyl]-3,4-tetrazolium}-bis(4-methoxy-6-nitro)benzenesulfonic acid hydrate (XTT) has the advantage of producing an orange-colored, water-soluble XTT-formazan product (Roslev and King, 1993). Because XTT-formazan is water soluble, its production can be quantified spectrophotometrically without the need to use solvent extraction. Another tetrazolium salt is 5-cyano-2,3-ditolyl tetrazolium chloride (CTC). The reduced form of CTC is a water-insoluble, red-fluorescent CTC-formazan, which forms deposits within the cell (Rodriguez *et al.*, 1992). The fluorescence of CTC-formazan allows respiring cells to be highly visible when viewed under epifluorescence microscopy. This provides for easier enumeration of respiring cells in environmental samples. Using CTC, Winding *et al.* (1994) determined that actively respiring bacterial cells accounted for only 2 to 6% of the total population in an agricultural soil, and Schuale *et al.* (1993) determined that between 1 and 10% of bacteria in samples of drinking water were actively respiring. The fluorescent properties of CTC-formazan also provide a means of determining the location of physio-

logically active cells within attached biofilm communities without having to disrupt the biofilm structure (Yu and McFeters, 1994a). Schuale *et al.* (1993) used CTC to examine the viability of thin biofilms formed by microorganisms present in drinking water (Fig. 11.20). Their results revealed that between 5 and 35% of total sessile bacteria were actively respiring. Yu and McFeters (1994b) reported that CTC provided a sensitive measure of the efficacy of biocidal compounds in disinfecting biofilms formed by a potentially pathogenic waterborne bacterium. A disadvantage of the use of CTC is that it is redox sensitive and in a low redox environment can be reduced by an abiotic chemi-

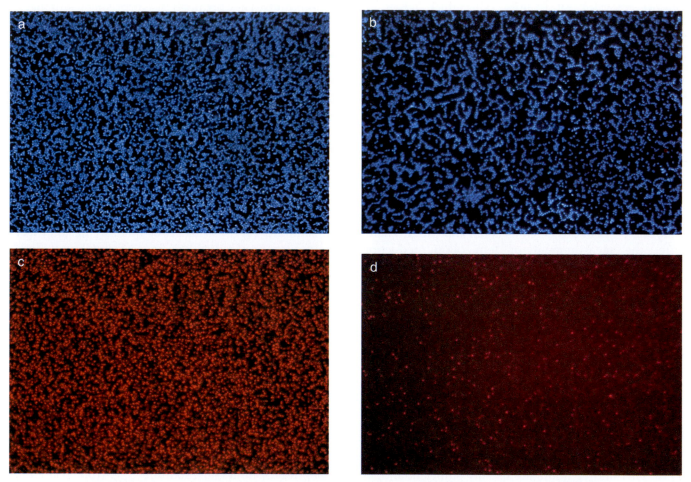

FIGURE 11.20 This figure shows how CTC-staining can be used to detect respiring cells in a biofilm. In this experiment, UV-sterilized, plastic, uncoated, microscope slides were placed in a sterile one-liter beaker containing 750 ml of 10% R2A medium. The beaker was inoculated with 2 ml of a *Pseudomonas putida* strain 54G cell suspension, and after 24 hours, the slides were taken out and placed into mineral salts medium (starved cells) or R2A medium (unstarved cells). After a further period of incubation each slide was stained with CTC and then counterstained with DAPI (see Chapter 9.6). The micrographs that are stained with CTC-DAPI (which stains all cells regardless of physiological state) shows that comparable biofilms were formed on both slides (A, unstarved cells) and (B, starved cells). Examination of the slides stained only with CTC show that nearly the entire biofilm of unstarved cells is respiring (C) while only a few of the starved cells are respiring (D). The results of this experiment were enumerated and showed that cells that were actively respiring comprised 76.8% of unstarved biofilm cells but only 9.6 % of starved biofilm cells. (Data and photos courtesy of G. Rodriguez and H. Ridgway, Orange County Water District, Fountain Valley, CA.)

cal reaction (Schuale *et al.*, 1993). Therefore, anaerobic pockets within a biofilm formation or within the soil environment may interfere with the detection of respiring cells, although these are issues for further research.

11.7.2 Esterase Assay

Another enzyme assay system used to reflect general or total microbial activity in environmental samples is based on the hydrolysis of fluorescein diacetate (FDA). FDA is hydrolyzed by a wide variety of enzymes including esterases, proteases, and lipases (Schnürer and Rosswall, 1982). The product of FDA hydrolysis is the highly fluorescent molecule fluorescein, which is water soluble and can be detected spectrophotometrically or fluorometrically. FDA hydrolysis has been used to examine microbial activity in soil (Schnürer and Rosswall, 1982), subsurface samples (Federle *et al.*, 1986), and activated sewage sludge (Fontvieille *et al.*, 1992). The general procedure is to incubate an environmental sample with FDA in a buffer solution for a matter of minutes to hours. The supernatant is then separated from the cells or environmental debris by centrifugation and the fluorescein production quantified. Fluorescein can also be extracted from the sample using acetone to improve detection. Interpretation of the results requires careful consideration because the enzymes involved in FDA hydrolysis can be intracellular or extracellular, which means that the assay determines enzymatic potential and not necessarily the activity of proliferating microorganisms. Several studies appear to validate the use of FDA by correlating fluorescein production with other measures of microbial activity or biomass. For example, Federle *et al.* (1986) determined changes in microbial biomass and FDA hydrolysis with depth in several soils. Microbial biomass declined at least an order of magnitude from the surface to approximately 100 cm depth in each soil. FDA hydrolysis correlated well with biomass measurements, indicating a decline in microbial activity as a function of depth in this study.

QUESTIONS AND PROBLEMS

1. You are given a pure culture of four different benzoate-degrading microbes and asked to evaluate which has the fastest degradation rate (μ_{max}). Design an experiment to answer this question. In your experimental design be sure to specify the culture medium to be used and how you will analyze benzoate degradation.

2. Next you are given 4 different soils and asked to evaluate the capacity of each soil to degrade the herbicide 2,4-dichlorophenoxy acetic acid (2,4-D). Design an experiment to answer this question. Be sure to specify how the experiment will be setup and how you will analyze 2,4-D degradation.

3. One of the soils that you tested in problem 2 shows rapid rates of 2,4-D degradation. How would you determine whether there is one or more than one different populations degrading the 2,4-D?

4. You are given an assignment to determine the impact of agricultural practices on microbial activity in soil. You collect a series of soil samples from plots that are undisturbed (no crops, no fertilizer), cropped with normal tillage, cropped with reduced tillage, and cropped with no tillage. In addition, a high and low fertilizer application rate was used. Thus, there is a combination of 7 different treatments. What microbial activity tests would you choose to run and why. How would you set up these tests?

5. You are assigned to restore a site in a National Park that has been severely disturbed by overuse. The site currently has no visible plant growth. You would like to add an organic amendment to stimulate microbial activity. You test several amendments for their effect on activity and for their longevity. These include a manure amendment, wood chips, and a manure-wood chip mixture. Describe how you would determine the effect of these different amendments on 1) microbial biomass, and 2) microbial activity including nitrification and heterotrophic activity.

References and Recommended Readings

Atlas, R. M., and Bartha, R. (1993) "Microbial Ecology." Benjamin/Cummings, Redwood City, CA.

Anderson, J. P. E. (1982) Soil respiration. *In* "Methods of Soil Analysis," Part 2, "Chemical and Microbiological Properties," 2nd ed. Agron. Monogr. 9. ASA and SSSA, Madison, WI, pp. 831–871.

Anderson, J. P. E., and Domsch, K. H. (1978) A physiological method for the quantitative measurement of microbial biomass in soils. *Soil Biol. Biochem.* **10**, 215–221.

APHA, AWWA, WEF American Public Health Association, American Water Works Association, and Water Environment Federation. (1995) "Standard Methods for the Examination of Water and Wastewater," 19th ed. American Public Health Association, Washington, DC.

Azam, F., and Fuhrman, J. A. (1984) Measurement of bacterioplankton growth in the sea and its regulation by environmental conditions. *In* "Heterotrophic Activity in the Sea" (J. E. Hobbie, and P. J. LeB. Williams, eds.) Plenum, New York, pp. 179–196.

Beare, M. H., Neely, C. L., Coleman, D. C., and Hargrove, W. L.

(1990) A substrate-induced respiration (SIR) method for measurement of fungal and bacterial biomass on plant residues. *Soil. Biol. Biochem.* **22,** 585–594.

Beloin, R. M., Sinclair, J. L., and Ghiorse, W. C. (1988) Distribution and activity of microorganisms in subsurface sediments of a pristine study site in Oklahoma. *Microb. Ecol.* **16,** 85–97.

Blenkinsopp, S. A., and Lock, M. A. (1990) The measurement of electron transport system activity in river biofilms. *Water Res.* **24,** 441–445.

Bowman, G. T., and Delfino, J. J. (1980) Sediment oxygen demand techniques: A review and comparison of laboratory and *in situ* systems. *Water Res.* **14,** 491–499.

Bradford, M. M. (1976) A rapid and sensitive method for the quantitation of microgram quantities of protein utilizing the principle of protein–dye binding. *Anal. Biochem.* **72,** 248–254.

Brookes, P. C., Tate, K. R., and Jenkinson, D. J. (1983) The adenylate energy charge of the soil microbial biomass. *Soil Biol. Biochem.* **15,** 9–16.

Brooks, P. D., and Paul, E. A. (1987) A new automated technique for measuring respiration in soil samples. *Plant Soil* **101,** 183–187.

Burns, R. G., ed. (1978) "Soil Enzymes." Academic Press, New York.

Chin-Leo, G., and Kirchman, D. (1988) Estimating bacterial production in marine waters from the simultaneous incorporation of thymidine and leucine. *Appl. Environ. Microbiol.* **54,** 1934–1939.

Ciardi, C., Ceccanti, B., and Nannipieri, P. (1991) Method to determine the adenylate energy charge in soil. *Soil. Biol. Biochem.* **23,** 1099–1101.

Daniels, L., Hanson, R. S., and Phillips, J. A. (1994) Chemical analysis. *In* "Methods for General and Molecular Bacteriology" (P. Gerhardt, R. G. E. Murray, W. A. Wood, and N. R. Krieg, eds.) American Society for Microbiology, Washington, DC, pp. 512–554.

Diáz-Raviña, M., Bååth, E., and Frostegård, Å. (1994) Multiple heavy metal tolerance of soil bacterial communities and its measurement by a thymidine incorporation technique. *Appl. Environ. Microbiol.* **60,** 2238–2247.

Dulohery, C. J., Morris, L. A., and Lowrance, R. (1996) Assessing forest soil disturbances through biogenic gas fluxes. *Soil Sci. Soc. Am. J.* **60,** 291–298.

Federle, T. W., Dobbins, D. C., Thornton-Manning, J. R., and Jones, D. D. (1986) Microbial biomass, activity, and community structure in subsurface soils. *Ground Water* **24,** 365–374.

Fontvieille, D.A., Outaguerouine, A., and Thevenot, D. R. (1992) Fluorescein diacetate hydrolysis as a measure of microbial activity in aquatic systems: Application to activated sludge. *Environmental Technology* **13,** 531–540.

Ghiorse, W. C. (1994) Iron and manganese oxidation and reduction. *In* "Methods of Soil Analysis," Part 2, "Microbiological and Biochemical Properties." Soil Science Society of America, Madison, WI, pp. 1079–1096.

Hinchee, R. E., and Ong, S. K. (1992) A rapid *in situ* respiration test for measuring aerobic biodegradation rates of hydrocarbons in soil. *J. Air Waste Manage. Assoc.* **42,** 1305–1312.

Inubushi, K., Brookes, P. C., and Jenkinson, D. S. (1989) Adenosine 5'-triphosphate and adenylate energy charge in waterlogged soil. *Soil Biol. Biochem.* **21,** 733–739.

Jenkinson, D. S. (1988) Determination of microbial biomass carbon and nitrogen in soil. *In* "Advances in Nitrogen Cycling in Agricultural Ecosystems" (J. R. Wilson, ed.) C.A.B. International, Wallingford, Oxon, UK pp. 368–386.

Johnson, C. K., Vigil, M. F., Doxtader, K. G., and Beard, W. E. (1996) Measuring bacterial and fungal substrate-induced respiration in dry soil. *Soil Biol. Biochem.* **28,** 427–432.

Kelly, D. P., and Wood, A. P. (1998) Microbes of the sulfur cycle. *In* "Techniques in Microbial Ecology" (R. S. Burlage, R. Atlas, D.

Stahl, G. Geesey, and G. Sayler, eds.) Oxford University Press, New York, pp. 31–57.

Kieft, T. L., Fredrickson, J. K., McKinley, J. P., Bjornstad, B. N., Rawson, S. A., Phelps, T. J., Brockman, F. J., and Pfiffner, S. M. (1995) Microbiological comparisons within and across contiguous lacustrine, paleosol, and fluvial subsurface sediments. *Appl. Environ. Microbiol.* **61,** 749–757.

Kieft, T. L, and Rosacker, L. L. (1991) Application of respiration- and adenylate-based soil microbiological assays to deep subsurface terrestrial sediments. *Soil Biol. Biochem.* **23,** 563–568.

Kirchman, D., K'ness, E., and Hodson, R. (1985) Leucine incorporation and its potential as a measure of protein synthesis by bacteria in natural aquatic systems. *Appl. Environ. Microbiol.* **49,** 599–607.

Ladd, T. I., Costerton, J. W., and Geesey, G. G. (1979) Determination of the heterotrophic activity of epilithic microbial populations. *In* "Native Aquatic Bacteria: Enumeration, Activity, and Ecology" (J. W. Costerton and R. R. Colwell, eds.) ASTM STP 695. American Society for Testing and Materials, Pittsburgh, pp. 180–195.

Lowry, O. H., Rosebrough, N. J., Farr, A. L., and Randall, R. J. (1951) Protein measurement with the Folin phenol reagent. *J. Biol. Chem.* **193,** 265–275.

Martens, R. (1987) Estimation of microbial biomass in soil by the respiration method: Importance of soil pH and flushing methods for the measurement of respired carbon dioxide. *Soil Biol. Biochem.* **19,** 77–81.

Moriarty, D. J. W. (1986) Measurement of bacterial growth rates in aquatic systems from rates of nucleic acid synthesis. *Adv. Microb. Ecol.* **9,** 245–292.

Morra, M. J. (1997) Assessment of extracellular enzymatic activity in soil. *In* "Manual of Environmental Microbiology, " (C. J. Hurst, G. R. Knudsen, M. J. McInerney, L. D. Stetzenbach, and M. V. Walter, eds.) American Society for Microbiology Press, Washington, DC., pp. 459–465.

Nannipieri, P., Grego, S., and Ceccanti, B. (1990) Ecological significance of the biological activity in soil. *In* "Soil Biochemistry," Vol. 6 (J.-M. Bollag, and G. Stotsky, eds.) Marcel Dekker, New York, pp. 293–355.

Nielsen, L. P., Christensen, P. B., Revsbech, N. P., and Sørensen, J. (1990) Denitrification and oxygen respiration in biofilms: Studies with a microsensor for nitrous oxide and oxygen. *Microb. Ecol.* **19,** 63–72.

Paerl, H. W. (1998) Microbially mediated nitrogen cycling. *In* "Techniques in Microbial Ecology" (R. S. Burlage, R. Atlas, D. Stahl, G. Geesey, and G. Sayler, eds.) Oxford University Press, New York, pp. 3–30.

Pomeroy, L. R., Sheldon, J. E., and Sheldon, W. M. Jr. (1994) Changes in bacterial numbers and leucine assimilation during estimations of microbial respiratory rates in seawater by the precision Winkler method. *Appl. Environ. Microbiol.* **60,** 328–332.

Posch, T., Pernthaler, J., Alfreider, A., and Psenner, R. (1997) Cell-specific respiratory activity of aquatic bacteria studied with the tetrazolium reduction method, cyto-clear slides, and image analysis. *Appl. Environ. Microbiol.* **63,** 867–873.

Revsbech, N. P. (1994) Analysis of microbial mats by use of electrochemical microsensors: Recent advances. *In* "Microbial Mats: Structure, Development, and Environmental Significance." (L. J. Stal and P. Coumette, eds.) Springer-Verlag, New York, pp. 135–147.

Revsbech, N. P., and Jørgensen, B. B. (1986) Microelectrodes: Their use in microbial ecology. *Adv. Microb. Ecol.* **9,** 293–352.

Robarts, R. D., and Zohary, T. (1993) Fact or fiction—bacterial growth rates and production as determined by [*methyl*-3H]-thymidine? *Adv. Microb. Ecol.* **13,** 371–425.

Rodriguez, G. G., Phipps, D., Ishiguro, K., and Ridgway, H. F. (1992) Use of a fluorescent redox probe for direct visualization of actively respiring bacteria. *Appl. Environ. Microbiol.* **58,** 1801–1808.

Rosacker, L. L., and Kieft, T. L. (1990) Biomass and adenylate energy charge of a grassland soil during drying. *Soil Biol. Biochem.* **22,** 1121–1127.

Roslev, P., and King, G. M. (1993) Application of a tetrazolium salt with a water-soluble formazan as an indicator of viability in respiring bacteria. *Appl. Environ. Microbiol.* **59,** 2891–2896.

Schnürer, J., and Rosswall, T. (1982) Fluorescein diacetate hydrolysis as a measure of total microbial activity in soil and litter. *Appl. Environ. Microbiol.* **43,** 1256–1261.

Schramm, A., Larsen, L. H., Revsbech, N. P., Ramsing, N. B., Amann, R., and Schleifer, K.-H. (1996) Structure and function of a nitrifying biofilm as determined by *in situ* hybridization and the use of microelectrodes. *Appl. Environ. Microbiol.* **62,** 4641–4647.

Schuale, G., Flemming, H.-C., and Ridgway, H. F. (1993) Use of 5-cyano-2,3-ditolyl tetrazolium chloride for quantifying planktonic and sessile respiring bacteria in drinking water. *Appl. Environ. Microbiol.* **59,** 3850–3857.

Smith, P. K., Krohn, R. I., Hermanson, G. T., Mallia, A. K., Gartner, R. H., Provenzano, M. D., Fujimoto, E. K., Goeke, N. M., Olson, B. J., and Klenk, D. C. (1985) Measurement of protein using bichinchoninic acid. *Anal. Biochem.* **150,** 76–85.

Songster-Alpin, M. S., and Klotz, R. L. (1995) A comparison of electron transport system activity in stream and beaver pond sediments. *Can. J. Fish. Aquat. Sci.* **52,** 1318–1326.

Sparling, G. P., West, A. W., and Whale, K. N. (1985) Interference from plant roots in the estimation of soil microbial ATP, C, N, and P. *Soil Biol. Biochem.* **17,** 275–278.

Tabatabai, M. A. (1994a) Sulfur oxidation and reduction in soils. *In* "Methods of Soils Analysis," Part 2, "Microbiological and Biochemical Properties." Soil Science Society of America, Madison, WI, pp. 1067-1078.

Tabatabai, M. A. (1994b) Soil enzymes. *In* "Methods of Soil Analysis," Part 2, "Microbiological and Biochemical Properties." Soil Science Society of America, Book Series 5. SSSA, Madison, WI, pp. 775–834.

Thorn, P. M., and Ventullo, R. M. (1988) Measurement of bacterial growth rates in subsurface sediments using the incorporation of tritiated thymidine. *Microb. Ecol.* **16,** 3–16.

Tibbles, B. J., and Harris, J. M. (1996) Use of radiolabelled thymidine and leucine to estimate bacterial production in soils from continental Antarctica. *Appl. Environ. Microbiol.* **62,** 694–701.

Trevors, J. T. (1984) Electron transport system activity in soil, sediment, and pure cultures. *Crit. Rev. Microbiol.* **11,** 83–100.

Trevors, J. T., Mayfield, C. I., and Inniss, W. E. (1982) Measurement of electron transport system (ETS) activity in soil. *Microb. Ecol.* **8,** 163–168.

Van Es, F. B., and Meyer-Reil, L.-A. (1982) Biomass and metabolic activity of heterotrophic marine bacteria. *Adv. Microb. Ecol.* **6,** 111–170.

Webster, J. J., Hall, S. M., and Leach, F. R. (1992) ATP and adenylate energy charge determinations on core samples from an aviation fuel spill site at the Travers City, Michigan airport. *Bull. Environ. Contam. Toxicol.* **49,** 232–237.

Wetzel, R. G., and Likens, G. E. (1991) "Limnoligical Analyses," 2nd ed. Springer-Verlag, NewYork.

Winding, A., Binnerup, S. J., and Sorrensen, J. (1994) Viability of indigenous soil bacteria assayed by respiratory activity and growth. *Appl. Environ. Microbiol.* **60,** 2869–2875.

Wright, R. T., and Burnison, B. K. (1979) Heterotrophic activity measured with radiolabelled organic substrates. *In* "Native Aquatic Bacteria: Enumeration, Activity, and Ecology," (J. W. Costerton and R. R. Colwell, eds.) ASTM STP 695. American Society for Testing and Materials, Pittsburgh, pp. 140–155.

Yu, F. P., and McFeters, G. A. (1994a) Rapid *in situ* assessment of physiological activities in bacterial biofilms using fluorescent probes. *J. Microbiol. Methods* **20,** 1–10.

Yu, F. P., and McFeters, G. A. (1994b) Physiological responses of bacteria in biofilms to disinfection. *Appl. Environ. Microbiol.* **60,** 2462–2466.

Zimmermann, R., Iturriaga, R., and Becker-Birck, J. (1978) Simultaneous determination of the total number of aquatic bacteria and the number thereof involved in respiration. *Appl. Environ. Microbiol.* **36,** 926–935.

Zinder, S. H. (1998) Methanogens. *In* "Techniques in Microbial Ecology" (R. S. Burlage, R. Atlas, D. Stahl, G. Geesey, and G. Sayler, eds.) Oxford University Press, New York, pp. 113–136.

12

Immunological Methods

SCOT E. DOWD   RAINA M. MAIER

12.1 INTRODUCTION

Immunology is the study of the immune system of higher organisms in relation to disease. Specifically, immunology can be defined as the branch of biology that is concerned with the structure and function of the immune system, the bodily distinction of self from nonself, and the use of antibody based laboratory techniques or immunoassays. In general, the **immune system** of higher organisms can be broken down into two primary response systems that work together to create immunity. The two primary response systems are the cell-mediated and the antibody-mediated responses. The **cell-mediated response** is produced when sensitized white blood cells or **lymphocytes** directly attack material, which has been determined to be foreign to the body. The **antibody-mediated response** involves the transformation of a subset of lymphocytes into cells that produce and secrete specific antibodies against these foreign objects. These two immune responses are triggered when foreign material is introduced into the host as depicted in Fig. 12.1.

Environmental Microbiology does not deal with all aspects of immunology or the immune responses per se but instead adapts immunology-based research technologies or **immunoassays** for the study of microorganisms in association with the environment. The primary immunologic-based tool used in environmental microbiology is the antibody. In this chapter an introduction to antibodies is given, with respect to structure of antibodies, the various classes of antibodies [immunoglobulin G (IgG), IgA, etc.], and the interaction of antibodies with foreign objects **(antigens).** Following this introduction, several of the basic immunological methodologies (immunoassays), that are widely used in environmental microbiology, are discussed. These immunoassays include fluorescent immunolabeling, enzyme-linked immunosorbent assay (ELISA), magnetic bead antigen capture, Western immunoblotting, immunoaffinity chromatography, and immunoprecipitation. Finally, in order to provide perspective and illustrate how these immunoassays can be used in the field of environmental microbiology, an example of each immunoassay is provided in relation to current research topics such as bioremediation and pathogen detection.

12.2 WHAT IS AN ANTIBODY?

Antibodies are protein complexes produced by the immune system of higher life forms that help defend the host against foreign invasion. When a host is challenged by foreign material (bacteria, virus, toxins, etc.) the first response of certain host immune cells called **macrophages** is to engulf these invaders **(antigens)** and process them biochemically. This biochemical processing essentially creates a blueprint that is used for the development of an immune response that results in the production of antibodies (Fig. 12.2). The unique feature of antibodies produced in response to an antigen is that they are synthesized in such a way that they are highly specific for that antigen. Thus, they can chemically react and bind only with that particular antigen, neutralize it, and aid in its destruction and removal from the body.

There are five different classes of antibodies or **immunoglobulins** (Igs): IgA, IgD, IgE, IgG, and IgM. These

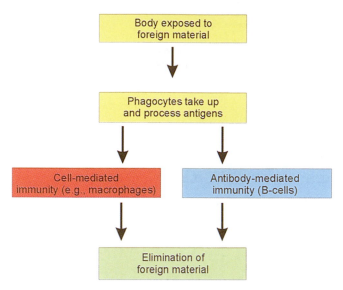

FIGURE 12.1 Flow chart showing the two primary immune response systems that comprises the host response to foreign materials. Together these two branches of the immune system work together to create immunity. The cell-mediated response is designed to directly attack and destroy material determined by the body to be non-self. The antibody-mediated response is the branch of the immune system involved in the formation of antibodies.

immunoglobulins (antibodies) differ in many ways including their overall structures (Fig. 12.3). The most common type of antibody used for immunoassays is the IgG class of immunoglobulins (Fig. 12.4). IgG antibodies are Y-shaped proteins composed of four protein chains that are joined together by disulfide linkages. There are two major structural fragments or regions of Igs called the Fc and Fab regions. The antigen binding fragments **(Fabs)** of the IgG immunoglobulins are the two regions at the top of the molecule, which, as indicated, are the sites of antibody-antigen interaction. The **Fc** region is the tail of the antibody and is the fragment, that is recognized by the host as "self." Because antibodies are relatively large proteins they can also act as antigens, so the ability to recognize a particular antibody as being self prevents the host from responding against its own antibodies. For example, all rabbits and humans have the same classes of antibodies (IgG, IgM, etc.) but do not have the same subclasses (IgG1, IgG2, etc.) or the same host recognition sites on the immunoglobulins. This species-specific difference makes the immune system of one species recognize another species' antibodies as being foreign. For example, if you immunize a rabbit (purposefully expose it to a foreign antigen) with a human's antibody, the rabbit will produce antibodies against them. These antibodies that are directed against another animal species' or another individuals' antibodies are termed **antiglobulins.**

12.2.1 Antibody Diversity

A very important characteristic of the antibody response in animals is that B cells must be able to produce a wide variety of antibodies that can interact with a diverse range of antigens. In fact, the vast populations of B cells have been estimated to have the potential to produce up to 1×10^{10} structurally different IgG antibodies, which in theory could recognize 1×10^{10} different antigens. This enormous diversity has been exploited in the field of environmental microbiology.

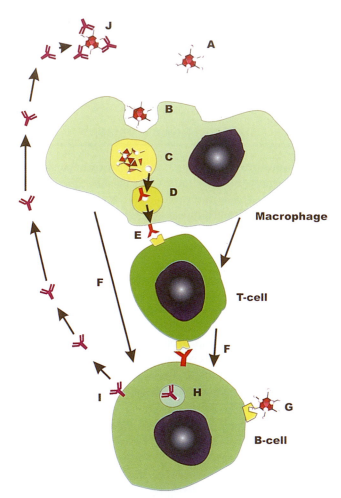

FIGURE 12.2 Schematic representation is showing the processes that lead to the formation of antibodies. (A) Foreign material represented here by a virus that has gained access to the body; (B) The virus is phagocytosed by a macrophage; (C) The virus is broken down into subunits by enzymes contained in the phagocytic vacuole; (D) An antigen presenting molecule escorts antigenic subunits from the virus to the macrophage surface; (E) The antigenic molecule derived from the virus is presented to a T-cell; (F) The macrophage and the T-cell release chemicals that stimulate B-cells; (G) The B-cell then encounters the virus and produces antibodies specific to the viral epitope; (I) Antibodies are released into circulation; (J) Antibodies neutralize the virus in circulation by binding to the viral epitopes inhibiting viral attachment to target cells.

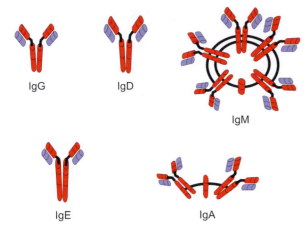

FIGURE 12.3 Schematic representation of the 5 classes of immunoglobulins. IgG (molecular weight 146,000), IgD (184,000) and IgE (188,000) are all comprised of one basic subunit. IgG is the major immunoglobulin involved in the humeral response. IgM, which consists of 5 subunits (970,000), is the second major antibody involved in the humeral response. IgA (385,000), is commonly found in body secretions such as saliva, milk and intestinal fluid. IgD is found in low concentrations in plasma but is not found in serum, and IgE is involved in allergic reactions. IgG and IgM are the antibody classes most commonly utilized in immunoassays.

An animal's immune system can adapt and produce an immune response against many different antigens. Therefore, essentially any bacterium, virus, protein, or pollutant that can stimulate an immune response can be used as an antigen to create a specific antibody. This antibody can then be used to design immunoassays to aid in the study of that bacterium, virus, or pollutant.

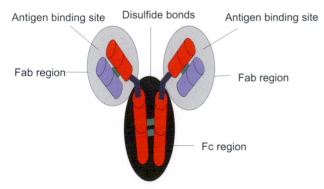

FIGURE 12.4 Schematic representation of an IgG antibody, showing the various regions associated with the antigen/antibody interaction. There are two antigen-binding fragments (Fab), which interact with the antigen. There is also one crystal fragment (Fc), which is the part of the antibody recognized by the host immune system as self. There are two light chains (in purple) joined to two heavy chains (red/orange) by disulfide bonds and the two heavy chains are in turn joined to each other in a similar fashion.

12.2.2 Antibody Specificity

Specificity for a particular antigen is one characteristic that makes immunology-based methodologies such valuable tools. In essence, once they are produced, antibodies are very precise in recognition of the particular antigen. This discrimination is based on the molecular structure of the antigen binding sites located on the Fab portion of the antibody and on the epitopes or chemically reactive sites of the antigen (Fig. 12.5). Antigen–antibody binding is the result of specific chemical interactions (i.e., charge–charge, dipole-dipole, hydrogen bonding, and van der Waals) that occur between the antigen and amino acid residues of the antibody that are located on the Fab region. This reaction is so specific that even a small change, such as an alteration of one amino acid in the binding site of the antibody, may weaken or nullify the antigen–antibody binding. It should be noted, however, that even with this specificity there are still instances in which the antibody may react with more than one antigen. For example, antibodies directed against protozoan parasites, such as *Giardia lamblia*, may recognize epitopes on the surface of some species of algae (Dowd and Pillai, 1997). The protozoa and the algae are obviously not closely related and yet they both have epitopes on their surfaces that interact with the antibody designed to recognize the protozoan. This phenomenon in which an antibody reacts with two unrelated epitopes is termed **cross-reactivity.**

12.2.3 Antibody Affinity

Affinity is defined as the attraction between an antibody and an antigen. More specifically, affinity is a

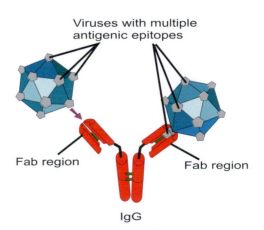

FIGURE 12.5 Schematic representation of the antigen antibody interaction. This figure shows an IgG antibody binding to an enterovirus. Not that the Fab regions of the antibody are chemically formed to fit perfectly with the antigenic epitopes of the virus.

measure of the strength of this interaction and is usually expressed as an interaction or association constant. Quantitatively, affinity is the sum of the chemical bonds that form between the antigen and the antibody. These are usually relatively weak interactions such as hydrophobic interactions and hydrogen bonds. Even though individually such chemical bonds are relatively weak, collectively they form very strong and tight interactions. Thus, the strongest binding occurs between epitopes and antibodies only when their shapes are complimentary (Fig. 12.6). Affinity can be described as the reversible formation of the antigen (Ag)–antibody (Ab) complex by the equation

$$Ag + Ab \Leftrightarrow AgAb \qquad (Eq.\ 12.1)$$

The affinity constant (K) can thus be determined by the mass balance equation, which is expressed in the form:

$$K=[AgAb]/[Ag][Ab] \qquad (Eq.\ 12.2)$$

where [AgAb] is the concentration of the antigen–antibody complex at equilibrium, [Ag] is the concentration of free antigen binding sites at equilibrium, [Ab] is the concentration of free antibody binding sites at equilibrium, and K is the affinity constant or the measure of strength of the bond formation between Ag and Ab. Essentially, the higher the K value, the stronger the affinity of an antibody for an antigen. This is a direct consequence of a stronger molecular interaction. This is important in the development of monoclonal antibodies for immunoassays, especially in the selection of the best hybridoma to use for monoclonal antibody production. The affinity constant not only gives an indication of which of the antibodies might be best for a certain assay, but it can also indicate the concentration of antibody required for the assay and the potential for cross-reactivity with other antigens.

12.2.4 Polyclonal and Monoclonal Antibodies

In the past, injecting a laboratory animal with the antigen of interest produced particular antibodies needed for environmental and other research purposes. In response to immunization, the animal produces antibodies that can be collected (in serum) directly from the blood of the animal (Fig. 12.7). These blood-derived antibodies are termed **polyclonal antibodies** because they are not derived from a single B lymphocyte but are instead a product of many different B cells reacting in slightly different ways to the same antigen. As a result, this method yields a very impure product that contains a wide variety of antibodies, proteins, and blood factors.

To avoid problems associated with use of such antibody mixtures, scientists developed the technology to produce monoclonal antibodies. A **monoclonal antibody** is an antibody that is the product of a single B cell. Production of monoclonal antibodies involves the *in vitro* combination of two types of cells. The first type of cell is a B cell that produces a single, unique antibody. The second type of cell is an immortalized myeloma cell: a cancer cell that is able to thrive and multiply *in vitro*. The specific antibody-producing B cell is fused with the myeloma cell to form a hybrid cell called a hybridoma. This hybridoma combines the characteristic of "immortality" with the ability to produce the desired specific antibody in high concentrations and in pure form. The procedure for the production of monoclonal antibodies is schematically diagramed in Figure 12.8. As a result of the development of monoclonal antibody technology, scientists are able to produce large amounts of pure and highly specific antibodies.

What are the advantages and disadvantages of using polyclonal or monoclonal antibodies? In general, monoclonal antibodies have higher specificity and lower cross-reactivity than polyclonal antibodies. They can also be produced indefinitely and in relatively large concentrations (>13 mg/ml). Polyclonal antibodies, on the other hand, can be produced more rapidly and much less expensively because they are prepared directly from the serum of immunized animals. Thus polyclonal and monoclonal antibodies each have benefits and drawbacks that must be considered during the design of any immunoassay.

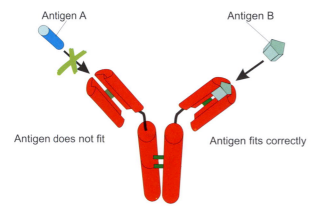

FIGURE 12.6 Schematic representation of an IgG antibody interacting with two different antigens. In the first case, antigen A does not have the correct biochemical conformation so the antibody is unable to interact and bind it. Antigen B has the correct conformation and the antibody is able to interact and bind this antigen tightly.

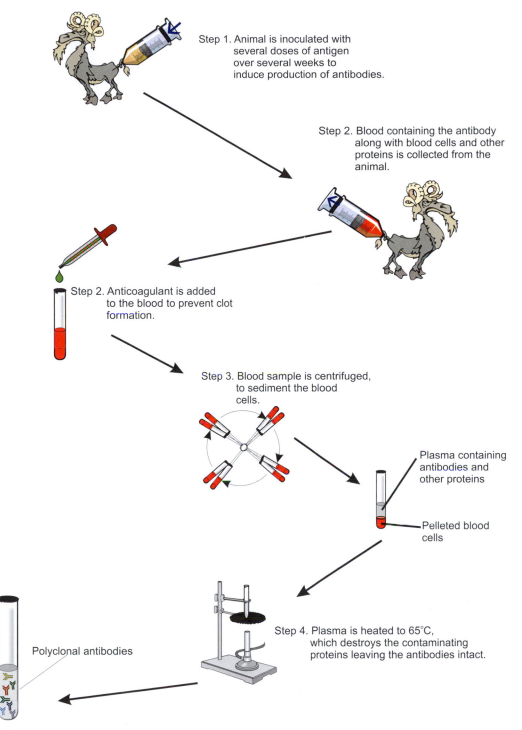

Step 1. Animal is inoculated with several doses of antigen over several weeks to induce production of antibodies.

Step 2. Blood containing the antibody along with blood cells and other proteins is collected from the animal.

Step 2. Anticoagulant is added to the blood to prevent clot formation.

Step 3. Blood sample is centrifuged, to sediment the blood cells.

Plasma containing antibodies and other proteins

Pelleted blood cells

Step 4. Plasma is heated to 65°C, which destroys the contaminating proteins leaving the antibodies intact.

Polyclonal antibodies

FIGURE 12.7   General outline of steps used to produce polyclonal antibodies. The primary step is the immunization of the animal and the collection of the blood containing the antibodies. Following collection the blood cells are separated by centrifugation and heating denatures other proteins such as complement. This results in a stable suspension of antibodies that can be used in many types of immunoassays.

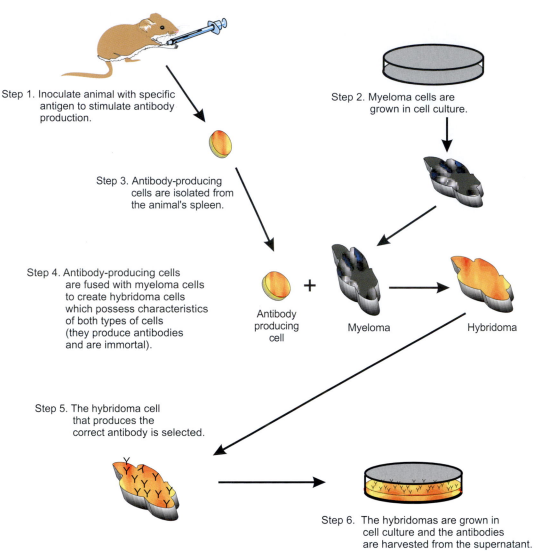

Step 1. Inoculate animal with specific
        antigen to stimulate antibody
        production.

Step 2. Myeloma cells are
        grown in cell culture.

Step 3. Antibody-producing
        cells are isolated from
        the animal's spleen.

Step 4. Antibody-producing cells
        are fused with myeloma cells
        to create hybridoma cells
        which possess characteristics
        of both types of cells
        (they produce antibodies
        and are immortal).

Antibody
producing
cell

Myeloma

Hybridoma

Step 5. The hybridoma cell
        that produces the
        correct antibody is selected.

Step 6. The hybridomas are grown in
         cell culture and the antibodies
         are harvested from the supernatant.

FIGURE 12.8   General outline of the steps used to produce monoclonal antibodies. Essentially the spleen cells from the immunized mouse are removed and combined with myeloma cells using polyethylene glycol to fuse the two. After this the cells that produce the specific antibody needed are selected and cultured in order to produce large quantities of highly purified monoclonal antibody.

12.2.5 Antiglobulins

Antiglobulins, as indicated earlier, are antibodies that are specific (usually targeting the Fc portion) for another individual antibody. Usually, antiglobulins are developed to recognize a whole antibody class for a specific organism e.g., a mouse. Because antibodies are large proteins with complex structures, they have the potential to be seen as antigens if they do not have the "self" recognition sites common to the host that produced the antibody. For example, an antiglobulin can be raised in a goat by immunizing the goat with a mouse IgG antibody. Because the goat does not recog-nize the mouse antibody as self, it produces its own set of antibodies against the mouse's IgG antibody. The antibodies produced by the goat are specific for the mouse-derived IgG antibody. Using monoclonal methods for production of specific antibodies, goat antibodies can be derived that are specific against all mouse IgGs or against specific IgGs, for example IgG2a. Thus, as depicted in Figure 12.9, if you can develop mouse monoclonal IgG2 antibodies that are specific for the protozoan parasite microsporidia (mouse–antimicrosporidia IgG2), these mouse antibodies can then be used to immunize a goat to produce goat–antimouse antiglobulins. These antiglobulins

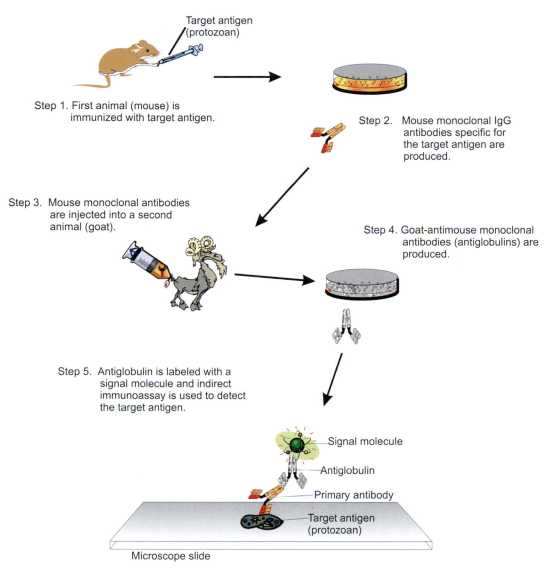

Step 1. First animal (mouse) is immunized with target antigen.

Target antigen (protozoan)

Step 2. Mouse monoclonal IgG antibodies specific for the target antigen are produced.

Step 3. Mouse monoclonal antibodies are injected into a second animal (goat).

Step 4. Goat-antimouse monoclonal antibodies (antiglobulins) are produced.

Step 5. Antiglobulin is labeled with a signal molecule and indirect immunoassay is used to detect the target antigen.

Signal molecule
Antiglobulin
Primary antibody
Target antigen (protozoan)

Microscope slide

FIGURE 12.9 General outline of the steps involved in the production of monoclonal antiglobulins. The mouse is inoculated with protozoan antigens and allowed to produce monoclonal antibodies and their use as described in Figure 11.8. The goat is then inoculated with the mouse anti-protozoan antibodies. The goat produces antibodies against the mouse antibody. These goat anti-mouse monoclonal antibodies are then cloned and used as secondary labeling antibodies.

can then be attached to a signal molecule and used as secondary or indirect detection molecules in immunoassays.

12.3 IMMUNOASSAYS

Immunoassays are analytical methods used for the detection and/or quantitation of the antigen–antibody interaction. For the most part, the types of immunoassays used in environmental microbiology are based on quantitation or detection of antigens as opposed to

characterization of the antigens. That is, we are usually interested in using immunoassays to determine how much antigen is in an environmental sample and not in characterizing an antibody/antigen interaction or the role of the antigen in disease or in the immune response. However, in order to quantitate or detect the antigen there must be a way to visualize the antigen-antibody interaction. This visual signal is produced by the attachment of specific signal molecules to the antibodies or antiglobulins used to detect the antigen within an environmental sample.

For almost all types of immunoassays, attachment

of a signal molecule to the antibody and/or antigen is very important. Many types of signal molecules are used in immunoassays, including iodine, enzymes, fluorochromes, and radioisotopes. These signal molecules produce a visual signal that allows quantitation of the specific antibody–antigen interaction being investigated. The signal is usually indicated by the production of some type of color change. For example, enzymes such as horseradish peroxidase and alkaline phosphatase act by enzymatically cleaving colorless substrates to produce a colored product. This signal is then detected qualitatively by the naked eye or quantitatively using an instrument such as a spectrophotometer. For fluorochrome signal molecules, the antigen–antibody interaction can be detected by exciting the fluorochrome with a particular wavelength of light. The fluorochrome will emit energy (light) at a second wavelength, which can then be detected visually or instrumentally. The fluorescent dye most commonly used as an antibody label is fluorescein isothiocyanate (FITC). Other examples of fluorescing chemicals used in this type of assay includes R-phycoerythrin, rhodamine, and Texas red. Radioisotopes are quantitated by liquid scintillation counting (see Fig. 11.12) or by exposing the sample to a photographic emulsion (x-ray film), which produces a qualitative signal on the film that can be visualized as a dark spot.

As just mentioned, attachment of an antibody or antigen to a signal molecule is an almost universal way to allow for visual detection via immunoassays. There are also two universal formats for immunoassays. These are termed direct labeling and sandwich (often termed indirect) labeling (Fig. 12.10). Essentially with **direct labeling,** the primary antibody (antibody specific for the target) has the signal molecule attached to it allowing for one-step detection. **Sandwich** or **indirect labeling** involves two steps. The first is the attachment of a primary antibody to the target, and the second is the attachment of a secondary (antiglobulin) antibody to the primary antibody. In indirect labeling, the secondary antibody has the signal molecule attached. Both methods work well though both have advantages and disadvantages. With direct labeling, the binding and signal are usually more specific because there is a smaller signal to background noise ratio. However the use of sandwich labeling allows for one labeled antibody to be used with many different primary antibodies provided they are all of the same type, meaning that each primary antibody does not have to be labeled separately. For instance, if you use a mouse to produce monoclonal IgG primary antibodies against 4 different protozoa (*Microsporidia, Giardia, Cryptosporidium,* and *Entamoeba*), then you can use the same la-

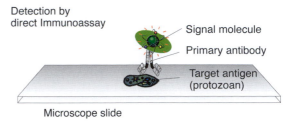

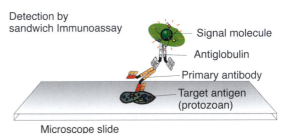

FIGURE 12.10 Direct immunoassay versus sandwich immunoassay.

beled antiglobulin to bind to each of these. Because conjugating signal molecules to antibodies is tedious and often difficult, and because a wide range of antiglobulins conjugated to various signal molecules are available commercially, this may be the format of choice. However if you desire a one-step assay that is slightly more specific, then the use of direct labeling is often preferred.

With this brief introduction to the technique of immunoassay, the following sections focus on descriptions of the main types of immunoassays routinely used in environmental microbiology.

12.3.1 Fluorescent Immunolabeling

12.3.1.1 Technique

Fluorescent immunolabeling (immunofluorescence) is the use of fluorescent signal molecules conjugated to antibodies to interact with and subsequently indicate the presence of a particular antigen by the production of fluorescent light. The basic procedure for immunofluorescence microscopy, for example, is to attach the sample antigen to a microscope slide, add a fluorescent chemical/antibody conjugate specific to the antigen, and view the sample under a microscope equipped with a fluorescent light source. When viewing fluorescence-labeled samples under the microscope, the labeled antigen appears bright green against a dark background. Immunofluorescence microscopy is one of the easiest and most widely used immunoassays. Immunofluorescence microscopy was first used in environmental microbiology for the study of Rhizobia in soil.

12.3.1.2 Application

One of the main uses for immunofluorescence is in the detection of protozoan parasites in water. The protozoan parasites *Giardia* sp. and *Cryptosporidium* sp. are two of the main causes of diarrhea in humans and many animals and are transmitted by the fecal–oral route, often through the contamination of surface waters (see Chapter 19). The hypothesis of a study by Dowd *et al.* (1997) was that freshwater microarthropods, such as *Daphnia* sp., bivalve freshwater mollusks, and mosquito larvae, act as natural predators for these protozoa in surface waters. The study was carried out in the laboratory by creating microcosms (feeding troughs) for the microarthropods out of petri dishes. Microarthropods were placed in the petri dishes containing *Giardia* cysts and *Cryptosporidium* oocysts. After feeding, the microarthropods were carefully removed from these microcosms and the remaining cysts and oocysts were enumerated by direct immunofluorescence. This was done using a cocktail of two different fluorescein-labeled monoclonal antibodies, one specific for the cyst and one specific for the oocyst walls of *Giardia* sp. and *Cryptosporidium* sp., respectively (Fig. 12.11). The microarthropods, after removal from the feeding troughs, were subsequently kept overnight to determine whether they passed live cysts or oocysts out in their feces. The same fluorescein-labeled monoclonal antibodies were used once again to screen the feces to see whether intact and/or viable cysts or oocysts were present. It was shown that the microarthropods were very efficient at removing the cysts and oocysts from the feeding troughs and that they did not pass live or intact cysts or oocysts in their feces. Finally, fluorescein-labeled antibodies were also used to stain the internal organs of the microarthropods to determine whether they had cysts or oocysts within their bodies (Fig. 12.12). The internal organs including the digestive tracts of the microarthropods fluoresced very strongly in comparison with control microarthropods that had not fed on cysts and oocysts. But, it was noted that no intact cysts or oocysts could be seen within the microarthropods. It was concluded that microarthropods may act as predators of these protozoa and that they digest the protozoa after ingestion rather than shedding viable entities. This study shows the versatility of fluorescence immunoassays, which can be utilized in a wide variety of samples (water, feces, internal structures of the arthropods) to detect microscopically specific antigens, such as those represented by the protozoan parasites in this example.

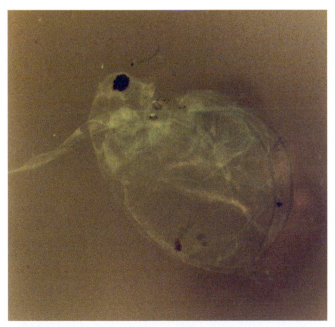

FIGURE 12.12 This is a photomicrograph of the microarthropod (*Daphnia* spp.) also known as the water flea. The water flea was stained with antibodies specific for cyst and oocysts of *Giardia* sp. and *Cryptosporidium* sp., respectively. This was done to ascertain if there were any of these protozoa present within the internal organs of the water flea. The internal organs of the *Daphnia* sp. fluoresced green while control *Daphnia* stained in the same manner did not fluoresce. This experiment indicates that such microarthropods may be natural predators or biological control agents for waterborne pathogenic protozoa. (Photo courtesy S. E. Dowd)

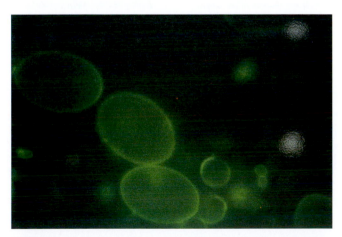

FIGURE 12.11 This is a photomicrograph of *Cryptosporidium parvum* and *Giardia lamblia* stained with fluorescene labeled monoclonal antibodies. There are 4 *Cryptosporidium* oocysts located above and to the left of the *Giardia* cyst (magnification × 1000). (Photo courtesy S. E. Dowd)

12.3.1.3 Advantages and Disadvantages

One of the primary advantages of fluorescent antibody-based techniques over other detection methodologies is the ease of use. Individuals with a minimal amount of training can perform immunofluorescence assays. Another obvious advantage is that results can be obtained usually within a few hours. In many cases immunofluorescence is also highly specific and sensitive. On the other hand there are several problems which must be considered when performing immunofluorescence assays. One serious disadvantage is the potential for crossreaction between the antibody and non-target antigens. Such crossreaction can provide false positive results using immunofluorescence assays. False negative results are also possible. This could occur if antigenic epitopes on a target cell were damaged or genetically altered thus preventing antibody recognition, binding, and subsequent detection.

12.3.2 Enzyme-Linked Immunosorbent Assays

12.3.2.1 Technique

The enzyme-linked immunosorbent assay (ELISA) is a very sensitive laboratory method used to detect the presence of antigens. There are many different approaches to ELISAs. They are typically performed as a direct or an indirect sandwich method (Fig. 12.10) but can also be performed as a competitive assay (Section 12.3.3). For any ELISA procedure the antigens of interest are concentrated (if necessary) and solubilized in an appropriate buffer. At this point a direct or indirect assay can be performed, depending on the type of antigen and the antibodies. An outline of indirect and direct ELISAs is given in Figure 12.13. For a direct sandwich ELISA, a primary antibody is attached to a microtiter plate, microcentrifuge tube, or other solid support. The antigen is then added and allowed to incubate in order to bind with the antibody **(antigen capture).** After the antigen is bound, a second antibody called the signal antibody is added. These secondary antibodies are conjugated to a signal molecule, for example, the enzyme alkaline phosphatase. This secondary antibody then binds to the antigen **(antibody capture),** which is already bound to the primary antibody. Substrate is then added that causes a color change in response to the presence of the signal molecule. This color change is usually in proportion to the amount of antigen present; thus, the assay becomes quantitative. This makes it possible to quantify the amount of antigen present in a given sample. Once a signal is produced it can be used to visually score the results based on the color change or an automated

Direct ELISA	Indirect ELISA

Well is coated with antibody	Well is coated with antigen
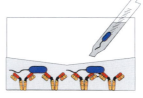	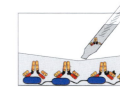
Sample containing antigen is added	Sample containing antibody is added
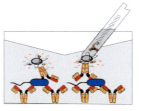	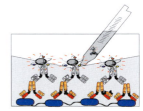
Second enzyme linked antibody is added and bound to antigen	Enzyme linked antiglobulin is added
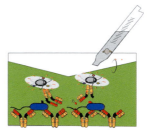	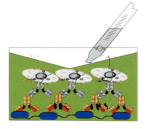
Enzyme substrate is added to produce color reaction	Enzyme substrate is added to produce color reaction

FIGURE 12.13 These figures are schematic representations of direct and indirect sandwich ELISAs. These reactions are usually carried out in a microtiter plate and the color change shown can be detected and quantitated using a plate reader.

plate reader can be used. Plate readers provide highly sensitive detection of low-level signals and can determine accurately the strength of a given signal in comparison with a standard curve.

12.3.2.2 Application

In many environmentally relevant processes such as the degradation of pollutants in water treatment plants and bioreactors, the concentration of certain bacterial species within biofilms is very important. Biofilms are specialized environments where microorganisms are firmly attached to surfaces and to one another by exopolymeric substances. Previously, the enumeration of specific organisms within biofilms was done using cultural counts, most probable number (see Chapter 10), or immunofluorescence microscopy. However, the inefficiency of these methods and the tendency of cells to clump as a result of production of exopolymers leads to significant underestimation of the actual numbers. ELISAs have proved to be powerful tools for quantifying specific biomass within biofilms (Gorris *et al.* 1994; Lee *et al.* 1990). In a study by Bouer-Kreisel *et al.* (1996), an ELISA was used to quantify populations of *Dehalospirillum multivorans* (*D. multivorans* are bacteria associated with the biodegradation of organic pollutants) in mixed culture biofilms. A standard curve was developed using ELISAs and immunofluorescence microscopy to quantify signal from and enumerate serial dilutions of pure cultures of *D. multivorans*. The standard curve related the amount of signal provided by the ELISA to the direct counts provided by microscopic enumeration. The two assays were found to be directly proportional. ELISAs were then performed on actual biofilms, and the results were used to quantify the total biomass of specific organisms responsible for the degradation of organic pollutants in mixed culture biofilms. Thus, the rates of degradation of these pollutants could be directly correlated with biomass using ELISA methodology.

12.3.2.3 Advantages and Disadvantages

As mentioned, there are many advantages of using ELISA over other detection or quantification methods. ELISA is sensitive and can be quantitative when used in conjunction with standard curves. Disadvantages are similar to all antibody-based methods and are related to crossreactivity and non-specific signal production. ELISAs must also be optimized to provide consistent results especially when using environmental samples. ELISAs, as described above, are also poorly suited for detection of extremely low concentrations of antigens. Other types of ELISAs have been described that are capable of detecting low target concentrations. One of these assays is known as competitive ELISA.

12.3.3 Competitive ELISA

12.3.3.1 Technique

In competitive ELISA, both a labeled control antigen and a sample, that contains an unknown quantity of unlabeled antigen are added to a sample well coated with antibody (Fig. 12.14). In this assay, the sample is added first and the bound antibody captures the unlabeled antigen. The labeled antigen is then added and is captured by any remaining antibody binding sites. If all sites are taken, no labeled antigen will bind and no signal will be emitted. If there is no antigen in the sample, only labeled antigen will bind and the signal will be maximized. Because the standard curve in a competitive ELISA exhibits the maximum signal at the lowest concentrations of antigen, this assay is very sensitive.

Competitive ELISA

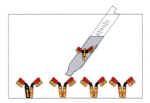

Well is coated with antibody

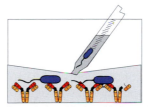

Sample is added and is bound by the antibody

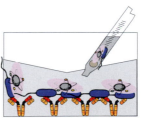

Enzyme linked antigen is added and binds to the remaining unoccupied sites on the antibody

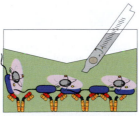

Enzyme substrate is added to produce color reaction

FIGURE 12.14 This is a schematic representation of the procedure used for competitive ELISA. This variation of a typical ELISA shown in Figure 12.12 can be used to detect very low concentrations of antigen.

12.3.3.2 Application

Mycotoxins such as T-2 toxin and zearalenone (also referred to as F-2 toxin) are frequent contaminants of cereal grains and other corn-based food and fodder. These compounds can cause serious health problems in humans and other animals in extremely low (nanogram) quantities. Barna-Vetro *et al.* (1994) developed a competitive ELISA for the detection of minute quantities of these mycotoxins in cereals. In order to conduct a competitive ELISA for environmental samples, purified toxin controls were first conjugated to horseradish peroxidase. Microtiter plates were coated with monoclonal antibodies specific for the mycotoxins T-2 and F-2. Standard curves were determined using known toxin standards with conjugate labels. For environmental sample analysis, 50 μl samples of toxin standards and extracted samples were co-incubated in wells with peroxidase (enzyme signal molecule). Results were determined by adding tetramethylbenzidine, which is the substrate for the enzyme, and measuring the degree of the resulting color reaction. This competitive ELISA assay was capable of detecting nanogram concentrations of toxins in foods.

12.3.3.3 Advantages and Disadvantages

The advantage of competitive ELISA lies in its ability to detect extremely low antigen concentrations. This is related to the inverse relationship between target concentration and signal strength described previously. The disadvantage is that these assays require a great deal of optimization and must be used in conjunction with a plate reader.

12.3.4 Immunomagnetic Separation Assays

12.3.4.1 Technique

Magnetic immunoseparation is an antigen capture methodology that uses antibodies conjugated to paramagnetic beads to attach to, concentrate, and purify antigens. Immunomagnetic separation is rapidly becoming one of the more popular approaches to the specific manipulation of microorganisms, proteins, and nucleic acids. The increase in popularity of this approach is due to its ease, mild conditions, low cost, and the ability to automate the process. In its most simplistic form, an immunomagnetic separation is accomplished with antibody-coated magnetite beads and a magnet. Essentially, the antibodies coated on the magnetic beads bind with antigens in solution and are then separated from the solution using a magnet (Fig. 12.15). The beads used for magnetic separation are small, ranging from 100 nm to 20 μm, and are typically made of iron oxide (magnetite). Such particles react strongly in magnetic field but do not retain any magnetism when the magnetic field is removed.

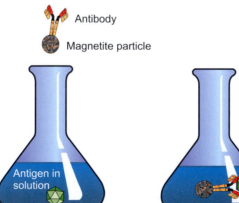   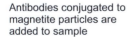

Antibodies conjugated to magnetite particles are added to sample

Antibody/magnetite complex binds to the target antigen

Magnet is used to separate the magnetite/antibody/antigen complex from the remaining solution

FIGURE 12.15 This is a schematic representation of the principle of immunomagnetic separation. In the first flask is a solution containing the antigen (virus). Antibodies specific to the virus are added to this solution. These antibodies, which have been conjugated (attached chemically) to a paramagnetic particle such as magnetite, then bind the virus. When a magnet is applied to the solution the magnetic particle is attracted to the magnet along with the attached antigen (virus). Immunomagnetic separation is a valuable tool for concentrating and purifying antigens from complex solutions.

12.3.4.2 Application

Immunomagnetic separation was used to recover thermophilic sulfate-reducing bacteria from oil field waters below oil production platforms in the North Sea (Christensen *et al.*, 1992). These bacteria can proliferate in oil field waters and cause considerable problems for oil companies during oil recovery, so the ability to detect their presence is of great value to the petroleum industry. In this example, immunobeads specific against cell wall antigens of the thermophilic *Thermodesulfobacterium mobile* captured several different isolates from oil-containing strata. Only one of the bacteria isolated by this method was serologically and morphologically identical to the bacterium (*T. mobile*) for which the antibodies were designed. Two others species of bacteria isolated using the immunomagnetic beads were spore-forming and similar to *Desulfotomaculum* sp., a sulfate-reducing bacterium that had previously been isolated from oil fields. However, western blots (see Section 12.3.5) of whole cells showed that the isolates were serologically different from *Desulfotomaculum* sp. This is a good illustration of the fact that an antibody that is designed against one organism, can cross-react with other organisms that are serologically different.

12.3.4.3 Advantages and Disadvantages

Immunomagnetic separation is an efficient means of target separation and purification from heterogenous matrices. In most cases immunomagnetic separation is also one of the easiest methods available for specific target isolation. Though crossreactivity issues are also evident, as with any antibody-oriented methodology, in most cases the problems associated with nonspecific signal production are not as problematic. As with any microbiology protocol such separations require optimization and proper choice of format (i.e., microcentrifuge separation or column matrix separation).

12.3.5 Western Immunoblotting Assays

12.3.5.1 Technique

Western immunoblotting is a three step, binding assay used to identify the presence of target antigens in a complex mixture of many other nontarget antigens such as might be found in environmental samples. This assay can be done with simple dot blot hybridization with a labeled antibody or an electrophoretic separation followed by hybridization with the labeled antibody. In dot blot hybridization, environmental samples

are added directly to an immobilizing nitrocellulose membrane, followed by immunolabeling and signal detection. In the second technique a sample of antigen is added to a gel and separated by size using electrophoresis. After electrophoretic separation, the sample is transferred to an immobilizing nitrocellulose membrane. This membrane is then incubated with enzyme-labeled or radiolabeled antibodies, that specifically bind to the antigen. After incubation, a substrate for enzyme-labeled or photographic film for radiolabeled substrate is used to detect the presence of the target antigen. Either method (dot blot or electrophoretic separation) indicates the presence and relative quantity of an antigen. If a separation step is used, this also allows molecular size determination of the antigen, which aids in confirming its identity. A schematic diagram of the electrophoretic separation and detection process is shown in Fig. 12.16.

12.3.5.2 Application

Methylosinus trichosporium OB3b is a methanotrophic bacterium that has been studied for bioremediation of trichloroethylene (TCE). TCE is a common environmental pollutant (see Chapter 16.6.2.1.3). The first step in the degradation pathway of TCE is the enzymatic cleavage of TCE by an enzyme called methane monooxygenase (MMO). Because little is known about the environmental factors that influence the rates of TCE degradation by *M. trichosporium*, studies to optimize the cellular expression of the enzyme (MMO) are common. In order to determine the amounts of MMO produced by a microbe, Western blotting can be used (Fitch *et al.*, 1993, 1996). The amount of signal produced by the Western blot analysis is compared with a standard curve. The maximum amount of signal is then used as an index of optimized MMO expression conditions, which have been correlated with optimal rates of TCE degradation. Thus, Western blotting is a useful technique for determining the presence and amounts of environmentally relevant proteins such as MMO produced by microorganisms such as *M. trichosporium* OB3B.

12.3.5.3 Advantages and Disadvantages

Obvious advantages of immunoblotting lie in its ability to specifically detect and quantify a particular antigen (target) within heterogeneous matrices. Further, immunoblotting can detect extremely low levels of target antigen. However, this procedure can be very time consuming and is subject to the problems inherent to all antibody-based methods.

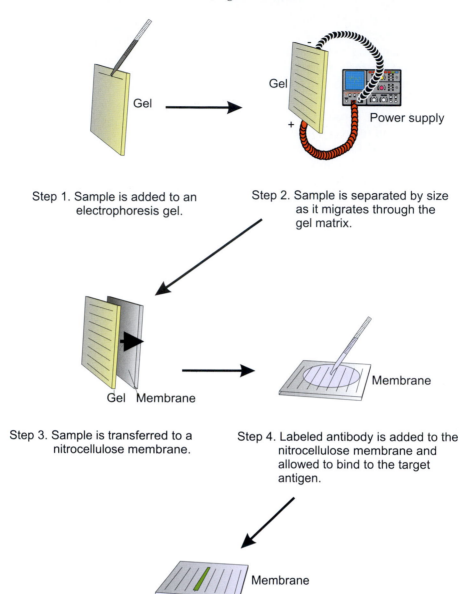

Step 1. Sample is added to an
electrophoresis gel.

Step 2. Sample is separated by size
as it migrates through the
gel matrix.

Step 3. Sample is transferred to a
nitrocellulose membrane.

Step 4. Labeled antibody is added to the
nitrocellulose membrane and
allowed to bind to the target
antigen.

Target antigen is detected
by color production

FIGURE 12.16   This is a schematic representation showing the basic steps involved in Western blot immunoassay.

12.3.6 Immunoaffinity Chromatography Assays

12.3.6.1 Technique

Affinity chromatography is a very powerful method used in purification and concentration of antigens. In affinity chromatography, the antibody is chemically bound to an inert support matrix (usually a glass, latex, or plastic bead) in a chromatography column. The sample containing the antigen is eluted through the column, and the antigen is selectively retained within the column while the sample passes through. After the sample is run through the column, the purified antigen is eluted, usually by changing the pH of the column, which causes the antigen to detach from the antibody. The antigen can then pass out of the column and be collected in highly purified form (Fig. 12.17). This process provides a very efficient means of both concentration and purification. Immunoaffinity chromatography offers several advantages compared with conventional purification techniques. Not only is the process selective and efficient, it also enables the processing of large-volume samples with relatively few steps.

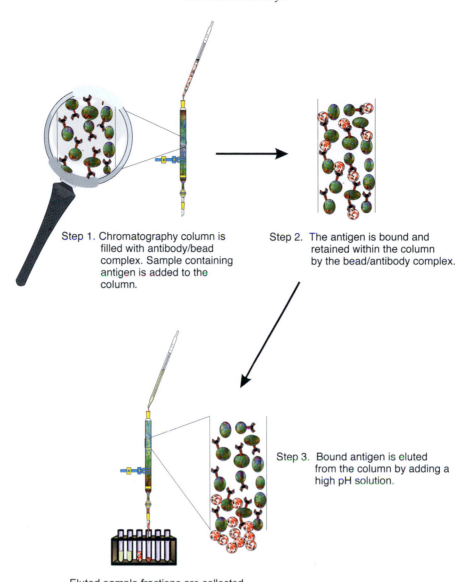

Step 1. Chromatography column is filled with antibody/bead complex. Sample containing antigen is added to the column.

Step 2. The antigen is bound and retained within the column by the bead/antibody complex.

Step 3. Bound antigen is eluted from the column by adding a high pH solution.

Eluted sample fractions are collected

FIGURE 12.17 This is a schematic representation showing the basic steps involved in antibody-mediated chromatographic separation.

12.3.6.2 Application

Pyranose oxidase is an enzyme made by many types of fungi. It catalyzes the oxidation of D-glucose in the presence of oxygen to form 2-dehydro-D-glucose and hydrogen peroxide (H_2O_2). These types of enzymes are important in many industrial processes to aid in catalysis of reactions that are used to synthesize different carbohydrate products. For this reason, efficient purification methods are needed. Schafer *et al.* (1996) described the use of immunoaffinity chromatography for the highly efficient purification of a pyranose oxidase produced by *Phlebiopsus gigantea*. They used antibodies specific for other pyranose oxi-

dases to construct an immunoaffinity column to purify this enzyme from mycelial extracts. The extracts were obtained by first harvesting the fungal mycelia by filtration. The extracts were homogenized, and then partially purified by heating the extracts and sedimentation of dense impurities. The mycelial extracts were then added to the immunoaffinity columns to allow binding of the pyranose oxidase enzyme. The researchers were able to get yields of 71% of highly purified enzyme. Thus, immunoaffinity chromatography is one of the most powerful methodologies for the isolation, purification and concentration of antigens including active enzymes from complex samples.

12.3.6.3 Advantages and Disadvantages

Immunoaffinity chromatography is a very efficient method for rapid purification of specific targets from a heterogeneous sample. Some problems encountered with this technique include a limited degree of sample concentration and the possibility for column clogging if the sample applied is too turbid. Sample concentration is problematic because usually several column volumes are required for target elution after purification. Further, the column matrix can often retain targets even after elution.

12.3.7 Immunocytochemical Assays

12.3.7.1 Technique

An immunocytochemical assay is used for the detection and determination of the cellular localization of target antigens. The purpose of an immunocytochemical assay is to determine where target antigens are localized within a particular cell. For instance, you can determine whether the target antigen (a particular protein for example) is localized in the cytoplasm or on the cell surface. In many cases, light microscopes are used to determine the location of antigens within a eukaryotic cell. However, immunocytochemical assays more often involve the use of electron microscopes to increase resolution and magnification of the area being studied. Many types of antibody labels can be used for electron microscopy, including heavy metals such as colloidal gold, enzymes such as horseradish peroxidase, and proteins such as ferritin. These labels are visualized by the electron microscope as electron dense regions (Bozzola and Russell, 1992). The process involved in the preparation of a sample for immunocytological analysis is complicated but can be summarized as follows: The sample is fixed with a preservative to maintain the original localization and antigenicity of the target. The sample is then embedded in plastic, sectioned, placed on an electron microscope grid, immunolabeled, poststained, and finally viewed under the electron microscope.

12.3.7.2 Application

An example of the use of immunocytochemical techniques in environmental microbiology is in the study of the parasitism of certain amoebea by *Legionella pneumophila*. *L. pneumophila* is an intracellular parasite and is the causative agent of legionnaires' disease. This bacterium is pervasive in aquatic environments and has been shown to parasitize and multiply within certain protozoa. The sequence of events in the intracellular infection of the amoeba *Hartmannella vermiformis* by *L. pneumophila* was examined by Kwaik (1996) using an

immunocytochemical assay. The goal of this study was to compare the intracellular infection of the amoebae with the infection of human alveolar macrophages that occurs during onset of legionnaires' pneumonia to aid in understanding the environmental life cycle of this human pathogen. Specifically, these researchers wanted to determine whether the accumulation of ribosomes and rough endoplasmic reticulum (RER) around amoebic phagosomes containing *L. pneumophila* was similar to that observed in human phagosomes. To accomplish this, monolayers of *H. vermiformis* amoeba cells were infected with *L. pneumophila*. The infected cells were harvested, fixed, and embedded in plastic. The samples were then thin sectioned, collected on sample grids, and incubated with an antibody specific for Bip, a heat shock protein associated with RER. This was followed by secondary incubation with an antiglobulin labeled with colloidal gold particles (indirect immunoassay) and examination by transmission electron microscopy. The results demonstrated that there is considerable similarity in the ultrastructure of phagosomes containing *L. pneumophila* in both amoeba and humans. Using immunocytochemical assay it was also found that similar to the human system, RER-specific protein (Bip) was present in the phagosomal membrane of the amoebae. Further, the localization of the RER and ribosomes in the amoebic phagosome was identical to the localization seen in human macrophage phagosomes. This study helped indicate a possible role of the RER in the protection and growth of the *L. pneumophila* within the phagosomes of both hosts.

12.3.7.3 Advantages and Disadvantages

The major advantage of immunocytochemical assay is the high resolution that can be achieved using electron microscopy. However, this technique is very specialized and requires a great deal of training for sample preparation and performing the microscopy. There is also a great deal of expense involved, compared to other immunology-based assays, but the information gathered is often very important in elucidation of actual cellular processes such as the example described, infection of a eukaryotic cell by a disease-causing bacterium.

12.3.8 Immunoprecipitation Assays

12.3.8.1 Technique

Immunoprecipitation is a methodology that uses the antigen–antibody reaction in solution to semiquantitatively determine the amount of antigen or antibody in a sample by determining the amount of precipitation or clumping of the antigen–antibody complex. Immunoprecipitation can be used to determine the concentration of low levels of antigen or can be used to quantify or titer antibodies or antigens. Immunoprecipitation can also be used to determine the optimal concentration ratio for an antibody and antigen. Most commonly, a series of reaction tubes are set up, each of which contains a constant titer of antibodies. Antigen is then added in increasing concentration to consecutive tubes. In the initial tubes, where the lowest concentration of antigen has been added, there is no obvious precipitation. As the antigen concentration is increased, the formation of antigen–antibody complexes increases until a visible precipitate is formed. As the antigen concentration is further increased it will eventually exceed the concentration of antibody present and the amount of precipitate will decrease again (Fig. 12.18).

This may seem strange until one entirely understands the nature of antibodies and antigens. Most antibodies are considered to be bivalent, whereas antigens are considered to be multivalent, in other words they have multiple antibody reactive sites or **epitopes** as shown in Fig. 12.19a. This means that antibodies are able to bind at least two antigens at once and anti-

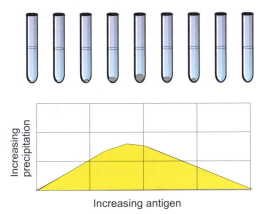

FIGURE 12.18 This is a schematic representation of an immunoprecipitation assay. The test tubes at the top all have constant concentrations of antibody and increasing concentrations of antigen starting from the left. In the middle tubes there is optimum antigen–antibody interaction and precipitation of the antibody and antigen occur. As indicated by the graph on the bottom there is a point where the highest amount of precipitation forms. This type of assay is used to determine optimum antibody to antigen ratios for immunoassays. It is also useful for quantitating either antibody or antigen concentrations in solutions.

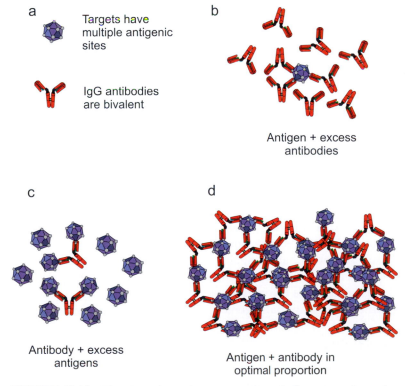

FIGURE 12.19 This is a schematic representation of what occurs in an immunoprecipitation assays. (a) Targets can have multiple antigenic sites. An IgG antibody has two binding sites (bivalent). (b) In this case there is excess antibody so little antibody–antigen binding occurs resulting in no precipitation. (c) In this case there is excess antigen and while antibody–antigen binding occurs, there is little cross-linking and so precipitation does not occur. (d) In this case the amount of antibody and antigen are optimal, there is extensive cross-linking of antibody–antigen complexes and this results in maximum precipitation as shown in Figure 12.18.

gens can be bound by more than one antibody at a time. So, in the case in which there is excess antibody (Fig. 12.19b), multiple antibodies bind to each antigen and no cross-linking or complex interactions take place. In this case, there is no visible precipitation or coagulation and considerable antibody is still found in solution. When antibody and antigen are present in optimal proportions, complex interactions and antigen cross-linking occur (Fig. 12.19d). This results in clumping and precipitation from solution. As the antigen concentration is increased past this optimal antigen–antibody proportion, it reaches a point where no precipitation occurs and excess antigen is found in the supernatant (Fig. 12.19c).

12.3.8.2 Application

This technology has been utilized in environmental microbiology to determine the mechanism for the inhibition of certain economically important fungal plant pathogens by another nonpathogenic fungus. It was hypothesized that the fungus *Talaromyces flavus* can control the proliferation of several fungal plant pathogens including *Sclerotinia sclerotionum*, *Rhizoctonia solani*, and *Verticillium dahliae* by the production of hydrogen peroxide. *T. flavus* produces hydrogen peroxide as a product of glucose metabolism in the presence of the enzyme glucose oxidase. To determine whether hydrogen peroxide was responsible for the inhibition, cultures containing the pathogens were incubated with and without glucose oxidase. The cultures without glucose oxidase showed a high percentage of germination, whereas cultures containing glucose oxidase exhibited very low germination (inhibition). To confirm these results, the cultures with glucose oxidase were subjected to immunoprecipitation. Antibody to the glucose oxidase was added to the cultures, effectively removing the glucose oxidase from the culture. After immunoprecipitation of the glucose oxidase, the fungal pathogens recovered and showed high levels of germination. Germination was subsequently halted again when more glucose oxidase was added. This research effectively showed that glucose oxidase is the enzyme produced by *T. flavus* that controls the plant pathogens.

12.3.8.3 Advantages and Disadvantages

In addition to the type of application described, this method is important in the development of immunoassays and the characterization of antigen–antibody interactions. This assay is simple to perform though its use requires careful optimization. Disadvantages vary depending on what type of assay and application is being considered. As with all immunoassays, the possible non-specific interactions with non-target antigens are always an issue.

QUESTIONS AND PROBLEMS

1. Describe an immunological assay for the accurate quantitation of *Rhizobium* around root nodules (hint: see Chapter 18).

2. Justify your choice of assay in question 1.

3. Immunomagnetic purification methods are becoming popular in environmental microbiology. Sediment slurries can contain paramagnetic particles, which are copurified along with target antigens, making subsequent assay procedures difficult. Describe the problems that would arise with immunomagnetic analysis in this case, and devise potential methods for solving these issues.

4. How could you determine whether an enzyme is intercellular or extracellular using immunoassay techniques?

5. You have two monoclonal antibodies available from two different commercial companies. Both cost the same and both are specific for the rhizobia you are studying in question 1. After ordering both antibodies, how would you determine which antibody was the best for the assay you designed in question 1? How would you perform these evaluations?

6. Rapid detection of biological warfare agents is an emerging area of research (Chapter 5). Design a rapid detection method using an immunoassay that can aid in the detection of *Bacillus anthracis*.

7. What immunoassays described offer quantitative results?

8. What immunoassays described allow for purification of antigens from heterogeneous samples?

References and Recommended Readings

Barna-Vetro, I., Gyongyosi, A., and Solti, L. (1994) Monoclonal antibody-based enzyme-linked immunosorbent assay of fusarium T-2 and zearalenone toxins in cereals. *Appl. Environ. Microbiol.* **60**, 729–731.

Bouer-Kreisel, P., Eisenbeis, M., and Scholz-Muramatsu, H. (1996) Quantification of *Dehalospirillum multivorans* in mixed-culture biofilms with an enzyme-linked immunosorbent assay. *Appl. Environ. Microbiol.* **62**, 3050–3052.

Bozzola, J. J., and Russell, L. D. (1992) "Electron Microscopy: Principles and Techniques for Biologists." Jones and Bartlett, Boston.

Christensen, B., Torsvik, T., and Lien, T. (1992) Immunomagnetically captured thermophilic sulfate-reducing bacteria from North Sea oil field waters. *Appl. Environ. Microbiol.* **58,** 1244–1248.

Collins, W. P. (1985) "Alternate Immunoassays." John Wiley & Sons, New York.

Dowd, S. E., and Pillai, S. D. (1997) A rapid viability assay for *Cryptosporidium* oocysts and *Giardia* cysts for use in conjunction with indirect fluorescent antibody detection. *Can. J. Microbiol.* **43,** 658–662.

Dowd, S. E., and Pillai, S. D. (1997) Microarthropods as natural predators of the protozoan parasites *Cryptosporidium* sp. and *Giardia* sp. Manuscript in preparation.

Edwards, R. (1996) "Immunoassays: Essential Data." John Wiley & Sons, New York.

Fitch, M. W., Graham, D. W., Arnold, R. G., Speitel, G. E., Jr., Agarwal, S. K., Phelps, P., and Georgiou, G. (1993) Phenotypic characterization of copper-resistant mutants of *Methylosinus trichosporium* OB3b. *Appl. Environ. Microbiol.* **59,** 2771–2776.

Fitch, M. W., Speitel, G. E., Jr., and Georgiou, G. (1996) Degradation of trichloroethylene by methanol-grown cultures of *Methylosinus trichosporium* OB3b PP358. *Appl. Environ. Microbiol.* **62,** 1124–1128.

Gorris, M. T., Alarcon, B., Lopez, M. M., and Cambra, M. (1994) Characterization of monoclonal antibodies specific for *Erwinia carotovora* subsp. *atroseptica* and comparison of serological methods for its sensitive detection on potato tubers. *Appl. Environ. Microbiol.* **60,** 2076–2085.

Kwaik, Y. A. (1996) The phagosome containing *Legionella pneumophila* within the protozoan *Hartmannella vermiformis* is surrounded by the rough endoplasmic reticulum. *Appl. Environ. Microbiol.* **62,** 2022–2028.

Lee, H. A., Wyatt, G. M., Bramham, S., and Morgan, M. R. A. (1990) Enzyme-linked immunosorbent assay for *Salmonella typhimurium* in food: Feasibility of 1-day *Salmonella* detection. *Appl. Environ. Microbiol.* **56,** 1541–1546.

Price, C. P., and Newman, D. J. (1991) "Principles and Practice of Immunoassay." Stockton Press, New York.

Schafer, A., Bieg, S., Huwig, A., Kohring, G., and Giffhorn, F. (1996) Purification by immunoaffinity chromatography, characterization, and structural analysis of a thermostable pyranose oxidase from the white rot fungus *Phlebiopsis gigantea*. *Appl. Environ. Microbiol.* **62,** 2586–2592.

Stosz, S. D., Fravel, D. R., and Roberts, E. P. (1996) *In vitro* analysis of the role of glucose oxidase from *Talaromyces flavus* in biocontrol of the plant pathogen *Verticillium dahliae*. *Appl. Environ. Microbiol.* **62,** 3183–3186.

Tizard, I. R. (1995) "Immunology: An Introduction," 4th ed. Harcourt Brace College Publishers, New York.

13

Nucleic Acid-Based Methods of Analysis

ELIZABETH M. MARLOWE   KAREN L. JOSEPHSON   IAN L. PEPPER

The advent of molecular biology has created a new array of methodologies for examining microorganisms in the environment. The premise of these methodologies relies on basic concepts that have created some very powerful tools. Techniques such as gene probing and polymerase chain reaction (PCR) have made possible a very specific and sensitive evaluation of the microbial world. Microbiologists are now able to use a small sample of microbial nucleic acids to identify unculturable bacteria, track genes, and evaluate genetic activity in the environment. The theories and applications of these concepts and techniques are discussed in this chapter.

13.1 STRUCTURE AND COMPLEMENTARITY OF NUCLEIC ACIDS

Specific molecular analyses are based on nucleic acid structures and on the intricate mechanisms microbes use to synthesize nucleic acids. Nucleic acids are either **deoxyribonucleic acid (DNA)** or **ribonucleic acid (RNA)**. DNA is made up of four deoxynucleotide bases: guanine (G), cytosine (C), adenine (A), and thymine (T). Guanine and adenine are purine bases, and cytosine and thymine are pyrimidine bases (Fig. 13.1a). It is more convenient simply to refer to each of these nucleotide bases by its abbreviation (G, C, A, or T). DNA consists of these bases linked to the sugar, deoxyribose, and a phosphate moiety. Thus, DNA has a phosphate sugar backbone, which always forms a link between the 5' carbon atom of one deoxyribose

and the 3' carbon atom of the adjacent deoxyribose. DNA polymerase, the enzyme used to replicate DNA, can add a new nucleotide only to a free hydroxyl at the 3' carbon of the last sugar on the growing DNA chain, and thus DNA is synthesized only in the 5' to 3' direction.

Structurally, DNA consists of two strands of deoxyribonucleic acids combined together to form a double helix. One strand of DNA is oriented 5' to 3' while the complementary strand is oriented 3' to 5'. These two strands are linked by hydrogen bonds between corresponding pairs of bases. Specifically, G binds only to C (i.e., G-C), and A binds only to T (i.e., A-T). For each G-C pairing, there are three hydrogen bonds, whereas for the A-T pairing there are only two hydrogen bonds (Fig. 13.1b). Because these bases are specific in their ability to bind together, bases are said to be **complementary** to each other. Single-stranded pieces of DNA (ssDNA) would be expected to form a double-stranded DNA helix (dsDNA) only if the sequences of the two strands are complementary to each other. In other words, a DNA strand containing the nine bases A-T-T-C-G-G-A-A-T will anneal only with the complementary strand T-A-A-G-C-C-T-T-A with the resulting dsDNA being nine base pairs (bp) in length. The double-stranded, complementary nature of DNA is the basis for two extremely important molecular analyses that will be discussed later, **gene probes** and **polymerase chain reaction (PCR)** (see Information Box 1).

In contrast to DNA, RNA is single stranded and consists of the three bases adenine (A), guanine (G), and cytosine (C), but contains uracil (U), which re-

A

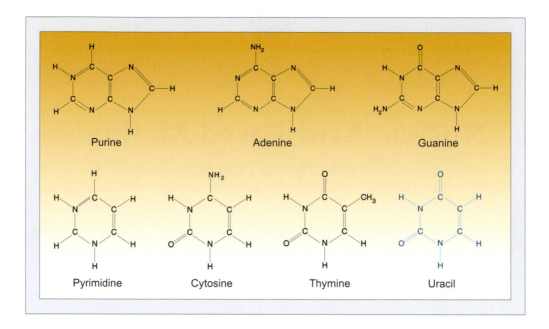

B

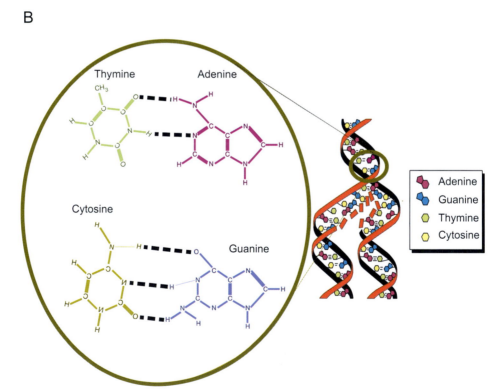

FIGURE 13.1   (a) Structures of purine and pyrimidine bases used in nucleic acid sequences. (b) Hydrogen bonding between purine and pyrimidine bases.

places the thymine (T) found in DNA. An additional difference is that in RNA, the sugar component is a ribose rather than a deoxyribose. In RNA, G still binds to C, but A binds to U. In all organisms there are three forms of RNA: **ribosomal RNA (rRNA), messenger RNA (mRNA), and transfer RNA (tRNA).** rRNA is a

---

Information Box 1
Hybridization and Denaturation

When two DNA strands combine together because they are complementary, the process is known as **DNA–DNA hybridization,** because the resulting dsDNA is a hybrid of the two separate strands. The reverse process, in which dsDNA melts into two single strands, is called **denaturation**. This can be done chemically or simply by heating the DNA to 94°C. Upon cooling, the two single strands automatically hybridize back into a double-stranded molecule, a process known as **reannealing.** Complementarity is an important concept because if one strand sequence is known, the sequence of the other strand is easily deduced. This concept is the basis of many of the nucleic acid-based methodologies discussed in this chapter.

---

class of RNA molecules that play a role in protein translation. mRNA is a class of RNA molecules that are transcribed from an encoded gene consisting of DNA. The transcribed "message" can then be translated into a protein by ribosomes. tRNA is a small population of RNA molecules that transfer specific amino acids to the ribosome during mRNA translation. The specific amino acid is inserted into the growing polypeptide chain based on specific tRNA and mRNA sequences, which in turn are based on the original DNA sequence. Two types of RNA, rRNA and mRNA, are used extensively in molecular genetic analyses, including phylogenetic studies and estimates of metabolic activity. In all cases, the base sequence of RNA is **transcribed** from the original conserved DNA sequences. For more in-depth information on the structure and function of DNA and RNA see Watson *et al.* (1983, 1987).

13.2 OBTAINING MICROBIAL NUCLEIC ACIDS FROM THE ENVIRONMENT

13.2.1 Extracting Community Bacterial DNA from Soil Samples

This concept is discussed in detail in Chapter 8.1.3.1.1. Here a brief review of the concepts is presented. Because less than 1% of soil microflora are culturable, traditional cultural techniques tend to underestimate the true soil microbial population. Depending on the environment under investigation, several methods have been developed to obtain a more representative portion of the soil microflora. These are based on the total nucleic acid content of the community or **community DNA.** In the soil environment one way to obtain community DNA is to separate all cells from the soil and subsequently lyse the cells and extract the nucleic acids (see Chapter 8.1.3.1.2). Intact cells may be lysed by heat, chemical or enzymatic treatment. Following lysis, it is essential that the sample undergo

one or more purification steps for successful subsequent molecular analyses. There are problems with this approach that were discussed earlier. However, cell extraction followed by DNA analysis is still likely to result in a more representative estimate of the total bacterial community than a simple plate dilution on growth medium.

A second approach for extraction of community DNA from soil is to lyse the bacterial cells *in situ* (within the soil matrix) and subsequently extract and purify the DNA. Several protocols for the extraction of community DNA have been published (Moré *et al.,* 1994) with most relying on the use of lysozyme or sodium dodecyl sulfate as the lysing agent. Once extracted, the DNA is purified by cesium chloride density centrifugation or the use of commercial purification kits. An alternative ap-proach is to use phenol extraction followed by ethanol precipitation as a means of DNA purification. Once the DNA has been purified, it is ready for molecular analysis.

In aquatic environments, cells are concentrated by filtering a representative portion of water through the appropriate filter (see Chapter 8.2.3). Bacteria are captured by membrane filtration with a 0.2-μm filter. Viruses are too small to be filtered by size exclusion and thus require an electronegative or electropositive filter for collection. In contrast, the larger parasites such as *Giardia* and *Cryptosporidium* are trapped in a simple fiber filter cartridge that captures the parasites via size exclusion. As with soil samples, once the organisms are collected from the water the total biomass can be lysed, purified, and subjected to further molecular analysis. Air samples are collected as described in Chapter 5.

DNA extraction methods are not without their limitations. Drawbacks include the loss of DNA during the purification process. In addition, the presence of metals or humic residues from environmental samples can interfere with further analysis even after careful purification. Finally, it should be noted that community DNA extracts may contain eukaryotic DNA as well as bacterial DNA.

Overall, it is generally agreed that the community-based extraction methods (cells or DNA) obtain a higher proportion of the total bacterial population than dilution and plating alone. They are also likely to represent a different population of bacteria than that obtained by cultural methods.

13.3 NUCLEIC ACID–BASED METHODS

13.3.1 Gene Probes and Probing

Gene probes are an application of nucleic acid hybridization. Typically, probes are small pieces of DNA known as **oligonucleotides,** which are complementary to the target sequence of interest, that are marked or labeled in some way in order to make them detectable. Probes have been used in environmental microbiology to examine soil microbial diversity, to identify a particular genotype, and to test for virulence genes of suspected pathogens that have been isolated from water. Construction of gene probes, and specific application methodologies are outlined in detail in manuals such as that by Sambrook *et al.* (1989). Here we present the general concepts involved in the construction of gene probes along with some of their practical applications. These concepts are also critical to the understanding of some of the more sophisticated techniques that are discussed later.

Gene probe methodology takes advantage of the fact that DNA can be denatured and reannealed. To make a gene probe, the DNA sequence of the gene of interest must be known. This gene may be unique to a particular microbial species, in which case the sequence may be useful for the specific detection of that organism. Alternatively, the gene may code for the production of an enzyme unique to some metabolic pathway, and a gene probe constructed from such a sequence may indicate the potential activity of a group of bacteria in a soil or water sample. This kind of probe can be defined as a **functional gene probe.** For example, gene probes can be made from sequences that code for enzymes involved in nitrogen fixation, and such a probe could then be used to estimate whether a particular soil contained any bacteria with nitrogen-fixing genes. Additional probes could later be constructed to determine whether such organisms were in fact nitrogen-fixing *Rhizobium* or *Azospirillum* spp. or even cyanobacteria. The gene could even be universal to all bacteria, thus allowing detection of all known

bacteria. Many gene sequences are now known and are readily accessible via the gene databases, which will be discussed later in this chapter. There are also computer software packages designed to search for sequences that are unique to a particular bacterium or, conversely, sequences that are conserved within particular groups of bacteria. The strategy that one uses to select a particular target sequence will depend on the intended use of the gene probe.

The basic strategy in the construction of a gene probe is to obtain the sequence of the target gene and then to select a portion of this sequence for use as a probe. The size of the probe can range from 18 base pairs to as many as several hundred base pairs. The probe is then synthesized and labeled in such a way that it can be detected after it hybridizes to the target sequence.

A researcher has several options when constructing and labeling a probe. In the early days of molecular biology the only labeling option was radioactivity, but several nonradioactive methods are now available. Radioactive labeling of a probe is typically done by labeling the sequence with a radioactive chemical, such as ^{32}P, which is incorporated into the sugar–phosphate backbone of DNA. Later, when the probe or labeled sequence is hybridized to the target molecule, it can be detected by autoradiography (see Information Box 2). Nonradioactive alternatives include probes labeled with **digoxigenin (DIG), biotin,** or **fluorescein,** which can be incorporated into the sequence by chemical synthesis. The different labels are detected by binding the respective antibody or streptavidin-alkaline phosphate conjugate, which, when reacted with the appropriate substrate, will give a signal (Fig. 13.2).

The probing process is explained using the following example. A gene probe is constructed for a particular intestinal bacterium such as *Salmonella*. A well water sample suspected to contain the pathogen is collected and then filtered onto a membrane. The cells captured on the membrane are lysed, and the released bacterial DNA is denatured into two single strands (Fig 13.3) The gene probe is similarly denatured into single strands and is subsequently added to the membrane with the lysed cells. The DNA is allowed to reanneal, and in some cases the single strand of the gene probe will anneal with the complementary target DNA sequence from the *Salmonella* cells. After washing the membrane free of unhybridized probe, the filter is exposed to x-ray film and the radioactivity of the ^{32}P associated with the hybridized DNA results in a photographic image. This process is called **autoradiography.**

A positive image indicates that the ^{32}P probe annealed to target DNA on the filter, implying that hy-

Information Box 2
Autoradiography

During autoradiography a photon of light, beta particle, or gamma ray emitted by the filter activates silver bromide crystals on the film. When the film is developed, the silver bromide is reduced to silver metal and forms a visible grain or black spot on the film. Film images obtained by this technique may be quantified using microdensitometry.

bridization between the probe and the sample DNA has occurred. Such hybridization, and hence detection, can occur only because the gene probe has annealed to the specific *Salmonella* sequence. In contrast no signal would indicate that the target *Salmonella* sequence was not present. The probe should not hybridize to the DNA from other cells such as *Bacillus* or *Pseudomonas*. This process could be repeated using another probe specific for an indicator organism such as *Escherichia coli*. This example illustrates how a probe can be used to detect a target organism within a sample containing a mixture of organisms. The probe could similarly be used to confirm the identity of a pure culture isolate.

13.3.1.1 Colony Lifts or Hybridization

Gene probes can also be used to detect a specific gene sequence within bacterial colonies on a petri plate containing a mixed population of bacteria through use of a process termed **colony hybridization** or **lifts.** To perform a colony hybridization, a piece of filter paper is lightly pressed onto the petri plate so that some bacterial cells from each colony adhere to the paper. The cells are lysed directly on the filter paper, and the DNA is fixed to the filter. The filter is then probed and detected as described above. After this procedure, only the colonies that contain the specific

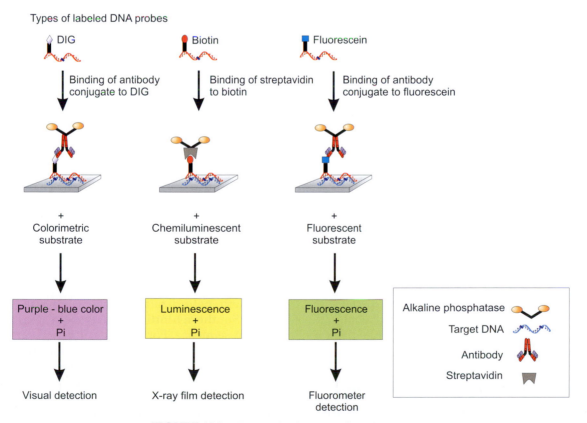

FIGURE 13.2   Gene probe detection of a DNA sequence.

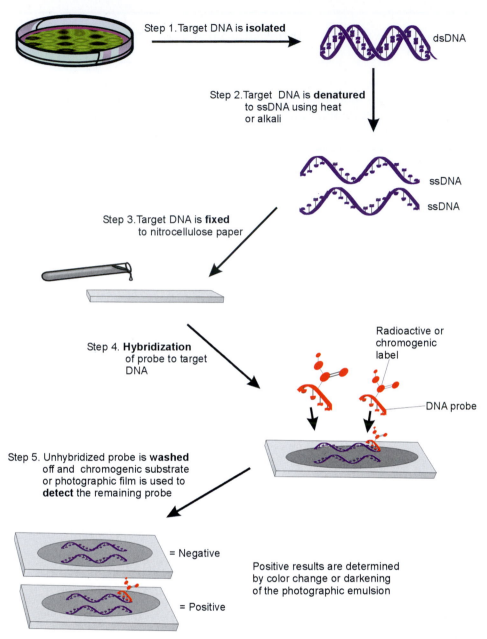

Step 1. Target DNA is **isolated**

dsDNA

Step 2. Target DNA is **denatured** to ssDNA using heat or alkali

ssDNA

ssDNA

Step 3. Target DNA is **fixed** to nitrocellulose paper

Step 4. **Hybridization** of probe to target DNA

Radioactive or chromogenic label

DNA probe

Step 5. Unhybridized probe is **washed** off and chromogenic substrate or photographic film is used to **detect** the remaining probe

= Negative

= Positive

Positive results are determined by color change or darkening of the photographic emulsion

FIGURE 13.3 Alternative probe labeling and detection systems. Probes are labeled with either radioactivity, digoxigenin, biotin, or fluorescein. These labels are then detected by autoradiography or by binding the appropriate antibody or streptavidin-alkaline phosphate conjugate, which gives a signal when combined with the respective substrate. Detection of these signals can be by fluorescence, chemiluminescence, or a colorimetric assay.

DNA sequence give a radioactive signal. Because the original petri plate contains all the intact colonies, the viable colony of interest can now be identified and retained for further study (Fig. 13.4).

13.3.1.2 Southern and Northern Hybridizations

Another application of gene probe technology is to identify a target sequence of interest via Southern or Northern hybridizations, which are also known as blots. **Southern blotting** is a technique used to identify a DNA sequence. For example, it may be important to know whether a gene is plasmid or chromosomally borne. To determine whether the sequence is plasmid borne, all the plasmids within the strain can be extracted and separated by gel electrophoresis (see Information Box 3). The plasmid DNA is then transferred onto a special membrane by blotting, and the

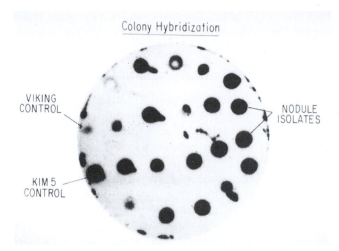

FIGURE 13.4 A colony lift from a petri plate containing a mixed population of the nitrogen-fixing bacteria rhizobia isolated from root nodules. The gene probe was constructed from a unique plasmid associated with the isolate known as KIM 5. (Reprinted from *Soil Biol. Chem.,* 21, I. L. Pepper *et. al.,* "Strain identification of highly competitive bean rhizobia isolated from root nodules: use of fluorescent antibodies, plasmid profiles and gene probes, " pp. 749–753, © 1989 with permission from Elsevier Science.)

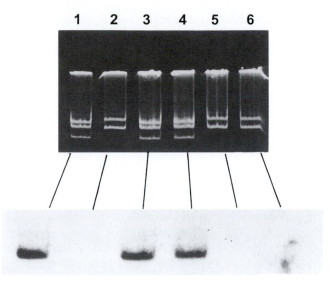

FIGURE 13.5 Southern blot of a DNA sequence associated with a unique plasmid contained within the rhizobial strain KIM 5. The upper portion of the figure shows the original plasmid profiles of two different rhizobial isolates. The gene probe was constructed from the smallest plasmid associated with this strain. Isolates that do not contain the small plasmid are not detected. The bottom portion of this figure shows the Southern blot. (Reprinted from *Soil Biol. Chem.,* 21, I. L. Pepper *et. al.,* "Strain identification of highly competitive bean rhizobia isolated from root nodules: use of fluorescent antibodies, plasmid profiles and gene probes, " pp. 749–753, © 1989 with permission from Elsevier Science.)

membrane is subsequently probed. Once again, only the DNA molecules that contain the target sequence hybridize with the probe, thus allowing detection of those plasmids containing the target sequence (Fig. 13.5).

Northern blotting is the analogous process used to analyze RNA. Total RNA that has been extracted from an environmental sample can be run on a gel and/or transferred to a membrane, as described earlier, and specific RNA molecules can be detected by the appropriate probe. Although extraction and stability of RNA are problematic, this technique can be used in gene expression studies to show induction of a specific gene. Thus, detection of DNA sequences can give information about the presence of a gene in a population, whereas detection of RNA provides information about the expression of the gene in the given population.

Dot blotting or **dot hybridization** is a technique used to evaluate the presence of a specific nucleic acid sequence without the need to run a gel. Rather, identical amounts of nucleic acid are spotted on a nitrocellulose filter and subsequently probed. Dot blots can be used to indicate the presence or absence of a sequence or can be used to quantitate the amount of sequence present. For quantitation, the relative amount or intensity of hybridization will give an estimate of the quan-

---

Information Box 3
Agarose Gel Electrophoresis

Agarose gel electrophoresis is a fundamental tool in nucleic acid analysis. It is a simple and effective technique for viewing and sizing DNA molecules such as plasmids or DNA fragments. The DNA samples are loaded into wells in an agarose gel medium. Voltage is applied to the gel, causing the DNA to migrate toward the anode because of the negatively charged phosphates along the DNA backbone. The gel is stained with a dye such as ethidium bromide, allowing visualization of the DNA when viewed under ultraviolet (UV) light. The smaller DNA fragments migrate faster through the gel matrix, and the larger fragments migrate more slowly. The molecular weight in base pairs (bp) of the DNA determines the rate of migration through the gel and is estimated from standards of known size that are run in parallel on the gel (see Fig. 13.7).

tity of sequences in a sample when compared with similarly spotted known standards (Fig. 13.6). The intensity of signal is determined using an instrument called a densitometer.

13.3.1.3 Microarrays

Recent developments in high throughput screening of DNA sequences has resulted in the development of microarray technology. A microarray refers to a set of DNA probes, usually purified PCR products from cDNA or genomic clones, deposited on a solid support (generally a glass microscope slide) with a spot density of several hundred individual spots per cm^2. Arrays have been used primarily for gene expression profiling. For example, RNA is extracted from cells which have been exposed to stress and cells that have not been exposed to stress. A cDNA copy of the mRNA is labeled enzymatically by the inclusion of a fluorochrome labeled nucleotide during an amplification step (see Section 13.3.2). The array is hybridized with the target, the labeled cDNA copy of the mRNA, and the hybridization intensity is acquired by laser scanning and CCD camera detection (Jordan, 1998). Since it is known where on the array that the individual DNA probes are located, the result is a global fingerprint or signature of the genes that are expressed as a result of a specific stress. This technology has been used in the pharmaceutical industry to look at drug responses. However the applications of this technology seems limitless. In environmental microbiology microarrays

have the potential to allow screening of a drinking water supply for several hundred pathogens with one assay.

13.3.2 Polymerase Chain Reaction

The **PCR**, a technique used to amplify the amount of target DNA up to 10^6-fold or more, has revolutionized molecular biology methodologies. First discovered in 1985, PCR has become a key protocol in many biological laboratories, including those concerned with environmental microbiology. The ability to amplify small amounts of DNA that may be present in the environment makes detection very sensitive. For example, the detection of 100 *Salmonella* cells in 1 g of soil using traditional cultural techniques would be very difficult. However, amplifying the 100 cells to the equivalent of 10^6 cells makes detection much easier. PCR is a relatively simple enzymatic reaction that uses a DNA polymerase enzyme to copy a target DNA sequence repeatedly during a series of 25-30 cycles. During each cycle of PCR, the amount of target DNA is doubled, resulting in an exponential increase in the amount of DNA. In theory, 25 cycles result in amplification by 2^{25}, but in practice one obtains approximately a 10^6-fold increase in the amount of target sequence present. This is because the efficiency of amplification is not perfect. The PCR product is typically visualized by agarose gel electrophoresis, and the size is estimated by comparison with DNA standards of known size (Fig. 13.7).

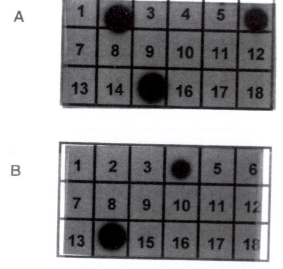

FIGURE 13.6 Dot blot hybridization showing detection of two bacterial isolates. *Arthrobacter* isolates that degraded toluene, ethylbenzene, and xylene (TEX) and *Sphingomonas* isolates that degraded xylene (X) were detected by the use of two individual gene probes. (Photos courtesy E. M. Jutras.)

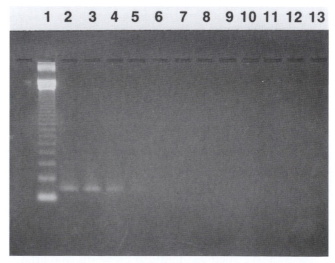

FIGURE 13.7 Agarose gel electrophoresis of PCR-amplified product DNA. The internal lanes show DNA products of different sizes. The external lane shows a DNA ladder of known size fragments (123 bp), which can be used to estimate the size of the amplified DNA. (Photo courtesy I. L. Pepper.)

13.3.2.1 The Steps of PCR

A typical cycle of PCR has three steps (Fig. 13.8a and b). The first step involves the denaturation of ds-DNA into two single strands of **target** or **template DNA** (ssDNA). Added to the reaction mixture are two different short pieces of single-stranded DNA called **primers** that have been carefully chosen and commercially synthesized. In specialized PCR reactions, such as arbitrarily primed PCR (AP-PCR), only a single primer is added (Section 13.3.2.6). Primers are **oligonucleotides** that have a sequence complementary to the target ssDNA template, so they can hybridize or anneal to this DNA, defining the region of amplification. One primer is described as the upstream primer and the other is the downstream primer. This description simply defines the locations where the primers anneal on the ssDNA template. The second step in the PCR cycle is **primer annealing,** which consists of the primers hybridizing to the appropriate target sequence. The third and final step is **extension**. Here, a DNA polymerase synthesizes a strand complementary to the original ssDNA by the addition of appropriate bases to the primers that have already hybridized to the target. The net result at the end of a cycle is two double-stranded molecules of DNA identical to the original double-stranded molecule of DNA. Repeating the process results in PCR amplification of the DNA and an exponential increase in the number of copies of the original DNA present.

The key to each PCR cycle is that each of the three steps of PCR amplification occur at different but defined temperatures for specific time intervals. Generally, these three steps are repeated 25 to 30 times to obtain sufficient amplification of the target DNA. These cycles are conducted in an automated, self-contained temperature cycler or **thermocycler** (Fig. 13.9). Thermocyclers range in cost from $2500 to $7000 and allow precise temperature control for each step. The actual amplification reaction takes place in small (200-500 μl) microfuge tubes, and commercial kits are available that provide all the nucleotides, enzymes, and buffers required for the reaction (Fig. 13.10).

Temperature is a critical part of the PCR process. Denaturation of the target sequence or template occurs at a temperature greater than the melting temperature of the DNA. For most PCR reactions this is standardized at 94°C for 1.5 minutes, because it guarantees complete denaturation of all DNA molecules. Primer annealing occurs at a lower temperature, typically between 50 and 70°C for 1 minute, depending on the base composition of the primer. It is possible for a primer to anneal to a DNA sequence that is similar to the correct target sequence but which contains a few incorrect bases. This results in incorrect amplification of the DNA, which is termed **nonspecific amplification,** and gives a false-positive result. The higher the temperature of annealing, the more specific the annealing is, and thus the extent of nonspecific amplification is reduced. However, as the annealing temperature increases, PCR sensitivity normally decreases.

The final step of PCR is extension. The essential component of this reaction is a polymerase enzyme, such as *Taq* **polymerase,** which sequentially adds bases to the primers. This enzyme was obtained from the thermophilic bacterium *Thermus aquaticus* and is uniquely suited for PCR because it is heat stable, withstanding temperatures up to 98°C, and can therefore be reused for many cycles. The extension step is normally 1 minute in length and is performed at 72°C. A 25-cycle PCR reaction may take about 3 hours, including ramp times between steps, although the time varies depending on the type of thermocycler. Ramp time is simply the time interval it takes the thermocycler to go from one temperature to the next. It is important to keep in mind that for each primer pair, the researcher must initially optimize the conditions of temperature, incubation time, and concentration of the various reaction components to obtain the desired results. This can be a labor intensive process.

Design of Primers

The choice of the primer sequences is critical for successful amplification of a specific DNA sequence. As in the case of gene probes, primer sequences can be deduced from known DNA sequences. The overall choice of primers is guided by the objectives of the researcher. If detection of a target DNA that is specific to a given species or genus of a bacterium is required, then only sequences unique to that bacterium are appropriate for the design of the primers. For example, the *lamB* gene encodes the production of an outer membrane protein in *E. coli*, and primers designed from this gene sequence specifically detect *E. coli* (Josephson *et al.*, 1991). However, some objectives may require primers from sequences conserved across the eubacterial kingdom. **Conserved sequences** are those that are similar and are found in many bacterial species. Some conserved sequences are **universal** and found in all known species of a genus. For example, primers designed from common *nod* (nitrogen fixation) gene sequences (*nod abc*) are conserved and can detect all species of *Rhizobium*. Other sequences are universal, such as some regions of 16S rDNA, and primers derived from these sequences anneal to all known bacterial species. Therefore, the experimental objective is an important consideration in the choice of the sequences from which the primers will be designed.

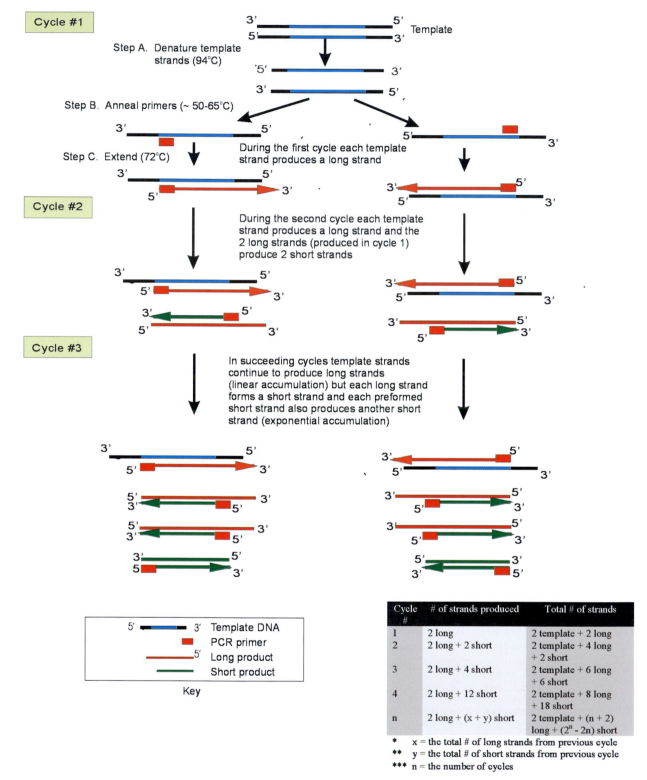

Cycle #	# of strands produced	Total # of strands
1	2 long	2 template + 2 long
2	2 long + 2 short	2 template + 4 long + 2 short
3	2 long + 4 short	2 template + 6 long + 6 short
4	2 long + 12 short	2 template + 8 long + 18 short
n	2 long + (x + y) short	2 template + (n + 2) long + (2^n - 2n) short

*   x = the total # of long strands from previous cycle
**   y = the total # of short strands from previous cycle
***   n = the number of cycles

FIGURE 13.8   (a) The PCR cycle. Step A, denaturation of the DNA template; step B, annealing of primers to the single-stranded template; Step C, extension of the primer to make a complementary copy of the DNA template. These three steps make up one cycle of PCR. (b) Steps involved in PCR amplification.

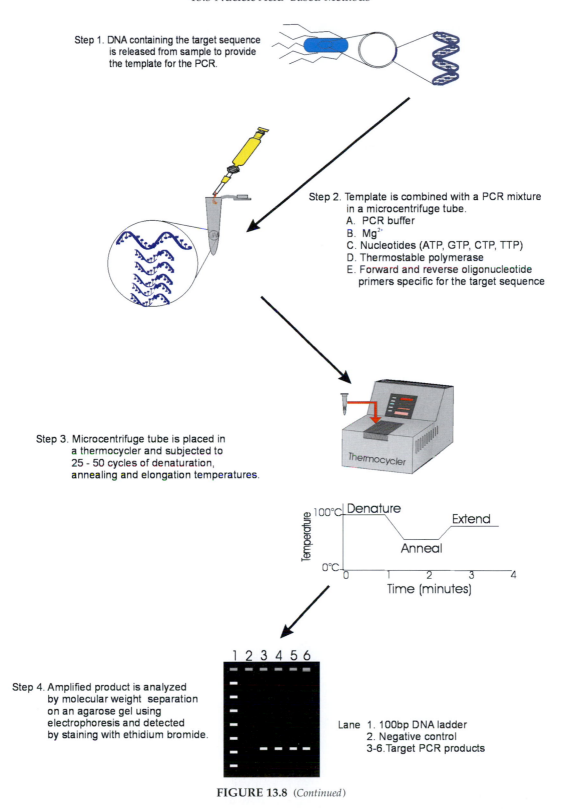

Step 1. DNA containing the target sequence is released from sample to provide the template for the PCR.

Step 2. Template is combined with a PCR mixture in a microcentrifuge tube.
A. PCR buffer
B. Mg^{2+}
C. Nucleotides (ATP, GTP, CTP, TTP)
D. Thermostable polymerase
E. Forward and reverse oligonucleotide primers specific for the target sequence

Step 3. Microcentrifuge tube is placed in a thermocycler and subjected to 25 - 50 cycles of denaturation, annealing and elongation temperatures.

Thermocycler

Step 4. Amplified product is analyzed by molecular weight separation on an agarose gel using electrophoresis and detected by staining with ethidium bromide.

Lane 1. 100bp DNA ladder
2. Negative control
3-6. Target PCR products

FIGURE 13.8 (*Continued*)

In addition, there are other criteria that determine the ultimate choice of primer sequences. In general, most primers are 17 to 30 bp in length and are separated by as few as a hundred base pairs or as many as several thousand base pairs. The distance between the primers determines the size of the amplification product. Because the location of the primers within the genome defines the size of the amplification product,

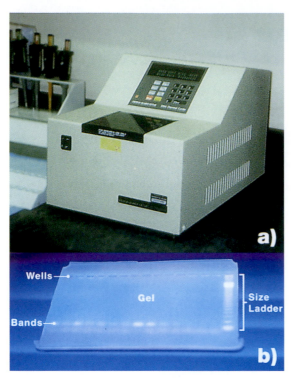

FIGURE 13.9 (a) An automated PCR thermocycler that is used to amplify target DNA. (b) A gel stained with ethidium bromide. (From *Pollution Science*, © 1996, Academic Press, San Diego, CA.)

this theoretical size can be compared with the actual size of the product obtained on an electrophoresis gel containing DNA standards. Further confirmation that the amplified product is the DNA sequence of interest can be obtained by the use of a gene probe (see Section 13.3.1). Southern blot hybridizations are used to confirm that the PCR product is the gene target of interest by using a probe internal to the PCR product sequence. The probe will hybridize to the complemen-

FIGURE 13.10 A commercial kit that contains aliquots of buffer, individual nucleotides (A, T, G, C), and the enzyme *Taq* polymerase. (Photo courtesy S. E. Dowd.)

tary region of the PCR product only if successful amplification of the correct sequence including internal regions has occurred. Probing can also increase the sensitivity of PCR detection. It is also critical that the primers contain sequences that are different from each other, so that they anneal at different sites within the chromosome. If primers contain complementary sequences, they can hybridize to each other, producing a **primer dimer**. Most primers should have a high G-C content (>50%), which results in a higher melting temperature of the primers (recall that G-C pairs have three hydrogen bonds and A-T pairs have only two).

13.3.2.2 PCR Detection of a Specific Gene

As explained earlier, appropriately designed primers can be used to detect a gene of interest such as the *nod* gene found in all species of *Rhizobium*. Another such example is the work by Marlowe *et al.* (1997), who used primers specific for the *lamB* gene to detect *E. coli* in marine waters. Tsai *et al.* (1993) used primers to amplify a *uidA* gene fragment (the *E. coli* gene that codes for β-D-glucuronidase) to detect *E. coli* in sewage and sludge. As discussed earlier, PCR can also be used to detect the universal 16S ribosomal gene sequence conserved in all eubacteria. External parts of this sequence are very highly conserved, but internal sequences are unique to a particular bacterium. Following amplification using primers designed from the universal sequences, the internal sequence of the product can be used in phylogenetic analyses that allow identification of a specific isolate. Analysis of 16S rDNA sequences has resulted in an explosion of phylogenetic studies, because analyses can be made of total DNA extract obtained from environmental samples without prior culturing of bacterial isolates. Instead, a researcher can extract the community DNA of the bacterial population within the environmental sample, amplify the 16S region from the bacterial genomes within the extracted DNA, shotgun clone this PCR product (see Section 13.3.3.1), and subsequently analyze the sequences of the clones (Amann *et al.*, 1995). Alternatively, DGGE can be used to separate the different PCR products prior to sequencing (see Section 13.3.5).

13.3.2.3 RT-PCR

Since the discovery of PCR, scientists in many fields of microbiology have become very creative in developing new types and applications of PCR. The field of environmental microbiology is no exception. In addition to allowing the detection of a specific gene sequence, other PCR methods have been used in environmental microbiology. One of these is a PCR method for the detection of RNA which involves the use of the enzyme reverse transcriptase and is known as **reverse transcriptase-PCR (RT-PCR).** In RT-PCR, the first step is to make a DNA copy of the RNA sequence of interest (Fig. 13.11). This copy is known as **complementary DNA** or **cDNA.** The key enzyme in the reaction is reverse transcriptase, which is an RNA-dependent DNA polymerase used to synthesize DNA from an RNA template. The first step in RT-PCR is performed using either the downstream antisense primer, which has the sequence complementary to the RNA, or random hexamers (small, ~6 bp random primers) to make a complete cDNA copy of the RNA molecule. Thus, the

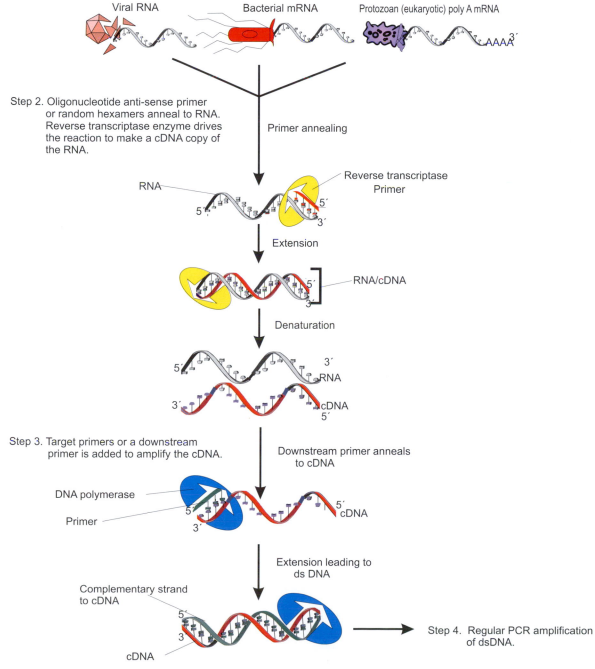

FIGURE 13.11 RT-PCR amplification of RNA. RNA is reverse transcribed to synthesize cDNA by random hexamers or a specific antisense primer. PCR can then be performed on the cDNA template.

cDNA molecule is the complement of the RNA strand. Normal PCR can now be conducted. During the first cycle, a strand of DNA complementary to the cDNA is synthesized. Following this, PCR proceeds as before from the double-stranded DNA template.

RT-PCR has been used in environmental microbiology to detect RNA viruses, mRNA transcripts, and rRNA sequences. For example, the detection of enteric viruses in water has classically been dependent on animal cell cultures. This methodology is time consuming, labor intensive and expensive. Also, there is no one cell line that will allow the detection of all enteric viruses. To overcome these problems, water microbiologists have employed RT-PCR to detect a wide range of enteric viruses. Abbaszadegan *et al.* (1993) used RT-PCR to amplify the conserved *VP1* gene to detect enteric viruses in groundwater. A second useful application of RT-PCR is in the analysis of mRNA, which allows estimation of metabolic activity. In a sense, normal PCR has the ability to detect the genetic potential within an organism, whereas RT-PCR can detect actual gene activity. For example, Selvaratnam *et al.* (1995) used RT-PCR to monitor gene expression of the transcripts of phenol degradative genes (*dmpN*) in a sequence batch reactor. It was found that phenol concentration and aeration influenced transcriptional activity. These findings could then be used to optimize reactor performance. Finally, the amplification of rRNA and subsequent cloning and comparative sequence analysis have proved particularly powerful in studies of the distribution of specific phylogenetic groups in soil, hot springs, and sediment (Amann *et al.*, 1995).

13.3.2.4 ICC PCR

A new procedure identified as **integrated cell culture–PCR** (ICC-PCR) was introduced by Reynolds *et al.* (1996). Here, normal cell culture flasks are inocu-lated with pure cultures of virus or aliquots of environmental samples and incubated for 2–3 days. Following incubation, the flasks are frozen, which lyses the cells, releasing virus particles. RT-PCR is then applied to the cell culture lysate. In essence, this is a biological amplification followed by an enzymatic amplification, and the procedure can identify virus in 3 days as opposed to the 10–15 days required for cell culture alone. Thus, this new technique greatly enhances the speed of detection of virus (Table 13.1). It has other important advantages as well. Because growth of the virus precedes PCR amplification, only infectious viruses are detected. In addition, the sensitivity of ICC-PCR is better than that of direct RT-PCR because larger sample volumes can be added to the cell culture. Dilution of sample within the cell culture also dilutes any PCR inhibitory substances that may be present in the environmental sample. Finally, because ICC-PCR–positive samples are confirmed by PCR, there is no need to run an additional cell culture assay to confirm positives, as in the case in cell culture positives. Thus, overall costs are reduced by approximately 50%.

13.3.2.5 Multiplex and Seminested PCR

Sometimes more than two primers are used in a PCR reaction. **Seminested PCR** uses three primers. The first two primers are used to detect a particular gene (Fig. 13.12A and B). Then a second round of PCR is done with replenished reagents, one original downstream primer and one new internal primer that is complementary to the internal sequence of the original PCR product. This second round of PCR not only increases the sensitivity of the reaction but also allows confirmation of the first amplification product by generating a second product that is smaller than the original product and based on internal sequences.

Multiplex PCR uses several sets of primers in one reaction to generate several diagnostic bands (Fig.

TABLE 13.1   Comparison of Methods for Detection of Virus

Issue	Method of detection		
	Cell culture	RT-PCR	ICC-PCR
Reduced time of detection	No	Yes	Yes
Infectious virus detected	Yes	Yes/No	Yes
Increased sensitivity	Yes	No	Yes
Affected by PCR inhibitory substances	No	Yes	No
Reduced costs	No	Yes	Yes
Detects only viable organisms	Yes	No	Yes
Detects viable but nonculturable virus	No	Yes	Yes

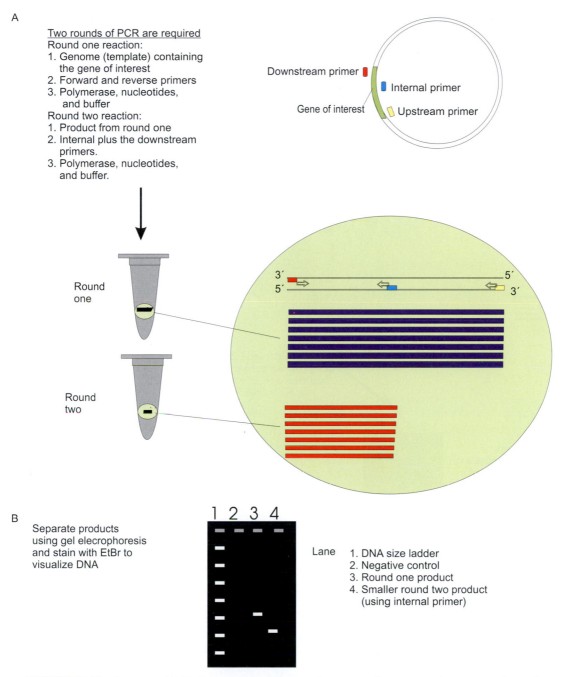

A

Two rounds of PCR are required
Round one reaction:
1. Genome (template) containing the gene of interest
2. Forward and reverse primers
3. Polymerase, nucleotides, and buffer
Round two reaction:
1. Product from round one
2. Internal plus the downstream primers.
3. Polymerase, nucleotides, and buffer.

Downstream primer
Internal primer
Gene of interest
Upstream primer

Round one

Round two

B

Separate products using gel elecrophoresis and stain with EtBr to visualize DNA

1 2 3 4

Lane   1. DNA size ladder
2. Negative control
3. Round one product
4. Smaller round two product (using internal primer)

FIGURE 13.12   Seminested PCR. Two external primers and one internal primer result in two products of different size.

13.13a and b). One objective of using more than one set of primers is to allow detection of multiple areas on the genome (Way *et al.*, 1993). For example, to confirm that an *Escherichia coli* detected with the *lamB* primers also carries toxin genes, a PCR reaction could be set up with primers for both the *lamB* gene and the toxin gene. Multiplex PCR can also be used to detect multiple genomes of multiple organisms in one sample if, for example, a water supply was to be evaluated for more than one pathogen (Tsai *et al.*, 1994). However, it is important to note that an established multiplex reaction must have a balanced ratio of primer sets in order to have valid and specific results. Essentially, these PCR methods allow a researcher to answer more than one question in a single experiment. Seminested PCR allows identification and confirmation of the PCR

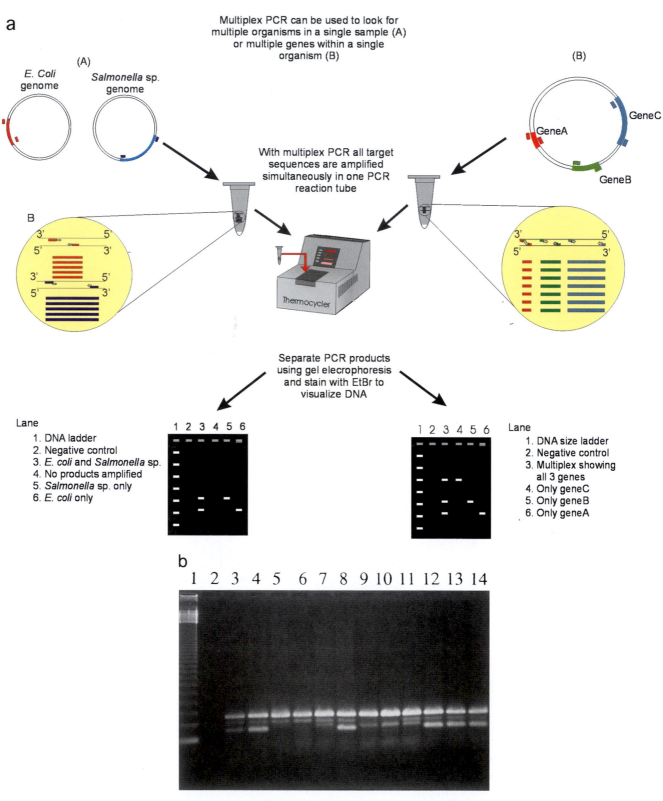

FIGURE 13.13 (a) Schematic representation of multiplex PCR in which multiple sets of primers allow simultaneous amplifications of more than one sequence. (b) A gel illustrating multiplex PCR amplifications. Here three primer sets were used to amplify three sequences associated with *Salmonella* spp., allowing detection of the pathogen. (Photo reprinted from Way et al., 1993.)

product in one assay, and multiplex PCR allows identification of more than one target sequence in a single reaction.

13.3.2.6 PCR Fingerprinting

Multiple fragments of DNA of variable size associated with a particular bacterial isolate can be amplified, separated by electrophoresis, and stained, resulting in a DNA fingerprint. The fingerprinting of a species can help identify and also distinguish between microbial isolates in the environment. Several PCR fingerprinting techniques have been applied to environmental samples and allow rapid analysis of whole genomes or mixtures of genomes. These techniques produce patterns of multiple anonymous DNA fragments of unpredicted length (Burr *et al.*, 1997). In essence, PCR fingerprinting relies on primers that anneal to sequences that are repeated at multiple locations throughout the chromosome. One PCR fingerprinting method is **arbitrarily primed PCR (AP-PCR)**,

also referred to as **random amplified polymorphic DNA (RAPD)** (Welsh and McClelland, 1990, 1993). AP-PCR uses one random primer of length 10-20 bp, which is annealed at a low temperature, resulting in nonspecific amplification of a genome (Fig. 13.14). Thus, this reaction requires no prior sequence information and will generate a fingerprint based on the uniqueness of the genome of a species. Depending on the random primer used, fingerprint patterns can be simple (1–2 bands) or complex (>10 bands). Other fingerprinting methods include **repetitive extragenic palindromic sequence PCR (REP-PCR)** which targets REP sequences, **ERIC (enterobacterial repetitive intergenic consensus)** sequences (de Bruijn, 1992), or **BOX sequences** (Burr *et al.*, 1997). These sequence elements are present throughout prokaryotic genomes as short (<160 bp), inverted repeats in noncoding regions, possibly associated with the organization of the nucleoide (Moat and Foster, 1995). ERIC PCR differs from AP-PCR in that two primers are required. Although REP, ERIC, and BOX sequences are unrelated to each other

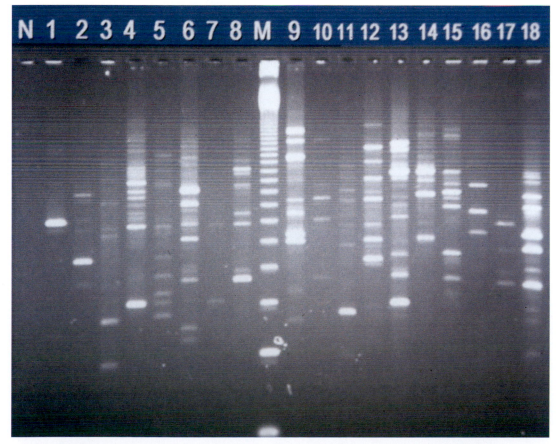

FIGURE 13.14 Illustration of AP-PCR. Here a single arbitrary primer (682) generates unique DNA fingerprints from 18 different bacterial genomes (lanes 1–18). Lane N is a negative control. Lane M is a 123- bp DNA size ladder in which the size of each successive band increases by 123 bp. (Photo courtesy I. L. Pepper.)

because of their high copy number and variable spacing in the genome, they have a mutual use as targets for PCR fingerprinting. An example of an ERIC fingerprint is shown in Fig. 13.15. PCR fingerprinting is simple and rapid compared with restriction fragment length polymorphism (RFLP) fingerprinting, which is discussed later in this chapter (Section 13.3.4).

13.3.2.7 Advantages and Disadvantages of PCR

PCR has many advantages over traditional cultural techniques. It is more sensitive and detects all organisms regardless of their physiological state. Once a PCR assay has been optimized in a laboratory, it is relatively fast and easy to get results within 24 hours, compared to the several days or weeks that may be necessary for cultural methods. PCR also has some disadvantages. In some cases it is not advantageous to detect both viable and non-viable organisms (Josephson *et al.*, 1993). For example, while PCR will detect pathogenic viruses in as little as a one day, PCR will detect both infectious and noninfectious virus leaving open the question of how harmful a sample may be. In contrast, cell culture of the same virus may take several weeks but will give information as to the infectivity of the sample. Additionally, PCR is subject to inhibition due to humic substances and metals that may be present in environmental samples (Abbaszadegan *et al.*, 1993). Contamination can also be a problem when using PCR resulting in false positive results. Even given its disadvantages, PCR has allowed environ-

mental microbiologists to examine questions that were formerly impossible to address. In particular, PCR used in conjunction with traditional techniques is a powerful tool that is allowing the next generation of microbiologists to address important questions related to environmental microbiology.

13.3.3 Recombinant DNA Techniques

13.3.3.1 Cloning

Recombinant DNA technology or DNA cloning has been widely used in environmental microbiology to examine the genetic structure of bacterial populations. Cloning has enabled scientists to find new or closely related genes, as well as characterize and identify unculturable or unknown isolates. Cloning may also be used to examine the activity of specific genes.

A **clone** is a foreign DNA fragment that is replicated in a host organism after being shuttled in by a **cloning vector** such as a plasmid. Cloning basically involves three steps: (1) choosing the source of DNA for cloning, (2) producing a collection of DNA fragments that can be inserted into a vector and in many cases ultimately creating a cDNA library, and (3) screening for the desired sequence of interest. Thus, cloning results in a population of organisms that contain recombinant DNA molecules. Following screening for the target sequence of interest, a particular clone can be propagated to amplify the recombinant molecules, resulting in a large mass of the DNA sequence of interest.

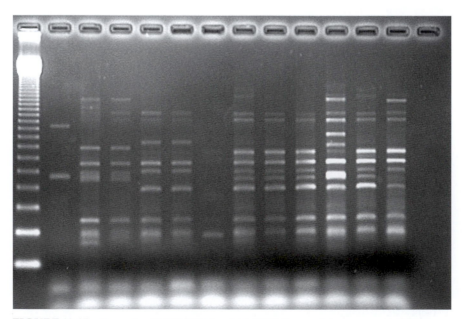

FIGURE 13.15   Example of ERIC PCR, in which DNA fingerprints are generated from several different bacterial isolates. Lane 1 is a 123 bp ladder. (Photo courtesy I. L. Pepper.)

A cloning vector or vehicle is a self-replicating DNA molecule, such as a plasmid or phage, that transfers a DNA fragment between host cells (Fig. 13.16). A useful cloning vector has three properties: (1) it must be able to replicate; (2) it must be able to introduce vector DNA into a cell; and (3) there must be a means of detecting its presence. For example, some cloning vectors screen for recombinant isolates via the *lacZ* gene. This gene codes for β-galactosidase, the en-zyme that catalyzes the hydrolysis of lactose into glucose and galactose. β-galactosidase also cleaves the compound 5-bromo-4-chloro-3-indolyl-β-d-galac-toside (X-Gal), which results in blue colonies. The cloning vector in this screening system carries the first half of the *lacZ* gene, and the host cell carries the carboxyl terminal regions of the β-galactosidase gene, which together allow coding that, when expressed, results in an active protein. This phenomenon is called

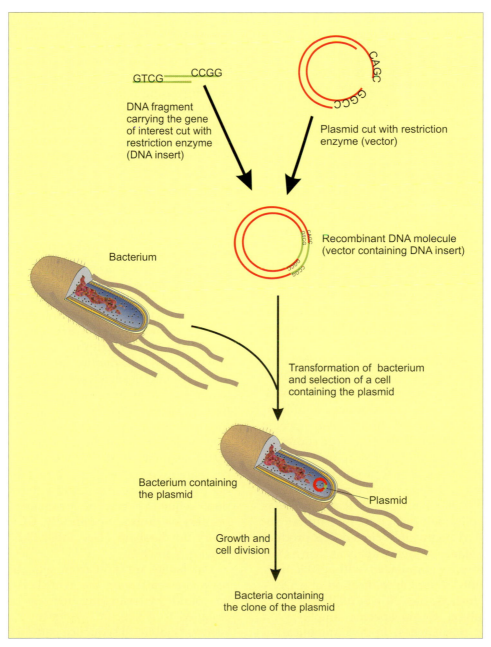

FIGURE 13.16 A foreign piece of DNA is inserted into a plasmid cloning vector. The vector can then transform a bacterial host, which will grow and divide. Subsequently, all progeny bacteria will carry the foreign DNA by plasmid replication.

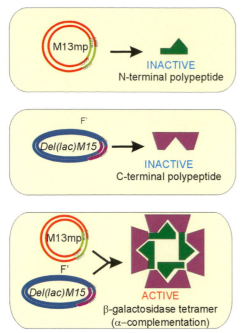

FIGURE 13.17 Complementation of the *lacZ* gene. (Adapted with permission from Maloy, Cronan, and Freifelder, 1994.)

α-complementation (Fig. 13.17). Within the vector coding region is the **multiple cloning site (MCS),** which is a short sequence that has been constructed with cleavage sites for several restriction enzymes (Fig. 13.18). Insertion of the foreign DNA into the MCS disrupts α-complementation. Thus, clones from the cDNA library can be selectively screened because white clones contain the insert but blue colonies do not.

PCR has had a dramatic impact on the steps involved in the sometimes labor-intensive cloning of DNA fragments. As discussed earlier, PCR is an enzymatic reaction that amplifies a specific fragment a million-fold in hours, expediting the hunt for a specific piece of DNA. Thus, PCR produces a large source of the DNA fragment for cloning and simplifies the first two steps of cloning. A single PCR product can then be directly inserted into a cloning vector and replicated in a host, typically *E. coli*. Cloning of PCR products has become a very applicable research tool for further sequencing of the DNA fragment. A cloned PCR product results in a much cleaner sequence analysis than a PCR product that is directly sequenced without cloning. This is because with *in vitro* amplification of a DNA fragment, the polymerase has a lower fidelity of replication and is subject to nonspecific amplification caused by primer mismatches. *In vivo* amplification via cloning is much more precise, because the fidelity of the polymerase is higher and there is a lack of random mismatched products that can interfere with the sequencing. Also, sequence analysis becomes much

simpler, because the cloned fragment is replicated in a known vector with known sites of insertion. Commercially available kits for PCR cloning are now widely available.

13.3.3.2 Sequence Analysis

Automated DNA sequencing has made the task of sequencing very routine and economical. Most laboratories send DNA samples to a commercial sequencing laboratory to be analyzed rather than perform the sequencing process themselves. Computerized sequence databases have been compiled to catalog this information for the entire scientific community (see Table 13.2). Cataloged sequences carry an accession number for identification within the database, the citation of the published research, as well as the DNA, RNA, and protein coding regions for a given sequence. Several software programs allow researchers to utilize these databases to identify sequences, translate a DNA sequence into its potential protein coding regions, and look for homology or relatedness based on sequences. The most common uses of databases in the field of environmental microbiology are in the identification of isolates by their 16S rDNA sequence, and the retrieval and evaluation of gene sequences for the design of probes or PCR primers.

Several computer software sequence analysis programs are available to aid in sequences searches. Software tools include **Genetics Computer Group (GCG)** and several Internet sites such as FASTA and BLAST that are provided by the **National Center for Biotechnology Information (NCBI).** These programs allow researchers to perform sequence comparisons whereby one can identify homologous or unique regions in nucleic acid sequences of interest. These search programs are used when researchers want to find out what sequences in the databases are similar to their unknown sequences and identify distal relationships to look at common ancestry. For example, a 16S rDNA sequence from an unknown soil isolate can be compared with other sequences in the database in order to characterize and identify the organism. In addition, sequences can be evaluated to design PCR primers or probes. Specifically, GCG can be used to query the databanks (see Table 13.2) to examine a sequence of interest.

Databanks are built from sequences submitted by researchers all over the world. Sequence data can be found by searching for an accession number assigned to a specific sequence, a character pattern such as "globin", a sequence pattern such as "GGTAC-CTTGAG", a file name such as "lambdga", or a pattern of another known sequence. A number of algorithms are available for a wide scope of nucleic acid and protein sequence comparisons. FASTA (Pearson,

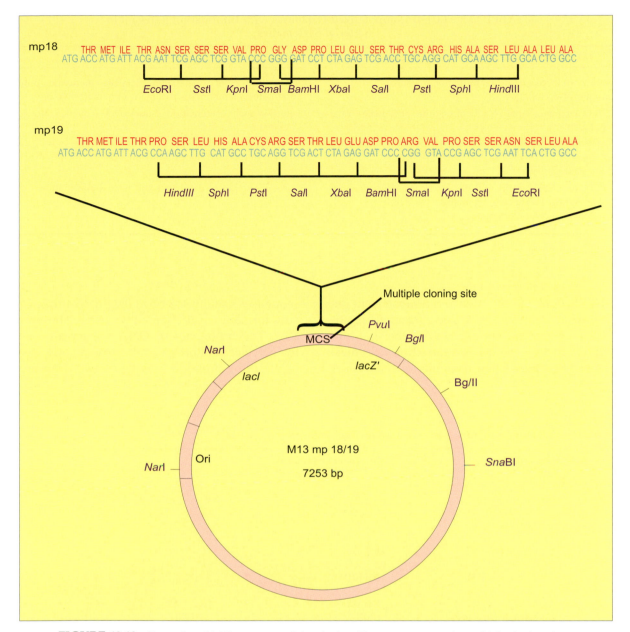

FIGURE 13.18 Examples of M13 vectors used for cloning. These vectors contain multiple cloning sites. Only restriction enzymes that cut the vectors at a single site are shown on the restriction maps. The two vectors mp18 and mp19 are identical except that their multiple cloning sites are in opposite orientations to facilitate cloning DNA fragments in both directions. The multiple cloning sites of mp18 and mp19 are shown at the top of the figure. The amino acids in the peptide are shown with the amino acids encoded by the polylinker. The following restriction enzymes do not cut mp18 or mp19: *Aat* II, *Apa* I, *Asu* II, *Bcl* I, *BssH* II, *BstE* II, *BstX* I, *Eco* RV, *Hpa* I, *Mlu* I, *Mst* I, *Nco* I, *Nhe* I, *Not* I, *Nru* I, *Nsi* I, *Sst* I, *Sca* I, *Sfi* I, *Sin* I, *Spe* I, *Stu* I, *Xho* I, *Xma* I. (Adapted with permission from Maloy, Cronan, and Freifelder, 1994.)

1990) and BLAST (Altuschul *et al.*, 1990) are two such rapid searching programs that facilitate effective comparisons of unidentified sequences with those contained in the current databanks. Alternatively, PILEUP, part of the GCG program, is a program that creates multiple sequence alignments using progressive pairwise alignments. GCG can also plot a phylogenetic tree.

13.3.4 Restriction Fragment Length Polymorphism Analysis

13.3.4.1 Theory and Concept

Restriction fragment length polymorphism (RFLP) analysis is frequently used on bacterial isolates (Tenover and Unger, 1993). The method consists of first extracting total DNA from a pure culture of a bacterial

TABLE 13.2   List of Computerized Databases with Information Relevant to Bacterial Systematics

DATABASE producer and e-mail address	Scope and comments
DBIR, Directory of Biotechnology Information Resources E-mail: request@atcc.nih.gov	Online directory of organizations, databases, networks, publications, and nomenclature resources relevant to biotechnology
DDBJ, DNA Databank of Japan, National Institute of Genetics, Yata, Mishima, Japan E-mail: tgojobor@ddbj.nih.ac.ip	Collection and dissemination of nucleotide sequence data generated in Japan
GenBank E-mail: info@nibi.nim.gov	Database of reported nucleotide sequences.
EMBL, EMBL-New, EMBL-Daily UEMBL: European Molecular Biology Laboratory, Heidelberg, Germany E-mail: datalib@embl-heidelberg.de	Nucleotide sequence databanks and related information

Data from Canhos *et al.* (1993).

isolate. This is done via cell lysis and a subsequent genomic preparation of the DNA using standard methods (Wilson, 1994). The genomic DNA is then cut into smaller fragments by the use of **restriction enzymes**. These are endonucleases that recognize specific DNA sequences usually 4–6 bp in length. Examples of target sites for some restriction enzymes are shown in Table 13.3. The fragments of DNA are usually separated by gel electrophoresis and subsequently visualized by ethidium bromide staining and UV illumination. If

TABLE 13.3   Target Sites for Some Restriction Endonucleases[a]

Organism	Restriction endonuclease	Site
Anabaena variabilis	*Ava* I	$C \downarrow {C \choose T} CG {A \choose G} G$
Bacillus amyloliquefaciens H	*Bam* HI	$G \downarrow GATCC$
Bacillus globigii	*Bgl* II	$A \downarrow GATCT$
Escherichia coli RY13	*Eco* RI	$G \downarrow A\overset{*}{A}TTC$
Escherichia coli R245	*Eco* RII	$\downarrow CC {A \choose T} GG$
Haemophilus aegyptius	*Hae* III	$GG \downarrow \overset{*}{C}C$
Haemophilus gallinarum	*Hga* I	$GACGC$
Haemophilus haemolyticus	*Hha* I	$G\overset{*}{C}G \downarrow C$
Haemophilus influenzae Rd	*Hind* II	$GT {C \choose T} \downarrow {A \choose G}\overset{*}{A}C$
Haemophilus parainfluenzae	*Hind* III	$\overset{*}{A} \downarrow AGCTT$
	Hpa I	$GTT \downarrow AAC$
	Hpa II	$C \downarrow \overset{*}{C}GG$
Klebsiella pneumoniae	*Kpn* II	$GGTAC \downarrow C$
Moraxella bovis	*Mbo* I	$\downarrow GATC$
Providencia stuartii	*Pst* I	$CTGCA \downarrow G$
Serratia marcescens	*Sma* I	$CCC \downarrow GGG$
Streptomyces stanford	*Sst* I	$GAGCT \downarrow C$
Xanthomonas malvacearum	*Xma* I	$C \downarrow CCGGG$

Data from Old and Primrose (1989).

[a] Recognition sequences are written from 5′ to 3′, only one strand being given, and the point of cleavage is indicated by an arrow. Bases written in parentheses signify that either base may occupy that position. Where known, bases modified by a specific methylase, are indicated by an asterisk. $\overset{*}{A}$ is N^6-methyladenine, and $\overset{*}{C}$ is 5-methylcytosine.

the restriction enzymes cut at many sites within the genomic DNA, several or even hundreds of DNA fragments can result, and the pattern of these fragments following electrophoresis and probing produces a fingerprint characteristic of the original bacterial isolate.

13.3.4.2 RFLP Analysis of Whole Genomes

Following cell lysis, bacterial DNA consisting of a single chromosome can be isolated and purified as described previously. Restriction enzymes can then be used to cut at specific recognition sites. Because a single base change resulting from an insertion, deletion, or substitution can create or eliminate a site, the locations of restriction sites can vary even among closely related bacteria. Distantly related bacteria would be expected to have many differently located restriction sites. These subtle changes are known as **polymorphisms.** DNA polymorphisms result in different recognition sites, and thus different sizes of the restriction fragments from different isolates. The recognition sequences of known restriction enzymes have been published, but enzyme selection for RFLP analysis is usually empirical because sequence information on target isolates is often unavailable. Restriction digestion of a few micrograms of DNA is usually sufficient, following the manufacturer's recommendations or a standard protocol (Block, 1995). After the enzyme digestions, polymorphisms are detected by ethidium bromide staining following gel electrophoresis.

A restriction digest of genomic bacterial DNA contains potentially hundreds of fragments of different sizes. Sometimes the observed pattern is photographed and used as the actual fingerprint. However, it is often difficult to interpret the large number of restriction fragments, and in some cases there may be so many fragments that a smear is all that can be observed (Hansen *et al.*, 1993). To simplify the pattern, the DNA fragments can be transferred to a membrane and only fragments that contain a specific sequence detected using a gene probe. Many gram-negative enteric bacteria contain repetitive DNA sequences that are located at different points within the bacterial chromosome. Because these sequences are evolutionarily conserved, gene probes utilizing portions of these sequences can be designed. If a genomic DNA preparation of a particular isolate is cut with specific restriction enzymes, some DNA fragments will contain the conserved repetitive DNA sequence. The use of gene probes can identify these fragments, and two different strains of the same species will produce different patterns of fingerprints. Insertion sequences have also been used as probes.

Insertion sequences are genetic elements that can insert copies of themselves into different sites in a bacterial genome. For example, the 708-bp insertion sequence IS200 has been used as a probe for *Salmonella* isolates (Threlfall *et al.*, 1994). Copy numbers of IS200 can range from zero to 18, depending on serotype. Therefore, up to 18 restriction fragments can potentially be identified, resulting in a relatively easily interpreted fingerprint.

13.3.4.3 RFLP Analysis of PCR Sequences

RFLP analysis can also be used to identify or confirm specific gene sequences associated with a specific bacterium rather than using the whole bacterial chromosome. A new technique relies on PCR amplification of a specific sequence, such as a 16S rDNA sequence, followed by a restriction digest. All bacteria contain 16S rDNA sequences that code for ribosomal RNA; however, portions of these 16S rDNA sequences are highly conserved, whereas other portions of the sequence are unique to each organism. Primers designed from the conserved sequences are known as "universal" primers and theoretically will amplify 16S rDNA sequences from all known bacteria.

The amplified DNA between the two external universal primers contains unique sequences that can be cut by restriction enzymes and result in a fingerprint of the bacterial isolate. Navarro *et al.* (1992) used PCR-RFLP analysis to characterize natural populations of *Nitrobacter* spp. This approach is also known as **PCR ribotyping.** Ribotyping, then, is yet another way to identify a specific bacterial isolate. Use of a specific enzyme normally results in two to four fragments, but in some instances ribotyping will not differentiate isolates because of insufficient discrimination. In other words, a single restriction enzyme might cut at similar sites for two different isolates. Normally, the amplified DNA is cut with at least two enzymes to ensure discrimination between isolates.

13.3.4.4 Pulsed Field Gel Electrophoresis

Pulsed field gel electrophoresis (PFGE) was first described by Schwartz and Cantor in 1984. More recently, protocols for PFGE were provided by Hood (1995). PFGE is another way to detect polymorphisms using restriction enzymes. However, in contrast to normal RFLP, there is no transfer of DNA to a membrane, and there is no detection of specific fragments using a gene probe. The restriction digest is separated by gel electrophoresis of alternately pulsed, perpendicularly oriented electrical fields and stained, resulting in the fingerprint. PFGE is used to detect fragments of higher molecular weight than those in

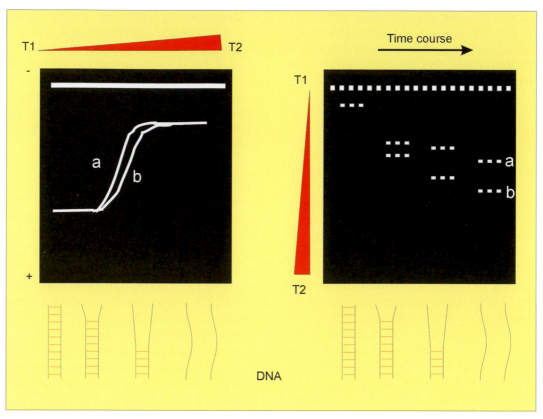

FIGURE 13.19 Schematic description of TGGE analysis of a 16S rDNA fragment amplified from a mixture of two strains (e.g., a, *Pseudomonas fluorescens*; b, *Pantoea agglomerans*) in a perpendicular gel (left) and in a parallel gel (right). (Adapted from Heuer and Smalla, 1997, p. 355, by courtesy of Marcel Dekker, Inc.)

normal RFLP, often greater than 40 kb (Burr *et al.*, 1997). These larger fragments are often generated by the use of rare cutting enzymes (e.g., *Not* I). Overall cell lysis and restriction digestion occur in plugs made from bacterial broth cultures and molten agarose. The plugs are melted into wells in the gel prior to electrophoresis. The process reduces shearing of the DNA.

13.3.4.5 Advantages and Disadvantages of RFLP Analysis

RFLP analysis is most often used to identify specific bacterial isolates. The greatest single advantage of RFLP coupled with gene probe analysis is that the banding patterns produced are unambiguous. This is because relatively large amounts of DNA are being cut, and the resultant fragments are easily visualized following electrophoresis and staining. In contrast, PCR-generated fingerprints such as those from AP-PCR or ERIC PCR can be difficult to reproduce and may contain "faint" or "ghost" bands, making interpretation difficult (Burr and Pepper, 1997). However,

the choice of restriction enzymes used is usually empirical, and normally multiple enzymes must be used. For example, if a single enzyme used on two chromosomal preparations and subsequent gene probe analysis results in two different fingerprints, then assuredly the isolates are different. If, however, the fingerprints are the same, it does not necessarily mean that the isolates are the same, because one enzyme may cut two preparations at the same locations. This situation is exacerbated when the gene probe used detects only one or two DNA fragments. Finally, note that RFLP analysis is more tedious than PCR fingerprinting, particularly when coupled with gene probe analysis.

13.3.5 Denaturing/Temperature Gradient Gel Electrophoresis

13.3.5.1 Theory and Concept

A new approach to the study of the diversity of microbial communities is the analysis of PCR-amplified DNA fragments by **denaturing gradient gel elec-**

trophoresis (DGGE) or **temperature gradient gel electrophoresis (TGGE).**

This technique was originally developed in medical research to detect point mutations in genetic linkage studies but was introduced in microbial ecology by Muyzer *et al.* (1993). In essence, DGGE or TGGE can be used for analysis of PCR-amplified 16S rDNA sequences obtained from community DNA extractions (see Section 13.3.2.2) or mixtures of different bacterial isolates. The use of universal primers derived from conserved 16S rDNA sequences results in amplified DNA fragments of nearly identical lengths, but with variable sequence composition.

Separation is based on changes in the electrophoretic mobility of different DNA fragments migrating in a gel containing a linearly increasing gradient of DNA denaturants (urea/formaldehyde or temperature). Changes in fragment mobility result from partial melting of the double-stranded DNA in discrete regions; in other words, the fragments remain double stranded until they reach the conditions that cause melting of domains less resistant to denaturants or temperature. In the case of DGGE, this occurs because the temperature of the gel is held constant, so that the melting of domains each varies according to the concentration of the denaturant and therefore position in the gel. When the DNA enters a region of the gel containing sufficient denaturant, a transition from helical to partially melted molecules occurs, with a resultant branching of the molecule that sharply decreases the mobility of the DNA fragment. Sequence variation within particular domains alters their melting behavior; thus different PCR amplification products stop migrating at different positions in the denaturing gradient. The use of a GC-rich sequence or **GC clamp,** which consists of 40–45 bases of a GC-rich sequence, acts as a high-temperature-melting domain and prevents complete melting of the PCR product. The GC clamp is normally attached to the 5' end of the forward primer (Sheffield *et al.,* 1992). Using these techniques, fragments with only a few base pair variants can be separated efficiently in a linearly increasing denaturing gradient of urea and formaldehyde at 60°C. In contrast, a linearly increasing temperature gradient in the presence of a high constant concentration of urea and formaldehyde is used for separation of PCR products in TGGE.

The use of DGGE or TGGE is illustrated in Figure 13.19, which shows a perpendicular gel (left) and a parallel time course gel (right). The perpendicular gel has an increasing gradient of the denaturants (either temperature or chemical) perpendicular to the electrophoresis direction. The DNA is loaded in a slot across the entire width of the gel. At the left side of the gel (low denaturant concentration) the PCR products have high mobility because the DNA remains double stranded. At the right side of the gel (high denaturant concentration), migration is almost zero. At intermediate concentrations of denaturant, the DNAs have different degrees of melting and variable mobilities. In parallel gels, the denaturing gradient increases from the top to the bottom of the gel, i.e., parallel to the electrophoresis direction. Generally, in this case, a time course experiment is run in which the same sample is loaded repeatedly to different lanes of a parallel gradient gel, with a defined time interval between sample loadings. This allows estimation of the optimum conditions for separating different fragments.

13.3.5.2 Advantages and Disadvantages of DGGE/TGGE

Once separated, specific fragments can be excised from the gel and subjected to further analysis such as sequencing. This strategy applied to community DNA extractions allows estimates to be made of the dominant 16S rDNAs present without prior culturing of isolates. Problems with the technique include the errors associated with community DNA extractions (described in Section 13.2.1). Finally, note that sequences between rRNA operons of an individual organism can vary significantly, so individual organisms could potentially produce multiple bands on a gradient gel. Finally, even though this approach provides information about what organisms are present, it does not provide information about their activity.

13.3.6 Plasmid Analysis

13.3.6.1 Theory and Concept

Many bacterial isolates contain one or more plasmids of variable size (see Chapter 2.2.5). In some cases, detection of these plasmids allows identification of specific bacterial isolates. The size and number of plasmids associated with a given bacterium are often unique. Note, however, that plasmids can be transferred to other species or in some cases lost from a bacterium if selective pressure is not present to maintain the plasmid. **Curing** is the process whereby a plasmid is removed from a bacterium, and it can often be achieved by the application of a stress such as heat shock. For example, pJP4 is an 80-kb plasmid that encodes some of the enzymes used to degrade 2,4-dichlorophenoxyacetic acid as well as sequences responsible for mercury resistance. This plasmid has been shown to undergo horizontal gene transfer from *Alcaligenes eutrophus* to other soil organisms (see Chapter 3, Case Study) and is not stable within the *Alcali-*

genes isolate if grown in laboratory culture without selective pressure (Di Giovanni *et al.*, 1996). Other plasmids, particularly large megaplasmids, cannot be cured easily.

Despite the potential for plasmid transfer or loss, plasmids have been used to detect specific bacteria or genes. Specific gene sequences associated with a plasmid or homologous gene sequences among plasmids can be detected using specific gene probes. Bacteria that are associated with specific plasmids can be identified by plasmid profiles or plasmid fingerprints. **Plasmid profiles** are prepared by electrophoresis and staining of whole plasmids (Fig. 13.5), whereas **plasmid fingerprints** are produced by restriction digestion of plasmid DNA, followed by electrophoresis and staining of restriction fragments. For gene probe analysis, bacterial isolates can be lysed and probed via colony lifts using labeled plasmid DNA sequences (Fig. 13.4). Alternatively, plasmid extracts free of chromosomal DNA can be prepared, electrophoresed, and probed, in essence becoming a Southern hybridization (Fig. 13.5).

Crude plasmid extracts can be prepared via an alkaline lysis (Birnboim and Doly, 1979) and used for Southern or plasmid profiles. For megaplasmids larger than 20 kb, bacteria can be lysed directly in the gel wells (Eckhardt, 1978). This *in situ* lysis reduces the shearing of megaplasmids that can easily occur if other methods are employed.

The large variety of plasmid functions is reflected in the variety of assays in which plasmid analyses have been utilized. For example, plasmid analyses have been used to identify *Salmonella* spp. in epidemiological studies (Burr *et al.*, 1997). They have also been used to identify *Rhizobium* spp. involved in symbiotic relations with plants (Shishido and Pepper, 1990), in biodegradation studies, and in studies of antibiotic and metal resistance.

13.3.6.2 Advantages and Disadvantages of Plasmid Analyses

Regardless of whether plasmid profiles or plasmid fingerprints are generated, this analysis provides a relatively quick method for identification of specific DNA sequences associated with a bacterium. Often these DNA sequences can be correlated with specific phenotypic characteristics such as resistance to antibiotics or metals or the ability to degrade specific organic contaminants. In addition, plasmids have been used to identify specific bacterial isolates. However, in this case, the greatest disadvantage of plasmid analysis is the potential loss of the plasmid from the bacterium of interest. This can occur through a plasmid

curing process, which could result in false-negative results, or the transfer of the plasmid to other isolates through horizontal gene transfer, which could result in false-positive results.

13.3.7 Reporter Genes

13.3.7.1 Theory and Concept

Nucleic acid sequences can be utilized not only to detect specific organisms but also to demonstrate specific activity. One recently developed method involves the use of **reporter genes,** which signal the activity of an associated gene of interest. A reporter gene is a gene (or set of genes) that is inserted into a target gene of interest and subsequently signals or reports the activity of that gene. Specifically, a reporter gene system is composed of a gene lacking its endogenous promoter, which will be transcribed only if placed downstream from an exogenous promoter (Loper and Lindow, 1997).

The theory behind a reporter gene is that most genes encode products that are not easily detected, whereas the inserted reporter gene produces a product that is easy to detect and quantify. Because it is linked to the promoter of the gene of interest, it is in essence turned on only when the target gene of interest is turned on. Reporter genes are typically inserted into an operon of interest, creating a **gene fusion.** These gene fusions are often created using transposons to insert reporter genes randomly in the bacterial genome. The process produces mutants with many different gene fusions. This collection of mutants must be screened; first of all, to ensure that the mutant still has the activity of interest, and second, to ensure that the activity of interest occurs concomitantly with the expression of the reporter gene. Table 13.4 shows some of the important characteristics of reporter genes. Reporter genes are most useful in assessing *in situ* gene expression by bacteria in natural habitats such as soil, water, or plant habitats (Lindow, 1995).

13.3.7.2 Specific Reporter Gene Systems

The **lacZ** gene codes for the enzyme β-galactosidase, which cleaves the disaccharide lactose into glucose and galactose. These products can be quantified colorimetrically by the development of a blue color in colonies grown on agar supplemented with X-Gal and quantified by colorimetric assay. In addition, activity can also be quantified in assays generating fluorescent and chemiluminescent products (Loper and Lindow, 1997). The activity of *lacZ* fusions is detectable with cells exhibiting 1 molecule of β-galactosidase per cell, and samples must contain at least 10^5-10^6 cells to

TABLE 13.4   Characteristics of Reporter Genes Useful in Environmental Microbiology

Characteristics	Issues
Detectability	Reporter gene product must be uncommon in the environment.
	Must be easy to detect, e.g., soil particles can obscure pigmented products.
Sensitivity	Single cell detection allows evaluation of temporal and spatial heterogeneity of gene expression in natural environments.
Ability to quantify	Ideally, should be able to quantify gene product and relate to population size and population activity
Stability of reporter gene product	If stable, the gene product represents cumulative transcriptional activity rather than current level of activity.
	If labile, sensitivity of assay decreases.

Data from Loper and Lindow (1997).

quantify the activity of *lacZ*. Because it is easily detected and because there are many vectors available for making fusions, *lacZ* is probably the most common reporter gene used in studies of gene regulation in bacterial cultures. *LacZ* fusions have been used by researchers in a variety of ways, including investigations of the expression of biosynthetic genes and gene expression in the rhizosphere (Arsène *et al.*, 1994). Unfortunately, *LacZ* has limited use for assessing *in situ* gene expression because of significant background levels of β-galactosidase activity in many indigenous bacteria. Pigmented and fluorescent compounds in plant tissue and soil also interfere with the colorimetric and fluorometric assay of β-galactosidase activity.

The β-glucuronidase gene (**gus A**) from *E. coli* has become a very useful marker, partly because of the range of fluorescent substrates available for the enzyme β-glucuronidase and its rare occurrence in environmental isolates. This gene is also known as *uidA*, and the enzyme catalyzes the hydrolysis of a wide range of glucuronides (Jefferson, 1989). Many substrates are available that produce fluorescence upon cleavage, such as 5-bromo-4-chloro-3-indolyl-β-d-glucuronide (X-Gluc) (Bronstein *et al.*, 1994). Fusions of *gusA* have been used extensively in studies of gene expression by viral plant pathogens, plants, and fungi, all of which lack indigenous β-glucuronidase activity. These conditions allow sensitive gus activity to be measured in as little as a single cell (Jefferson, 1989). However, its use with bacterial pathogens has been limited, because some bacteria do contain β-glucuronidase (Loper and Lindow, 1997).

The *xylE* gene from the TOL plasmid of *Pseudomonas putida* encodes the enzyme catechol 2,3-dioxygenase, which converts catechol to 2-hydroxymuconic semialdehyde. This reaction produces a yellow pigment that can be measured spectrophotometrically (Loper and Lindow, 1997). This reporter gene has been used successfully in a variety of gram-negative

bacteria. The gene is rarely found in microorganisms that have not previously been exposed to aromatic hydrocarbon contaminants, which makes it useful for *in situ* assays of gene expression. The system has been used successfully to study *in situ* expression by *P. putida* inhabiting the rhizosphere (Buell and Anderson, 1993). However, the stability of the reporter product is influenced by oxygen and the physiology of the cell. Thus, such factors must be taken into account when quantifying of *xylE* activity.

Luminescent Reporter Genes

Bacterial enzymes known as **luciferases** cause light emission in the presence of the substrate **luciferin** (Fig. 13.20). Luciferase and its aldehyde substrate are coded by *lux* (*lux* CDABE) genes and are active in the presence of oxygen and a source of reducing power such as flavin mononucleotide ($FMNH_2$). The *luxAB* genes encode the active form of luciferase, and the *lux*CDE genes encode the synthesis of the aldehyde. The basis of the *lux* system is that *lux* genes are inserted into the operon being studied and when that operon is induced, luminescence is given off. The *lux* system is widely used because it is uncommon to find large numbers of indigenous microorganisms that emit light. Most *lux* systems require $\sim 10^5$ cells per sample for the detection of light emission, and detection can be done directly in a nondestructive manner. The biochemistry and genetics of bacterial bioluminescence have been reviewed by Hill *et al.* (1993). Most commonly the light emitted is quantified by scintillation counters used in the single-photon counting mode. Luminescent reporter genes are extremely useful for evaluating gene expression and have been used to evaluate bioavailability of xenobiotics such as naphthalene (Heitzen *et al.*, 1992). In a further application of the lux system, a fiber optic detection system is being used to study the *in situ* response of microbes to addition of salicylate in a soil column system. In this system, an optic fiber is emplaced

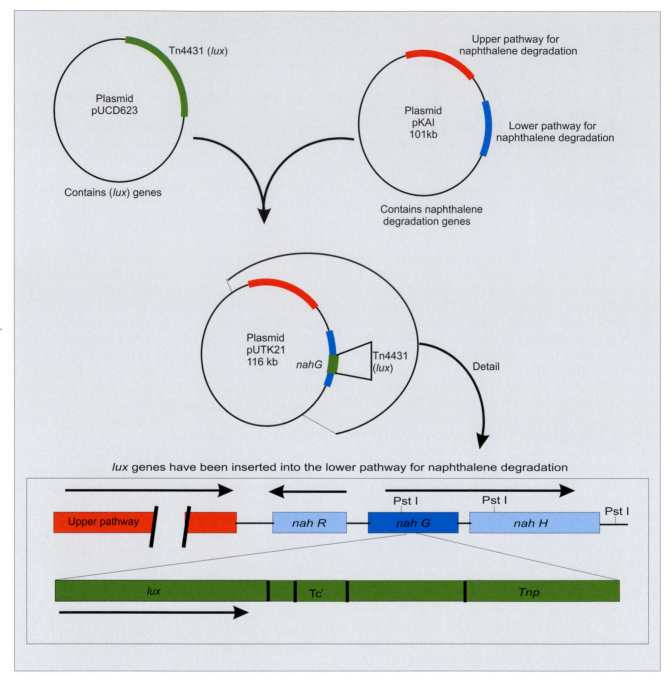

FIGURE 13.20   Example of a reporter gene. The *lux* genes, which code for a luminescence reaction, have been inserted into the naphthalene degradation pathway. As a result, an organism with the pUTK21 plasmid will luminesce while degrading naphthalene. (Adapted with permission from King *et al.*, 1990.)

into the soil column and used to collect luminescence that occurs in response to salicylate addition (Neilson *et al.*, 1999). This system is being used to better understand microbial metabolism in an undisturbed soil including the lag phase, and the effect of oxygen depletion, substrate concentration and cell number on degradation activity.

The **Green fluorescent protein (GFP)** is a relatively new bioreporter that relies on fluorescent light production instead of luminescence. Luminescence is different from fluorescence in that luminescence is emitted and detected without any light input. In contrast, fluorescence requires the fluorescent molecule to be excited by exposure to a certain wavelength of light. This causes an emission of fluorescence at a different wavelength that can be detected and measured. The gene for GFP is found in the jellyfish *Aequorea victoria*. GFP itself converts the blue bioluminescent light of the jellyfish to a green color, for reasons that are not yet known. When the GFP gene is expressed in a cell, which can be prokaryotic or eukaryotic, it forms a cyclic structure (Fig. 13.21). Excitation of GFP with UV light (395 nm) results in a bright green fluorescence (509 nm), which can subsequently be measured (Chalfie *et al.*, 1994). GFP has several advantages over *lacZ* and *xylE* fusions. First, as with the *lux* system, because of

the abundance of GFP in a single cell, transcription of a gene in individual cells rather than the average transcription of a population of cells can be measured. However, unlike the case of the *lux* system, cellular metabolism is not required for fluorescence of GFP, thus allowing *in situ* gene expression by cells that are not actively growing, which is likely to be the case in many environments. GFP fluorescence is also stable at 37°C, whereas the luciferase from *Vibrio fischeri* is heat labile (Burlage, 1997). Finally, both GFP and *lux* are affected by oxygen levels. Although the GFP system functions in the absence of O_2, the GFP protein must be oxidized in order to fluoresce. The *lux* system is affected when O_2 levels drop below 5 mg/l; however, luminescence occurs even as low as 0.5 mg/l O_2. Thus, in low O_2, the *lux* system must be calibrated for the amount of O_2 present. One other important difference between GFP and *lux* is that luminesance stops as soon as gene expression stops. However, the GFP protein will keep on fluorescing for as long as it remains intact. Thus, *lux* is a more reliable indicator of real-time gene activity.

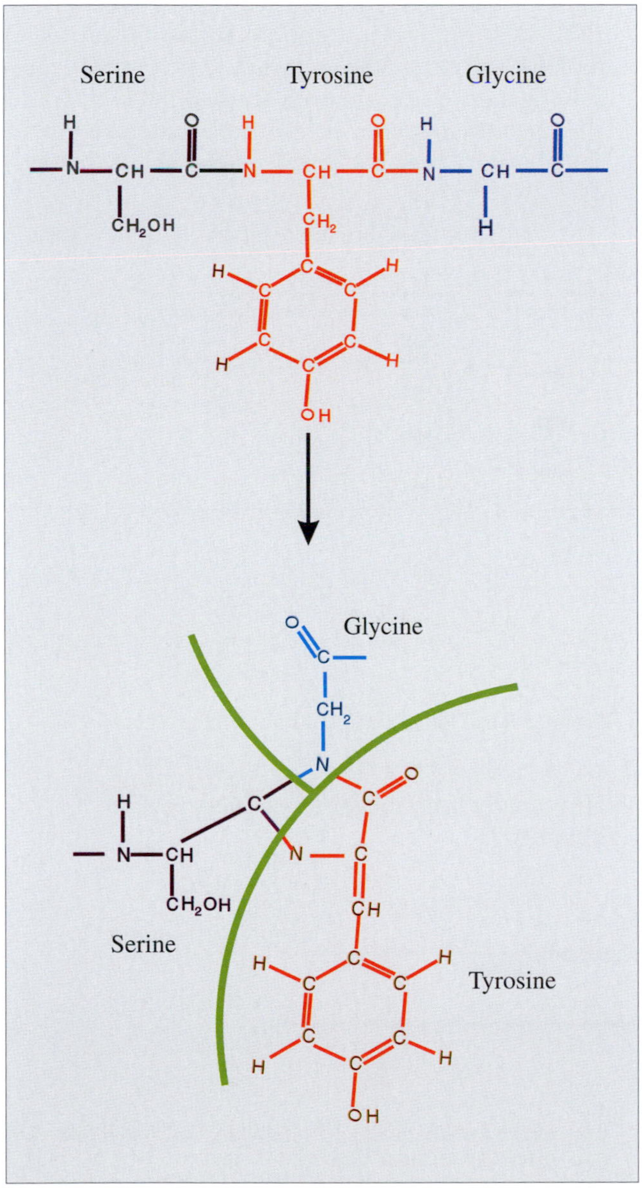

13.3.7.3 Advantages and Disadvantages of Reporter Genes

Reporter genes are unique in that they allow evaluation of microbial activity in natural environments through detection of gene expression. The perfect reporter gene system would take into account desirable characteristics including detectability, sensitivity, the ability to quantify, and a reasonable degree of stability (see Table 13.4). In reality, of course, there are always limitations. The *lacZ* system suffers from the presence of background levels of β-galactosidase activity expressed by many indigenous bacteria. The *lux* luminescent reporter systems are sensitive to O_2 levels, and the emitted light is not always proportional to metabolic activity in substrate-limited environments. In addition, the emitted light depends on growth stage and substrate concentration. Finally, the advantage of the GFP system is that it is less sensitive to cell energy levels than *lux*. However, the GFP system reflects previous and current activity while the *lux* represents only current activity. Despite these problems, reporter genes represent an innovative approach to environmental microbiology studies, and new reporter gene systems are constantly being evaluated.

FIGURE 13.21 The chromophore of *Aequorea* GFP. Amino acids 65, 66, and 67 of GFP form a cyclic structure by an autocatalytic reaction. This chromophore is the source of the bright fluorescence seen with this protein. The solid green lines delineate the separate amino acids in the chromophore. (Adapted with permission from Burlage, 1997.)

QUESTIONS AND PROBLEMS

1. What are the major advantages of PCR when it is applied to environmental samples?

2. What are the major disadvantages of PCR when it is applied to environmental samples?

3. What nucleic acid–based methods are best to detect microbial activity?

4. At the beginning of PCR there are 2 template DNA strands (T). As amplification proceeds, short strands (S) and long strands (L) are generated. After 5 cycles of PCR, how many total strands each of T, S and L are there?

5. Compare and contrast ERIC & AP-PCR methods of fingerprinting.

6. (a) You want to identify, at the genus level, several bacterial strains isolated from the soil. Explain how you would do this using molecular techniques.

6. (b) How would you identify a soil bacterium without prior cultivation using molecular techniques?

7. Compare and contrast the *lux* reporter system and the GFP reporter system. Based on these characteristics which would be more useful in evaluating contaminant transport in a soil column study?

8. You are trying to characterize isolates that you suspect have the ability to degrade 2,4-D. You isolate 2 strains (strain 1 & strain 2) and begin to characterize them using molecular techniques. You screen for the *tfd-b* gene (one of the genes that encodes for an enzyme responsible for 2,4-D degradation) using seminested PCR, generate ERIC PCR fingerprints and do a plasmid profile to look for the large 80kb plasmid which carries the 2,4-D genes. Using the well studied 2,4-D degrader *Alcaligenes eutrophous* as a control, what can you say about each isolate based on the results below?

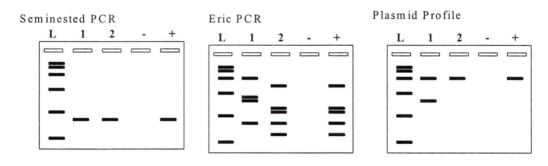

L = 123 bp ladder
1 = strain 1
2 = strain 2
- = negative control
+ = positive control (*Alcaligenes eutrophus*)

9. The schematic below shows the position and orientation of 4 primers used in nested PCR.

How many amplification products will be obtained?

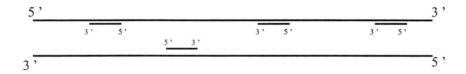

References and Recommended Readings

Abbaszadegan, M., Huber, M. S., Gerba, C. P., and Pepper, I. L. (1993) Detection of enteroviruses in groundwater with the polymerase chain reaction. *Appl. Environ. Microbiol.* **59**, 1318–1324.

Altuschul, S. F., Gish, W., Miller, W., Myers, E. W., and Lipman, D. J. (1990) Basic local alignment search tool. *J. Mol. Biol.* **215**, 403–410.

Amann, R. I., Ludwig, W., and Schleifer, K. (1995) Phylogenetic iden-tification and *in situ* detection of individual microbial cells without cultivation. *Microbiol. Rev.* **59**, 143–169.

Arsène, F., Katupitiya, S., Kennedy, I. R., and Elmerich, C. (1994) Use of *lacZ* fusions to study the expression of *nif* genes of *Azospirillum brasilense* in association with plants. *Mol. Plant Microbe Interact.* **7**, 748–757.

Birnboim, H. C., and Doly, J. (1979) A rapid alkaline extraction method for screening recombinant plasmid DNA. *Nucleic Acids Res.* **7**, 1513–1523.

Bloch, K. D. (1995) Digestion of DNA with restriction enzymes. *In* "Current Protocols in Molecular Biology" (F. M. Ausubel, ed.) John Wiley & Sons, New York, Section 3.3.2.

Bronstein, I., Fortin, J., Stanley, P. E., Stewart, G. S. A. B., and Kricka, L. J. (1994) Chemiluminescent and bioluminescent reporter gene assays. *Anal. Biochem.* **219,** 169–181.

Buell, C. R., and Anderson, A. J. (1993) Expression of the agg A locus of *Pseudomonas putida* is *in vitro* and in plants as detected by the reporter gene, *xylE. Mol. Plant Microbe Interact.* **6,** 331–340.

Burlage, R. S. (1997) Emerging technologies: Bioreporters, biosensors, and microprobes. *In* "Manual of Environmental Microbiology" (C. J. Hurst, G. R. Knudsen, M. J. McInerney, L. D. Stetzenbach, and M. V. Walter, eds.) American Society for Microbiology Press, Washington, DC. pp 115–123.

Burr, M. D., and Pepper. I. L. (1997) Variability in presence–absence scoring of AP PCR fingerprints affects computer matching of bacterial isolates. *J. Microbiol. Methods* 29, 63–68.

Burr, M. D., Josephson, K. L., and Pepper, I. L. (1997) An evaluation of DNA based methodologies for subtyping *Salmonella. Crit. Rev. Environ. Sci. Technol.,* **28,** 283–323.

Canhos, V. P., Manifo, G. P., and Blaine, L. D. (1993) Software tools and databases for bacterial systematics and their dissemination via global networks. *Antonie Leeuwenhoek.* **64,** 205–229.

Chalfie, M., Tu, Y., Euskirchen, G., Ward, W. W., and Prasher, D. C. (1994) Green fluorescent protein as a marker for gene expression. *Science* 263, 802–805.

de Bruijn, F. J. (1992) Use of repetitive (repetitive extragenic palindromic and enterobacterial repetitive intergenic consensus) sequences and the polymerase chain reaction to fingerprint the genomes of *Rhizobium meliloti* isolates and other soil bacteria. *Appl. Environ. Microbiol.* 58, 2180–2187.

Di Giovanni, G. D., Neilson, J. W., Pepper, I. L., and Sinclair, N. A. (1996) Gene transfer of *Alcaligenes eutrophus* JMP134 plasmid pJP4 to indigenous soil recipients. *Appl. Environ. Microbiol.* **62,** 2521–2526.

Eckhardt, T. (1978) A rapid method for the identification of plasmid deoxyribonucleic acid in bacteria. *Plasmid* **1,** 584–588.

Hansen, L. M., Jang, S. J., and Hirsh, D. C. (1993) Use of random fragments of chromosomal DNA to highlight restriction site heterogeneity for fingerprinting isolates of *Salmonella typhimurium* from hospitalized animals. *Am. J. Vet. Res.* **54,** 1648–1652.

Heitzen, A., Webb, O. F., Thonnard, J. E., and Saylor, G. S. (1992) Specific and quantitative assessment of naphthalene and salicylate bioavailability by using a bioluminescent catabolic reporter bacterium. *Appl. Environ. Microbiol.* **58,** 1839–1846.

Heuer, H. and Smalla, K. (1997) Application of denaturing gradient gel electrophoresis and temperature gradient gel electrophoresis for studying soil microbial communities. *In* "Modern Soil Microbiology" (J. D. van Elsa, J. T. Trevors, and E. M. H. Wellington, eds.) Marcel Dekker Inc., New York.

Hill, P. J., Rees, C. E. D., Winson, M. K., and Stewart, G. S. A. B. (1993) The application of *lux* genes. *Biotechnol. Appl. Biochem.* **17,** 3–14.

Hood, D. W. (1995) PFGE in the study of a bacterial pathogen (*Haemophilus influenzae*). *In* "Pulsed Field Gel Electrophoresis—A Practice Approach" (A. P. Monaco, ed.) Oxford University Press, Oxford.

Jefferson, R. A. (1989) The *gus* reporter gene system. *Nature* **342,** 837–838.

Jordan, B. R. (1998) Large-scale expression measurement by hybridization methods: From high density membranes to "DNA Chips." *J. Biochem.* **124,** 251–258.

Josephson, K. L., Pillai, S. D., Way, J., Gerba, C. P., and Pepper, I. L. (1991) Fecal coliforms in soil: detection by polymerase chain reaction and DNA–DNA hybridizations. *Soil Sci. Soc. Am. J.* **55,** 1326–1332.

Josephson, K. L., Gerba, C. P., and Pepper, I. L. (1993) Polymerase chain reaction detection of nonviable bacterial pathogens. *Appl. Environ. Microbiol.* **59,** 3513–3515.

King, J. M., DiGazia, P. M., Applegate, B., Burlage, R., Sanseverino, J., Dunbar, P., Larimer, F., and Sayler, G. S. (1990) Rapid sensitive bioluminescent reporter technology for naphthalene exposure and biodegradation. *Science,* 242, 778–781.

Lindow, S. E. (1995) The use of reporter genes in the study of microbial ecology. *Mol. Ecol.* **4,** 555–566.

Loper, J. E., and Lindow, S. E. (1997) Reporter gene systems useful in evaluating *in situ* gene expression by soil and plant-associated bacteria. *In* "Manual of Environmental Microbiology" (C. J. Hurst, G. R. Knudsen, M. J. McInerney, L. D. Stetzenbach, and M. V. Walter, eds.) American Society for Microbiology Press, Washington, DC. pp 482–492.

Maloy, S. R., Cronan, J. E. Jr., Freifelder, D. (1994) "Microbial Genetics, 2nd ed." Jones and Bartlett Publishers, Boston, MA.

Marlowe, E. M., Josephson, K. L., Miller, R. M., and Pepper, I. L. (1997) Quantitation of polymerase amplified products from environmental samples by high performance liquid chromatography. *J. Microbiol. Methods* **28,** 45–53.

Moat, A. G., and Foster, J. W. (1995) "Microbial Physiology," 3rd ed. John Wiley & Sons, Inc., New York.

Moré, M. I., Herrick, J. B., Silva, M. S., Ghiorse, W. C., and Madsen, E. L. (1994) Quantitative cell lysis of indigenous microorganisms and rapid extraction of microbial DNA from sediment. *Appl. Environ. Microbiol.* **60,** 1572–1580.

Muyzer, G., de Waal, E. C., and Uitterlinden, A. (1993) Profiling of complex microbial populations by denaturing gradient gel electrophoresis analysis of polymerase chain reaction-amplified genes coding for 16S rRNA. *Appl. Environ. Microbiol.* **59.** 695–700.

Navarro, E., Simonet, P., Normand, P., and Bardin, R. (1992) Characterization of natural populations of *Nitrobacter* spp. using PCR/RFLP analysis of the ribosomal intergenic spacer. *Arch. Microbiol.* **157,** 107–115.

Neilson, J. W., Pierce, S. A., and Maier, R. M. (1999) Factors Influencing the Expression of *lux* CDABE and NAH7 Genes in Pseudomonas putidaRB1353 (NAH7, pUTK9) in Dynamic Systems. *Appl. Environ. Microbiol.,* August issue.

Old, R. W., and Primrose, S. B. (1989) "Principles of Gene Manipulation–An Introduction to Genetic Engineering." Blackwell Scientific, London.

Pearson, W. R. (1990) Rapid and sensitive sequence comparison with FASTP and FASTA. *Methods Enzymol.* **183,** 63–98.

Reynolds, K. A., Gerba, C. P., and Pepper, I. L. (1996) Detection of infectious enteroviruses by an integrated cell culture–PCR procedure. *Appl. Environ. Microbiol.* **62,** 1424–1427.

Sambrook, J., Maniatis, T., and Frisch, E. F. (1989) "Molecular Cloning: A Laboratory Manual." Cold Spring Harbor Laboratory Press, Cold Spring Harbor, NY.

Schwartz, D. C., and Cantor, C. R. (1984) Separation of yeast chromosome–sized DNAs by pulsed-field gradient gel electrophoresis. *Cell* **37,** 67–75.

Selvaratnam, S., Schoedel, B. A., McFarland, B. L., and Kulpa, C. F. (1995) Application of reverse transcriptase PCR for monitoring expression of the catabolic *dmpN* gene in a phenol-degrading sequence batch reactor. *Appl. Environ. Microbiol.* **61,** 3981–3985.

Sheffield, V. C., Beck, J. S., Stone, E. M., and Myers, R. M. (1992) A simple and efficient method for attaching a 40bp, GC-rich sequence to PCR amplified DNA. *Biotechniques* **12,** 386–387.

Shishido, M., and Pepper, I. L. (1990) Identification of dominant indigenous *Rhizobium meliloti* by plasmid profiles and intrinsic antibiotic resistance. *Soil Biol. Biochem.* **22,** 11–16.

Tenover, F. C., and Unger, R. R. (1993) Nucleic acid probes for detection and identification of infectious agents. *In* "Diagnostic Molecular Microbiology—Principles and Applications" (D. H. Persing, T. F. Smith, F. C. Tenover, and T .J. White, eds.) American Society for Microbiology, Washington, DC. pp 223–228.

Threlfall, E. J., Torre, E., Ward, L. R., Davolos-Perez, A., Rowe, B., and Gibert, I. (1994) Insertion sequence IS200 fingerprinting of *Salmonella typhi*: An assessment of epidemiological applicability. *Epidem. Infect.* **112**, 253–261.

Tsai, Y., Palmer, C., and Sangermano, L. R. (1993) Detection of *Escherichia coli* in sewage and sludge by polymerase chain reaction. *Appl. Environ. Microbiol.* **59**, 353-357.

Tsai, Y., Tran, B., Sangermano, L. R., and Palmer, C. J. (1994) Detection of poliovirus, hepatitis A virus, and rotavirus from sewage and ocean water by triplex reverse transcriptase PCR. *Appl. Environ. Microbiol.* **60**, 2400–2407.

Watson, J. D., Tooze, J., and Kurtz, D. T. (1983) "Recombinant DNA: A Short Course." W. H. Freeman, New York.

Watson, J. D., Hopkins, N. H., Roberts, J. W., Steitz, J. A., and Weiner, A. M. (1987) "Molecular Biology of the Gene." Benjamin/Cummings, Menlo Park, CA.

Way, J. S., Josephson, K. L., Pillai, S. D., Abbaszadegan, M., Gerba, C. P., and Pepper, I. L. (1993) Specific detection of *Salmonella* spp. by multiplex polymerase chain reaction. *Appl. Environ. Microbiol.* **59**, 1473–1479.

Welsh, J., and McClelland, M. (1990) Fingerprinting genomes using PCR with arbitrary primers. *Nucleic Acids Res.* **19**, 861–866.

Welsh, J., and McClelland, M. (1993) Characterization of pathogenic microorganisms by genomic fingerprinting using arbitrarily primed PCR. *In* "Diagnostic Molecular Microbiology—Principles and Applications." (D. H. Pershing, T. F. Smith, F. C. Tenover, and T. J. White, eds.) ASM Press, Washington, DC. pp. 331–342.

Wilson, K. (1994) Preparation of genomic DNA from bacteria. *In* "Current Protocols in Molecular Biology" (F. M. Ausubel, ed.) John Wiley & Sons, New York. Section 2.4.5.

14

Biogeochemical Cycling

RAINA M. MAIER

14.1 INTRODUCTION

14.1.1 Biogeochemical Cycles

What happens to the vast array of organic matter that is produced on the earth during photosynthetic processes? This material does not keep accumulating rather, it is consumed and degraded, and a delicate global balance of carbon is maintained; carbon dioxide is removed from the atmosphere during photosynthesis and released during respiration. This balance is a result of the biologically driven, characteristic cycling of carbon between biotic forms such as sugar or other cellular building blocks and abiotic forms such as carbon dioxide. Cycling between biotic and abiotic forms is not limited to carbon. All of the major elements found in biological organisms (see Table 14.1), as well as some of the minor and trace elements, are similarly cycled in predictable and definable ways.

Taken together, the various element cycles are called the **biogeochemical cycles.** Understanding these cycles allows scientists to understand and predict the development of microbial communities and activities in the environment. There are many activites that can be harnessed in a beneficial way, such as for remediation of organic and metal pollutants or for recovery of precious metals such as copper or uranium from low-grade ores. There are also detrimental aspects of the cycles that can cause global environmental problems such as the formation of acid rain and acid mine drainage, metal corrosion processes, and formation of nitrous oxide, which can deplete the earth's ozone layer (see Chapter 15). As these examples illustrate, the microbial activities that drive biogeochemical cycles are highly relevant to the field of environ-

mental microbiology. Thus, the knowledge of these cycles is increasingly critical as the human population continues to grow and the impact of human activity on the earth's environment becomes more significant. In this chapter, the biogeochemical cycles pertaining to carbon, nitrogen, and sulfur are delineated while our discussion will be limited to these three cycles, it should be noted that there are a number of other cycles. These include the phosphorus cycle, the iron cycle, the calcium cycle and more (Dobrovolsky, 1994).

14.1.2 Gaia Hypothesis

In the early 1970s, James Lovelock theorized that Earth behaves like a superorganism, and this concept developed into what is now known as the **Gaia hypothesis.** To quote Lovelock (1995), "Living organisms and their material environment are tightly coupled. The coupled system is a superorganism, and as it evolves there emerges a new property, the ability to self-regulate climate and chemistry." The basic tenet of this hypothesis is that the earth's physicochemical properties are self-regulated so that they are maintained in a favorable range for life. As evidence for this, consider that the sun has heated up by 30% during the past 4–5 billion years. Given the earth's original carbon dioxide-rich atmosphere, the average surface temperature of a lifeless earth today would be approximately 290°C (Table 14.2). In fact, when one compares Earth's present-day atmosphere with the atmospheres found on our nearest neighbors Venus and Mars, one can see that something has drastically affected the development of Earth's atmosphere. According to the Gaia hypothesis, this is the development and continued presence of life. Micro-

TABLE 14.1   Chemical Composition of an *E. coli* Cell

Elemental breakdown	% dry mass of an *E. coli* cell
Major elements	
Carbon	50
Oxygen	20
Hydrogen	8
Nitrogen	14
Sulfur	1
Phosphorus	3
Minor elements	
Potassium	2
Calcium	0.05
Magnesium	0.05
Chlorine	0.05
Iron	0.2
Trace elements	
Manganese	all trace elements
Molybdenum	combined comprise 0.3%
Cobalt	of dry weight of cell
Copper	
Zinc	

Adapted from Neidhardt *et al.* (1990).

bial activity, and later the appearance of plants, have changed the original heat-trapping carbon dioxide–rich atmosphere to the present oxidizing, carbon dioxide-poor atmosphere. This has allowed Earth to maintain an average surface temperature of 13°C, which is favorable to the life that exists on Earth.

How do biogeochemical activities relate to the Gaia hypothesis? These biological activities have driven the response to the slow warming of the sun resulting in the major atmospheric changes that have occurred over the last 4–5 billion years. When Earth was formed 4–5 billion years ago, a reducing (anaerobic) atmosphere existed. The initial reactions that mediated the formation of organic carbon were abiotic, driven by large influxes of ultraviolet (UV) light. The resulting reservoir of organic matter was utilized by early anaerobic heterotrophic organisms. This was followed by the development of the ability of microbes to fix car-

bon dioxide photosynthetically. Evidence from **stromatolites** suggests that the ability to photosynthesize was developed at least 3.5 billion years ago. Stromatolites are fossilized laminated structures that have been found in Africa and Australia. These structures were formed primarily by cyanobacteria that grew in mats and entrapped or precipitated inorganic material as they grew. The evolution of photosynthetic organisms tapped into an unlimited source of energy, the sun, and provided a mechanism for carbon recycling, i.e., the first carbon cycle (Fig. 14.1). This first carbon cycle was maintained for approximately 1.5 billion years. Geologic evidence then suggests that approximately 2 billion years ago, photosynthetic microorganisms developed the ability to produce oxygen. This allowed oxygen to accumulate in the atmosphere, resulting, in time, in a change from reducing to oxidizing conditions. Further, oxygen accumulation in the atmosphere created an ozone layer, which reduced the influx of harmful UV radiation, allowing the development of higher forms of life to begin.

At the same time that the carbon cycle evolved, the nitrogen cycle emerged because nitrogen was a limiting element for microbial growth. Although molecular nitrogen was abundant in the atmosphere, microbial cells could not directly utilize nitrogen as N_2 gas. Cells required organic nitrogen compounds or reduced inorganic forms of nitrogen for growth. Therefore, under the reducing conditions found on early Earth, some organisms developed a mechanism for fixing nitrogen using the enzyme nitrogenase. Nitrogen fixation remains an important microbiological process, and to this day, the nitrogenase enzyme is totally inhibited in the presence of oxygen.

When considered over this geologic time scale of several billion years, it is apparent that biogeochemical activities have been unidirectional. This means that the predominant microbial activities on earth have evolved over this long period of time to produce changes and to respond to changes that have occurred in the atmosphere, i.e., the appearance of oxygen and

TABLE 14.2   Atmosphere and Temperatures Found on Venus, Mars, and Earth

Gas	Venus	Mars	Earth without life	Earth with life
Carbon dioxide	96.5%	95%	98%	0.03%
Nitrogen	3.5%	2.7%	1.9%	9%
Oxygen	trace	0.13%	0.0	21%
Argon	70 ppm	1.6%	0.1%	1%
Methane	0.0	0.0	0.0	1.7 ppm
Surface temperature (°C)	459	−53	290 ± 50	13

Adapted from Lovelock (1995).

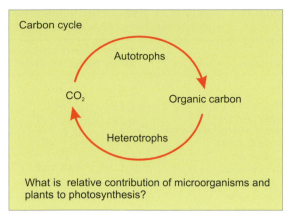

FIGURE 14.1 The carbon cycle is dependent on autotrophic organisms that fix carbon dioxide into organic carbon and heterotrophic organisms that respire organic carbon to carbon dioxide.

TABLE 14.3 Global Carbon Reservoirs

Carbon reservoir	Metric tons carbon	Actively cycled
Atmosphere		
$\quad CO_2$	6.7×10^{11}	Yes
Ocean		
$\quad$ Biomass	4.0×10^{9}	No
$\quad$ Carbonates	3.8×10^{13}	No
$\quad$ Dissolved and	2.1×10^{12}	Yes
$\quad\quad$ particulate organics		
Land		
$\quad$ Biota	5.0×10^{11}	Yes
$\quad$ Humus	1.2×10^{12}	Yes
$\quad$ Fossil fuel	1.0×10^{13}	Yes
$\quad$ Earth's crust[a]	1.2×10^{17}	No

[a] This reservoir includes the entire lithosphere found in either terrestrial or ocean environments. (Data from Dobrovolsky, 1994.)

the decrease in carbon dioxide content. Presumably these changes will continue to occur, but they occur so slowly that we do not have the capacity to observe them. However, one can also consider biogeochemical activities on a more contemporary time scale, that of tens to hundreds of years. On this much shorter time scale, biogeochemical activities are regular and cyclic in nature, and it is these activities that are addressed in this chapter. On the one hand, the presumption that earth is a superorganism and can respond to drastic environmental changes is heartening when one considers that human activity is effecting unexpected changes in the atmosphere, such as ozone depletion and buildup of carbon dioxide. However, it is important to point out that the response of a superorganism is necessarily slow (thousands to millions of years), and as residents of earth we must be sure not to overtax Earth's ability to respond to change by artificially changing the environment in a much shorter time frame.

14.2 CARBON CYCLE

14.2.1 Carbon Reservoirs

A **reservoir** is a sink or source of an element such as carbon. There are various global reservoirs of carbon, some of which are immense in size and some of which are relatively small (Table 14.3). The largest carbon reservoir is carbonate rock found in the earth's sediments. This reservoir is four orders of magnitude larger than the carbonate reservoir found in the ocean and six orders of magnitude larger than the carbon reservoir found as carbon dioxide in the atmosphere. If one considers these three reservoirs, it is obvious that the carbon most available for photosynthesis is in the smallest of the reservoirs, the at-

mosphere. Therefore, it is the smallest reservoir that is most actively cycled. It is small, actively cycled reservoirs such as atmospheric carbon dioxide that are subject to perturbation from human activity. In fact, since global industrialization began in the late 1800s, humans have affected several of the smaller carbon reservoirs. Utilization of fossil fuels (an example of a small, inactive carbon reservoir) and deforestation (an example of a small, active carbon reservoir) are two activities that have reduced the amount of fixed organic carbon in these reservoirs and added to the atmospheric carbon dioxide reservoir (Table 14.4).

The increase in atmospheric carbon dioxide has not been as great as expected. This is because the reservoir of carbonate found in the ocean acts as a buffer between the atmospheric and sediment carbon reservoirs through the equilibrium equation shown below.

$$H_2CO_3 \rightleftarrows HCO_3^- \rightleftarrows CO_2$$

Thus, some of the excess carbon dioxide that has been released has been absorbed by the oceans. However, there has still been a net efflux of carbon dioxide into the atmosphere of approximately 7×10^9 metric tons/year. The problem with this imbalance is that be-

TABLE 14.4 Net Carbon Flux between Selected Carbon Reservoirs

Carbon source	Flux (metric tons carbon/year)
Release by fossil fuel combustion	7×10^9
Land clearing	3×10^9
Forest harvest and decay	6×10^9
Forest regrowth	-4×10^9
Net uptake by oceans (diffusion)	-3×10^9
Annual flux	9×10^9

cause atmospheric carbon dioxide is a small carbon reservoir, the result of a continued net efflux over the past 100 years or so has been a 28% increase in atmospheric carbon dioxide from 0.026% to 0.033%. A consequence of the increase in atmospheric carbon dioxide is that it may contribute to global warming through the **greenhouse effect.** The greenhouse effect is caused by gases in the atmosphere that trap heat from the sun and cause the earth to warm up. This effect is not solely due to carbon dioxide; other gases such as methane, chlorofluorocarbons (CFCs), and nitrous oxide add to the problem.

14.2.2 Carbon Fixation and Energy Flow

The ability to photosynthesize allows sunlight energy to be trapped and stored. In this process carbon dioxide is fixed into organic matter (Fig. 14.1). Photosynthetic organisms, also called **primary producers,** include plants and microorganisms such as algae, cyanobacteria, some bacteria, and some protozoa. As shown in Fig. 14.2, the efficiency of sunlight trapping is very low; less than 0.1% of the sunlight energy that hits the earth is actually utilized. As the fixed sunlight energy moves up each level of the food chain, up to 90% or more of the trapped energy is lost through respiration. Despite this seemingly inefficient trapping, photoautotrophic primary producers support most of the considerable ecosystems found on the earth. Productivity varies widely among different ecosystems depending on the climate, the type of primary producer, and whether the system is a managed one (Table 14.5). For example, one of the most productive natural areas is the coral reefs. Managed agricultural systems such as corn and sugarcane systems are also very productive, but it should be remembered that a

TABLE 14.5 Net Primary Productivity of Some Natural and Managed Ecosystems

Description of ecosystem	Net primary productivity (g dry organic matter/m^2/year)
Tundra	400
Desert	200
Temperate grassland	Up to 1500
Temperate or deciduous forest	1200–1600
Tropical rain forest	Up to 2800
Cattail swamp	2500
Freshwater pond	950–1500
Open ocean	100
Coastal seawater	200
Upwelling area	600
Coral reef	4900
Corn field	1000–6000
Rice paddy	340–1200
Sugarcane field	Up to 9400

Adapted from Atlas and Bartha (1993).

significant amount of energy is put into these systems in terms of fertilizer addition and care. The open ocean has much lower productivity, but covers a majority of the earth's surface and so is a major contributor to primary production. In fact, aquatic and terrestrial environments contribute almost equally to global primary production. Plants predominate in terrestrial environments, but with the exception of immediate coastal zones, microorganisms are responsible for most primary production in aquatic environments. It follows that microorganisms are responsible for approximately one half of all primary production on the earth.

14.2.3 Carbon Respiration

Carbon dioxide that is fixed into organic compounds as a result of photoautotrophic activity is available for consumption or respiration by animals and heterotrophic microorganisms. This is the second half of the carbon cycle shown in Fig. 14.1. The end-products of respiration are carbon dioxide and new cell mass. An interesting question to consider is the following: if respiration were to stop, how long would it take for photosynthesis to use up all of the carbon dioxide reservoir in the atmosphere? Based on estimates of global photosynthesis, it has been estimated that it would take 30 to 300 years. This illustrates the importance of both legs of the carbon cycle in maintaining a carbon balance.

The following sections discuss the most common organic compounds found in the environment and the microbial catabolic activities that have evolved in response. These include organic polymers, humus, and

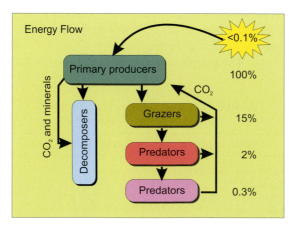

FIGURE 14.2 Diagram of the efficiency of sunlight energy flow from primary producers to consumers.

C1 compounds such as methane (CH_4). It is important to understand the fate of these naturally occurring organic compounds. This is because degradative activities that evolved for these compounds form the basis for degradation pathways that may be applicable to organic contaminants that are spilled in the environment (see Chapter 16). But before looking more closely at the individual carbon compounds, it should be pointed out that the carbon cycle is actually not quite as simple as depicted in Fig. 4.1. This simplified figure does not include anaerobic processes, which were predominant on the early earth and remain important in carbon cycling even today. A more complex carbon cycle containing anaerobic activity is shown in Fig. 14.3. Under anaerobic conditions, which predominated for the first several billion years on earth, some cellular components were less degradable than others (Fig. 14.4). This is especially true for highly reduced molecules such as cellular lipids. These components were therefore left over and buried with sediments over time and became the present-day fossil fuel reserves. Another carbon compound produced under anaerobic conditions is methane. Methane is produced in soils as an end product of anaerobic respiration (see Eq. 3.20, Chapter 3). Methane is also produced under the anaerobic conditions found in ruminants such as cows.

14.2.3.1 Organic Polymers

What are the predominant types of organic carbon found in the environment? They include plant polymers, polymers used to build fungal and bacterial cell walls, and arthropod exoskeletons (Fig. 14.5). Because these polymers constitute the majority of organic carbon, they are the basic food supply available to support heterotrophic activity. The three most common polymers are the plant polymers cellulose, hemicellu-

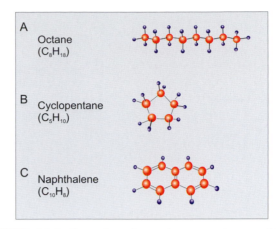

FIGURE 14.4 Examples of petroleum constituents: (A) an alkane, (B) an alicyclic, and (C) an aromatic compound. A crude oil contains some of each of these types of compounds but the types and amounts vary in different petroleum reservoirs.

lose, and lignin (Table 14.6) (Wagner and Wolf, 1998). There are also various other polymers including starch, chitin, and peptidoglycan. These various polymers can be divided into two groups on the basis of their structures: the carbohydrate-based polymers, which include the majority of the polymers found in the environment, and the phenylpropane-based polymer, lignin.

Carbohydrate-based Polymers

Cellulose is not only the most abundant of the plant polymers, it is also the most abundant polymer found on the earth. It is a linear molecule containing β-1,4 linked glucose subunits (Fig. 14.5a). Each molecule contains 1000 to 10,000 subunits with a resulting molecular weight of up to 1.8×10^6. These linear molecules are arranged in microcrystalline fibers that help make up the woody structure of plants. Cellulose is not only a large molecule, it is also insoluble in water. How then do microbial cells get such a large, insoluble molecule across their walls and membranes? The answer is that they have developed an alternative strategy, which is to make and release enzymes, called **extracellular enzymes,** that can begin the polymer degradation process outside the cell (Deobald and Crawford, 1997). There are two extracellular enzymes that initiate cellulose degradation. These are β-1,4-endoglucanase and β-1,4-exoglucanase. The endoglucanase hydrolyzes cellulose molecules randomly within the polymer, producing smaller and smaller cellulose molecules (Fig. 14.6). The exoglucanase consecutively hydrolyzes two glucose subunits from the reducing end of the cellulose molecule, releasing the disaccharide cellobiose. A third enzyme, known as β-glucosidase or cellobiase, then hydrolyzes cellobiose to glucose. Cellobiase can be found as both an extracellular and an intracellular enzyme.

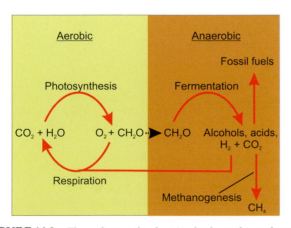

FIGURE 14.3 The carbon cycle, showing both aerobic and anaerobic contributions.

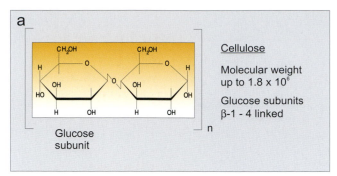

a

Cellulose

Molecular weight
up to 1.8 x 10^6

Glucose subunits
β-1 - 4 linked

Glucose
subunit

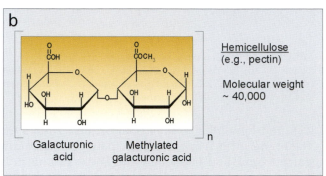

b

Hemicellulose
(e.g., pectin)

Molecular weight
~ 40,000

Galacturonic   Methylated
acid   galacturonic acid

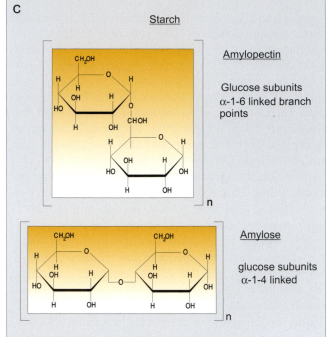

c

Starch

Amylopectin

Glucose subunits
α-1-6 linked branch
points

Amylose

glucose subunits
α-1-4 linked

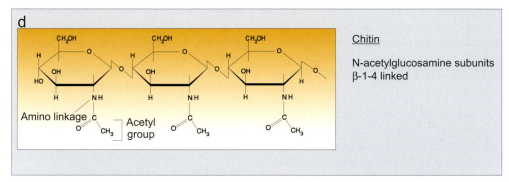

d

Chitin

N-acetylglucosamine subunits
β-1-4 linked

Amino linkage   Acetyl
group

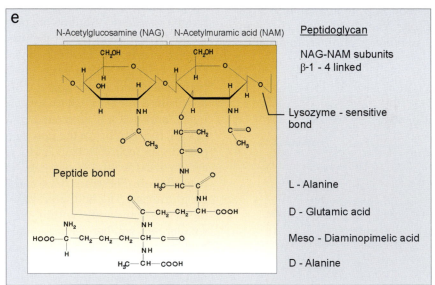

e

N-Acetylglucosamine (NAG)   N-Acetylmuramic acid (NAM)

Peptidoglycan

NAG-NAM subunits
β-1 - 4 linked

Lysozyme - sensitive
bond

Peptide bond

L - Alanine

D - Glutamic acid

Meso - Diaminopimelic acid

D - Alanine

Table 14.6 Major Types of Organic Components of Plants

Plant component	% dry mass of plant
Cellulose	15–60
Hemicellulose	10–30
Lignin	5–30
Protein and nucleic acids	2–15

Both cellobiose and glucose can be taken up by many bacterial and fungal cells.

Hemicellulose is the second most common plant polymer. This molecule is more heterogeneous than cellulose, consisting of a mixture of several monosaccharides including various hexoses and pentoses as well as uronic acids. In addition, the polymer is branched instead of linear. An example of a hemicellulose polymer is the pectin molecule shown in Fig. 14.5b, which contains galacturonic acid and methylated galacturonic acid. Degradation of hemicellulose is similar to the process described for cellulose except that, because the molecule is more heterogeneous, many more extracellular enzymes are involved.

In addition to the two major plant polymers, several other important organic polymers are carbohydrate based. One of these is **starch,** a polysaccharide synthesized by plants to store energy (Fig. 14.5c). Starch is formed from glucose subunits and can be linear (α-1,4 linked), a structure known as amylose, or can be branched (α-1,4 and α-1,6 linked), a structure known as amylopectin. Amylases (α-1,4–linked exo- and endoglucanases) are extracellular enzymes produced by many bacteria and fungi. Amylases produce the disaccharide maltose, which can be taken up by cells and mineralized. Another common polymer is **chitin,** which is formed from β-1,4–linked subunits of N-acetylglucosamine (Fig. 14.5d). This linear, nitrogen-containing polymer is an important component of fungal cell walls and of the exoskeleton of arthropods. Finally, there is **peptidoglycan,** a polymer of N-acetylglucosamine and N-acetylmuramic acid, which is an important component of bacterial cell walls (Fig. 14.5e).

Lignin

Lignin is the third most common plant polymer and is strikingly different in structure from all of the carbohydrate-based polymers. The basic building blocks of lignin are the two aromatic amino acids tyrosine and phenylalanine. These are converted to phenylpropene subunits such as coumaryl alcohol, coniferyl alcohol, and sinapyl alcohol. Then 500 to 600 phenylpropene subunits are randomly polymerized, resulting in the formation of the amorphous aromatic polymer known as lignin. In plants lignin surrounds cellulose microfibrils and strengthens the cell wall. Lignin also helps make plants more resistant to pathogens.

Biodegradation of lignin is slower and less complete than degradation of other organic polymers. This is shown experimentally in Figure 14.7a, and visually in Figure 14.7b. Lignin degrades slowly because it is constructed as a highly heterogeneous polymer and in addition contains aromatic residues rather than carbohydrate residues. The great heterogeneity of the molecule precludes the evolution of specific degradative enzymes comparable to cellulose. Instead, a nonspecific extracellular enzyme, H_2O_2-dependent lignin peroxidase, is used in conjunction with an extracellular oxidase enzyme that generates H_2O_2. The peroxidase enzyme and H_2O_2 system generate oxygen-based free radicals that react with the lignin polymer to release phenylpropene residues (Morgan *et al.*, 1993). These residues are taken up by microbial cells and degraded as shown in Fig. 14.8. Biodegradation of intact lignin polymers occurs only aerobically, which is not surprising because reactive oxygen is needed to release lignin residues. However, once residues are released, they can be degraded under anaerobic conditions.

Phenylpropene residues are aromatic in nature, similar in structure to several types of organic pollutant molecules such as the BTEX (benzene, toluene, ethylbenzene, xylene) and polyaromatic hydrocarbon compounds found in crude oil and gasoline and creosote compounds found in wood preservatives (see Fig. 16.13). These naturally occurring aromatic biodegrada-

FIGURE 14.5 Common organic polymers found in the environment. (A) Cellulose is the most common plant polymer. It is a linear polymer of β-1,4-linked glucose subunits. Each polymer contains 1000 to 10,000 subunits. (B) Hemicellulose is the second most common polymer. This molecule is more heterogeneous, consisting of hexoses, pentoses, and uronic acids. An example of a hemicellulose polymer is pectin. (C) Starch is a polysaccharide synthesized by plants to store energy. Starch is formed from glucose subunits and can be linear (α-1,4-linked), a structure known as amylose, or can be branched (α-1,4 and α-1,6 linked), known as amylopectin. (D) Chitin is formed from subunits of N-acetylglucosamine linked β-1,4. This polymer is found in fungal cell walls. (E) Bacterial cell walls are composed of polymers of N-acetylglucosamine and N-acetylmuramic acid connected by β-1,4 linkages.

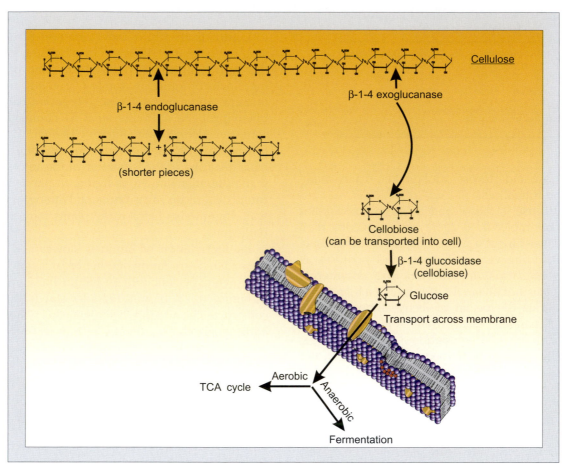

FIGURE 14.6 The degradation of cellulose begins outside the cell with a series of extracellular enzymes called cellulases. The resulting smaller glucose subunit structures can be taken up by the cell and metabolized.

tion pathways are of considerable importance in the field of bioremediation. In fact, a comparison of the pathway shown in Fig. 14.8 with the pathway for degradation of aromatics presented in Chapter 16 shows that they are very similar. Lignin is degraded by a variety of microbes including fungi, actinomycetes, and bacteria. The best studied organism with respect to lignin degradation is the white rot fungus *Phanerochaete chrysosporium*. This organism is also capable of degrading several pollutant molecules with structures similar to those of lignin residues (see Case Study Chapter 16).

14.2.3.2 Humus

Humus was introduced in Chapter 4.2.2 and its structure is shown in Fig. 4.8. How does humus form? It forms in a two-stage process that involves the formation of reactive monomers during the degradation of organic matter, followed by the spontaneous polymerization of some of these monomers into the humus

molecule. Although the majority of organic matter that is released into the environment is respired to form new cell mass and carbon dioxide, a small amount of this carbon becomes available to form humus. To understand this spontaneous process, consider the degradation of the common organic polymers found in soil that were described in the preceding sections. Each of these polymers requires the production of extracellular enzymes that begin the polymer degradation process. In particular, for lignin these extracellular enzymes are nonspecific and produce hydrogen peroxide and oxygen radicals. It is not surprising, then, that some of the reactive residues released during polymer degradation might repolymerize and result in the production of humus. In addition, nucleic acid and protein residues that are released from dying and decaying cells contribute to the pool of molecules available for humus formation. This process is illustrated in Fig. 14.9. Considering the wide array of residues that

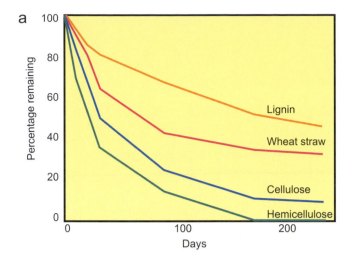

FIGURE 14.7 (a) Decomposition of wheat straw and its major constituents in a silt loam. The initial composition of the wheat straw was 50% cellulose, 25% hemicellulose, and 20% lignin. (Adapted from Wagner and Wolf, 1998.) (b) Leaves from a rain forest in different stages of decomposition. Cellulose and hemicellulose are degraded first, leaving the lignin skeleton. (Photo taken in Henry Pittier National Park, Venezuela, courtesy C. M. Miller.)

can contribute to humus formation, it is not surprising that humus is even more heterogeneous than lignin. Table 14.7 compares the different properties of these two complex molecules.

Humus is the most complex organic molecule found in soil, and as a result it is the most stable organic molecule. The turnover rate for humus ranges from 2 to 5% per year, depending on climatic conditions (Wagner and Wolf, 1998). This can be compared with the degradation of lignin shown in Fig. 14.7a, where approximately 50% of lignin added to a silt loam was degraded in 250 days. Thus, humus provides a slowly released source of carbon and energy for indigenous autochthonous microbial populations. The release of humic residues most likely occurs in a manner similar to release of lignin residues. Because the humus content of most soils does not change, the

rate of formation of humus must be similar to the rate of turnover. Thus, humus can be thought of as a molecule that is in a state of dynamic equilibrium (Haider, 1992).

14.2.3.3 Methane

Methanogenesis

The formation of methane, **methanogenesis,** is predominantly a microbial process, although a small amount of methane is generated naturally through volcanic activity (Table 14.8) (Ehrlich, 1981). Methanogenesis is an anaerobic process and occurs extensively in specialized environments including water-saturated areas such as wetlands and paddy fields; anaerobic niches in the soil; landfills; the rumen; and termite guts. Methane is an end-product of anaerobic degradation (see Chapter 3.4) and as such is associated with petroleum, natural gas, and coal deposits.

At present a substantial amount of methane is released to the atmosphere as a result of energy harvesting and utilization. A second way in which methane is released is through landfill gas emissions. It is estimated that landfills in the United States alone generate 10×10^6 metric tons of methane per year. Although methane makes a relatively minor carbon contribution to the global carbon cycle (compare Table 14.8 with Table 14.3), methane emission is of concern from several environmental aspects. First, like carbon dioxide, meth-ane is a greenhouse gas and contributes to global warming. In fact, it is the second most common greenhouse gas emitted to the atmosphere. Further, it is 22 times more effective than carbon dioxide at trapping heat. Second, localized production of methane in landfills can create safety and health concerns. Methane is an explosive gas at concentrations as small as 5%. Thus, to avoid accidents, the methane generated in a landfill must be managed in some way. If methane is present in concentrations higher than 35%, it can be collected and used for energy. Alternatively, the methane can be burned off at concentrations of 15% percent or higher. However, most commonly, it is simply vented to the atmosphere to prevent it from building up in high enough concentrations to ignite. Although venting landfill gas to the atmosphere does help prevent explosions, it clearly adds to the global warming problem.

The organisms responsible for methanogenesis are a group of obligately anaerobic archaebacteria called the **methanogens.** The basic metabolic pathway used by the methanogens is:

$$4H_2 + CO_2 \rightarrow CH_4 + 2H_2O$$
$$\Delta G^{\circ\prime} = -130.7 \text{ kJ} \tag{Eq. 14.1}$$

This is an exothermic reaction where CO_2 acts as the TEA and H_2 acts as the electron donor providing en-

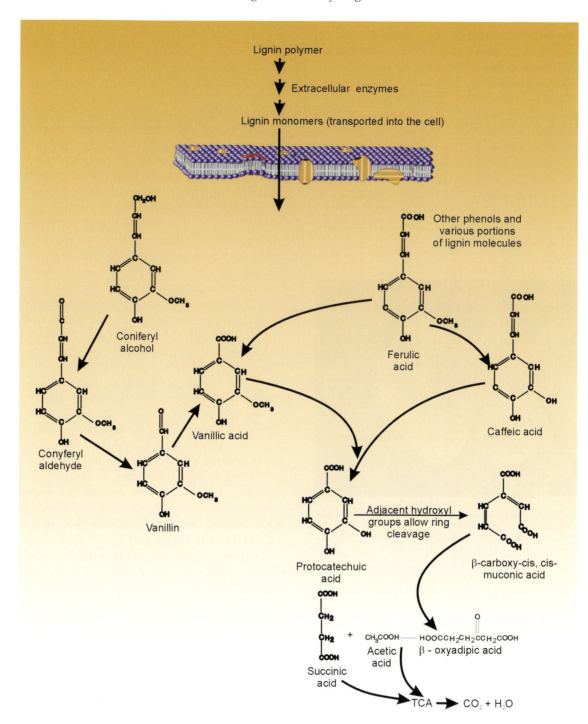

FIGURE 14.8   Lignin degradation. (Adapted from Wagner and Wolf, 1998.)

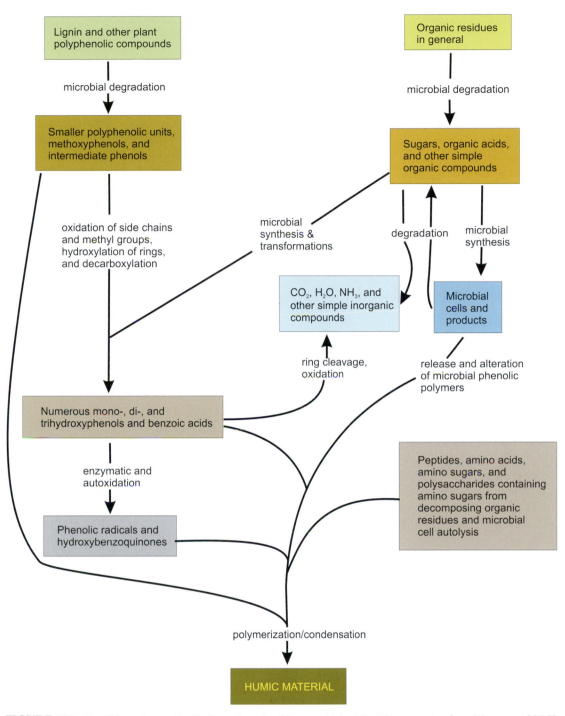

FIGURE 14.9   Possible pathways for the formation of soil humus. (Adapted with permission from Wagner and Wolf, 1998.)

ergy for the fixation of carbon dioxide. Methanogens that utilize CO_2/H_2 are therefore autotrophic. In addition to the autotrophic reaction shown in Eq. 14.1, methanogens can produce methane during heterotrophic growth on a limited number of other C_1 and C_2 substrates including acetate, methanol, and formate. Since there are very few carbon compounds that can be used by methanogens, these organisms are dependent on the production of these compounds by other microbes in the surrounding community. As such an interdependent community of microbes typically develops in anaerobic environments. In this community, the more complex organic molecules are catabolized by populations that ferment or respire anaerobically, gen-

TABLE 14.7   Chemical Properties of Humus and Lignin

Characteristic	Humic material	Lignin
Color	Black	Light Brown
Methoxyl (-OCH$_3$) content	Low	High
Nitrogen content	3–6%	0%
Carboxyl and phenolic hydroxyl content	High	Low
Total exchangeable acidity (cmol/kg)	≥150	≤0.5
α Amino nitrogen	Present	0
Vanillin content	<1%	15–25%

Data from Wagner and Wolf (1998).

erating C$_1$ and C$_2$ carbon substrates as well as CO$_2$ and H$_2$ that are then used by methanogens.

Methane Oxidation

Clearly, methane as the end product of anaerobiosis is found extensively in nature. As such, it is an available food source, and a group of bacteria called the **methanotrophs** have developed the ability to utilize methane as a source of carbon and energy. The methanotrophs are chemoheterotrophic and obligately aerobic. They metabolize methane as shown in Eq. 14.2.

$$\underset{\text{methane}}{CH_4} + O_2 \quad \xrightarrow{\underset{\text{monooxygenase}}{\text{methane}}} \quad \underset{\text{methanol}}{CH_3OH} \rightarrow \qquad \text{(Eq. 14.2)}$$

$$\underset{\text{formaldehye}}{HCHO} \rightarrow \underset{\text{formic acid}}{HCOOH} \rightarrow \underset{\text{carbon dioxide}}{CO_2 + H_2O}$$

The first enzyme in the biodegradation pathway is called **methane monooxygenase.** Oxygenases in general incorporate oxygen into a substrate and are important enzymes in the intial degradation steps for hydrocarbons (see Section 16.6.2.1.1). However, methane monooxygenase is of particular interest because it was the first of a series of enzymes isolated that can cometabolize highly chlorinated solvents such as trichloroethylene (TCE) (see Section 16.6.2.1.3). Until this discovery it was believed that biodegradation of highly chlorinated solvents could occur only under anaerobic conditions as an incomplete reaction. The application of methanogens for cometabolic degradation of TCE is a strategy under development for bioremediation of groundwater contaminated with TCE. This example is a good illustration of the way in which naturally occurring microbial activities can be harnessed to solve pollution problems.

14.2.3.4 Carbon Monoxide and Other C1 Compounds

Bacteria that can utilize C1 carbon compounds other than methane are called **methylotrophs.** There

TABLE 14.8 Estimates of Methane Released into the Atmosphere

Source	Methane emission (10^6 metric tons/year)	
Biogenic		
Ruminants	80–100	
Termites	25–150	
Paddy fields	70–120	
Natural wetlands	120–200	
Landfills	5–70	
Oceans and lakes	1–20	
Tundra	1–5	
Abiogenic		
Coal mining	10–35	
Natural gas flaring and venting	10–35	
Industrial and pipeline losses	15–45	
Biomass burning	10–40	
Methane hydrates	2–4	
Volcanoes	0.5	
Automobiles	0.5	
Total	350–820	
Total biogenic	302–665	81–86% of total
Total abiogenic	48–155	13–19% of total

Adapted from Madigan *et al.* (1997).

are a number of important C1 compounds produced from both natural and anthropogenic activities (Table 14.9). One of these is carbon monoxide. The annual global production of carbon monoxide is $3–4 \times 10^9$ metric tons/year (Atlas and Bartha, 1993). The two major carbon monoxide inputs are abiotic. Approximately 1.5×10^9 metric tons/year result from atmospheric photochemical oxidation of carbon compounds such as methane, and 1.6×10^9 metric tons/year results from burning of wood, forests, and fossil fuels. A small proportion, 0.2×10^9 metric tons/year, results from biological activity in ocean and soil environments. Carbon monoxide is a highly toxic molecule because it has a strong affinity for cytochromes, and in binding to cyctochromes, it can completely inhibit the activity of the respiratory electron transport chain.

Destruction of carbon monoxide can occur abiotically by photochemical reactions in the atmosphere. Microbial processes also contribute significantly to its destruction, even though it is a highly toxic molecule. The destruction of carbon monoxide seems to be quite efficient because the level of carbon monoxide in the atmosphere has not risen significantly since industrialization began, even though CO emissions have increased. The ocean is a net producer of carbon monoxide and releases CO to the atmosphere. In contrast, the terrestrial environment is a net sink for carbon monoxide and absorbs approximately 0.4×10^9 metric tons/year. Key microbes found in terrestrial environments that can metabolize carbon monoxide include both aerobic and anaerobic organisms. Under aerobic conditions *Pseudomonas carboxydoflava* is an example of an organism that oxidizes carbon monoxide to carbon dioxide:

$$CO + H_2O \rightarrow CO_2 + H_2 \qquad (Eq.\ 14.3)$$

$$2H_2 + O_2 \rightarrow 2H_2O \qquad (Eq.\ 14.4)$$

This organism is a chemoautotroph and fixes the CO_2 generated in Eq. 14.3 into organic carbon. The oxidation of the hydrogen produced provides energy for CO_2 fixation (see Eq. 14.4). Under anaerobic conditions, methanogenic bacteria can reduce carbon monoxide to methane:

$$CO + 3H_2 \rightarrow CH_4 + H_2O \qquad (Eq.\ 14.5)$$

A number of other C1 compounds support the growth of methylotrophic bacteria (Table 14.9). Many, but not all, methylotrophs are also methanotrophic. Both types of bacteria are widespread in the environment because these C1 compounds are ubiquitous metabolites, and in response to their presence, microbes have evolved the capacity to metabolize them under either aerobic or anaerobic conditions.

14.3 NITROGEN CYCLE

In contrast to carbon, elements such as nitrogen and sulfur are taken up in the form of mineral salts and cycle oxidoreductively. For example, nitrogen can exist in numerous oxidation states, from -3 in ammonium (NH_4^+) to $+5$ in nitrate (NO_3^-). These element cycles are referred to as mineral cycles. The best studied and most complex of the mineral cycles is the nitrogen cycle. There is great interest in the nitrogen cycle because nitrogen is the mineral nutrient most in demand by microorganisms and plants. It is the fourth most com-

TABLE 14.9   C1 Compounds of Major Environmental Importance

Compound	Formula	Comments
Carbon dioxide	CO_2	Combustion, respiration, and fermentation end product, a major reservoir of carbon on earth.
Carbon monoxide	CO	Combustion product, common pollutant. Product of plant, animal and microbial respiration, highly toxic.
Methane	CH_4	End product of anaerobic fermentation or respiration.
Methanol	CH_3OH	Generated during breakdown of hemicellulose, fermentation by-product.
Formaldehyde	$HCHO$	Combustion product, intermediate metabolite.
Formate	$HCOOH$	Found in plant and animal tissues, fermentation product.
Formamide	$HCONH_2$	Formed from plant cyanides.
Dimethyl ether	CH_3OCH_3	Generated from methane by methanotrophs, industrial pollutant.
Cyanide ion	CN^-	Generated by plants, fungi, and bacteria. Industrial pollutant, highly toxic.
Dimethyl sulfide	$(CH_3)_2S$	Most common organic sulfur compound found in the environment, generated by algae.
Dimethyl sulfoxide	$(CH_3)_2SO$	Generated anaerobically from dimethyl sulfide.

mon element found in cells, making up approximately 12% of cell dry weight and includes the microbially catalyzed processes of nitrogen fixation, ammonium oxidation, assimilatory and dissimilatory nitrate reduction, ammonification, and ammonium assimilation.

14.3.1 Nitrogen Reservoirs

Nitrogen in the form of the inert gas, dinitrogen (N_2), has accumulated in the earth's atmosphere since the planet was formed. Nitrogen gas is continually released into the atmosphere from volcanic and hydrothermal eruptions, and is one of the major global reservoirs of nitrogen (Table 14.10). A second major reservoir is the nitrogen that is found in the earth's crust as bound, nonexchangeable ammonium. Neither of these reservoirs is actively cycled; the nitrogen in the earth's crust is unavailable and the N_2 in the atmosphere must be fixed before it is available for biological use. Because nitrogen fixation is an energy-intensive process and is carried out by a limited number of microorganisms, it is a relatively slow process. Smaller reservoirs of nitrogen include the organic nitrogen found in living biomass and in dead organic matter and soluble inorganic nitrogen salts. These small reservoirs tend to be actively cycled, particularly because nitrogen is often a limiting nutrient in the environment. For example, soluble inorganic nitrogen salts in terrestrial environments can have turnover rates greater than one per day. Nitrogen in plant biomass turns over approximately once a year, and nitrogen in organic matter turns over once in several decades.

TABLE 14.10   Global Nitrogen Reservoirs

Nitrogen reservoir	Metric tons nitrogen	Actively cycled
Atmosphere		
N_2	3.9×10^{15}	No
Ocean		
Biomass	5.2×10^8	Yes
Dissolved and particulate organics	3.0×10^{11}	Yes
Soluble salts (NO_3^-, NO_2^-, NH_4^+)	6.9×10^{11}	Yes
Dissolved N_2	2.0×10^{13}	No
Land		
Biota	2.5×10^{10}	Yes
Organic matter	1.1×10^{11}	Slow
Earth's crust[a]	7.7×10^{14}	No

[a] This reservoir includes the entire lithosphere found in either terrestrial or ocean environments. (Adapted from Dobrovolsky, 1994.)

TABLE 14.11   Relative Inputs of Nitrogen Fixation from Biological Sources

Source	Nitrogen fixation (metric tons/year)
Terrestrial	1.35×10^8
Aquatic	4.0×10^7
Fertilizer manufacture	3.0×10^7

14.3.2 Nitrogen Fixation

Ultimately, all fixed forms of nitrogen, NH_4^+, NO_3^-, and organic N, come from atmospheric N_2. The relative contributions to **nitrogen fixation** of microbes in terrestrial and aquatic environments are compared with human inputs in Table 14.11. Approximately 65% of the N_2 fixed annually is from terrestrial environments, including both natural systems and managed agricultural systems. Marine ecosystems account for a smaller proportion, 20%, of N_2 fixation. As already mentioned, nitrogen fixation is an energy-intensive process and until recently was performed only by selected bacteria and cyanobacteria. However, a substantial amount of nitrogen fixation, 15% of the total N_2 fixed, now occurs through the manufacture of fertilizers. Because this is an energy-driven process, fertilizer prices are tied to the price of fossil fuels. As fertilizers are expensive, management alternatives to fertilizer addition have become attractive. These include rotation between nitrogen-fixing crops such as soybeans and nonfixing crops such as corn. Wastewater reuse is another alternative that has become especially popular in the desert southwestern United States for nonfood crops and uses, such as cotton and golf courses, where both water and nitrogen are limiting (see Chapter 21).

Nitrogen is fixed into ammonia (NH_3) by over 100 different free-living bacteria, both aerobic and anaerobic, as well as some actinomycetes and cyanobacteria (Table 18.2). For example, *Azotobacter* (aerobic), *Beijerinckia* (aerobic), *Azospirillum* (facultative), and *Clostridium* (anaerobic) can all fix N_2. Because fixed nitrogen is required by all biological organisms, nitrogen-fixing organisms occur in most environmental niches. The amount of N_2 fixed in each niche depends on the environment (Table 14.12). Free-living bacterial cells that

TABLE 14.12   Rates of Nitrogen Fixation

N_2-fixing system	Nitrogen fixation (kg N/hectare/year)
Rhizobium–legume	200–300
Anabaena–*Azolla*	100–120
Cyanobacteria–moss	30–40
Rhizosphere associations	2–25
Free–living	1–2

are not in the vicinity of a plant root fix small amounts of nitrogen (1 to 2 kg N/hectare/year). Bacterial cells associated with the nutrient-rich rhizosphere environment can fix larger amounts of N_2 (2 to 25 kg N/hectare/year). Cyanobacteria are the predominant N_2-fixing organisms in aquatic environments, and because they are photosynthetic, N_2 fixation rates are one to two orders of magnitude higher than for free-living nonphotosynthetic bacteria. An evolutionary strategy developed collaboratively by plants and microbes to increase N_2 fixation efficiency was to enter into a symbiotic or mutualistic relationship to maximize N_2 fixation. The best studied of these symbioses is the *Rhizobium*-legume relationship, which can increase N_2 fixation to 200 to 300 kg N/hectare/year. This symbiosis irrevocably changes both the plant and the microbe involved but is beneficial to both organisms. Both the enzymology of N_2 fixation and the development of symbiotic N_2-fixation are presented in detail in Chapter 18.2.

As the various transformations of nitrogen are discussed in this section, the objective is to understand how they are interconnected and controlled. As already mentioned, N_2 fixation is limited to bacteria and is an energy-intensive process. Therefore, it does not make sense for a microbe to fix N_2 if sufficient amounts are present for growth. Thus, one control on this part of the nitrogen cycle is that ammonia, the end product of N_2-fixation, is an inhibitor for the N_2-fixation reaction. A second control in some situations is the presence of oxygen. Nitrogenase is extremely oxygen sensitive, and some free-living aerobic bacteria fix N_2 only at reduced oxygen tension. Other bacteria such as *Azotobacter* and *Beijerinckia* can fix N_2 at normal oxygen tension because they have developed mechanisms to protect the nitrogenase enzyme.

14.3.3 Ammonium Assimilation (Immobilization) and Ammonification (Mineralization)

The end product of N_2 fixation is ammonium. The ammonium produced is assimiliated by cells into amino acids to form proteins, cell wall components such as *N*-acetylmuramic acid, and purines and pyrimidines to form nucleic acids. This process is known as **ammonium assimilation** or **immobilization.** Nitrogen can also be immobilized by the uptake and incorporation of nitrate into organic matter, a process known as assimilatory nitrate reduction (Section 14.3.5.1). Because nitrate must be reduced to ammonium before it is incorporated into organic molecules, most organisms prefer to take up nitrogen as ammonium if it is available. The process that reverses immobilization, the release of ammonia from dead and decaying cells, is called **ammonification** or **ammonium mineralization.** Both immobilization and mineralization of nitrogen occur under aerobic and anaerobic conditions.

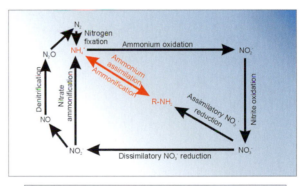

Summary:
Assimilation and ammonification cycles ammonia between its organic and inorganic forms
Assimilation predominates at C:N ratios > 20
ammonification predominates at C:N ratios < 20

14.3.3.1 Ammonium Assimilation (Immobilization)

There are two pathways that microbes use to assimilate ammonium. The first is a reversible reaction that incorporates or removes ammonia from the amino acid glutamate (Fig. 14.10A). This reaction is driven by ammonium availability. At high ammonium concentrations, (>0.1 mM or >0.5 mg N/kg soil), in the presence of reducing equivalents (reduced nicotinamide adenine dinucleotide phosphate, NADPH₂), ammonium is incorporated into α-ketoglutarate to

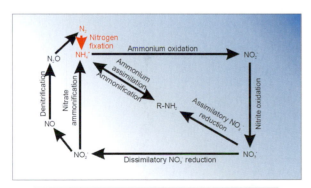

Summary:
N_2 fixation is energy intensive
End product of N_2 fixation is ammonia
N_2 fixation is inhibited by ammonia
Nitrogenase is O_2 sensitive, some free-living N_2 fixers require reduced O_2 tension

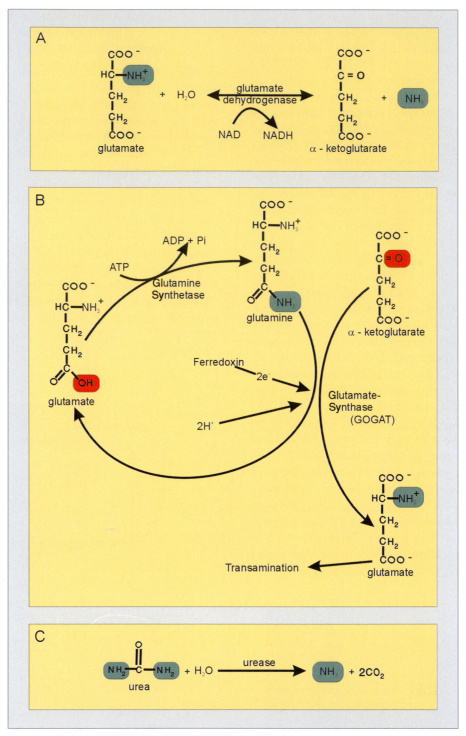

FIGURE 14.10 Pathways of ammonium assimilation and ammonification. Assimilation: (A) The enzyme glutamate dehydrogenase catalyzes a reversible reaction that immobilizes ammonium at high ammonium concentrations. (B) The enzyme system glutamine synthase–glutamate synthetase (GOGAT) that is induced at low ammonium concentrations. This ammonium uptake system requires ATP energy. Ammonification: Ammonium is released from the amino acid glutamate as shown in (A). (C) Ammonium is also released from urea, a molecule that is found in animal waste and is a common component of fertilizer. Note that the first of these reactions (A and B) occurs within the cell. In contrast, urease enzymes are extracellular enzymes resulting in release of ammonia to the environment.

form glutamate. However, in most soil and many aquatic environments, ammonium is present at low concentrations. Therefore microbes have a second ammonium uptake pathway that is energy dependent. This reaction is driven by ATP and two enzymes, glutamine synthase and glutamate synthetase (Fig. 14.10B). The first step in this reaction adds ammonium to glutamate to form glutamine, and the second step transfers the ammonium molecule from glutamine to α-ketoglutarate resulting in the formation of two glutamate molecules.

14.3.3.2 Ammonification (Mineralization)

Ammonium mineralization can occur intracellularly by the reversible reaction shown in Fig. 14.10A. Mineralization reactions can also occur extracellularly. As discussed in Section 14.2.4.1, microorganisms release a variety of extracellular enzymes that initiate degradation of plant polymers. Microorganisms also release a variety of enzymes including proteases, lysozymes, nucleases, and ureases that can initiate degradation of nitrogen-containing molecules found outside the cell such as proteins, cell walls, nucleic acids, and urea. Some of these monomers are taken up by the cell and degraded further, but some of the monomers are acted upon by extracellular enzymes to release ammonium directly to the environment, as shown in Fig. 14.10C for urea and the extracellular enzyme urease.

Which of these two processes, immobilization or mineralization, predominates in the environment? This depends on whether nitrogen is the limiting nutrient. If nitrogen is limiting, then immobilization will become the more important process. For environments where nitrogen is not limiting, mineralization will predominate. Nitrogen limitation is dictated by the carbon/nitrogen (C/N) ratio in the environment. Generally, the C/N ratio required for bacteria is 4 to 5 and for fungi is 10. So a typical average C/N ratio for soil microbial biomass is 8 (Myrold, 1998). It would then seem logical that at the C/N ratio of 8, there would be a net balance between mineralization and immobilization. However, one must take into account that only approximately 40% of the carbon in organic matter is actually incorporated into cell mass (the rest is lost as carbon dioxide). Thus, the C/N ratio must be increased by a factor of 2.5 to account for the carbon lost as carbon dioxide during respiration. Note that nitrogen is cycled more efficiently than carbon, and there are essentially no losses in its uptake. In fact, a C/N ratio of 20 is not only the theoretical balance point but also the practically observed one. When organic amend-

ments with C/N ratios less than 20 are added to soil, net mineralization of ammonium occurs. In contrast, when organic amendments with C/N ratios greater than 20 are added, net immobilization occurs.

There are numerous possible fates for ammonium that is released into the environment as a result of ammonium mineralization. It can be taken up by plants or microorganisms and incorporated into living biomass, or it can become bound to nonliving organic matter such as soil colloids or humus. In this capacity, ammonium adds to the cation-exchange capacity (CEC) of the soil. Ammonium can become fixed inside clay minerals, which essentially traps the molecule and removes the ammonium from active cycling. Also, because ammonium is volatile, some mineralized ammonium can escape into the atmosphere. Finally, ammonium can be utilized by chemoautotrophic microbes in a process known as nitrification.

14.3.4 Nitrification

Nitrification is the microbially catalyzed conversion of ammonium to nitrate. This is predominantly an aerobic chemoautotrophic process, but some methylotrophs can use the methane monooxygenase enzyme to oxidize ammonium and a few heterotrophic fungi and bacteria can also perform this oxidation. The autotrophic nitrifiers are a closely related group of bacteria (Table 14.13). The best studied nitrifiers are from

TABLE 14.13 Chemoautotrophic Nitrifying Bacteria

Genus	Species
Ammonium oxidizers	
Nitrosomonas	europaea
	eutrophus
	marina
Nitrosococcus	nitrosus
	mobilis
	oceanus
Nitrosospira	briensis
Nitrosolobus	multiformis
Nitrosovibrio	tenuis
Nitrite oxidizers	
Nitrobacter	winogradskyi
	hamburgensis
	vulgaris
Nitrospina	gracilis
Nitrococcus	mobilis
Nitrospira	marina

the genus *Nitrosomonas*, which oxidizes ammonium to nitrite, and *Nitrobacter*, which oxidizes nitrite to nitrate. The oxidation of ammonium is shown in Eq. 14.6:

$$\begin{aligned} &\text{ammonium}\\ &\text{monooxygenase}\\ NH_4^+ + O_2 + 2H^+ &\rightarrow NH_2OH\\ + H_2O &\rightarrow NO_2^- + 5H^+\ \Delta G = -66\ kcal \end{aligned}$$ (Eq. 14.6)

This is a two-step energy-producing reaction and the energy produced is used to fix carbon dioxide. There are two things to note about this reaction. First, it is an inefficient reaction requiring 34 moles of ammonium to fix 1 mole of carbon dioxide. Second, the first step of this reaction is catalyzed by the enzyme ammonium monooxygenase. This first step is analogous to Eq. 14.2, where the enzyme methane monooxygenase initiates oxidation of methane. Like methane monooxygenase, ammonium monooxygenase has broad substrate specificity and can be used to oxidize pollutant molecules such as TCE cometabolically (see Chapter 16.6.2.1.1).

The second step in nitrification, shown in Eq. 14.7, is even less efficient than the first step, requiring approximately 100 moles of nitrite to fix 1 mole of carbon dioxide.

$$NO_2^- + 0.5O_2 \rightarrow NO_3^-\ \Delta G = -18\ kcal$$ (Eq. 14.7)

These two types of nitrifiers, i.e., those that carry out the reactions shown in Eqs. 14.6 and 14.7, are generally found together in the environment. As a result, nitrite does not normally accumulate in the environment. Nitrifiers are sensitive populations. The optimum pH for nitrification is 6.6 to 8.0. In environments with pH < 6.0, nitrification rates are slowed, and below pH 4.5, nitrification seems to be completely inhibited.

Heterotrophic microbes that oxidize ammonium include some fungi and some bacteria. These organisms gain no energy from nitrification, so it is unclear why they carry out the reaction. The relative importance of autotrophic and heterotrophic nitrification in the environment has not yet been clearly determined. Although the measured rates of autotrophic nitrification in the laboratory are an order of magnitude higher than those of heterotrophic nitrification, some data for acidic forest soils have indicated that heterotrophic nitrification may be more important in such environments (Myrold, 1998).

Nitrate does not normally accumulate in natural, undisturbed ecosystems. There are several reasons for this. One is that nitrifiers are sensitive to many environmental stresses. But perhaps the most important reason is that natural ecosystems do not have much excess ammonium. However, in agricultural systems that have large inputs of fertilizer, nitrification can be-

come an important process resulting in the production of large amounts of nitrate. Other examples of managed systems that result in increased nitrogen inputs into the environment are feedlots, septic tanks, and landfills. The nitrogen released from these systems also becomes subject to nitrification processes. Because nitrate is an anion (negatively charged), it is very mobile in soil systems, which also have an overall net negative charge. Therefore, nitrate moves easily with water and this results in nitrate leaching into groundwater and surface waters. There are several health concerns related to high levels of nitrate in groundwater, including methemoglobinemia and the formation of nitrosamines. High levels of nitrate in surface waters can also lead to eutrophication and the degradation of surface aquatic systems. These consequences as well as the control of nitrification in managed systems are discussed in detail in Chapter 15.6.

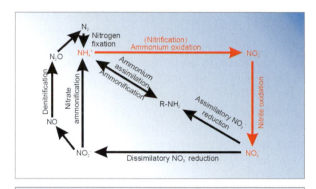

Summary:
Nitrification is an aerobic process
Nitrification is sensitive to a variety of chemical
   inhibitors and is inhibited at low pH
Nitrification processes in managed systems can
   result in nitrate leaching and groundwater contamination

14.3.5 Nitrate Reduction

What are the possible fates of nitrate in the environment? We have just discussed nitrate leaching into groundwater and surface waters as one possible fate. In addition, nitrate can be taken up and incorporated into living biomass by plants and microorganisms. The uptake of nitrate is followed by its reduction to ammonium, which is then incorporated into biomass. This process is called **assimilatory nitrate reduction** or **nitrate immobilization.** Finally, microorganisms can utilize nitrate as a terminal electron acceptor in anaerobic respiration to drive the oxidation of organic compounds. There are two separate pathways for this dissimilatory process, one called **dissimilatory nitrate reduction to ammonium,** where ammonium is the end product, and one called **denitrification,** where a mixture of gaseous products including N_2 and N_2O is formed.

14.3.5.1 Assimilatory Nitrate Reduction

Assimilatory nitrate reduction refers to the uptake of nitrate, its reduction to ammonium, and its incorporation into biomass (see Fig. 14.10A and B). Most microbes utilize ammonium preferentially, when it is present, to avoid having to reduce nitrate to ammonium, a process requiring energy. So if ammonium is present in the environment, assimilatory nitrate reduction is suppressed. Oxygen does not inhibit this activity. In contrast to microbes, for plants that are actively photosynthesizing and producing energy, the uptake of nitrate for assimilation is less problematic in terms of energy. In fact, because nitrate is much more mobile than ammonium, it is possible that in the vicinity of the plant roots, nitrification of ammonium to nitrate makes nitrogen more available for plant uptake. Because this process incorporates nitrate into biomass, it is also known as nitrate immobilization (see Section 14.3.3).

ergy to drive the oxidation of organic compounds. The end product of DRNA is ammonium:

$$NO_3^- + 4H_2 + 2H^+ \rightarrow NH_4^+ + 3H_2O$$
$$\Delta G = -144 \text{ kcal/8e}^- \text{ transfer} \qquad \text{(Eq. 14.8)}$$

The first step in this reaction, the reduction of nitrate to nitrite, is the energy-producing step. The further reduction of nitrite to ammonium is catalyzed by an NADH-dependent reductase. This second step provides no additional energy, but it does conserve fixed nitrogen and also regenerates reducing equivalents through the reoxidation of $NADH_2$ to NAD. These reducing equivalents are then used to help in the oxidation of carbon substrates. In fact, it has been demonstrated that under carbon-limiting conditions nitrite accumulates (denitrification predominates), while under carbon-rich conditions ammonium is the major product (DNRA predominates). A second environmental factor that selects for DNRA is low levels of available electron acceptors. It is not surprising therefore that this process is found predominantly in saturated, carbon-rich environments such as stagnant water, sewage sludge, some high-organic-matter sediments, and the rumen. Table 14.14 lists a variety of bacteria that perform DNRA. It is interesting to note that most of the bacteria on this list have fermentative rather than oxidative metabolisms.

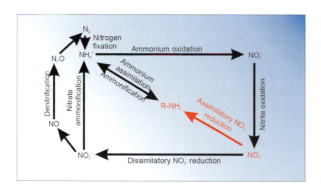

Summary:
Nitrate taken up must be reduced to ammonia before it is assimilated
Ammonia inhibits this process
O_2 does not inhibit this process

14.3.5.2 Dissimilatory Nitrate Reduction

Dissimilatory Nitrate Reduction to Ammonium

There are two separate dissimilatory nitrate reduction processes both of which are used by facultative chemoheterotrophic organisms under microaerophilic or anaerobic conditions. The first process, called dissimilatory nitrate reduction to ammonia (DNRA), uses nitrate as a terminal electron acceptor to produce en-

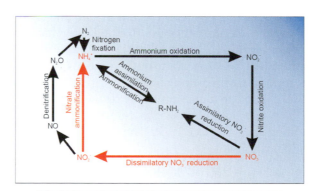

Summary:
Anaerobic respiration using nitrate as TEA
Inhibited by O_2
Not inhibited by ammonia
Found in a limited number of carbon rich, TEA poor environments
Fermentative bacteria predominate

Denitrification

The second type of dissimilatory nitrate reduction is known as **denitrification.** Denitrification refers to the microbial reduction of nitrate, through various gaseous

TABLE 14.14 Bacteria That Utilize Dissimilatory Nitrate or Nitrite to Ammonium (DRNA)

Genus	Typical habitat
Obligate anaerobes	
Clostridium	Soil, sediment
Desulfovibrio	Sediment
Selenomonas	Rumen
Veillonella	Intestinal tract
Wolinella	Rumen
Facultative anaerobes	
Citrobacter	Soil, wastewater
Enterobacter	Soil, wastewater
Erwinia	Soil
Escherichia	Soil, wastewater
Klebsiella	Soil, wastewater
Photobacterium	Seawater
Salmonella	Sewage
Serratia	Intestinal tract
Vibrio	Sediment
Microaerophiles	
Campylobacter	Oral cavity
Aerobes	
Bacillus	Soil, food
Neisseria	Mucous membranes
Pseudomonas	Soil, water

Adapted from Tiedje (1988).

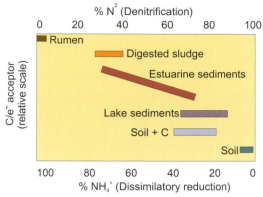

FIGURE 14.11 Partitioning of nitrate between denitrification and DNRA as a function of available carbon/electron (C/e⁻) acceptor ratio. (Adapted from J. M. Tiedje in *Biology of Anaerobic Microorganisms*, A. J. B. Zehnder, ed. © 1988. Reprinted by permission of John Wiley & Sons, Inc.)

inorganic forms, to N_2. This is the primary type of dissimilatory nitrate reduction found in soil, and as such is of concern because it cycles fixed nitrogen back into N_2. This process removes a limiting nutrient from the environment. Further, some of the gaseous intermediates formed during denitrification, e.g., nitrous oxide (N_2O), can cause depletion of the ozone layer and can also act as a greenhouse gas contributing to global warming (see Section 15.5). The overall reaction for denitrification is

$$NO_3^- + 5H_2 + 2H^+ \rightarrow N_2 + 6H_2O$$
$$\Delta G = -212 \text{ kcal/8e}^- \text{ transfer} \qquad \text{(Eq. 14.9)}$$

Denitrification, when calculated in terms of energy produced for every eight-electron transfer, provides more energy per mole of nitrate reduced than DNRA. Thus, in a carbon-limited, electron acceptor–rich environment, denitrification will be the preferred process because it provides more energy than DNRA. The relationship between denitrification and DNRA is summarized in Fig. 14.11.

The four steps involved in denitrification are shown in more detail in Fig. 14.12. The first step, reduction of nitrate to nitrite, is catalyzed by the enzyme nitrate reductase. This is a membrane-bound molybdenum–

iron–sulfur protein that is found not only in denitrifiers but also in DNRA organisms. Both the synthesis and the activity of nitrate reductase are inhibited by oxygen. Thus, both denitrification and DNRA are inhibited by oxygen. The second enzyme in this pathway is nitrite reductase, which catalyzes the conversion of nitrite to nitric oxide. Nitrite reductase is unique to denitrifying organisms and is not present in the DNRA process. It is found in the periplasm and exists in two forms, a copper-containing form and a heme form, both of which are distributed widely in the environment. Synthesis of nitrite reductase is inhibited by oxygen and induced by nitrate. Nitric oxide reductase, a membrane-bound protein, is the third enzyme in the pathway, catalyzing the conversion of nitric oxide to nitrous oxide. The synthesis of this enzyme is inhibited by oxygen and induced by various nitrogen oxide forms. Nitrous oxide reductase is the last enzyme in the pathway and converts nitrous oxide to dinitrogen gas. This is a periplasmic copper-containing protein. The activity of the nitrous oxide reductase enzyme is inhibited by low pH and is even more sensitive to oxygen than the other three enzymes in the denitrification pathway. Thus, nitrous oxide is the final product of denitrification under conditions of high oxygen (in a relative sense, given a microaerophilic niche) and low pH. In summary, both the synthesis and activity of denitrification enzymes are controlled by oxygen. Enzyme activity is more sensitive to oxygen than enzyme synthesis as shown in Fig. 14.13. The amount of dissolved oxygen in equilibrium with water at 20°C and 1 atm pressure is 9.3 mg/l. However, as little as 0.5 mg/l or less inhibits the activity of denitrification enzymes. As already stated, nitrous oxide re-

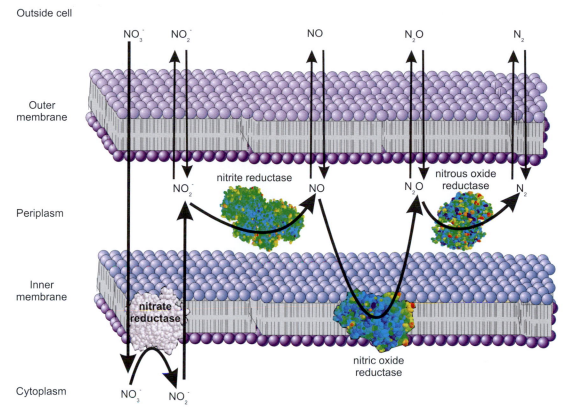

FIGURE 14.12   The denitrification pathway. (Adapted from Myrold, 1998.)

ductase is the most sensitive denitrification enzyme and it is inhibited by dissolved oxygen concentrations of less than 0.2 mg/l.

Whereas the denitrification pathway is very sensitive to oxygen, neither it nor the DNRA pathway is in-hibited by ammonium as is the assimilatory nitrate reduction pathway. However, the initial nitrate level in an environmental system can help determine the extent of the denitrification pathway. Low nitrate levels tend to favor production of nitrous oxide as the end

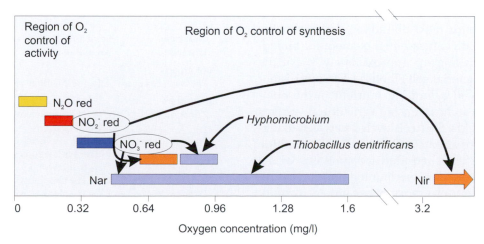

FIGURE 14.13   Approximate regions of oxygen concentration that inhibit the enzyme activity and synthesis for three steps in the denitrification pathway. (Adapted from J. M. Tiedje in *Biology of Anaerobic Microorganisms,* A. J. B. Zehnder, ed. © 1988. Reprinted by permission of John Wiley & Sons, Inc.)

product. High nitrate levels favor production of N_2 gas, a much more desirable end product.

Organisms that denitrify are found widely in the environment and display a variety of different characteristics in terms of metabolism and activities. In contrast to DNRA organisms, which are predominantly heterotrophic using fermentative metabolism, the majority of denitrifiers also are heterotrophic but use respiratory pathways of metabolism. However, as shown in Table 14.15, some denitrifiers are autotrophic, some are fermentative, and some are associated with other aspects of the nitrogen cycle; for example, they can fix N_2.

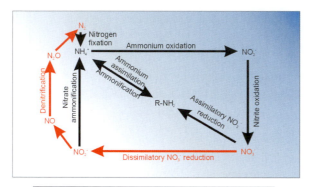

Summary:
Anaerobic respiration using nitrate as TEA
Inhibited by O_2
Not inhibited by ammonia
Produces a mix of N_2 and N_2O
Many heterotrophic bacteria are denitrifiers

14.4 SULFUR CYCLE

Sulfur is the tenth most abundant element in the crust of the earth. It is an essential element for biological organisms, making up approximately 1% of the dry weight of a bacterial cell (Table 14.1). Sulfur is not generally considered a limiting nutrient in the environment except in some intensive agricultural systems with high crop yields. Sulfur is cycled between oxidation states of +6 for sulfate (SO_4^{2-}) and −2 for sulfide (S^{2-}). In cells, sulfur is required for synthesis of the amino acids cysteine and methionine and is also required for some vitamins, hormones, and coenzymes. In proteins, the sulfur-containing amino acid cysteine is especially important because the formation of disulfide bridges between cysteine residues helps govern protein folding and hence activity. All of these compounds contain sulfur in the reduced or sulfide form. Cells also contain organic sulfur compounds in which

the sulfur is in the oxidized state. Examples of such compounds are glucose sulfate, choline sulfate, phenolic sulfate, and two ATP–sulfate compounds that are required to sulfate assimilation and can also serve to store sulfur for the cell. Although the sulfur cycle is not as complex as the nitrogen cycle, the global impacts of the sulfur cycle are extremely important, including the formation of acid rain, acid mine drainage, and corrosion of concrete and metal (see Chapter 15.2 and 15.3).

14.4.1 Sulfur Reservoirs

Sulfur is outgassed from the earth's core through volcanic activity. The sulfur gases released, primarily

TABLE 14.15   Genera of Denitrifying Bacteria

Genus	Interesting characteristics
Organotrophs	
Alcaligenes	Common soil bacterium
Agrobacterium	Some species are plant pathogens
Aquaspirillum	Some are magnetotactic, oligotrophic
Azospirillum	Associative N_2 fixer, fermentative
Bacillus	Spore former, fermentative, some species thermophilic
Blastobacter	Budding bacterium, phylogenetically related to *Rhizobium*
Bradyrhizobium	Symbiotic N_2 fixer with legumes
Branhamella	Animal pathogen
Chromobacterium	Purple pigmentation
Cytophaga	Gliding bacterium; cellulose degrader
Flavobacterium	Common soil bacterium
Flexibacter	Gliding bacterium
Halobacterium	Halophilic
Hyphomicrobium	Grows on one-C substrates, oligotrophic
Kingella	Animal pathogen
Neisseria	Animal pathogen
Paracoccus	Halophilic, also lithotrophic
Propionibacterium	Fermentative
Pseudomonas	Commonly isolated from soil, very diverse genus
Rhizobium	Symbiotic N_2 fixer with legumes
Wolinella	Animal pathogen
Phototrophs	
Rhodopseudomonas	Anaerobic, sulfate reducer
Lithotrophs	
Alcaligenes	Uses H_2, also heterotrophic, common soil isolate
Bradyrhizobium	Uses H_2, also heterotrophic, symbiotic N_2 fixer with legumes
Nitrosomonas	NH_3 oxidizer
Paracoccus	Uses H_2, also heterotrophic, halophilic
Pseudomonas	Uses H_2, also heterotrophic, common soil isolate
Thiobacillus	S-oxidizer
Thiomicrospira	S-oxidizer
Thiosphaera	S-oxidizer, heterotrophic nitrifier, aerobic denitrification

(From Myrold, 1998.)

sulfur dioxide (SO_2) and hydrogen sulfide (H_2S), become dissolved in the ocean and aquifers. Here the hydrogen sulfide forms sparingly soluble metal sulfides, mainly iron sulfide (pyrite), and sulfur dioxide forms metal sulfates with calcium, barium, and strontium as shown in Eqs. 14.10 and 14.11.

$$2S^{2-} + Fe^{2+} \rightarrow FeS_2 \text{ (pyrite)} \qquad \text{(Eq. 14.10)}$$

$$SO_2^{2-} + Ca^{2+} \rightarrow CaSO_4 \text{ (gypsum)} \qquad \text{(Eq. 14.11)}$$

This results in a substantial portion of the outgassed sulfur being converted to rock. Some of the gaseous sulfur compounds find their way into the upper reaches of the ocean and the soil. In these environments, microbes take up and cycle the sulfur. Finally, the small portions of these gases that remain after precipitating and cycling find their way into the atmosphere. Here, they are oxidized to the water-soluble sulfate form, which is washed out of the atmosphere by rain. Thus, the atmosphere is a relatively small reservoir of sulfur (Table 14.16). Of the sulfur found in the atmosphere, the majority is found as sulfur dioxide. Currently, one third to one half of the sulfur dioxide emitted to the atmosphere is from industrial and automobile emissions as a result of the burning of fossil fuels. A smaller portion of the sulfur in the atmosphere is present as hydrogen sulfide and is biological in origin.

The largest reservoir of sulfur is found in the earth's crust and is composed of inert elemental sulfur deposits, sulfur–metal precipitates such as pyrite (FeS_2) and gypsum ($CaSO_4$), and sulfur associated with buried fossil fuels. A second large reservoir that is slowly cycled is the sulfate found in the ocean, where it is the second most common anion (Dobrovolsky, 1994). Smaller and more actively cycled reservoirs of sulfur include sulfur found in biomass and organic matter in the terrestrial and ocean environments. Two recent

TABLE 14.16 Global Sulfur Reservoirs

Sulfur reservoir	Metric tons sulfur	Actively cycled
Atmosphere		
SO_2/H_2S	1.4×10^6	Yes
Ocean		
Biomass	1.5×10^8	Yes
Soluble inorganic ions		
(primarily SO_4^{2-})	1.2×10^{15}	Slow
Land		
Living biomass	8.5×10^9	Yes
Organic matter	1.6×10^{10}	Yes
Earth's crust[a]	1.8×10^{16}	No

[a] This reservoir includes the entire lithosphere found in either terrestrial or ocean environments. (Adapted from Dobrovolsky, 1994.)

practices have caused a disturbance in the global sulfur reservoirs. The first is strip mining, which has exposed large areas of metal–sulfide ores to the atmosphere, resulting in the formation of acid mine drainage. The second is the burning of fossil fuels, a sulfur reservoir that was quite inert until recently. This has resulted in sulfur dioxide emissions into the atmosphere with the resultant formation of acid rain. These processes are discussed further in Section 15.3.

14.4.2 Assimilatory Sulfate Reduction and Sulfur Mineralization

The primary soluble form of inorganic sulfur found in soil is sulfate. Whereas plants and most microorganisms incorporate reduced sulfur (sulfide) into amino acids or other sulfur-requiring molecules, they take up sulfur in the oxidized sulfate form and then reduce it internally (Widdel, 1988). This is called **assimilatory sulfate reduction.** Cells assimilate sulfur in the form of sulfate because it is the most available sulfur form, and because sulfide is toxic. Sulfide toxicity occurs because inside the cell sulfide reacts with metals in cytochromes to form metal–sulfide precipitates, destroying cytochrome activity. However, under the controlled conditions of sulfate reduction inside the cell, the sulfide can be removed immediately and incorporated into an organic form (see Information Box 1). Although this process does protect the cell from harmful effects of the sulfide, it is an energy-consuming reaction. After sulfate is transported inside the cell, ATP is used to convert the sulfate into the energy-rich molecule adenosine 5'-phosphosulfate (APS) (Eq. 14.13). A second ATP molecule is used to transform APS to 3'-phosphoadenosine-5'-phosphosulfate (PAPS) (Eq. 14.14). This allows the sulfate to be reduced to sulfite and then sulfide in two steps (Eq. 14.15 and 14.16). Most commonly, the amino acid serine is used to remove sulfide as it is reduced, forming the sulfur-containing amino acid cysteine (see Eq. 14.17).

The release of sulfur from organic forms is called **sulfur mineralization.** The release of sulfur from organic molecules occurs under both aerobic and anaerobic conditions. The enzyme serine sulfhydrylase can remove sulfide from cysteine in the reverse of the reaction shown in Eq. 14.17, or a second enzyme, cysteine sulfhydrylase, can remove both sulfide and ammonia as shown in Eq. 14.18.

$$\text{cysteine} \xrightarrow{\text{cysteine sulfhydrylase}} \text{serine} + H_2S \qquad \text{(Eq. 14.18)}$$

In marine environments, one of the major products of algal metabolism is the compound dimethylsulfoniopropionate (DMSP), which is used in osmoregulation of the cell. The major degradation product of DMSP is

---

Information Box 1

$$\text{sulfate (outside cell)} \xrightarrow{\text{active transport}} \text{sulfate (inside cell)} \tag{Eq. 14.12}$$

$$\text{ATP} + \text{sulfate} \xrightarrow{\text{ATP sulfurylase}} \underset{\text{adenosine phosphosulfate}}{\text{APS}} + \text{Ppi} \tag{Eq. 14.13}$$

$$\text{ATP} + \text{APS} \xrightarrow{\text{APS phosphokinase}} \underset{\text{3-phosphoadenosine-5-phosphosulfate}}{\text{PAPS}} \tag{Eq. 14.14}$$

$$\underset{\substack{\text{thioredoxin}\\\text{(reduced)}}}{2\text{RSH}} + \text{PAPS} \xrightarrow{\text{PAPS reductase}} \text{sulfite} + \underset{\text{AMP-3-phosphate}}{\text{PAP}} + \underset{\substack{\text{thioredoxin}\\\text{(oxidized)}}}{\text{RSSP}} \tag{Eq. 14.15}$$

$$\text{sulfite} + 3\text{NADPH} \xrightarrow{\text{sulfite reductase}} \text{H}_2\text{S} + 3\text{NADP} \tag{Eq. 14.16}$$

$$O\text{-acetyl-L-serine} + \text{H}_2\text{S} \xrightarrow{\substack{O\text{-acetylserine}\\\text{sulfhydrylase}}} \text{L-cysteine} + \text{acetate} + \text{H}_2\text{O} \tag{Eq. 14.17}$$

---

dimethylsulfide (DMS). Both H_2S and DMS are volatile compounds and therefore can be released to the atmosphere. Once in the atmosphere, these compounds are photooxidized to sulfate (Eq. 14.19).

$$\text{H}_2\text{S/DMS} \xrightarrow{\text{UV light}} \text{SO}_4^{2-} \xrightarrow{+\text{H}_2\text{O}} \underset{\text{sulfuric acid}}{\text{H}_2\text{SO}_4} \tag{Eq. 14.19}$$

Normal biological release of reduced volatile sulfur compounds results in the formation of approximately 1 kg SO_4^{2-}/hectare/year. The use of fossil fuels, which all contain organic sulfur compounds, increases the amount of sulfur released to the atmosphere to up to 100 kg SO_4^{2-}/hectare/year in some urban areas. Exacerbating this problem is the fact that reserves of fossil fuels that are low in sulfur are shrinking, forcing the use of reserves with higher sulfur content. Burning of fossil fuels produces sulfite as shown in Eq. 14.20.

$$\text{Fossil fuel combustion} \rightarrow \text{SO}_2 \xrightarrow{+\text{H}_2\text{O}} \underset{\text{sulfurous acid}}{\text{H}_2\text{SO}_3} \tag{Eq. 14.20}$$

Thus, increased emission of sulfur compounds to the atmosphere results in the formation of sulfur acid compounds. These acidic compounds dissolve in rainwater and can decrease the rainwater pH from neutral to as low as pH 3.5, a process also known as the formation of **acid rain.** Acid rain damages plant foliage, causes corrosion of stone and concrete building surfaces, and can affect weakly buffered soils and lakes (see Chapter 15.2).

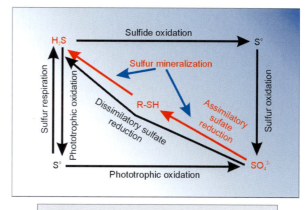

Summary:
Sulfur is taken up as sulfate due to H_2S toxicity
Assimilation and mineralization cycles sulfur between its organic and inorganic forms
Mineralization predominates at C:S ratios < 200, assimilation predominates at C:S > 400

14.4.3 Sulfur Oxidation

In the presence of oxygen, reduced sulfur compounds can support the growth of a group of chemoautotrophic bacteria under strictly aerobic conditions and a group of photoautrophic bacteria under strictly anaerobic conditions (Table 14.17). In addition, a number of aerobic heterotrophic microbes, including both bacteria and fungi, oxidize sulfur to thiosulfate or to

TABLE 14.17   Sulfur-Oxidizing Bacteria

Group	Sulfur conversion	Habitat requirements	Habitat	Genera
Obligate or facultative chemoautotrophs	$H_2S \rightarrow S°$ $S° \rightarrow SO_4^{2-}$ $S_2O_3^{2-} \rightarrow SO_4^{2-}$	H_2S-O_2 interface	Mud, hot springs, mining surfaces, soil	*Thiobacillus* *Thiomicrospira* *Achromatium* *Beggiatoa* *Thermothrix*
Anaerobic phototrophs	$H_2S \rightarrow S°$ $S° \rightarrow SO_4^{2-}$	Anaerobic, H_2S, light	Shallow water, anaerobic sediments meta- or hypolimnion, anaerobic water	*Chlorobium* *Chromatium* *Ectothiorhodospira* *Thiopedia* *Rhodopseudomonas*

Adapted from Germida (1998).

sulfate. The heterotrophic sulfur oxidation pathway is still unclear, but apparently no energy is obtained in this process. Chemoautotrophs are considered the predominant sulfur oxidizers in most environments. However, because many chemoautotrophic sulfur oxidizers require a low pH for optimal activity, heterotrophs may be more important in some aerobic, neutral to alkaline soils. Further, heterotrophs may initiate sulfur oxidation, resulting in a lowered pH that is more amenable for chemoautotrophic activity.

14.4.3.1 Chemoautotrophic Sulfur Oxidation

Of the chemoautotrophs, most oxidize sulfide to elemental sulfur, which is then deposited inside the cell as characteristic granules (Eq. 14.21).

$$H_2S + \tfrac{1}{2} O_2 \rightarrow S° + H_2O$$
$$\Delta G = -50 \text{ kcal} \qquad \text{(Eq. 14.21)}$$

The energy provided by this oxidation is used to fix CO_2 for cell growth. In examining Eq. 14.21, it is apparent that these organisms require both oxygen and sulfide. However, reduced compounds are generally abundant in areas that contain little or no oxygen. So these microbes are **microaerophilic;** they grow best under conditions of low oxygen tension (Fig. 14.14). Characteristics of marsh sediments that contain these organisms are their black color due to sulfur deposits and their "rotten egg" smell due to the presence of H_2S. Most of these organisms are filamentous and can easily be observed by examining a marsh sediment under the microscope and looking for small white filaments.

Some chemoautotrophs, most notably *Thiobacillus thiooxidans*, can oxidize elemental sulfur as shown in Eq. 14.22.

$$S° + 1.5O_2 + H_2O \rightarrow H_2SO_4$$
$$\Delta G = -150 \text{ kcal} \qquad \text{(Eq. 14.22)}$$

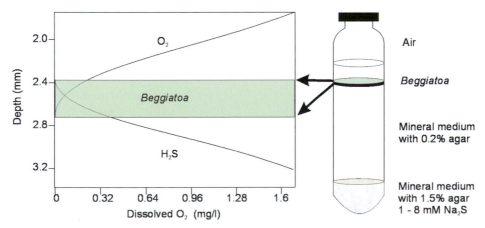

FIGURE 14.14   Cultivation of the sulfur-oxidizing chemolithotroph *Beggiatoa*. At the right is a culture tube with sulfide agar overlaid with initially sulfide-free soft mineral agar. The airspace in the closed tube is the source of oxygen. Stab-inoculated *Beggiatoa* grows in a narrowly defined gradient of H_2S and oxygen as shown. (Adapted from *Microbial Ecology* by R. M. Atlas and R. Bartha. © 1993 by Benjamin Cummings. Adapted by permission.)

This reaction produces acid, and as a result, *T. thiooxidans* is extremely acid tolerant with an optimal growth pH of 2. It should be noted that there are various *Thiobacillus* species, and these vary widely in their acid tolerance. However, the activity of *T. thiooxidans* in conjunction with the iron-oxidizing, acid-tolerant, chemoautotroph *T. ferrooxidans* is responsible for the formation of acid mine drainage, an undesirable consequence of sulfur cycle activity (see Chapter 15.3). It should be noted that the same organisms can be harnessed for the acid leaching and recovery of precious metals from low-grade ore, also known as **biometallurgy** (see Chapter 17). Thus, depending on one's perspective, these organisms can be very harmful or very helpful.

Although most of the sulfur-oxidizing chemoautotrophs are obligate aerobes, there is one exception, *Thiobacillus denitrificans*, a facultative anaerobic organism that can substitute nitrate as a terminal electron acceptor for oxygen (Eq. 14.23).

$$S° + NO_3^- + CaCO_3 \rightarrow CaSO_4 + N_2 \qquad \text{(Eq. 14.23)}$$

In the above equation, the sulfate formed is shown as precipitating with calcium to form gypsum. *T. denitrificans* is not acid tolerant but has an optimal pH for growth of 7.0.

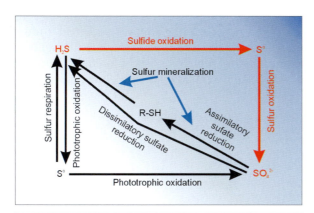

Summary:
Sulfur oxidation is a chemoautotrophic, aerobic process
Most sulfur oxidizers are microaerophilic
Some heterotrophs oxidize sulfur but obtain no energy
This process can result in the formation of acid mine
    drainage
This process can be used in metal recovery

14.4.3.2 Photoautotrophic Sulfur Oxidation

Photoautotrophic oxidation of sulfur is limited to green and purple sulfur bacteria (Table 14.17). This group of bacteria evolved on early Earth when the atmosphere contained no oxygen. These microbes fix carbon using light energy, but instead of oxidizing water to oxygen, they use an analogous oxidization of sulfide to sulfur.

$$CO_2 + H_2S \rightarrow S° + \text{fixed carbon} \qquad \text{(Eq. 14.24)}$$

These organisms are found in mud and stagnant water, sulfur springs, and saline lakes. In each of these environments, both sulfide and light must be present. Although the contribution to primary productivity is small in comparison with aerobic photosynthesis, these organisms are important in the sulfur cycle. They serve to remove sulfide from the surrounding environment, effectively preventing its movement into the atmosphere and its precipitation as metal sulfide.

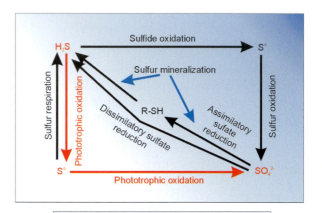

Summary:
Photoautotrophic, anaerobic process
Limited to purple and green sulfur bacteria
Responsible for a very small portion of
    photosynthetic activity

14.4.4 Sulfur Reduction

There are three types of sulfur reduction. The first, already discussed in Section 14.4.2, is performed to assimilate sulfur into cell components (Widdel, 1988). Assimilatory sulfate reduction occurs under either aerobic or anaerobic conditions. In contrast, there are two dissimilatory pathways, both of which use an inorganic form of sulfur as a terminal electron acceptor. In this case sulfur reduction occurs only under anaerobic conditions. The two types of sulfur that can be used as terminal electron acceptors are elemental sulfur and sulfate. These two types of metabolism are differentiated as **sulfur respiration** and **dissimilatory sulfate reduction.** *Desulfuromonas acetooxidans* is an example of a bacterium that grows on small carbon compounds such as acetate, ethanol, and propanol using elemental sulfur as the terminal electron acceptor (Eq. 14.25).

$$CH_3COOH + 2H_2O + 4S° \rightarrow$$
$$\text{acetate} \qquad 2CO_2 + 4S^{2-} + 8H^+ \qquad \text{(Eq. 14.25)}$$

However, the use of sulfate as a terminal electron acceptor seems to be the more important environmental process. The following genera, all of which utilize sulfate as a terminal electron acceptor, are found widely distributed in the environment, especially in anaerobic sediments of aquatic environments, water-saturated soils, and animal intestines: *Desulfobacter*, *Desulfobulbus*, *Desulfococcus*, *Desulfonema*, *Desulfosarcina*, *Desulfotomaculum*, and *Desulfovibrio*. Together these organisms are known as the **sulfate-reducing bacteria (SRB)**. These organisms can utilize H_2 as an electron donor to drive the reduction of sulfate (Eq. 14.26):

$$4H_2 + SO_4^{2-} \rightarrow S^{2-} + 4H_2O \qquad (Eq.\ 14.26)$$

Thus, SRB compete for available H_2 in the environment, as H_2 is also the electron donor required by methanogens. It should be noted that this is not usually a chemoautotrophic process because most SRB cannot fix carbon dioxide. Instead they obtain carbon from low-molecular-weight compounds such as acetate or methanol. The overall reaction for utilization of methanol is shown in Eq. 14.27.

$$\underset{\text{methanol}}{4CH_3OH} + 3SO_4^{2-} \rightarrow 4CO_2 + 3S^{2-} + 8H_2O \qquad (Eq.\ 14.27)$$

Both sulfur and sulfate reducers are strict anaerobic chemoheterotrophic organisms that prefer small carbon substrates such as acetate, lactate, pyruvate, and low-molecular-weight alcohols. Where do these small carbon compounds come from in the environment? They are by-products of fermentation of plant and microbial biomass that occurs in anaerobic regions. Thus, the sulfate reducers are part of an anaerobic consortium of bacteria including fermenters, sulfate reducers, and methanogens that act together to completely mineralize organic compounds to carbon dioxide and methane (see Chapter 3). More recently, it has been found that some SRB can also metabolize more complex carbon compounds including some aromatic compounds and some longer chain fatty acids. These organisms are being looked at closely to determine whether they can be used in remediation of contaminated sites that are highly anaerobic and that would be difficult to oxygenate.

The end product of sulfate reduction is hydrogen sulfide. What are the fates of this compound? It can be taken up by chemoautotrophs or photoautotrophs and reoxidized, it can be volatilized into the atmosphere, or it can react with metals to form metal sulfides. In fact, the activity of sulfate reducers and the production of hydrogen sulfide are responsible for the corrosion of underground metal pipes. In this process, the hydrogen sulfide produced reacts with ferrous iron metal to more iron sulfide (see Chapter 15).

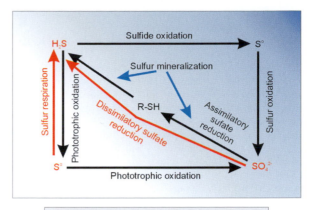

Summary:
Anaerobic respiration using SO_4^{2-} or S^0 as a TEA
Completely inhibited by O_2
Produces H_2S which can cause metal corrosion

QUESTIONS AND PROBLEMS

1. Give an example of
   a. a small actively cycled reservoir
   b. a large actively cycled reservoir
   c. a large inactively cycled reservoir

2. Describe how the ocean has reduced the expected rate of increase of CO_2 in the atmosphere since industrialization began.

3. What strategy is used by microbes to initiate degradation of large plant polymers such as cellulose?

4. Define what is meant by a greenhouse gas and give 2 examples. For each example describe how microorganisms mediate generation of the gas, and then describe how human activity influence generation of the gas.

5. Both autotrophic and heterotrophic activities are important in element cycling. For each cycle discussed in this chapter (carbon, nitrogen, and sulfur), name the most important heterotrophic and autotrophic activity. Justify your answer.

6. What would happen if microbial nitrogen fixation suddenly ceased?

References and Recommended Readings

Atlas, R. M., and Bartha, R. (1993) "Microbial Ecology." Benjamin Cummings, New York.

Deobald, L. A., and Crawford, D. L. (1997) Lignocellulose biodegradation. *In* "Manual of Environmental Microbiology" (C. J. Hurst, G. R. Knudsen, M. J. McInerney, L. D. Stetzenbach, and

M. V. Walter, eds.) American Society for Microbiology, Washington, DC, pp. 730–737.

Dobrovolsky, V. V. (1994) "Biogeochemistry of the World's Land." CRC Press, Boca Raton, FL.

Ehrlich, H. L. (1981) "Geomicrobiology." Marcel Dekker, New York.

Germida, J. J. (1998) Transformations of sulfur. *In* "Principles and Applications of Soil Microbiology" (D. M. Sylvia, J. J. Fuhrmann, P. G. Hartel, and D. A. Zuberer, eds.) Prentice-Hall, Upper Saddle River, NJ, pp. 346–368.

Haider, K. (1992) Problems related to the humification processes in soils of temperate climates. *In* "Soil Biochemistry", Vol. 7 (G. Stotzky, and J.-M. Bollag, eds.) Marcel Dekker, New York, pp. 55–94.

Killham, K. (1994) "Soil Ecology." Cambridge University Press, New York.

Lovelock, J. (1995) "The Ages of Gaia." W. W. Norton, New York.

Madigan, M. T., Martinko, J. M., and Parker, J. (1997) "Brock Biology of Microorganisms," 8th ed. Prentice Hall, Upper Saddle River, NJ.

Morgan, P., Lee, S. A., Lewis, S. T., Sheppard, A. N., and Watkinson, R. J. (1993) Growth and biodegradation by white-rot fungi inoculated into soil. *Soil Biol. Biochem.* **25,** 279–287.

Myrold, D. D. (1998) Transformations of nitrogen. *In* "Principles and Applications of Soil Microbiology" (D. M. Sylvia, J. J. Fuhrmann, P. G. Hartel, and D. A. Zuberer, eds.) Prentice-Hall, Upper Saddle River, NJ, pp. 218–258.

Neidhardt, F. C., Ingraham, J. L., Schaechter, M. (1990) "Physiology of the Bacterial Cell: A Molecular Approach." Sinauer Associates, Sunderland, MA.

Paul, E. A., and Clark, F. E. (1989) "Soil Microbiology and Biochemistry." Academic Press, New York.

Tiedje, J. M. (1988) Ecology of denitrification and dissimilatory nitrate reduction to ammonium. *In* "Biology of Anaerobic Microorganisms" (A. J. B. Zehnder, ed.) John Wiley & Sons, New York, pp. 179–244.

Wagner, G. H., and Wolf, D. C. (1998) Carbon transformations and soil organic matter formation. *In* "Principles and Applications of Soil Microbiology" (D. M. Sylvia, J. J. Fuhrmann, P. G. Hartel, and D. A. Zuberer, eds.) Prentice-Hall, Upper Saddle River, NJ, pp. 259–294.

Widdel, F. (1988) Microbiology and ecology of sulfate- and sulfur-reducing bacteria. *In* "Biology of Anaerobic Microorganisms" (A. J. B. Zehnder, ed.) John Wiley & Sons, New York, pp. 469–585.

15

Consequences of Biogeochemical Cycles Gone Wild

DAVID C. HERMAN   RAINA M. MAIER

15.1 INTRODUCTION

In Chapter 14 the basic biogeochemical cycles for carbon, nitrogen, and sulfur were introduced. Without interference, these cycles have remained stable for thousands to millions of years and the changes that have occurred did so very slowly. The increasing need for food and energy by a growing human population has recently interfered with these natural cycles, in some cases accelerating or in some cases slowing part of a cycle. As a result, there have been detrimental impacts on a global scale. One example is microbial activities that accelerate the formation of "greenhouse" gases such as carbon dioxide (Chapter 14.2.1), methane (Chapter 14.2.4.3), and nitrous oxide (Chapter 15.5), resulting in concern about global warming. This has occurred because of direct human action, namely the mining and utilization of fossil fuels, waste disposal in landfills, and fertilizer application. A second example is the accelerated formation of acid leachates in metal-mining areas that have affected the pH and water quality of receiving rivers and streams (Section 15.3.1). In this case, the human activity is mining, which in turn results in the enhanced microbial activity of sulfur oxidation (see Chapter 14.4.3). In some instances the impact of a biogeochemical activity may be more localized but still costly in terms of damage and repair, an example being corrosion of metal, stone, and concrete (Fig. 15.1, Section 15.2). This is not to say that all impacts of changing these cycles are detrimental. In fact, some activities of the biogeochemical cycles have been used to our benefit. An example is the use of

sulfur-oxidizing bacteria to aid in the recovery of metals such as gold, copper, and uranium from ore (Section 15.3.2). Yet another example is the use of microbial degradation activities in composting (Section 15.8) or remediation of contaminated sites (see Chapter 16). The objective of this chapter is to examine some of the detrimental and beneficial aspects of accelerated biogeochemical cycles that pose or help us solve modern pollution problems.

15.2. MICROBIALLY INFLUENCED CORROSION

15.2.1 Metal Corrosion

Microbially mediated corrosion damage can occur on metal structures submersed in water, such as the hulls of ships, and in pipelines that carry water and oil products. It is thought that microbially influenced metal corrosion accounts for 15 to 30% of the corrosion failures in the gas and nuclear industries. It is a major cause of pipeline failures in water treatment and chemical industries and is also associated with corrosion failures and souring in gas and oil production and storage (Dowling and Guezennec, 1997). Resulting damage is costly, with one 1990 estimate being $1.6 to $5.0 billion for underground iron pipes in the United States alone. Although the mechanisms and scale of microbially influenced corrosion are still not fully understood, a common mechanism by which sulfate-reducing bacteria (SRB) are thought to influence corrosion has been described (Dowling and

FIGURE 15.1 Microbially influenced deterioration of an angel statue above the "Peters" Portal on the cathedral of Cologne (Germany); documented by the original object in 1880 (photo by Anselm Schmitz, Cologne) and the respective weathered statue in 1993 (photo by Dombaumeister Prof., Dr. A. Wolff, Cologne). (Used with permission from Warscheid and Krumbein, 1996.)

Guezennec, 1997). SRB are active under strictly anaerobic conditions, utilizing sulfate instead of oxygen in the respiration of organic compounds (Chapter 14.4.4). SRB-mediated corrosion requires anaerobic conditions, but oxygen is often present in the environment surrounding a metal surface. However, when biofilms (Chapter 6.2.4.1) form on the metal surface, anaerobic microsites can develop within the biofilm. These sites occur as a result of rapid oxygen utilization by bacteria in the outer layers of the biofilm coupled with a

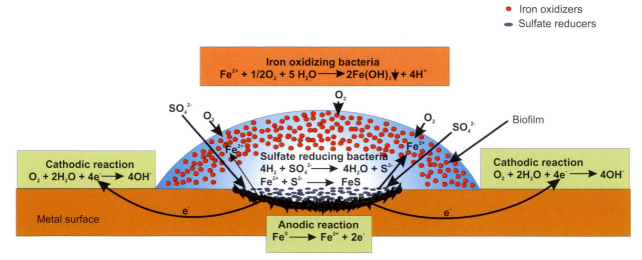

FIGURE 15.2 General representation of microbially influenced corrosion of a metal surface. (Adapted from Hamilton, 1995.)

slow rate of oxygen diffusion through the exopolymer matrix. These anaerobic biofilm microsites support the growth and activity of SRB, which ultimately cause the metal surface to corrode. Because SRB are common constituents of natural environments, the use of natural water sources in cooling towers or the pumping of oil from underground reservoirs serves to inoculate metal pipeline surfaces with SRB.

Metal corrosion is initiated by two spontaneous electrochemical reactions. In the first reaction (Eqs. 15.1 and 15.2), a differential aeration cell is set up in which the metal surface acts as the anode to produce metal ions, and in the corresponding cathodic reaction oxygen accepts the electrons produced from the oxidation of elemental iron:

Anodic reaction: $\quad Fe^{\circ} \rightarrow Fe^{2+} + 2e^{-}$ $\quad$ (Eq. 15.1)

Cathodic reaction:
$$\tfrac{1}{2}O_2 + H_2O + 2e^{-} \rightarrow 2(OH^{-}) \quad \text{(Eq. 15.2)}$$

In the second reaction (Eqs. 15.3 and 15.4) a concentration cell is created under anaerobic conditions in which the anodic reaction remains the same but the cathodic reaction produces H_2:

Anodic reaction: $\quad Fe^{\circ} \rightarrow Fe^{2+} + 2e^{-}$ $\quad$ (Eq. 15.3)

Cathodic reaction: $\quad 2H^{+} + 2e^{-} \rightarrow 2H \rightarrow H_2$ $\quad$ (Eq. 15.4)

Microbially influenced corrosion refers to the involvement of biofilms in stimulating these electrochemical reactions (Hamilton, 1995). How do microorganisms, specifically SRB, participate in this process? First of all, as already mentioned, a biofilm is formed on the metal surface that facilitates the establishment of these separate electrochemical reactions by establishing an anaerobic environment at the metal surface (Lappin-Scott and Costerton, 1989; Hamilton, 1995) (Fig. 15.2). Second, SRB utilize H_2 as an electron donor (see Chapter 14.4.4), thereby removing it from the environment and providing a driving force for the anodic reaction. Finally, the end product of sulfate reduction is sulfide (S^{2-}), which reacts with Fe^{2+} to form metal sulfide precipitate:

Step 1 $\quad Fe^{2+} + 2H_2O \rightarrow Fe(OH)_2 + H_2$
$\qquad\qquad\qquad\qquad\qquad\qquad$ (spontaneous)

Step 2 $\quad 4H_2 + SO_4^{2-} \rightarrow H_2S + 2OH^{-} + 2H_2O$
$\qquad\qquad$ (SRB, e.g., *Desulfovibrio desulfuricans*)

Step 3 $\quad H_2S + Fe^{2+} \rightarrow FeS \downarrow + H_2$ $\quad$ (spontaneous)

Overall $\quad 2Fe^{2+} + SO_4^{2-} + 2H_2$
$\qquad\qquad\qquad \rightarrow FeS \downarrow + Fe(OH)_2 \downarrow + 2OH^{-}$

Other mechanisms also contribute to microbially influenced corrosion. For example, the biofilm matrix is thought to contribute to corrosion by trapping corrosion products including organic acids and metal species at the metal–biofilm interface. As a result of all of these corrosion processes, pits and cracks appear on the metal surface that eventually compromise the integrity of the metal structure.

How can microbially influenced corrosion be controlled? There are basically two strategies. The first is to coat the metal surface with bactericidal chemicals. These include quaternary ammonia compounds, phenolic compounds, surface-active substances, and metals such as copper. However, these compounds can leach into the surrounding environment, raising concern about their use. The second strategy is to disrupt the organization of surface biofilms, thus removing the microenvironment that supports the activity of SRB. Disinfectant chemicals, such as chlorine (see Chapter 23), and addition of cationic surfactants are used to control microbial growth in pipelines, but these chemical are more effective in killing planktonic cells than in destroying biofilms. Further, the high chlorine concentrations required to destroy biofilms may have undesirable side effects. The oil industry uses an additional step besides chemical biofilm disruption, mechanical scraping (pigging) of the inside surface of pipes. Regular treatment is required because of regrowth of biofilms.

15.2.2 Concrete Corrosion

Microorganisms have also been shown to participate in the corrosion of concrete structures through the production of acidic metabolites (Diercks *et al.*, 1991). The binding components of concrete are acid sensitive, and therefore microbes that produce acidic metabolites contribute to this type of corrosion as well as to corrosion of other acid-sensitive ceramic and natural stone objects. Different forms of acidic metabolites can be produced by microorganisms, including organic acids produced by fermentative bacteria. Also, the chemoautotrophic sulfur oxidizing bacteria and the nitrifying bacteria can produce sulfuric and nitric acids, respectively (Chapter 14.4.3.1). These and other microorganisms have been found to colonize the surface of concrete structures, and their level of activity is strongly influenced by the amount of moisture that is available.

A well-documented consequence of biogenic acid production is the corrosion of concrete sanitary sewer pipes. There are thousands of miles of sewer pipes in any major city, and corrosion of these pipes has cost hundreds of millions of dollars in damage and replacement costs (Sydney *et al.*, 1996). Sewer pipes are corroded from the inside out by a two-step process involving both sulfate-reducing and sulfur-oxidizing bacterial populations that cycle sulfur as described in

Chapters 14.4.3.1 and 14.4.4 (Fig. 15.3). Pipes carrying sewage can contain two distinct environments, namely the liquid sewage and the headspace area. In the liquid environment, anaerobic conditions are created by the high rate of microbial activity in the organic-rich sewage. Under anaerobic conditions, the SRB generate sulfide, which is converted to the volatile H_2S form and exchanged across the liquid–headspace interface. In the aerobic and moist environment of the headspace, sulfide-oxidizing bacteria colonizing the concrete wall surface oxidize H_2S to sulfuric acid. Moisture condensation along the inside of the sewer pipe walls improves the habitat for microbial activity.

The actual corrosion process occurs when the sulfuric acid produced by sulfide oxidation reacts with the calcium hydroxide binder in the concrete to form calcium sulfate, which is a soft, expansive compound with no binding capability (Sydney *et al.*, 1996):

$$H_2SO_4 + Ca(OH)_2 \rightarrow CaSO_4 + 2H_2O \qquad \text{(Eq. 15.5)}$$

Mori *et al.* (1992) found that maximum corrosion occurred just above the sewage liquid level and that corrosion on the crown of the pipe was also evident. Near the surface of corroded concrete (0 to 4 mm depth), the population of a common sulfur oxidizer, *Thiobacillus thiooxidans*, was found to reach almost 100,000 cells per gram of concrete, and their activity created a highly acidic environment. The pH of water moisture in corroded concrete is reduced to around 2. In the zones of highest activity, the corrosion rate was estimated to be between 4.3 and 4.7 mm per year. Thus, over a 12-year period of use, the thickness of a sewer pipe can be reduced from 88 mm to between 32 and 36 mm. This rate of corrosion effectively reduces the life expectancy of concrete sewer pipes to approximately 20 years.

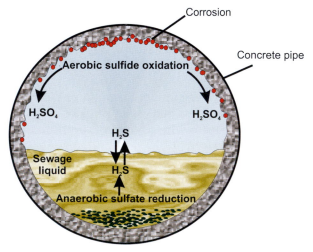

FIGURE 15.3   Cross section showing microbial involvement in the corrosion of a concrete sewer pipe. (Adapted from Sydney *et al.*, 1996.)

Two approaches have been used to control sewer pipeline corrosion (Sydney *et al.*, 1996). One approach is to treat the sewage liquid to prevent the production of H_2S. This type of treatment may include the use of caustic soda to inhibit microbial growth or the addition of alternative electron acceptors, such as iron (e.g., ferric chloride) or oxygen, to inhibit SRB activity competitively. An alternative approach is to neutralize the surface of the concrete above the sewage liquid level by spraying the concrete with a high-pH solution, such as a magnesium hydroxide slurry (Sydney *et al.*, 1996).

15.3 ACID MINE DRAINAGE AND METAL RECOVERY

15.3.1 Acid Mine Drainage

As discussed in Chapter 14.4.1, pyrite (FeS_2) is a major source of sulfur in the lithosphere. During mining of ore deposits and of bituminous coal, pyrite is exposed to oxygen and moisture and becomes the source of an acidic, iron-rich leachate known as acid mine drainage. Acid mine drainage can have a pH of less than 2 and can seriously impair the quality of receiving waters, such as surface streams or rivers. Two mining practices have exacerbated the formation of acid mine drainage. The first is strip mining, an activity that uncovers and exposes large surface areas to the atmosphere. Mining also generates large amounts of pyrite-containing rock or mine tailings. Tailings are the less valuable rock that remains after the material of interest, such as metal-bearing minerals, has been removed. Deposition of these tailings in open impoundments where they are exposed to air and rainwater is a second practice that results in the formation of acid mine drainage. As a result of past mining activities and millions of tons of tailings remaining from abandoned mining operations, acid mine drainage currently affects 10,000 miles of waterways on the East Coast of the United States alone.

The formation of acid from pyrite ore is a complex mechanism that involves the oxidation of both iron and sulfur. The initial reaction leading to the formation of acid mine drainage is the spontaneous chemical oxidation of pyrite:

$$4FeS_2 \text{ (pyrite)} + 15O_2 + 14H_2O$$
$$\rightarrow 4Fe(OH)_3{}^- + 8H_2SO_4 \qquad \text{(Eq. 15.6)}$$

This process is initiated spontaneously, but as the local pH drops, a sulfur- and iron-oxidizing bacterium, *Thiobacillus ferrooxidans*, also begins to participate in the reaction. *T. ferrooxidans* is a chemoautotrophic organism that derives energy for carbon fixation and growth from the oxidation of inorganic sulfur- and iron-containing compounds, such as pyrite (Chapter 14.4.3.1). This organism is unusual in that its pH optimum is around 2. *T. ferrooxidans* is a commonly occurring microbe and has been intensely studied because of its role in acid mine drainage formation and for its use in metal recovery. It is thought to attach directly to the pyrite crystal lattice and there utilize FeS_2 as an electron donor (Ehrlich, 1996). Other microbes, such as the thermophilic archaebacteria *Sulfolobus* and *Acidianus*, are also associated with acid mine drainage. These are acidophilic chemolithotrophs capable of maximal growth around pH 1.5 to 2.5 (Ehrlich, 1996). Because oxidation of pyrite is an exothermic reaction, temperatures within tailings impoundments can build until they exceed 60°C (Brierley and Brierley, 1997). *T. ferrooxidans* is a mesophile, oxidizing inorganic substrates in a temperature range around 10 to 40°C. Temperatures above this range create a selective advantage for the thermophilic archaebacteria, which then begin to predominate in the formation of acid mine drainage.

The ferrous iron (Fe^{2+}) formed by the oxidation of pyrite will, in a neutral aerobic environment, spontaneously oxidize to the ferric iron form (Fe^{3+}):

$$2Fe^{2+} + \frac{1}{2}O_2 + 2H^+ \rightarrow 2Fe^{3+} + H_2O \qquad \text{(Eq. 15.7)}$$

This is the rate-limiting step in the formation of acid leachate. The autoxidation of Fe^{2+} is a very slow process. Alternatively, *T. ferrooxidans* can oxidize Fe^{2+}. Because its pH optimum is 2, at first the microbial contribution to iron oxidation is small. But as the pH decreases, the microbial contribution grows and the reaction begins to occur more rapidly. The oxidized iron formed in Eq. 15.7 can have two fates. It can be precipitated as iron oxide, a reaction that generates more acid:

$$Fe^{3+} + 3H_2O \rightarrow Fe(OH)_3 \downarrow + 3H^+ \qquad \text{(Eq. 15.8)}$$

Alternatively, the ferric iron can aid in the further chemical oxidation of pyrite:.

$$FeS_2 \text{ (pyrite)} + 14Fe^{3+} + 8H_2O$$
$$\rightarrow 15Fe^{2+} + 2SO_4^{2-} + 16H^+ \qquad \text{(Eq. 15.9)}$$

Note that this reaction produces acid and regenerates reduced or ferrous iron, which can then be reoxidized by *T. ferroxidans* (Eq. 15.7). Thus, the combination of a microbially mediated reaction and a chemical oxidation reaction creates a reaction loop that speeds the oxidation of pyrite. Overall, these reactions can be summarized as

$$4FeS_2 \text{ (pyrite)} + 15O_2 + 14H_2O$$
$$\rightarrow 4Fe(OH)_3^- \downarrow + 8H_2SO_4 \qquad \text{(Eq. 15.10)}$$

Thus, the leachate produced is highly acidic and contains high levels of a dark brown ferrous hydroxide precipitate. Ferric iron remaining in the leachate can, in the presence of sulfate and a monovalent cation such as potassium, precipitate as a complex sulfate mineral, $KFe_3(SO_4)_2(OH)_6$, which is a yellow–brown product characteristic of acid mine drainage (Brierley and Brierley, 1997). In addition, the leachate can contain other acid-soluble elements including aluminum, which can be toxic to aquatic organisms. The leachate is carried by surface water and groundwater flow into receiving stream and river waters.

As already mentioned, acid mine drainage can have a pH of less than 2. The actual pH depends on the rate of growth of the pyrite-oxidizing bacteria, which in turn depends on environmental factors such as the amount of rainfall, nitrogen availability, temperature, and pH conditions (Ehrlich, 1996). The amount of acid produced is also dependent on the mineral composition of the tailings. Acid-consuming reactions can occur, which neutralize the pH to some extent. These reactions include the dissolution of carbonate minerals, such as calcite and dolomite; the formation of gypsum; and the bacterial oxidation of certain metal sulfides, such as copper and zinc sulfides (Brierley and Brierley, 1997). The balance between acid-generating and acid-consuming reactions influences the final pH of tailing leachate.

Abatement actions to remediate acid mine drainage are best applied as the tailings are being produced to prevent the initiation of pyrite oxidation. Prevention methods focus on restricting the exposure of tailings to oxygen and moisture, such as capping or encapsulating tailings with low-permeability material, such as clay. Other methods include mixing tailings with acid-consuming rocks to neutralize acid production and depositing tailings on impermeable material so that acid leachate can be collected and treated. It may also be possible to treat acid mine drainage using engineered environments such as wetlands that promote the activity of the sulfate-reducing bacteria (Southam *et al.*, 1995). The SRB can neutralize drainage waters by utilizing sulfate as an electron acceptor during anaerobic respiration of a supplied carbon source, such as acetate. The sulfides produced will also contribute to acid drainage abatement by the precipitation of divalent metal ions as metal sulfides.

15.3.2 Metal Recovery and Desulfurization of Coal

The reactions just described for formation of acid mine drainage that are so environmentally damaging can be harnessed in a controlled way for use in metal recovery and desulfurization of coal. As high-grade ore deposits have been increasingly mined, the recovery of metals from remaining low-grade deposits becomes more important. As it is less economical to smelt low-grade ores, microbially mediated metal recovery has become an attractive alternative to smelting. In recent years, metal recovery using the organism *T. ferrooxidans* has become a well-understood, efficient, and cost-effective process. As of 1989, more than 30% of U.S. copper and uranium production was microbially mediated. *T. ferrooxidans* can participate in both the **direct** and **indirect leaching** of metals such as copper from a variety of ores. The reactions for removal of copper from ores such as chalcopyrite ($CuFeS_2$), chalcocite (CuS_2), and covellite (CuS) are shown next:

Direct leaching: $MS + 2O_2 \rightarrow MSO_4$ (Eq. 15.11)

where M represents the metal being leached. An example of direct leaching is the recovery of copper from chalcocite as shown in Eq. 15.12. Note that energy is provided to the microbe by the oxidation of Cu^+ to Cu^{2+}, and there is no change in the valence of sulfur (Atlas and Bartha, 1993).

$$\underset{\text{chalcocite}}{2Cu_2S} + O_2 + 4H^+ \rightarrow \underset{\text{covellite}}{2CuS}$$
$$+ \underset{\text{leached copper}}{2Cu^{2+}} + 2H_2O \quad \text{(Eq. 15.12)}$$

More important overall are the indirect metal leaching reactions:

$$Cu_2S + 2Fe_2(SO_4)_3 \rightarrow 2CuSO_4 + 4FeSO_4 + S°$$
$$\text{(Eq. 15.13)}$$

$$2FeSO_4 + \tfrac{1}{2}O_2 + H_2SO_4 \rightarrow Fe_2(SO_4)_3 + H_2O$$
$$\text{(Eq. 15.14)}$$

where copper is spontaneously oxidized by the presence of the ferric ion (Fe^{3+}) and acid (Eq. 15.13), and then the resulting ferrous iron (Fe^{2+}) is reoxidized biologically by *T. ferrooxidans* (Eq. 15.14). Finally, the copper ions (Cu^{2+}) can be recovered from solution by spontaneous precipitation in the presence of scrap iron:

$$Cu^{2+} + Fe° \rightarrow Fe^{2+} + Cu° \quad \text{(Eq. 15.15)}$$

The related reactions for recovery of uranium and gold are described in Chapter 17.9, which also describes common approaches to microbial recovery of metals from ore.

A second controlled process that utilizes these reactions is the microbial desulfurization of coal. Coal reserves can contain high levels of sulfur-containing compounds. As cleaner coal reserves are utilized, sources of coal that contain higher levels of sulfur re-

main. The industrial burning of coal releases sulfur compounds as sulfur dioxide (SO_2). In the atmosphere, SO_2 combines with water to form sulfuric acid (H_2SO_4), which lowers the pH of the water, forming acid rain. Sulfur is found as two forms in coal; it can be in an organic form, most commonly as dibenzothiophene (Fig. 15.4), or as inorganic sulfur. The most important inorganic sulfur compound in coal is pyrite (Bos *et al.*, 1992). Thus, the burning of coal contributes not only to the greenhouse effect, which is connected with the use of all fossil fuels, but also to the formation of acid rain. However, the extent of SO_2 emissions from coal burning is determined by the sulfur content of coal. One strategy that is being investigated is to use the microbial leaching processes shown in Eqs. (15.7) to (15.10) to remove inorganic sulfur from coal. Note that although this process will remove inorganic sulfur, it does not remove organic sulfur forms. Other forms of microbial desulfurization being investigated include the bacterial oxidation of the organic sulfur molecule dibenzothiophene.

15.4 BIOMETHYLATION OF METALS AND METALLOIDS

Some metals and metalloids are environmental pollutants that can inhibit the activity of microorganisms and can also be toxic to higher organisms. Bioalkylation is the biologically mediated linking of an alkyl group, e.g., $-CH_3$, to a metal or metalloid element, thus forming an organometal(loid) compound. Metal(loid)s known to be methylated by microorganisms are listed in Table 15.1. Methylation, the linking of a methyl group to a metal(loid), occurs extensively in nature, with mercury methylation being the most intensively studied process. Microbes can also bind alkyl groups with longer carbon chains to the metalloids arsenic and selenium. Bioalkylation greatly alters the physical and chemical properties of a metal(loid) and has a major effect on the fate and biological impact of metal(loid)s

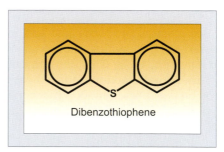

FIGURE 15.4 Dibenzothiophene, the most common organic form of sulfur found in coal.

TABLE 15.1   Elements Undergoing Biomethylation in the Natural Environment

Nickel (Ni)

Tin (Sn)

Antinomy (Sb)

Mercury (Hg)

Lead (Pb)

Arsenic (As)

Selenium (Se)

Germanium (Ge)

From Thayer (1993).

in the environment. Methylation also alters the toxicity of the element because the methylated form is more volatile and more soluble in lipids (Thayer, 1993). A consequence of increased lipid solubility is that methylated metal(loid)s are less easily excreted and thus accumulate in living organisms causing toxicity (Chapter 17.8).

Mercury is among the most common of the metal pollutants present in the environment. Trace levels of mercury are present in soil and sediments, and higher concentrations can be released into the environment through industrial activities. The most common form of mercury released to the environment is the divalent form, Hg^{2+}. In the aquatic environment, the organic-rich sediment becomes a natural sink for Hg^{2+}. Biomethylation of mercury occurs in the sediments of lakes, rivers, estuaries, and other water bodies. Microorganisms can methylate mercury in both aerobic and anaerobic environments, although the rate of mercury methylation is much greater under anaerobic conditions. The primary generators of methylmercury in the environment are believed to be the sulfate-reducing bacteria, although a variety of microorganisms are capable of methylating mercury. The most important intracellular agent of mercury methylation is believed to be methylcobalamine (CH_3CoB_{12}), a derivative of vitamin B_{12}. Methylation reactions can be summarized as follows (Gadd, 1993):

$$\underset{\text{methylcobalamine}}{CH_3CoB_{12}} + Hg^{2+} + H_2O \rightarrow CH_3Hg^+ + H_2OCoB_{12}^+$$
$$\underset{}{\qquad\qquad\qquad\qquad\qquad\qquad\text{methylmercury}}$$

(Eq. 15.16)

$$\underset{\text{methylcobalamine}}{CH_3CoB_{12}} + CH_3Hg^+ + H_2O \rightarrow$$

(Eq. 15.17)

$$\underset{\text{dimethylmercury}}{(CH_3)_2Hg} + H_2OCoB_{12}^+$$

The dominant product formed is the salt of the methylmercuric ion, CH_3Hg^+ (methylmercury) because the volatile dimethylmercury, $(CH_3)_2Hg$, forms at a much slower rate. Microbially mediated reactions affecting the fate of Hg^{2+} are shown in Fig. 15.5.

The reason for the methylation of mercury by bacteria is not fully understood, although the methylation of certain metal(loid)s may be a mechanism of detoxification. For example, the methylation of arsenic to methylarsenicals results in a compound that is less toxic, less reactive, and more readily excreted from the cell (Vahter and Marafante, 1993). Thus, the transfor-

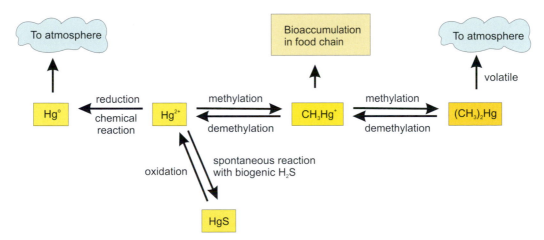

FIGURE 15.5   Microbially-mediated reactions with Hg^{2+} in the environment. Hg^{2+} can be reduced to elemental $Hg°$ by chemical reaction with humic acids or by microbially-mediated reactions which are believed to be a detoxification mechanism. Hg^{2+} can be precipitated by reaction with H_2S produced under sulfate-reducing conditions but can also be released by microbial oxidation of HgS. Methylation of Hg^{2+} produces organometals, which can accumulate in the tissue of living organisms. The production of organometals may to some extent be balanced by demethylation reactions occurring in both aerobic and anaerobic environments. (Based on Gadd, 1993, and Ehrlich, 1996.)

mation of the metal(loid) is one mechanism by which an organism can be protected from the toxicity of the element. Methylation of mercury may provide some degree of resistance to the toxicity of Hg^{2+}. Methylmercury is still toxic to the cell, although it may diffuse more easily from the cell than Hg^{2+} (Gadd, 1993). It can also be immobilized in the sediment matrix by interaction with negatively charged soil particles or by partitioning into soil organic debris (Ehrlich, 1996). The formation of the volatile dimethylmercury can remove mercury from the environment of the cell by its release into the atmosphere. However, in the case of mercury, the most common mechanism used by microorganisms to resist Hg^{2+} toxicity is not methylation but the enzymatic reduction of Hg^{2+} to elemental mercury (Hg°) by cytoplasmic mercuric reductase. Elemental mercury is less toxic than Hg^{2+} and is also volatile and thus can be removed from the environment of the cell.

The environmental significance of mercury methylation is the fate of methylmercury within the aquatic food chain. Methylation converts sediment-associated Hg^{2+} into an organometal that has greater lipid solubility. Methylated mercury can be absorbed by living organisms and can accumulate to high levels in fish and shellfish. On consumption of contaminated fish and shellfish, the small concentrations of mercury present in the fish are transferred to the consumer and, with continued consumption of fish, can gradually accumulate to toxic levels.

Mercury is a strong neurotoxin and if ingested in high enough levels by humans, can be fatal. Incidents of mercury poisoning due to the consumption of contaminated fish have occurred in Minamata Bay in Japan, in isolated Native American populations in North America, and elsewhere. In Japan, industrial activity released mercury into the environment and contaminated the aquatic food chain in an area where the residents were dependent on fishing. Several thousand became poisoned by mercury and hundreds died. The symptoms of mercury accumulation in the central nervous system include tremors, inability to coordinate body movements, impairment of vision and speech, as well as liver and kidney damage. The combined symptoms of mercury poisoning are known as Minamata disease but have also been called cat's dance disease, which refers to the tremors experienced by cats when they consumed contaminated fish. In the case of Minamata Bay, the source of the mercury was industrial pollution and the extent to which biological methylation of inorganic mercury contributed to the contamination of the food chain is unclear. However, the incidents of mercury poisoning provide compelling reasons to monitor closely the fate of mercury in natural systems.

One area of concern is the flooding of terrestrial environments in order to create water reservoirs (Bodaly et al., 1997). Fish in certain newly created reservoirs, particularly in subarctic environments, have shown unexpectedly high levels of mercury in their tissue. The source of the mercury in fish in these sites is the accumulation of methylmercury in the aquatic food chain. While mercury occurs naturally in the soil and vegetation that was flooded, this mercury was methylated by microbial activity in the anoxic sediments created when the vegetation-rich terrain was flooded. Thus, although very low levels of available mercury were present in the environment before flooding, the methylation of mercury and the process of bioaccumulation in the food chain resulted in mercury levels in fish that exceeded accepted concentrations for safe human consumption.

Methylation of other metal(loid)s may also be of environmental significance. Tin and possibly lead may undergo methylation in the environment to form more lipophilic organometals that, like mercury, can bioaccumulate in the food chain. In the case of selenium, selenium-rich agricultural drainage water was impounded in reservoirs in Southern California. Evaporation of the water increased selenium concentration to a level that was toxic to wildlife (Thompson-Eagle and Frankenberger, 1992). Selenium in the water is mostly in the water-soluble selenate form (SeO_4^{2-}), and the microbially-mediated methylation of selenium produces a highly volatile methylated product. Therefore, promotion of methylation is seen as a means of removing selenium from the water and dispersing it into the atmosphere, thus greatly reducing the level of exposure to wildlife.

15.5 NITROUS OXIDE AND EARTH'S ATMOSPHERE

Nitrous oxide (N_2O) gas is released to the atmosphere from both industrial and biogenic sources. Concern over the fate of N_2O arises from the fact that N_2O can contribute to both global warming and the destruction of the protective ozone layer in the earth's stratosphere. Increasing emissions of CO_2 and other greenhouse gases, such as chlorofluorocarbons (CFCs), CH_4, and N_2O, to the atmosphere are expected to result in a gradual increase in global temperature (Matthais, 1996). The greenhouse gases are of concern because they absorb long-wave radiation from the sun after it hits the earth and is reflected back into space. This effectively traps heat in the atmosphere. N_2O is of concern in two respects. It has a long residence time (>100 years) in the atmosphere, and it is highly efficient in absorbing long-wave radiation. One molecule of N_2O is equivalent in heat-trapping ability to about 200 molecules of CO_2. Therefore, small increases in atmospheric N_2O concentration can have a large impact on warming trends.

A second concern about N_2O is that in the upper atmosphere, solar radiation can photolytically convert N_2O to nitric oxide (NO), which is a contributor to the depletion of the protective ozone layer (Fig. 15.6). The ozone layer acts as a filter to remove biologically harmful ultraviolet (UV) light. Stratospheric ozone depletion occurs through a chemical interaction between sunlight, ozone, and certain reactive chemical species, including nitrogen oxides and organohalogens such as CFCs. Depletion of the protective ozone layer can have serious ecological and human health consequences. Increased levels of UV radiation may be inhibitory to certain microorganisms, such as phytoplankton, and may also increase the incidence of skin cancer in humans. Ozone is depleted in the series of reactions shown in Eqs. 15.18 to 15.21, where light energy begins

the reaction by splitting nitrous oxide into N_2 and singlet oxygen, in which one of the electrons is in a high-energy state. This singlet oxygen can react with nitrous oxide to form two molecules of nitric oxide. Nitric oxide in turn reacts with ozone (O_3) to produce nitrogen dioxide and oxygen. The nitrogen dioxide then reacts with singlet oxygen (O^*) to produce oxygen and regenerate nitric oxide. The fact that nitric oxide is regenerated in this series of reactions means that for every nitrous oxide molecule released to the atmosphere, a large number of ozone molecules can be destroyed.

$$N_2O + h\nu \rightarrow N_2 + O^* \qquad \text{(Eq. 15.18)}$$

$$N_2O + O^* \rightarrow 2NO \qquad \text{(Eq. 15.19)}$$

$$NO + O_3 \rightarrow NO_2 + O_2 \qquad \text{(Eq. 15.20)}$$

$$NO_2 + O^* \rightarrow NO + O_2 \qquad \text{(Eq. 15.21)}$$

A major source of N_2O released to the atmosphere is the soil environment. N_2O is produced during denitrification, which is the utilization of nitrate (NO_3^-) as a

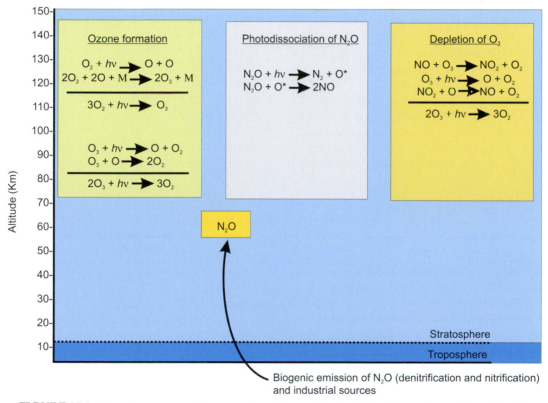

FIGURE 15.6 Equations summarizing ozone formation and the depletion of ozone by reaction with nitric oxide (NO). Solar UV radiation ($h\nu$) photodissociates molecular oxygen (O_2) into two oxygen atoms (O), which recombine with undissociated O_2 (in the presence of another chemical species, M) to form ozone (O_3). Ozone is then photodissociated back to molecular oxygen (O_2). The constant cycling between ozone and oxygen is important because it consumes harmful UV radiation in the stratosphere. Nitrous oxide (N_2O) emitted to the atmosphere is photodissociated to nitrogen and an oxygen atom in an electronically excited state, O^*, which reacts with N_2O to produce nitric oxide (NO). Nitric oxide can react with ozone and, in a series of reactions, produce O_2 resulting in a net depletion of ozone. (From Matthias, 1996.)

terminal electron acceptor during anaerobic respiration of organic compounds (Chapter 14.3.6). N_2O is an intermediate in the reduction of NO_3^- to N_2, and it may also be the end point of NO_3^- reduction under certain environmental conditions. N_2O production is also associated with nitrification, which is a two-step process in which ammonium is oxidized to nitrite and then nitrite is oxidized to nitrate. It has been suggested that N_2O production is associated with low-oxygen conditions, in which nitrifying bacteria utilize NO_2 as an electron acceptor in place of oxygen, as a result reducing NO_2 to N_2O (Conrad, 1990). Thus, N_2O is produced as an intermediate in denitrification and as a by-product of nitrification. There has been intense interest in determining the relative contribution of denitrification and nitrification processes to N_2O production and to determine the environmental factors that control each process. In simplistic terms, N_2O production from nitrification is favored in aerated, moist soils, and N_2O production from denitrification is favored in wet soils where aeration is restricted (Mosier et al., 1996).

The agricultural practice of added nitrogen sources, either as chemical fertilizers or as animal manure, to soil is a major contributor to the gradual increase in N_2O emissions to the atmosphere (Mosier et al., 1996). The source of nitrogen in fertilizers is mainly ammonia or ammonium-producing compounds such as urea. However, only about half the total nitrogen applied to a field is assimilated by the crop (Delgado and Mosier, 1996). The remaining nitrogen is lost through leaching, erosion, and gaseous emissions. Millions of acres of agricultural land are also fertilized with animal ma-

nure, with the result that about 1% of the nitrogen applied is released to the atmosphere (Lessard et al., 1996). Thus, agricultural practices account for a large percentage of the N_2O that is released by human activities, although other sources of N_2O in the atmosphere include burning of biomass, combustion of fossil fuels, and chemical manufacturing of nylon.

In agriculture, certain strategies can be implemented to reduce biogenic N_2O emissions (Mosier et al., 1996). A primary factor in N_2O emissions is the low efficiency of utilization of nitrogen fertilizers, and improved management practices can be applied to maximize plant assimilation of nitrogen. Decreased loss of added nitrogen will decrease the nitrogen available for denitrification. A reduction in N_2O emissions from agricultural soils also can be achieved through the use of specific nitrification inhibitors (Table 15.2).

15.6 NITRATE CONTAMINATION OF GROUNDWATER

Nitrate does not normally accumulate in the environment. However, in agricultural systems the use of ammonia fertilizers and the production of large amounts of animal waste in dairy or feedlot operations result in the addition of excess nitrogen to the soil and groundwater environment. Septic tanks are another source of elevated nitrogen addition. Although nitrogen is normally added to crops or produced by animals as ammonia or an ammonia-containing compound, the conversion of ammonia to nitrate by aerobic, chemoautotrophic nitrifying bacteria results

TABLE 15.2 Nitrification Inhibitors

Common name(s)	Chemical name	% Inhibition (day 14)	(day 28)
—	2-Ethynylpyridine	97	87
—	Phenylacetylene[a]	92	55
Dwell	Etridiazole	90	75
N-serve, nitrapyrin	2-Chloro-6-(trichloromethyl)pyridine	85	65
ATC	4-Amino-1,2,4-triazole	87	60
DCD, dicyan	2,4-Diamino-6-trichloromethyltriazine	76	41
—	Dicyandiamide	61	15
AM	2-Amino-4-chloro-6-methylpyrimidine	60	37
ST	Sulfathiazole	52	17
Tu	Thiourea	2	0

From Myrold (1988).
[a] Average of two soils after 10 and 30 days incubation.

in the accumulation of nitrate in many agricultural systems (Chapter 14.3.4). Because nitrate is an anion, it is very mobile in soil and, when produced in excess, is easily transported with water flow into groundwater supplies. In fact, according to a recent survey by the Environmental Protection Agency, most states have nitrate levels that exceed drinking water standards in 1 to 10% of the groundwater wells tested (Fig. 15.7).

As a result, there is concern about potential health effects on humans and animals that use these water supplies. One of these effects is a condition called **methemoglobinemia** or "blue baby syndrome" that affects infants less than 6 months old as well as young animals. The stomachs of these young are not yet acidic enough to prevent the growth of denitrifying bacteria that convert nitrate to nitrite. The nitrite produced binds to hemoglobin and prevents the hemoglobin from carrying oxygen throughout the body. This can result in cyanosis, which causes the infant's skin to turn a bluish color. At its worst methemoglobinemia can result in brain damage or death, although few such cases have been reported. When high nitrate levels in drinking water are suspected and detected, it is a problem that is easy to solve by substituting bottled water. The allowable level of nitrate-nitrogen (NO_3-N) in water for children 6 months or less is 10 mg/l. A second concern about elevated nitrate levels in groundwater is that in adults denitrifi-cation of nitrate to nitrite in the stomach can lead to the formation of nitrosamines, which are highly carcinogenic. Although there is no proven link between nitrate consumption and human cancer, in laboratory studies, nitrites have been shown to interact with amino compounds in the stomach to form N-nitrosamines. Many of these N-nitrosamine compounds have been shown to be carcinogenic in test animals.

Several measures can be taken to prevent or minimize nitrate contamination of groundwater. The most economical approach is simply to manage the amount and time of fertilizer application to a crop (Thompson, 1996). Enough fertilizer must be added to meet crop needs, but overfertilization will result in increased nitrate formation and leaching. A second way to minimize nitrogen losses in irrigated croplands is through control of the timing and amount of irrigation. Remember, plants can take up either ammonia or nitrate, but nitrate is easily removed from the plant root area by water leaching. Many states in the U.S. have studied and adopted **Best Management Practices (BMPs)** for fertilizer application and irrigation. These BMPs are region specific since climate and soil types change dramatically from region to region. Finally, two other approaches to minimizing nitrate formation and leaching are the use of slow-release fertilizers, which allow more controlled release of ammonia into the environment, and the application of nitrification inhibitors

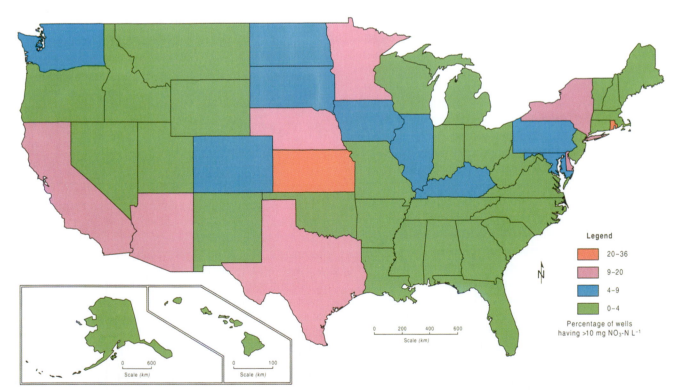

FIGURE 15.7 Distribution of nitrate-N concentrations in groundwater in the United States. (From *Pollution Science* © 1996, Academic Press, San Diego, CA.)

(Table 15.2), which suppress the formation of nitrate. However, application of slow-release fertilizers or nitrification inhibitors is costly and so neither approach is widely used.

15.7 BIOGENESIS OF HALOMETHANES AND OTHER ORGANOHALOGENS

Microbial reactions resulting in the production of methylated halogens are of interest in terms of global biogeochemical cycles. Halomethanes and other organohalogens are natural products of the marine environment and have been shown to be synthesized by marine algae and macroalgae, such as kelp and other seaweeds (Manley et al., 1992; Nightingale et al., 1995). Examples include chloromethane (CH_3Cl), chloroform ($CHCl_3$), bromoform (CHB_3), methyl iodide (CH_3I), and a wide variety of other chlorinated, brominated, and iodinated organic compounds. The production of halomethane metabolites is believed to be the result of reactions with **haloperoxidase.** These enzymes are of biochemical importance in terms of plant hormone production and other metabolic pathways. Halomethanes may be formed at random by the activity of relatively nondiscriminating haloperoxidases or other oxidase reactions. It has also been suggested that the production of toxic organohalogen metabolites may be of benefit to marine macroalgae by providing a deterrent to grazing herbivores or discouraging the colonization of their surfaces by microorganisms (Gschwend et al., 1985).

Halogenated organic compounds have numerous industrial applications, particularly as fumigants and solvents, and are manufactured by chemical synthesis to meet industrial demand. There has been intense scientific interest in volatile halogenated compounds because of the role of reactive organohalogens in the destruction of the ozone layer in the earth's stratosphere (Matthias, 1996). International treaties have been signed to phase out production of chlorofluorocarbons (CFCs), which are volatile halogenated compounds that have been used in refrigerators, air conditioners, and industrial processes. Other organohalogens, such as methyl bromide, may also play a role in ozone destruction.

There has been increasing interest in identifying and quantifying all sources of volatile halogenated compounds in our environment and in determining the relative contribution of natural and anthropogenic sources to the global cycling of halogenated gases. In this vein, it is known that the oceans can be a significant source of certain halomethanes, such as bromomethanes (Gschwend et al., 1985; Manley et al.,

1992) and chloroform (Nightingale et al., 1995). The small concentrations of halomethanes synthesized and released to the atmosphere by macroalgae are magnified to global significance by vast areas of productive macroalgae in the world's oceans.

The biological origin of organohalogen compounds will also contribute to the presence of halocarbons in "unpolluted" freshwater and groundwater environments where there is no known source of anthropogenic organohalogen pollution (Grimvall and de Leer, 1995). Some of the organohalgens present in unpolluted environments can be accounted for by long-range deposition of halocarbons from industrial sources that can be located hundreds of miles away. However, there can be significant levels of organohalogens from biogenic sources. Uncertainty exists with regard to the mechanism of biological origin, although it is known that organohalogen metabolites may be formed by many types of algae, fungi, and plants and that natural halogenation of humic substances produces halogenated organic compounds. One area of concern is the production of trihalomethanes in water supply reservoirs. Trihalomethanes, including chloroform, bromoform, and mixed trihalomethanes, were formed as a result of chlorine application combined with algal growth or, more specifically, the release of dissolved organic compounds from algae in a water supply reservoir (Karimi and Singer, 1991). Chlorination of the water in open reservoirs is used to control algal growth and is also necessary to prevent the spread of waterborne diseases in drinking water supplies. However, the small level of biogenic halocarbon production in water may be of concern because of the classification of certain trihalomethanes as possible human carcinogens. Government regulations mandate strict control of the level of halocarbons from all sources present in drinking water.

15.8 COMPOSTING

Composting is a solid waste disposal technique that has been practiced since ancient times. It is a technique that can be used to treat a variety of wastes including biosolids, municipal solid waste, food and agricultural wastes, and even hazardous wastes (Table 15.3). Essentially, the composting process turns waste products into an organic soil amendment by taking advantage of the normal microbes found in soil and optimizing their carbon cycling activities. In fact, the spatial characteristics and physical–chemical gradients found in a compost system are very similar to those found in a soil system. But compost and soil systems are distinct in several other respects. For example, compost is primarily

TABLE 15.3   Potential Compost Substrates

Type of compost substrate	Examples
Biosolids	8 million dry metric tons produced per year (1989 estimate for United States)
Industrial sludge	Food, textile, pulp and paper, pharmaceutical, and petroleum industries
Manure	Feedlots, dairies, poultry production
Yard waste	Second largest component of U.S. solid waste stream, primarily grass clippings, leaves, and brush
Food and agricultural waste	Fruit and vegetable processing wastes, agricultural residues (e.g. rice and nut hulls, corn cobs, cotton gin residues)
Municipal solid waste	Production is 4.3 lb/day per capita in U.S., includes paper, food, and yard wastes
Special wastes	Hazardous wastes such as TNT, petroleum sludge, and pesticide residues that are incorporated into compost materials

Adapted from Haug (1993).

organic in composition, and the rate of microbial activity is extremely high during compost formation and after it is added to soil. In contrast, in a soil, both organic matter content and microbial activity levels are much lower. In soils, physical factors including temperature, moisture, and bulk density are imposed externally, but in composting systems these factors are controlled internally. Finally, the soil matrix is a very stable one and undergoes change slowly. The compost matrix is organic substrate that undergoes rapid physical change as a result of degradative activities.

As an ecosystem, a compost pile is characterized by high substrate density and diverse and highly interactive microbial populations that go through a succession of populations dominated first by mesophilic organisms and then by thermophiles. As shown in Table 15.4, the microorganisms found in a compost system are a mixed population of bacteria, actinomycetes, and fungi. The numbers of actinomycetes and fungi peak during the thermophilic stage of the composting process. Bacteria peak in numbers as the compost system starts to cool. All of these microbes are important components of the system in terms of carbon assimilation.

Temperature is one of the most critical factors in composting. The elevated temperatures reached in composting are fundamental to the high rate of decomposition but are also important from several other aspects. High temperatures in composting sludge wastes help to kill any viruses and coliform bacteria present, especially those that are added as a part of biosolids or municipal solid waste. High temperatures also help to kill weed seeds, which are an undesirable component of the final compost product that will be used on lawns and gardens. The optimal temperature for composting is around 60°C. Above 60°C, the temperature optimum for thermophilic growth is exceeded. Above 82°C, activity will be totally inhibited until the compost pile cools down.

The overall process of composting is to build a pile of organic matter that can be easily aerated to maintain aerobic conditions and to provide a way to cool the compost system. Details of the construction of compost systems are given in Section 22.9.2. Once the compost system is constructed, the organic matter in the compost pile acts as an insulator and allows heat to build up as the degradation proceeds. As the temperature increases and reaches 40°C, usually after 2 to 3 days, the initial mesophilic populations die off and are replaced by thermophilic microbes. Once all of the usable organic material is degraded, decomposition ac-

TABLE 15.4   Microbial Populations during Aerobic Composting

Microbe	Mesophilic, initial temperature <40°C	Thermophilic, 40–70°C	Mesophilic, 70°C to cooler	Number of species identified
Bacteria				
Mesophilic	10^8	10^6	10^{11}	6
Thermophilic	10^4	10^9	10^7	1
Actinomyces				
Thermophilic	10^4	10^8	10^5	14
Fungi				
Mesophilic	10^6	10^3	10^5	18
Thermophilic	10^3	10^7	10^6	16

CFU/g compost (wet weight)

From Haug (1993), p. 135.

tivity slows, as does the generation of heat. After composting is complete, which usually takes 3 to 4 weeks, the compost pile cools to ambient temperature.

The success of the composting process depends on the construction of the compost pile. If it is too large, the organic matter will be compressed and the system will become anaerobic. If it is too small, a majority of the heat generated will escape and the rate of decomposition will not be maximized. It is also important to maintain sufficient oxygen levels. If oxygen is depleted, the rate of activity and hence heat generation will slow. Also, under anaerobic conditions, undesirable gaseous degradation products are formed, including carbon compounds such as methane and volatile organic acids as well as sulfur and nitrogen compounds such as H_2S and NH_3.

QUESTIONS AND PROBLEMS

1. Think about the activities involved in the carbon, nitrogen, and sulfur cycles and develop your own idea for applying one of these activities in the commercial sector.

2. As city mayor, you are informed that the concrete sewer system is corroded and will need to be replaced within the next 3 years. You are running for reelection and the cost of replacing the sewer system means that taxes will have to be raised. How would you explain, in lay terms, to your constituents why the concrete pipes have corroded and justify raising taxes to replace the system.

3. A mine that supports much of the local economy is releasing acid leachates into a receiving stream that eventually feeds a fishery important to local tourism. The concern is that a significant decrease in pH of the fishery waters will affect fish survival. What suggestions would you make to the local mine to 1) prevent acid leachate formation, 2) treat acid leachates before their release so that this does not become an insurmountable problem.

4. Methylation of metals increases their volatility. Is it a good idea to use microbial methylation as a basis for removal of metals such as mercury and selenium from contaminated soils? Support your answer.

5. Municipal sludge or biosolids left after wastewater treatment can be landfilled, combusted, or composted. Discuss the advantages and disadvantages of each approach.

6. Groundwater under cattle feedlot operations is often found to have nitrate contamination. Explain the microbial basis for nitrate contamination of groundwater in these areas.

References and Recommended Readings

Atlas, R. M., and Bartha, R. (1993) "Microbial Ecology." Benjamin Cummings, New York.

Bodaly, R. A., St. Louis, V. L., Paterson, M. J., Fudge, R. J. P., Hall, B., Rosenberg, D. M., and Rudd, J. W. M. (1997) Bioaccumulation of mercury in the aquatic food chain in newly formed flooded areas. In "Metal Ions in Biological Systems," Vol. 34, "Mercury and Its Effect on Environment and Biology" (A. Sigel and H. Sigel, eds.) Marcel Dekker, New York, pp. 259–287.

Bos, P., Boogerd, F. C., and Kuenen, J. G. (1992) Microbial desulfurization of coal. In "Environmental Microbiology" (R. Mitchell, ed.) Wiley-Liss, New York, pp. 375–403.

Brierley, C. L., and Brierley, J. A. (1997) Microbiology of the metal mining industry. In "Manual of Environmental Microbiology" (C. J. Hurst, G. R. Knudsen, M. J. McInerney, L. D. Stetzenback and M. V. Walter, eds.) American Society for Microbiology (ASM) Press, Washington, DC, pp. 830–841.

Conrad, R. (1990) Flux of NO_x between soil and atmosphere: Importance and soil microbial metabolism. In "Denitrification in Soil and Sediment" (N. P. Revsbech and J. Sørensen, eds.) Plenum, New York, pp. 105–128.

Delgado, J. A., and Mosier, A. R. (1996) Mitigation alternative to decrease nitrous oxides emissions and urea-nitrogen loss and their effect on methane flux. J. Environ. Qual. 25, 1105–1111.

Diercks, M., Sand, W., and Block, E. (1991) Microbial corrosion of concrete. Experientia 47, 514–516.

Dowling, N. J. E., and Guezennec, J. (1997) Microbiologically influenced corrosion. In "Manual of Environmental Microbiology" (C. J. Hurst, G. R. Knudsen, M. J. McInerney, L. D. Stetzenbach and M. V. Walter, eds.) American Society for Microbiology (ASM) Press, Washington, DC, pp 842–855.

Ehrlich, H. L. (1996) "Geomicrobiology," 3rd ed. Marcel Dekker, New York.

Gadd, G. M. (1993) Microbial formation and transformation of organometallic and organometalloid compounds. FEMS Microbiol. Rev. 11, 297–316.

Grimvall, A., and de Leer, E. W. B. (1995) "Naturally-Produced Organohalgens." Kluwer Academic Publishers, Boston.

Gschwend, P. M., MacFarlane, J. K., and Newman, K. A. (1985) Volatile halogenated organic compounds released to seawater from temperate marine macroalgae. Science 227, 1033–1035.

Hamilton, W. A. (1995) Biofilms and microbially influenced corrosion. In "Microbial Biofilms" (H. M. Lappin-Scott and J. W. Costerton, eds.) Cambridge University Press, Cambridge, pp. 171–182.

Haug, R. T. (1993) "The Practical Handbook of Compost Engineering." Lewis, CRC Press, Boca Raton, FL.

Heitz, E. (1996) Electrochemical and chemical mechanisms. In "Microbially Influenced Corrosion of Materials" (E. Heitz, H.-C. Flemming, and W. Sand, eds.) Springer-Verlag, New York, pp. 27–38.

Karimi, A. A., and Singer, P. C. (1991) Trihalomethane formation in open reservoirs. J. Am. Water Works Assoc. 83, 84–88.

Lappin-Scott, H. M., and Costerton, J. W. (1989) Bacterial biofilms and surface fouling. Biofouling 1, 323–342.

Lessard, R., Rochette, P., Gregorich, E. G., Pattey, E., and Desjardins, R. L. (1996) Nitrous oxide fluxes from manure-amended soil under maize. J. Environ. Qual. 25, 1371–1377.

Manley, S. L., Goodwin, K., and North, W. J. (1992) Laboratory production of bromoform, methylene bromide, and methyl iodide by macroalgae and distribution in nearshore southern California waters. *Limnol. Oceanogr.* **37,** 1652–1659.

Matthias, A. D. (1996) Atmospheric pollution. *In* "Pollution Science" (I. L. Pepper, C. P. Gerba and M. L. Brusseau, eds.) Academic Press, San Diego, pp. 171–188.

Mori, T., Nonaka, T., Tazaki, K., Koga, M., Hikosaka, Y., and Noda, S. (1992) Interactions of nutrients, moisture and pH on microbial corrosion of concrete sewer pipes. *Water Res.* **26,** 29–37.

Mosier, A. R., Duxbury, J. M., Freney, J. R., Heinemeyer, O., and Minami, K. (1996) Nitrous oxide emissions from agricultural fields: Assessment, measurement, and mitigation. *Plant Soil* **181,** 95–108.

Myrold, D. D. (1998) Transformations of nitrogen. *In* "Principles and Applications of Soil Microbiology" (D. M. Sylvia, J. J. Fuhrmann, P. G. Hartel, D. A. Zuberer, eds.) Prentice-Hall, Upper Saddle River, NJ, pp. 218–258.

Nightingale, P. D., Malin, G., and Liss, P. S. (1995) Production of chloroform and other low-molecular-weight halocarbons by some species of macroalgae. *Limnol. Oceanogr.* **40,** 680–689.

Southam, G., Ferris, F. G., and Beveridge, T. J. (1995) Mineralized bacterial biofilms in sulphide tailings and in acid mine drainage systems. *In* "Microbial Biofilms" (H. M. Lappin-Scott and J. W. Costerton, eds.) Cambridge University Press, Cambridge, pp. 148–170.

Sydney, R., Esfandi, E., and Surapaneni, S. (1996) Control concrete sewer corrosion via the crown spray process. *Water Environ. Res.* **68,** 338–347.

Thayer, J. S. (1993) Global bioalkylation of the heavy metals. *In* "Metal Ions in Biological Systems," Vol. 29, "Biological Properties of Metal Alkyl Derivatives" (H. Segel and A. Sigel, eds.) Marcel Dekker, New York, pp. 1–36.

Thompson, T. L. (1996) Agricultural fertilizers as a source of pollution. *In* "Pollution Science" (I. L. Pepper, C. P. Gerba and M. L. Brusseau, eds.) Academic Press, San Diego, pp. 211–223.

Thompson-Eagle, E. T., and Frankenberger, W. T. Jr. (1992) Bioremediation of soils contaminated with selenium. *In* "Advances in Soil Science," Vol. 17 (R. Lal and B. A. Stewart, eds.) Springer-Verlag, New York, pp. 262–310.

Vahter, M., and Marafante, E. (1993) Metabolism of alkylarsenic and antimony compounds. *In* "Metal Ions in Biological Systems," Vol. 29, "Biological Properties of Metal Aklyl Derivatives" (H. Segel and A. Sigel, eds.) Marcel Dekker, New York, pp. 162–184.

Warscheid, T., and Krumbein, W. E. (1996) General aspects and selected cases. *In* "Microbially Influenced Corrosion of Materials" (E. Heitz, H.-C. Flemming and W. Sand, eds) Springer-Verlag, New York, pp. 237–295.

16

Microorganisms and Organic Pollutants

RAINA M. MAIER

16.1 INTRODUCTION

Since before the beginning of the century, we have dumped enormous amounts of waste products into the environment following the principle of "out of sight, out of mind." Until World War II (WWII), these waste products and their effects went largely unnoticed. However, after WWII, the severity of pollution problems, resulting from careless waste disposal, has steadily increased. Today, more than ever, the general public is aware of these problems.

Historically, sources of waste products have been both industrial and agricultural. For instance, the energy production industry generates huge amounts of waste during the processing of coal and oil and also nuclear energy production. In the past, these wastes have been buried, sometimes not so carefully, and as a result waste constituents have migrated through the soil into groundwater supplies. This type of contamination is said to be from a **point source.** On the other hand, the agricultural sector is the nation's largest user of pesticides and fertilizers. The application of pesticides and fertilizers over vast land areas is responsible for what is called **nonpoint source** contamination. Studies have shown that these agricultural chemicals are being found in our surface water and groundwater supplies in increasing amounts. A 1984 EPA report on nonpoint source pollution identified 44 states where agriculture is an identified nonpoint pollution problem. In 1988, EPA reported that 26 states had confirmed various amounts of 46 different pesticides in their groundwater resulting from normal agricultural practices.

Eighteen of these pesticides (in 24 out of the 26 states) were discovered at levels equal to or greater than health advisory levels established or proposed by EPA. Groundwater contamination is a critical issue because groundwater is a major source of potable water. In fact, groundwater constitutes 96% of all available fresh water in the United States (USDA, 1990). Fifty percent of the general population and at least 95% of rural residents obtain drinking water from groundwater sources.

The objective of this chapter is to examine microbial interactions with organic pollutants that can be harnessed to help prevent contamination and clean up contaminated sites. As will be seen in the following section, the United States has passed a series of environmental laws mandating the cleanup of such sites. However, the cost of cleanup has been estimated to be in the trillions of dollars. Therefore, we as a society are reexamining the cleanup issue from several perspectives. The first is related to the cleanup target and the question "How clean is clean?" As you can imagine, the stricter the cleanup provision (e.g., lower contaminant concentration), the greater the attendant cleanup costs. It may require tens to hundreds of millions of dollars for complete cleanup of a large, complex hazardous waste site. In fact, it may be impossible to clean many sites completely. It is very important, therefore, that the physical feasibility of cleanup and the degree of potential risk posed by the contamination be weighed against the economic impact and the future use of the site. The consideration of the future use of the site will help focus our scarce resources on the sites that pose the greatest current and future risk. The second perspective being considered is whether natural microbial activities in the environment can aid in the cleanup of contaminated sites, and whether these activities will occur rapidly enough naturally, or whether they can and should be enhanced. These two perspectives are closely tied together because although microbial activities can reduce contamination significantly, they often do not remove contamination entirely.

16.2 ENVIRONMENTAL LAW

Society began responding to environmental concerns long ago, beginning with the recognition that our environment is fragile and human activities can have a great impact on it. This led to the creation of Yellowstone National Park in 1872 and the assignment of the care of forested public domain lands to the U.S. Forest Service in 1897. After World War II, as the pollution impacts of industrialization began to be apparent, Congress began to legislate in the area of pollution control. This legislation culminated in the major federal pollution control statutes of the 1970s that now constitute a large body of law called **environmental law.** Federal environmental law consists of laws in the conservation and pollution control areas, along with key planning and coordination statutes. Environmental law is constantly changing and evolving as we try to respond to shifting priorities and pressures on resources. Sometimes changes are made to the law to allow further contamination or risk of contamination to occur for the "good of society." An example is oil exploration in Alaska and off the coast of California. On the other hand, some laws can be made more stringent. Again, using an example from the oil industry, whereas refinery wastes are heavily regulated for disposal, the same types of wastes generated in an oil field were not regulated and were routinely buried without treatment. When attention was drawn to this practice, laws were enacted to require the oil industry to implement proper oil field disposal practices. As these examples suggest, environmental law comprises a complex body of laws, regulations, and decisions now established in the United States. This body of law, which evolved quite quickly in comparison with labor, tax, banking, and communications laws, already ranks with these other areas in size and complexity (Arbuckle et al., 1987).

The term "environmental law" came into being with the enactment of the **National Environmental Policy Act (NEPA)** in early 1970. Although environmental law can vary considerably from state to state

and even from city to city, a series of major federal environmental protection laws have been enacted that pertain to the generation, use, and disposition of hazardous waste. NEPA requires each agency, government, or industry that proposes a major action which may have a significant effect on the human environment to prepare an **environmental impact statement (EIS).** The EIS must address the environmental impact of the proposed action and any reasonable alternatives that may exist. The types of projects that NEPA covers are landfills, roads, dams, building complexes, research projects, and any private endeavor requiring a federal license that may affect the environment. *NEPA does not mandate particular results and it does not require a federal agency to adopt the least environmentally damaging alternative.* Because of this, NEPA can be thought of as an "environmental full disclosure law," which requires the applicant to take a "hard look" at the environmental consequences of its action. Thus, an EIS allows environmental concerns and planning to be integrated into the early stages of project planning. Unfortunately, an EIS is often done as an afterthought and becomes a rationale for a project that may be a poor alternative.

Table 16.1 shows a series of environmental laws that have been passed since NEPA to protect our natural resources. These include laws such as the **Clean Air Act** and the **Clean Water Act** that protect air and water resources. There are also laws that govern the permitting of the sale of hazardous chemicals and laws that mandate specific action to be taken in the cleanup of contaminated sites. When the **Comprehensive Environmental Response, Compensation and Liability Act (CERCLA),** more commonly known as **Superfund,** was enacted, it became clear that technology for cleaning up hazardous waste sites was needed. Early remedial actions for contaminated sites consisted of excavation and removal of the contaminated soil to a landfill. Very soon, it became apparant that this was simply moving the problem around, not solving it. As a result, the **Superfund Amendments and Reauthorization Act (SARA)** was passed in 1986. This act added several new dimensions to CERCLA. SARA stipulates cleanup standards and mandates the use of the **National Contingency Plan** to determine the most appropriate action to take in site cleanup.

Two types of responses are available within Superfund: (1) removal actions in response to immediate threats, e.g., removing leaking drums, and (2) remedial actions, which involve cleanup of hazardous sites. The Superfund provisions can be used when a hazardous substance is released or there is a substantial threat of a release that poses imminent and substantial endanger-

TABLE 16.1 History of Environmental Law

Law	Year passed	Goals
Clean Air Act (CAA)	1970	Sets nationwide ambient air quality standards for conventional air pollutants. Sets standards for emissions from both stationary and mobile sources (e.g., motor vehicles).
Clean Water Act (CWA)	1972	Mandates "fishable/swimmable" waters wherever attainable. Provides for (1) a construction grants program for publicly owned water treatment plants and requires plants to achieve the equivalent of secondary treatment; (2) a permit system to regulate point sources of pollution; (3) areawide water quality management to reduce nonpoint sources of pollution; (4) wetlands protection, sludge disposal and ocean discharges; (5) regulation of cleanup of oil spills.
Surface Mining Control and Reclamation Act (SMCRA)	1977	Regulates coal surface mining on private lands and strip mining on public lands. Prohibits surface mining in environmentally sensitive areas.
Resource Conservation and Recovery Act (RCRA)	1976	Provides a comprehensive management scheme for hazardous waste disposal. This includes a system to track the transportation of wastes and federal performance standards for hazardous waste treatment, storage, and disposal facilities. Open dumps are prohibited.
Toxic Substances Control Act (TOSCA)	1976	Requires premarket notification of EPA by the manufacturer of a new chemical. Based on testing information submitted by the manufacturer or premarket test ordered by EPA (including biodegradability and toxicity), a court injunction can be obtained barring the chemical from distribution or sale. EPA can also seek a recall of chemicals already on the market. It is this act that prohibits all but closed-circuit uses of PCBs.
Comprehensive Environmental Response, Compensation and Liability Act (CERCLA)	1980	Commonly known as Superfund, this act covers the cleanup of hazardous substance spills, from vessels, active, or inactive facilities. Establishes a Hazardous Substances Response Trust Fund, financed by a tax on the sale of hazardous chemicals, to be used for removal and cleanup of hazardous waste releases. Cleanup costs must be shared by the affected state. Within certain limits and subject to a few defenses, anyone associated with the release is strictly liable to reimburse the fund for cleanup costs, including damage to natural resources.
Superfund Amendments and Reauthorization Act (SARA)	1986	SARA provides cleanup standards and stipulates rules through the National Contingency Plan for the selection and review of remedial actions. It strongly recommends that remedial actions use on-site treatments that "permanently and significantly reduce the volume, toxicity, or mobility of hazardous substances" and requires remedial action that is "protective of human health and the environment, that is cost-effective, and that utilizes permanent solutions and alternative treatment technologies or resource recovery technologies to the maximum extent practicable."
National Contingency Plan (NCP)	1988	A five-step process to use in evaluation of contaminated sites and suggest the best plan for remediation.

ment to public health and welfare. The process by which Superfund is applied to a site is illustrated in Fig. 16.1. The first step is to place the potential site in the **Superfund Site Inventory.** After a preliminary assessment and site inspection, the decision is made as to whether or not the site will be placed on the **National Priority List (NPL).** Sites placed on this list are those deemed to require a remedial action. Currently, there are more than 1200 sites on the NPL. **The National Contingency Plan (NCP)** is the next component. The purpose of the NCP is to characterize the nature and extent of risk posed by contamination and to evaluate potential remedial options. The investigation and feasibility study components are normally conducted concurrently and with a "phased" approach. This allows

feedback between the two components. A diagram of the NCP procedure is provided in Fig. 16.2. The selection of the specific remedial action to be used at a particular site is a very complex process. The goals of the remedial action are that it be protective of human health and the environment, that it maintain protection over time, and that it maximize waste treatment.

16.3 THE OVERALL PROCESS OF BIODEGRADATION

The remainder of this chapter deals with biodegradation of organic contaminants and ways in which these processes can be harnessed to remediate contam-

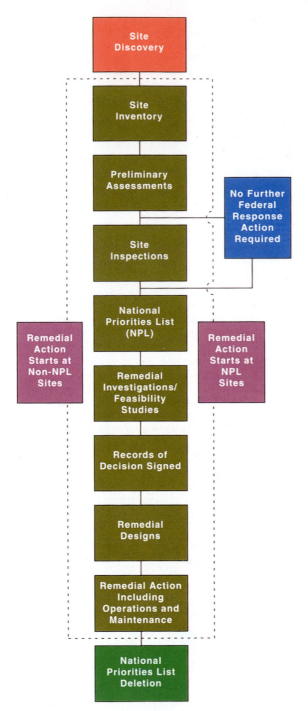

FIGURE 16.1 The Superfund process for treatment of a hazardous waste site. (From *Pollution Science* © 1996, Academic Press, San Diego, CA.)

inated sites. **Biodegradation** is the breakdown of organic contaminants that occurs due to microbial activity. As such, these contaminants can be considered as the microbial food source or **substrate.** Biodegradation of any organic compound can be thought of as a series of biological degradation steps or a pathway that ultimately results in the oxidation of the

parent compound. Often, the degradation of these compounds results in the generation of energy as described in Chapter 3.

Complete biodegradation or **mineralization** involves oxidation of the parent compound to form carbon dioxide and water, a process that provides both carbon and energy for growth and reproduction of cells. Figure 16.3 illustrates the mineralization of an organic compound under either aerobic or anaerobic conditions. The series of degradation steps constituting mineralization is similar whether the carbon source is a simple sugar such as glucose, a plant polymer such as cellulose, or a pollutant molecule. Each degradation step in the pathway is catalyzed by a specific **enzyme** made by the degrading cell. Enzymes are most often found within a cell but are also made and released from the cell to help initiate degradation reactions. Enzymes found external to the cell are known as **extracellular enzymes.** Extracellular enzymes are important in the degradation of macromolecules such as the plant polymer cellulose (see Chapter 14.2.4.1). Macromolecules must be broken down into smaller subunits outside the cell to allow transport of the smaller subunits into the cell. Degradation by either internal or extracellular enzymes will stop at any step if the appropriate enzyme is not present (Fig. 16.4). Lack of appropriate biodegrading enzymes is one common reason for persistence of organic contaminants, particularly those with unusual chemical structures that existing enzymes do not recognize. Thus, contaminant compounds that have structures similar to those of natural substrates are normally easily degraded. Those that are quite dissimilar to natural substrates are often degraded slowly or not at all.

Some organic contaminants are degraded partially but not completely. This can result from absence of the appropriate degrading enzyme as mentioned earlier. A second type of incomplete degradation is **cometabolism,** in which a partial oxidation of the substrate occurs but the energy derived from the oxidation is not used to support microbial growth. The process occurs when organisms coincidentally possess enzymes that can degrade a particular contaminant. Thus, such enzymes are nonspecific. Cometabolism can occur during periods of active growth or can result from interaction of resting (nongrowing) cells with an organic compound. Cometabolism is difficult to measure in the environment but has been demonstrated for some environmental contaminants. For example, the industrial solvent trichloroethylene (TCE) can be oxidized cometabolically by methanotrophic bacteria that grow on methane as a sole carbon source. TCE is currently of great interest for several reasons. It is one of the most frequently reported contaminants at hazardous waste sites, it is a suspected carcinogen, and it is generally re-

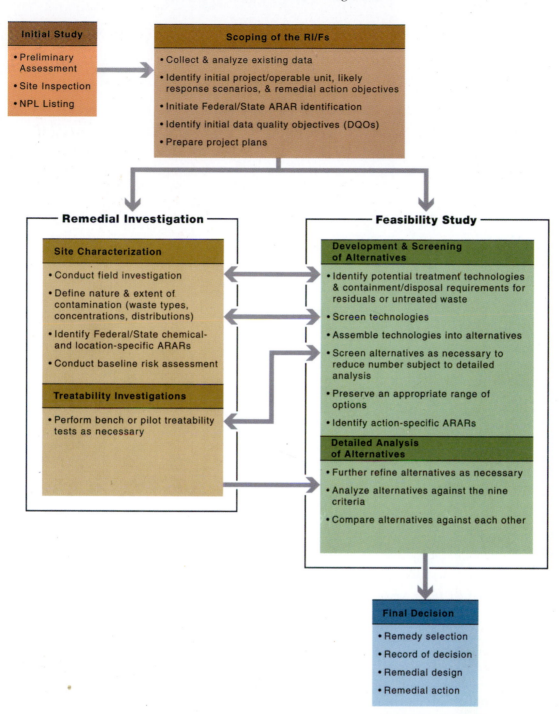

FIGURE 16.2   The remedial investigation–feasibility study (RI/FS) process. ARAR(s) = Applicable or relevant and appropriate requirements. (From *Pollution Science* © 1996, Academic Press, San Diego, CA.)

sistant to biodegradation. As shown in Fig. 16.5, the first step in the oxidation of methane by **methanotrophic bacteria** is catalyzed by the enzyme **methane monooxygenase.** This enzyme is so nonspecific that it can also cometabolically catalyze the first step in the oxidation of TCE when both methane and TCE are present. The bacteria receive no energy benefit from this cometabolic

degradation step. The subsequent degradation steps shown in Figure 16.5 may be catalyzed spontaneously, by other bacteria, or in some cases by the methanotroph. This is an example of a cometabolic reaction that may have great significance in remediation. Research is currently investigating the application of these methanotrophs to TCE-contaminated sites. Other cometaboliz-

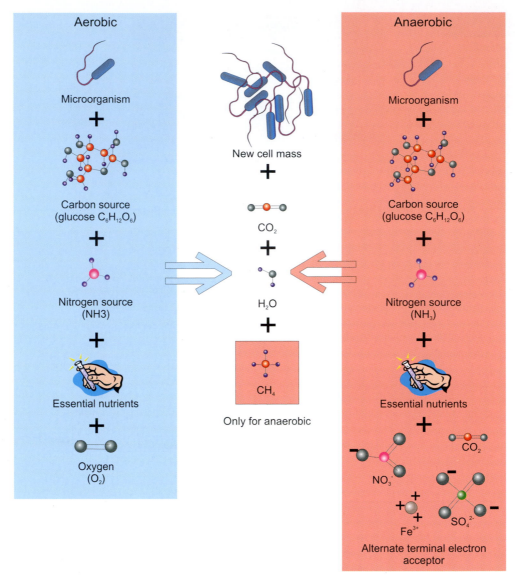

FIGURE 16.3   Aerobic (blue) or anaerobic (red) mineralization of an organic compound.

ing microorganisms that grow on toluene, propane, and even ammonia are also being evaluated for use in TCE bioremediation.

Partial or incomplete degradation can also result in **polymerization** or synthesis of compounds more complex and stable than the parent compound. This occurs when initial degradation steps, often catalyzed by extracellular enzymes, create reactive intermediate compounds. These highly reactive intermediate compounds can then combine with each other or with other organic matter present in the environment. This is illustrated in Fig. 16.6, which shows some possible polymerization reactions that occur with the herbicide propanil during biodegradation. These include formation of dimers or larger polymers, which are quite stable in the environment. Stability is due to low

bioavailability (high sorption and low solubility), lack of degrading enzymes, and the fact that some of these residues become chemically bound to the soil organic matter fraction.

16.4 RELATIONSHIP BETWEEN CONTAMINANT STRUCTURE, TOXICITY, AND BIODEGRADABILITY

The vast majority of the organic carbon available to microorganisms in the environment is material that has been photosynthetically fixed (plant material). Of concern are environments that receive large additional inputs of carbon from agriculture or industry (petroleum products, organic solvents, pesticides).

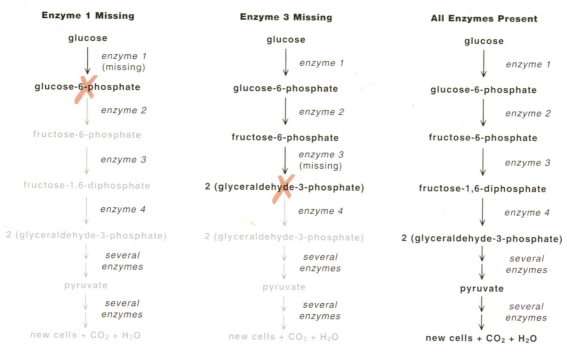

FIGURE 16.4 Stepwise degradation of organic compounds. A different enzyme catalyzes each step of the biodegradation pathway. (From *Pollution Science* © 1996, Academic Press, San Diego, CA.)

Although many of these chemicals can be readily degraded because of their structural similarity to naturally occurring organic carbon, the amounts added may exceed the existing **carrying capacity** of the environment. Carrying capacity is defined here as the maximum level of microbial activity that can be expected under existing environmental conditions. Microbial activity may be limited by both biological and

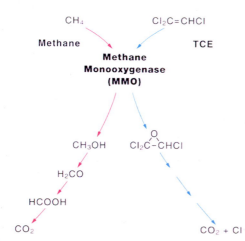

FIGURE 16.5 The oxidation of methane by methanotrophic bacteria is catalyzed by the enzyme methane monooxygenase. Subsequent degradation steps may be catalyzed spontaneously, by other bacteria, or in some cases by the methanotroph. (From *Pollution Science* © 1996, Academic Press, San Diego, CA.)

physical–chemical factors. These factors include low numbers of microbes, insufficient oxygen or nutrient availability, as well as suboptimal temperature and water availability. These factors are discussed further in Section 16.5. Microbial activity, whether degradation occurs and the rate of degradation, also depends on several factors related to the structure and physical–chemical properties of the contaminant (Miller and Herman, 1997). These factors include

1. Genetic potential, or the presence and expression of appropriate degrading genes by the indigenous microbial community.
2. Bioavailability, or the effect of limited water solubility and sorption on the rate at which a contaminant is taken up by a microbial cell.
3. Contaminant structure, including both steric and electronic effects. Steric effects involve the extent to which substituent groups on a contaminant molecule sterically hinder recognition by the active site of the degrading enzyme. Electronic effects involve the extent to which substituent groups electronically interfere with the interaction between the active site of the enzyme and the contaminant. Electronic effects can also alter the energy required to break critical bonds in the molecule.
4. Toxicity, or the inhibitory effect of the contaminant on cellular metabolism.

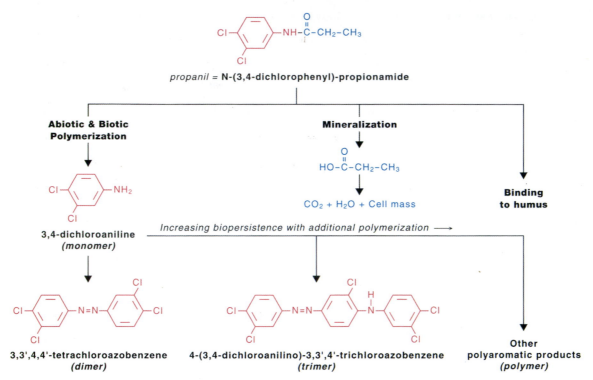

FIGURE 16.6 Polymerization reactions that occur with the herbicide propanil during biodegradation. Propanil is a selective postemergence herbicide used in growing rice. It is toxic to many annual and perennial weeds. The environmental fate of propanil is of concern because it, like many other pesticides, is toxic to most crops except for cereal grains. It is also toxic to fish. Care is used in propanil application to avoid contamination of nearby lakes and streams. (From *Pollution Science* © 1996, Academic Press, San Diego, CA.)

16.4.1 Genetic Potential

The onset of contaminant biodegradation generally follows a period of adaptation or acclimation of indigenous microbes, the length of which depends on the contaminant structure. The efficient cycling of plant-based organic matter by soil microorganisms can promote the rapid degradation of organic contaminants that have a chemical structure similar to those of natural soil organic compounds. Previous exposure to a contaminant through repeated pesticide applications or through frequent spills will create an environment in which a biodegradation pathway is maintained within an adapted community. Adaptation of microbial populations most commonly occurs by induction of enzymes necessary for biodegradation followed by an increase in the population of biodegrading organisms (Leahy and Colwell, 1990).

Naturally occurring analogues of certain contaminants may not exist, and previous exposure may not have occurred. Degradation of these contaminants requires a second type of adaptation that involves a genetic change such as a mutation or a gene transfer (Boyle, 1992; Fulthorpe and Wyndham, 1992; Van der Meer *et al.*, 1992). This results in the development of new metabolic capabilities. The time needed for an adaptation requiring a genetic change or for the selection and development of an adapted community is not yet predictable, but it may require weeks to years or may not occur at all.

16.4.2 Bioavailability

For a long time biodegradation was thought to occur if the appropriate microbial enzymes were present. As a result, most research focused on the actual biodegradative process, specifically the isolation and characterization of biodegradative enzymes and genes. There are, however, two steps in the biodegradative process. The first is the uptake of the substrate by the cell, and the second is the metabolism or degradation of the substrate. Assuming the presence of an appropriate metabolic pathway, degradation of a contaminant can proceed rapidly if the contaminant is available in a water-soluble form. However, degradation of contaminants with limited water solubility or that are strongly sorbed to soil or sediments can be limited due to their low bioavailability (Miller, 1995).

Growth on an organic compound with limited water solubility poses a unique problem for microorganisms because the compound is not freely available in the aqueous phase. Most microorganisms require high water activity (>0.96) for active metabolism (Atlas and Bartha, 1993), and thus the contact between the degrading organism and an organic compound with low water solubility is limited. The compound may be present in a liquid or solid state, both of which can form a two-phase system with water. Liquid hydrocarbons can be less or more dense than water, forming a separate phase above or below the water surface. For example, polychlorinated biphenyls (PCBs) and chlorinated solvents such as TCE are denser than water and form a separate phase below the water surface. Solvents less dense than water, such as benzene and other petroleum constituents, form a separate phase above the water surface. There are three possible modes of microbial uptake of a liquid organic (Fig. 16.7):

1. Utilization of the solubilized organic compound (Fig. 16.7A).
2. Direct contact of cells with the organic compound. This can be mediated by cell modifications such as fimbriae (Rosenberg et al., 1982) or cell surface hydrophobicity (Zhang and Miller, 1994), which increase attachment of the cell to the organic compound (Fig. 16.7B).
3. Direct contact with fine or submicrometer size substrate droplets dispersed in the aqueous phase (Fig. 16.7C).

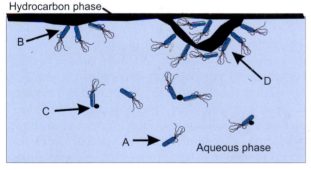

FIGURE 16.7 A water environment with a oil phase floating on the surface. This is typical of what might occur when oil is spilled in the ocean. There are several ways in which microbes reach the oil phase in this type of situation. (A) Microbes can take up hydrocarbons dissolved in the aqueous phase surrounding degrading cells. This uptake mode becomes limiting as the aqueous solubility of the hydrocarbon decreases. (B) Uptake via direct contact of degrading cells at the aqueous–hydrocarbon interface of large oil drops in water. This uptake mode is limited by the interfacial area between the water and hydrocarbon phase. (C) Uptake through direct contact of degrading cells with fine or submicrometer-size oil droplets dispersed in the aqueous phase. This uptake mode is limited by the formation of such droplets. In the ocean environment, wave action can create substantial dispersion of oil. In a soil environment such dispersion is more limited. (D) Enhanced uptake as a result of production of biosurfactants or emulsifiers that effectively increase the apparent aqueous solubility of the hydrocarbon or allow better attachment of cells to the hydrocarbon.

The mode that predominates depends largely on the water solubility of the organic compound. In general, direct contact with the organic compound plays a more important role (modes 2 and 3) as water solubility decreases.

Some microbes can enhance the rate of uptake and biodegradation as a result of production of **biosurfactants** or emulsifiers (Fig. 16.7D). There are two effects of biosurfactants. First, they can effectively increase the aqueous solubility of the hydrocarbon through formation of micelles or vesicles (Fig. 16.8) that associate with hydrocarbons. Second, they can facilitate attachment of cells to the hydrocarbon by making the cell surface more hydrophobic and, thus, better able to stick to a separate oil phase (Fig. 16.9). This makes it possible to achieve greatly enhanced biodegradation rates in the presence of biosurfactants (Herman et al., 1997).

For organic compounds in the solid phase, e.g., waxes, plastics, or polyaromatic hydrocarbons (PAHs), there are only two modes by which a cell can take up the substrate:

1. Direct contact with the substrate
2. Utilization of solubilized substrate

Available evidence suggests that for solid-phase organic compounds, utilization of solubilized substrate is most important. Thus, low water solubility has a greater impact on degradation of solid- phase organic compounds than on liquid-phase organics.

Another factor that affects bioavailability of an organic compound is sorption of the compound by soil or sediment (Novak et al., 1995). Depending on the sorption mechanism, organic compounds can be weakly (hydrogen bonding, van der Waals forces, hydrophobic interactions) or strongly (covalent binding) bound to soil. Sorption of weakly bound or labile residues is reversible, and when a sorbed residue is released back into solution it becomes available for microbial utilization (Scow, 1993). Bioavailability can also be reduced by the diffusion of contaminants into soil matrix microsites that are inaccessible to bacteria because of pore size exclusion (Alexander, 1995). There is evidence that the proportion of labile residues made available by desorption decreases with the length of time the residues are in the soil. Thus, as contaminants age and become sequestered more deeply within inaccessible microsites (Fig. 4.3), bioavailability, and therefore biodegradation, can be expected to decrease.

Finally, some contaminants may be incorporated into soil organic matter by the catalytic activity of a wide variety of oxidative enzymes that are present in the soil matrix. The incorporation of contaminants into

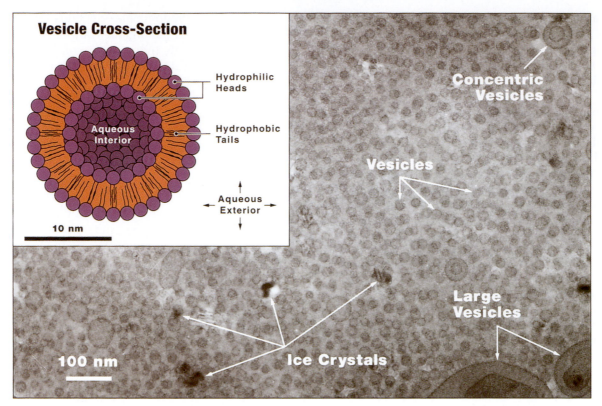

FIGURE 16.8 Cryo–transmission electron micrograph of a microbially produced surfactant, rhamnolipid. In water, this compound spontaneously forms aggregates such as the vesicles shown here. Hydrocarbons like to associate with the lipid-like layer formed by the hydrophobic tails of the surfactant vesicles. These tiny surfactant–hydrocarbon structures are soluble in the aqueous phase. (From *Pollution Science* © 1996, Academic Press, San Diego, CA.)

soil organic matter is called **humification,** a process that is usually irreversible and that may be considered as one factor in the aging process (Bollag, 1992). These bound or humified residues are released and degraded only very slowly as part of the normal turnover of humic material in soil (see Chapter 14.2.4.2).

16.4.3 Contaminant Structure

16.4.3.1 Steric Effects

What type of contaminant structures can lead to low degradation rates even if the contaminant structure is similar to naturally-occurring molecules? It is the presence of branching and functional groups that often slows degradation by changing the chemistry of the degradation **reaction site.** The reaction site is where a degradative enzyme comes into contact with a contaminant substrate causing a transformation step to occur. When the reaction site is blocked by branching or a functional group, contact between the contaminant and enzyme at the reaction site is hindered. This is known as a steric effect and is illustrated in Fig. 16.10, which compares two structures, an eight-carbon

n-alkane (A) and the same eight-carbon backbone with four methyl branches (B). Whereas octane is readily degradable by the pathway shown in Fig. 16.14, the four methyl substituents in structure B inhibit degradation at both ends of the molecule. Branching or functional groups can also affect transport of the substrate across the cell membrane, especially if the transport is enzyme-assisted. Steric effects usually increase as the size of the functional group increases (Pitter and Chudoba, 1990).

16.4.3.2 Electronic Effects

In addition to steric effects, functional groups may contribute electronic effects that hinder biodegradation by affecting the interaction between the contaminant and the enzyme. Functional groups can be electron-donating (e.g., CH_3) or electron-withdrawing (e.g., Cl), and therefore, can change the electron density of the reaction site. In general, functional groups that add to the electron density of the reaction site increase biodegradation rates, and functional groups that decrease the electron density of the reaction site decrease biodegradation rates. To illustrate the relation-

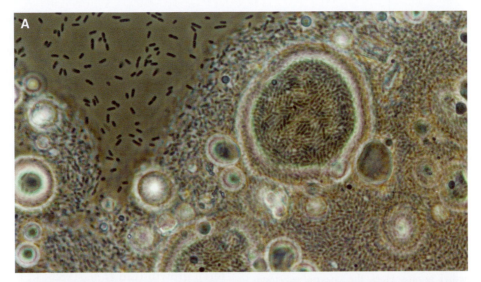

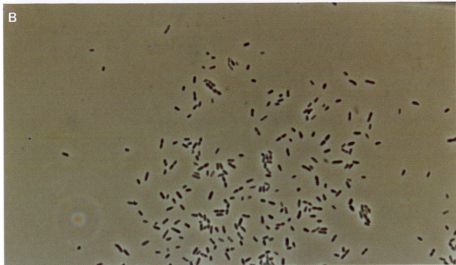

FIGURE 16.9   Phase-contrast micrographs showing the effect of a biosurfactant on the ability of *Pseudomonas aeruginosa* ATCC 15524 to stick to hexadecane droplets (magnification × 1000). (A) Addition of rhamnolipid biosurfactant (0.1 mM) causes cells to clump and to stick to oil droplets. (B) No biosurfactant is present and individual cells do not clump and do not stick to oil droplets in the solution. (Photos courtesy D. C. Herman.)

ship between functional group electronegativities and rate of biodegradation, Pitter and Chudoba (1990) compared the electronegativity of a series of ortho-substituted phenols with their biodegradation rates. Five different functional groups were tested, and it was found that as the electronegativity of the substituents increased, biodegradation rates decreased (Fig. 16.11).

16.4.4 Toxicity

Contaminant spills and engineered remediation projects, such as landfarming of petroleum refinery sludges, can involve extremely high contaminant concentrations. In these cases, toxicity of the contaminant to microbial populations can slow the remediation process. The toxicity of nonionized organic contaminants such as petroleum hydrocarbons or organic solvents is to a large extent due to a nonspecific narcotic-type mode of action. Nonspecific narcotic-type toxicity is based on the partitioning of a dissolved contaminant into the lipophilic layer of the cell membrane, which causes a disruption of membrane integrity (Sikkema *et al.*, 1995). The cell membrane is believed to be the major site of organic contaminant accumulation in mi-

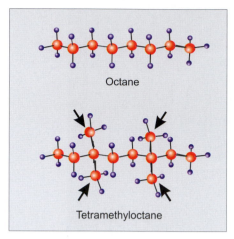

FIGURE 16.10 The structure of (A) octane, which is readily degradable, and (B) a tetramethyl-substituted octane that is not degraded because the methyl groups block the enzyme–substrate catalysis site.

croorganisms. One result of membrane exposure to contaminants is increased permeability of the membrane. The disruption of membrane permeability has been assayed in several ways: using fluorescent dyes (Herman *et al.*, 1991); by the loss of intracellular components, such as potassium from algae (Hutchinson *et al.*, 1981) or bacteria (Bernheim, 1974); or by the loss of ^{14}C-labeled photosynthates from algal cells (Stratton, 1989).

Models have been developed that relate bioconcentration (the accumulation of a hydrophobic contaminant by a cell or organism) and toxicity to physicochemical attributes of the organic contaminant. These models are referred to as **quantitative structure–activ-**

ity relationship (QSAR) models. QSAR models allow physical descriptors to be used to predict the toxicity of a wide variety of nonionized organics to aquatic organisms and have also been used to predict the inhibition of microbial activity (Blum and Speece, 1990). Such descriptors include hydrophobicity, as determined from the octanol–water partition coefficient (K_{ow}), and molecular connectivity, which represents the surface topography of a compound. An example of a QSAR model was provided by Warne *et al.* (1989a) who found a strong correlation between growth inhibition of a mixed culture of marine bacteria and the K_{ow} of shale oil components, including alkyl-substituted benzenes, naphthalenes, pyridines, and phenols. Similar correlations were found for the toxicity of mono-, di-, tri-, tetra-, and pentachlorophenols to bacterial growth (Liu *et al.*, 1982). In this case, increasing chlorination of phenols produced a more lipophilic (hydrophobic) compound (greater K_{ow}) that was more toxic to bacterial growth.

Many QSAR models are based on the hydrophobic partitioning of a nonionized organic compound, leading to a nonspecific, narcotic-type mode of action. Exceptions can occur if a compound has a more specific mode of inhibition. For example, steric or electronic factors can increase the toxicity of a compound over that attributed to hydrophobicity alone. This was demonstrated in a comparison of isomers of mono-, di-, tri-, and tetrachlorophenols (Ruckdeschel *et al.*, 1987) and of di- and trichlorobenzenes (Liu and Thomson, 1983). Some isomers with similar K_{ow} values differed greatly in their ability to inhibit microbial growth. Another example is aromatic compounds that can be microbially transformed to highly reactive pheno-

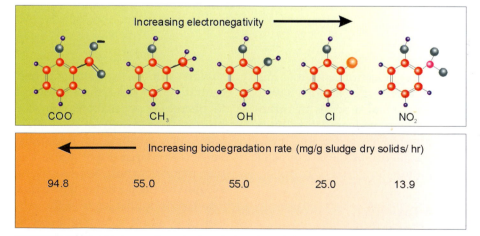

FIGURE 16.11 Various ortho-substituted phenols and their respective biodegradation rates. (Adapted from Pitter and Chudoba, 1990.)

lic or quinone derivatives that have greater toxicity than would be predicted on the basis of hydrophobic partitioning alone (Narro *et al.,* 1992).

One limitation of hydrophobicity-based QSAR modeling is the condition that the organic contaminant is partitioned from an aqueous (solubilized) phase into a lipid phase. This necessitates that a toxic aqueous concentration is achievable within the water solubility limits of a particular compound. Thus, certain organic contaminants, such as high-molecular-weight aliphatic or asphaltene components of crude oil, that are strongly hydrophobic are not thought to be toxic to microorganisms. In contrast, low-molecular-weight aliphatic and aromatic components with higher water solubilities are strongly associated with the toxicity of crude oil to microorganisms.

16.5 ENVIRONMENTAL FACTORS AFFECTING BIODEGRADATION

A number of parameters influence the survival and activity of microorganisms in any environment. One factor that has great influence on microbial activity is organic matter, the primary source of carbon for heterotrophic microorganisms in most environments. Surface soils have a relatively high and variable organic matter content and therefore are characterized by high microbial numbers and diverse metabolic activity (see Chapter 4). In contrast, the subsurface unsaturated (vadose) zone and saturated zone usually have a much lower content and diversity of organic matter, resulting in lower microbial numbers and activity. Exceptions to this rule are some areas of the saturated zone that have high flow or recharge rates, which can lead to numbers and activities of microorganisms similar to those found in surface soils.

Occurrence and abundance of microorganisms in an environment are determined not only by available carbon but also by various physical and chemical factors. These include oxygen availability, nutrient availability, temperature, pH, salinity, and water activity. Inhibition of biodegradation can be caused by a limitation imposed by any one of these factors, but the cause of the persistence of a contaminant is sometimes difficult to determine. Perhaps the most important factors controlling contaminant biodegradation in the environment are oxygen availability, organic matter content, nitrogen availability, and contaminant bioavailability. Interestingly, the first three of these factors can change considerably depending on the location of the contaminant. Figure 16.12 shows the relationship between organic carbon, oxygen, and microbial activity in a profile of the terrestrial ecosystem including surface soils, the vadose zone, and the saturated zone.

16.5.1 Oxygen

Oxygen is very important in determining the extent and rate of contaminant biodegradation. In general, aerobic biodegradation is much faster than anaerobic biodegradation. For example, petroleum-based hydrocarbons entering the aerobic zones of freshwater lakes and rivers are generally susceptible to microbial degradation, but oil accumulated in anaerobic sediments can be highly persistent (Cooney, 1984). It follows that as oxygen is depleted, a reduction in the rate of hydrocarbon degradation can be expected. Oxygen is especially important for degradation of highly reduced hydrocarbons such as the alkane hexadecane ($C_{16}H_{34}$). Biodegradation of hexadecane was found to occur only in the presence of oxygen, although an oxygen tension as low as 1% of full oxygen saturation was enough to allow degradation to occur (Michaelsen *et al.,* 1992).

Benzene, oxygenated aromatics such as benzoate or phenols, and alkylated aromatics such as toluene have been shown to be biodegraded under anaerobic conditions when nitrate, iron, and sulfate are available for use as terminal electron acceptors (Evans and Fuchs, 1988). Nevertheless, it is still accepted that the biodegradation of such compounds is much slower under anaerobic conditions and requires much longer adaptation periods than degradation in an aerobic environment. Some studies have shown that anaerobic biodegradation by specifically adapted consortia can achieve rapid loss of contaminants (Lovley *et al.,* 1995), indicating that anaerobic degradation may become more important as a bioremediation tool as more is understood about this process.

16.5.2 Organic Matter Content

Surface soils have large numbers of microorganisms. Bacterial numbers are generally 10^6 to 10^9 organisms per gram of soil. Fungal numbers are somewhat lower, 10^4 to 10^6 per gram of soil. In contrast, microbial populations in deeper regions such as the deep vadose zone and groundwater region are often lower by two orders of magnitude or more (see Chapter 4). This large decrease in microbial numbers with depth is due primarily to differences in organic matter content. Both the vadose zone and the groundwater region have low amounts of organic matter. One result of low total numbers of microorganisms is that a low population of contaminant degraders may be present initially. Thus, biodegradation of a particular contaminant may be slow until a sufficient biodegrading population has been built up. A second reason for slow biodegradation in the vadose zone and groundwater region is that because a low amount of organic matter is present, the organisms in this region are often dormant. This can cause their response to an added carbon source to be slow, especially if the carbon source is a contaminant molecule that has low bioavailability or to which the organisms have not had prior exposure.

Because of these trends in oxygen availability and organic matter content, several generalizations can be made with respect to surface soils, the vadose zone, and the groundwater region (Fig. 16.12):

1. Biodegradation in surface soils is primarily aerobic and rapid.
2. Biodegradation in the vadose zone is also primarily aerobic, but significant acclimation times may be necessary for significant biodegrading populations to build up.
3. Biodegradation in the deep groundwater region is also initially slow because of low numbers, and can rapidly become anaerobic because of lack of available oxygen. Biodegradation in shallow groundwater regions is initially more rapid because of higher microbial numbers but is similarly slowed by low oxygen availability.

16.5.3 Nitrogen

Microbial utilization of organic contaminants, particularly hydrocarbons composed primarily of carbon and hydrogen, creates a demand for essential nutrients such as nitrogen and phosphorus. Ward and Brock (1976) monitored seasonal variations in hydrocarbon degradation in a temperate lake and found that variations in the available forms of nitrogen and phosphorus limited degradation. Maximum rates of hydrocarbon degradation were evident in early spring, when available nitrogen and phosphorus levels were high, but rapid consumption of these nutrients reduced the rate of hydrocarbon degradation during the summer months. Thus, biodegradation can often be improved simply by the addition of nitrogen fertilizers. This is particularly true in the case of biodegradation of petroleum oil spills, in which nitrogen shortages can be acute. In general, microbes have an average $C:N$ ratio within their biomass of about 5:1 to 10:1 depending on the type of microorganism. Therefore a ratio of approximately $100:10:1$ ($C:N:P$) is often used in such sites. However, in some instances, quite different ratios have been used. For example, Wang and Bartha (1990) found that effective remediation of hydrocarbons in soil required the addition of nitrogen and phosphorus additions to maintain a $C:N$ ratio of 200:1 and a $C:P$ ratio of 1000:1. Why were the $C:N$ and $C:P$ ratios maintained at levels so much higher than the cell $C:N$ and $C:P$ ratios? As discussed in Chapter 14.3.3.2, it is because much of the hydrocarbon carbon that is metabolized is released as carbon dioxide so that much of the carbon is lost from the system. In contrast, almost all of the nitrogen and phosphorus metabolized is incorporated into microbial biomass and thus is conserved in the system.

16.5.4 Other Environmental Factors

16.5.4.1 Temperature

Hydrocarbon degradation has been reported to occur at a range of temperatures between close to freezing and more than 30°C. Bacteria can adapt to temperature extremes in order to maintain metabolic activity; however, seasonal temperature fluctuations in the natural environment have been shown to affect the rate at which degradation occurs (Palmisano et al., 1991). For example, the degradation rates of hexadecane and naphthalene in a river sediment were reduced approximately 4.5-fold and 40-fold, respectively, in winter (0–4°C) compared with summer (8–21°C) samples (Wyndham and Costerton, 1981).

16.5.4.2 pH

In soils, the rate of hydrocarbon degradation is often higher in alkaline conditions than in acidic conditions. In acidic soils, fungi are more competitive than bacteria, which prefer a neutral environment. There-

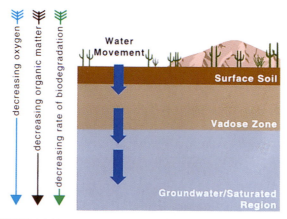

FIGURE 16.12 There are three major locations where contamination can occur in terrestrial ecosystems: surface soils, the vadose zone, and the saturated zone. The availability of both oxygen and organic matter varies considerably in these zones. As indicated, oxygen and organic matter both decrease with depth, resulting in a decrease in biodegradation activity with depth. (From *Pollution Science* © 1996, Academic Press, San Diego, CA.)

fore, at lower soil pH, fungi become more important in hydrocarbon degradation. Acidic soils favor the growth of fungi, which degrade hydrocarbons but usually at a slower rate than soil bacteria, which prefer a more neutral to alkaline environment. Hambrick *et al.* (1980) examined the effect of pH on hydrocarbon degradation in a salt marsh sediment. The pH of the sediment ranged from 6.5 to 8.0, but when incubated at different pH levels, lower rates of hydrocarbon degradation were evident at pH 5.0 and 6.5 than at pH 8.0.

16.5.4.3 Salinity

Hydrocarbon degradation has been shown to occur in saline environments. Samples of freshwater sediment incubated under saline conditions showed a reduced rate of hydrocarbon degradation. In contrast, hydrocarbon degradation in estuarine sediments incubated with increasing levels of salinity were little affected (Kerr and Capone, 1988), although hypersaline conditions were reported to reduce the rate of hydrocarbon degradation in sediments sampled from a saline lake (Ward and Brock, 1978).

16.5.4.4 Water Activity

Optimal conditions for activity of aerobic soil microorganisms occur between 38% and 81% of the soil pore space (also referred to as % saturation) because in this range of water contents, water and oxygen availability are maximized. At higher water contents, the slow rate of oxygen diffusion through water limits oxygen replenishment, thereby limiting aerobic activity. At lower water contents, water availability becomes limiting. Why is the optimal % saturation range so broad? It is because optimal activity really depends upon a combination of factors including water content and available pore space. Available pore space is measured as bulk density which is defined as the mass of soil per unit volume (g/cm^3). This means that in any given soil, increasing bulk density indicates increasing compaction of the soil. In a soil that is loosely compacted (lower bulk density), a water saturation of 70% represents more water (more filled small pores and pore throats) than in a highly compacted soil. Thus, for soils of low bulk density, oxygen diffusion constraints become important at lower water saturation than for highly compacted soils.

For example, Neilson and Pepper (1990) showed that respiration in a clay loam soil was maximal at different % saturation depending on bulk density. At a bulk density of 1.1 g/cm^3, respiration was optimal between 38 and 45% water saturation. At a bulk density of 1.6 g/cm^3, respiration was optimal at a higher water saturation, 81%.

16.6 BIODEGRADATION OF ORGANIC POLLUTANTS

16.6.1 Pollutant Sources and Types

In 1981, the United States used almost 6 billion barrels of oil for heating, generation of electricity, and gasoline. Other sources of energy are coal, natural gas, and nuclear energy. The paper, transportation, electronics, defense, and metals industries all produce large amounts of waste, including solvents, acids, bases, and metals. It is estimated that in 1985, 3 million tons ($15.9 billion market value) of pesticides were used worldwide, with the United States using approximately one fourth of this total. As these figures demonstrate, both industry and agriculture produce large amounts of chemicals. Inevitably, some of these find their way into the environment as a result of normal handling procedures and spills. Figure 16.13 shows various contaminant molecules that are added to the environment in significant quantities by anthropogenic activities. The structure of most contaminant molecules is based on the one of the three structures shown in Fig. 16.13: aliphatic, alicyclic, or aromatic. By combining or adding to these structures, a variety of complex molecules can be formed that have unique properties useful in industry and agriculture. The objective of this section is to become familiar with these structures and their biodegradation pathways so that given a structure, a reasonable biodegradation pathway can be predicted.

16.6.2 Aerobic Conditions

16.6.2.1 Aliphatics

There are several common sources of aliphatic hydrocarbons that enter the environment as contaminants. These include straight-chain and branched-chain structures found in petroleum hydrocarbons, the linear alkyl benzenesulfonate (LAS) detergents, and the one- and two-carbon halogenated compounds such as TCE that are commonly used as industrial solvents. Some general rules for aliphatic biodegradation are presented in Information Box 1, and specific biodegradation pathways are summarized in the following sections for alkanes, alkenes, and chlorinated aliphatics.

Alkanes

Because of their structural similarity to fatty acids and plant paraffins, which are ubiquitous in nature, many microorganisms in the environment can utilize *n*-alkanes (straight-chain alkanes) as a sole source of carbon and energy. In fact, it is easy to isolate alkane-

Hydrocarbon Type	Structure	Name	Physical state at room temp.	Source and uses
Aliphatics		propane n=1	gas	Petroleum contains both linear and branched aliphatics. The gasoline fraction of crude oil is 30 - 70% aliphatic depending on the source of the crude oil.
		hexane n=4	liquid	
		hexatriacontane n=34	solid	
Alicyclics		cyclopentane	liquid	Petroleum contains both unsubstituted and alkyl substituted alicyclics. The gasoline fraction of crude oil is 20-70% alicyclic depending on the source of the crude oil
		cyclohexane	liquid	
Aromatics		benzene	liquid	Petroleum contains both unsubstituted and alkyl substituted aromatics. The gasoline fraction of crude oil is 10-15% depending on the source of the crude oil.
		naphthalene	solid	
		phenanthrene	solid	
Substituted aliphatics		chloroform	liquid	Anthropogenically manufactured and used as solvents, degreasing agents, and in organic syntheses.
		trichloroethylene (TCE)	liquid	
Substituted aromatics		phenol	liquid	Found in coal tar or manufactured and used as a disinfectant, and in manufacture of resins, dyes and industrial chemicals.
		pentachlorophenol	liquid	Manufactured and used as an insecticide, defoliant, and wood preservative.
		toluene	liquid	Found in tar oil, used in manufacture of organics, explosives, and dyes. Also used as a solvent.
		benzoate	liquid	Found in plants and animals and manufactured for use as a food preservative, dye component, and in curing tobacco.

378

Category	Compound	State	Description
Biaryl hydrocarbons	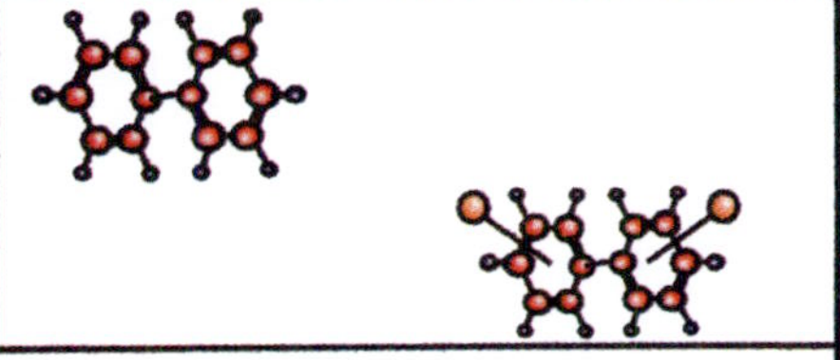 biphenyl	solid	Biphenyl is the parent compound of variously chlorinated biphenyl mixtures known as the PCBs. PCBs are used as transformer oils and plasticizers.
	polychlorinated biphenyls (PCBs)	liquid	
Heterocyclics	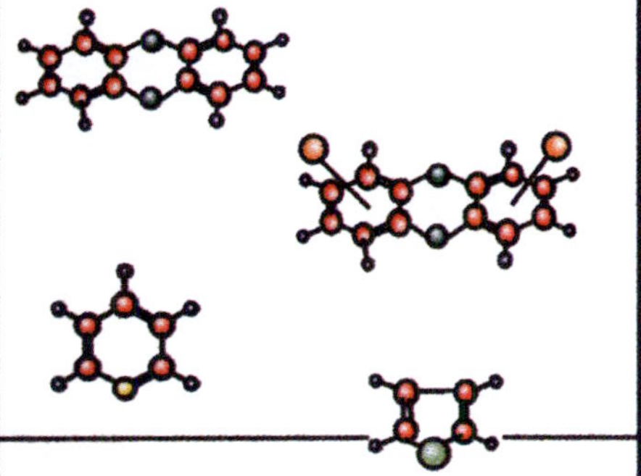 dibenzodioxin	solid	Dioxins are created during incineration processes and are contaminants associated with the manufacture of herbicides including 2,4-D and 2,4,5-T.
	chlorinated dioxins	solid	
	pyridine	liquid	Found in coal tar. Used as a solvent and synthetic intermediate.
	thiophene	liquid	Found in coal tar, coal gas and crude oil. Used as a solvent and in manufacture of resins, dyes, and pharmaceuticals.
Pesticides Organic acids	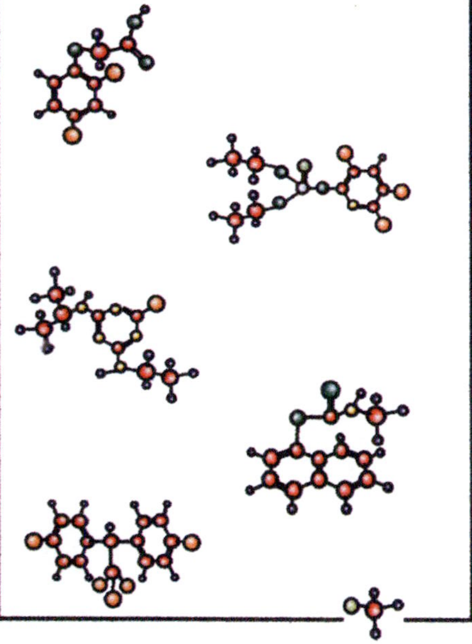 2,4-dichlorophenoxy acetic acid	solid	Broadleaf herbicide
Organophosphates	chlorpyrifos	solid	Used as an insecticide and an acaricide
Triazenes	atrazine	solid	Selective herbicide
Carbamates	carbaryl	solid	Contact insecticide
Chlorinated hydrocarbons	1,1,1-trichloro-2,2-bis-(4-chlorophenyl)-ethane (DDT)	solid	Contact insecticide
	methyl bromide	gas	Used to degrease wool, extract oil from nuts, seeds and flowers, used as an insect and soil fumigant.

LEGEND

Carbon, Hydrogen, Oxygen, Chlorine, Nitrogen, Sulfur, Phosphorus, Bromine

FIGURE 16.13 Representative pollutant structures.

Information Box 1

General rules for degradation of aliphatic compounds:

1. Midsize straight-chain aliphatics (*n*-alkanes C_{10} to C_{18} in length) are utilized more readily than *n*-alkanes with either shorter or longer chains.
2. Saturated aliphatics and alkenes are degraded similarly.
3. Hydrocarbon branching decreases biodegradability.
4. Halogen substitution decreases biodegradability.

---

degrading microbes from any environmental sample. As a result, alkanes are usually considered to be the most readily biodegradable type of hydrocarbon. Biodegradation of alkanes occurs with a high biological oxygen demand (BOD) using one of the two pathways shown in Fig. 16.14. The more common pathway is the direct incorporation of one atom of oxygen onto one of the end carbons of the alkane by a **monooxygenase enzyme**. This results in the formation of a primary alcohol (see pathway I). Alternatively, a **dioxygenase enzyme** can incorporate both oxygen atoms into the alkane to form a hydroperoxide (pathway II). The end result of both pathways is the production of a primary fatty acid. There are also examples in the literature of diterminal oxidation, with both ends of the alkane oxidized, and of subterminal oxidation, with an interior carbon oxidized (Britton, 1984).

Fatty acids are common metabolites found in all cells. They are used in the synthesis of membrane phospholipids and lipid storage materials. The common pathway used to catabolize fatty acids is known as *β*-**oxidation**, a pathway that cleaves off consecutive two-carbon fragments (Fig. 16.14). Each two carbon fragment is removed by coenzyme A as acetyl-CoA, which then enters the tricarboxylic acid (TCA) cycle for complete mineralization to CO_2 and H_2O. If you think about this process, it becomes apparent that if one starts with an alkane that has an even number of carbons, the two-carbon fragment acetyl-CoA will be the last residue. If one starts with an alkane with an odd number of carbons, the three-carbon fragment propionyl-CoA will be the last residue. Propionyl-CoA is then converted to succinyl-CoA, a four-carbon molecule that is an intermediate of the TCA cycle.

What types of alkanes do microbes most prefer? In general, midsize straight-chain aliphatics (*n*-alkanes C_{10} to C_{18} in length) are utilized more readily than *n*-alkanes with either shorter or longer chains. Long-chain *n*-alkanes are utilized more slowly because of low bioavailability resulting from extremely low water

solubilities (Miller and Bartha, 1989). For example, the water solubility of decane (C_{10}) is 0.052 mg/l, and the solubility of octadecane (C_{18}) is almost 10-fold less (0.006 mg/l). Solubility continues to decrease with increasing chain length. In contrast, short-chain *n*-alkanes have higher aqueous solubility, e.g., the water solubility of butane (C_4) is 61.4 mg/l, but they are toxic to cells. Short-chain alkanes are toxic to microorganisms because their increased water solubility results in increased uptake of the alkanes, which are then dissolved in the cell membrane. The presence of these short alkanes within the cell membrane can alter the fluidity and integrity of the cell membrane.

The toxicity of short-chain *n*-alkanes can be mediated in some cases by the presence of free phase oil droplets. Protection occurs because the short-chain alkanes partition into the oil droplets. This results in reduced bioavailability because the aqueous phase concentration is decreased. Thus, *n*-alkane degradation rates will differ depending on whether the substrate is present as a pure compound or in a mixture of compounds.

Biodegradability of aliphatics is also negatively influenced by branching in the hydrocarbon chain. The degree of resistance to biodegradation depends on both the number of branches and the positions of methyl groups in the molecule. Compounds with a quaternary carbon atom (four carbon–carbon bonds) such as that shown in Fig. 16.10B are extremely stable because of steric effects, as discussed in Section 16.4.3.1.

Alkenes

Alkenes are hydrocarbons that contain one or more double bonds. The majority of alkene biodegradability studies have used 1-alkenes as model compounds (Britton, 1984). These studies have shown that alkenes and alkanes have comparable biodegradation rates. As illustrated in Fig. 16.15, the initial step in 1-alkene degradation can involve attack at the terminal (1) or a

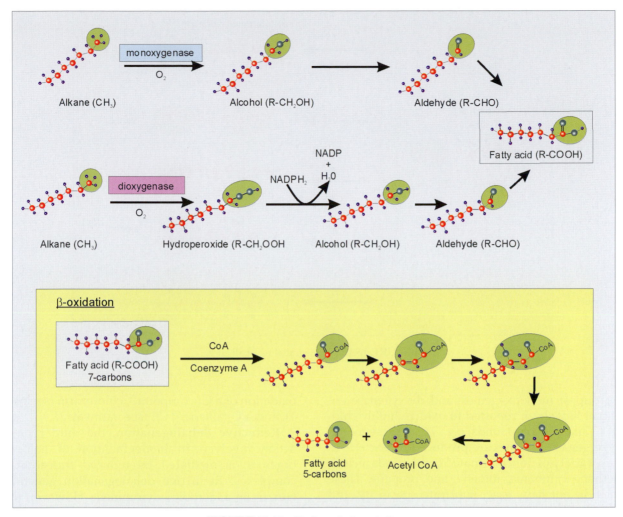

FIGURE 16.14　Biodegradation of alkanes.

subterminal (2) methyl group as described for alkanes. Alternatively, the initial step can be attack at the double bond, which can yield a primary (3) or secondary alcohol (4) or an epoxide (5). Each of these initial degradation products is further oxidized to a primary fatty acid, which is degraded by β-oxidation as shown in Fig. 16.14 for alkanes.

Halogenated Aliphatics

Chlorinated solvents such as trichloroethylene (TCE) have been extensively used as industrial solvents. As a result of improper use and disposal, these solvents are among the most frequently detected types of organic contaminants in groundwater (U.S. EPA, 1984). The need for efficient and cost-effective remediation of solvent-contaminated sites has stimulated interest in the biodegradation of these C_1 and C_2 halogenated aliphatics. Halogenated aliphatics are generally degraded more slowly than aliphatics with-

out halogen substitution. For example, although 1-chloroalkanes ranging from C_1 to C_{12} are degraded as a sole source of carbon and energy in pure culture, they are degraded more slowly than their nonchlorinated counterparts. The presence of two or three chlorines bound to the same carbon atom inhibits aerobic degradation (Jannsen *et al.*, 1990). Further, degradation rates of 1-chloroalkanes, ranging from C_3 to C_{12}, increased with increasing alkyl chain length (Okey and Bogan, 1965). These results can be explained by the decreasing electronic effects of the chlorine atom on the enzyme–carbon reaction center as the alkane chain length increases (see Section 16.4.3.2).

Biodegradation of halogenated aliphatics occurs by three basic types of reactions. **Substitution** is a nucleophilic reaction (the reacting species brings an electron pair) in which the halogens on a mono- or dihalogenated compound are substituted by a hydroxy group (Fig. 16.16). **Oxidation** reactions are catalyzed by a

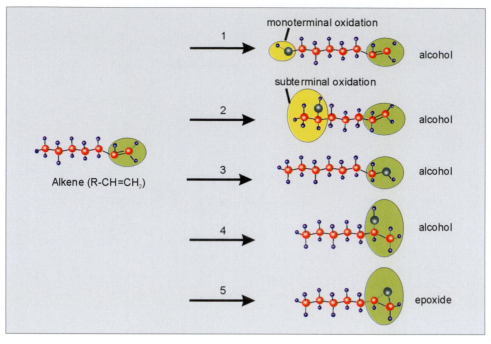

FIGURE 16.15   Biodegradation of alkenes.

select group of monooxygenase and dioxygenase enzymes that have been reported to oxidize highly chlorinated C_1 and C_2 compounds, e.g., TCE. These mono- and dioxygenase enzymes are produced by bacteria to oxidize a variety of nonchlorinated compounds including methane, ammonia, toluene, and propane. These enzymes do not have exact substrate specificity, and thus they can also participate in the cometabolic degradation of chlorinated aliphatics (Section 16.3, Fig. 16.5). Usually, a large ratio of substrate to chlorinated aliphatic is required to achieve cometabolic degradation of the chlorinated aliphatic. Reported examples of oxygenase–substrate systems that cometabolically degrade chlorinated hydrocarbons include methane monooxygenase produced by methanotrophic bacteria during growth on carbon sources such as methane or formate, toluene dioxygenase produced during growth of some bacteria on toluene, ammonia monooxygenase produced by *Nitrosomonas europaea* during growth on ammonia (Vannelli *et al.*, 1990), and propane monooxygenase produced by *Mycobacterium vaccae* JOB5 during growth on propane (Wackett *et al.*, 1989). Figure 16.17 shows an example of cometabolic

oxidation of a C_1 compound, chloroform (see pathway 1), and a C_2 alkene, TCE (see pathway 2).

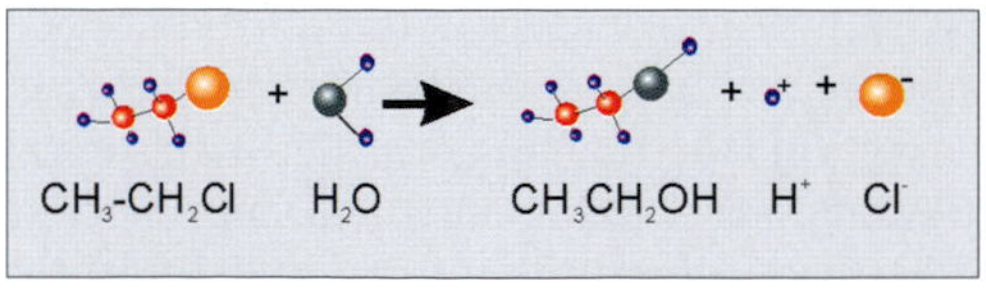

Reductive dehalogenation, the third type of reaction involved in biodegradation of halogenated organics, is mediated by reduced transition metal complexes. Reductive dehalogenation generally occurs in an anaerobic environment. However, the second of the reactions shown in Fig. 16.18, formation of alkenes, can occur aerobically for a limited number of chlorinated compounds that have a higher reducing potential than O_2, e.g., hexachloroethane and dibromoethane (Vogel *et al.*, 1987). In the first step of reductive dehalogenation, electrons are transferred from the reduced metal to the halogenated aliphatic, resulting in an alkyl radical and free halogen. The alkyl radical can either scavenge a hydrogen atom (1) or lose a second halogen to form an alkene (2).

Generally, aerobic conditions favor the biodegradation of compounds with fewer halogen substituents, and anaerobic conditions favor the biodegradation of compounds with a high number of halogen substituents. However, complete biodegradation of highly halogenated aliphatics under anaerobic conditions often does not take place. Therefore, some researchers have proposed the use of a sequential anaerobic and aerobic treatment (Fathepure and Vogel, 1991). Initial incubation under anaerobic conditions would be used to decrease the halogen content, and subsequent addition of oxygen would create aerobic conditions to allow complete degradation to proceed aerobically.

$$CH_3\text{-}CH_2Cl \quad H_2O \quad \longrightarrow \quad CH_3CH_2OH \quad H^+ \quad Cl^-$$

FIGURE 16.16   Dechlorination by substitution.

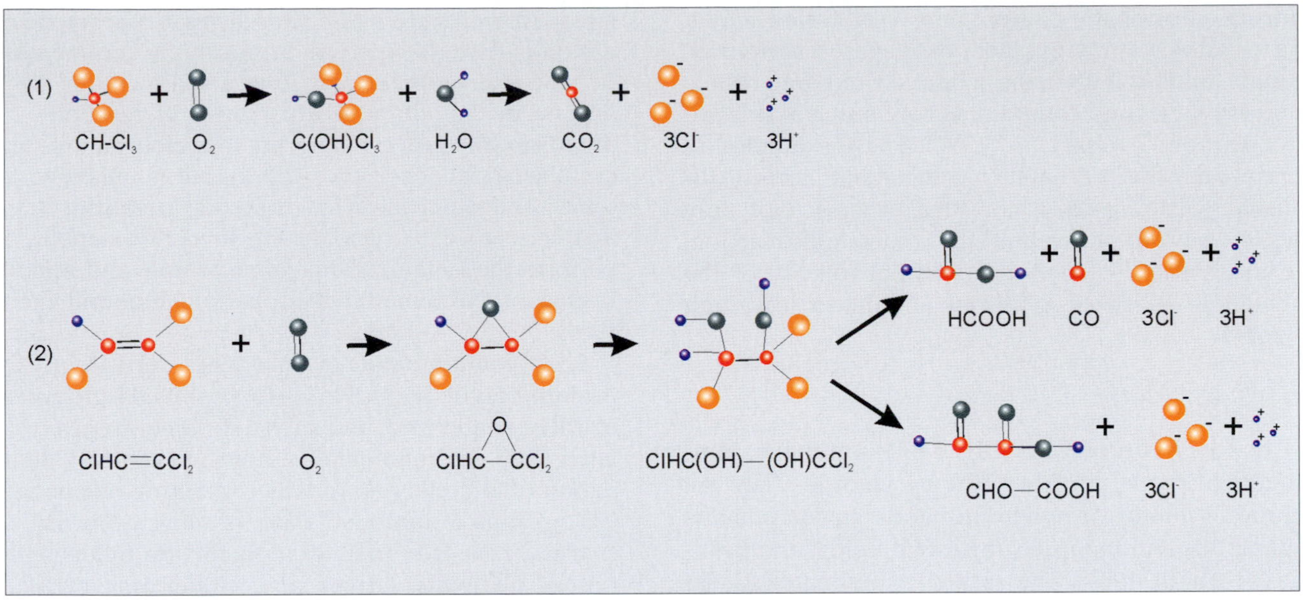

FIGURE 16.17 Dehalogenation by oxidation.

One important consideration in the bioremediation of sites containing chlorinated aliphatics is the potential toxicity of such compounds. For example, the effect of solvent concentration on inhibition of a propylene-grown *Xanthobacter* strain capable of oxidizing several chlorinated alkenes, including TCE, was determined (Ensign *et al.*, 1992). It was found that alkene monooxygenase activity was up to 90% inhibited by chlorinated alkene concentrations between 25 and 330 μM, depending on the compound. Similarly, it was found that toluene used to support the aerobic metabolism of TCE in soil could be inbitory (Mu and Scow, 1994). Specifically, increasing the TCE concentration from 1 to 20 μg/ml decreased the numbers of toluene and TCE degraders in the soil and decreased the rate of TCE degradation. Also of concern is that TCE transformations can result in oxidized derivatives more toxic than TCE itself. This toxicity has been attributed to a nonspecific binding of TCE transformation products to cellular proteins. For example, TCE transformation by methane monooxygenase results in a transformation product that can bind to a protein subunit of the monooxygenase enzyme, resulting in inhibition of enzyme activity (Oldenhuis *et al.*, 1991). Recovery of oxidizing activity is possible by replacement of the enzyme following *de novo* protein synthesis. Similar behavior was found for ammonia monoxy-genase. In this experiment, a wide variety of

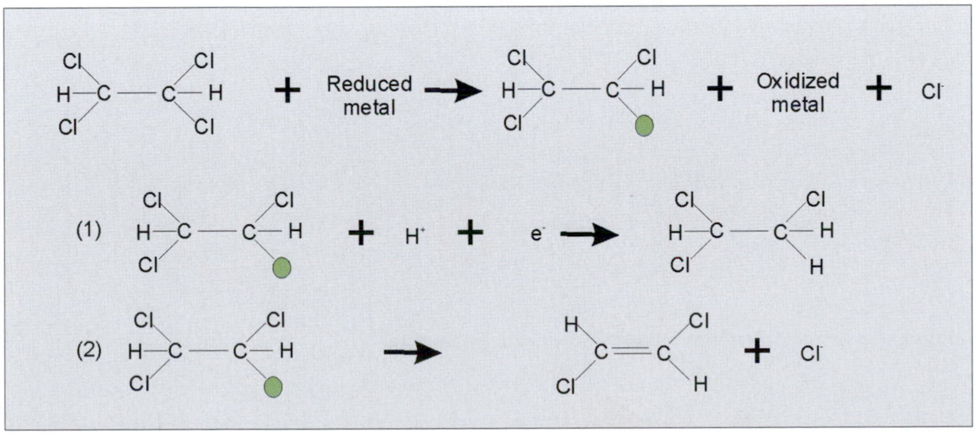

FIGURE 16.18 Reductive dehalogenation of tetrachloroethane to trichloroethane (1) or dichloroethylene (2).

chlorinated aliphatic compounds were tested and it was found that ammonia monooxygenase activity was strongly inhibited after the oxidation of some chlorinated ethylenes and compounds containing a dichlorinated carbon (Rasche *et al.,* 1991). These examples all demonstrate that harnessing cometabolic activity for unusual substrates is a complex process. Not only does one have to understand the cometabolic reaction, one must also consider the potential toxicity of the cometabolic substrate and its metabolites to degrading microbes.

16.6.2.2 Alicyclics

Alicyclic hydrocarbons (Fig. 16.13) are major components of crude oil, 20 to 70% by volume. They are commonly found elsewhere in nature as components of plant oils and paraffins, microbial lipids, and pesticides (Trudgill, 1984). The various components can be simple, such as cyclopentane and cyclohexane, or complex, such as trimethylcyclopentane and various cycloparaffins. The use of alicyclic compounds in the chemical industry, and the release of alicyclics to the environment through industrial processes, other than oil processing and utilization, is more limited than for aliphatics and aromatics. Consequently, the issue of health risks associated with human exposure to alicyclics has not reached the same level of importance as for the other classes of compounds, especially the aromatics. As a result, far less research has focused on the study of alicyclic biodegradation.

It is known that there is no correlation between the ability to utilize *n*-alkanes and the ability to oxidize cycloalkanes fully. Further, it is difficult to isolate pure cultures that degrade alicyclic hydrocarbons using enrichment techniques. Although microorganisms with complete degradation pathways have been isolated (Trower *et al.*, 1985), alicyclic hydrocarbon degradation is thought to occur primarily by commensalistic and cometabolic reactions as shown for cyclohexane in Fig. 16.19. In this series of reactions, one organism converts cyclohexane to cyclohexanone via a cyclohexanol (step 1 and step 2), but is unable to lactonize and open the ring. A second organism that is unable to oxidize cyclohexane to cyclohexanone can perform the lactonization, ring opening, and mineralization of the remaining aliphatic compound (Perry, 1984).

Cyclopentane and cyclohexane derivatives that contain one or two OH, C=O, or COOH groups are readily metabolized, and such degraders are easily isolated from environmental samples. In contrast, degradation of alicyclic derivatives containing one or more CH_3 groups is inhibited. This is reflected in the decreasing rate of biodegradation for the following series of alkyl derivatives of cyclohexanol: cyclohexanol > methylcyclohexanol > dimethylcyclohexanol (Pitter and Chudoba, 1990).

16.6.2.3 Aromatics

Unsubstituted Aromatics

Aromatic compounds contain at least one unsaturated ring system with the general structure C_6R_6, where R is any functional group (Fig. 16.13). Benzene (C_6H_6) is the parent hydrocarbon of this family of unsaturated cyclic compounds. Compounds containing two or more fused benzene rings are called polyaromatic hydrocarbons (PAH). Aromatic hydrocarbons are natural products; they are part of lignin and are formed as organic materials are burned, for example, in forest fires. However, the addition of aromatic compounds to the environment has increased dramatically through activities such as fossil fuel processing and

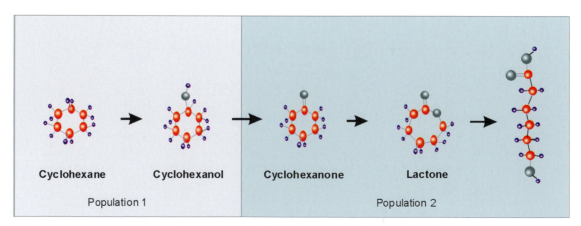

FIGURE 16.19   Degradation of cyclohexane.

utilization and burning of wood and coal (Cerniglia and Heitkamp, 1989). The quantity and composition of the aromatic hydrocarbon component in petroleum products are of major concern when evaluating a contaminated site because several components of the aromatic fraction have been shown to be carcinogenic to humans. Aromatic compounds also have demonstrated toxic effects toward microorganisms. For example, the toxicity of the water-soluble fraction of refined oil is more toxic to the growth of heterotrophic microorganisms than the water soluble fraction of crude oil. This is attributed to the greater proportion of aromatic compounds, particularly naphthalene and alkylnaphthalenes, present in refined oils compared with crude oils. For example, the total naphthalene concentration of the refined oil was 1800 μg/l, compared with only 45 μg/l in the crude oil tested (Hodson *et al.*, 1977). Autotrophs are also affected adversely. When the impact of the aliphatic, aromatic, and asphaltic fractions from five petroleum oils on photosynthesis and respiration of a representative cyanobacterium was determined, it was found that growth inhibition was most strongly associated with the aromatic fraction (Singh and Guar, 1990).

Because of the potential human health impacts of aromatic compounds, their biodegradation has been extensively studied (Gibson and Subramanian, 1984). The results of this work showed that a wide variety of bacteria and fungi can carry out aromatic transformations, both partial and complete, under a variety of environmental conditions. Under aerobic conditions, the most common initial transformation is a hydroxylation that involves the incorporation of molecular oxygen. The enzymes involved in these initial transformations are either monooxygenases or dioxygenases. In general, prokaryotic microorganisms transform aromatics by an initial dioxygenase attack to *cis-* dihydrodiols. The *cis*-dihydrodiol is rearomatized to form a dihydroxylated intermediate, catechol. The catechol ring is cleaved by a second dioxygenase either between the two hydroxyl groups **(ortho pathway)** or next to one of the hydroxyl groups **(meta pathway)** and further degraded to completion (Fig. 16.20).

Eukaryotic microorganisms initially attack aromatics with a cytochrome P-450 monooxygenase, incorporating one atom of molecular oxygen into the aromatic and reducing the second to water, resulting in the formation of an arene oxide. This is followed by the enzymatic addition of water to yield a *trans*-dihydrodiol (Fig. 16.21). Alternatively, the arene oxide can be isomerized to form phenols, which can be conjugated

with sulfate, glucuronic acid, and glutathione. These conjugates are similar to those formed in higher organisms and seem to aid in detoxification and elimination of aromatic compounds. The exception to this is the white-rot fungi which under certain conditions are able to completely mineralize aromatic compounds (see Case Study 1).

Thus far, six families of genes responsible for PAH degradation have been identified. Often the capacity for aromatic degradation, especially chlorinated aromatics, is plasmid mediated (Ghosal *et al.*, 1985). Plasmids can carry both individual genes and operons encoding partial or complete biodegradation of an aromatic compound. An example of a plasmid that carries a family of genes involved in the degradation of aromatic compounds is the NAH7 plasmid, which codes for the degradation of naphthalene. The NAH7 plasmid was obtained from *Pseudomonas putida* and contains genes that encode the enzymes for the first 11 steps of naphthalene oxidation. This plasmid or closely related plasmids are frequently found in sites that are contaminated with PAHs (Ahn *et al.*, 1999). As discussed in Chapter 13, this plasmid has also been used to construct a luminescent bioreporter gene system (Fig. 13.20). Here the lux genes that cause luminescence have been inserted into the *nah* operon in the NAH plasmid. When the *nah* operon is induced by the presence of naphthalene, both naphthalene-degrading genes and the lux gene are expressed. As a result, naphthalene is degraded and the reporter organism luminesces. Such reporter organisms are currently being used to study the effect of oxygen and substrate level on degradation of naphthalene in soil systems.

There is also interest in construction of bacterial strains with a broad aromatic biodegradation potential. Although it is possible to manipulate the chromosome, plasmids offer an easy mechanism by which new genes can be introduced into a bacterial cell.

In general, aromatics composed of one, two, or three condensed rings are transformed rapidly and often completely mineralized, whereas aromatics containing four or more condensed rings are transformed much more slowly, often as a result of cometabolic attack. This is due to the limited bioavailability of these high-molecular-weight aromatics. Such PAHs have very limited aqueous solubility and sorb strongly to particle surfaces in soil and sediments. However, it has been demonstrated that chronic exposure to aromatic compounds will result in increased transformation rates because of adaptation of an indigenous population to growth on aromatic compounds.

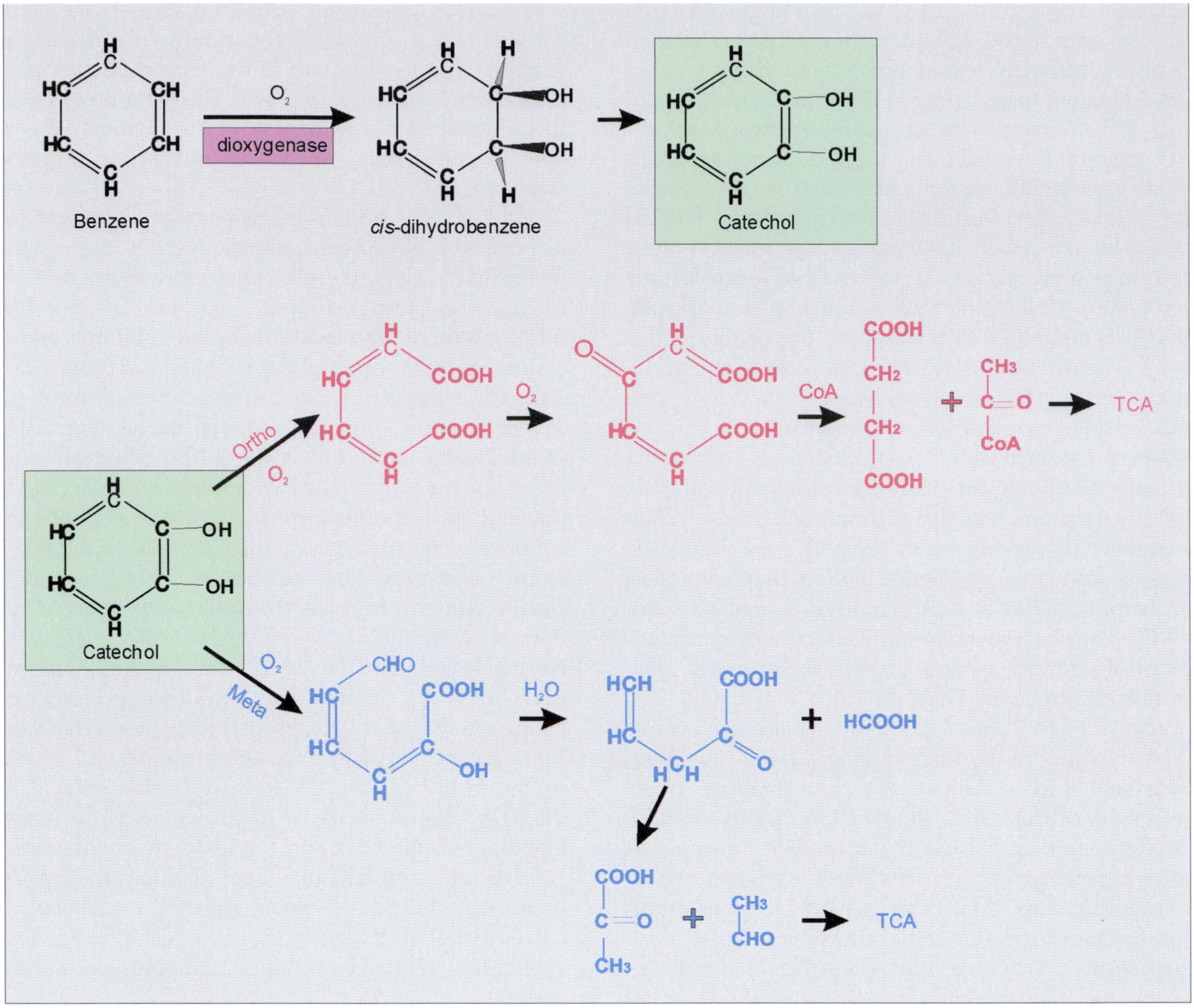

FIGURE 16.20   Incorporation of oxygen into the aromatic ring by the dioxygenase enzyme, followed by meta or ortho ring cleavage.

Substituted Aromatics

One group of aromatics of special interest is the **chlorinated aromatics.** These compounds have been used extensively as solvents and fumigants (e.g., dichlorobenzene), and wood preservatives (e.g., pentachlorophenol (PCP)) and are parent compounds for pesticides such as 2,4-dichlorophenoxyacetic acid (2,4-D) and DDT. The difficulty for microbes in the degradation of chlorinated organics is that the carbon–chlorine bond is very strong and requires a large input of energy to break. Second, the common intermediate in aromatic degradation is catechol or dihydroxybenzene (see Fig. 16.20). This molecule requires two adjacent unsubstituted carbons so that

hydroxyl groups can be added. Chlorine substituents can block these sites. Some strategies for degradation of chlorinated aromatics are shown in Fig. 16.22 for pentachlorophenol and dichlorobenzene.

Chlorinated phenols are particularly toxic to microorganisms. In fact, phenol itself is very toxic and is used as a disinfectant. Chlorination adds to toxicity, which increases with the degree of chlorination (Chaudhry and Chapalamadugu, 1991). For example, Van Beelen and Fleuren-Kemilä (1993) quantified the effect of PCP and several other pollutants on the ability of soil microorganisms to mineralize [14C]acetate in soil. The amount of PCP required to reduce the initial rate of acetate mineralization by 10% ranged between

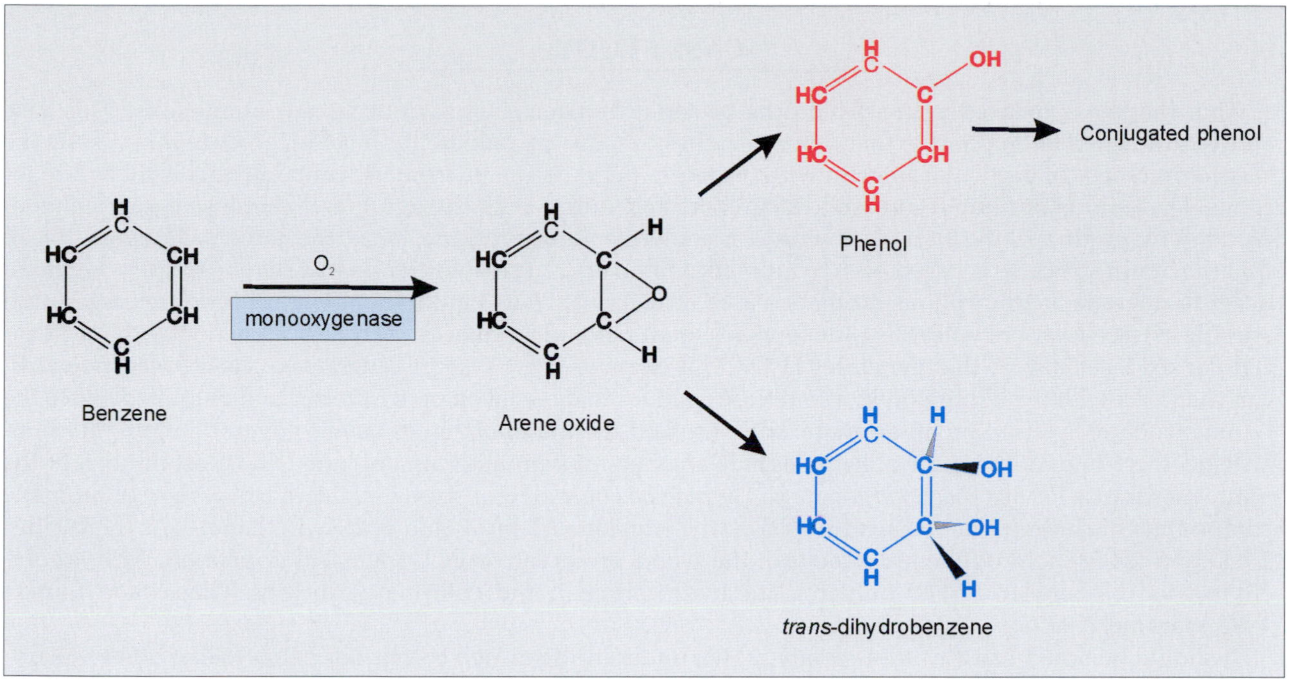

FIGURE 16.21   Fungal monooxygenase incorporation of oxygen into the aromatic ring.

0.3 and 50 mg/kg dry soil, depending on the soil type. High concentrations of PCP also have inhibitory effects on PCP-degrading microorganisms. For example, Alleman *et al.* (1992) investigated the effect of PCP on six species of PCP-degrading fungi. They showed that increasing the PCP concentration from 5 to 40 mg/l decreased fungal growth and decreased the ability of the fungi to degrade PCP.

Methylated aromatic derivatives, such as toluene, constitute another common group of substituted aromatics. These are major components of gasoline and are commonly used as solvents. These compounds can initially be attacked either on the methyl group or directly on the ring as shown in Fig. 16.23. **Alkyl** derivatives are attacked first at the alkyl chain, which is shortened by β-oxidation to the corresponding benzoic acid or phenylacetic acid, depending on the number of carbon atoms. This is followed by ring hydroxylation and cleavage (Fig. 16.23).

Dioxins

Dioxins and dibenzofurans are created during waste incineration and are part of the released smoke stack effluent. Once thought to be one of the most potent carcinogens known, 2,3,7,8-tetrachlorodibenzo-*p*-dioxin (TCDD) is associated with the manufacture of 2,4-D and 2,4,5-trichlorophenoxy acetic acid (2,4,5-T), hexacholorophene, and other pesticides that have

2,4,5-T as a precursor. Current thinking is that TCDD is less dangerous in terms of carcinogenicity and teratogenicity than once thought, but that noncancer risks including diabetes, reduced IQ, and behavioral impacts may be more important. The structure of TCDD (Fig. 16.13) and its low water solubility, 0.002 mg/l, result in great stability of this molecule in the environment. Although bacterial and fungal biodegradation of TCDD has been demonstrated, the extent of biodegradation is very minimal. For example, a mixture of six bacterial strains isolated from TCDD-contaminated soil obtained from Seveso, Italy was able to produce a metabolite presumed to be 1-hydroxy-TCDD (Phillippi *et al.,* 1982). However, less than 1% of the original TCDD was degraded in 12 weeks. In addition, the fungus *Phanerochaete chrysosporium* (see Case Study 1) was only able to mineralize 2% of the parent compound to CO_2 (Bumpus *et al.,* 1985). More recent work has focused on the reductive dechlorination of TCDD and more highly chlorinated isomers.

Heterocyclic Compounds

Heterocyclics are cyclic compounds containing one or more heteroatoms (nitrogen, sulfur, or oxygen) in addition to carbon atoms. The dioxins already discussed as well as other compounds shown in Fig. 16.13 fall into this category. In general, heterocyclic compounds are more difficult to degrade than analogous

CASE STUDY 1

This chapter is focused primarily on the bacterial transformations of organic contaminants. This case study is included to emphasize that other microbes can also participate in the biodegradation process. One class of microbes of particular interest with respect to degradation of aromatic contaminants is the white rot fungi. These fungi are important wood degraders and utilize both the lignin and cellulose components of wood. A by-product of the degradation process is a white fibrous residue, hence the name "white rot". Recall that the lignin structure is based on two aromatic amino acids, tyrosine and phenylalanine (Figure 14.5F). In order to degrade an amorphous aromatic-based structure such as lignin, the white rot fungi release a non-specific extracellular enzyme, H_2O_2-dependent lignin peroxidase, that is used in conjunction with an extra-cellular oxidase enzyme that generates H_2O_2. This enzyme-H_2O_2 system generates oxygen-based free radicals that react with the lignin polymer to release residues that are taken up by the cell and degraded. Since the lignin structure is based on an aromatic structure and the initial enzymes used to degrade lignin are non-specific, the white rot fungi are able to degrade a variety of aromatic contaminants. The most famous of the white rot fungi is *Phanerochaete chrysosporium,* which has been demonstrated to degrade a variety of aromatic compounds including the pesticide DDT (1,1,1-trichloro-2,2-bis(4-chlorophenyl)ethane), the herbicides 2,4-D and 2,4,5-T (chlorophenoxyacetates), the wood preservative PCP (pentachlorophenol), PCBs (poly-chlorinated biphenyls), PAHs (polyaromatic hydrocarbons), and chlorinated dioxins (Glaser and Lamar, 1995; Hammel, 1995).

It should be noted that *P. chrysosporium* grown under nitrogen-rich conditions behaves like other eukaryotic microbes, it produces trans-dihydrodiols (see Figure 16.21) and various conjugates that are excreted. Under these conditions aromatic contaminant molecules are not degraded. However, under nitrogen-limiting conditions, which induce production of lignin-degrading enzymes, aromatic contaminants are mineralized. The application of white-rot fungi and in particular *P. chrysosporium* to problems in Environmental Microbiology is being investigated from several aspects. These include the use of the fungus to delignify lignocellulosic materials that would normally be considered as waste, e.g. straw, to produce cattle feed (Carlile and Watkinson, 1994). These fungi can also be used to delignify wood pulp that is used in paper manufacture. In this case a biological process replaces one that is normally done mechanically. A further application of these fungi is the degradation of aromatic components found in wastestreams associated with the paper industry (a variety of unchlorinated and chlorinated aromatic components), wood treatment industry (PCP and various creosote components), and the munitions industry (TNT (2,4,6-trinitrotoluene) and DNT (2,4-dinitrotoluene)).

aromatics that contain only carbon. This is probably due to the higher electronegativity of the nitrogen and oxygen atoms compared with the carbon atom, leading to deactivation of the molecule toward electrophilic substitution. Heterocyclic compounds with five-membered rings and one heteroatom are readily biodegradable, probably because five-membered ring compounds exhibit higher reactivity toward electrophilic agents and are hence more readily biologically hydroxylated. The susceptiblity of heterocyclic compounds to biodegradation decreases with increasing number of heteroatoms in the molecules.

16.6.2.4 Pesticides

Pesticides are the biggest nonpoint source of chemicals added to the environment. The majority of the currently used organic pesticides are subject to exten-sive mineralization within the time of one growing season or less. Synthetic pesticides show a bewildering variety of chemical structures, but most can be traced to relatively simple aliphatic, alicyclic, and aromatic base structures already discussed. These base structures bear a variety of halogen, amino, nitro, hydroxyl, carboxyl, and phosphorus substituents. For example, the chlorophenoxyacetates, such as 2,4-D and 2,4,5-T, have been released into the environment as herbicides over the past 40 years. Both of these structures are biodegradable and pathways are presented in Fig. 16.24.

As an exercise, examine the pesticide structures presented in Fig. 16.25. For each set of pesticides, predict which is more easily degraded. You are correct if you predicted 2,4-D for the first set. It is rapidly degraded by soil microorganisms. Although 2,4,5-T is also degraded, the degradation is much slower. For the second set of pesticides, propham is more degradable. In

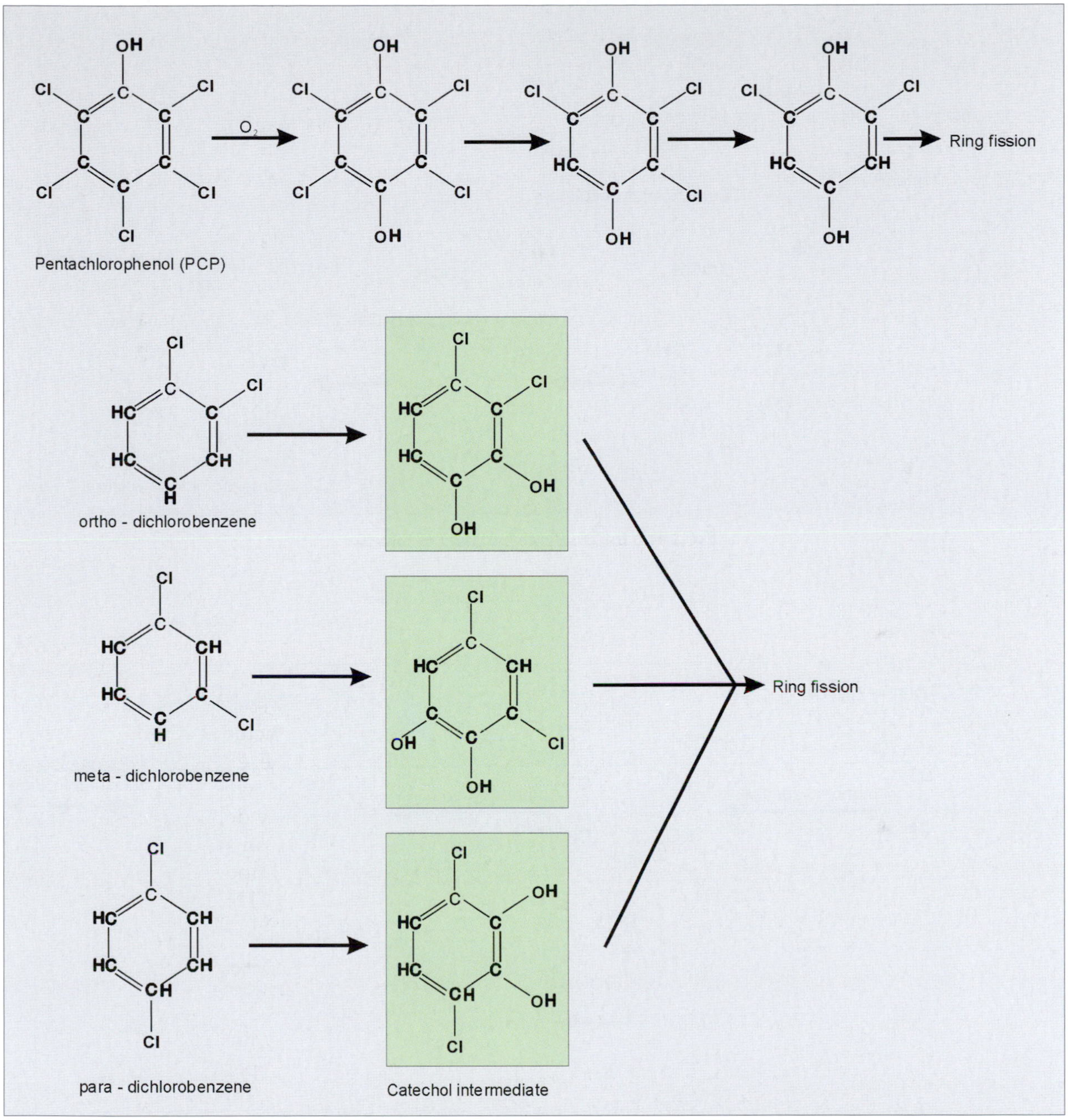

FIGURE 16.22   Initial steps in the aerobic degradation of pentachlorophenol and three dichlorobenzenes.

propachlor, the extensive branching so close to the ring structure blocks biodegradation. In the third set, carbaryl is more degradable because of the extensive chlorination and complex ring structures of aldrin. In fact, the estimated **half-life** of carbaryl in soil is 30 days, compared with 1.6 years for aldrin. Half-life is a term used to express the time it takes for 50% of the compound to disappear. Generally, 5½ half-lives are believed sufficient for the compound to be completely degraded. Finally, in the fourth set, methoxychlor is more degradable than DDT. In this case, the half-lives

are even longer, 1 year for methoxychlor and 15.6 years for DDT.

16.6.3 Anaerobic Conditions

Anaerobic conditions are not uncommon in the environment. Most often, such conditions develop in water or saturated sediment environments. But, even in well-aerated soils there are microenvironments with little or no oxygen. In all of these environments, **anaerobiosis** occurs. Anaerobiosis occurs when the rate of

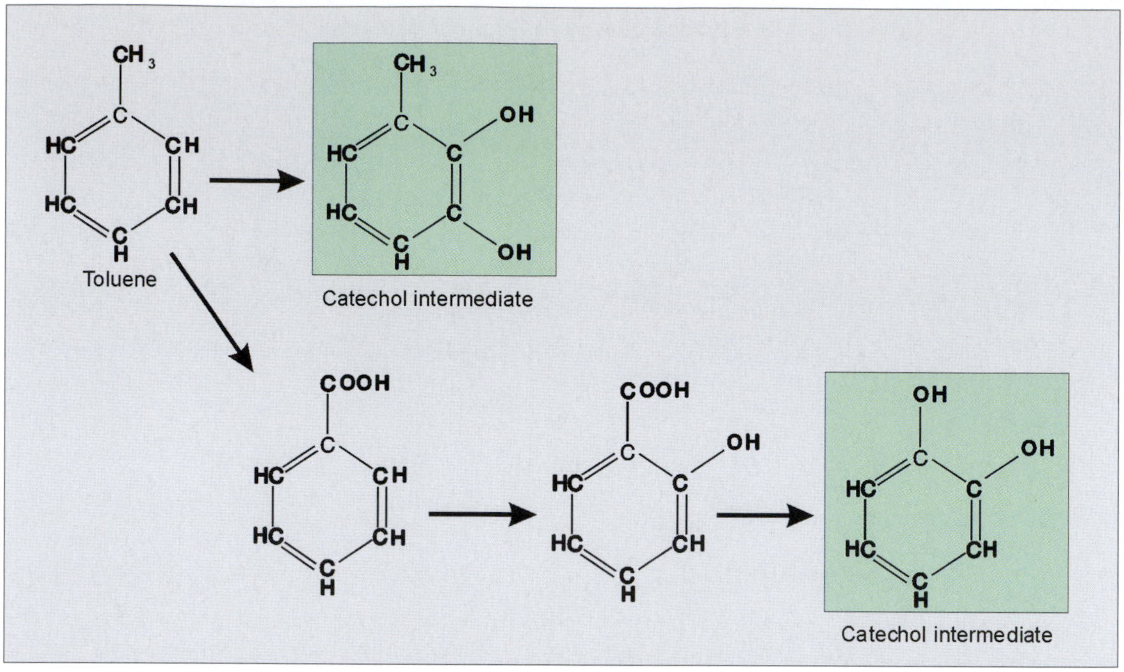

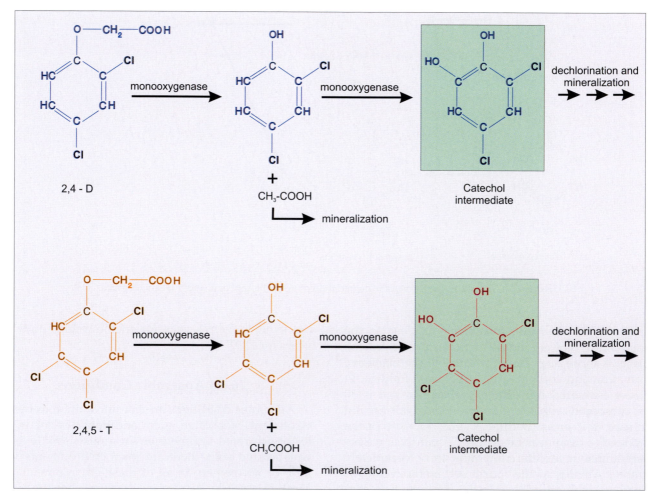

FIGURE 16.23 Biodegradation of toluene.

FIGURE 16.24 Biodegradation of 2,4-D and 2,4,5-T.

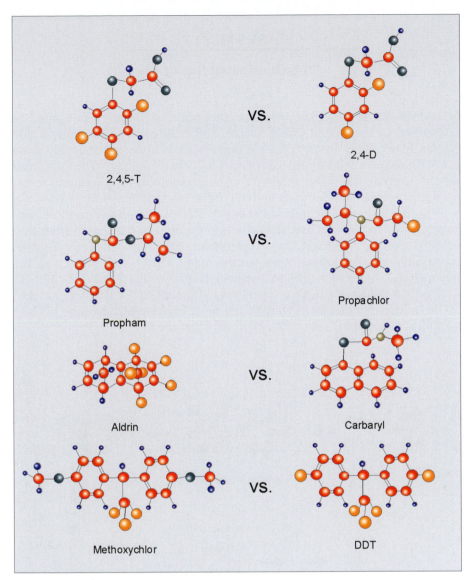

FIGURE 16.25 Comparison of four sets of pesticides. Can you predict which of each set is more easily biodegraded?

oxygen consumption by microorganisms is greater than the rate of oxygen diffusion through either air or water. In the absence of oxygen, organic compounds can be mineralized through anaerobic respiration using a terminal electron acceptor other than oxygen (see Chapter 3.4). There is a series of alternative terminal electron acceptors in the environment including iron, nitrate, manganese, sulfate, and carbonate. These alternative electron acceptors have been listed in the order of most oxidizing to most reducing, and are usually utilized in this order, because the amount of energy generated for growth depends on the oxidation potential of the electron acceptor. Because none of these electron acceptors are as oxidizing as oxygen, growth under anaerobic conditions is never as efficient as growth under aerobic conditions.

16.6.3.1 Aliphatics

Saturated aliphatic hydrocarbons are degraded slowly if at all under anaerobic conditions. This is supported by the fact that hydrocarbons in natural underground reservoirs of oil (which are under anaerobic conditions) are not degraded despite the presence of microorganisms. However, both unsaturated aliphatics and aliphatics containing oxygen (aliphatic alcohols and ketones) are readily biodegraded anaerobically. The suggested pathway of biodegradation for unsaturated hydrocarbons is hydration of the double bond to an alcohol, with further oxidation to a ketone or aldehyde and, finally, formation of a fatty acid (Fig. 16.26).

As previously discussed, halogenated aliphatics can be partially or completely degraded under anaerobic

CASE STUDY 2

Polychlorinated Biphenyls

Biphenyl is the unchlorinated analogue or parent compound of the polychlorinated biphenyls (PCBs), which were first described in 1881 (Waid 1986a; 1986b). The PCBs consist of variously chlorine-substituted biphenyls of which, in theory, there are 209 possible isomers. Only approximately 100 actually exist in commercial formulations. The aqueous solubility of biphenyl is 7.5 mg/l, and any chlorine substituent decreases the water solubility. In general, the water solubility of monochlorobiphenyls ranges from 1 to 6 mg/l, for dichlorobiphenyls the range is 0.08–5 mg/l, and for hexachlorobiphenyl the aqueous solubility is just 0.00095 mg/l. By 1930, because of their unusual stability, PCBs were widely used as nonflammable heat-resistant oils in heat transfer systems, as hydraulic fluids and lubricants, as transformer fluids, in capacitors, as plasticizers in food packaging materials, and as petroleum additives. PCBs were used as mixtures of variously chlorinated isomers and marketed under various trade names, e.g., Aroclor (U.S.), Clophen (West Germany), Phenoclor (Italy), Kanechlor (Japan), Pyralene (France), and Soval (U.S.S.R.). The U.S. domestic sales of Aroclors 1221–1268 (where the last two numbers indicated the % chlorination) went from 32 million pounds in 1957 to 80 million pounds in 1970, with the most popular blend being 1242. PCB use in transformers and capacitors accounted for about 50% of all Aroclors used.

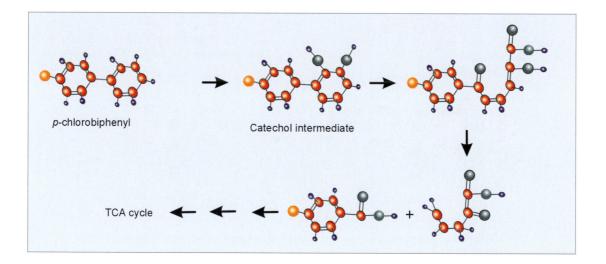

A rice oil factory accident in Japan in 1968 brought PCBs international attention. In the factory, the heat exchanger pipes used to process rice oil contained PCBs as the heat exchange fluid. Unnoticed, a heat exchange pipe broke and leaked PCBs into a batch of rice oil, which was then packaged and consumed by the local population. The contaminated rice oil poisoned over 1000 people, producing a spectrum of symptoms including chloracne, gum and nail bed discoloration, joint swelling, emission of waxy secretions from eyelid glands, and lethargy. As a result, the U.S. Food and Drug Administration (FDA) issued tolerance levels for PCBs in food and packaging products, and the Environmental Protection Agency (EPA) under TOSCA, issued rules governing the use of PCBs. This has drastically reduced domestic production and use of PCBs.

Despite decreased use, PCBs still pose an environmental problem because they are not only chemically stable, they also resist biodegradation. Because past use of PCBs has been high, PCBs have accumulated in the environment. There are several ways in which PCBs make their way into the environment. One of these is in runoff from industrial waste dumps and spills. Other sources include points of PCB manufacture and processing into other products. Because PCB input into fresh water has been high in the past, PCBs have accumulated in sediments. Even if PCB input were stopped completely, previously contaminated sediments could continue to release PCBs into freshwater systems for years to come.

Although some PCB degradation occurs, it is limited by low bioavailability. The case study figure shows a typical degradation pathway. Note that the pathway includes the familiar catechol intermediate. In this example only one of the biaryl rings is chlorinated, so degradation begins with the unchlorinated ring. Not surprisingly, the number and location of chlorines help determine the rate and extent of biodegradation. This is illustrated in Table 16.2, which shows that chlorination of one of the biaryl rings does not generally inhibit degradation with the exception of the 2,6 isomer. However, chlorines on both rings inhibit degradation to some extent in most cases. This table is from an early study performed to evaluate PCB degradation. As illustrated in Table 16.2, the two degraders studied vary somewhat in their preference for certain PCB isomers. Subsequent work confirmed this finding and further found that there are bacterial isolates that actually prefer more highly chlorinated isomers to the less chlorinated ones.

The extensive research that has been performed to understand PCB degradation has suggested several strategies for promoting biodegradation. These include the use of a sequential anaerobic-aerobic process to remove chlorines and then allow mineralization of the less chlorinated isomers. A second approach is the addition of a co-substrate, such as the parent PCB compound biphenyl, to stimulate PCB-degrading populations. The rationale for this strategy is that one problem with PCB degradation is that their low bioavailability simply does not allow induction of biodegradation pathways. The addition of the biphenyl, which has a higher aqueous solubility (7.5 mg/l), can induce PCB biodegradation. Unfortunately, the addition of biphenyl to the environment is problematic because of cost and toxicity. However, it has been demonstrated that some plant compounds can also induce PCB biodegradation. One such compound is l-carvone from spearmint (Gilbert and Crowley, 1997). A third approach is **bioaugmention** or addition of specific PCB-degrading microorganisms to contaminated areas that may not contain indigenous degraders.

TABLE 16.2   Effect of Chlorine Substitution on Biodegradability of Various Polychlorinated Biphenyl Isomers

PCB chlorine position	Degradation rate (nmol/ml/hr)		PCB chlorine position	Degradation rate (nmol/ml/hr)	
	Alcaligenes sp.	*Acinetobacter sp.*		*Alcaligenes sp.*	*Acinetobacter sp.*
2	>50	>50	2,4,6	3.1	46.0
3	>50	>50	2,5,2′	1.6	5.1
4	>50	>50	2,5,3′	42.1	41.3
2,3	>50	46.4	2,5,4′	21.8	30.4
2,4	>50	>50	2,4,4′	41.3	40.2
2,5	>50	>50	3,4,2′	15.6	38.6
2,6	0	4.1	2,3,4,5	25.8	19.1
3,4	>50	>50	2,3,5,6	0	0
3,5	>50	>50	2,3,2′,3′	8.7	7.3
2,2′	6.3	14.0	2,4,2′,4′	0	0
2,4′	48.2	49.1	2,4,3′,4′	0	0
3,3′	—	18.5	2,5,2′,5′	0	3.5
4,4′	16.2	25.2	2,6,2′,6′	0	0
2,3,4	35.1	32.0	3,4,3′,4′	0	0
2,3,6	0	0	2,4,5,2′,5′	0.6	0
2,4,5	46.0	32.4			

From Furukawa, *et al.* (1978).

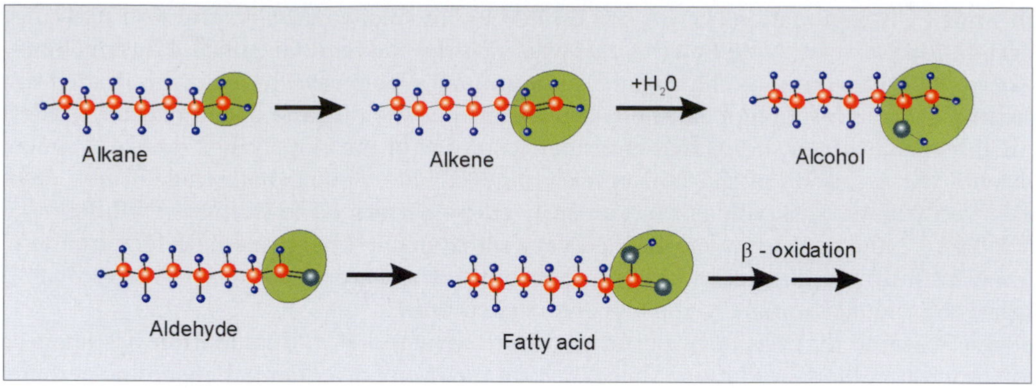

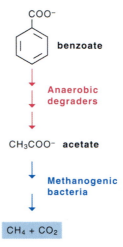

FIGURE 16.26   Anaerobic biodegradation of aliphatic compounds.

conditions by reductive dehalogenation, a cometabolic degradation step that is mediated by reduced transition metal complexes. As shown in Fig. 16.18, in the first step, electrons are transferred from the reduced metal to the halogenated aliphatic, resulting in an alkyl radical and free halogen. Then the alkyl radical can either scavenge a hydrogen atom (1) or lose a second halogen to form an alkene (2). In general, anaerobic conditions favor the degradation of highly halogenated compounds, whereas aerobic conditions favor the degradation of mono- and di-substituted halogenated compounds.

16.6.3.2 Aromatics

Like aliphatic hydrocarbons, aromatic compounds can be completely degraded under anaerobic conditions if the aromatic is oxygenated. There is also evidence that even nonsubstituted aromatics are degraded slowly under anaerobic conditions. Anaerobic mineralization of aromatics often requires a mixed microbial community that works together even though each of the microbial components requires a different redox potential. For example, mineralization of benzoate can be achieved by growing an anaerobic benzoate degrader in co-culture with a methanogen and sulfate reducer. The initial transformations in such a system are often carried out fermentatively, and this results in the formation of aromatic acids, which in turn are transformed to methanogenic precursors such as acetate, carbon dioxide, and formate. These small molecules can then be utilized by methanogens (Fig. 16.27). Such a mixed community is called a consortium. It is not known how this consortium solves the problem of requiring different redox potentials in the same vicinity in a soil system. Clearly, higher redox potentials are required for degradation of the more complex substrates such as benzoate, leaving smaller organic acid or alcohol molecules that are degraded at lower redox potentials. To ultimately achieve degradation may require that the organic acids and alcohols formed at higher redox potential be trans-

ported by diffusion or by movement with water (advection) to a region of lower redox potential. On the other hand, it may be that biofilms form on the soil surface and that redox gradients are formed within the biofilm allowing complete degradation to take place. Environmental microbiologists are actively exploring how anaerobes operate in the soil and vadose environments. Such an understanding is expected to aid in developing technologies to enhance degradation processes in anaerobic contaminated environments.

16.7 BIOREMEDIATION

The objective of bioremediation is to exploit naturally occurring biodegradative processes to clean up contaminated sites (National Research Council, 1993). There are several types of bioremediation. *In situ* **bioremediation** is the in-place treatment of a contaminated site. *Ex situ* **bioremediation** may be implemented to treat contaminated soil or water that

FIGURE 16.27   Anaerobic biodegradation of aromatic compounds by a consortium of anaerobic bacteria. (From *Pollution Science* © 1996, Academic Press, San Diego, CA.)

is removed from a contaminated site. **Intrinsic bioremediation** or **natural attenuation** is the indigenous level of contaminant biodegradation that occurs without any stimulation or treatment. All of these types of bioremediation continue to receive increasing attention as viable remediation alternatives for several reasons. These include generally good public acceptance and support, good success rates for some applications, and a comparatively low cost of bioremediation when it is successful. As with any technology, there are also drawbacks. Success can be unpredictable because a biological system is being used. A second consideration is that bioremediation rarely restores an environment completely. Often the residual contamination left after treatment is strongly sorbed and not available to microorganisms for degradation. Over a long period of time (years), these residuals can be slowly released.

There is little research concerning the fate and potential toxicity of such released residuals, and therefore there is both public and regulatory concern about the importance of residual contamination.

Although it is often not thought of as bioremediation, domestic sewage waste has been treated biologically for many years with resounding success. In application of bioremediation to other environmental problems, it must be kept in mind that biodegradation is dependent on the pollutant structure and bioavailability. Therefore, application of bioremediation to other pollutants depends on the type of pollutant or pollutant mixtures present and the type of microorganisms present. The first successful application of bioremediation outside sewage treatment was the cleanup of oil spills, and success in this area is now well documented (see Case Study 3). In the past few

---

CASE STUDY 3

Exxon *Valdez*

In March 1989, 33,000 tons of crude oil were spilled from the Exxon *Valdez* in the Prince William Sound, Alaska. This oil was subsequently spread by a storm and coated the shores of the islands in the Sound. Conventional cleanup, primarily by physical methods, did not remove all of the oil on the beaches, particularly under rocks and in the beach sediments. Exxon reached a cooperative agreement with the Environmental Protection Agency (EPA) in late May 1989 to test bioremediation as a cleanup strategy.

The approach followed was to capitalize on existing conditions as much as possible. For example, for oil-degrading microorganisms, scientists realized that the beaches most likely contained indigenous organisms that were already adapted to the cold climate. In fact, preliminary studies showed that, indeed, the polluted beaches contained such organisms as well as a plentiful carbon source (spilled oil) and sufficient oxygen. In fact, intrinsic biodegradation was already occurring in the area (Button, 1992). Further studies showed that intrinsic degradation rates were limited by availability of nutrients such as nitrogen, phosphorus, and other trace elements. Time was of the essence in treating the contaminated sites because the relatively temperate Alaskan season was rapidly advancing. Therefore, it was decided to attempt to enhance the intrinsic rates of biodegradation by amendment of the contaminated areas with nitrogen and phosphorus.

In early June 1989, field demonstration work was started. Several different fertilizer nutrient formulations and application procedures were tested. One problem encountered was the tidal action, which tended to quickly wash away added nutrients. Therefore a liquid oleophilic fertilizer (Inipol EAP-22) that adhered to the oil-covered surfaces and a slow-release water-soluble fertilizer (Customblen) were tested as nutrient sources. Within about 2 weeks after application of the fertilizers, there was a visible decrease in the amount of oil on rock surfaces treated with Inipol EAP-22. Subsequent independent scientific studies of the effectiveness of bioremediation in Prince William Sound have concluded that bioremediation enhanced removal from three- to eightfold over the intrinsic rate of biodegradation without any adverse effects on the environment. Bioremediation in Prince William Sound was considered a success because the rapid removal of the petroleum contamination prevented further spread of the petroleum to other uncontaminated areas and restored the contaminated areas. This well-documented and highly visible case study has helped gain attention for the potential of bioremediation. It should be noted that while the bioremediation was considered successful; today, more than 10 years later, weathered oil can still be found at dozens of sites in the Prince William Sound even hundreds of miles from the spill site. Thus, bioremediation is not a perfect solution, and the question discussed in Chapter 16.1 remains; "How clean should clean be?"

years, many new bioremediation technologies have emerged that are being used to address other types of pollutants (Table 16.3). In fact, every other year, beginning in 1991, a meeting entitled "International *In Situ* and On-Site Bioreclamation Symposium" has been held so that the latest technologies in bioremediation can be presented, discussed, and published.

Several key factors are critical to successful application of bioremediation: environmental conditions, contaminant and nutrient availability, and the presence of degrading microorganisms. If biodegradation does not occur, the first thing that must be done is to isolate the factor limiting bioremediation, and this can sometimes be a very difficult task. Initial laboratory tests using soil or water from a polluted site can usually determine whether degrading microorganisms are present and whether there is an obvious environmental factor that limits biodegradation, for example, extremely low or high pH or lack of nitrogen and/or phosphorus. However, sometimes the limiting factor is not easy to identify. Often pollutants are present as mixtures and one component of the pollutant mixture can have toxic effects on the growth and activity of degrading microorganisms. Low bioavailability due to sorption and aging is another factor that can limit bioremediation and can be difficult to evaluate in the environment.

Actual application of bioremediation is still limited in practice but is rapidly gaining in popularity. Most of the developed bioremediation technologies are based on two standard practices: addition of oxygen and addition of other nutrients (Norris, 1994).

16.7.1 Addition of Oxygen or Other Gases

One of the most common limiting factors in bioremediation is availability of oxygen. Oxygen is an element required for aerobic biodegradation. In addition, oxygen has low solubility in water and a low rate of diffusion (movement) through both air and water. The combination of these three factors makes it easy to understand that inadequate oxygen supplies will slow bioremediation. Several technologies have been developed to overcome a lack of oxygen. A typical bioremediation system used to treat a contaminated aquifer as well as the contaminated zone above the water table is shown in Fig. 16.28a. This system contains a series of injection wells or galleries and a series of recovery wells that comprise a two-pronged approach to bioremediation. First, the recovery wells remove contaminated groundwater, which is treated above ground, in this case using a **bioreactor** containing microorganisms that are acclimated to the contaminant. This would be considered *ex situ* treatment. Followng bioreactor treatment, the clean water is supplied with oxygen and nutrients, and then it is reinjected into the site. The

reinjected water provides oxygen and nutrients to stimulate *in situ* biodegradation. In addition, the reinjected water flushes the vadose zone to aid in removal of the contaminant for above-ground bioreactor treatment. This remediation scheme is a very good example of a combination of physical, chemical, and biological treatments being used to maximize the effectiveness of the remediation treatment.

Bioventing is a technique used to add oxygen directly to a site of contamination in the vadose zone (unsaturated zone). Bioventing is a merging of soil vapor extraction technology and bioremediation. The bioventing zone is highlighted in red in Fig. 16.28b and includes the vadose zone and contaminated regions just below the water table. As shown in this figure, a series of wells have been constructed around the zone of contamination. To initiate bioventing, a vacuum is drawn on these wells to force accelerated air movement through the contamination zone. This effectively increases the supply of oxygen throughout the site, and thus the rate of contaminant biodegradation. In some cases, depending on the type of pollutant in the site, pollutant volatility becomes an issue, e.g., in gasoline spills. In this case, some of the pollutants will be removed as air is forced through this system. This contaminated air can also be treated biologically by passing the air through aboveground soil beds in a process called **biofiltration** as shown in Fig. 16.29 (Jutras *et al.*, 1997).

In contrast, **air sparging** is used to add oxygen to the saturated zone (Fig. 16.28c). In this process, an air sparger well is used to inject air under pressure below the water table. The injected air displaces water in the soil matrix, creating a temporary air-filled porosity. This causes oxygen levels to increase, resulting in enhanced biodegradation rates. In addition, volatile organics will volatilize into the airstream and be removed by a vapor extraction well.

Methane is another gas that can be added with oxygen in extracted groundwater and reinjected into the saturated zone. Methane is used specifically to stimulate methanotrophic activity and cometabolic degradation of chlorinated solvents. As described in Chapter 14.2.3.3, methanotrophic organisms produce the enzyme methane monooxygenase to degrade methane, and this enzyme also cometabolically degrades several chlorinated solvents. Cometabolic degradation of chlorinated solvents is presently being tested in field trials to determine the usefulness of this technology.

16.7.2 Nutrient Addition

Perhaps the second most common bioremediation treatment is the addition of nutrients, in particular nitrogen and phosphorus. Many contaminated sites con-

TABLE 16.3 Current Status of Bioremediation

Chemical class	Frequency of occurrence	Status of bioremediation	Evidence of future success	Limitations
Hydrocarbons and derivatives				
Gasoline, fuel oil	Very frequent	Established		Forms nonaqueous phase liquid
PAHs	Common	Emerging	Aerobically biodegradable under a narrow range of conditions	Sorbs strongly to subsurface soilds
Creosote	Infrequent	Emerging	Readily biodegradable under aerobic conditions	Sorbs strongly to subsurface soilds; forms nonaqueous phase liquid
Alcohols, ketones, esters	Common	Established		
Ethers	Common	Emerging	Biodegradable under a narrow range of conditions using aerobic or nitrate-reducing microbes	
Halogenated aliphatics				
Highly chlorinated	Very frequent	Emerging	Cometabolized by anaerobic microbes; cometabolized by aerobes in special cases	Forms nonaqueous phase liquid
Less chlorinated	Very frequent	Emerging	Aerobically biodegradable under a narrow range of conditions; cometabolized by anaerobic microbes	Forms nonaqueous phase liquid
Halogenated aromatics				
Highly chlorinated	Common	Emerging	Aerobically biodegradable under a narrow range of conditions; cometabolized by anaerobic microbes	Sorbs strongly to subsurface solids; forms nonaqueous phase either liquid or solid
Less chlorinated	Common	Emerging	Readily biodegradable under aerobic conditions	Forms nonaqueous phase either liquid or solid
Polycholorinated biphenyls				
Highly chlorinated	Infrequent	Emerging	Cometabolized by anaerobic microbes	Sorbs strongly to subsurface solids
Less chlorinated	Infrequent	Emerging	Aerobically biodegradable under a narrow range of conditions	Sorbs strongly to subsurface solids
Nitroaroamatics	Common	Emerging	Aerobically biodegradable; converted to innocuous volatile organic acids under anaerobic conditions	
Metals (Cr, Cu, Ni, Pb, Hg, Cd, Zn, etc.)	Common	Possible (see Chapter 17)	Solubility and reactivity can be changed by a variety of microbial processes	Availability highly variable and controlled by solution and solid-phase chemistry

Adapted from National Research Council (1993).

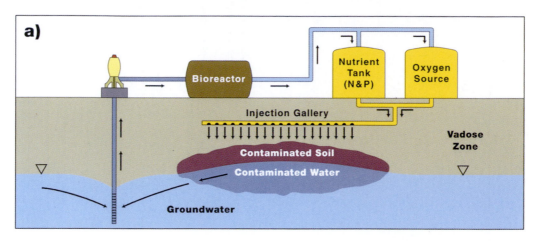

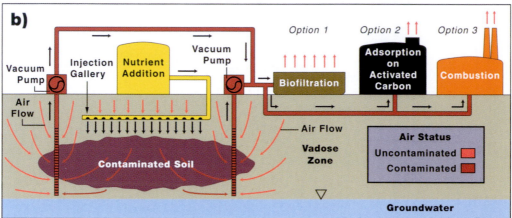

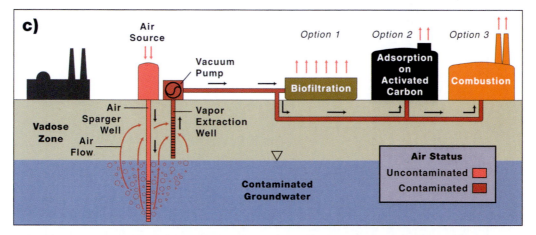

FIGURE 16.28 (a) *In situ* bioremediation in the vadose zone and groundwater. Nutrients and oxygen are pumped into the contaminated area to promote *in situ* processes. This figure also shows *ex situ* treatment. *Ex situ* treatment is for water pumped to the surface and uses an aboveground bioreactor, as shown, or other methods, e.g., air stripping, activated carbon, oil-water separation, or oxidation. An injection well returns treated water to the aquifer. (b) Bioventing and biofiltration in the vadose zone. Air drawn through the contaminated site (bioventing) stimulates *in situ* aerobic degradation. Volatile contaminants removed with the air are treated in a biofilter, by adsorption on activated carbon, or by combustion. (c) Bioremediation in the groundwater by air sparging. Air pumped into the contaminated site stimulates aerobic degradation in the saturated zone. Volatile contaminants brought to the surface are treated by biofiltration, activated carbon, or combustion. (From *Pollution Science* © 1996, Academic Press, San Diego, CA.)

a) Soil vapor extraction unit

b) Side view of biofilter

FIGURE 16.29 Bioremediation of petroleum vapors from a leaking underground storage tank. (a) Red arrows denote the direction of airflow out of the ground and through the manifold, which controls the airflow. The air then passes through the humidifier, a tank containing water, and into the biofilter. The large red vertical cylinders serve as protective barriers. (b) Air flows into the biofilter through a pipe running lengthwise along the bottom of the filter and into the soil via perforations in the pipe. (From *Pollution Science* © 1996, Academic Press, San Diego, CA.)

tain organic wastes that are rich in carbon but contain minimal amounts of nitrogen and phosphorus. Nutrient addition is illustrated in the bioremediation schemes shown in Fig. 16.28a and b. Injection of nutrient solutions takes place from an above-ground batch feed system. The goal of nutrient injection is to optimize the ratio of carbon, nitrogen, and phosphorus (C:N:P) in the site to approximately 100:10:1. However, sorption of added nutrients can make it difficult to achieve the optimal ratio accurately.

16.7.3 Stimulation of Anaerobic Degradation Using Alternative Electron Acceptors

Until recently, anaerobic degradation of many organic compounds was not even considered feasible (Grbíc-Galíc, 1990). Now it is being proposed as an al-

ternative bioremediation strategy even though aerobic degradation is generally considered a much more rapid process. This is because it is difficult to establish and maintain aerobic conditions in some groundwater sites. Several alternative electron acceptors have been proposed for use in anaerobic degradation, including nitrate, sulfate, iron (Fe^{3+}), and carbon dioxide. There has even been a limited number of field trials using nitrate that show promise for this approach. This is a relatively new area in bioremediation that will undoubtedly receive increased attention in the next few years.

16.7.4 Addition of Surfactants

Surfactant addition has been proposed as a technique for increasing the bioavailability and hence biodegradation of contaminants (see Section 16.4.2). Surfactants can be synthesized chemically and are also produced by many microorganisms, in which case they are called biosurfactants. Surfactants work similarly to industrial and household detergents that effectively remove oily residues from machinery, clothing, or dishes. As shown in the cartoon in Fig. 16.8, individual contaminant molecules can be "solubilized" inside surfactant micelles. These micelles range from 5 to 10 nm in diameter. Alternatively, surfactant molecules can coat oil droplets and emulsify them into solution. In addition, biosurfactants seem to enhance the ability of microbes to stick to oil droplets. In laboratory tests, synthetic surfactants and biosurfactants can be used to increase the apparent aqueous solubility of organic contaminants. However, field tests have been attempted only with synthetic surfactants and results have been mixed (Miller, 1995).

16.7.5 Addition of Microorganisms or DNA

If appropriate biodegrading microorganisms are not present in soil or if microbial populations have been reduced because of contaminant toxicity, specific microorganisms can be added as "introduced organisms" to enhance the existing populations. This process is known as **bioaugmentation.** Scientists are now capable of creating "superbugs," organisms that can degrade pollutants at extremely rapid rates. Such organisms can be developed through successive adaptations under laboratory condition, or can be **genetically engineered.** In terms of biodegradation, these superbugs are far superior to organisms found in the environment. The problem is that introduction of a microorganism to a contaminated site may fail for two reasons. First, the introduced microbe often cannot establish a niche in the environment. In fact, these introduced organisms often do not survive in a new environment beyond a few weeks. Second, there are difficulties in delivering the introduced organisms to the site of contamination, because microorganisms, like contaminants, can be strongly sorbed by solid surfaces. Currently, very little is known about microbial transport and establishment of environmental niches. These are areas of active research, and in the next few years scientists may gain a further understanding of microbial behavior in soil ecosystems. However, until we discover how to successfully deliver and establish introduced microorganisms, their addition to contaminated sites will not be a viable bioremediation option.

One way to take advantage of the superbugs that have been developed is to use them in bioreactor systems under controlled conditions. Extremely efficient biodegradation rates can be achieved in bioreactors that are used in aboveground treatment systems.

A second bioaugmentation strategy is to add specific genes that can confer a specific degradation capability to indigenous microbial populations. The addition of degradative genes relies on the delivery and uptake of the genetic material by indigenous microbes. There are two approaches that can be taken in delivery of genes. The first is to use microbial cells to deliver the DNA via conjugation. The second is to add "naked" DNA to the soil to allow uptake via transformation. This second approach may reduce the difficulty of delivery since DNA alone is much smaller than a whole cell. However, little is known as yet about these two approaches. As discussed in the Chapter 3 Case Study, Di Giovanni *et al.* (1996) demonstrated that gene transfer can occur in soil resulting in 2,4-D degradation activity. However, whether such transfer is common, and conditions that are conducive or inhibitory to such transfer are not yet defined.

QUESTIONS AND PROBLEMS

1. Why is there concern about the presence of organic contaminants in the environment?

2. Describe the different factors that can limit biodegradation of organic contaminants in the environment.

3. Draw and name an aliphatic, alicyclic, and aromatic structure, each with 6 carbons.

4. Outline the biodegradation pathway for each of the structures that you just drew under aerobic conditions.

5. Why are aerobic conditions usually preferred for biodegradation of organic contaminants? Under what conditions might anaerobic biodegradation be preferred?

6. Compare the advantages and disadvantages of intrinsic, *ex situ*, and *in situ* bioremediation.

7. You have been hired to bioremediate a site in which the groundwater is contaminated with petroleum. Groundwater samples have a strong sulfide smell and gas chromatographic analysis of the samples show negligible biodegradation of the petroleum has occurred. What is your recommendation?

8. Kleen Co. is in charge of a site in Nevada that was used for pesticide preparation. As a result of years of operation, the groundwater below this site has elevated levels of pesticides (up to 20 mg/l). Your initial investigation shows that 1) the pesticide-containing plume is neither growing nor shrinking in size, 2) there are pesticide degraders in the plume, 3) and the dissolved oxygen levels in the plume range from 2 to 4 mg/l. This site is not being used presently, and the groundwater is not used for drinking water purposes. What is your best recommendation based on these site characteristics and on your knowledge of cost of remediation?

References and Recommended Readings

Ahn, Y., Sanseverino, J., and Sayler, G. (1999) Analyses of polycyclic aromatic hydrocarbon degrading bacteria isolated from contaminated soils. *Biodegradation* **10**, 149–157.

Alexander, M. (1995) How toxic are toxic chemicals in soil? *Environ. Sci. Technol.* **29**, 2713–2717.

Alleman, B. C., Logan, B. E., and Gilbertson, R. L. (1992) Toxicity of pentachlorophenol to six species of white rot fungi as a function of chemical dose. *Appl. Environ. Microbiol.* **58**, 4048–4050.

Arbuckle, J. G., Bryson, N. S., Case, D. R., Cherney, C. T., Hall, R. M. Jr., Martin, H. C., Miller, J. G., Miller, M. L., Pedersen W. F. Jr., Randle, R. V., Stoll, R. G., Sullivan, T. F. P., and Vanderver, R. A. Jr. (1987) "Environmental Law Handbook," 9th ed. Government Institutes, Rockville, MD.

Atlas, R. M., and Bartha, R. (1993) "Microbial Ecology" Benjamin Cummings, New York.

Bernheim, F. (1974) Effect of aromatic hydrocarbons, alcohols, ketones, and aliphatic alcohols on cell swelling and potassium efflux of *Pseudomonas aeruginosa*. *Cytobios* **11**, 91–95.

Blum, D. J., and Speece, R. E. (1990) Determining chemical toxicity to aquatic species. *Environ. Sci. Technol.* **24**, 284–293.

Bollag, J.-M. (1992) Decontaminating soil with enzymes. *Environ. Sci. Technol.* **26**, 1876–1881.

Boyle, M. (1992) The importance of genetic exchange in degradation of xenobiotic chemicals. *In* "Environmental Microbiology" (R. Mitchell, ed.) Wiley-Liss, Inc., New York, NY pp. 319–333.

Britton, L. N. (1984) Microbial Degradation of Aliphatic Hydrocarbons. *In* "Microbial Degradation of Organic Compounds" (D. T. Gibson, ed.) Marcel Dekker Inc., New York, NY pp. 89–129.

Bumpus, J. A., Tien, M., Wright, D., and Aust, S. D. (1985) Oxidation of persistent environmental pollutants by a white rot fungus. *Science* **228**, 1434–1436.

Button, D. K., Robertson, B. R., McIntosh, D., and Jüttner, F. (1992) Interactions between marine bacteria and dissolved-phase and beached hydrocarbons after the Exxon *Valdez* Oil Spill. *Appl. Environ. Microbiol.* **58**, 243–251.

Carlile, M. J., and Watkinson, S. C. (1994) The Fungi. Academic Press, San Diego, CA, pp. 251–306.

Cerniglia, C. E., and Heitkamp, M. A. (1989) Microbial degradation of polyaromatic hydrocarbons (PAH) in the aquatic environment. *In* "Metabolism of Polycyclic Aromatic Hydrocarbons in the Aquatic Environment" (U. Varanasi, ed.) CRC Press, Ann Arbor, MI.

Chaudhry, G. R., and Chapalamadugu, S. (1991) Biodegradation of halogenated organic compounds. *Microbiol. Rev.* **55**, 59–79.

Cooney, J. J. (1984) The fate of petroleum pollutants in freshwater ecosystems. *In* "Petroleum Microbiology," (R. M. Atlas, ed.) Macmillian Publishing, New York, NY, pp. 399–434.

Ensign, S. A., Hyman, M. R., and Arp, D. J. (1992) Cometabolic degradation of chlorinated alkenes by alkene monooxygenase in a propylene-grown *Xanthobacter* strain. *Appl. Environ. Microbiol.* **58**, 3038–3046.

Evans, W. C., and Fuchs, G. (1988) Anaerobic degradation of aromatic compounds. *Ann. Rev. Microbiol.* **42**, 289–317.

Fathepure, B. Z., and Vogel, R. M. (1991) Complete degradation of polychlorinated hydrocarbons by a two-stage biofilm reactor. *Appl. Environ. Microbiol.* **57**, 3418–3422.

Fulthorpe, R. S., and Wyndham, R. C. (1992) Involvement of a chlorobenzoate-catabolic transposon, Tn5271, in community adaptation to chlorobiphenyl, chloroaniline, and 2,4-dichlorophenoxyacetic acid in a freshwater ecosystem. *Appl. Environ. Microbiol.* **58**, 314–325.

Furukawa, K., Tonomura, K., and Kamibayashi, A. (1978) Effect of chlorine substitution on biodegradability of polychlorinated biphenyls. *Appl. Environ. Microbiol.* **35**, 223–227.

Ghosal, D., You, I.-S., Chatterjee, D. K., and Chakrabarty, A. M. (1985) Plasmids in the degradation of chlorinated aromatic compounds. *In* "Plasmids in Bacteria" (D. R. Helinski, S. N. Cohen, D. B. Clewell, D. A. Jackson, A. Hollaender , eds.) Plenum Press, New York, NY, pp. 667–686.

Gibson, D. T., and Subramanian, V. (1984) Microbial degradation of aromatic hydrocarbons. *In* "Microbial Degradation of Organic Compounds" (D. T. Gibson, ed.) Marcel Dekker, Inc., New York, NY, pp. 181–252.

Gilbert, E. S., and Crowley, D. E. (1997) Plant compounds that induce polychlorinated biphenyl biodegradation by *Arthrobacter* sp. strain B1B. *Appl. Environ. Microbiol.* **63**, 1933–1938.

Glaser, J. A., and Lamar, R. T. (1995) Lignin-degrading fungi as degraders of pentachlorophenol and creosote in soil. *In* "Bioremediation: Science and Applications" (H. D. Skipper, and R. F. Turco, eds.) *Soil Science Society of America*, Madison, WI, pp. 117–133.

Grbic-Galic, D. (1990) Anaerobic microbial transformation of nonoxygenated aromatic and alicyclic compounds in soil, subsurface, and freshwater sediments. *In* "Soil Biochemistry, Vol. 6" (J.-M. Bollag and G. Stotsky, eds.) Marcel Dekker, New York.

Hambrick, G. A., DeLuane, R. D., and Patrick, W. H. (1980) Effect of estuarine sediment pH and oxidation-reduction potential on microbial hydrocarbon degradation. *Appl. Environ. Microbiol.* **40**, 365–369.

Hammel, K. E. (1995) Organopollutant degradation by ligninolytic fungi. *In* "Microbial Transformation and Degradation of Toxic Organic Chemicals" (L. Y. Young and C. E. Cerniglia, eds.), Wiley and Sons, New York, pp. 331–346.

Herman, D. C., Inniss, W. E., and Mayfield, C. I. (1991) Toxicity testing of aromatic hydrocarbons utilizing a measure of their impact on the membrane integrity of the green alga *Selenastrum capricornutum*. *Bull. Environ. Contam. Toxicol.* **47**, 874–881.

Herman, D. C., Lenhard, R. J., and Miller, R. M. (1997) Formation and removal of hydrocarbon residual in porous media: effects of bacterial biomass and biosurfactants. *Environ. Sci. Technol.* **31**, 1290–1294.

Hodson, R. E., Azam, F., and Lee, R. F. (1977) Effects of four oils on marine bacterial populations: controlled ecosystem pollution experiment. *Bull. Mar. Sci.* **27**, 119–126.

Hutchinson, T. C., Hellebust, J. A., and Soto, C. (1981) Effects of naphthalene and aqueous crude oil extracts on the green flagellate *Chlamydomonas angulosa*. IV. Decrease in cellular manganese and potassium. *Can. J. Bot.* **59**, 742–749.

Janssen, D. B., Oldenhuis, R., and van den Wijngarrd, A. J. (1990) Hydrolytic and oxidative degradation of chlorinated aliphatic compounds by aerobic microorganisms. *In* "Biotechnology and Biodegradation" (D. Kamely, A. Chakrabarty, and G. S. Omenn, eds.) Gulf Publishing Company, Houston, TX.

Jutras, E. M., Smart, C. M., Rupert, R., Pepper, I. L., and Miller, R. M. (1997) Field scale biofiltration of gasoline vapors extracted from beneath a leaking underground storage tank. *Biodegradation* **8**, 31–42.

Kerr, R. P., and Capone, D. G. (1988) Effect of salinity on the microbial mineralization of two polycyclic aromatic hydrocarbons in estuarine sediments. *Mar. Environ. Res.* **26**, 181–198.

Leahy, J. G., and Colwell, R. R. (1990) Microbial degradation of hydrocarbons in the environment. *Microbiol. Rev.* **54**, 305–315.

Liu, D., and Thomson, K. (1983) Toxicity assessment of chlorobenzenes using bacteria. *Bull. Environ. Contam. Toxicol.* **31**, 105–111.

Liu, D., Thomson, K., and Kaiser, K. L. E. (1982) Quantitative structure-toxicity relationship of halogenated phenols on bacteria. *Bull. Environ. Contam. Toxicol.* **29**, 130–136.

Lovley, D. R., Coates, J. D., Woodward, J. C., and Phillips, E. J. P. (1995) Benzene oxidation coupled to sulfate reduction. *Appl. Environ. Microbiol.* **61**, 953–958.

Michaelsen, M., Hulsch, R., Hopper, T., and Berthe-Corti, L. (1992) Hexadecane mineralization in oxygen-controlled sediment-seawater cultivations with autochthonous microorganisms. *Appl. Environ. Microbiol.* **58**, 3072–3077.

Miller, R. M. (1995) Surfactant-Enhanced Bioavailability of Slightly Soluble Organic Compounds. *In* "Bioremediation—Science & Applications" (H. Skipper, and R. Turco, eds.) Soil Science Society of America Special Publication Number 43, Madison, WI, pp. 33–45.

Miller, R. and Bartha, R. (1989) Evidence from liposome encapsulation for transport-limited microbial metabolism of solid alkanes. *Appl. Environ. Microbiol.* **55**, 269–274.

Miller, R. M., and Herman, D. H. (1997) Biotransformation of organic compounds—remediation and ecotoxicological implications. In "Soil Ecotoxicology" (J. Tarradellas, G. Bitton, and D. Rossel, eds.) Lewis Publishers, Boca Raton, FL, pp. 53–84.

Mu, Y. D., and Scow, K. M. (1994) Effect of trichloroethylene (TCE) and toluene concentrations on TCE and toluene biodegradation and the population density of TCE and toluene degraders in soil. *Appl. Environ. Microbiol.* **60**, 2661–2665.

Narro, M. L., Cerniglia, C. E., van Baalen, C., and Gibson, D. T. (1992) Metabolism of phenanthrene by the marine cyanobacterium *Agmenellum quadruplicatum* PR-6. *Appl. Environ. Microbiol.* **58**, 1351–1359.

National Research Council. (1993) "*In Situ* Bioremediation, When Does It Work?" National Academy Press, Washington, DC.

Neilson, J. W., and I. L. Pepper. 1990. Soil respiration as an index of soil aeration. *Soil Science Society of America Journal* **54**, 428–432.

Norris, R. D. (1994) "Handbook of Bioremediation" Lewis Publishers, Boca Raton, FL.

Novak, J. M., Jayachandran, K., Moorman, T. B., and Weber, J. B. (1995) Sorption and binding of organic compounds in soils and their relation to bioavailability. *In* "Bioremediation—Science & Applications" (H. Skipper and R. F. Turco, eds.) Soil Science Society of America Special Publication Number 43, Soil Science Society of America, Madison, WI., pp. 13–32.

Oldenhuis, R., Oedzes, J. Y., van der Waarde, J. J., and Janssen, D. B. (1991) Kinetics of chlorinated hydrocarbon degradation by *Methylosinus trichosporium* OB3b and toxicity of trichloroethylene. *Appl. Environ. Microbiol.* **57**, 7–14.

Palmisano, A. C., Schwab, B. S., Maruscik, D. A., and Ventullo, R. M. (1991) Seasonal changes in mineralization of xenobiotics by stream microbial communities. *Can. J. Microbiol.* **37**, 939–948.

Perry, J. J. (1984) Microbial metabolism of cyclic alkanes. *In* "Petroleum Microbiology" (R. Atlas, ed.) Macmillan, New York, NY, pp. 61–97.

Pitter, P., and Chudoba, J. (1990) "Biodegradability of Organic Substances in the Aquatic Environment." CRC Press, Ann Arbor, MI.

Rasche, M. E., Hyman, M. R., and Arp, D. J. (1991) Factors limiting aliphatic chlorocarbon degradation by Nitrosomonas europaea: cometabolic inactivation of ammonia monooxygenase and substrate specificity. *Appl. Environ. Microbiol.* **57**, 2986–2994.

Rosenberg, E., Bayer, E. A., Delarea, J., and Rosenberg, E. (1982) Role of thin fimbriae in adherence and growth of *Acinetobacter calcoaceticus* RAG-1 on hexadecane. *Appl. Environ. Microbiol.* **44**, 929–937.

Ruckdeschel, G., Renner, G., and Schwarz, K. (1987) Effects of pentachlorophenol and some of its known and possible metabolites on different species of bacteria. *Appl. Environ. Microbiol.* **53**, 2689–2692.

Scow, K. M. (1993) Effect of sorption-desorption and diffusion processes on the kinetics of biodegradation of organic chemicals in soil. *In* "Sorption and Degradation of Pesticides and Organic Chemicals in Soil" (D. M. Linn, T. H. Carski, M. L. Brusseau, F-H. Chang, eds.) Soil Science Society of America Special Publication, Madison, WI, pp. 73–114.

Sikkema, J., de Bont, J. A. M., Poolman, B. (1995) Mechanisms of membrane toxicity of hydrocarbons. *Microbiol. Rev.* **59**, 201–222.

Singh, A. K., and Gaur, J. P. (1990) Effects of petroleum oils and their paraffinic, asphaltic, and aromatic fractions on photosynthesis and respiration of microalgae. *Ecotoxicol. Environ. Saf.* **19**, 8–16.

Song H.-G., and Bartha, R. (1990) Effects of jet fuel spills on the microbial community of soils. *Appl. Environ. Microbiol.* **56**, 646–651.

Stratton, G. W. (1989) Effect of the solvent acetone on membrane integrity in the green alga *Chlorella pyrenoidosa. Bull. Environ. Contam. Toxicol.* **11**, 437–445.

Trower, M. K., Buckland, R. M., Higgins, R., and Griffin, M. (1985) Isolation and characterization of a cyclohexane-metabolizing *Xanthobacter* sp. *Appl. Environ. Microbiol.* **49**, 1282–1289.

Trudgill, P. W. (1984) Microbial degradation of the alicyclic ring. *In* "Microbial Degradation of Organic Compounds" (D. T. Gibson, ed.) Marcel Dekker, Inc., New York, NY, pp. 131–180.

U.S. Environmental Protection Agency, Summary Report: Remedial Response at Hazardous Waste Sites, EPA–540/2–84–002a, March 1984.

Van Beelen, P., and Fleuren-Kemilä, A. K. (1993) Toxic effects of pentachlorophenol and other pollutants on the mineralization of acetate in several soils. *Ecotox. Environ. Saf.* **26**, 10–17.

Van der Meer, J. R., de Vos, W. M., Harayama, S., and Zehnder, A. J. B. (1992) Molecular mechanisms of genetic adaptation to xenobiotic compounds. *Microbiol. Rev.* **56**, 677–694.

Vannelli, T., Logan, M., Arciero, D. M., and Hooper, A. B. (1990) Degradation of halogenated aliphatic compounds by the ammonia-oxidizing bacterium *Nitrosomonas europaea. Appl. Environ. Microbiol.* **56**, 1169–1171.

Vogel, T. M., Criddle, C. S., and McCarty, P. L. (1987) Transformations of halogenated aliphatic compounds. *Environ. Sci. Technol.* **21**, 722–732.

Wackett, L. P., Brusseau, G. A., Householder, S. R., and Hanson, R. S. (1989) Survey of microbial oxygenases: trichloroethylene degradation by propane-oxidizing bacteria. *Appl. Environ. Microbiol.* **55**, 2960–2964.

Waid, J. S., ed. (1986a) "PCBs and the Environment," Vol. I. CRC Press, Boca Raton, FL.

Waid, J. S., ed. (1986b) "PCBs and the Environment," Vol. II. CRC Press, Boca Raton, FL.

Wang, X., and Bartha, R. (1990) Effects of bioremediation on residues, activity and toxicity in soil contaminated by fuel spills. *Soil Biol. Biochem.* **22**, 501–505.

Ward, D. M., and Brock, T. D. (1976) Environmental factors influencing the rate of hydrocarbon oxidation in temperate lakes. *Appl. Environ. Microbiol.* **31**, 764–772.

Ward, D. M., and Brock, T. D. (1978) Hydrocarbon biodegradation in hypersaline environments. *Appl. Environ. Microbiol.* **35**, 353–359.

Warne, M. St. J., Connell, D. W., Hawker, D. W., and Schüürmann, G. (1989b) Prediction of the toxicity of mixtures of shale oil components. *Ecotoxicol. Environ. Saf.* **18**, 121–128.

Wyndham, R. C., and Costerton, J. W. (1981) Heterotrophic potentials and hydrocarbon biodegradation potentials of sediment microorganisms within the Athabasca oil sands deposit. *Appl. Environ. Microbiol.* **41**, 783–790.

Zhang, Y., and Miller, R. M. (1994) Effect of a *Pseudomonas* Rhamnolipid Biosurfactant on Cell Hydrophobicity and Biodegradation of Octadecane. *Appl. Environ. Microbiol.* **60**, 2101–2106.

17

Microorganisms and Metal Pollutants

TIMBERLEY M. ROANE   IAN L. PEPPER

17.1 CAUSE FOR CONCERN

Metal pollution is a global concern. The levels of metals in all environments, including air, water and soil, are increasing, in some cases to toxic levels, with contributions from a wide variety of industrial and domestic sources. For example, anthropogenic emissions of lead, cadmium, vanadium, and zinc exceed those from natural sources by up to 100-fold.

Metal-contaminated environments pose serious health and ecological risks. Metals, such as aluminum, antimony, arsenic, cadmium, lead, mercury, and silver, cause conditions including hypophosphatemia, heart disease and liver damage, cancer, neurological and cardiovascular disease, central nervous system damage, encephalopathy, and sensory disturbances. Historians have speculated that the decline of the Roman Empire may have been due in part to a decrease in the mental skills of the ruling class as a result of lead poisoning from wine stored in pottery lined with lead and from lead water pipes. The problem of mercury pollution came into focus after the discovery of high levels of methylmercury in fish and shellfish in Minamata Bay, Japan, that resulted in 46 deaths (see Chapter 15.4). The mercury contamination originated from an acetaldehyde factory that had dumped relatively nontoxic elemental mercury into Minamata Bay. However, microbial activity in the sediment converted the elemental mercury into the highly toxic and bioavailable methylmercury, which accumulated in fish. Lead is a second metal of concern especially because lead poisoning of children is common and leads to retardation and semi-permanent brain damage. The Centers for Disease Control (CDC) and the Agency for Toxic Sub-

stances and Disease Registry (ASTDR) estimate that 15–20% of U.S. children have blood lead levels greater than 15 mg/dl, which is considered a potentially toxic level. Although contamination of drinking water supplies and concentration of metals in edible fish are of particular concern, soils and sediments are the major sinks for metals.

Because of their toxic nature, metals are not as amenable to bioremediation as organics (see Chapter 16). Unlike organics, metals are persistent in the environment and cannot be degraded through biological, chemical, or physical means to an innocuous by-product. The chemical nature and, thus, bioavailability of a metal can be changed through oxidation or reduction; however, the elemental nature remains the same because metals are neither thermally decomposable nor microbiologically degradable. Consequently, metals are difficult to remove from the environment. In addition, total metal concentrations in the environment do not necessarily reflect the degree of biological metal toxicity or bioavailability, making it difficult to assess accurately the extent of risk posed by metals. Sposito (1989) defined a metal as bioavailable if it is present as, or can be transformed readily to, the free-ion species. Only recently have investigators been trying to elucidate the ecological significance of bioavailable metal concentrations.

Because metal availability is strongly dependent on environmental components, such as pH, redox, and organic content, there is often a discrepancy between the total and the bioavailable amount of a metal present. In environmental samples, a bioavailable metal is generally soluble and not sorbed to colloids or soil surfaces. A current research focus in Environmental Microbiology is how to determine and relate metal toxicity and bioavailability.

Because of the toxicity and the ubiquity of metals in the environment, microbes have developed unique and sometimes bizarre ways of dealing with unwanted metals. Some microorganisms have mechanisms to sequester and immobilize metals, whereas others actually enhance metal solubility in the environment. The goal of this chapter is to demonstrate how the presence of metals influences both the degree and type of metal resistance mechanisms expressed and how microbial resistance, in turn, can influence the fate of metals in the environment. The first part of the chapter introduces metals and their interaction with the physical-chemical components of the environment. The remainder of the chapter then focuses on specific microbe-metal interactions including mechanisms of metal resistance, positive and negative effects of metal-microbe interactions, and applications of microorganisms to metal mining and remediation of metal-contaminated sites.

17.2 METALS DEFINED

There are three classes of metals: metals, metalloids, and heavy metals. Metals, in general, are a class of chemical elements that form lustrous solids that are good conductors of heat and electricity. However, not all metals fit this definition; for example, mercury is a liquid. Metals such as arsenic, boron, germanium, and tellurium are generally considered metalloids or semimetals in that their properties are intermediate between those of metals and those of nonmetals. Heavy metals are defined in a number of ways based on cationic–hydroxide formation, a specific gravity greater than 5 g/ml, complex formation, hard–soft acids and bases, and, more recently, association with eutrophication and environmental toxicity. This chapter will focus on the metals most commonly associated with metal pollution: arsenic (As), cadmium (Cd), copper (Cu), chromium (Cr), mercury (Hg), lead (Pb), and zinc (Zn). The metals most commonly associated with severe pollution at Superfund sites throughout the United States are shown in Table 17.1.

As a result of the complexity of the chemical definitions, metals have also been classified into three additional classes on the basis of their biological functions and effects: (1) the essential metals with known biological functions, (2) the toxic metals, and (3) the nonessential, nontoxic metals with no known biological effects.

The following metals are currently known to have essential functions in microorganisms: Na, K, Mg, Ca,

TABLE 17.1 Common Metal Contaminants Found at Superfund Sites

Chemical	Occurrence (%)
Lead (Pb)	70.9
Arsenic (As)	60.0
Zinc (Zn)	56.6
Nickel (Ni)	49.6
Mercury (Hg)	47.1
Barium (Ba)	45.8
Cadmium (Cd)	29.8

From http://www.epa.gov/superfund/oerr/atsdr/index.com

V, Mn, Fe, Co, Ni, Cu, Zn, Mo, and W (Fig. 17.1). Chromium is also thought to be essential, although this is still in dispute. Metals such as Na, K, Mg, and Ca are required by all organisms. Tungsten (W), on the other hand, appears to be essential only in hyperthermophilic bacteria, such as *Pyrococcus furiosus*, found in hydrothermal vents (Adams, 1993). Tungstate is thought to replace molybdate in these environments. In general, essential metals are required for enzyme catalysis, molecule transport, protein structure, charge neutralization, and the control of osmotic pressure (Hughes and Poole, 1989). These metals are usually transported into the cell via membrane transport systems. An important point to note is that although these metals are essential for microbial growth and metabolism, at high concentrations even essential metals can become toxic, e.g., Cu.

The toxic metals include those with no known biological function. These include Ag, Cd, Sn, Au, Hg, Tl, Pb, Al, and the metalloids Ge, As, Sb, and Se. The metalloids exert different toxic effects than the metals because they have different chemistries. Metals are predominantly present as cationic species and metalloids are predominantly present as anionic species. Radionuclides are also toxic and are of increasing concern as a result of careless disposal in the past. Toxic metals exert their toxicity in a number of ways including the displacement of essential metals from their normal binding sites on biological molecules (e.g., arsenic and cadmium compete with phosphate and zinc, respectively), inhibition of enzymatic functioning, and disruption of nucleic acid structure. Intuitively, one would expect the toxicity of these metals to increase as concentration increases; however, recent studies have found that, in some cases, higher metal concentrations activate aggressive resistance mechanisms that increase microbial tolerance of these metals. It is important to note that the toxicity of a metal depends to a

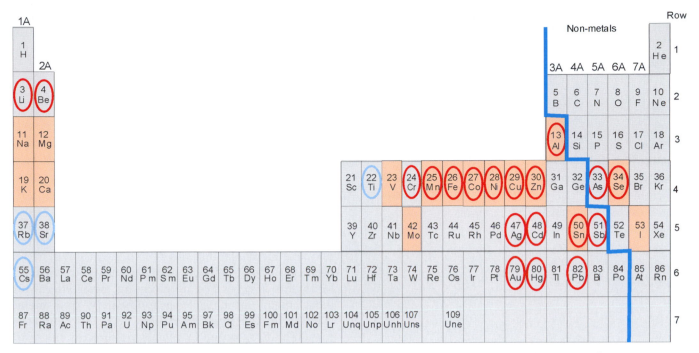

FIGURE 17.1   Periodic table showing the essential metals (pink squares), the toxic metals (circled in red), and the nonessential, nontoxic metals (circled in blue). Chromium is thought to be essential but is often found to be highly toxic.

large extent on its speciation which in turn influences metal bioavailability.

The nontoxic nonessential metals include Rb, Cs, Sr, and Ti. These metals are sometimes found accumulated in cells as a result of nonspecific sequestration and transport (Gadd, 1988). Cation replacement is the general biological effect, for example, Cs^+ replacement of K^+ but there seems to be no apparent effect on the cell (Avery *et al.*, 1991). In general, the appearance of nontoxic metals in the cell results from elevated environmental levels of the metals.

17.3 SOURCES OF METALS

Metal pollution results when human activity disrupts normal biogeochemical activities or results in disposal of concentrated metal wastes. Sometimes a single metal is involved but more often mixtures of metals are present. Mining; ore refinement; nuclear processing; and the industrial manufacture of batteries, metal alloys, electrical components, paints, preservatives, and insecticides are examples of processes that produce metal by-products. Examples of specific metal contaminants include copper and zinc salts that are used extensively as pesticides in agricultural settings; silver salts that are used to treat skin burns; lead which is utilized in the production of batteries, cable sheath-

ing, pigments, and alloys. Other examples include mercury compounds that are used in electrical equipment, paints, thermometers, and fungicides and as preservatives in pharmaceuticals and cosmetics; and triorganotin compounds, such as tributyltin chloride and triphenyltin chloride, which can be used as antifouling agents in marine paints because of their toxicity to plankton and bacteria. The extent of metal pollution becomes even more obvious when one considers the amount of waste generated in metal processing. For example, for every kilogram of copper produced in the United States, 198 kg of copper-laden waste is produced (Debus, 1990).

Thus while metals are ubiquitous in nature (Table 17.2), human activities have caused metals to accumulate in soil. Such contaminated soils provide a metal sink from which surface waters, groundwaters, and the vadose zone can become contaminated. Metal contamination has occurred for centuries since metals have been mined and used extensively throughout human history. Archeological evidence unearthed in Timma, Israel, indicates that mining and smelting of ores has been carried out in Western civilization since at least 4500 B.C.E. (Debus, 1990). Roman lead–zinc mines in Wales are still a source of contamination nearly 2000 years after they were first used. Atmospheric metal concentrations have also increased. Contaminated soil contributes to high metal concen-

TABLE 17.2   Typical Background Levels of Metals in Soil and Aquatic Systems

Metal	Fresh water[a] (μm)	Seawater[b] (μm)	Soil[c] (μm)
Gold (Au)	ND[d]	0.0028	ND
Aluminum (Al)	Trace[e]	0.37	2.63×10^7
Arsenic (As)	Trace	0.040	660
Barium (Ba)	ND	0.22	31,623
Cadmium (Cd)	0.00053	0.00098	5.37
Cobalt (Co)	0.012	0.0068	1,349
Chromium (Cr)	Trace	0.00096	19,054
Cesium (Cs)	Trace	0.0023	447
Copper (Cu)	0.010	0.047	4.667
Mercury (Hg)	Trace	0.0010	1.48
Manganese (Mn)	0.18	0.036	1.10×10^5
Nickel (Ni)	ND	0.12	6.761
Lead (Pb)	0.00029	0.00014	478
Tin (Sn)	Trace	0.0067	851
Zinc (Zn)	0.30	0.153	7.585

[a] From Goldman and Horne (1983), Leppard (1981), and Sigg (1985).
[b] From Bidwell and Spotte (1985).
[c] From Lindsay (1979).
[d] ND, no data reported.
[e] Trace, levels below detection.

trations in the air through metal volatilization. In addition, industrial emissions and smelting activities cause release of substantial amounts of metals to the atmosphere. For example, in 1973, a lead smelter in northern Idaho released an estimated 27,215 kg of lead per 1.6 km^2 within a 6-month period (Keely *et al.*, 1976). Naturally high metal concentrations can also occur as a result of weathering of parent materials containing high levels of metals. For example Stone and Timmer (1975) found a natural copper concentration as high as 10% in surface peat that was filtering copperrich spring water in New Brunswich, Canada; Forgeron (1971) described a natural surface soil with up to 3% lead and zinc at a site on Baffin Island, Canada; and Warren *et al.* (1966) reported a mercury concentration of 1–10 mg/Kg in soil overlying a cinnabar (HgS) deposit in British Columbia, compared with <0.1 mg/Kg in a reference soil. Sewage sludges containing Cd, Zn, Cr, Cu, Pb, Co, Ni, and Hg can be significant sources of soil and water metal contamination. In some regions in the United States, sludge with low metal concentrations has been applied to nonfood crops, such as cotton, for many years. Sludge is an excellent source of water, nitrogen, phosphorus and other plant-essential nutrients and reduces the need for fertilization. Of the

many contaminants that may be present in such wastes, toxic metals are of primary concern because their nonbiodegradability can result in metal accumulation over time.

17.4 METAL BIOAVAILABILITY IN THE ENVIRONMENT

Metals in the environment can be divided into two classes: (1) bioavailable (soluble, nonsorbed, and mobile) and (2) nonbioavailable (precipitated, complexed, sorbed, and nonmobile). It is the bioavailable metal concentration that is taken up and is thus toxic to biological systems (Sposito, 1989). Much of the research on metal bioavailability has been done in soil systems because understanding the fate of metals in soil and sediments is crucial to determining metal effects on biota, metal leaching to groundwaters, and metal transfer up the food chain (Fig. 17.2). The environmental hazards posed by metals are directly linked to their mobility and thus to their concentrations in the soil solution. High metal concentrations in the soil solution results in greater plant uptake and/or leaching of metals; while metals that are retained in the soil solid phase pose a greatly reduced environmental hazard.

Soils usually exhibit higher concentrations of metals than waters because metals are more likely to accumulate in soil versus being diluted or carried elsewhere in water, and soils are composed of minerals which can naturally contain high concentrations of metals. The

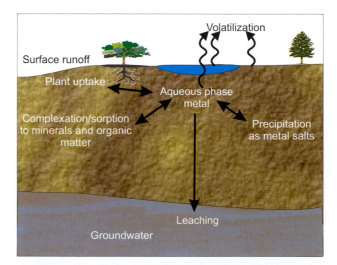

FIGURE 17.2   Potential fates and transformations of metals in the soil environment. The ultimate fate of metals in soil may be dissolved in groundwater. Metal transformations may also occur at the surface of a soil particle or colloid or in the soil solution.

cation exchange capacity of soils allows metals to attach to soil particles in response to ionic attractions and accumulate.

Several abiotic and biotic factors can affect the chemical speciation of metals in soil, and, thus, affect the bioavailability and toxicity of metals to microbial populations. These factors include metal chemistry, sorption to clay minerals and organic matter, pH, redox potential, and the microorganisms present. All of these factors interact to influence metal speciation, bioavailability, and the overall toxicity of metals in the environment. Thus, it must be emphasized that determination of the total concentration of a metal in a soil is not enough to predict toxicity in biological systems. It is the bioavailable amount that is important.

17.4.1 Metal Chemistry

Whether a metal is cationic or anionic in nature determines its fate and bioavailability in an environment. Most metals are cationic which means they exhibit a positive charge when in their free ionic state and are most reactive with negatively charged surfaces. Thus, in soil, cationic metals such as Pb^{2+} or Ca^{2+} strongly interact with the negative charges on clay minerals and with anionic salts, such as phosphates and sulfates. Positively charged metals are additionally attracted to negatively charged functional groups such as hydroxides and thiols on humic residues. Unfortunately, cationic metals are also attracted to negatively charged cell surfaces where they can be taken up and cause toxicity. Cationic metals sorb to both soil particles and cell surfaces with varying strengths, termed adsorption affinity. For example, of the common soil cations, aluminum binds more strongly than calcium or magnesium.

$$Al^{3+} > Ca^{2+} = Mg^{2+} > K^+ > Na^+$$

As this affinity series shows, the size and charge of the cationic metal determines the strength of adsorption. Thus, metal cations compete for the limited number of cation exchange sites present in a soil with the larger multivalent cations replacing smaller monovalent cations such as Na^+. For example, Al^{3+} has such strong affinity for clay surfaces that it is primarily found as $Al(OH)_3$ and therefore has extremely low bioavailability. Since many of the toxic metals are large, divalent cations, they have high adsorption affinities and thus are not readily exchanged. In contrast, negatively charged or anionic metals such as AsO_4^{3-} (arsenate) are attracted to positively charged surfaces. In soils, anions can be sorbed to negatively charged clays by divalent cation bridging, using cations such as Ca^{2+} or Mg^{2+}. Consequently, the fate of a metal in the environment is strongly influenced by the nature and intensity of its charge.

Components in the soil solution also affect metal solubility. Specifically phosphates, sulfates and carbonates in the soil solution form sparingly soluble metal-salt compounds. Polyvalence tends to increase the strength of the interactions between metals and soil components.

17.4.2 Cation Exchange Capacity

One of the most important factors affecting metal bioavailability is the soil cation exchange capacity (CEC), which is dependent on both the organic matter and clay content of the soil (see Chapter 4.2.1). Cation exchange reflects the capacity for a soil to sorb metals. Thus, the toxicity of metals within soils with high CEC (organic and clay soils) is often low even at high total metal concentrations. In contrast, sandy soils with low CEC, and therefore low metal binding capacity, show decreased microbial activity at comparatively low total metal concentrations, indicative of metal toxicity.

Clay minerals provide a wide range of negative adsorption sites for cationic metals. These include planar sites associated with isomorphous replacements, edge sites derived from partly dissociated Si–OH groups at the edges of clay minerals, and interlayer sites located between clay platelets. Permanently charged sites on clay minerals interact with metallic ions by means of non-specific electrostatic forces. Metallic oxides and hydrated metal oxides offer surface sites for the sorption of metals. Iron, aluminum and manganese oxides are an important group of minerals that form colloidal size particles, which, in the presence of water, assume various hydrated forms able to strongly retain most metals in the soil.

Soil organic matter, contains non-humic substances, such as carbohydrates, proteins and nucleic acids, that are normally readily and quickly degraded, as well as less readily degraded substances including lignin, cellulose and hemicellulose. The non-humic organics in soil are relatively short-lived, and thus they have little influence on the long-term fate of metals. In the short-term, however, organic acids readily dissociate ($COOH \rightarrow COO^- + H^+$) and the resultant carboxylate anion can form soluble complexes with metal cations thereby lowering their bioavailability. This effect is pH dependent because at low pH this complexation does not occur. Thus, metal bioavailability can vary considerably in microsite environments depending on pH and the presence of nonhumic substances.

In contrast, humic substances are relatively stable and patchily coat the particle surfaces of natural soils. Humic substances contain a variety of organic functional groups that are able to interact with metals.

These functional groups include carboxyl, carbonyl, phenyl, hydroxyl, amino, imidazole, sulfhydryl and sulfonic groups. Metals complexed with humic substances are generally not bioavailable, and therefore less toxic to biological systems.

17.4.3 Redox Potential

Metal bioavailability changes in response to changing redox conditions. Under oxidizing or aerobic conditions ($+800$ to 0 mV), metals are usually found as soluble cationic forms, e.g., Cu^{2+}, Cd^{2+}, Pb^{2+}, and Ca^{2+}. In contrast, reduced or anaerobic conditions (0 to -400 mV), such as those commonly found in sediments or saturated soils, often result in metal precipitation. For example, in areas rich in sulfur and sulfate-reducing bacteria, the sulfide that is generated is available to form nontoxic, insoluble sulfide deposits, e.g., CuS, and PbS. As another example, for soils rich in carbonates, metals are precipitated as metal carbonates, e.g., $CdCO_3$.

17.4.4 pH

For cationic metals the pH of a system can have an appreciable affect on metal solubility, and, hence, metal bioavailability. At high pH, metals are predominantly found as insoluble metal mineral phosphates and carbonates while at low pH they are more commonly found as free ionic species or as soluble organometals. The pH of a system also affects metal sorption to soil surfaces. The effect of pH on metal sorption is principally the result of changes in the net charge on soil and organic particles. As the pH increases, the electrostatic attraction between a metal and soil constituents is enhanced by increased pH-dependent CEC. In addition as already mentioned, as pH increases, metal solubility decreases. Thus the net effect of increased soil pH is to decrease metal bioavailability. In contrast, as soil pH decreases, metal solubility is increased and pH dependent charge decreases, enhancing metal bioavailability. For anionic metals in a soil with low pH, these metals may become increasingly sorbed as the positive charge on soil particles increases. As the pH increases, the fate of anionic metals is highly dependent on other environmental factors including redox and metal speciation.

The influence of pH, redox, and organic and inorganic minerals on the chemical form and nature of metals demonstrates how important and difficult it is to define clearly metal speciation and bioavailability. As discussed in the following section, metal bioavailability is further complicated by the complexity of microbiological interactions with metals.

17.5 METAL TOXICITY EFFECTS ON THE MICROBIAL CELL

Metals are toxic because, as a result of their strongly ionic nature, they bind to many cellular ligands and displace native essential metals from their normal binding sites (Fig. 17.3). For example, arsenate can replace phosphate in the cell. Metals also disrupt proteins by binding to sulfhydryl groups and nucleic acids by binding to phosphate or hydroxyl groups. As a result protein and DNA conformation are changed and function is disrupted. For example, cadmium competes with cellular zinc and nonspecifically binds to DNA, inducing single-strand breaks. Metals may also affect oxidative phosphorylation and membrane permeability, as seen with vanadate and mercury. Microorganisms generally use specific transport pathways to bring essential metals across the cell membrane into the cytoplasm. Unfortunately, toxic metals can also cross membranes, via diffusion or via pathways designed for other metals. For instance, Cd^{2+} transport occurs via the Mn^{2+} active transport system in *Staphylococcus aureus*. Figure 17.4 illustrates various mechanisms used to transport metals into the cell.

These metal–microbe interactions result in decreased growth, abnormal morphological changes, and inhibition of biochemical processes in individual cells. The toxic effects of metals can been seen on a community level as well. In response to metal toxicity, overall community numbers and diversity decrease. However, few studies have addressed community resistance. While individual microbial populations may be metal resistant, how do microbial populations interact with each other when metals are present. Further, is it possible for metal resistant populations to interact in such a way as to confer resistance on the consortium of organisms? Likewise, are there symbiotic relationships between metal resistant and metal sensitive populations such that the metal sensitive organism receives protection from metal toxicity while providing the metal resistant organism with some essential nutrient or carbon source? The answer to questions like these is the goal of future research in microbial metal resistance.

17.6 MECHANISMS OF MICROBIAL METAL RESISTANCE AND DETOXIFICATION

Some microorganisms are believed to have evolved metal resistance because of their exposure to toxic metals shortly after life began (Information Box 1). Others are believed to have evolved metal resistance in response to recent exposure to metal pollution over the

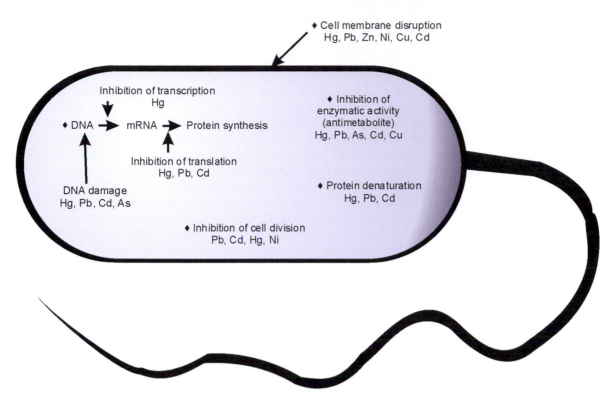

FIGURE 17.3 Summary of the various toxic influences of metals on the microbial cell demonstrating the ubiquity of metal toxicity. Metal toxicity generally inhibits cell division and metabolism. As a result of this ubiquity, microorganisms have to develop "global" mechanisms of resistance that protect the entire cell from metal toxicity.

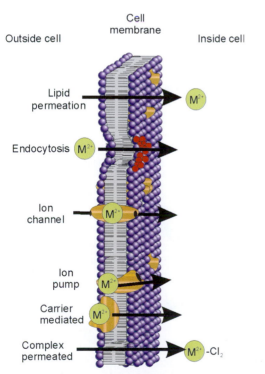

FIGURE 17.4 Mechanisms of metal (M) flux across the microbial cell membrane. (Adapted from Simkiss and Taylor, 1989.)

past 50 years. The evolutionarily recent pollution of the environment with anthropogenic sources of metals has increased the need for research concerning microbial metal resistance as well as remediation. Microorganisms directly influence the fate of metals in the environment and so may provide the key to decreasing already existing contamination.

In response to metals in the environment, microorganisms have evolved ingenious mechanisms of metal resistance and detoxification (Fig. 17.5). Some resistance mechanisms are plasmid encoded and tend to be specific for a particular metal. Others are general conferring resistance to a variety of metals. Microbial metal resistance may be divided into three categories. These include resistance mechanisms that: (1) are general and do not require metal stress; (2) are dependent on a specific metal for activation; and (3) are general and are activated by metal stress.

General mechanisms of metal resistance often serve other functions. For example, slime layer production, while effectively providing a barrier against metal entry into the cell, also serves in surface adhesion and protection against desiccation and predation. The sole purpose of **metal-dependent mechanisms,**

Information Box 1

A question plaguing environmental microbiologists is how do microorganisms become metal resistant? The earth is estimated to be 4.7 Ga (Ga = giga anna, 10^9 years) old and microbial life is thought to have appeared approximately 3.5 Ga ago. Since microorganisms were the first life forms on earth, microorganisms may have developed their resistance in response to the toxic metal gases that existed when the earth was created and early life began. Another possible scenario is that microorganisms have recently developed metal resistance in response to increasing anthropogenic pollution with various metals. Metal concentrations in the atmosphere, in surface soils, and in both surface and groundwaters have increased with industrialization. With an average mutation occurring every 10^5 base pairs, microorganisms have the capability to rapidly respond to metal-contaminated environments and evolve metal resistance. The answer to the origins of microbial metal resistance is probably a combination of both early and recent exposure to toxic metals. A comprehensive study of the physiological and the genetic diversity in microbial metal resistance mechanisms may provide an answer to this intriguing question.

---

both specific and general, is cell protection from metal toxicity. Examples of such mechanisms include efflux pumps and metallothionein production.

17.6.1 General Mechanisms of Metal Resistance

Binding of metals to extracellular materials immobilizes the metal and prevents its entry into the cell.

Metal binding to anionic cell surfaces occurs with a large number of cationic metals, including cadmium, lead, zinc and iron. For example, algal surfaces contain carboxylic, amino, thio, hydroxo and hydroxy–carboxylic groups that strongly bind metals. Phosphoryl groups and phospholipids in bacterial lipopolysaccharides in the outer membrane also strongly interact with cationic metals. Binding of metals by microbial cells is important both ecologically and practically.

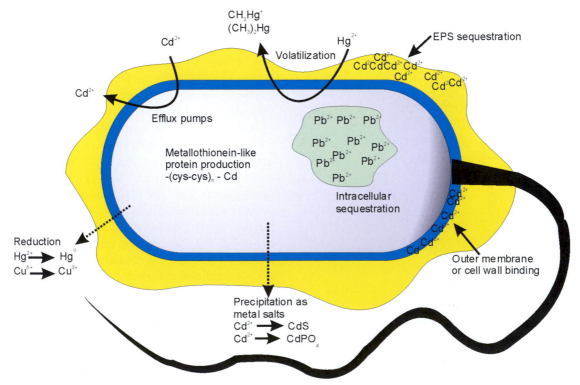

FIGURE 17.5 In response to metal toxicity, many microorganisms have developed unique mechanisms to resist and detoxify harmful metals. These mechanisms of resistance may be intracellular or extracellular and may be specific to a particular metal, or a general mechanism able to interact with a variety of metals.

Ecologically, cell surface metal binding plays a dominant role in the distribution of metals, especially in aquatic environments. In practical terms, the capacity of cells to sorb metals is being used to remove metals from contaminated wastestreams.

Extracellular binding is usually to slime layers or **exopolymers** composed of carbohydrates, polysaccharides, and sometimes nucleic acids and fatty acids. Offering protection against desiccation, phagocytosis, and parasitism, exopolymers or extracellular polymeric substances (EPSs) are common in natural environments. Microbial exopolymers are particularly efficient in binding heavy metals, such as lead, cadmium and uranium. Exopolymer functional groups are generally negatively charged, and consequently, the efficiency of metal:exopolymer binding is pH-dependent. Metal detoxification through EPS production results in metal immobilization and prevention of metal entry into the cell. For example, the immobilization of lead by exopolymers has been demonstrated in several bacterial genera, including *Staphylococcus aureus, Micrococcus luteus,* and *Azotobacter* spp. In fact, extracellular polymeric metal binding is the most common resistance mechanism against lead.

A second extracellular molecule produced microbially that complexes metals is the **siderophore.** Siderophores are iron-complexing, low-molecular-weight organic compounds. Their biological function is to concentrate iron in environments where concentration is low and to facilitate its transport into the cell. Siderophores may interact with other metals that have chemistry similar to that of iron, such as aluminum, gallium, and chromium (which form trivalent ions similar in size to iron). By binding metals, siderophores can reduce metal bioavailability and thereby metal toxicity. For example, siderophore complexation reduces copper toxicity in cyanobacteria.

Biosurfactants are a class of compounds produced by many microbes that in some cases are excreted. Recently biosurfactants have been investigated for their ability to complex metals such as cadmium, lead, and zinc (Fig. 17.6) (Miller, 1995). Biosurfactant complexation can actually increase the apparent solubility of metals, however, the biosurfactant-complexed metal is not toxic to cells. It is not clear why some microorganisms produce biosurfactants but they may play a role in reducing metal toxicity. Recent evidence shows that biosurfactant-producing microorganisms can be isolated in greater diversity from metal-contaminated environments than from uncontaminated ones. More information on biosurfactants is available in Chapter 16.4.2 and 16.7.5.

Finally, metal bioavailability can be influenced by common metabolic by-products that result in metal reduction. In this case, soluble metals are reduced to less soluble metal salts, including sulfidic and phosphidic precipitates. For example, under aerobic conditions *Citrobacter* spp. can enzymatically produce phosphate which results in the precipitation of lead and copper. Under anaerobic conditions, high H_2S concentrations from sulfate reducing bacteria, e.g., *Desulfovibrio* spp., readily cause metal precipitation.

17.6.2 Metal-Dependent Mechanisms of Resistance

Intracellular metal resistance mechanisms in bacteria are not clearly understood. Possibly the best-known mechanism involves metal binding or sequestration by **metallothioneins** or similar proteins. Primarily documented in higher microorganisms, plants, algae, yeast, and some fungi, metallothioneins are low molecular weight, cysteine-rich proteins with a high affinity for cadmium, zinc, copper, silver and mercury metals. Their production is induced by the presence of metals, and their primary function is metal detoxification. Metallothionein-like proteins have been isolated from the cyanobacterium *Synechococcus* spp.,

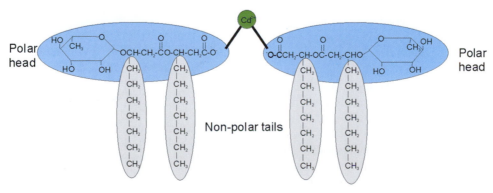

FIGURE 17.6   Chemical structure of rhamnolipid, a microbial surfactant, binding cadmium.

as well as *Escherichia coli,* and *Pseudomonas putida* (Gupta *et al.,* 1993). However, it is not known how widespread the production of metallothionein-like proteins is in bacteria. Metal binding by metallo-thioneins can result in cellular accumulations visible as electron dense areas within the cell matrix. Suspected deposits are confirmed using electron dispersive spectroscopy (discussed below) which can identify the metal.

Some microorganisms use plasmid-encoded energy-dependent **metal efflux** systems to remove metals from the cell. Some efflux systems involve ATPases, and others are chemiosmotic ion/proton pumps. These mechanisms effectively pump toxic ions that have entered the cell back out of the cell via active transport (ATPase pump) or diffusion (chemiosmotic ion/proton pump). Arsenate, chromium and cadmium are the three metals most commonly associated with efflux resistance. Arsenic resistance in a number of bacteria involves the enzymatic reduction of arsenate to arsenite followed by a plasmid-mediated arsenite efflux (Fig. 17.7). Arsenate reduction and subsequent arsenite efflux are common in both Gram-negative and Gram-positive microorganisms. Gram-negative bacteria, such as *Escherichia coli,* use a two-component membrane-bound ATPase complex, ArsA and ArsB, responsible for the active efflux of arsenite from the cell. ArsC is an NADPH-dependent cytoplasmic protein responsible for enzymatically reducing arsenate to arsenite. Gram-positive bacteria lack the ArsA protein (Cervantes *et al.,* 1994).

Another example of efflux-based resistance is cadmium resistance. The toxicity of cadmium is primarily due to its binding to sulfhydryl groups of proteins and single-stranded DNA breaks. Several camium-resistant organisms have been studied, including *Staphylo-coccus aureus, Bacillus subtilis, Listeria* spp., *Alcaligenes eutrophus, Escherichia coli, Pseudomonas putida,* some cyanobacteria, fungi and algae. While Gram-positive bacteria use an ATP-mediated efflux pump for cadmium resistance, Gram-negative bacteria use chemiosmotic antiporter pumps (Fig. 17.8).

The plasmid-borne *cad*A gene encodes a cadmium specific ATPase in several bacterial genera, including *Staphylococcus, Pseudomonas, Bacillus* and *Escherichia.* In *Alcaligenes eutrophus* CH34, the *czc* operon is responsible for cadmium resistance. Another cadmium resistance operon (*ncc*) has recently been identified in *Alcaligenes xylosoxidans* (Schmidt and Schlegel, 1994).

The **methylation** of metals is considered a metal-dependent mechanism of resistance because only certain metals are involved. Methylation generally in-

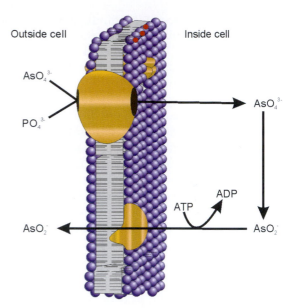

FIGURE 17.7   Schematic of arsenic resistance demonstrating both the influx and efflux systems for arsenic. Arsenate enters the cell via a phosphate specific transport pathway. Once in the cell, arsenate is reduced to arsenite and an efflux mechanism then pumps arsenite out of the cell via an anion pump and an F_1F_0-ATPase. Arsenite efflux is fueled by ATP. Notice that arsenic is not detoxified by this mechanism because arsenite can still be toxic.

creases metal toxicity as a result of increased lipophilicity, thus increased permeation across cell membranes. However, metal volatilization facilitates metal diffusion away from the cell, and in this way effectively decreases metal toxicity. Metal volatilization has been ob-

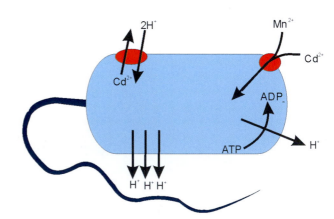

FIGURE 17.8   A proposed model for cadmium influx and efflux in a bacterial cell. Cadmium enters the cell via a manganese transport pathway, which relies on membrane potential. The cadmium efflux system excretes cadmium via a $Cd^{2+}/2H^+$ antiport. (Adapted with permission from Silver, 1983.)

served with lead, mercury, tin, selenium, and arsenic. For example, mercury (Hg^{2+}) is readily oxidized to the volatile and very toxic forms methylmercury and dimethylmercury, which can then diffuse away from the cell. Methylation of metals has been known to remove significant amounts of metal from contaminated surface waters, sewage, and soils.

Mercuric compounds are highly toxic owing to their affinity for thiol groups in proteins and their lipophilic tendencies. Mercury resistance may involve the enzymatic reduction of Hg^{2+} to elemental mercury (Hg^0) in both gram-positive and gram-negative bacteria. Often plasmid mediated, two additional pathways of mercury resistance involve the detoxification of organomercurial compounds via cleavage of C-Hg bonds by an organomercurial lyase (MerB) followed by reduction of Hg^{2+} to Hg^0 by a flavin adenine dinucleotide (FAD)-containing, NADPH-dependent mercuric reductase (MerA). Specific to inorganic mercury, the MerP protein in the periplasmic space shuttles Hg^{2+} to the membrane-bound MerT protein, which releases Hg^{+2} to the cytoplasm. Once in the cytoplasm, Hg^{2+} is reduced to Hg^0 by mercuric reductase (Fig. 17.9).

17.7 METHODS FOR STUDYING METAL–MICROBIAL INTERACTIONS

Unique considerations apply when studying metal–microorganism interactions. Metal concentrations on a macroscale poorly reflect the toxic influences of metals on the microorganismal scale. Recall that total metal concentrations do not accurately assess the biologically toxic concentration. Because metals are not biologically degradable, it is difficult to determine whether and how a metal is being detoxified when the total metal concentration does not change. However, new and exciting technical and analytical developments in metal chemistry, microbiology, and molecular biology are now making it possible to expand our understanding of how microorganisms influence metal fates in the environment. An extensive review of such methodologies is available in Beveridge *et al.* (1997). A few of these approaches are highlighted here.

The **culturing** of metal-resistant microorganisms in the laboratory often occurs in either nutrient-rich or chemically defined media, which may contain yeast

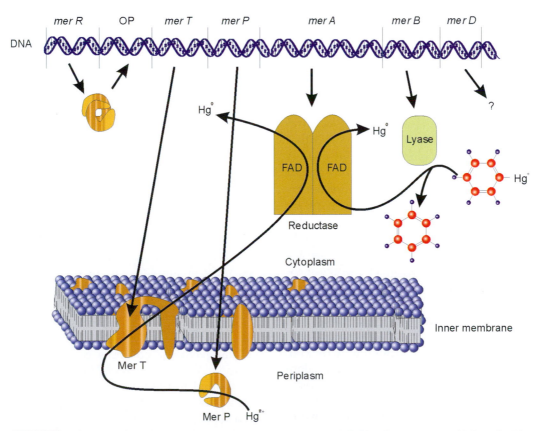

FIGURE 17.9 Proposed model for bacterial mercury resistance encoded by the *mer* operon. (Adapted with permission from Silver *et al.*, 1986.)

extract, phosphate buffers, and amino acids that bind metal ions. Neutral medium pH is an additional factor increasing metal binding in culture media. The presence and amount of these reagents strongly influence metal bioavailability, thereby influencing metal toxicity to microorganisms. Thus, depending on the growth medium, metal toxicity will vary. For example, it has been shown that *Tetrahymena pyriformis* can tolerate a 100-fold higher concentration of copper(II) in a rich organic medium compared with a defined one (Nilsson, 1981). Consequently, several factors need to be taken into consideration when choosing a culture medium to assess microbial metal resistance. Medium nutrients need to be defined and chosen in such a way as to minimize metal binding. This applies to both carbon substrates and to buffers. For example, phosphate buffers strongly precipitate metals. Recall that phosphate production in some microorganisms affords them protection from certain metals. Non-metal-binding buffers, including the sulfonic acids such as MES [2-(*N*-morpholino)ethanesulfonic acid; $C_6H_{13}NO_4SCH_2O$], pKa = 6.15, and PIPES (1,4-piperazinediethanesulfonic acid; $C_8H_{18}N_2O_6S_2$), pKa = 6.80, optimize metal bioavailability in culture media. Finally, pH strongly influences metal bioavailability. Metals readily precipitate as carbonic salts at pH > 7.0. Therefore, medium pH should be kept slightly acidic (~pH 6.0) to maintain metal solubility.

To experimentally determine the relationship between metal bioavailability and toxicity, and to determine the rate and extent that microbes sequester metals, one must be able to determine both total and soluble metal concentrations. Total metal concentration is determined by digestion of the sample with a hot acid such as nitric or perchloric acid. This process dissolves soil particles and releases even tightly bound metals. Soluble metal determination often involves extraction with the weak acid DTPA (diethylenetriamine pentaacetic acid) or extraction with deionized water to release loosely bound, readily exchangeable metals. In either case, metal concentrations in the extract are determined using atomic absorption spectroscopy or inductively coupled plasma atomic emission spectroscopy.

When studying metal resistance, it is assumed that a decrease in the soluble metal concentration corresponds to the amount of metal sequestered by the cell. However, care must be taken in interpreting these data since metals often precipitate with culture medium components and in addition, may bind to the walls of flasks and test tubes. Metal controls with no inoculum are therefore crucial in distinguishing between biological and chemical metal removal. It should be noted

that while it is relatively easy to determine a macro-scale estimate of bioavailability in the environment, such an estimate does not necessarily reflect micro-scale metal concentrations. So even if very low soluble levels of metal are measured, it is likely that in some micropores (where most soil microorganisms live) substantial levels of bioavailable metal may be encountered. This would explain why microorganisms in metal-contaminated environments with no detectable soluble metal may exhibit extreme resistance.

Flame or flameless **atomic absorption spectroscopy (AAS)** and **inductively coupled plasma atomic emission spectroscopy (ICPAES)** are easy, efficient techniques for determination of a metal ion in solution. For AAS, metal is determined by aspirating a metal solution into an air-acetylene flame to atomize the metal. A metal-specific lamp is placed into the AAS and is used to determine the difference in light absorbance between a reference source and the metal solution. This difference reflects the amount of metal present. ICPAES determination is based on light emitted from metal atom electrons in the excited state. An argon plasma is used to produce the excited state atoms. Matrix interference can be a significant problem with these techniques, and some samples need to be acid digested before analysis. Spectroscopy can be used for any metal, however detection limits vary for each metal. For example, for AAS, the limit of detection is 1 μg/l (8.9×10^{-3} μm) for cadmium and 700 μg/l (2.9 μm) for uranium. For comparison the detection limits for ICPAES are 1 μg/l and 75 μg/l (0.32 μm) for cadmium and uranium respectively.

Ion-selective electrodes (ISEs), available for some metals such as cadmium, lead, and arsenic, provide a quick way to determine metal concentrations in solution to approximately 10^{-6} M. The electrodes are easy to use, relatively inexpensive, and not strongly influenced by sample color or turbidity. They are, however, influenced by the presence of other ions in solution. ISEs are different from AAS and ICPAES in that they allow the determination of extracellular free metal ion in solution following cellular interactions. In this case the ISE will only measure free metal ions, and does not measure complexed metal even if the complexed metal is soluble. In some senses this can be considered an advantage since the ISE measures only metal that is truly bioavailable.

Ion-exchange columns can be used to chromatographically separate metal species in a sample. Metals are bound to the ion-exchange resin and then are removed using a metal-complexing solution. The eluant fractions are analyzed by conductivity or AAS or ICPAES spectroscopy. This is a highly sensitive and rapid (several metal species can be identified within minutes) technique that can be automated and sample pre-treatment is often unnecessary. Both gas and liquid chromatographic techniques can also be used to separate mixtures of metal species.

Whereas the methods discussed so far can detect the presence of a metal, **transmission electron microscopy** (TEM) provides an effective means of locating and visualizing suspected metal deposits associated with microorganisms (Fig. 17.10). This technique is particularly useful for determining whether microbes sequester metals inside or outside of the cell. When TEM is coupled with **energy-dispersive x-ray spectroscopy (EDS),** the metal element can be identified. Metals emit characteristic x-rays as the electron beam interacts with deposits. Elements can be identified by signature spectral lines (Fig. 17.11). High- and low-magnification TEM micrographs can help distinguish between intracellular and extracellular metal interactions, depending on the location of metal deposition.

Much of the progress in studying metal–microbial interactions has been made using molecular tools, including cloning, polymerase chain reaction (PCR), and DNA sequencing. For example, PCR techniques, such as enterobacterial repetitive entergenic consensus (ERIC) PCR and RNA arbitrarily primed (RAP) PCR, provide ways to characterize and identify new metal-resistant isolates and the genes responsible for resistance. As another example, DNA primers specific for the *mer* operon can be used to amplify mercury resistance genes (see Chapter 13).

17.8 ADVERSE EFFECTS OF MICROBIAL METAL TRANSFORMATIONS

In geological formations, metals are found in a reduced state as for example, pyrite (FeS_2). Pyrite is often associated with metal ore deposits and as ex-

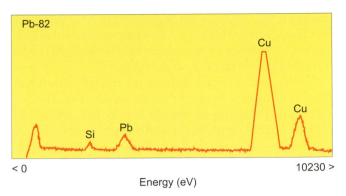

FIGURE 17.11 X-ray analysis of the suspected lead (Pb) deposit in Fig. 17.10. Copper and silica were present from the copper grid and embedding medium, respectively, used in mounting the sample for analysis. (Courtesy T. M. Roane.)

plained in Chapter 15.3.1, when exposed to oxygen, the combination of autoxidation and the microbial oxidation of iron and sulfur results in the production of large amounts of acid. Acid, in turn, facilitates metal solubilization, resulting in a metal-rich leachate called **acid mine drainage.** Contaminating groundwater and over 10,000 miles of rivers in the United States alone, acid mine drainage is highly toxic to plants and animals, often resulting in widespread fish kills. Acid mine drainage is a problem associated with any type of mining activity including subsurface mining, where metal deposits become exposed to atmospheric oxygen; strip mining, where large expanses of land are exposed to oxygen; and mine tailings, where metal deposits are brought to the surface.

Similar in theory to the microbial production of acid mine drainage, the microbially induced **corrosion** of metal pipes is a significant concern (see Chapter 15.2.1). Fuel and storage tanks are likewise susceptible to this biological corrosion. Corrosion can be the result of acid production by bacteria or fungi and/or attack by sulfate-reducing bacteria.

The production of sulfuric acid by *Thiobacillus* spp. and other sulfur oxidizers is one cause of acid corrosion of ferrous metals. For nonferrous metals, the production of organic acids by for example, the fungus *Cladosporium resinae*, is more important. This organism is often the culprit in the corrosion of aluminum fuel tanks in aircraft. In some cases, microbial organic acids destroy protective metal coatings, exposing the metal to attack by other microorganisms. As discussed in Chapter 15.2.1, sulfate-reducing bacteria are primarily responsible for the destruction of buried and water-immersed iron or steel structures and also participate in aerobic corrosion by forming biofilms on the metal surface. Typical of the sulfate-reducing genera are *Desulfovibrio* and *Desulfotomaculum*, which use sulfate as the terminal sulfate-reducing electron acceptor, re-

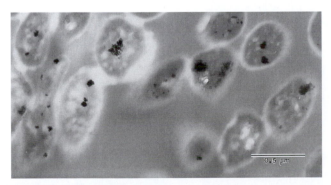

FIGURE 17.10 Transmission electron micrograph of a bacillus exhibiting an intracellular accumulation (dark material) of lead in response to the production of a metallothionein-like protein. (Photo courtesy of T. M. Roane.)

ducing it to sulfide (a black precipitate visible on the metal surface). The precise mechanism for sulfate-reducing corrosion is not clear but is thought to be linked to the production of sulfide.

Under certain circumstances, the microbial transformation of metals results in increased toxicity. The microbial **methylation of metals** results not only in increased metal mobility because organometals are volatile (see Chapter 15.4 for more detail) but also increased toxicity. Since methylation involves the transfer of methyl groups (CH_3) to metals and metalloids, e.g., mercury, arsenic, and selenium, the resulting organometal is more lipophilic than the metal species. This results in the potential of bioaccumulation and biomagnification in food webs (Fig. 17.12).

Mercury is the best studied of the metals that are methylated. It is used extensively in the electrical industry, instrument manufacturing, electrolytic processes, and chemical catalysis. Mercury salts and phenylmercury compounds are also used as fungicides and disinfectants. Approximately 10,000 metric tons of mercury are produced worldwide annually. Fossil fuel burning releases an additional 3000 metric tons. Methylmercury compounds are highly lipophilic and neurotoxic. Several outbreaks of mercury poisoning have occurred throughout history. In Minimata Bay, Japan, release of mercury-containing effluents by a chemical processing plant resulted in serious illness in people who consumed fish with elevated levels of mercury. Another example is the Great Lakes in the United States which until the 1970's had relatively uncontrolled releases of polychlorinated

biphenyls (PCBs), dioxins, and mercury. For a period of time parts of the lakes were closed to fishing. But the problem has improved due to restricted use and release of the organic and metal pollutants. At this time health advisories are in effect that make recommendations about the type and amounts of fish that can be safely consumed.

Arsenic is another example of a metal that is methylated. It is methylated by some fungi such as *Scopulariopsis brevicaulis* to mono-, di-, and trimethylarsenes, highly toxic forms of arsenic. Arsenic poisonings have occurred in the past when fungi growing on damp wallpaper converted and volatilized arsenate (used as a coloring agent) in the wallpaper. Illness occurred upon inhalation of the resulting methylarsines.

17.9 THE BENEFITS OF METAL–MICROBIAL INTERACTIONS

Although microorganisms can sometimes increase metal toxicity, solubility, and mobilization, microorganisms can also be used to reduce metal waste and bioavailability. For example, microbially catalyzed **oxidation** of minerals is used in the commercial recovery of copper, uranium and gold from low-grade (<0.4% metal content) ores by exploiting the same activities that cause formation of acid mine drainage (see Chapter 15.3.2 for more details).

Controlled leaching or metal recovery is done using one of several approaches including: heap leaching, vat leaching or in-place leaching (Fig. 17.13). **Heap**

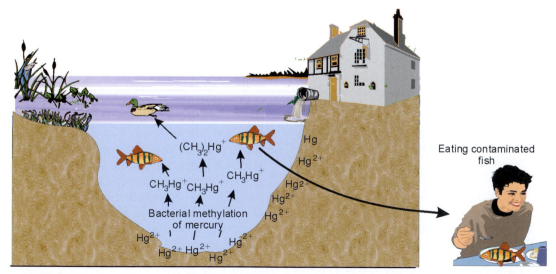

FIGURE 17.12 Schematic demonstrating the potential for the bioaccumulation of mercury as a result of mercury methylation. Once in the food chain, methylmercury poses serious health risks to the human population.

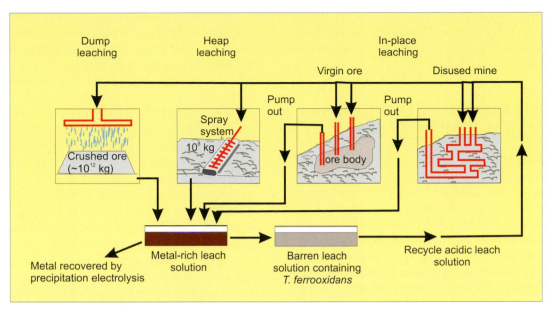

FIGURE 17.13   Various applications of micrometallurgy in the recovery of metals from crushed ores. Metals can be recovered from ores either in place, in heaps, or in dump leaching. In each method, an acidic leach solution containing *Thiobacillus ferrooxidans* is flushed through the ore, leaching out the metal to be recovered via precipitation or electrolysis.

leaching usually involves excavating a long narrow ditch about 30 cm in depth and lining it with polyethylene. Air lines are evenly placed along the ditch to provide aeration. The ditch is filled with crushed ore to a height of 2 to 3 m and then saturated with acidified water (pH ranging from 2 to 3) containing a bacterial inoculum of *Thiobacillus ferrooxidans*. The leachate is collected at the bottom and recycled through the heap. Like heap leaching, **dump leaching** occurs on an impermeable surface. However, there is no artificial aeration so that set up costs are lower. In this case, efficiency of extraction is considerably decreased.

Vat leaching involves the continuous stirring of an ore slurry in large tanks. Again, the ore is mixed with acidified water and *T. ferrooxidans*. The advantage of vat leaching over heap leaching is the ability to control the leaching conditions efficiently (discussed further later). **In-place leaching** is carried out underground in spent mines. Up to 39% of the metal can remain in these mines. The mine is flooded with the acid–bacterial solution, resulting in an acid-rich, metal-rich leachate several months later from which metal is extracted and recovered. There are several advantages to in-place leaching. In-place leaching has lower associated energy costs and has a less negative environmental impact when the production and collection of leachate is properly controlled. The in-place leaching of uranium has received considerable attention as it involves less risk of human exposure, and thus better control of radiation exposure. Uranium leaching relies

on the bacterially facilitated oxidation of insoluble UO_2 to the acid-soluble U^{6+} (Eq. 17.1). *T. ferrooxidans* reoxidizes Fe^{2+} to Fe^{3+} facilitating the uranium oxidation using the indirect method of leaching discussed in Chapter 15.3.2 (Eq. 17.2).

$$UO_2 + 2Fe^{3+} \rightarrow 6\ UO_2^{2+} + 2Fe^{2+} \qquad \text{(Eq. 17.1)}$$

$$4Fe^{2+} + O_2 + 4H^+ \rightarrow 6\ 4Fe^{3+} + 2H_2O \qquad \text{(Eq. 17.2)}$$

Each of these approaches produces a metal-containing leachate. A number of methods can be used to recover the metal from the leachate, including solvent extraction, increasing the pH to precipitate the metals, and electrolysis, in which an electrical current is passed through the metal solution to accumulate cationic metals at the anode and anionic metals at the cathode. In the case of gold recovery, cyanidation is used, where free gold (Au^{2+}) is precipitated as a cyanogold(I) species (Eq. 17.3).

$$2Au + 4CN^- + O_2 \rightarrow 2Au(CN)_2^- + O_2^{2-} \qquad \text{(Eq. 17.3)}$$

While uncontrolled leaching and chelation of metals can result in metal-contaminated environments, in some cases increased metal solubility can be beneficial, e.g., when essential nutrients are solubilized and delivered to plants. Metals such as zinc, copper, and manganese are necessary for proper plant growth. Yet, like the toxic metals, essential metals may be present in a nonbioavailable state so that plants cannot readily absorb them. Microbial activity through acid production or the presence of microbially produced chelators can increase the solubility, and hence the bioavailability, of these metals. This can be particularly important in the rhizosphere (see Chapter 18).

17.10 PHYSICAL/CHEMICAL METHODS OF METAL REMEDIATION

The remedial methods used to treat contaminated soil or sediments may be broadly divided into two main categories: (1) methods aimed at preventing movement of metals to the immediate surroundings also called immobilization; and (2) methods aimed at physically removing or destroying the contamination. Removal of metals from soils and sediments is difficult because metals cannot be degraded. Metals have to be physically removed from soils, either by excavation or by chemical-washing of the soil. Alternatively, reducing the solubility of metals in soils serves to reduce the mobility of the metal, thus reduce contaminant spreading. While metal removal by excavation may be appropriate when the area of contamination is small or there is immediate risk to human health, increasing cost and shrinking landfill space emphasize the need for cheaper, environmentally friendly alternatives.

The goal in **metal immobilization** is to reduce metal solubility. Two immobilization strategies include pH alteration and addition of organic matter. Since metal solubility decreases with increasing pH, metal solubility should be reduced when site pH is raised. Liming is sometimes used to increase soil pH causing precipitation of contaminating metals as calcic and phosphoric metal-containing minerals. Amendment with organic matter can also aid in metal immobilization as a result of the electrostatic attraction between metals and organic particles. The addition of organic matter may involve the addition of highly organic waste material, such as sludge. Often sites containing high levels of toxic metals have little or no vegetation. Revegetation of such sites, while sometimes difficult to achieve, is a good way to increase organic matter content.

Metals can be **physically removed** from soils or sediments via excavation or by using soil washing techniques. Soil washing methods rely on chemicals to facilitate metal removal. Washing with acidic solutions or chelating agents, e.g., ethylenediaminetetraacetic acid (EDTA) or nitrilotriacetic acid (NTA), solubilizes metals, enhancing removal from the system. For example, 60% of the lead in a soil containing 10,000 mg lead/kg was reportedly removed by washing with 0.1 M EDTA (Peters and Shem, 1992). One problem with the use of these chemical agents is the residual

toxicity left by the washing agent after treatment. Researchers are looking at biological alternatives to these chemicals. For example, a rhamnolipid biosurfactant that complexes cadmium was able to remove >80% of cadmium from an experimentally contaminated soil.

Physical **excavation** is another approach to use in dealing with metal-contaminated soils. Excavation is simply a removal technique and neither destroys nor detoxifies contaminating metals. Contaminated soils can then be stored in a hazardous waste containment facility or incinerated. In some situations, however, excavation can exacerbate the problem. For example, excavation of sediments, called **dredging,** can actually result in increased metal toxicity. Metal sediments are often anaerobic and the metals within the sediment exist in an immobile, reduced state. Exposure to oxidizing conditions results in metal oxidation and increased metal solubility, increasing both bioavailability and transport. There is current discussion about whether physically removing metal-containing sediments is more detrimental than leaving them in place.

Incineration of soils can be used to remove metals from soils. However, incineration not only is expensive and impractical for large volumes of soil but also releases metals to the atmosphere only to be deposited elsewhere. Unfortunately, such thermal treatment of soil also destroys important soil properties, destroying soil structure and soil biota.

The nonbiological remediation of **aquatic systems,** including surface water, groundwater, and wastewater, is fairly straightforward, albeit costly. Metals are removed and concentrated from contaminated waters through flocculation, complexation, and/or precipitation. Lime addition precipitates metals as metal hydroxides. Chelating agents complex metals and can be recovered with a change in pH. Electroreclamation methods include ion exchange, reverse osmosis, and electrochemical recovery of metals.

17.11 INNOVATIVE MICROBIAL APPROACHES IN THE REMEDIATION OF METAL-CONTAMINATED SOILS AND SEDIMENTS

The goals of microbial remediation of metal-contaminated soils and sediments are to (1) immobilize the metal *in situ* to reduce metal bioavailability and mobility or (2) remove the metal from the soil (Fig. 17.14). A problem with metal immobilization (both biological and nonbiological) as a remediation technology is that it is difficult to predict whether the metals will remain immobilized indefinitely. On the other hand,

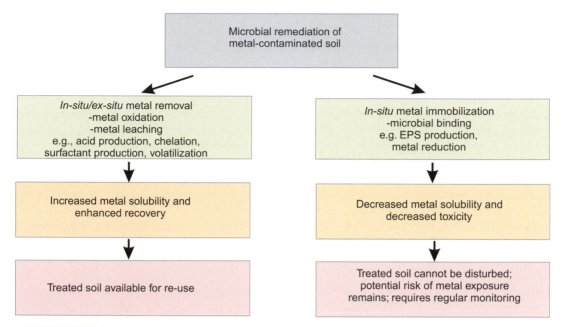

FIGURE 17.14 Microbial metal remediation in metal-contaminated soils relies on either metal removal or, more commonly, metal immobilization. Metal removal is ideal because following treatment the soil is available for reuse. In metal immobilization, soil reuse is limited because of the continued potential risk of exposure.

metal removal is difficult due to the heterogeneous nature of soil and is expensive, and so may not be practical for some metal-contaminated sites. However such microbial approaches are of interest because physical/chemical methods for metal removal including incineration and excavation are not practical or cost-effective when large volumes of soil are involved. There are several proposed methods for microbial remediation of metal contaminated soils including microbial leaching, microbial surfactants, microbially induced metal volatilization, and microbial immobilization and complexation.

Certain microorganisms, such as *T. ferrooxidans,* can facilitate the removal of metals from soil through **metal solubilization** or **leaching.** Generally used in the recovery of economically valuable metals from ores, bioleaching is used to recover copper, lead, zinc, and uranium from tailings. Metals recovered by leaching can be concentrated by complexation with chelating agents or precipitation with lime. Bioleaching also has potential in the removal of metals from contaminated soils and metal-containing sludges. Unfortunately, this aspect of microbial leaching has received little attention.

Microorganisms can also increase metal solubility for recovery through the production of **surfactants.** Because of their small size, biosurfactants are a potentially powerful tool in metal remediation. Bacterial surfactants are water-soluble, low-molecular-weight molecules (<1500) that can move freely through soil pores. In addition to their small size, biosurfactants have a high affinity for metals so that, once complexed, contaminating metals can be removed from the soil by soil flushing. Some surfactants, such as the rhamnolipid produced by *Pseudomonas aeruginosa,* show specificity for certain metals, such as cadmium and lead (Torrens *et al.,* 1998). Biosurfactant specificity allows the optimization of a particular metal. Related to biosurfactants, the higher molecular weight ($\sim 10^6$) bioemulsifiers such as emulsan, produced by *Acinetobacter calcoaceticus,* can also aid in metal removal. Metal removal by bioemulsifiers has been demonstrated, and removal efficiencies range between 12 to 91% (Chen *et al.,* 1995).

Like leaching, **volatilization** of metals, specifically their methylation, increases metal bioavailability and hence toxicity. Methylated metals are more lipophilic than their nonmethylated counterparts. In spite of the increased toxicity, many microorganisms volatilize metals to facilitate their removal from the immediate environment. Because methylation enhances metal removal, methylation of certain metals has been used as a remediation strategy. The most famous example is the removal of selenium from contaminated soil in San Joaquin, California (see Case Study) by selenium-volatilizing microorganisms. Mercury is another metal commonly methylated by microorganisms. However, mercury is susceptible to bioaccumulation in the food chain, posing serious health risks to the human popu-

CASE STUDY

Selenium Bioremediation in San Joaquin Valley, California

Selenium is known to bioaccumulate and can cause death and deformities in waterfowl. Agriculture is the primary cause of selenium contamination in the San Joaquin Valley, but other anthropogenic sources of selenium include petroleum refining, mining, and fossil fuel combustion. Once in soil, selenium exists as selenate (SeO_4^{2-}, Se^{6+}), selenite (SeO_3^{2-} Se^{4+}), elemental selenium ($Se°$), dimethylselenide [DMSe; $(CH_3)_2Se$], and/or dimethyldiselenide [DMDSe; $(CH_3)_2Se_2$]. The most toxic forms of selenium are selenate and selenite; elemental selenium is considered insoluble and least toxic of the selenium ions. In this case, the methylated forms of selenium are 500–700 times less toxic than inorganic forms.

The nonbiological treatment of selenium-contaminated soils includes soil washing or leaching; however, because of a large, low-solubility pool of selenium in soils, these are not effective treatments. Excavation is another option, but there are extensive costs associated with soil excavation, making it impractical for the large contaminated areas.

As a result of agricultural practices, largely irrigation, soils found in the San Joaquin Valley generally have elevated selenium levels (400 to 1000 mgSe/kg soil). Drainage waters in the valley can contain up to 4200 mg Se/l. Consequently, in some areas, selenium has concentrated to hazardous levels in evaporation ponds and in soils. The Kesterson Reservoir, located within the valley, is one such area and has resulted in extensive birdkills in the past.

The bioremediation of the San Joaquin Valley and specifically the Kesterson Reservoir is based on the ability of a large number of microorganisms to reduce selenium oxyanions to the insoluble, elemental selenium or to the volatile, methylated forms, e.g., DMSe. Under anaerobic conditions, organisms such as *Wolinella succinogenes* and *Desulfovibrio desulfuricans* can reduce SeO_4^{2-} and SeO_3^{2-} to elemental selenium. However, reduction to insoluble forms of selenium is not sufficient to stabilize the selenium pool within the soil matrix. Both microbial reoxidation and resolubilization are facilitated by organisms including *Thiobacillus ferrooxidans* and *Bacillus megaterium*. Consequently, long-term remediation of contaminated soils requires selenium removal, hence the role of selenium-volatilizing microorganisms becomes important.

Laboratory experiments initially conducted with Kesterson sediments found that certain environmental conditions influence microbial selenium volatilization. Researchers found that increased soil moisture (-33 kPa), soil mixing, increased temperature ($35°C$), and application of an organic amendment increased selenium volatilization. In greenhouse experiments, sediment samples containing 60.7 mg Se/kg sediment) were treated with various carbon sources to enhance microbial selenium methylation (Karlson and Frankenberger 1990). The highest amount of selenium volatilization was seen upon the addition of citrus peel (~44%), compared with manure (19.5%), pectin (16.4%), and straw plus nitrogen (8.8%). Without amendments, selenium volatilization was 6.1%.

On the basis of promising laboratory and greenhouse results, field plots (3.7×3.7 m²) were set up at the Kesterson Reservoir. Plots were treated with different carbon amendments, including manure, gluten, citrus, and casein, in an attempt to stimulate microbial activity and selenium methylation. With periodic tilling and irrigation, approximately 68–88% of the total amount of selenium was removed from the top 15 cm of soil within 100 months (Flury *et al.*, 1997). The highest rates of selenium removal were in soils amended with casein.

The remediation of selenium-contaminated soils is a successful example of how metal- transforming microorganisms can be used to detoxify and remove metals from affected systems. Microbial selenium remediation is a prime example of how laboratory experimentation has led to a viable approach to remediating selenium-contaminated environments. The effective microbial remediation of other metals will need development of strategies similar to those used in this case study.

lation and therefore removal of mercury by volatilization would not be an acceptable approach.

The amount of soil involved and the type of metal present limit the use of the preceding technologies in remediating metal-contaminated soils. Not all metals are readily volatilized, and large tracts of contaminated soil may not be amenable to surfactant application. **Metal sequestration** provides an alternative approach. Metal sequestration relies on the ability of some microorganisms to produce metal-complexing polymers (both extracellular and intracellular). Recall that exopolymers have high affinities for various metals. The overall approach in microbial metal sequestration is to introduce the polymer-producing microorganism into the contaminated soil and allow the organism to grow and replicate, thereby increasing the amount of polymer present in the soil and increasing the number of organisms producing the polymer. Microbial metal sequestration has been shown to be effective in laboratory studies but has yet to be proven effective in the field.

17.12 INNOVATIVE MICROBIAL APPROACHES IN THE REMEDIATION OF METAL-CONTAMINATED AQUATIC SYSTEMS

Microbially facilitated removal of metals from water is based on the ability of microorganisms to complex and precipitate metals, resulting in both detoxification and removal from the water column. Specific interactions for metal removal include metal binding to microbial cell surfaces and exopolymer layers, intracellular uptake, metal volatilization, and metal precipitation via microbially facilitated metal redox reactions (Fig. 17.15). Although these microbial mechanisms can effectively remove metals from contaminated aquatic systems, it is important to note that the metals are not destroyed and still have to be properly disposed of. The use of wetlands and microbial biofilms are readily available treatments for metal-polluted waters.

Wetland treatment is a cost-effective and efficient method for removal of metals from contaminated waters, such as acid mine drainage. Metal reductions are often greater than 90% (Wildeman *et al.*, 1994). Wetland remediation is based on microbial adsorption of metals, metal bioaccumulation, bacterial metal oxidation, and sulfate reduction. The high organic matter content of wetlands provided by high plant and algal growth encourages both the growth of sulfate-reducing microorganisms and metal sorption to the organic material. Although these various processes together contribute to the removal of toxic metals from the water column, the metals are not destroyed. Consequently, wetlands are constantly monitored for any environmental change that may adversely affect metal removal. For example, a decrease in pH may solubilize precipitated metals, or a disturbance of the wetland sediment may change the redox conditions and oxidize reduced metals. Wetlands are resilient systems,

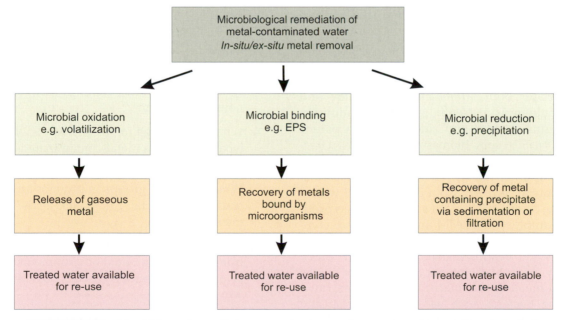

FIGURE 17.15 Microbial metal remediation approaches for metal-contaminated waters. In each method, the treated water is safe to release into the environment. Both metals and microorganisms can easily be recovered during treatment for proper disposal.

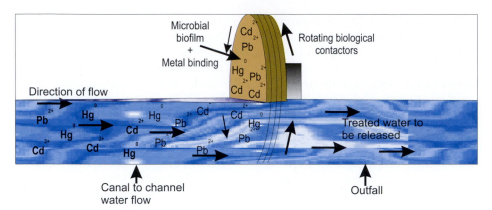

FIGURE 17.16 Schematic demonstrating how microbial biofilms are used in removing metals from contaminated wastestreams. The biofilm located on the rotating drum accumulates metals as the water passes through the drum. The treated water can be safely released. The biofilm may either be viable or nonviable. When viable, the biofilm rarely needs to be replaced; however, non-living biofilms need to be replaced periodically for their metal removal efficiency will decrease with time.

and as long as new vegetative growth and organic inputs occur, wetlands can effectively remove metals for an indefinite period of time.

The most common treatment for metal-contaminated waters is with **microbial biofilms.** As previously discussed, many microorganisms, including *Bacillus, Citrobacter, Arthrobacter, Streptomyces,* and the yeasts *Saccharomyces* and *Candida,* produce exopolymers as part of their growth regime. Metals have high affinities for these anionic exopolymers. Microbial biofilms may be viable or nonviable when used in remediation. In general, the biofilm is immobilized on a support as contaminated water is passed through the support (Fig. 17.16). Often, a mixture of biofilm-producing organisms grow on these supports, providing a constant supply of fresh biofilm. For example, live *Citrobacter* spp. biofilms are used to remove uranium from contaminated water. Both *Arthrobacter* spp. biofilms and biomass (nonliving) are used in recovery of cadmium, lead, chromium, copper, and zinc from wastewaters. Nonliving *Bacillus* spp. biomass preparations effectively bind cadmium, chromium, copper, mercury, and nickel, among other metals. The success of microbial biomass in metal recovery from contaminated waters has led to the commercial sale of several biomass products. For example, BIOCLAIM and BIO-FIX are commercially available immobilized, nonliving microbial preparations for treating metal-contaminated water. Interestingly, microbial biofilms are also used in the treatment of metal-contaminated marine waters; however, marine bacteria such as *Deleya venustas* and *Moraxella* sp. are used. Microbial biofilms are likewise used in the removal of metals from domestic wastewater. In domestic waste treatment, the important biofilm-producing organisms include *Zoogloea, Kleb-*

siella, and *Pseudomonas* spp. Complexed metals are removed from the wastewater via sedimentation before release from the sewage treatment plant.

QUESTIONS AND PROBLEMS

1. Address the structural differences between prokaryotic and eukaryotic cells and how these differences influence cell resistance or sensitivity to metal toxicity.

2. In a metal contaminated lake, discuss metal bioavailability throughout the water column including the sediment.

3. Is dredging a viable option for removal of metal contamination in an aquatic system? What are the potential problems?

4. What factors need to be considered when bioaugmenting a metal-contaminated site with a metal-resistant microorganism?

5. Which chemical groups in proteins are most reactive with metals? In nucleic acids? In membranes?

6. Metal-resistant microorganisms are often isolated from non-contaminated environments (with no prior metal exposure). Discuss possible reasons why.

7. A metal-contaminated soil has been remediated using a metal-complexing microorganism. What factors need to be considered to ensure that the metal does not become "re-available"?

8. Summarize the possible mechanisms of metal resistance in microorganisms and discuss which mechanism would be most effective in remediating a metal-contaminated surface soil, metal-contaminated soil in the vadose zone, a metal-contaminated stream, metal-contaminated groundwater?

9. Discuss the fate (both chemical and biological) of lead in (a) an acid soil, pH 4.0; (b) a neutral soil; (c) a basic soil, pH 8.5; (d) an anaerobic soil; and (e) an aerobic/anaerobic soil interface.

References and Recommended Readings

Adams, M. W. W. (1993) Enzymes and proteins from organisms that grow near and above 100°C. *Annu. Rev. Microbiol.* **47**, 627–658.

Anderson, G. L., Williams, J., and Hille, R. (1992) The purification and characterization of arsenite oxidase from *Alcaligenes faecalis*, a molybdenum-containing hydroxylase. *J. Biol. Chem.* **267**, 23674–23682.

Avery, S. V., Codd, G. A., and Gadd, G. M. (1991) Caesium accumulation and interactions with other monovalent cations in the cyanobacterium *Synechocystis* PCC 6893. *J. Gen. Microbiol.* **137**, 405–413.

Beveridge, T. J., Hughes, M. N., Lee, H., Leung, K. T., Poole, R. K., Savvaidis, I., Silver, S., and Trevors, J. T. (1997) Metal–microbe interactions: Contemporary approaches. *Adv. Microbiol. Phys.* **38**, 177–243.

Bidwell, J. P., and Spotte, S. (1985) "Artificial Seawaters: Formulas and Methods" Jones and Barlett Publishers, Inc., Boston, MA.

Cervantes, C., Ji, G., Ramirez, J. L., and Silver, S. (1994) Resistance to arsenic compounds in microorganisms. *FEMS Microbiol. Rev.* **15**, 355–367.

Chen, J.-H., Lion, L. W., Ghiorse, W. C., and Shuler, M. L. (1995) Mobilization of adsorbed cadmium and lead in aquifer material by bacterial extracellular polymers. *Water Res.* **29**, 421–430.

Debus, K. H. (1990) Mining with microbes. *Technol. Rev.* **93**, 50–57.

Flury, M., Frankenberger, W. T., Jr., and Jury, W. A. (1997) Long-term depletion of selenium from Kesterson dewatered sediments. *Sci. Total Environ.* **198**, 259–270.

Forgeron, F. D. (1971) Soil geochemistry in the Canadian Shield. *Can. Min. Metall.* **64**, 37–42.

Gadd, G. M. (1988) Accumulation of metals by microorganisms and algae. *In* "Biotechnology—A Comprehensive Treatise. Special Microbial Processes" (H. J. Rehm, ed.) VCH Verlagsgesellschaft, Weinheim, pp. 401–433.

Goldman, C. R., and Horne, A. J. (1983) "Limnology" McGraw-Hill, New York.

Gupta, A., Morby, A. P., Turner, J. S., Whitton, B. A., and Robinson, N. J. (1993). Deletion within the metallothionein locus of cadmium-tolerant *Synechococcus* PCC 6301 involving a highly iterated palindrome (HIP1). *Mol. Microbiol.* **7**, 189–195.

Hughes, M. N., and Poole, R. K. (1989) "Metals and Micro-Organisms." Chapman & Hill, New York.

Karlson, U., and Frankenberger, W. T., Jr. (1990) Volatilization of selenium from agricultural evaporation pond sediments. *Sci. Total Environ.* **92**, 41–54.

Keely, J. F., Hutchinson, F. I., Sholley, M. G., and Wai, C. M. (1976) Heavy metal pollution in the Coeur d'Alene mining district. Project Technical Report to National Science Foundation, University of Idaho, Moscow.

Leppard, G. G. (1981) "Trace Element Speciation in Surface Waters" Plenum Press, New York.

Lindsay, W. L. (1979) "Chemical Equilibria in Soils" John Wiley and Sons Inc., New York.

Miller, R. M. (1995) Biosurfactant-facilitated remediation of metal-contaminated soils. *Environ. Health Perspec.* **103**, 59–62.

Nilsson, J. R. (1981) Effects of copper on phagocytosis in *Tetrahymena*. *Protoplasma* **109**, 359–370.

Peters, R. W., and Shem, L. (1992) Use of chelating agents for remediation of heavy metal contaminated soil. *In* "Environmental Remediation." American Chemical Society, Washington, DC, pp. 70–84.

Schmidt, T., and Schlegel, H. G. (1994) Combined nickel-cobalt-cadmium resistance encoded by the *ncc* locus of *Alcaligenes xylosoxidans* 31A. *J. Bacteriol.* **176**, 7045–7054.

Sigg, L. (1985) Metal transfer mechanisms in lakes; the role of settling particles. *In* "Chemical Processes in Lakes" (W. Stumm, ed.) John Wiley, New York.

Silver, S. (1983) *In* "Biomineralization and Biological Metal Accumulation. Biological and Geological Perspectives" (P. Westbroek, and E. W. de Jong, eds.) Reidel, Dordrecht, pp. 439–457.

Silver, S., Rosen, B. P., and Misra, T. K. (1986) *In* "Fifth International Symposium on the Genetics of Industrial Micro-Organisms" (M. Alacevic, D. Hranueli, and Z. Toman, eds.), pp. 357-376.

Simkiss, K., and Taylor, M. G. (1989) Metal fluxes across the membranes of aquatic organisms. *Rev. Aquat. Sci.* **1**, 173.

Sposito, G. (1989) "The Chemistry of Soils." Oxford Press, New York.

Stone, E. L., and Timmer, V. R. (1975) Copper content of some northern conifers. *Can. J. Bot.* **53**, 1453–1456.

Torrens, J. L., Herman, D. C., and Miller-Maier, R. M. 1998. Biosurfactant (rhamnolipid) sorption and the impact on rhamnolipid-facilitated removal of cadmium from various soils. *Environ. Sci. Technol.* **32**, 776–781.

Warren, H. V., Delevault, R. E., and Barakso, J. (1966) Some observations on the geochemistry of mercury as applied to prospecting. *Econ. Geol. Ser. Can.* **61**, 1010–1028.

Wildeman, T. R., Updrgraff, D. M., Reynolds, J. S., and Bolis, J. L. (1994) Passive bioremediation of metals from water using reactors or constructed wetlands. *In* "Emerging Technology for Bioremediation of Metals" (J. L. Means and R. E. Hinchee, eds.) Lewis, London, pp. 13–25.

18

Beneficial and Pathogenic Microbes in Agriculture

IAN L. PEPPER

18.1 OVERVIEW

18.1.1 The Soil–Plant–Microorganism System

As we have seen, most normal soils do not contain abundant microbial nutrients, because microbial communities utilize any nutrients that are available (see Chapter 4). In contrast, the **rhizosphere** is a unique soil environment found in close proximity to plant roots, where nutrients are more abundant because of the influence of the plant itself. Increased nutrient availability in turn results in enhanced microbial activity and numbers. Thus the rhizosphere exists because of **soil–plant–microorganism interactions** (Fig. 18.1). Ultimately microbial gene expression in the rhizosphere is controlled by these interactions, which in turn are influenced by direct or indirect environmental factors. Overall, the microbial populations within the soil–plant–microorganism system can affect plant growth in beneficial or detrimental ways (Fig. 18.2). In this chapter we will examine the role of these microbes in the agricultural arena.

18.1.2 The Rhizosphere Environment

The term rhizosphere was coined by Hiltner in 1904 to describe the part of the soil that is influenced by plant roots. Originally the rhizosphere was thought to extend 2 mm outward from the root surface. Now it is recognized that the rhizosphere can extend 5 mm or more as a series of gradients of organic substrate, pH, O_2, CO_2, and H_2O (Fig. 18.3). Essentially two regions of the rhi-

zosphere are now recognized: (1) the rhizosphere soil and (2) the soil in direct contact with the plant root, which is the **rhizoplane.** Microorganisms also inhabit the root itself and are known as **endophytes.** The portion of the root occupied by microbes was formerly known as the **endorhizosphere** but this term is no longer used by soil microbiologists. Finally, note that the rhizosphere effect occurs almost as soon as a seed is planted, with the area of increased microbial activity around a seed bring known as the **spermosphere.**

The rhizosphere effect is caused by the release of organic and inorganic compounds from the plant roots. In particular, the rhizosphere is influenced by living root border cells that are released by the root (see Section 18.3.4.3). Because of these releases and because of the influence of the plant roots themselves, rhizosphere soil is thought to be quite different from non-rhizosphere or bulk soil. However, despite hundreds of different studies, very little can actually be said with certainty about rhizosphere soil. Part of the problem lies in the methods used to sample rhizosphere soil. Despite even the most sophisticated of analyses performed on "rhizosphere soil," most studies are restricted due to the historically crude method of obtaining such a sample. Typically, this has involved extracting a plant from soil and shaking the roots until most of the soil particles fall off. To this day, this is still the method of choice. Other problems include the subtle interactions between specific plants and specific soils in specific environments. Thus, an infinite array of different "rhizosphere environments" is possible. This is evidenced by a perusal of rhizosphere litera-

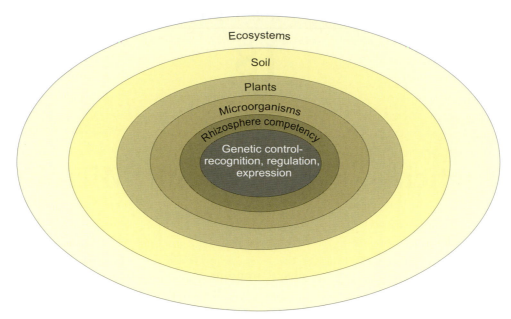

FIGURE 18.1   The soil–plant–microorganism system. Here, the influence of the environment on both soil and plant controls gene expression in the rhizosphere.

ture, which tends to be confusing and inconsistent. This is not an indictment of the scientists who have worked on the rhizosphere, rather it illustrates the difficulties of studying the complex rhizosphere ecosystem. As an example, rhizosphere soil can have pH values one unit higher or lower than those of the bulk soil, depending on nitrogen nutrition and other factors. It is, however, known that rhizosphere soil tends to be drier than bulk soil because of plant transpiration, and it also contains greater concentrations of organics because of plant-released compounds.

18.1.3 Organic Compounds Released by Plants

In natural vegetation systems, plant roots are in intimate contact with soil particles. Soil exists as a discontinuous environment with a matrix of organic and inorganic constituents combined in diverse conditions and is therefore a unique environment for many microorganisms. These organisms include viruses, bacteria, actinomycetes, fungi, algae, protozoa, and nematodes. Aerobic and anaerobic microsites exist in close proximity, allowing organic and inorganic substrates to be metabolized by organisms with different modes of nutrition. These conditions permit billions of organisms to coexist in soil. Populations vary with the soil, environment, and method of analysis, but reasonable values for "normal" soils are shown in Table 18.1.

Roots are therefore surrounded by organisms and exist as part of the soil–plant–microorganism system, which can be termed the rhizosphere. The complexity of the rhizosphere is shown in Fig. 18.1, where the two inner circles depict, respectively, the early events necessary for colonization. In addition, the figure shows subsequent factors that contribute to rhizosphere competence and the ability to metabolize and reproduce in the rhizosphere in the presence of other organisms. The components of this system (microorganisms, plant, and soil) interact with each other, which distinguishes the rhizosphere from the bulk soil. The activity of root microorganisms is affected by soil environmental factors or by environmental factors operating indirectly through the plant. Root microorganisms can affect the plant and plant nutrient uptake, directly by colonizing the root and modifying its structure or indirectly by modifying the soil environment around the root. The spokes of the wheel in Fig. 18.2 depict not only the significant processes that affect plant growth but also those that have potential for enhancement through improved cultural practices, genetic manipulation, and modeling. Substrates released from roots have many origins and were originally classified by Rovira *et al.* (1979) as

1. **Exudates**—compounds of low molecular weight that leak nonmetabolically from intact plant cells
2. **Secretions**—compounds metabolically released from active plant cells
3. **Lysates**—compounds released by the autolysis of older cells

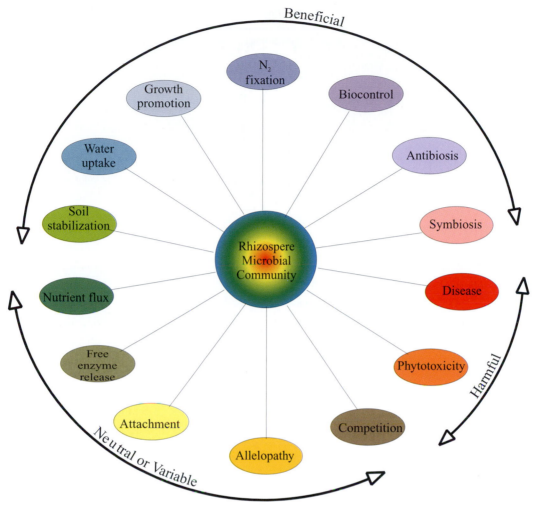

The beneficial, harmful, and neutral or variable effects of the
rhizosphere microbial community on plant growth.

FIGURE 18.2 Potential influences of the rhizosphere microbial community on plant growth. (Reprinted
with permission from *Principles and Applications of Soil Microbiology* by Sylvia, D., *et al.,* © 1991, Prentice-Hall,
Inc., Upper Saddle River, NJ.)

4. **Plant mucilages**—polysaccharides from the root
cap, root cap cells, primary cell wall, and other cells
5. **Mucigel**—gelatinous material of plant and
microbial origin

More recently, it has been demonstrated that living
root border cells affect the rhizosphere ecology more
than any other plant source of carbon substrate (see
Section 18.1.3.2). The terms "exudates" and "exudation" were sometimes used collectively and perhaps in-
correctly to include all of the organic compounds released from roots.

8.1.3.1 Exudates, Secretions, and Lysates

The release of soluble organic compounds (loosely
known as root exudates) is also responsible for some
of the rhizosphere effect. Loss of substrates from roots

can change the pH, the structure of rhizosphere soil,
the availability of inorganic nutrients, and can induce
toxic or stimulatory effects on soil microorganisms
(Hale *et al.,* 1978). The major mechanisms are leakage
and secretion. **Leakage** involves simple diffusion of
compounds because of the higher concentrations of
compounds within the root as compared with the soil.
Secretion can occur against concentration gradients
but requires the expenditure of metabolic energy.
Polysaccharides in particular are susceptible to secretion. Almost any plant metabolite has the potential to
be exuded including carbohydrates, amino acids, organic acids and lipids, growth factors, enzymes, and
miscellaneous compounds. Of these, the carbohydrates and the amino acids, which also represent a
source of nitrogen, are particularly important as substrates. Organic acids and lipids reduce the pH of

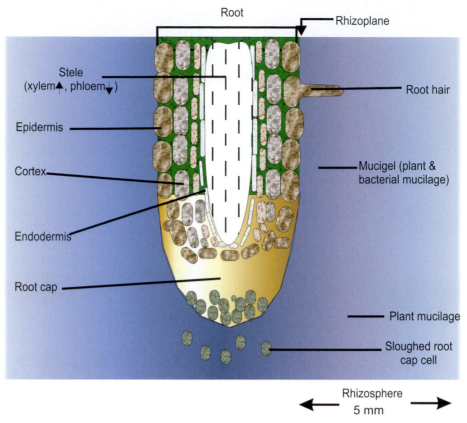

FIGURE 18.3 Structure of a root and corresponding rhizosphere.

the rhizosphere and also have a role in the chelation of metals (Curl and Truelove, 1985). Growth factors, including vitamins, and enzymes stimulate microbial activity and low growth of organisms with complex heterotrophic requirements. Miscellaneous compounds including volatiles can physiologically stimulate or inhibit organisms. When viewed collectively, it is apparent that the rhizosphere is a unique ecosystem in soil that provides a constant supply of substrate and growth factors for organisms.

TABLE 18.1   Numbers of Microorganisms in the Rhizosphere (R) of Wheat (*Triticum aestivum*) and Nonrhizosphere Soil (S) and Their Resultant *R/S* Ratio

Microorganisms	Rhizosphere	Nonrhizosphere	R/S ratio
	CFU[a] g soil		
Bacteria	120×10^7	5×10^7	24.0
Fungi	12×10^5	1×10^5	12.0
Protozoa	2.4×10^3	1×10^3	2.4
Ammonifiers	500×10^6	4×10^6	125.0
Denitrifiers	1260×10^5	1×10^5	1260.0

From Rouatt *et al.* (1960).
[a] CFU, colony-forming units.

18.1.3.2 Root Border Cells and Mucigel

As the root cap extends through soil, viable root border cells and some nonviable material (sloughed cells) are released into the soil. The role of border cells in controlling the rhizosphere ecosystem is discussed later in this chapter (see Section 18.3.4.3). The amount of sloughed material can be considerable. In solution culture, peanut plants released 0.15% of the plant's carbon, nitrogen, and hydrogen per week (Griffin *et al.*, 1976). One would predict that much more material would be lost in soil because of its abrasive nature. From the root tip to the root hair zone, the root is frequently covered with a layer composed of sloughed root border cells and polysaccharides of plant and microbial origin, which is termed mucigel (Miki *et al.*, 1980). These plant products are excellent substrates for microbial growth, in particular soil bacteria, which are extremely competitive at metabolizing simple sugars. Thus mucigel is in intimate contact with bacteria that consume the material, as well as bacteria that contribute bacterial polysaccharides to the mucigel. The amount of mucigel on a particular root depends on the net production and consumption of the material, so that in some instances parts of the root may have no mucigel. Mucigel may protect the root rip from injury and desiccation as well as play a role in nutrient up-take through its pH-dependent cation-exchange capacity (COO^- groups).

18.1.3.3 Factors Affecting the Release of Compounds

Major factors affecting release of organic compounds include plant species and cultivar, age and stage of plant development, light intensity and temperature, soil factors, plant nutrient, plant injury, and soil microorganisms (Pepper and Bezdicek, 1990). Because so many factors affect the release of compounds, generalizations are difficult, including the actual rate of exudation (Kennedy, 1997). However, it is known that plant genes control the release of root border cells (Hawes *et al.*, 1996).

18.1.4 Rhizosphere Populations

Rhizosphere populations are influenced by many plant, soil, and environmental factors. Crop plant roots tend to have greater rhizosphere populations than tree roots (Dangerfield *et al.*, 1978). Different cultivars of the same plant species may have different rhizosphere populations.

Soils directly affect the growth and vigor of plants and therefore influence shoot growth, photosynthesis, and the amount of exudation into the rhizosphere. The concentration of oxygen in the rhizosphere is usually lower than in nonrhizosphere soil as a result of its utilization by large rhizosphere populations. Hence, in heavy-textured soils oxygen may become limiting, resulting in reduced rhizosphere populations compared with coarser-textured soils.

The physical environment around the plant and its roots also affects rhizosphere populations by affecting the amount of organic material released into the soil. Factors such as light, moisture, and temperature can all cause changes in plant metabolism and the rhizosphere effect. In summary, rhizosphere populations are dependent on many diverse interacting factors, and care must be taken when interpreting different studies.

Overall, a vast number of different kinds of microorganisms are found in the rhizosphere, and their numbers generally decrease from the rhizoplane outward toward bulk soil. The rhizosphere effect is often evaluated in terms of **R/S ratios,** where R = the number of microbes in the rhizosphere and S = the number of similar microbes in bulk soil. Thus the greater the R/S ratio, the more pronounced the rhizosphere effect (Table 18.1).

18.1.4.1 Microflora

Bacteria including actinomycetes are the most numerous inhabitants of the rhizosphere, and R/S ratios

can typically be 20:1 (Table 18.1). Pseudomonads and other gram-negative bacteria are especially competitive in the rhizosphere. Typical actinomycete R/S ratios are 10:1 (Rouatt *et al.*, 1960). Overall R/S ratios are useful in delineating the rhizosphere effect, but they are only estimates and vary with different crop plants and different soil environments. The mechanisms that allow rhizosphere competence are discussed later (see Section 18.3.4.2).

Fungal plate counts are generally less than bacterial counts and in any case are often biased toward spore-forming species. However, fungal inhabitants of the rhizosphere are prevalent and can be extremely important because they can be beneficial, as in the case of mycorrhizal fungi, or harmful when they are pathogenic to plants.

18.1.4.2 Microfauna

Most research on the microfauna has centered on protozoa, nematodes, and the microarthropods. Soil protozoa are mostly rhizopods and flagellates, with smaller numbers of ciliates. Protozoan populations tend to mimic bacterial populations, because bacteria are their major food supply. Thus the rhizosphere should contain large populations of protozoa. Rouatt *et al.* (1960) reported R/S ratios for protozoa of 2:1 in wheat rhizospheres (Table 18.1). Darbyshire (1966) reported even higher protozoan populations in ryegrass rhizospheres. Many soil nematodes, including *Heterodera* and *Fylenchus*, are plant parasites that feed on underground roots. Little research has been conducted on nematodes in the rhizosphere, but populations have been reported to be higher in rhizosphere than in nonrhizosphere soil (Henderson and Katznelson, 1961). The Acari (mites) and Collembola (springtails) are important members of soil microarthropods. Mites are predatory on nematodes, and the rhizosphere would be expected to be a favorable habitat for the Acari, but studies in the rhizosphere have been limited in scope. Springtails have been shown to be abundant in cotton rhizospheres with R/S ratios of 4:1 in a sandy loam soil (Wiggins *et al.*, 1979), but the reasons for their attraction to roots are not clear.

18.2 BENEFICIAL ROOT–MICROBIAL INTERACTIONS

The fact that there are many beneficial root–microbial interactions can easily be demonstrated by growing plants in the laboratory in sterilized soil and comparing plant growth with that achieved in non-sterilized soil. Inevitably, growth in the nonsterilized system is superior to that in the sterile system. There are many ways that microorganisms can beneficially influence plant growth, but two of the predominant mechanisms involve the macroelements nitrogen and phosphorus. The prokaryotic bacteria can enhance plant nitrogen uptake through the process of biological nitrogen fixation, whereas eukaryotic fungi enhance plant phosphorus uptake through mycorrhizal associations. Because of the importance of these two processes in agriculture, each will now be discussed in detail.

18.2.1 Biological Dinitrogen Fixation

Nitrogen is critical for plant growth and is often applied to plants as organic or inorganic fertilizers. However, organic forms of nitrogen can be converted to nitrate by the microbial processes of ammonification and nitrification and subsequently denitrified to an inorganic form, nitrogen gas or nitrous oxide (see Chapter 14.3). Clearly, some microbial process that converts nitrogen gas back to ammonia must be present, or all nitrogen would ultimately end up as nitrogen gas. This process, **biological dinitrogen fixation,** is mediated only by prokaryotes, including bacteria, cyanobacteria, and the actinomycete *Frankia.* These nitrogen-fixing organisms can exist as independent free-living organisms or as part of complex interactions with other microbes, plants, and animals. Organisms that can utilize atmospheric nitrogen gas as their sole source of nitrogen for growth are known as **diazotrophs.** In terms of benefits to agriculture, the following major systems can be delineated:

1. SYMBIOTIC RELATIONSHIPS

Symbiont	Host
Rhizobia (bacterium)	Legumes
Frankia (actinomycete)	Nonlegume
Anabaena (cyanobacterium)	Azolla (fern)

2. ASSOCIATIVE SYMBIOTIC RELATIONSHIPS INVOLVING FREE-LIVING DIAZOTROPHS

Microbe	Benefitting crop
Acetobacter	Sugarcane
Azotobacter	Tropical grasses

The preceding list is by no means inclusive of all of the possible associations between microbes and plants, but in terms of importance to agriculture, these are the major players. They are also the systems that have been most easily manipulated by human activity, including that of environmental microbiologists. Also note that there are about 100 true diazotrophs that can exist free living and that can contribute fixed nitro-

gen into the rhizosphere and other environments. Each of these nitrogen-fixing associations will be examined, with emphasis on the rhizobia–legume symbiosis, because it is the best studied system and is critical to agricultural crop production. First, however, we will examine the enzymatic process of biological nitrogen fixation, because many characteristics are common to all nitrogen-fixing associations.

18.2.1.1 The Process of Nitrogen Fixation

Nitrogen gas is a triple-bonded molecule that unites two nitrogen atoms. As such it is a very stable molecule that requires a large amount of energy (226 kcal per mole) to break bands and initiate nitrogen fixation. The overall reaction can be written as follows:

$$N_2 + 3H_2 \rightarrow 2NH_3$$

Energy for this process arises from the oxidation of carbon sources in the case of heterotrophs or from light in the case of photosynthetic diazotrophs. Central to biological nitrogen fixation is the enzyme complex **nitrogenase.** Initially it was thought that only one type of nitrogenase existed, but now it is clear that at least three different enzyme complexes are involved in different nitrogen systems. The classical nitrogenase complex was described by Evans and Burris (1992) and is illustrated in Fig. 18.4.

The overall nitrogenase complex consists of two protein components, which in turn consist of multiple subunits (Fig. 18.4). The iron protein termed **dinitrogenase reductase** is thought to function in the reduction of the molybdenum–iron protein **dinitrogenase,** which reduces nitrogen gas to ammonia. In the 1980s, it was shown that some nitrogenase enzymes did not contain molybdenum and that vanadium and perhaps other metals could substitute for molybdenum (Bishop and Premakumar, 1992).

A schematic of the nitrogen fixation process is shown in Fig. 18.5. Two Mg ATPs are required for each electron transferred from dinitrogenase reductase to dinitrogenase. Thus, under optimal conditions at least 16 molecules of ATP are required, illustrating the energy-intensive nature of the reaction. In practice perhaps 30 ATPs are needed because the overall process is not 100% efficient in the environment (Burris and Roberts, 1993). Initially the dinitrogenase reductase accepts electrons from a low-redox donor such as reduced ferredoxin (Fd_{red}) and binds two Mg ATPs. The dinitrogenase reductase and dinitrogenase form a complex during which an electron is transferred and the two Mg ATPs are hydrolyzed to Mg ADP + inorganic phosphate (P_i). The two proteins then dissociate

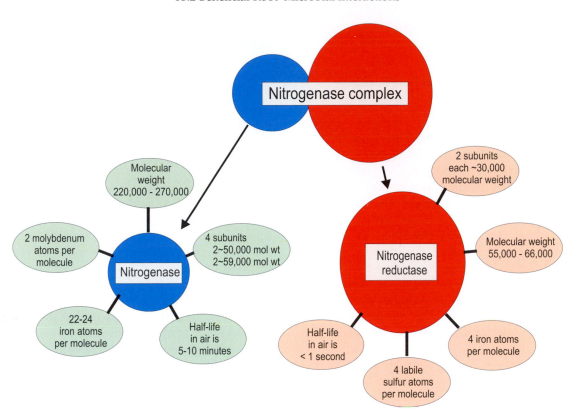

FIGURE 18.4 Characteristics of dinitrogenase (the Mo–Fe protein) and dinitrogenase reductase (the Fe protein).

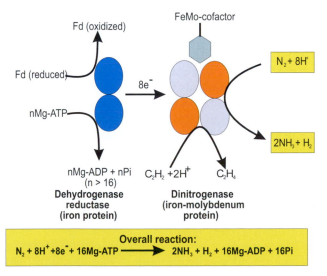

FIGURE 18.5 The nitrogen fixation process. (Adapted from Sylvia *et al.*, 1997.)

and the process repeats. After the dinitrogenase protein has collected sufficient electrons, it binds a molecule of nitrogen gas and reduces it, producing ammonia and hydrogen gas. Thus, during reduction of one N_2 molecule, the two proteins must complex and then dissociate a total of eight times. This is the rate-limiting step of the process and takes considerable time. In fact, it takes 1.25 seconds for a molecule of enzyme to reduce one molecule of N_2 (Zubener, 1997). This is why nitrogen-fixing bacteria require a great deal of the nitrogenase enzyme, which can constitute 10–40% of the bacterial cell's proteins (Postgate, 1994).

Figure 18.5 also shows that one H_2 molecule is released for each N_2 reduced to $2NH_3$. Thus, of the 16 theoretical ATPs required, 4 are spent on the formation of H_2 (25%). Some diazotrophs contain an uptake hydrogenase that reoxidizes some of the H_2 and regains some of the 25% of energy lost during nitrogen fixation (Zuberer, 1997).

An important point to note is that the nitrogenase enzyme is inactivated by molecular oxygen because of the sensitivity of dinitrogenase reductase. Thus, many organisms have developed unique strategies to protect the nitrogenase enzyme. Finally, note that the enzyme can reduce many substrates including H^+, N_2, N_2O, cyanide, carbon monoxide, and acetylene. The reduction of acetylene to ethylene and subsequent measurement of the ethylene led to the development of the **acetylene reduction assay** for nitrogenase activity, which has been used extensively as an indicator of nitrogen fixation (Burris, 1974).

18.2.2 Free-Living Dinitrogen Fixation

Nitrogen fixation occurs in a diverse array of prokaryotic organisms including bacteria, actinomycetes, and cyanobacteria. These organisms include heterotrophic and autotrophic organisms that utilize different terminal electron acceptors for generation of energy. Table 18.2 identifies typical members of the various types of nitrogen-fixing groups. In general, the amount of nitrogen fixed by these free-living diazotrophs is small, in the neighborhood of 2–25 kg per hectare per year. This is due in part to the energy-intensive nature of the conversion, which is about 15 mg of nitrogen fixed per gram of carbon metabolized. Oxygen is an additional problem for many of the nitrogen fixers. The extent of the problem is maximal for strict aerobes, but even anaerobic microbes need to keep the nitrogenase enzyme free of oxygen. This can be done by a variety of mechanisms, including existence in the microaerophilic conditions found in the rhizosphere or production of extracellular polysaccharides to reduce free oxygen diffusion into the cell. Other organisms use high rates of respiration to reduce oxygen concentrations or undergo conformational changes of the nitrogenase enzyme when oxygen is present. During this conformational change the enzyme is protected but cannot fix nitrogen. Finally, some of the cyanobacteria produce thick-walled cells known as **heterocysts.** Fixation is also limited by the presence of free available nitrogen as nitrate or ammonium. Although the amounts of nitrogen fixed are small on an individual basis, almost all environments show some fixation. At a global level, the amount fixed is significant at about 50 Tg (10^{12} g) annually (Paul and Clark, 1989).

18.2.3 Associative Dinitrogen Fixation

Some of the free-living diazotrophs have developed the ability to form associations with plants in which the organisms are established on or in plant cells, where they have available carbon reserves supplied to them by the plant. In return they fix nitrogen, which is taken up by the plant. The casual or **associative symbioses** do not appear to require genetic interactions between the plant and the microbe, and no morphological modifications occur to either partner. Examples of associative symbioses include tropical grasses such as *Paspalum notatum* with *Azotobacter paspali,* in which the microbe exists within the grass root rhizosphere. *Azospirillum* spp. are also found associated with a diverse range of plant hosts including sugarcane, rye, and sorghum. Besides colonizing the rhizosphere, some diazotrophs occupy the outer root cell layers or even internal root tissues. Presumably, the more intimate the association, the more exudates are available for the microbe. *Acetobacter diazotrophicus* has been shown to fix nitrogen from within the internal root cells of sugarcane (Kennedy, 1997).

Attempts have been made to enhance crop production by inoculation with free-living diazotrophs such as *Azospirillum* or *Azotobacter,* but results have been inconsistent and many reports anecdotal. This may be due in part to the generally low rates of nitrogen that have been shown to be fixed within associative symbioses (Table 18.3). These have usually been reported as up to 20 kg per hectare per year. In addition, inoculant bacteria can have additional beneficial or detrimental effects on plants by mechanisms other than nitrogen fixation (see Section 18.3.4.2). The reasons for the low amounts of nitrogen fixed are likely to be similar to those limiting free-living fixation, namely high energy requirements, low oxygen requirement, and the inhibitory effect of nitrogen applied as fertilizer.

Overall, associative symbioses are unlikely to be significant for major crop production but may be highly significant for range and prairie grasses and a few tropical crops.

One example of an important association is that between tropical sugarcane and *A. diazotrophicus.* This organism exists inside the root cells and appears to be reasonably protected against oxygen inhibition of fixa-

TABLE 18.2　Representative Genera of Free-Living Nitrogen Fixers

Status with respect to oxygen	Mode of energy generation	Genus
Aerobe	Heterotrophic	*Azotobacter* *Beijerinckia* *Acetobacter* *Pseudomonas*
Facultative anaerobe	Heterotrophic	*Klebsiella* *Bacillus*
Microaerophile	Heterotrophic	*Xanthobacter* *Azospirillum*
Strict anaerobe	Autotrophic	*Thiobacillus*
	Heterotrophic	*Clostridium* *Desulfovibrio*
Aerobe	Phototrophic (cyanobacteria)	*Anabaena* *Nostoc*
Facultative anaerobe	Phototrophic (bacteria)	*Rhodospirillum*
Strict anaerobe	Phototrophic (bacteria)	*Chlorobium* *Chromatium*

TABLE 18.3 Estimated Average Rates of Biological N₂ Fixation for Specific Organisms and Associations

Organism or system	Dinitrogen fixed (kg/ha/yr)
Free-living microorganisms	
Cyanobacteria ("blue–green algae")	25
Azotobacter	0.3
Clostridium pasteurianum	0.1–0.5
Grass–bacteria associative symbioses	5–25
Plant–cyanobacterial associations	
Gunnera	12–21
Azolla	313
Lichens	39–84
Legumes	
Soybeans (*Glycine* max L. Merr.)	57–97
Cowpeas (*Vigna, Lespedeza, Phaseolus,* and others)	84
Clover (*Trifolium hybridum* L.)	104–160
Alfalfa (*Medicago sativa* L.)	128–300
Lupines (*Lupinus* sp.)	150–169
Nodulated nonlegumes	
Alnus (alders, e.g., red and black alders)	40–300
Hippophae (sea buckthorn)	2–179
Ceanothus (snow brush, New Jersey tea, California lilac)	60
Coriaria ("tutu" in New Zealand)	60–150
Casuarina (Australian pine)	58

tion. Some sugarcane cultivars have been estimated to fix 100 to 150 kg N per hectare per year (Zuberer, 1997). The other significant example is rice production, which can be enhanced by free-living fixation as well as the associative symbiosis between the aquatic fern *Azolla* and the cyanobacterium *Anabaena*. In the flooded conditions required for rice production, algal growth is promoted while oxygen levels around the root systems are reduced. In this situation rice may gain up to 50 kg of fixed nitrogen per hectare per year from combined fixation activities.

18.2.4 The Legume–Rhizobia Symbioses

In contrast to the free-living nitrogen-fixing organisms, the legume-rhizobia association involves a formal symbiosis in which both partners benefit. Here, gram-negative heterotrophic bacteria originally classified within the genus *Rhizobium* interact with leguminous plants causing profound physiological changes in both organisms. These bacteria are known colloquially as rhizobia and are characterized as fixing nitrogen for a plant host in return for carbon sources supplied by the plant as photosynthates. The symbiosis occurs within newly formed root organs called root nodules that develop in response to the presence of specific soilborne

rhizobia (Fig. 18.6). The rhizobia themselves undergo physiological changes, are known as **bacteroids,** and actually conduct the process of nitrogen fixation. As the plant host matures, ultimately the root nodules lyse and rhizobia are released back into the soil.

Many legumes such as peas, beans, and alfalfa are important agricultural crops, and many of them can be grown commercially with reduced inputs of fertilizer nitrogen because of the potential for symbiotically fixed nitrogen. Free-living nitrogen fixation associations often result in about 25 kg per hectare per year, but symbiotic fixation can be much more dramatic. Grain legumes such as peas, beans, and soybeans can fix about 50% of their total nitrogen requirements, with rates of fixation up to 100 kg per hectare per crop (Table 18.3). The remaining nitrogen required by the plant must be supplied from soil sources. Forage legumes such as alfalfa or clover can fix a greater fraction of their total nitrogen requirements, and fixation rates can be as high as 200 to 300 kg per hectare per year. Clearly, then, symbiotic nitrogen fixation has an economic impact globally, and environmental microbiologists have learned to maximize the efficiency of fixation (see Section 18.2.4.2).

Originally all rhizobia were classified within the genus *Rhizobium*, and species were identified on the basis of the legume host that each *Rhizobium* sp. nodulated. However, it became evident that some rhizobia could nodulate more than one host and different classification schemes evolved. The original genus *Rhizobium* is now divided into four genera and 16 species as illustrated in Information Box 1.

18.2.4.1 Root Nodule Initiation and Development

The formation of root nodules on the plant host root system is the result of subtle interactions between the host and the rhizobial endosymbiont. Here we describe the genetic exchanges that initiate nodule formation followed by the physiological manifestations of these interactions. Overall, rhizobia infect the growing nodules and ultimately inhabit the nodules as nitrogen-fixing bacteroids. The success of the symbiosis is due to both plant and bacterial genes that are turned on sequentially during nodule initiation and development. These interactions are highly specific and only specific genetic combinations lead to a successful symbiosis in which **effective nodules** result that can fix nitrogen. Other combinations can result in no nodule formation or nodules in which no nitrogen fixation occurs. Nodules that form but fix no nitrogen are termed **ineffective nodules.**

It has been shown that two different groups of microbial genes are necessary for infection. For *Rhizobium*

Specific nodule initiation

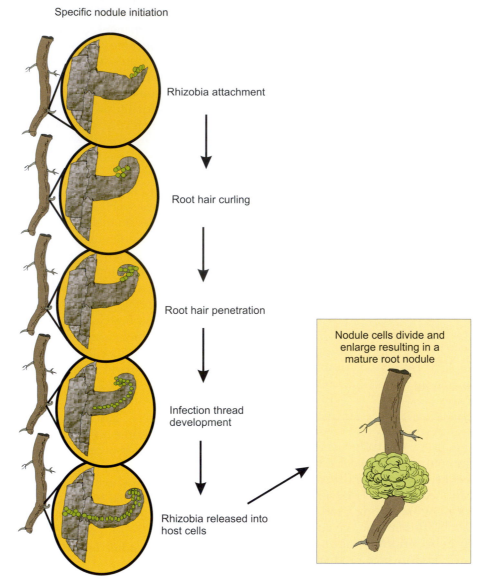

FIGURE 18.6 Nodule initiation and development.

spp., most of these genes are plasmid borne, whereas for *Bradyrhizobium* spp., *Azorhizobium* spp., and *R. loti* they are chromosomal. The **common nodulation genes** (*nodABC*) are found in all rhizobia. A fourth gene (*nodD*) is also common to all rhizobia but is the only *nod* gene expressed in the absence of a suitable host. Currently it is believed that the product of *nodD* interacts with the appropriate plant host and initiates nodule formation (Fig. 18.7). Specifically, **flavenoids,** which are complex phenolic compounds, are secreted by the plant host into the rhizosphere. If the correct rhizobial symbiont is present, the flavenoids interact with the *nodD* product and cause expression of the *nodABC* genes.

Host-specific nodulation genes differ depending on the type of rhizobia, and more than 50 genes have been defined in different rhizobia (Mergaert *et al.,* 1997). The roles of these different genes are now being researched vigorously as more information on the infection process becomes available. In essence, it is the expression of the common nodulation genes and the host-specific nodulation genes that controls successful nodule formation. Most of these nodulation genes are expressed from the initial encounter with the plant host until the release of rhizobia into the plant cells. Transcription is controlled by inducers from the plant, in particular flavenoids. One of the major host specificity determinants is the production of **lipo-**

Information Box 1
Genera and Species of the Root-Nodule Bacteria of Legumes

Genera in the square brackets refer to host legumes nodulated by each species of root-nodule bacteria. Common names are included for well-known legume genera. In several examples in this list, different species of root-nodule bacteria nodulate the same legume.*

Rhizobium[†]

      R. leguminosarum (three biovars: trifolii [*Trifolium*, clovers], viciae [*Pisum*, peas; *Vicia*, field beans; *Lathyrus*; and *Lens*, lentil], and phaseoli [*Phaseolus*, bean])

      R. loti [*Lotus*, trefoil]

      R. tropici [*Phaseolus*, bean; *Leucaena*, Ipil-Ipil, and *Macroptilium*]

      R. etli [*Phaseolus*]

      R. galegae [*Galega*, *Leucaena*]

      R. huakuii [*Astragalus*, milkvetch]

      R. ciceri [*Cicer*, chickpea]

      R. mediterraneum [*Cicer*, chickpea]

Sinorhizobium

      S. meliloti [*Melilotus*, sweetclover; *Medicago*, alfalfa; and *Trigonella*, fenugreek]

      S. fredii [*Glycine*, soybean]

      S. saheli [*Sesbania*]

      S. teranga [*Sesbania*, *Acacia*, wattle]

Bradyrhizobium

      B. japonicum [*Glycine*, soybean]

      B. elkanii [*Glycine*]

      B. liaoningense [*Glycine*]

Azorhizobium

      A. caulinodans [*Sesbania*]

Adapted from Sylvia *et al.* (1997).

* Other genus and species names exist in the literature. Some predate the present names. Others (e.g., *Photobacterium*) have not been accepted as valid.

[†] Strains of *Rhizobium* and *Bradyrhizobium* that do not belong in any named species are usually identified by the host from which they were isolated, e.g., *Rhizobium* spp. (*Acacia*) or *Bradyrhizobium* spp. (*Lupinus*).

---

chitooligosaccharide (LCO) molecules, which are also known as **Nod factors.** Nod factors from at least 13 rhizobial species have been characterized and their structures published (Dénarié *et al.*, 1996). All of these molecules have a similar basic structure composed of a chitooligosaccharide, which is a linear chain of β-1,4-linked *N*-acetylglucosamines, linked to an acyl chain (Fig. 18.8). Most rhizobial strains do not produce a single Nod factor but rather a population of factors with different combinations of features that may interact cooperatively to induce a specific nodulation response (Minami *et al.*, 1996). Originally it was thought that the host range of a specific strain of rhizobia was controlled by the number and diversity of the Nod factors, but it has been shown that even rhizobia with a narrow host range also produce mixtures of Nod factors (Mergaert *et al.*, 1997). However, even though total numbers of Nod factors are not correlated with the host range of a specific strain, it is clear that the Nod factors are the major determinants of host specificity, and that they are generated by transcription and translation of the *nodABC* genes plus host-specific genes. The common *nod* genes allow synthesis of the basic LCO structure, whereas the host-specific genes code for LCO-modifying transferases and enzymes for synthesis of precursors used by the transferases. Originally it was thought that the sequences of the common *nod* genes were functionally conserved, but it has been shown that different alleles of the *nodA* and *nodC* genes are present in different rhizobia and may also contribute to host range determination by specifying different degrees of polymerization of the LCOs (Kamst *et al.*, 1997).

The depth and range of information now available on the plant–microbe interactions involved in the legume–rhizobia symbiosis are remarkable and one of the triumphs of environmental microbiology in the

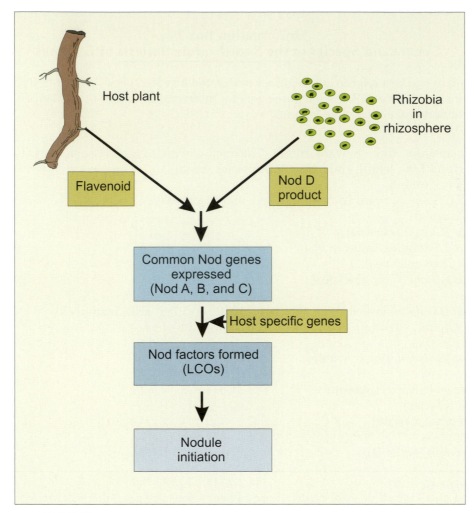

FIGURE 18.7   Plant–rhizobia genetic interactions that initiate nodule formation.

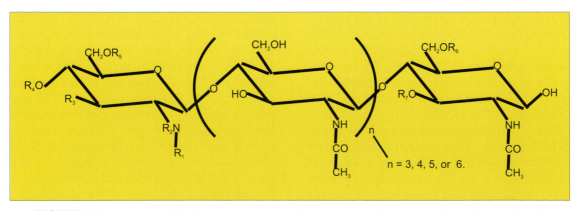

FIGURE 18.8   Overview of Nod factor structures. A chitooligosaccharide is shown. R_1 is a fatty acyl chain found in all Nod factors. R_2–R_7 are various functional groups that modify the basic structure. Nod factors of different rhizobia carry various combinations of these modifications and have different degrees of polymerization, resulting in a large diversity of structures.

molecular age. It is also apparent that it is an ongoing story, with more details still to be revealed. The end result of these genetic interactions is that nodule formation occurs (Fig. 18.6). Attachment of rhizobia to a root hair occurs very rapidly once plant and microbe are in close proximity. Rhizobia attach within minutes and after 4–5 hours, root hair curling begins. At this point, plant root cells are hydrolyzed, allowing rhizobia to enter the root hair. The rhizobia then travel along an **infection thread** manufactured by the plant, which separates the rhizobia from the rest of the root hair. As the infection thread penetrates the root cortex, the rhizobia are released, but remain enclosed by a plant-derived **peribacteroid membrane,** which again segregates the rhizobia from the plant host (Fig. 18.9).

18.2.4.2 Nodule Function and Fixation

As rhizobia are released from the infection thread, cell division occurs, and a visible nodule begins to be seen (1 to 2 weeks after infection). The size and shape of the nodules are in part determined by the plant host. **Determinate nodules** do not have continuous meri-

stematic growth and are usually round in shape, as typified by soybean and common green bean root nodules (Fig. 18.10). In contrast, indeterminate nodules have continuous meristematic growth over the plant growing season. This results in multi-lobed nodules, as evidenced by nodules associated with peas and clovers.

Within the root nodule, the rhizobia enlarge and elongate to perhaps five times the normal size of rhizobia and change physiologically to forms known as **bacteroids.** The nodule also contains leghemoglobin, which protects the nitrogenase enzyme within the bacteroids from the presence of oxygen. The leghemoglobin imparts a pink color to the interior of the nodule and is indicative of active nitrogen fixation. Thus, examination of the interior of a nodule allows instant determination of whether the nodule is active. Prior to the end of the growing season, nodules begin to break down or senesce, at which point they appear white, green, or brown.

In **indeterminate nodules,** nitrogen is fixed as ammonia and ultimately exported to the plant shoot as asparagine. In **determinate nodules,** fixation again occurs as ammonia, but the fixed nitrogen is ultimately exported to the shoot as a purine. Fixation in either nodule type usually begins after about 15 days. During the maturation of a nodule, several proteins are produced within the nodule that are not found in the plant or rhizobia alone. These so-called **nodulins** are produced during all stages of the infection process. They include leghemoglobin and enzymes such as nitrogenase and glutamine synthetase.

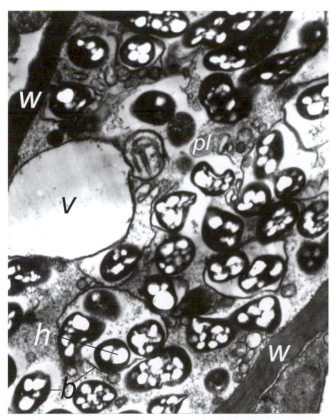

FIGURE 18.9 Transmission electron micrograph of bacteroids enclosed with the peribacteroid membrane. W; cell wall, V; vacuole, b; bacteroid (note that several bacteroids are enclosed within a membrane of plant origin) h; poly-β-hydroxy butyrate storage granule within a bacteroid, pl; plastid. (Photo courtesy I. L. Pepper)

FIGURE 18.10 Nodulated root system of *Phaseolus vulgaris* by strain KIM 5. Here nodules are determinate.

Overall, the amount of nitrogen fixed depends on both plant cultivar and rhizobial strain as well as a variety of environmental factors. In general, environmental factors that stress the plant such as extreme heat, cold, soil salinity, or lack of plant nutrients also limit nitrogen fixation, which is dependent on plant photosynthates (Graham, 1997).

18.2.4.3 Enhanced Fixation by the Legume–Rhizobia Symbiosis

Commercial legume crops are often aided in terms of nitrogen fixation through the application of rhizobial inoculants. This is particularly important when a new legume species is introduced into soils that are free of indigenous rhizobia. In this case, rhizobia introduced into the soil through the use of inoculants tend to establish themselves in the available ecological niche and are difficult to displace by any subsequent introduced rhizobia. Therefore, in such situations it is important that the originally introduced rhizobia are appropriate. The desired characteristics are outlined in Table 18.4. Usually rhizobia are impregnated into some kind of peat-based carrier with about 10^9 rhizobia per gram of peat. Production of commercial inoculants is now a big business in the United States, and internationally.

18.2.5 Mycorrhizal Associations

Clearly, mechanisms have evolved in soil microbial communities to enhance plant uptake of nitrogen as evidenced by the nitrogen-fixing bacteria. Of interest is the fact that microbial mechanisms have also evolved to enhance plant uptake of phosphates. The establishment and growth of most plants are enhanced by the presence of specialized fungi in soil that form close associations with their roots. These fungi are known as the **mycorrhizal fungi,** and they act as an extension of the plant root system. This aids in the uptake of almost all plant nutrients and is particularly important in the

uptake of phosphates, which typically have low solubility in the soil solution and therefore exist at low concentrations. Such fungi assist in the plant uptake of nutrients from dilute solutions by scavenging soil nutrients and utilizing active transport mechanisms to concentrate nutrients against steep concentration gradients. When released from fungal hyphae, such nutrients can be taken up by plant roots. In addition, when nutrients are stored within the fungus, the fungus can act as a reservoir of nutrients for future plant utilization. The mechanisms that cause the fungus to release its nutrients are not well understood. The plant supplying the fungus with carbon compounds, mostly as hexose sugars, completes the mutualistic association. Thus, each symbiont aids the other in terms of required nutrients.

Mycorrizal fungi become endemic in most soils and form extensive networks of fungal hyphae that can connect different plant species. In addition, on larger root systems, different fungi can infect the same root system. Mycorrhizal fungi naturally infect most plants, but in some commercial cropping systems such as the establishment of pine seedlings in pots, plants can be infected with known highly effective strains of fungi. There are several different types of mycorrhizal fungi, which are described next.

18.2.5.1 Vesicular–Arbuscular Mycorrhizas (VAMs)

These are the so-called **endomycorrhizal fungi,** which as the name implies are found mostly within the internal tissues of the root. This type of fungus is frequently found in fertile soils and is characterized by the presence of smooth **vesicles** and branched **arbuscules** that are involved in the storage and transfer of nutrients between the fungus and the plant (Fig. 18.11). About 90% of all vascular plants are associated with such fungal symbionts. The main group of fungi forming VAMs is within the order **Endoganales.** Six genera within this order are recognized, of which *Glomus* and *Gigaspora spp.* are typical.

18.2.5.2 Orchidaceous Mycorrhizas

These fungi are much more specific than other VAMs and infect only plants of the orchid family, which contains thousands of species, most of which are tropical. The physiological relationship between the orchid and the fungus is different because in this association it is the fungus that supplies the plant with a source of carbon. This is the only type of mycorrhizal association in which the carbon flow is into the plant from the fungus. In some cases, mature orchids can therefore live without conducting photosynthesis. It is also of interest that many orchids are associated with

TABLE 18.4   Desired Characteristics for Rhizobia Utilized as Commercial Legume Inoculants

Characteristic	Definition
Infective	Capable of causing nodule initiation and development
Effective	Capable of efficient nitrogen fixation
Competitive	Capable of causing nodule initiation in the presence of other rhizobia
Persistent	Capable of surviving in soil between crops in successive years

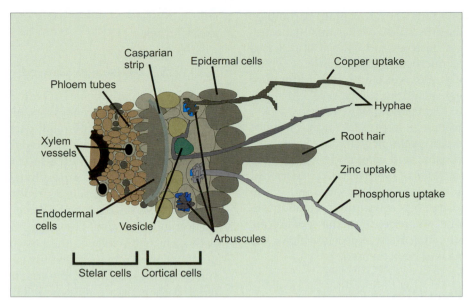

FIGURE 18.11 A typical endomycorrhiza showing hyphae extending beyond the root epidermis into the rhizosphere. Intracellular arbuscules and vesicles are also shown.

Rhizoctonia spp., including *R. solani*, which are common plant pathogens.

18.2.5.3 Ericaceous Mycorrhizas

These fungi are characterized by association with a specific group of plants known as the **Ericaceae,** which form important plant communities on moors, swamps, and peat. The plants involved include heathers, rhododendrons, and azaleas, which are often found on nutrient-poor, acid soil at high altitudes and at colder latitudes. The fungi involved are typical of the endomycorrhizal fungi in that they have intracellular hyphae. In this association, the fungus supplies the plant with nitrogen and the plant supplies the fungus with carbon substrate. The fungi also seem to be able to make the plants more tolerant of heavy metals and other soil contaminants. Most of the fungi involved seem to be members of the Ascomycetes or the Deuteromycetes.

18.2.5.4 Ectomycorrhizas

These associations are characterized by intercellular hyphae as opposed to the intracellular penetration of the VAMs. These mycorrhizas are formed on the roots of woody plants, with a thick fungal sheath developing around the terminal lateral branches of roots (Fig. 18.12). This is also known as the **mantle** and is connected to the network of intercellular hyphae found in the root cortex known as the **Hartig net.** The plants involved with these mycorrhizas are all trees or shrubs, whereas the fungi involved are often Basidiomycetes or Ascomycetes. Carbon substrate is supplied by the plant to the fungus, and minerals, in particular phosphates, are supplied by the fungus to the plant. The ectomycorrhizal fungi have been utilized as inoculants for pine seedlings in containers prior to use in reforestation projects.

18.3 PATHOGENIC MICROBES IN AGRICULTURE

The major source of detrimental microorganisms that affect plant growth is, of course, plant pathogens. The importance of these pathogens has led to the

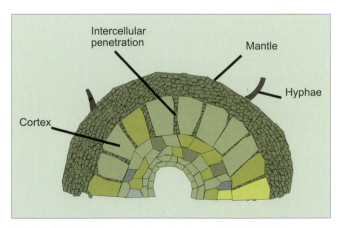

FIGURE 18.12 Cross section of ectomycorrhizal rootlet showing the exterior fungal sheath or mantle and intercellular penetration.

emergence of the discipline known as **plant pathology.** This can be defined as the study of the causes, mechanisms, environmental factors, and control of diseases of plants caused by microorganisms, many of which are soilborne. Clearly, a massive amount of literature exists on this subject including many excellent reference books such as "Plant Pathology" (Agrios, 1997). Therefore, the intent here is to present an overview of the important plant pathogens, which include viruses, bacteria, fungi, protozoa, and nematodes. The focus will be on *Agrobacterium tumefaciens*, a bacterial pathogen that has been extensively studied at the molecular level and exhibits a unique interaction between a prokaryotic microbe and the eukaryotic higher plants. Finally, important aspects of biological control of pathogens will be presented.

18.3.1 Plant Disease Caused by Fungi

Most plant pathogenic fungi have a filamentous structure known as a mycelium with individual branches known as hyphae (see Chapter 2). Almost all plant pathogenic fungi spend some of their time on the host plant and the remainder of their lives in soil or in plant debris within the soil. Thus, the survival and effects of the pathogen are controlled mainly by soil environmental factors including biotic (microbial) and abiotic factors such as temperature and moisture. The scope and diversity of plant fungal pathogens are extensive, and these organisms are responsible for billions of dollars of crop damage worldwide in all countries where agriculture is practiced. Some examples of important plant fungal pathogens and their plant hosts are shown in Table 18.5. Almost all commercial crops are subject to plant fungal attacks, which can re-

sult in diseases of seeds, roots, stems, leaves, fruit, or grain kernels.

18.3.2 Diseases Caused by Bacteria

Most plant pathogenic bacteria are rod shaped, with the exception of *Streptomyces,* which is a filamentous actinomycete. Almost all plant pathogenic bacteria occur within the host plant as parasites or on plant leaves as **epiphytes.** They also exist within plant debris or in soil as **saprophytes.** Some bacterial pathogens such as *Erwinia* predominate in the plant host, whereas others such as *Pseudomonas solanacearum* predominate in soil. However, most bacterial plant pathogens enter soil via the host tissue, and bacterial numbers often remain high only as long as the host tissue is still present within the soil. Important plant bacterial pathogens are shown in Table 18.6. Fruits and vegetables are particularly prone to pathogenic bacterial attack. Of all of the bacterial plant pathogens, *Agrobacterium tumefaciens* has perhaps been the best studied organism because of its mode of attack, which involves nucleic acids.

18.3.2.1 Crown Gall Disease– *Agrobacterium tumefaciens*

Crown gall disease is caused by the soilborne pathogen *Agrobacterium tumefaciens.* The disease manifests itself in uncontrolled cell division in the host plant, which results in the formation of a tumor or gall typically around the crown of the root. The disease is induced in a variety of dicotyledonous plants, particularly stone fruits, roses, and grapes. The majority of the bacterial genes necessary to induce the disease are plasmid borne. Specifically, a piece of the tumor-

TABLE 18.5 Examples of Important Fungal Plant Pathogens

Fungal pathogen	Plant host	Disease or symptom
Pythium	Almost all plants	Seed or root rot (damping off)
Phytophthora	Vegetables, fruit trees	Root rot
Plasmopara	Grapes	Downy mildew
Rhizopus	Fruits and vegetables	Soft rot of fruit or vegetables
Podosphaera	Fruit trees	Powdery mildewy growth
Alternaria	Vegetables	Leaf blight
Fusarium	Vegetables and field crops	Leaf wilting
Puccinia	Cereals and grains	Leaf and stem rusts
Ustilago	Cereals and grains	Corn smut
Rhizoctonia	Herbaceous plants	Root and stem rot
Armillaria	Fruit trees	Root rot

TABLE 18.6   Examples of Important Bacterial Plant Pathogens

Bacterial pathogen	Plant host	Disease or symptom
Pseudomonas syringae	Tobacco, vegetables	Leaf spots
Pseudomonas fluorescens	Potatoes	Soft rot
Xanthomonas campestris	Cereals, fruits	Several leaf spots (blights)
Xanthomonas campestris	Crucifers e.g., cabbage	Black rot
Erwinia tracheiphila	Cucumbers, melons	Vascular wilts
Erwinia carotovora	Fruits and vegetables	Soft rot
Agrobacterium tumefaciens	Fruit trees	Crown gall
Streptomyces scabies	Potatoes	Potato scab
Xylella fastidiosa	Grape	Pierce's disease

inducing **(Ti) plasmid** is transferred into the host plant cells, where it becomes integrated and functions within the plant. The overall features of the disease are shown in Figure 18.13. The transferred and integrated DNA (T-DNA) codes for the synthesis of two growth regulators, auxin and cytokinin, as well as for a group of amino acid derivatives known as "opines." The syn-thesis of these compounds, which is not regulated by the plant, gives rise to the symptoms of crown gall disease. The overall process also requires the expression of a variety of other genes on the Ti plasmid, which are termed virulence or *vir* **genes.** The *vir* genes are expressed in the presence of plant cell metabolites that are synthesized when a plant is wounded.

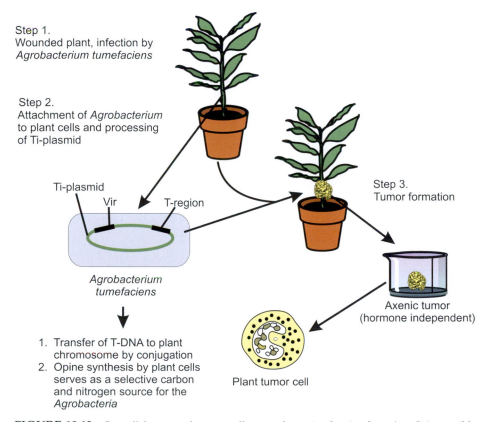

Step 1.
Wounded plant, infection by *Agrobacterium tumefaciens*

Step 2.
Attachment of *Agrobacterium* to plant cells and processing of Ti-plasmid

Ti-plasmid
Vir   T-region

Agrobacterium tumefaciens

1. Transfer of T-DNA to plant chromosome by conjugation
2. Opine synthesis by plant cells serves as a selective carbon and nitrogen source for the *Agrobacteria*

Step 3.
Tumor formation

Axenic tumor (hormone independent)

Plant tumor cell

FIGURE 18.13   Overall features of crown gall tumor formation by *Agrobacterium*. It is possible to delete the T-DNA and insert useful genes under the control of plant promoters. The expression of these genes integrated into the plant DNA confers desired properties on the plant. This technology forms the basis of genetic engineering of plants using *Agrobacterium*.

Attachment of Bacteria to Plant Cells

Chromosomal genes are also known to be important in the transformation of plants by *Agrobacterium*. Several genes including the chromosomally encoded virulence genes *chvA*, *chvB*, and *exoC* are involved in the attachment of the bacterium to plant cells. Mutations in these genes have been shown to result in avirulent phenotypes that are unable to attach to plant cells (Marks *et al.*, 1987). All of these genes are involved in the synthesis of β-1,2-glucan, although it is not clear how this molecule functions in the process of attachment.

Ti Plasmid

Most of the genes necessary for tumor induction are located on the large 180-kb Ti plasmid. This plasmid contains the virulence (*vir*) genes which are required for the processing and transfer of specific plasmid DNA known as **T-DNA**. The *vir* genes consist of about 35 kb of DNA and are essential for tumor formation, although they are not transferred into the plant. The *vir* genes consist of eight operons (Table 18.7). The induction of the *vir* genes occurs following exposure to plant signal molecules, which are synthesized by the plant upon wounding. This explains why crops that rely on root cuttings are particularly susceptible to crown gall disease. One of the signal molecules has been identified as the phenolic **acetosyringone**, which appears to be a precursor for lignin biosynthesis (Nester *et al.*, 1996). This molecule plus sugar monomers, which are precursors of the plant cell wall, are sensed by *Agrobacterium* through the *virA* and *virG* genes, which control expression of all other *vir* genes. The *virA* gene produces a protein that appears to sense the phenolic compound directly. This protein becomes activated and in turn activates a *virG* protein, which subsequently results in the expression of all other *vir* genes. Specifically, the *virG* protein binds to a 12-bp conserved sequence known as the "vir box."

Following activation of the *vir* genes, the T-DNA is processed for transfer into plant cells. The *virD* operon plays an important role in the early events in the processing of the T-DNA. **VirD1** codes for a topoisomerase that allows the conversion of the supercoiled Ti plasmid into a relaxed form. Following this, *virD2* produces an endonuclease that results in site-specific cleavage of the bottom strand of two 25-bp direct repeats at the right and left borders of the T-DNA. The *virD2* protein remains attached to the single-stranded displaced DNA at the 5' end and serves as a pilot protein. *VirC* also seem to be involved in the processing of T-DNA prior to its transfer into plant cells. Transport of the DNA is facilitated by products of the *virB* gene, which in conjunction with *virD4*, results in the formation of a pilus. Once inside the plant cell, the T strand is targeted to the plant cell nucleus, where both *virE* and *D* are necessary for optimal targeting.

Different T-DNAs code for different opines, which the bacterium can use as source of carbon, nitrogen, and energy, but the plant cannot. For example, two common opines are octopine and nopaline, and each is produced by different strains of *Agrobacterium*. Regardless of which opine is produced, the Ti plasmid of either strain contains *virA, G, B, C, D,* and *E* genes. However, there are differences in the *vir* gene composition. In addition, strains capable of octopine production contain genes concerned with the production of a cytochrome P-450 enzyme, which may detoxify inhibitory compounds in the plant cell environment. Octopine strains also contain the *virF* gene, which appears to be a host range determinant.

18.3.3 Plant Diseases Caused by Viruses

A variety of viruses can infect plants causing disease. Typically, plant viruses enter cells only through wounds made mechanically or perhaps via an infected pollen grain that is deposited in an ovule. Viruses which can contain RNA or DNA (see Chapter 2) typically result in leaf lesions. Although viral attacks can result in catastrophic crop losses, most virus diseases occur on crops year after year and cause small to moderate losses. Typical viral plant pathogens are shown in Table 18.8. Because of the small size of viral pathogens, their presence has been indicated primarily by the symptoms exhibited by the plant host. More recently, viruses are being identified by new molecular techniques including polymerase chain reaction (PCR) and reverse transcriptase (RT)-PCR analyses (see Chapter 13).

TABLE 18.7   Ti Plasmid-Encoded *vir* Genes of an Octopine Strain

vir	Inducibility	Size (kb)	Function
A	+	2.8	Plant signal sensor
G	+	1.0	Transcriptional activator
B	+	9.5	Transport of T-DNA
D	+	4.5	Processing of T-DNA, endonuclease, nuclear targeting
C	+	1.5	Processing of T-DNA
E	+	2.2	Single-stranded DNA binding protein, nuclear targeting
H	+	3.4	Cytochrome P-450 enzyme
F	+	0.6	Host range determinant

TABLE 18.8   Examples of Important Viral Plant Pathogens

Viral pathogen	Type of nucleic acid	Plant host	Disease or symptom
Tobamovirus (tobacco mosaic virus)	(ssRNA)[a]	Tobacco	Leaf chlorosis and distortion
Furovirus (wheat mosaic virus)	(ssRNA)	Wheat	Dwarf and mottled leaves
Potexvirus	(ssRNA)	Potato	Stunted plants
Potyvirus	(ssRNA)	Beans	Mottled, chlorotic leaves
Phytoreovirus	(dsRNA)	Rice	Galls or tumors
Caulimovirus	(dsDNA)	Cauliflower	General poor plant growth

[a]sS, single-stranded; ds, double-stranded.

18.3.4 Soil Biological Control of Plant Diseases

A relatively new approach to the control of plant pathogens is that of using microorganisms instead of chemicals. This can be done by either introducing microbes into a particular soil or manipulating the indigenous microflora. In either case, the objective is to reduce the numbers and activity of specific pathogens. **Biological control** can occur within the plant root itself, within the rhizosphere, or in the bulk soil in the vicinity of the root.

Antagonists are biological agents that reduce numbers or activities of pathogens through antibiosis, competition, or hyperparasitism. **Antibiosis** occurs when the pathogen is inhibited or lethally affected by metabolic products of the antagonist such as enzymes, acidic agents, or antibiotics. In contrast, **competition** can be for nutrients, growth factors, oxygen, or occasionally space. **Hyperparasitism** is due to the invasion of the parasite by the secretion of lytic enzymes. All of these mechanisms can result in decreased activities of pathogens, but because biological control acts by altering the biological equilibrium of the soil community, such control may take longer to act than chemical methods, and the efficacy of such methods may be more difficult to predict. On the other hand, when successful, biological control can last longer than chemical control. Finally, note that biological control methods are often most successful when used with integrated pest management strategies.

Pathogen-suppressive soils are soils in which a particular pathogen does not establish itself or persist, or if it does establish itself, it causes no damage or the disease becomes less severe with time (Cook and Baker, 1983). The main purpose of biological control is to maximize soil suppressiveness.

18.3.4.1 Maintenance of Suppressive Soils

Biological control through the manipulation of resident antagonists can be controlled by crop or soil management practices. **Soil management practices** that enhance suppressiveness include crop rotations and soil tillage, both of which reduce potential pathogen populations by reducing the incidence of specific crop residues that may harbor pathogens. Other practices include incorporation of organic amendments into soil, which apparently enhances the population of antagonists in the soil relative to pathogen populations. Crop management practices include the use of specific plant cultivars that select for specific rhizosphere antagonists. This may be due in part to the production of border cells that affect the ecology of the root system (Hawes *et al.*, 1998). In all of these cases, suppression of the activities of the pathogen occurs prior to the infection of the host root. In many cases, suppressive soils occur naturally (Agrios, 1997). However, in many instances, continuous cultivation of the same crop year after year results, at first, in a progressive increase of disease incidence, followed by years of reduced disease, presumably through a buildup of antagonists.

A well-documented example of soil suppression is the **"take-all disease"** of wheat, which is caused by the fungus *Gaeumannomyces graminis* var. *tritici*. General suppression of the pathogen is thought to be due to nonpathogenic, saprophytic *Fusarium* spp. In addition, specific suppression can occur due to antagonistic fluorescent pseudomonads. Specifically, *Pseudomonas fluorescens* has been shown to produce a phenazine antibiotic inhibitory to the pathogen. In an elegant piece of work, Thomashow and Weller (1988) showed that antibiotic-negative mutants produced by transposon mutagenesis were less suppressive to the fungus than the parent wild-type strain. Restoration of antibiotic production by complementation with a DNA fragment from the wild type restored the ability of the mutant to suppress take-all disease.

18.3.4.2 Introduced Biological Control Agents

A survey of the important plant pathogens shows that many act by attacking seeds or young root tissues. Because of this, attempts have been made to control pathogens by controlling the organisms within the rhizosphere through the use of introduced biological control agents. Successful agents were originally believed to be intrinsically **rhizosphere competent,** in other words, capable of colonizing the expanding root surface. However, it is now clear that the root itself, in many instances, controls the microbial populations within the rhizosphere through the production of border cells (see Section 18.3.4.3). This has made the performance of many biological control agents unpredictable when used with different plant hosts.

The history of biological control agents originated in the 1960s, when many scientists, particularly from the then Soviet Union, utilized introduced bacteria to increase crop yields. The so-called **bacterial fertilizers** were usually *Azotobacter* and *Bacillus* spp., and yield increases were believed to be due to associative nitrogen fixation and phosphate solubilization, respectively. In reality, the actual mechanisms may have been far more subtle and complex than those proposed and may have been the result of antagonistic interactions within the rhizosphere. In addition, many of the studies in the 1960s were not replicated and thus were disputed. In the 1980s, a flurry of new studies and concepts were introduced. Terms such as **plant growth–promoting rhizobacteria (PGPR)** (Kloepper *et al.,* 1980) and **deleterious rhizosphere microorganisms (DRMOs)** (Schippers *et al.,* 1987) were introduced. The PGPR were thought to improve plant growth by colonizing the root system and preventing the colonization of the DRMOs. The DRMOs in turn were defined as bacteria that reduced plant growth but were not parasitic. Deleterious activities were thought to include alterations in the supply of water, nutrients, and plant growth substances which altered root functions (Graham and Mitchell, 1997). More recently, PGPR have been shown to induce biocontrol through a mechanism known as **systematic resistance.** One example of this is the precolonization of roots by a *Pseudomonas* sp. that induced systematic resistance in the stem against *Fusarium* wilt (Liu *et al.,* 1995). Although many different mechanisms for rhizosphere competence have now been proposed, clearly the ultimate success of an organism in the rhizosphere depends on the interaction of the organism not only with other rhizosphere organisms but also with the plant root itself. This perhaps explains why many organisms that appear to be antagonistic to other pathogens in pure culture studies in the laboratory are later seen to be ineffective or inconsistent when used in field studies with different hosts. This concept of the plant root iself controlling the ecology of the root system has now been formulated in terms of root border cells controlling plant health.

Biological control agents produce a number of chemical metabolites that can participate in controlling plant disease. Microbes that produce these metabolites can be inoculated into the rhizosphere as just discussed, or the metabolites themselves can be added. Perhaps the best known example of a metabolite used as a biological control agent is the crystal toxin produced by *Bacillus thuringiensis*, a naturally-occurring soil microbe. *B. thuringiensis* produces a paracrystalline body during its growth that is toxic to specific groups of insects. The toxic crystal Bt protein is only effective when eaten by the insects which have an alkaline pH in their gut (recall that animal stomachs have an acidic pH). In the insect gut, the toxin binds to specific receptors and eventually leads to gut paralysis. The insect stops feeding and dies from a combination of tissue damage and starvation. Bt toxin is marketed world-wide as a commercial microbial insecticide and is sold under various trade names. Even more interesting is that the toxin genes have been moved into plants to confer self-protection against insects. As an example, the genes have been inserted into potato plants to allow protection against the Colorado Potato Beetle. Other metabolites that are useful in biological control include microbially-produced HCN, antibiotics, and siderophores (see Chapter 17.6.5). Most recently it has been discovered that a rhamnolipid biosurfactant (Fig. 17.6) is a very effective biological control agent against a class of pathogens called the zoosporic plant pathogens (see Fig. 18.14) (Stanghellini and Miller, 1997). One infamous zoosporic plant pathogen is *Phytophthora infestans* the causative agent of late blight of potato, the disease that caused the Irish Potato Famine in the 1840's.

FIGURE 18.14 *Plasmopara latacae-radeculus,* a zoosporic plant pathogen, infecting a root hair. (Photo courtesy M. E. Stanghellini.)

18.3.4.3 Function of Root Border Cells as Biocontrol Agents

Plant roots must move into new soil areas in order to obtain nutrients, and to do this, new root tissue must be generated by the root meristem. Evidence has suggested that the direction of root growth, and in fact the rhizosphere ecosystem, is controlled by **root border cells** (Hawes *et al.*, 1998). These have been shown to be what used to be called root cap cells, and they are actually living plant cells that are designed to be released from the root surface into the external environment. Under controlled conditions, border cells and their associated products can contribute up to 98% of carbon released as root exudates. The propagation and release of border cells are under the direction of the root and directly affect the behavior of rhizosphere bacteria and fungi. Many of the functions of these cells are still being elucidated, but they are thought to include chemotactic signals to beneficial microbes as well as plant pathogens. In the latter case, it may well be that the border cells act as decoys to the pathogens, protecting newly formed intact root tissue from attack. Border cells have also been shown to be a source of flavenoid-based *nod* gene–inducing signals that may affect nodulation of legumes by rhizobia. The ultimate role and scope of border cells are still being evaluated, but it is attractive to think that these agents are a vital link in controlling rhizosphere populations and their associated activities.

18.3.4.4 Biological Control of Crown Gall Disease

One very successful example of biological control worthy of mention is the control of crown gall disease by *Agrobacterium tumefaciens* strain K84. This commercially important biological control method was developed in Australia (Kerr, 1980). Inoculation of planting stock with the nonpathogenic strain K84 is often successful in preventing the disease. The mechanism of control involves the production of a **bacteriocin** that inhibits closely related bacteria—in this case other virulent strains of *A. tumefaciens*. The bacteriocin is known as agrocin 84 and is a fraudulent adenine nucleotide that inhibits DNA synthesis. It is taken up by *Agrobacterium* strains that synthesize nopaline or agrocinopine. Part of the agrocin molecule is similar to the structure of agrocinopine; thus strains that contain agrocinopine permease are capable of taking up agrocin 84. It is believed that bacterial colonization of the root surface may also be involved in the control mechanism. However, virulent bacterial mutant strains that do not take up agrocinopine are not sensitive to the bacteriocin.

QUESTIONS AND PROBLEMS

1. Discuss the usefulness and also the limitations of using R : S ratios as a means to evaluate the rhizosphere effect?

2. How does the rhizosphere differ physically, chemically and biologically from bulk soil?

3. How important is free living nitrogen fixation in i) production crop agriculture; and ii) natural ecosystems?

4. How important is symbiotic nitrogen fixation in crop production?

5. Describe your understanding of the genetic interactions between rhizobia and leguminous plants.

6. Compare and contrast the major mycorrhizal fungi/plant interactions.

7. Compare and contrast rhizobia and agrobacteria.

References and Recommended Readings

Agrios, G. N. (1997) "Plant Pathology," 4th ed. Academic Press, San Diego.

Bishop, P. E., and Premakumar, R. (1992) Alternative nitrogen fixation systems. *In* "Biological Nitrogen Fixation" (R. H. Burris and H. J. Evans, eds.) Chapman & Hall, New York, pp. 736–762.

Burris, R. H. (1974) Methodology. *In* "The Biology of Nitrogen Fixation" (A. Quispel, ed.) American Elsevier, New York, pp. 10–33.

Burris, R. H., and Roberts, G. P. (1993) Biological nitrogen fixation. *Annu. Rev. Nutr.* **13**, 317–335.

Cook, R. J., and Baker, K. F. (1983) "The Nature and Practice of Biological Control of Plant Pathogens." America Phytopathological Society Press, St. Paul, MN.

Curl, E. A., and Truelove, B. (1985) "The Rhizosphere." Springer-Verlag, Berlin.

Dangerfield, J. A., Westlake, D. W. S., and Cook, F. D. (1978) Characterization of the bacterial flora associated with root systems of *Pinus contorta* var. *Patifolia. Can. J. Microbiol.* **24**, 1520–1525.

Darbyshire, J. F., and Greaves, M. P. (1970) An improved method for the study of the inter-relationships of soil microorganisms and plant roots. *Soil Biol. Biochem.* **2**, 63–71.

Dénarié, J., Debellé, F., and Promé, J. C. (1996) *Rhizobium* lipochitooligosaccharide nodulation factors: Signaling molecules mediating recognition and morphogenesis. *Annu. Rev. Biochem.* **65**, 503–535.

Evans, H. J., and Burris, R. H. (1992) Highlights in biological nitrogen fixation during the last 50 years. *In* "Biological Nitrogen Fixation" (R. H. Burris and H. J. Evans, eds.) Chapman & Hall, New York, pp. 1–42.

Graham, P. H. (1997) Biological dinitrogen fixation: Symbiotic. *In* "Principles and Applications of Soil Microbiology" (D. M. Sylvia, J. J. Fuhrman, P. G. Hartel, and D. A. Zuberer, eds.) Prentice Hall, Upper Saddle River, NJ, pp. 322–345.

Graham, J. H., and Mitchell, D. J. (1997) Biological control of soilborne plant pathogens and nematodes. *In* "Principles and Applications of Soil Microbiology" (E. M. Sylvia, J. F. Fuhrman,

P. G. Hartel and D. A. Zuberer, eds.) Prentice Hall, Upper Saddle River, NJ, pp. 427–446.

Griffin, G. J., Hale, M. G., and Shay, F. J. (1976) Nature and quantity of sloughed organic matter produced by roots of axenic peanut plants. *Soil Biol. Biochem.* **8,** 29–32.

Hale, M. G., Moore, L. D., and Griffin, G. J. (1978) Root exudates and exudation. *In* "Interactions Between Non-Pathogenic Soil Microorganisms and Plants" (Y. R. Dommergues and S. V. Krupa, eds.) Elsevier, Amsterdam, pp. 163–203.

Hawes, M. C., Brigham, L A., Woo, H. H., Zhu, Y., and Wen, F. (1996) Root border cells. *Biol. Plant-Microbe Interact.* **8,** 509–514.

Hawes, M. C., Brigham, L. A., Wen, F., Woo, H. H., and Zhu, Y. (1998) Function of root border cells in plant health: Pioneers in the rhizosphere. *Annu. Rev. Phytopathol.* **36,** 311–327.

Henderson, V. E., and Katznelson, H. (1961) The effect of plant roots on the nematode population of the soil. *Can. J. Microbiol.* **7,** 163–167.

Kamst, E., Pilling, J., Raamsdonk, L. M., Lugtenberg, B. J. J., and Spaink, H. P. (1997) *Rhizobium* nodulation protein NodC is an important determinant of chitin oligosaccharide chain length in Nod factor biosynthesis. *J. Bacteriol.* **179,** 2103–2108.

Kennedy, A. C. (1997) The rhizosphere and spermosphere. *In* "Principles and Applications of Soil Microbiology" (D. M. Sylvia, J. J. Fuhrman, P. G. Hartel and D. A. Zuberer, eds.) Prentice Hall, Upper Saddle River, NJ, pp. 389–407.

Kerr, A. (1980) Biological control of crown gall through production of agrocin 84. *Plant Dis.* **64,** 25–30.

Kloepper, J. W., Shroth, M. N., and Miller, T. D. (1980) Effects of rhizosphere colonization by plant growth–promoting rhizobacteria on potato plant development and yield. *Phytopathology* **70,** 1078–1082.

Liu, L., Kloepper, J. W., and Tuzun, S. (1995) Induction of systematic resistance in cucumber against fusarium wilt by plant growth–promoting rhizobacteria. *Phytopathology* **85,** 843–847.

Marks, J. R., Lynch, T. J., Karlinsey, J. E., and Thomashow, M. F. (1987) *Agrobacterium tumefaciens* virulena locus *pscA* is related to the *Rhizobium meliloti exoC* locus. *J. Bacteriol.* **169,** 5835–5837.

Mergaert, P., Van Montagu, M., and Holsteres, M. (1997) Molecular mechanisms of Nod factor diversity. *Mol. Microbiol.* **25,** 811–817.

Miki, N. K., Clarke, K. J., and McCully, M. E. (1980) A histological and histochemical comparison of the mucilage on the root tips of several grasses. *Can. J. Bot.* **58,** 2581–2593.

Minami, E., Kouchi, H., Carlson, R. W., Cohn, J. R., Kolli, V. K., Day, R. B., Ogawa, T., and Stacey, G. (1996) Co-operative action of lipo-chitin nodulation signals on the indication of the early nodulin, ENOD2 in soybean roots. *Mol. Plant Microbe Interact.* **9,** 574–583.

Nester, E. G., Kemner, J., Deng, W., Lee, Y. W., Fullner, K., Liang, X., Pan, S., and Health, J. D. (1996) *Agrobacterium:* A natural genetic engineer exploited for plant biotechnology. *In* "Biology of Plant–Microbe Interactions" (G. Stacy, B. Mullin and P. M. Gresshoff, eds.) International Society for Molecular Plant–Microbe Interactions, St. Paul, MN, pp. 481–489.

Paul, E. A., and Clark, F. E. (1989) "Soil Microbiology and Biochemistry." Academic Press, San Diego.

Pepper, I. L., and Bezdicek, D. F. (1990) Root microbial interactions and rhizosphere nutrient dynamics. *In* "Crops as Enhancers of Nutrient Use" (V. C. Baligar and R. R. Duncan, eds.), Academic Press, San Diego, pp. 375–410.

Postgate, J. (1994) "The Outer Reaches of Life." Cambridge University Press, Cambridge.

Rouatt, J. W., Katznelson, H., and Payne, T. M. B. (1960) Statistical evaluation of the rhizosphere effect. *Soil Sci. Soc. Am. Proc.* **24,** 271–273.

Rovira, A. D., Foster, R. C., and Martin, J. K. (1979) Note on terminology: Origin, nature and nomenclature of the organic materials in the rhizosphere. *In* "The Soil–Root Interface" (J. L. Harley and R. Scott Russell, eds.) Academic Press, London, pp. 1–4.

Schippers, B., Bakker, A. W., and Bakker, P. A. H. M. (1987) Interactions of deleterious and beneficial rhizosphere microorganisms and the effect of cropping practice. *Annu. Rev. Phytopathol.* **25,** 339–358.

Stanghellini, M. E., and Miller, R. M. (1997) Biosurfactants: Their identity and potential efficacy in the biological control of zoosporic plant pathogens. *Plant Disease* **81,** 4–12.

Sylvia, D. M., Fuhrmann, J. J., Hartel, P. G., and Zuberer, D.A., eds. (1997) "Principles and Applications of Soil Microbiology." Prentice Hall, Upper Saddle River, NJ.

Thomashow, L. S., and Weller, D. M. (1988) Role of a phenazine antibiotic from *Pseudomonas fluorescens* in biological control of *Gaeumannomyces graminis* var. *tritici. J. Bacteriol.* **170,** 3497–3508.

Wiggins, E. A., Curl, E. A., and Harper, J. D. (1979) Effects of soil fertility and cotton rhizosphere on populations of Collembola. *Pedobiologia* **19,** 75–82.

Zuberer, D. A. (1997) Biological dinitrogen fixation: Introduction and nonsymbiotic. *In* "Principles and Applications of Soil Microbiology" (P. M. Sylvia, J. J. Fuhrmann, P. G. Hartel and D. A. Zuberer, eds.) Prentice Hall, Upper Saddle River, NJ, pp. 295–321.

CHAPTER

19

Environmentally
Transmitted Pathogens

PATRICIA RUSIN   CARLOS E. ENRIQUEZ   DANA JOHNSON   CHARLES P. GERBA

19.1 ENVIRONMENTAL
TRANSMITTED PATHOGENS

Although humans are continually exposed to a vast array of microorganisms in the environment, only a small proportion of these microbes are capable of interacting with the host in such a manner that infection and disease will result. Disease-causing microorganisms are called **pathogens. Infection** is the process in which the microorganism multiplies or grows in or on the host. Infection does not necessarily result in disease since it is possible for the organism to grow in or on the host but does not produce an illness (see Chapter 24). In the case of enteric infections (i.e., diarrhea) caused by *Salmonella*, only half of the individuals infected develop clinical signs of illness. A **frank pathogen** is a microorganism capable of producing disease in both normal healthy and immunocompromised persons. **Opportunistic pathogens** are usually capable of causing infections only in immunocompromised individuals (burn patients, patients taking antibiotics, those with impaired immune systems, elderly patients with diabetes, etc.). Opportunistic pathogens are common in the environment and may be present in the human gut or skin but normally do not cause disease.

To cause illness, the pathogen must usually first grow within or on the host. The time between infection and the appearance of clinical signs and symptoms (diarrhea, fever, rash, etc.) is the **incubation time** (Table 19.1). This may range from as short as 6–12 hours in the case of Norwalk virus diarrhea to 30–60 days for hepatitis A virus, which causes liver disease. At any

time during infection the pathogen may be released into the environment by the host in the feces, urine, respiratory secretions. Although the maximum release may occur at the height of the disease, it may also precede the first signs of clinical illness. In the case of hepatitis A virus, the maximum excretion in the feces occurs before the onset of signs of clinical illness. The concentration of organisms released into the environment varies with the type of organism and the route of transmission (Table 19.2). The concentration of enteric viruses during gastroenteritis may be as high as 10^{10}–10^{12} per gram of feces.

Pathogenic microorganisms usually originate from an infected host (either human or other animal) or directly from the environment. Many human pathogens can be transmitted only by direct or close contact with an infected person or animal. Examples include the herpes virus, *Neisseria gonorrhoeae* (gonorrhea), and *Treponema pallidum* (syphilis). This is because their survival time outside the host is very brief. Pathogens transmitted through the environment may survive from hours to years outside the host, depending on the organism and the environment. Pathogens may exit a host in respiratory secretions from the nose and mouth or be shed on dead skin or in feces, urine, saliva, or tears. Thus they may contaminate the air, water, food, or inanimate objects **(fomites).** When contaminated air is inhaled or food consumed, the organisms are effectively transmitted to another host, where the infection process begins again. Airborne transmission can occur via release from the host in droplets (i.e., coughing) or through natural (surf at a beach) or human activities

TABLE 19.1   Incubation Time for Common Enteric Pathogens

Agent	Incubation period	Modes of transmission	Duration of illness
Adenovirus	8–10 days	Fecal–oral–respiratory	8 days
Campylobacter jejuni	3–5 days	Food ingestion, direct contact	2–10 days
Cryptosporidium	2–14 days	Food or water ingestion, direct and indirect contact	Weeks, months
Escherichia coli			
ETEC	16–72 hr	Food or water ingestion	3–5 days
EPEC	16–48 hr	Food or water ingestion, direct and indirect contact	5–15 days
EHEC	72–120 hr	Food/ingestion, direct or indirect contact	2–12 days
Giardia lamblia	7–14 days	Food or water ingestion, direct and indirect contact	Weeks–months
Norwalk agent(s)	24–48 hr	Food or water ingestion, direct and indirect contact, aerosol?	1–2 days
Rotavirus	24–72 hr	Director and indirect contact	4–6 days
Hepatitis A	30–60 days	Hepatitis	2–4 weeks
Salmonella	16–72 hr	Food ingestion, direct and indirect contact	2–7 days
Shigella	16–72 hr	Food or water ingestion, direct and indirect contact	2–7 days
Yersinia enterocolitica	3–7 days	Food ingestion, direct contact	1–3 weeks

(cooling towers, showers) (see Chapter 5.4). Some organisms may be carried great distances, hundreds of meters or miles (e.g., legionnaires' disease and foot-and-mouth disease). Virus transmission by the airborne route may be both direct and indirect. Infection of a host may be by direct inhalation of infectious droplets or through contact with fomites on which the airborne droplets have settled. Hand or mouth contact with the organism on the surface of a fomite results in the transfer of the organism to the **portal of entry,** i.e., nose, mouth, or eye.

Microorganisms transmitted by the fecal–oral route are usually referred to as **enteric pathogens** because they infect the gastrointestinal tract. They are characteristically stable in water and food and, in the case of enteric bacteria, are capable of growth outside the host under the right environmental conditions.

TABLE 19.2   Concentration of Enteric Pathogens in Feces

Organism	Per gram of feces
Protozoan parasites	10^6–10^7
Helminths	
Ascarus	10^4–10^5
Enteric viruses	
Enteroviruses	10^3–10^7
Rotavirus	10^{10}
Adenovirus	10^{11}
Enteric bacteria	
Salmonella spp.	10^4–10^{10}
Shigella	10^5–10^9
Indicator bacteria	
Coliform	10^7–10^9
Fecal coliform	10^6–10^9

Waterborne diseases (Table 19.3) are those transmitted through the ingestion of contaminated water that serves as the passive carrier of the infectious agent. The classic waterborne diseases, cholera and typhoid fever, which frequently ravaged densely populated areas throughout human history, have been effectively controlled by the protection of water sources and by treatment of contaminated water supplies. In fact, the control of these classic diseases gave water supply treatment its reputation and played an important role in the reduction of infectious diseases. Other diseases caused by bacteria, viruses, protozoa, and helminths may also be transmitted by contaminated drinking water. However, it is important to remember that waterborne diseases are transmitted by the fecal–oral route, from human to human or animal to human, so that drinking water is only one of several possible sources of infection.

Water-washed diseases are those closely related to poor hygiene and improper sanitation. In this case, the availability of a sufficient quantity of water is generally considered more important than the quality of the water. The lack of water for washing and bathing contributes to diseases that affect the eye and skin, including infectious conjunctivitis and trachoma, as well as to diarrheal illnesses, which are a major cause of infant mortality and morbidity in developing countries. Diarrheal diseases may be directly transmitted through person-to-person contact or indirectly through contact with contaminated foods and utensils used by persons whose hands are fecally contaminated. When enough water is available for hand washing, the incidence of diarrheal diseases has been shown to decrease, as has the prevalence of enteric pathogens such as *Shigella*.

TABLE 19.3 Classification of Water-related Illnesses Associated with Microorganisms

Class	Cause	Example
Waterborne	Pathogens that originate in fecal material and are transmitted by ingestion	Cholera, typhoid fever
Water-washed	Organisms that originate in feces are transmitted through contact because of inadequate sanitation or hygiene	Trachoma
Water-based	Organisms that originate in the water or spend part of their life cycle in aquatic animals and come in direct contact with humans in water or by inhalation	Schistosomiasis Legionella
Water-related	Microorganisms with life cycles associated with insects that live or breed in water	Yellow fever

Modified from White and Bradley (1972).

Water-based diseases are caused by pathogens that either spend all (or essential parts) of their lives in water or depend on aquatic organisms for the completion of their life cycles. Examples of such organisms are the parasitic helminth *Schistosoma* and the bacterium *Legionella,* which cause schistosomiasis and legionnaires' disease, respectively.

Water-related diseases, such as yellow fever, dengue, filariasis, malaria, onchocerciasis, and sleeping sickness, are transmitted by insects that breed in water (e.g., mosquitoes that carry malaria) or live near water (e.g., the flies that transmit the filarial infection onchocerciasis). Such insects are known as **vectors.**

19.2 BACTERIA

19.2.1 *Salmonella*

In the United States, the concept of waterborne and foodborne disease was poorly understood until the late 19th century. During the Civil War (1860–1865), encamped soldiers often disposed of the waste upriver but drew drinking water from downriver. This practice resulted in widespread dysentery. In fact, dysentery, together with its sister disease typhoid fever (*Salmonella typhi*), was the leading cause of death among soldiers of all armies until the 20th century. It was not until the end of the 19th century that this state of affairs began to change. At that time, the germ theory was generally accepted, and steps were taken to treat wastes properly and protect drinking water and food supplies.

In 1890, more than 30 people out of every 100,000 in the United States died of typhoid. But by 1907, water filtration was becoming common in most U.S. cities, and in 1914 chlorination was introduced. Because of these new practices, the national typhoid death rate in the United States between 1900 and 1928 dropped from 36 to 5 cases per 100,000 people. The lower death toll was largely the result of a reduced number of outbreaks of waterborne diseases. In Cincinnati, for instance, the yearly typhoid rate of 379 per 100,000 people in the years 1905–1907 decreased to 60 per 100,000 people between 1908 and 1910, following the inception of sedimentation and filtration treatment of drinking water. The introduction of chlorination after 1910 decreased this rate even further.

Salmonella is a very large group of rod-shaped, gram-negative, bacteria comprising more than 2000 known serotypes that are members of the Enterobacteriaceae. All these serotypes are pathogenic to humans and can cause a range of symptoms from mild gastroenteritis to severe illness or death. *Salmonella* are capable of infecting a large variety of both cold- and warm-blooded animals. Typhoid fever, caused by *S. typhi*, and paratyphoid fever, caused by *S. paratyphi*, are normally found only in humans, although *S. paratyphi* is found in domestic animals on rare occasions. In the United States, salmonellosis is due primarily to foodborne transmission because the bacteria infect beef and poultry and are capable of growing in these foods. Salmonellosis is the second leading cause of foodborne illness in the United States. Since the route of transmission is fecal–oral, any food or water contaminated with feces may transmit the organism to a new host.

Surface waters receiving domestic sewage discharge, meat processing wastes, or stockyard wastes are likely to be contaminated with *Salmonella*. Nontyphi salmonellae were shown to have a half-life rate in fresh water at 9.5–12.5°C of 2.4–19.2 hours. In contrast, *S. typhi* were shown to have a half-life in fresh water at 9.5–12.5°C of 6.0 hours (McFeters *et al.,* 1974).

It is estimated that 2–4 million cases of salmonellosis (diarrhea) occur in the United States annually. The incidence appears to be rising both in the United States and in other industrialized nations. *S. enteritidis* isolations from humans have shown a dramatic rise in the past decade, particularly in the northeastern United States (six-fold or more), and the increase is spreading south and west, with sporadic outbreaks in other regions.

All age groups are susceptible, but symptoms are most severe in the elderly, infants, and the infirm. Patients with acquired immunodeficiency syndrome (AIDS) suffer from salmonellosis frequently (estimated 20-fold more that the general population) and suffer from recurrent episodes. The onset time is usually between 12 and 36 hours after ingestion of contaminated food or water. Intestinal disease occurs with the penetration of *Salmonella* organisms from the gut lumen into the lining of the intestines, where inflammation occurs and an enterotoxin is produced. The immediate symptoms include nausea, vomiting, abdominal cramps, diarrhea, fever, and headache. Acute symptoms may last for 1 to 2 days or may be prolonged, again depending on host factors, ingested dose, and individual strain characteristics.

S. typhi and *S. paratyphi* A, B, and C produce typhoid and typhoid-like (paratyphoid) fever in humans. Any of the internal organs may be infected. The fatality rate of untreated cases of typhoid fever is 10%, compared with less than 1% for most forms of salmonellosis. *Salmonella* septicemia (bacteria multiplying in the blood) has been associated with the subsequent infection of virtually every organ system.

Typhoid fever presents a very different clinical picture than salmonellosis. The onset of typhoid fever (1 to 3 weeks) is usually insidious, with fever, headache, anorexia, enlarged spleen, coughing, and constipation being more common than diarrhea in adults. Intestinal hemorrhage and perforation occur in about 1% of the cases. Paratyphoid presents a similar clinical picture but tends to be milder. A carrier state may follow infection (<1.0 to 3.9% of the population). A carrier is a person who is permanently infected and may transmit the organism, but does not demonstrate any signs or symptoms of disease. A chronic carrier state is most common among individuals infected during middle age, especially females. The organism is usually carried in the gallbladder and secreted in the bile.

19.2.2 *Escherichia coli* and *Shigella*

Escherichia coli is a gram-negative rod found in the gastrointestinal tract of all warm-blooded animals and is usually considered a harmless organism. However, several strains are capable of causing gastroenteritis; these are referred to as **enterotoxigenic (ETEC), enteropathogenic (EPEC), enteroinvasive (EIEC), or enterohemorrhagic (EHEC)** *E. coli.* All of the *E. coli* are spread by the fecal–oral route of transmission.

The enterotoxigenic *E. coli* are a major cause of traveler's diarrhea in persons from industrialized countries who visit less developed countries, and it is also an important cause of diarrhea in infants and children in less developed countries. Following an incubation period of 10–72 hours, symptoms including cramping, vomiting, diarrhea (may be profuse), prostration, and dehydration. The illness usually lasts less than 3 to 5 days. Disease is caused by two toxins, one the heat-labile toxin and the other the heat-stable toxin. The ETEC *E. coli* are usually species-specific; that is, humans are the reservoir for the strains causing diarrhea in humans. Only a few outbreaks have been documented in the United States. One resulted from the consumption of water contaminated with human sewage, another from the consumption of food prepared by an infected food handler, another in a hospital cafeteria, and the last aboard a cruise ship.

The EPEC is the oldest recognized category of diarrhea-causing *E. coli*. Diarrheal disease caused by this group of *E. coli* is virtually confined to infants less than 1 year of age. It is associated with infant summer diarrhea, outbreaks of diarrhea in nurseries, and community epidemics of infant diarrhea. Symptoms include watery diarrhea with mucus, fever, and dehydration. The diarrhea can be severe and prolonged with a high fatality rate (as high as 50%). Since the 1960s, EPEC has largely disappeared as an important cause of infant diarrhea in North America and Europe. However, it remains a major agent of infant diarrhea in South America, Africa, and Asia. Humans, cattle, and pigs can be infected by this organism. Thus, foods implicated in outbreaks are raw beef and chicken. Also it is often associated with contaminated infant formula, which is one reason that breast feeding is encouraged, particularly in developing countries.

The disease caused by EIEC closely resembles that caused by *Shigella*. Illness begins with severe abdominal cramps, watery stools, and fever. Disease is usually self-limiting with no known complications. Any age group is susceptible. This type of *E. coli* carries a plasmid that enables it to invade cells lining the gastrointestinal tract, resulting in a mild form of dysentery. EIEC infections are endemic in less developed countries, although occasional cases and outbreaks are reported in industrialized countries. It is not known which foods may transmit EIEC, but any food contaminated with human feces from an ill individual, either directly or via contaminated water, could cause disease in others. Outbreaks also have been associated with hamburger and unpasteurized milk.

The EHEC *E. coli* were described in 1982 when a multistate epidemic of hemorrhagic colitis occurred in the United States and was shown to be due to a specific serotype known as *E. coli* O157 : H7. This organism produces two toxins called verotoxins I and II. These toxins are closely related to or identical to the toxin produced by *Shigella dysenteriae*. The toxin production depends on the presence of a prophage. A prophage is a bacterial virus that inserts its DNA into a bacterial chromosome. A plasmid codes for a novel type of fimbriae (hairlike projections) that enable the organism to adhere to the intestinal lining and initiate disease.

EHEC infections are now recognized to be an important problem in North America, Europe, and some areas of South America. The illness usually includes severe cramping and diarrhea, which is initially watery but becomes grossly bloody. The illness is usually self-limiting and lasts for an average of 8 days. However, some victims, particularly the very young, have developed the **hemolytic–uremic syndrome (HUS),** resulting in renal failure and hemolytic anemia. This disease can result in permanent loss of kidney function. In older individuals, HUS plus two other symptoms, fever and neurologic symptoms, constitutes **thrombotic thrombocytopenic purpura (TTP).** This illness can have a mortality rate in the elderly as high as 50%. Although *E. coli* O157 : H7 has been isolated from 3.7% of beef, 1.5% of pork, 1.5% of poultry, and 2.0% of lamb samples in retail stores (Doyle and Schoeni, 1987), most outbreaks are associated with undercooked or raw hamburger. Raw milk, unpasteurized fruit juices, and vegetables contaminated with cow dung have also been implicated. Waterborne outbreaks have also occurred.

Shigella is closely related to *E. coli*. Four species have been described: *S. dysenteriae, S. flexneri, S. boydii,* and *S. sonnei. S. dysenteriae* causes the most severe disease and *S. sonnei* causes the mildest symptoms. Fortunately, *S. sonnei* is the serotype most often found in the United States (Lee *et al.,* 1991). *Shigella* very rarely occurs in animals. It is principally a disease of humans and rarely of other primates such as monkeys and chimpanzees. The organism is often found in water polluted with human sewage and is transmitted by the fecal–oral route. Surveillance data from the Centers for Disease Control and Prevention (CDC) between 1972 and 1985 showed that *Shigella* was the second most common cause of waterborne disease outbreaks of known origin, following only *Giardia lamblia* (a parasite). However, most cases of shigellosis are the result of person-to-person transmission through the fecal–oral route. Secondary attack rates are 20–40% of household contacts. *Shigella* spp. were shown to have a half-life in fresh water at 9.5–12.5°C of 22.4–26.8 hours (McFeters *et al.,* 1974). In well water, 50% of *S. flexneri* cells die off in 26.8 hours (Gerba *et al.,* 1975).

An estimated 300,000 cases of shigellosis occur annually in the United States. *Shigella* is associated with certain foods such as salads, raw vegetables, milk and dairy products, and poultry. After an onset time of 12 hours to one week, symptoms of abdominal pain, cramps, diarrhea, and vomiting may occur. The organisms multiply in the cells of the gastrointestinal tract and spread to neighboring cells, resulting in tissue destruction. Some infections are associated with ulceration, rectal bleeding, drastic dehydration, and fatality rates as high as 10-15%. Infants, the elderly, and the infirm are most susceptible.

19.2.3 *Campylobacter*

Campylobacter jejuni is a gram-negative curved rod. It is relatively fragile and sensitive to environmental stress (including an oxygen content of 21%, drying, heating, contact with disinfectants, or acidic conditions). Before 1972, when methods were developed for its isolation from feces, it was believed to be primarily an animal pathogen causing abortion and enteritis in sheep and cattle. Recent surveys have shown that *C. jejuni* is a leading cause of bacterial diarrheal illness in the United States. Although *C. jejuni* is not carried by healthy individuals in the United States or Europe, it is often isolated from healthy cattle, chickens, birds, and even flies. It is sometimes present in nonchlorinated water sources such as streams and ponds.

C. jejuni infections cause diarrhea with fever, abdominal pain, nausea, headache, and muscle pain. The illness usually occurs 2–5 days after the ingestion of the contaminated food or water. Illness generally lasts 7–10 days, and relapses are not uncommon. Although anyone can be infected with *C. jejuni*, children under 5 years of age and young adults (15–29 years old) are more frequently afflicted than other age groups.

Surveys show that 20 to 100% of retail chickens are contaminated with *C. jejuni*. This is not surprising because many healthy chickens carry these bacteria in their intestinal tracts. Nonchlorinated water may also be a source of infection. However, properly cooking chicken, pasteurizing milk, and chlorinating water will kill the bacteria (Blaser *et al.,* 1986).

C. jejuni has been isolated from 22% of coastal and estuary water samples in concentrations ranging from 10 to 230 campylobacters per 100 ml and from 28% of river samples in concentrations of 10 to 36 cells per 100 ml (Bolton *et al.,* 1982). Carter *et al.* (1987) found *Campylobacter* in 10–44% of pond water samples. However, it is thought that virtually all surface waters contain *Campylobacter*. Recovery rates from surface waters are highest in the fall and winter months and lowest during the spring and summer months. *Campylobacter* density does not show a significant correlation with the isolation of indicator bacteria such as fecal or total coliforms (Carter *et al.*, 1987).

This bacterium dies off rapidly in stream water at 37°C, showing a 9-log decrease within 3–12 days (Rollins and Colwell, 1986). However, the organism can remain viable for extended periods at cooler temperatures, surviving over 120 days at 4°C in stream water.

19.2.4 *Yersinia*

Yersinia enterocolitica and *Y. pseudotuberculosis* are small rod-shaped gram-negative bacteria. Both organisms have often been isolated from animals such as pigs, birds, beavers, cats, and dogs. Only *Y. enterocolitica* has been detected in environmental and food sources such as ponds, lakes, meats, ice cream, and milk. To date, no foodborne outbreaks caused by *Y. pseudotuberculosis* have been reported in the United States, but human infections transmitted via contaminated water and foods have been reported in Japan.

Symptoms usually begin 24 to 48 hours after ingestion of contaminated food or drink. Yersiniosis is frequently characterized by such symptoms as gastroenteritis with diarrhea and/or vomiting. However, fever and abdominal pain are the hallmark symptoms. Yersinia infections mimic appendicitis, but the bacteria may also cause infections of other sites such as wounds, joints, and the urinary tract.

Although strains of *Y. enterocolitica* can be found in meats, oysters, fish, and raw milk, the exact cause of food contamination is unknown. However, the prevalence of this organism in soil and water and in animals offers ample opportunities for it to enter our food supply. Poor and improper sanitation techniques of food handlers, including improper storage, cannot be overlooked as contributing to contamination. In addition, *Y. enterocolitica* is able to grow at refrigeration temperatures and under microaerophilic conditions posing an increased risk if uncured meat stored in evacuated plastic bags is undercooked.

Waterborne outbreaks have occurred in the United States but such documented outbreaks are very rare. One of these outbreaks occurred at a Montana ski resort from December 6, 1974 to January 17, 1975 (Craun, 1979). Approximately 750 cases of gastroenteritis caused by *Y. enterocolitica* occurred among 1550 guests and 350 employees. A significant association was found between drinking water and illness. Two 60-foot-deep wells developed in sand and gravel supplied water to the resort. A sewer line was found to pass near the wells, and samples collected from the

wells after the outbreak yielded *Y. enterocolitica* and coliform organisms. Routine bacteriological surveillance during the previous 3 years had not detected coliform contamination of the wells. Chlorination of the wells stopped the outbreak.

As stated earlier, *Yersinia* does not occur frequently in foods, unless a breakdown occurs in food processing techniques. The CDC estimates that about 17,000 cases occur annually in the United States. Yersiniosis is a far more common disease in Northern Europe, Scandinavia, and Japan. The greatest incidence of disease is seen during the cold season. The most susceptible populations for disease are the very young, the debilitated, the very old, and persons undergoing immune suppressive therapy.

19.2.5 *Vibrio*

London's Dr. John Snow (1813–1858) was one of the first to make a connection between certain infectious diseases and drinking water contaminated with sewage. In his famous study of London's Broad Street pump, published in 1854, he noted that people afflicted with cholera were clustered in a single area around the Broad Street pump, which he determined was the source of the infection. When, at his insistence, city officials removed the handle of the pump, Broad Street residents were forced to obtain their water elsewhere. Subsequently, the cholera epidemic in that area subsided.

The gram-negative genus *Vibrio* contains more than one member that is pathogenic to humans. The most famous member of the genus is still *V. cholerae*. Cholera is transmitted through the ingestion of fecally contaminated food and water. Cholera remains prevalent in many parts of Central America, South America, Asia, and Africa. Persons traveling in these areas should not eat food that is uncooked and should drink only beverages that are carbonated or made from boiled or chlorinated water.

V. cholerae serogroup O1 includes two biovars, cholerae (classical) and El Tor, each of which includes organisms of the Inaba and Ogawa serotypes. A similar enterotoxin is elaborated by each of these organisms, so the clinical pictures are similar. Asymptomatic infection is much more common than disease, but mild cases of diarrhea are also common. In severe untreated cases, death may occur within a few hours and the fatality rate without treatment may exceed 50%. This is due to a profuse watery diarrhea referred to as rice-water stools. The rice-water appearance is due to the shedding of intestinal mucosa and epithelial cells. With proper treatment, the fatality rate is below 1%. Humans are the only known natural host. Thus, the reservoir for *V. cholerae* is human, although environmental reservoirs may exist, apparently in association with copepods or other phytoplankton.

Vibrios that are biochemically indistinguishable, but do not agglutinate in *V. cholerae* serogroup O1 antiserum, were formerly known as nonagglutinable vibrios (NAGs) or noncholera vibrios (NCVs). They are now included in the species *V. cholerae*. Some of these strains produce enterotoxin but most do not. Thus reporting of non-O1 *V. cholerae* infections as cholera is inaccurate and leads to confusion. *V. mimicus* is a closely related species that can cause diarrhea, of which some strains elaborate an enterotoxin indistinguishable from that produced by *V. cholerae*.

The following example illustrates the possible devastating impacts of this disease. A cholera epidemic caused by *V. cholerae* O1 began in January 1991 in Peru and spread to Central America and South America. A total of 1,041,422 cases occurred with 9642 deaths (MMWR, 1995). The epidemic was believed to have been initiated by inadequate chlorination of drinking water. A second example is an outbreak in southern Asia, an epidemic caused by the newly recognized strain *V. cholerae* O139 which began in late 1992 and continued to spread to at least 11 countries. This latter strain of *V. cholerae* also produces severe watery diarrhea and dehydration that is indistinguishable from the illness caused by *V. cholerae* O1.

Another species, *V. parahaemolyticus*, is usually transmitted by contaminated food. The ingestion of inadequately cooked seafood or any food cross-contaminated by handling raw seafood or by rinsing with contaminated seawater may transmit this disease. A period of time at room temperature is generally necessary to allow multiplication of the organisms. Disease symptoms include watery diarrhea and abdominal cramps in the majority of cases, sometimes with nausea, vomiting, fever, and headache. Usually the disease is self-limiting and lasts 1 to 7 days. Most cases are reported during the warmer months, and marine coastal environments are a natural habitat. The organisms have been found in marine silt, in coastal waters, and in fish and shellfish.

Of all foodborne infectious diseases, infection with *V. vulnificus* is one of the most severe; the fatality rate for *V. vulnificus* septicemia exceeds 50% (Tacket *et al.*, 1984). Cases are most commonly reported during warm-weather months (April–November), and are often associated with eating raw oysters. Many patients are found to have had a preexisting liver disease, usually associated with alcohol use or viral hepatitis. All of these latter patients had eaten raw oysters 1–2 days before the onset of symptoms. These symptoms include thrombocytopenia, bullous skin lesions, hypotension, and shock. Cases have also been reported that arose through the contamination of a wound by seawater or seafood drippings.

19.2.6 *Helicobacter*

In 1982, a physician in Australia cultured a gram-negative, spiral-shaped bacterium observed in biopsied tissue of a stomach ulcer. Initially called *Campylobacter pylori* because of biochemical and morphological characteristics very similar to those of *Campylobacter*, the organisms are now named *Helicobacter pylori*.

The stomach mucosa contains cells that secrete proteolytic enzymes and hydrochloric acid. Other specialized cells produce a layer of mucus that protects the stomach itself from digestion. If this mucous layer is disrupted, the ensuing inflammation leads to an ulcer. *H. pylori* has been shown to bind to O^- blood group antigens on the gastric epithelial cells. People of this blood group are twice as likely to develop gastric ulcers. *H. pylori* produces large amounts of urease, an enzyme that converts urea to the alkaline product ammonia. This results in a high pH in the local region.

The reservoir of these organisms has not been identified with certainty. Humans are thought to be a natural host. Person-to-person spread is probably the dominant source of transmission, although controversy exists over whether the fecal–oral or oral–oral route of transmission predominates. *H. pylori* has been isolated from feces (Thomas *et al.*, 1992) and detected in saliva (Nguyen *et al.*, 1993). Infection may also be spread by ingesting contaminated food or water. In Lima, Peru, persons who drank from the municipal water system had a much higher risk of *H. pylori* infection than did those who drank from private wells (Klein *et al.*, 1991). Nonculturable *H. pylori* organisms have been detected in river water, suggesting that these bacteria can remain dormant in water for months (Shahamat *et al.*, 1993).

19.2.7 *Legionella*

Legionella is the causative agent of legionnaires' disease and Pontiac fever. Both syndromes are characterized initially by anorexia, malaise, myalgia, and headache. Within 24 hours, a fever ensues with chills. A nonproductive cough may occur and abdominal pain and diarrhea are seen in many patients. In legionnaires' disease, chest radiographs show areas of consolidation indicative of pneumonia. Respiratory failure may occur. This disease has a 15% fatality rate in hospitalized cases. Pontiac fever is not associated with pneumonia or death. Patients recover spontaneously in 2–5 days.

Legionella are poorly staining gram-negative bacilli that require cysteine and other nutrients when grown on artificial laboratory media. Over 28 species of *Legionella* have been shown to cause disease, but the most prominent pathogenic species is *L. pneumophila,* which first received extensive attention after an outbreak in 1976 in Philadelphia. The disease is found worldwide, with most sporadic cases occurring during the summer and fall months. The reservoir is primarily aquatic. Hot water systems, air-conditioning cooling towers), and evaporative condensers have all been implicated in outbreaks, as have decorative fountain and retail store misters (Fig. 19.1). The organism has also been found in creeks and ponds and the soil from their banks. The bacterium survives for months in tap and distilled water. The primary route of transmission is thought to be through the inhalation of aerosols. Exposure is not uncommon, as reflected in serologic assays that show that 1–20% of the general population have antibodies to *L. pneumophila.*

Concentrations of 1.4×10^4 to 1.7×10^5 cells/l of *Legionella* spp. have been detected in raw drinking water sources using direct fluorescent antibody (DFA) techniques (Tison and Seidler, 1983). The concentrations of *Legionella* found in distribution water samples by DFA were as follows [colony-forming units

FIGURE 19.1   (A) Sources of *Legionella* in the environment. (B) Cooling towers. Outbreaks of Legionnaires' disease have been commonly traced to cooling towers. (Photo by P. E. Gerba.)

455

(CFU)/l]: chlorinated water, $<8 \times 10^3$ to 1.4×10^4; water treated by slow sand filtration and chlorination, $<5.4 \times 10^3$ to 4.6×10^4; water treated by flocculation, filtration, and chlorination, $<8 \times 10^3$ to 2.2×10^4. Attempts to isolate the *Legionella* by plating techniques or guinea pig injection failed in all cases. The failure to recover *Legionella* may have been due to (1) nonviable and avirulent cells or (2) viable cells injured by the treatment process and rendered unable to survive.

Zacheus and Martikainen (1994) conducted one of the rare studies in which *L. pneumophila* was enumerated (by culture) in domestic potable water samples. They found that 30% of hot water systems in apartment buildings contained *Legionella*. The mean number of *L. pneumophila* was 2.7×10^3 CFU/ml with a range of <50 to 3.2×10^5 CFU/ml. For all positive hot water systems, *Legionella* was also isolated from the hot water tap and shower head. *Legionella pneumophila* was isolated from 12.5, 29.0, and 37.5% of the hot water distribution systems receiving chlorinated groundwater, unchlorinated groundwater, and chlorinated surface water, respectively.

Legionella was recovered from distribution water samples in 32.7% of homes in a survey in Quebec, Canada, most often associated with domestic hot water reservoirs (Alary and Joly, 1991). *L. pneumophila* was the species most frequently recovered, followed by unknown *Legionella* sp, *L. longbeachae*, and *L. micdadei*.

Stout *et al.*, (1992) isolated *L. pneumophila* from 6.4% of the distribution systems in private homes. The concentration of *L. pneumophila* isolated from hot water tank samples ranged from 1×10^4 to 6×10^5 CFU/l. Results of serological and urinary antigen testing showed that none of the subjects residing in *Legionella*-positive homes had any evidence of previous infection. As none of the subjects were immunosuppressed, this suggested that immunocompetent adults do not appear to be at high risk of acquiring legionnaires' disease even if residing in homes with contaminated water supplies.

A great deal of work has been carried out on the survival and growth of legionellae in potable water distribution systems and plumbing in hospitals and homes. Legionellae appear to be more resistant to chlorine than *E. coli*. For example, *Legionella* has been shown to exist in potable water systems even when exposed to 0.75 to 1.5 ppm free chlorine residual. Such protection is afforded by intracellular growth in protozoa such as *Tetrahymena pyriformis* and *Acanthamoeba castellani* (Fields *et al.*, 1984; Moffat and Tompkins, 1992).

Legionella survives well at 50°C and environmental isolates are able to grow in tap water at temperatures as high as 42°C. The enhanced survival and growth in these systems have been linked to stagnation stimu-

lated by rubber fittings in the plumbing system (Colbourne *et al.*, 1988) and trace concentrations of metals such as iron, zinc, and potassium (States *et al.*, 1989). Sediment promotes the growth of *Legionella*, as does stagnation in the hot water tank. *Legionella* can be removed from hot water heaters by raising the temperature to over 60°C near the heating element and to over 50°C at outlets, combined with regular flushing (Meenhorst *et al.*, 1985).

The overall attack rate of pneumonia in the United States is 12–15 cases per 1000 persons per year, resulting in approximately 3,957,000 cases annually. Pneumonia is the sixth leading cause of death in the United States, with an estimated annual cost of $23 billion. *L. pneumophila* causes 4.1 to 20.1% of community-acquired cases, many of which result in hospitalization (Marrie, 1994). These data suggest that *L. pneumophila* is a major cause of serious cases of pneumonia.

19.2.8 Opportunistic Bacterial Pathogens

An opportunistic pathogen is one that usually causes disease only in those whose immune system is compromised. The weakened immune system may be due to very young or old age, pregnancy, cancer therapy, immunosuppressive drugs, human immunodeficiency virus (HIV), and other causes. Opportunistic pathogens are numerous in the environment. Many opportunistic bacterial pathogens are found in surface and drinking waters.

Concern has been generated in the drinking water industry regarding the health effects of heterotrophic bacteria that are found in tap water, bottled water, and other sources of potable water. Heterotrophic bacteria are those that require organic carbon rather than carbon dioxide as a carbon source. All human bacterial pathogens are heterotrophic.

Most of the heterotrophic bacteria in drinking water are not human pathogens. However, some of the genera, including *Legionella*, *Mycobacterium*, *Pseudomonas*, *Acinetobacter*, *Stenotrophomonas*, and *Aeromonas*, include species that are opportunistic pathogens. See Section 19.2.7 for a discussion of the genus *Legionella*.

The most important opportunistic pathogen in the genus *Pseudomonas* is *P. aeruginosa*, which is primarily a **nosocomial** (hospital-acquired) pathogen responsible for 9.9% of nosocomial infections. This gram-negative microbe is the most frequent source of infection of burn patients, the second leading cause of nosocomial pneumonia, the third most common cause of nosocomial urinary tract infections, and the fourth leading cause of surgical wound infections. Persons in the community at the greatest risk of infection are those with cystic fibrosis and those who engage in a great deal of swimming ("swimmer's ear"). Human disease is often associated with water-related reservoirs such as swimming pools, whirl-pools, hot tubs, and contact lens solutions. The source of community-acquired infections of cystic fibrosis patients has not been clarified. It is important as the ultimate cause of death in these patients is often lung infections by *P. aeruginosa*. Although *P. aeruginosa* is found in drinking water, it is not ubiquitous. Results of surveys showed that 2% of bottled water and 2–3% of tap water samples contain *P. aeruginosa* at concentrations between 1 and 2300 organisms/ml (Allen and Geldreich, 1975).

Acinetobacter is the second most frequently isolated nonfermentative gram-negative rod in the clinical laboratory. However, it is generally considered to be of low virulence. *Acinetobacter* are isolated as commensals from the skin and respiratory tract of healthy individuals. Up to 25% of healthy adults carry this organism in the respiratory tract. The intestine does not seem to be an important reservoir. The human skin is the likely source for most outbreaks of hospital infections. As an opportunistic pathogen, *Acinetobacter* is involved in nosocomial urinary tract infections, bacteremia, wound infections, and pneumonia. *Acinetobacter* can also be a cause of community-acquired pneumonia and urinary tract infections. Predisposing factors include malignancy, burns, immunosuppression, major surgery, and very young or old age.

Acinetobacter has been isolated from 97% of natural surface water samples with a concentration of 0.1–100 cells/ml (Baumann, 1968). It has also been isolated from 42% of mineral water springs (Mosso *et al.*, 1994) and from two of three uncarbonated mineral springs (Gonzalez *et al.*, 1987). Bifulco *et al.*, (1989) isolated *Acinetobacter* from 38% of groundwater supplies at an arithmetic mean density of 8 CFU/100 ml in the positive samples. Mroz and Pillai (1994) found *Acinetobacter* in only 3 of 47 wells in the El Paso, Texas area.

Acinetobacter has been isolated from 5 to 92% of distribution water samples (LeChevallier *et al.*, 1980; Bifulco *et al.*, 1989). It comprised 1.0 to 5.5% of the heterotrophic plate count (HPC) flora in drinking water samples (Payment *et al.*, 1988) with concentrations of 6–21 CFU/ml. It has also been found in 5 to 35% of bottled water samples, at concentrations of 2–30 CFU/ml, and in 10% of cooler water samples at concentrations of 100–350 CFU/ml (Gonzalez *et al.*, 1987).

Pseudomomas maltophilia has been reclassified as *Stenotrophomonas maltophilia* and is the third most commonly isolated nonfermentative gram-negative rod in clinical laboratories. This organism can colonize the body and cause disease. Risk factors for infection include stays in intensive care units, mechanical ventilation, antibiotic treatment, and cancer. Diseases it can cause include septicemia, pneumonia, wound infections, and more rarely meningitis and endocarditis.

S. maltophilia constituted 5.7% of the HPC found in raw surface water samples and 0–1.2% of the flora in distribution water samples (LeChevallier *et al.*, 1980). Distribution water is drinking water that has exited the water treatment plant and is being distributed in pipes to the consumer. Payment *et al.* (1994) also isolated *S. maltophilia* from tap water but no quantitative information was included. The bacterium has also been found in 2% of bottled water samples at concentrations of 2–22 CFU/ml and in <5% of cooler water samples at 34 CFU/ml.

Aeromonas hydrophila, A. sobria, and *A. caviae* are very similar and were all referred to as *A. hydrophila* until 1976, when they were first split into separate species. Papers written before 1985 may use the term *A. hydrophila* including all three species and biochemical variants. The three gram-negative species are biochemically very similar and have all been implicated as diarrheal agents in humans. The exact mechanism for diarrhea has not been elucidated.

A strong association has been found between drinking untreated water and the occurrence of diarrhea with the isolation of *Aeromonas* species. Higher counts of *A. hydrophila* in distribution water correlate with greater frequency of diarrheal isolates; however, no waterborne outbreaks have been documented.

Children and the elderly tend to be affected most often. As with many diarrheal agents, outbreaks of diarrhea, in which *Aeromonas* has been implicated, have been associated with day care centers. *Aeromonas* is frequently found in environmental water samples (Table 19.4). It has been recovered from 0.6 to 18.2% of natural fresh water samples at a concentration range of 0.1–3,600 CFU/ml. It has also been recovered from 0.9 to 27% of distribution water samples at an average concentration of 0.022 CFU/ml. However, even at large oral doses of up to 10^{10} CFU, *A. hydrophila* failed to produce diarrhea in human volunteers (Morgan *et al.*, 1985).

The *Mycobacterium avium* complex (MAC) consists of 28 serovars of two distinct species, *M. avium* and *M. intracellulare*. It also included three serovars of *M. scrofulaceum* in the past, but its inclusion is no longer appropriate due to recent advances in mycobacterial systematics. Pulmonary disease caused by MAC is now as common as pulmonary tuberculosis in many areas of the United States. Evidence suggests the prevalence of this disease is growing. Predisposing factors include age, chronic lung disease, bronchogenic carcinoma, previous gastrectomy, and AIDS. Twenty to 40% of AIDS patients experience disseminated infections with MAC. However, MAC can also cause pulmonary

TABLE 19.4 Isolation of *Aeromonas* spp. from Water Sources

Lake or pond water	Well water	River water	Fresh water	Distribution water	References
0.1–205/ml (mean 20)[a]	—	0.4–3,600/ml (mean 161[a])	0.1–3600/ml (mean 130)[a]	—	Hazen *et al.*, 1978
—	—	2000/ml	—	—	Nishikawa and Kishi, 1988
16.7–18.2%	0.6–1.4%	—	—	0.9–3.0%	Clark *et al.*, 1982
—	—	—	15.9%	9.5%	LeChevallier, 1980
—	—	—	—	27%, 0.01–19.0/ml (average 0.022/ml)	LeChevallier, *et. al.*, 1982
—	—	—	—	22%[a]	Millership and Chattopadhyay, 1985
28–467/ml[a]	—	47/ml[a]	—	—	Rippey and Cabelli, 1979
10–300/ml (84%[a])	—	—	—	—	Seidler *et al.*, 1980
10^1–10^2ml[a]	—	—	—	—	Dan and Koppel, 1992
10^{-1}–10^1/ml	—	—	—	—	Baleux and Trousllier, 1989
10^{-1}–10^2/ml	—	—	—	—	Rippey and Cabelli, 1989
—	—	—	10/ml[a]	—	Biamon and Hazen, 1983

[a]*Aeromonas hydrophila.*

disease, osteomyelitis, and septic arthritis in people with no known predisposing factors (Jones *et al.*, 1995). In the United States, many infections are asymptomatic and occur early in life: 12% of the population has been infected by MAC (von Reyn *et al.*, 1993a). However, disease by MAC can be lethal and is difficult to treat because of resistance to many antimycobacterial agents.

Despite extensive investigation, the precise mode of transmission of MAC remains undetermined. Infection with MAC is thought to occur from colonization of the gastrointestinal and/or respiratory tract (Chin *et al.*, 1994). This suggests that exposure occurs by inhalation or ingestion.

Epidemiological investigations have associated water sources with infections by atypical mycobacteria (Burns *et al.*, 1991). These bacteria can multiply in water that is oligotrophic (Falcao *et al.*, 1993) or essentially free of nutrients, including dialysis water, and they are relatively resistant to disinfection by chlorination and chloramines (Collins *et al.*, 1984). Atypical mycobacteria are widespread in water environments (Table 19.5). Mycobacteria have been isolated from 11 to 38% of raw water samples at concentrations of <0.1–48 CFU/ml. It has also been found in up to 50% of municipal and private drinking water samples at concentrations of 0.01–5.2 CFU/ml.

Hospital water systems often harbor MAC and may be a source of nosocomial infection (von Reyn *et al.*, 1994). *M. avium* has primarily been associated with hot water systems (du Moulin *et al.*, 1988; von Reyn *et al.*, 1993b) as shown in Table 19.5. In some cases, a hot water system may be persistently colonized by the same strain of *M. avium* (von Reyn *et al.*, 1994).

19.2.9 Blue–Green Algae

Blue–green algae or Cyanobacteria occur in an enormous diversity of habitats, freshwater and marine, as plankton (free floating), mats, and periphyton (attached to surfaces). Hot spring mats of some *Oscillatoria* develop up to temperatures of 62°C. They have many beneficial functions such as nitrogen fixation and cycling of nutrients in the food chain.

Despite their beneficial roles in the environment, cyanobacteria sometimes become problematic. Occasionally, they increase rapidly resulting in cyanobacterial blooms (Fig. 19.2). Blooms are associated with eutrophic water, especially with levels of total phosphorus >0.01 mg/l and levels of ammonia- or

**TABLE 19.5   Isolation of *Mycobacterium avium* Complex (MAC) and Mycobacteria
from Water Sources**

Hot water	Cold water	Showers	Distribution water[a]	Raw water samples	References
33–50%[b]	0%		17–25%	33–38%	von Reyn *et al.*, 1993b
				11–35[c]	Haas and Fatal, 1990
	32%				Montecalvo *et al.*, 1994
0.4–4.2/ml		1.2–5.2/ml	0.2–4.2/ml		von Reyn *et al.*, 1994
			21.4%[b]	26.6%	Goslee and Wolinsky, 1976
			0.5/ml		du Moulin and Stottmeier, 1986
0.01–5/ml (25%)	0.01–.02/ml (13%)				du Moulin *et al.*, 1988
				<0.1–48/ml	Kirschner *et al.*, 1992

[a] Drinking water that has exited the water treatment plant and is in pipes for distribution to the consumers.
[b] Numbers indicate the % of samples that contained Mycobacteria.
[c] *Mycobacterium* species.

nitrate-nitrogen >0.1 mg/l. Optimal temperatures for blooms are 15–30°C, and optimal pH is 6–9. Calm or mild wind conditions sometimes allow blooms to cover the water surface, but the highest concentrations of cyanobacteria may occur at depths ranging from 2–9 m, which will not be visible from the shore. The offending bacteria may also grow in the sediment. These blooms can impart an off-taste and odor to the water and/or result in the production of toxins.

The most common complaints related to such blooms are of taste and odor. Geosmin and 2-methylisoborneol (MIB) can produce odors at levels as low as 1.3–10 and 6.3–29 ng/l, respectively (Young *et al.*, 1996). The odor produced by geosmin is described as earthy and that of MIB as musty or camphorous smelling. Concentrations of MIB and geosmin are usually highest in summer and fall. Several compounds produced by cyanobacteria can cause off-tastes and odors as shown in Table 19.6.

Geosmin is produced by several cyanobacteria including *Oscillatoria, Anabaena, Lyngbya, Phormidium, Symploca* (Narayan and Nunez, 1974), *Aphanizomenon*, and *Fischerella* (Wu and Juttner, 1988). *Lyngbya, Oscilla-*

FIGURE 19.2   Cyanobacterial bloom. (Photo courtesy C. P. Gerba.)

**TABLE 19.6   Cyanobacterial Compounds
Producing Off-Tastes and Odors**

Compound	Odor	Taste
Geosmin	Earthy, musty grassy	Musty, earthy, stale
MIB	Musty, earthy, peaty	Musty, earthy, stale
Isobutylmethoxypyrazine	Woody, stale, musty	Creosote, stale, dusty
Isoprophylmethoxypyrazine	Sooty, dusty, cabbage	Musty, vegetable water
Octa-1,3-diene	Musty	
Hexanal	Green apple-like	
Octan-1-ol	Rancid	
Octan-1-ol	Rancid	
β-Cyclocitral	Tobacco	

toria, and *Phormidium* (Izaguirre, 1992) are the most common genera producing MIB. Some strains of *Diplocystis* and *Schizothrix* can also cause off-tastes and odors. *Microcystis* release some odorous sulfur compounds, especially when they decay.

Many cyanobacteria found in algal blooms can produce toxins that cause liver damage, neural damage, and gastrointestinal (GI) disturbances. This has been well documented in many wild animal and livestock cases and implicated in human cases as well. *Microcystis* is the number one offender worldwide. Other toxin-producing genera include *Anabaena, Aphanizomenon, Alexandrium, Cylindrospermopsis, Nodularia, Nostoc,* and *Oscillatoria* (Turner *et al.,* 1990). Different types of toxins produced are shown in Tables 19.7 and 19.8. Most toxic species are associated with temperate rather than tropical climates.

In livestock and wild animals, the hepatotoxins cause weakness, anorexia, and liver damage. They can be lethal within minutes to a few days. Neurotoxins can cause twitching, muscle contraction, convulsions, and death. Signs and symptoms in humans associated with the ingestion of water with algal blooms are dizziness, headaches, muscle cramps, nausea, vomiting, gastroenteritis, and pneumonia (Phillip *et al.,* 1992). Long-term exposure to toxins is associated with liver cancer as well (Carmichael, 1994).

In mice, the median lethal dose fifty or LD_{50} (with intraperitoneal injection) of toxins varies. The LD_{50} of *Anabaena circinalis* neurotoxin is 36 ± 1.4 mg/kg mouse (Falconer *et al.,* 1989). The hepatotoxin of *Microcystis* had an LD_{50} of 14.8 ± 1.1 mg/kg mouse.

19.3 PARASITOLOGY

The study of parasitology embodies a large diversity of eukaryotic organisms. This group includes organisms that are unicellular, multicellular, and multi-

TABLE 19.7 Cyanobacteria and Types of Toxins Produced

Genus	Toxins produced
Anabaena	Anatoxin a, hepatotoxins
Aphanizomenon	Saxitoxin, neosaxitoxin, hepatotoxins
Alexandrium	Saxitoxin
Cylindrospermopsis	Hepatotoxin
Nodularia	Nodularins
Oscillatoria	Neurotoxins, hepatotoxins
Microcystis	Microcystins

TABLE 19.8 Characterization of Cyanobacterial Toxins

Toxin	Characterization of toxin
Anatoxins	Neurotoxins
Microcystins	Hepatotoxins
Nodularins	Hepatotoxins
Saxitoxins	Neurotoxins

nucleate; aerobic and anaerobic; motile and nonmotile; sexual and asexual. Generally, the study of viruses and bacteria is not included in parasitology as they each have their own discipline. For this chapter parasites will be grouped into two categories: protozoa and helminths. Protozoa are unicellular microorganisms in the Kingdom Protista and are classified according to their means of locomotion: flagella, cilia, pseudopodia, or no locomotion (Table 19.9). Some are parasitic, although the majority are nonparasitic. Some undergo a sexual stage, whereas others reproduce by asexual means: fission, budding, or schizogony. Helminths belong to Kingdom Animalia and include roundworms, flatworms, tapeworms, and flukes. They are multicellular complex organisms containing organs and tissue. They usually develop in the soil or in intermediate hosts

TABLE 19.9 Classification of Some Environmentally Transmitted Protozoa and Helminths

Protozoa
  Phylum Apicomplexa
    Cyclospora cayetanesis
    Cryptosporidium parvum
    Toxoplasma gondii
  Phylum Microspora
    Enterocytozoon bieneusi
    Encephalitozoon cuniculi
    Encephalitozoon hellem
    Encephalitozoon intestinalis
  Phylum Sarcomastigophora
    Entamoeba histolytica
    Giardia lamblia
    Naegleria fowleri
Helminths
  Phylum Nematoda
    Ascaris lumbricoides
    Necator americanus
    Trichuris trichiura
  Phylum Platyhelminthes
    Class Cestoidea
      Taenia saginata
    Class Trematoda
      Schistosoma mansoni

to complete their life cycle and have elaborate life cycles including larva, eggs, and adult stage. Parasitic protozoa and helminths can be acquired from the environment—water, soil, and contaminated food—and have had great health and economic impacts in many developing and developed countries. In the United States, there are two protozoa that are especially of concern: *Giardia* and *Cryptosporidium.* They form hardy cysts and oocysts that can survive water treatment disinfection and are one of the biggest concerns of water utilities today. Some of the characteristics of environmentally transmitted parasites are summarized in Table 19.10.

19.3.1 Protozoa

19.3.1.1 Giardia lamblia

Giardia lamblia was first described in 1681 by Antonie van Leeuwenhoek, who found them in his own feces. He called the trophozoites "animalcules." In 1859, Vilem Lambl rediscovered *Giardia* by finding the trophozoites in stools of young children with diarrhea. It was not until the early 20th century that physicians began associating diarrhea with the presence of *Giardia* in stools. In 1954 Robert Rendtorff confirmed infectivity in human volunteers with oral administration of *Giardia* cysts. *Giardia* is the most frequently identified

TABLE 19.10   Some Characteristics of Environmentally Transmitted Parasites

Organism	Infective form	Mechanism of transmission	Distribution	Reservoirs
Giardia lamblia	Cyst	Person–person Waterborne Foodborne	Worldwide	Humans, beavers, muskrats, voles
Cryptosporidium parvum	Sporulated oocyst	Person–person Waterborne Foodborne	Worldwide	Many vertebrates, especially cattle
Entamoeba histolytica	Cyst	Person–person Waterborne Foodborne	Areas of poor sanitation	Usually humans (potentially pigs, primates, and dogs)
Naegleria fowleri	Trophozoite	Trophozoite swims up nasal cavity	Worldwide	None: free living in aquatic or soil environment
Cyclospora cayetanesis	Sporulated oocyst	Waterborne foodborne	Asia, Caribbean, Mexico, and Peru	None known
Enterocytozoon bieneusi	Spore	Fecal–oral	Not known	None known
Encephalitozoon hellem	Spore	Urine–oral	Not known	None known
Encephalitozoon cuniculi	Spore	Fecal/urine–oral	Not known	Laboratory rabbits, rodents, dogs
Encephalitozoon intestinalis	Spore	Fecal/urine–oral	Not known	None known
Toxoplasma gondii	Sporulated oocyst	Oral ingestion from soil or litterbox (oocyst) Undercooked meats (tissue cysts)	Worldwide	Cats: definitive host humans, sheep, goats, pigs, cattle, birds intermediate hosts
Ascaris lumbricoides	Embroynated egg	Oral from soil contact	Worldwide	Humans
Tricuris trichiura	Embroynated egg	Waterborne foodborne	Worldwide especially tropics	Humans
Necator americanus	Filariform larva	Skin penetration	Tropical Africa, Asia, Central and	Humans
Ancylostoma duodenale	Filariaform larva	Skin penetration, Oral from soil contact	South America and Caribbean	Humans
Taenia saginata	Cysticercus	Ingestion of undercooked beef containing cysticerci, waterborne (?)	Worldwide	Intermediate host: cattle
Schistosoma mansoni	Cercariae	Penetrate skin	Arabia, Africa, South America, and Caribbean	Humans Intermediate host: snail

intestinal parasite in the United States (Adam, 1991). In the United States, the prevalence of *G. lamblia* infections has been estimated to range from <1% in midwestern middle-class adults (McHenry *et al.*, 1987) to as high as 10–13% in Oregon adults (Skeels *et al.*, 1986). In some settings, such as day care, the incidence can be as high as 33% in children (Ginsberg *et al.*, 1994). In the United States, it is one of the most frequent identified causes of waterborne disease (Craun, 1986).

Humans become infected with *G. lamblia* by ingesting the environmentally resistant stage, the cyst (Figs. 19.3 and 19.4). Once ingested, it passes through the stomach and into the upper intestine. The increase in acidity via passage through the stomach stimulates the cyst to excyst, which releases two trophozoites into the upper intestine. The trophozoites attach to the epithelial cells of the small intestine. It is believed that the trophozoites use their sucking disks to adhere to epithelial cells. The adherence to the cells flattens the villi, causing malabsorption and diarrhea by not allowing adsorption of water and nutrients across the intestine. It can cause both acute and chronic diarrhea within 1–4 weeks of ingestion of cysts resulting in foul-smelling, loose, and greasy stools. Once the trophozoites detach from the epithelial cells and travel down the intestine, cholesterol starvation is believed

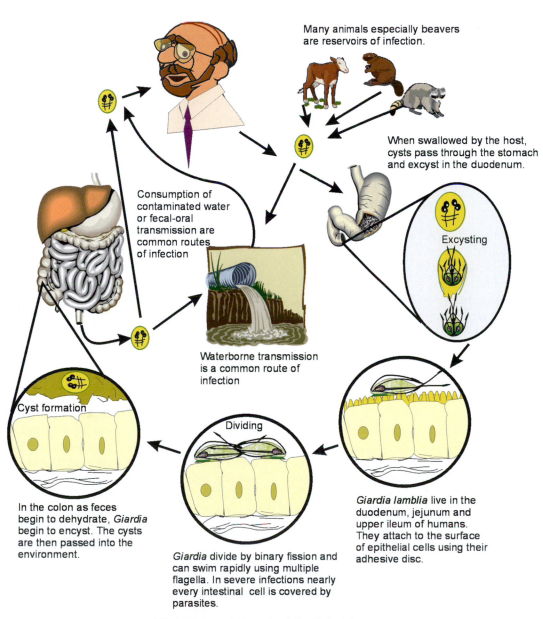

Many animals especially beavers are reservoirs of infection.

When swallowed by the host, cysts pass through the stomach and excyst in the duodenum.

Excysting

Consumption of contaminated water or fecal-oral transmission are common routes of infection

Waterborne transmission is a common route of infection

Cyst formation

Dividing

In the colon as feces begin to dehydrate, *Giardia* begin to encyst. The cysts are then passed into the environment.

Giardia divide by binary fission and can swim rapidly using multiple flagella. In severe infections nearly every intestinal cell is covered by parasites.

Giardia lamblia live in the duodenum, jejunum and upper ileum of humans. They attach to the surface of epithelial cells using their adhesive disc.

FIGURE 19.3   Life cycle of *Giardia lamblia*.

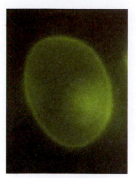

Giardia lamblia

Cryptosporidium parvum

FIGURE 19.4 Fluorescein isothiocyanate–labeled *Giardia lamblia* and *Cryptosporidium parvum* at 100× magnification. (Photos courtesy D. C. Johnson).

to stimulate the trophozoites to encyst and pass back into the environment as a cyst. Giaridiasis can be treated with metronidazole (Flagyl). Although in some cases the symptoms will spontaneously disappear without treatment, in most cases without treatment, the symptoms will wax and wane for many months. In symptomatic patients, more trophozoites than cysts are excreted into the feces, which cannot withstand the harshness outside the human body. In asymptomatic humans, mostly cysts are passed in stools; therefore *Giardia* carriers can serve as a source of cysts in the environment. The infectious dose of *Giardia* is quite low, which makes even small numbers in water a health threat.

When *Giardia* cysts enter the environment they can survive for prolonged periods. *G. lamblia* cysts have been documented to survive for up to 77 days at 8°C and 4 days at 37°C in distilled water (Bingham *et al.*, 1979). In another study, cysts of *Giardia muris* (a species that infects mice but is often used as a model for *G. lamblia*) were suspended in lake and river water and found to survive 28 days in lake water at a depth of 15 feet at 19.2 ± 1.3°C. At a 30-foot depth at a temperature of 6.6 ± 0.4°C, the cysts remained viable for 56 days. In cold (0–2°C) river water the cysts survived more than 56 days (DeRegnier *et al.*, 1989).

It is still controversial whether animals such as beavers produce a *G. lamblia* strain infectious for humans. Studies have shown that beavers can be infected with *G. lamblia* from humans, but the reverse has not

been demonstrated (Erlandsen *et al.*, 1988). In one study 40 to 45% of beavers in Colorado were found to be infected with *Giardia* and shedding up to 1×10^8 cysts/animal/day, making them a major source of *Giardia* in the environment (Hibler and Hancock, 1990). Other animals that may contribute to *Giardia* in the environment are muskrats, where 95% of the population is infected. Various other animals found to be infected are cattle, goats, sheep, pigs, cats, and dogs (Erlandsen, 1995). To date, no infections with *G. lamblia* in humans have been directly linked to an animal host, but there is evidence which suggests that animal-source *Giardia* could potentially infect humans. Studies based upon isoenzyme analysis and pulsed field gel electrophoresis (PFGE) banding patterns did not find a difference between cysts from beaver hosts and human hosts (Isaac-Renton *et al.*, 1993).

The cysts, which measure 8–16 μm in diameter, can survive certain levels of wastewater treatment. During primary settling only 0–53% of cysts are removed. Secondary treatment can reduce the number of cysts: secondary clarification can remove 98.6–99.7% of cysts (Casson *et al.*, 1990). Advanced tertiary treatment can further remove remaining cysts through physical filtration and chemical precipitation.

19.3.1.2 Cryptosporidium parvum

Cryptosporidium was first described by Tyzzer in 1907 when he identified the organism in the intestinal epithelium of a mouse. It was not identified as a human pathogen until 1976, when it was described in the stools of immunocompromised hosts (Meisel *et al.*, 1976). Since that time, there have been several waterborne outbreaks, the most notable being the Milwaukee outbreak in April 1993, which infected over 400,000 people, (MacKenzie *et al.*, 1994) and killed more than 50 (Hoxie *et al.*, 1997).

Cryptosporidium parvum has a complex life cycle involving both sexual and asexual stages (Fig. 19.5). The host ingests sporulated oocysts (ranging from 3 to 6 μm in diameter) from contaminated water, food, or direct contact (Fig 19.4). In the small intestine, the oocyst excysts, releasing four sporozoites, which attach to the epithelial cells of the mucosa. The sporozoite becomes enveloped by the microvilli, which fuse and elongate to cover the sporozoite. The sporozoite then matures into a trophozoite and into a schizont. The schizont, an asexual reproductive form in which multiple mitosis occurs followed by cytokinesis, results in eight first-generation merozoites. The cell ruptures, releasing the merozoites, which then infect neighboring epithelial cells, and schizogony occurs again, but forming only

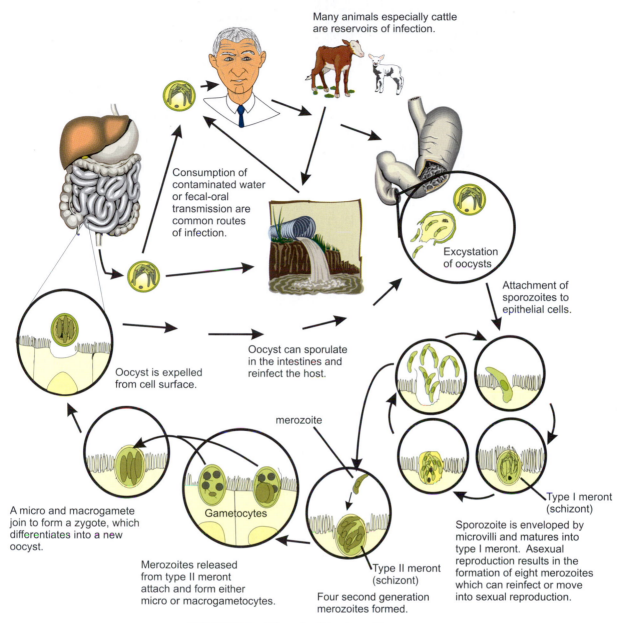

Many animals especially cattle
are reservoirs of infection.

Consumption of
contaminated water
or fecal-oral
transmission are
common routes
of infection.

Excystation
of oocysts

Attachment of
sporozoites to
epithelial cells.

Oocyst is expelled
from cell surface.

Oocyst can sporulate
in the intestines and
reinfect the host.

merozoite

Type I meront
(schizont)

A micro and macrogamete
join to form a zygote, which
differentiates into a new
oocyst.

Gametocytes

Type II meront
(schizont)

Four second generation
merozoites formed.

Sporozoite is enveloped by
microvilli and matures into
type I meront. Asexual
reproduction results in the
formation of eight merozoites
which can reinfect or move
into sexual reproduction.

Merozoites released
from type II meront
attach and form either
micro or macrogametocytes.

FIGURE 19.5   Life cycle of *Cryptosporidium parvum.*

four second-generation merozoites. When the cell ruptures, the merozoites attach to uninfected epithelial
cells and form either macrogametocytes or microgametocytes. The macrogametocytes and microgametocytes further divide and form macrogametes and microgametes, respectively. They join, forming a zygote,
which differentiates to form the unsporulated oocyst,
which is then expelled from the cell surface, sporulates, and is shed in the host's feces.

Within 3–10 days after ingestion of oocysts, nonbloody, voluminous watery diarrhea begins and lasts
for 10–14 days in most immunocompetent hosts. There
is no medical treatment and the disease is self-limiting;

however, immunocompromised hosts (e.g., AIDS patients) can succumb, and about 10–15% of AIDS patients die of complications related to cryptosporidiosis. The prevalence of cryptosporidiosis in the United
States is 0.3–4.3% (Ungar, 1990). The oocysts are very
infectious and the presence of low numbers of oocysts
in water or food poses a health threat (DuPont *et al.*,
1995).

Cryptosporidium oocysts can enter the environment
via human and animal wastes. They have been found in
marine water and bathing beaches in the vicinity of a
nearby sewage outfall (Johnson *et al.*, 1995). Cryptosporidiosis has been reported in many domestic an-

imals, especially cattle. An infected calf can excrete 10^{10} oocysts per day. In a study by Kemp *et al.* (1995), farm drains were found to contain 0.06 to 19.4 oocysts per liter. This can result in agricultural land runoff that can contaminate surface water.

C. parvum forms an extremely hardy oocyst that survives chlorine disinfection as commonly practiced at conventional water treatment plants. It has also been found to survive for weeks in surface waters (Johnson *et al.*, 1997).

19.3.1.3 Entamoeba histolytica

Entamoeba histolytica was discovered in 1873 by D. F. Lösch in St. Petersburg, Russia, although its life cycle was not determined until Dobell did so in 1928. It causes amebic dysentery (bloody diarrhea) and is the third most common cause of parasitic death in the world. The world prevalence exceeds 500 million infections with more than 100,000 deaths each year. There are two sizes of cysts, small (5–9 μm) and large (10–20 μm), with each cyst producing eight trophozoites in the host. Only the larger cyst has been associated with disease; the smaller cyst tends to be associated with a commensal lifestyle (the organism benefits from the host, while the host is unaffected). About 2–8% of people infected develop invasive amebic dysentery in which the trophozoites actively invade the intestinal wall, blood stream, and liver. It is unknown why this occurs. This organism is generally a problem in developing countries where sanitation is substandard and is transmitted via contaminated food and water. Flies and cockroaches can serve as mechanical vectors for transmission in developing countries. Humans are the main reservoir, although pigs, monkeys, and dogs have also been found to serve as reservoirs. No waterborne outbreaks have occurred in the United States for more than 30 years. *Entamoeba* is not as resistant to disinfectants as *Giardia* and *Cryptosporidium*.

19.3.1.4 Naegleria fowleri

Naegleria fowleri is an amoeboflagellate, changing between a cyst, ameba, and flagellate with the amoeba stage dominant. The free-living protozoa are ubiquitous and found throughout the world in fresh waters (John, 1982). Cysts are usually present in low numbers, but when the water temperature exceeds 35°C (hot springs and warm stagnant waters), the amoeba transforms to the flagellated form quite rapidly, which enables the microorganism to swim. Infections are usually associated with children swimming in natural springs or warm waters. The flagellate swims into the nose of a host and sheds its flagella (or it may be forced into the nose via diving). The amoeba then follows the nerves to the brain, producing a toxin that liquefies the brain. The organisms do not form cysts in the host. **Primary amoebic meningoencephalitis (PAM)** develops, causing severe, massive headaches. Death usually follows 4–6 days later. Diagnosis is most frequently postmortem upon brain examination. There is treatment (amphotericin B) if the diagnosis is made quickly enough, although permanent brain damage may already have occurred.

19.3.2 Nematodes

Nematodes are nonsegmented roundworms belonging to the phylum Nematoda. The majority of roundworms are free living in soil and fresh and salt water. The typical nematode has a flexible outer cuticle that protects the worm. They move via a muscular system and most lay eggs.

19.3.2.1 Ascaris lumbricoides

Ascaris lumbricoides is probably the most prevalent parasitic infection, with over 1 billion affected or about 22% of the world's population (Crompton, 1988). The prevalence is quite high in some regions of the world. Infection percentages range from 40–98% in Africa, 73% in Southeast Asia, 45% in Central America and South America, and 2% in the United States (Freedman, 1992; Stoll, 1947). The percentage actually infected may be higher in the United States, with one study finding about a 20% infection rate in the southeastern United States (Schultz, 1974).

Infection in humans occurs by ingestion of embryonated eggs (Figs. 19.6 and 19.7). There are no known animal reservoirs. The eggs are swallowed and the larvae hatch in the small intestine (Fig. 19.6). The larvae develop into second-stage larvae, which penetrate the lumen and enter into the blood stream and capillaries. They travel via pulmonary circulation to the liver and heart, where the larvae develop into third-stage organisms that can lodge in the alveolar space. This migration through the lungs can cause pneumonitis (Loeffler's syndrome). The immature worms leave the lungs and travel through the bronchi, trachea, and epiglottis. They are then swallowed and arrive in the small intestine. There they undergo two molts, mate, and produce eggs. There is medical treatment against the adult intestinal worm: mebendazole or pyrantel pamoate. Symptoms usually correspond to the worm load, and a heavy worm load can lead to intestinal blockage. The adult worm can reach more than 30 cm

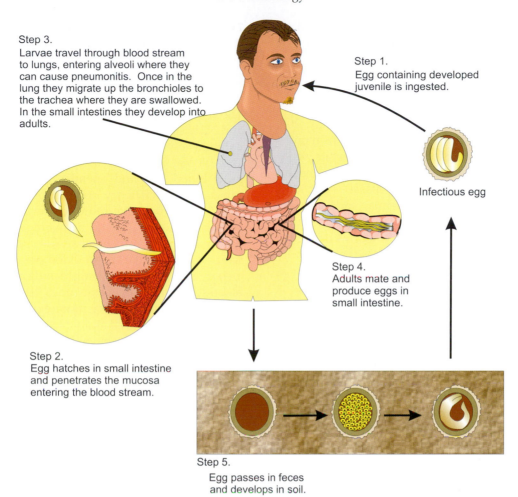

Step 3.
Larvae travel through blood stream to lungs, entering alveoli where they can cause pneumonitis. Once in the lung they migrate up the bronchioles to the trachea where they are swallowed. In the small intestines they develop into adults.

Step 1.
Egg containing developed juvenile is ingested.

Infectious egg

Step 4.
Adults mate and produce eggs in small intestine.

Step 2.
Egg hatches in small intestine and penetrates the mucosa entering the blood stream.

Step 5.
Egg passes in feces and develops in soil.

FIGURE 19.6   Life cycle of *Ascaris lumbricoides*.

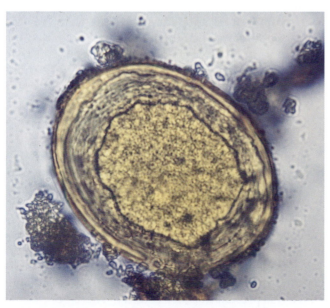

FIGURE 19.7   *Ascaris lumbricoides*. (Photo courtesy P. Watt.)

in size. Although most infections are mild, more than 20,000 people die annually with complications caused by intestinal blockage (Freedman, 1992).

An adult worm can produce more than 200,000 eggs per day. In 2–4 weeks after deposition in soil, they embryonate if the soil conditions are suitable (humid and warm) and are infectious. The eggs can survive months before embryonation if soil conditions are not appropriate. The eggs can survive freezing, chemicals, disinfectants, and sewage treatment. Because of their large size (35 × 55 μm), they accumulate in sewage sludge. This is especially a concern with sewage sludge-amended soils. In one study, 31–53% of eggs that had been deposited in soil were still viable 10 years later (Brudastov *et al.*, 1971).

19.3.2.2 *Trichuris trichiura*

Trichuris trichiura is a worm that measures about 30–50 mm in length and is referred to as a whipworm as the worm's shape resembles a whip. It has been estimated that there are over 355 million infections worldwide (Stoll, 1947), and it is the third most common nematode infection. It is common in the southeastern United States as the weather conditions are ideal for egg survival in the soil (Fig. 19.8). The egg must be deposited in the soil and requires 21 days in moist, shady, warm soil to embryonate. In one study, 20% of ova deposited in soil were viable after 18 months (Burden *et al.,* 1976). Infection occurs in humans via ingestion of contaminated water or soil. The worms can survive for years in a host, causing disease symptoms of diarrhea or constipation, anemia, inflamed appendix, vomiting, flatulence, and insomnia. The infection is diagnosed by identification of worms or eggs in the stool. The infection can be treated successfully with mebendazole. To prevent transmission, education on hand washing and sanitary feces disposal is necessary.

FIGURE 19.8 *Trichuris trichiura* eggs. (Photo courtesy P. Watt.)

19.3.2.3 Necator americanus and Ancylostoma duodenale

There are two major hookworm species that infect humans: *Necator americanus* (New World hookworm) and *Ancylostoma duodenale* (Old World hookworm). The species are differentiated by mouth parts of adults and body size. There are no known reservoirs. They inhabit the small intestine and feed on intestinal mucosa and blood. They secrete an anticoagulant, causing great blood loss and anemia. They are the leading cause of iron deficiency in the tropics.

N. americanus have round cutting teeth and are 7–10 mm in length. They lay an average of 10,000 eggs a day and enter humans via skin penetration. Each worm consumes 0.03 ml of blood per day. *A. duodenale* are larger (10–12 mm in length) and cause even greater blood loss. Each worm consumes 0.26 ml of blood per day. *A. duodenale* also lay more eggs (28,000) and are orally infective as well as able to penetrate the skin. In addition, *A. duodenale* have sharp cutting teeth.

The eggs embryonate once they are passed into the small intestine (Fig. 19.9). The eggs further develop into the rhabditiform larvae within 48 hours in warm, moist sandy or loamy soil. The larvae feed, grow, and molt twice and then transform to filariform larvae. The filariform do not eat. They seek out the highest point in the surroundings (e.g., top of grass blade) waiting for a host. Upon contact with skin, they penetrate the tissue and pass through a hair follicle or cut. They burrow through the subcutaneous tissue, then through the capillaries to the lungs. In the lungs they break out of the alveolar capillaries and migrate up the bronchi and trachea, where they are swallowed and enter the stomach and small intestine. They can live an average of 5 years but have been found to survive up to 15 years. In addition, the larvae can be vertically transmitted via breast milk. The larvae can survive up to 6 weeks in moist, shady sandy or loamy soil. They do not survive well in clay soil, dry conditions, or at temperatures below freezing or greater than 45°C.

19.3.3 Cestodes (*Taenia saginata*)

Cestodes are tapeworms consisting of a flat segmented body and a scolex (head) containing hooks and/or suckers and grooves for attachment. The segments are called proglottids and pregnant segments are called gravids. The adults are parasitic and live in the intestinal lumen of many vertebrates.

Taenia saginata is transmitted by infected beef products and is the most common tapeworm found in humans. It is present in every country where beef is con-

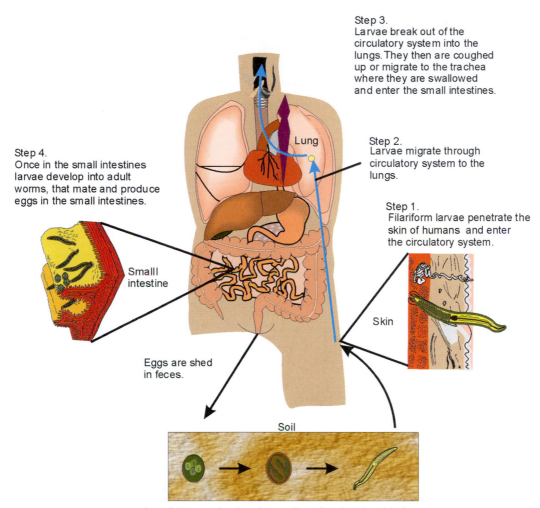

Step 3.
Larvae break out of the circulatory system into the lungs. They then are coughed up or migrate to the trachea where they are swallowed and enter the small intestines.

Step 4.
Once in the small intestines larvae develop into adult worms, that mate and produce eggs in the small intestines.

Lung

Step 2.
Larvae migrate through circulatory system to the lungs.

Step 1.
Filariform larvae penetrate the skin of humans and enter the circulatory system.

Small intestine

Skin

Eggs are shed in feces.

Soil

In soil the egg develops into embryo then hatches into larvae.

FIGURE 19.9   Life cycle of *Necator americanus*.

sumed (Fig. 19.10). Cattle become infected from eating grass or soil contaminated with human waste containing gravid proglottids. This occurs in areas where night soil (human waste) is used as fertilizer. The proglottids can survive in the environment for weeks. One study found that they could survive 71 days in liquid manure, 16 days in untreated sewage, and up to 159 days on grass (Jepsen and Roth, 1952). The eggs hatch in the duodenum, and hexacanths (tapeworm embryos containing six pairs of hooklets) are released. The hexacanth penetrate the mucosa, enter the intestine, and travel throughout the body. They then enter muscle and form a cysticercus (larval tapeworm enclosed in a cyst), which becomes infective in a few months. Humans become infected when they eat undercooked beef containing such cysticerci. The cys-

ticerci can be inactivated at 56°C or by freezing at −5°C for 1 week (Schmidt and Roberts, 1989). Within 2–12 weeks the worm begins shedding gravid proglottids. The average worm length is 10–15 feet. Symptoms of infection are abdominal pain, headache, nausea, diarrhea, intestinal blockage, and loss of appetite (contrary to the belief that tapeworms infections cause an increase in appetite). The infection can be diagnosed by examining a gravid proglottid or scolex. The best methods for prevention are sanitation (proper disposal of human wastes) and thorough cooking of beef.

19.3.4 Trematodes (*Schistosoma mansoni*)

Trematodes, or flukes, are bilaterally symmetric worms that have two deep suckers and flame cells (for

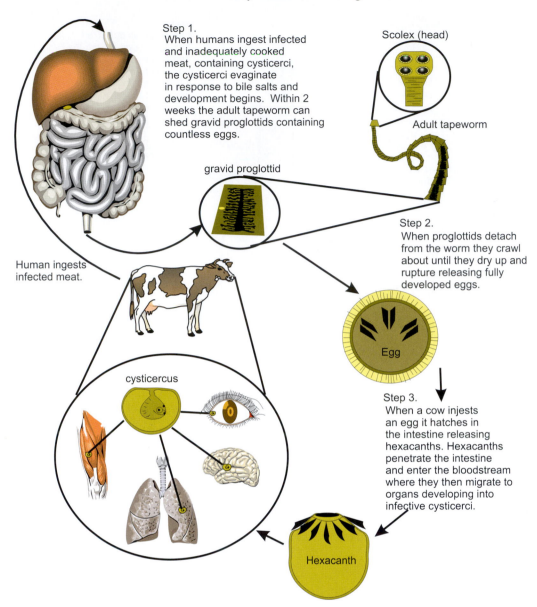

Step 1.
When humans ingest infected and inadequately cooked meat, containing cysticerci, the cysticerci evaginate in response to bile salts and development begins. Within 2 weeks the adult tapeworm can shed gravid proglottids containing countless eggs.

Scolex (head)

Adult tapeworm

gravid proglottid

Human ingests infected meat.

Step 2.
When proglottids detach from the worm they crawl about until they dry up and rupture releasing fully developed eggs.

Egg

cysticercus

Step 3.
When a cow injests an egg it hatches in the intestine releasing hexacanths. Hexacanths penetrate the intestine and enter the bloodstream where they then migrate to organs developing into infective cysticerci.

Hexacanth

FIGURE 19.10   Life cycle of *Taenia saginata*.

excretion). The suckers are used for both attachment and locomotion. The life cycles are complex, with trematodes being either hermaphrodites (adults have both female and male gonads) or schistosomes (separate sexes). Trematodes require an intermediate host (snail) to complete their life cycle; the human is the definitive host, excreting eggs in the feces.

Three species of *Schistosoma* are medically important. In the past, *S. japonicum* and *S. haematobium* were the main causes of schistosomiasis, but now *S. mansoni* is recognized to be the most widespread of the three. The genus *Schistosoma* is responsible for more than 200 million infections worldwide and causes up to 200,000 deaths annually (Hopkins, 1992). More than 400,000 of

those infected live in the United States (West and Olds, 1992). However, all these cases are imported as the intermediate host, the freshwater snail (*Biomphalaria sp.*), is not present in the United States. *S. mansoni* is distributed through most of Africa and the Middle East but is also found in parts of Central America and South America as well as some Caribbean Islands (Savioli *et al.*, 1997).

The larvae of schistosomes of bird and mammals can penetrate human skin, causing what is known as "swimmer's itch." These schistosomes do not mature in humans. This occurs in the Great Lakes region and along the coast of California.

The life cycle of *S. mansoni* is complicated, and human infection begins with the cercariae penetrating human skin. In addition, each parasitic stage involves different organs and thus different medical symptoms. The cercariae from infected snails are found in freshwater bodies. Once the cercariae penetrate the skin, they transform into the schistosomula, the first parasitic stage. This causes the disease cercarial dermatitis, which is immediate or delayed hypersensitivity to the penetration. After 1–2 days of living in the subcutaneous tissues, the schistomsomula migrate through the lungs to the liver. The organisms mature through this migration, developing into male or female adult worms. The adult worms then migrate to the circulatory system and to other organs. Each pair of adult worms live an average of 3–10 years and potentially up to 30 years (Lucey and Maguire, 1993). They also produce 300 eggs a day, only a percentage of which reach the lumen to be passed in the feces. Many of the eggs become trapped in the small blood vessels and tissues, causing granulomas and intestinal, visceral, liver, and fibroobstructive diseases. The acute form of disease, Katayama fever, can occur in previously uninfected persons 2–6 weeks after exposure to cercariae-contaminated water. Chemotherapy is available using praziquantel, although no vaccine is available.

No decrease of infection has been noted, and in fact the number of cases has increased over the past 50 years. The spread of this parasite has occurred as water development projects and population movements have introduced it into previously uninfected regions (UNDP, 1991). However, it has been controlled in some areas; the Caribbean, excluding Puerto Rico, and Brazil. Strategies that have been successful are chemotherapy, health education, water supply treatment, and sanitation. Very little success has been achieved by eradicating the intermediate host, the snail, with either mulluscicides or predator fish.

19.3.5 Emerging Pathogens

19.3.5.1 Cyclospora sp.

Cyclospora cayetanesis is an emerging waterborne and foodborne protozoan pathogen. It was first identified in the intestine of a mole in 1870 by Eimer. It was not recognized as a human pathogen until the early 1980s, and then it was believed to be a cyanobacterium (Soave and Johnson, 1995). In 1993, Ortega *et al.* determined through sporulation experiments and electron micrographs that it is a coccidian protozoan in the phylum Apicomplexa. It is a round spherical oocyst (Fig. 19.11) measuring $8 \times 10\ \mu$m and contains two sporocysts, each containing two sporozoites. Its sporu-

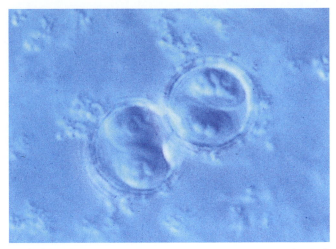

FIGURE 19.11 Oocyst of *Cyclospora cayotanesis*; 100× under DIC. (Photo courtesy H. Smith.)

lation life cycle is that typical of a coccidian protozoan parasite.

Cyclosporiasis causes voluminous, explosive, watery, nonbloody diarrhea in addition to abdominal cramps, nausea, and fatigue (Ooi *et al.*, 1995). However, cryptosporidiosis in the immunocompetent host has a duration of only 10 to 14 days, whereas illness associated with *Cyclospora* averages 43 days (Ortega *et al.*, 1993). Unlike cryptosporidiosis, it can be treated with Bactrim (trimethoprim–sulfamethoxazole), which appears to eliminate the parasite (Knight, 1995).

Acquisition of *Cyclospora* is not completely understood, as the host range and reservoirs are not known. However, water and food are believed to be a major route of infection, because for the organism to be infectious it must mature (sporulate) in the environment for 2 weeks (Ortega *et al.*, 1993). Water can potentially become contaminated with sporulated oocysts, but most infections come from contaminated produce. Outbreaks related to imported contaminated raspberries have been documented in Florida and Canada (Herwaldt and Ackers, 1997). Desiccation of the organism kills it, and therefore it must be in an aquatic environment during maturation.

Infections have been associated with individuals who live in or visit the Caribbean Islands, Central America and South America, Southeast Asia, and Eastern Europe (Knight, 1995). Two suspected waterborne outbreaks have been documented, one in Chicago and one in Nepal (Huang *et al.*, 1995; Rebold *et al.*, 1994).

19.3.5.2 Microsporidia

Microsporidia, the nontaxonomic name to describe organisms belonging to the phylum Microspora, were first described in 1857, when Nägeli identified *Nosema bombycis,* a microsporidium responsible for destruction of the silkworm industry (Nägeli, 1857). To date, over 1000 species of microsporidia infecting insects, invertebrates and all five phyla of vertebrate hosts have been described (Canning, 1993). Microsporidia are for the most part considered to be opportunistic pathogens in humans. There were only a handful of documented cases before the advent of the AIDS epidemic. Since then there have been hundreds of documented cases in immunocompromised patients. However, there have also been cases documented among the immunocompetent (Weber and Bryan, 1994). Recently, an *Encephalitozoon* sp. has been identified in the feces of immunocompetent persons in Mexico. So far, five genera have been associated with the majority of human infections: *Enterocytozoon bieneusi, Encephalitozoon hellem, Encephalitozoon council, Encephalitozoon intestinalis, Pleistophora* spp. *Nosema corneum,* and *Microsporidium* spp. (those currently not classified). The first four have the potential to be waterborne because they are shed in feces and urine. *E. bieneusi, E. hellem,* and *E. intestinalis* are the most common cause of microsporidian infections in patients with AIDS (Curry and Canning, 1993). In addition, they are much smaller (0.8 × 1.5 μm depending on species) than other parasites and potentially more difficult to remove by water treatment filtration.

The microsporidian spore has the potential of being transmitted by water. The life cycle of microsporidia contains three stages: the environmentally resistant spore, merogony, and sporogony. The spore is ingested by a host or possibly inhaled in some cases. Once in the body, it infects cells and goes through merogony followed by sporogony, which results in production of resistant infective spores. The spores are then shed via bodily fluids such as urine and excreta. Once in the environment they have a strong potential to enter water sources. *E. testinalis* spores were identified in sewage, surface, and groundwaters in the United States, supporting the notion of environmental transmission. The spores are highly resistant to heat inactivation and drying. Waller (1979) found that *E. cuniculi* survived 98 days at 4°C and 6 days at 22°C.

Enterocytozoon bieneusi, Encephalitozoon hellem, Encephalitozoon cuniculi, and *Encephalitozoon intestinalis* cause a variety of illnesses. *E. bieneusi* causes diarrhea and wasting disease. It is the most important cause of microsporidiosis in AIDS patients. Several surveys have determined that 7 to 30% of AIDS patients who have unexplained chronic diarrhea are infected with *E. bieneusi* (Weber *et al.,* 1994). This suggests that there could be a carrier state as with *Giardia,* which could lead to a constant source of microsporidia in the environment. *E. intestinalis* (Figs. 19.12 and 19.13) is similar

Encephalitozoon intestinalis

FIGURE 19.12 Florescein isothiocyanate–labeled *Encephalitzoor intestinalis* at 100× magnification. (Photo courtesy M. Panelli and D. C. Johnson.)

to *E. bieneusi* in that it infects the intestines and causes diarrhea, but it can also infect kidneys and bronchial and nasal cells. It can infect macrophages, which allows it to disseminate throughout the body. It is secreted in feces and urine, which supports the notion of water transmission. *E. cuniculi* is the most studied microsporidia and causes fulminating hepatitis; it is not an intestinal parasite. However, it can be shed in urine (Zeman and Baskin, 1985), and therefore environmental transmission is a possibility. It has also been described infecting many different mammals (Canning and Lom, 1986), which means that there could be many animal reservoirs that can contaminate the environment. *E. hellem* has been recognized and shown to cause eye infections (keratoconjunctivitis) and disseminated infections such as ureteritis and pneumonia. So far it has been identified only in humans. It does not

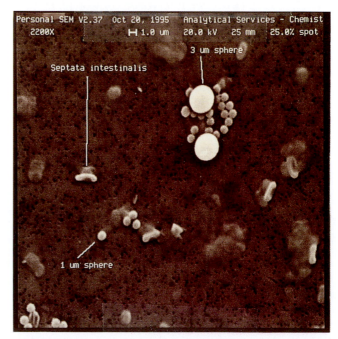

FIGURE 19.13 *Encaphalitzoon* intestinalis 1- and 3-μm microspheres under scanning electron microscope. (Photo courtesy R. Roth and P. Roessler.)

invade the intestine, but it can be shed in the urine (Schwartz *et al.*, 1994).

19.3.5.3 Toxoplasma gondii

Toxoplasma gondii, the causative agent of toxoplasmosis, is an intestinal coccidium of felines with a very wide range of intermediate hosts. *T. gondii* causes a wide range of clinical conditions including brain damage in children, lymphadenopathy, ocular disease, and encephalitis. In the immunocompetent adult it causes symptoms that could be mistaken for influenza. It is estimated that 13% of the world's population is infected with *T. gondii* (Hughes, 1985). Acquisition of the organism can be through contact with infected undercooked meat containing a bradyzoite (tissue cyst) and through contact with the environmental stage, the oocysts. The oocyst is excreted only in cat feces (domestic and wild) (Fig. 19.14). Congenital infection occurs when a pregnant woman becomes infected for the first time. Once

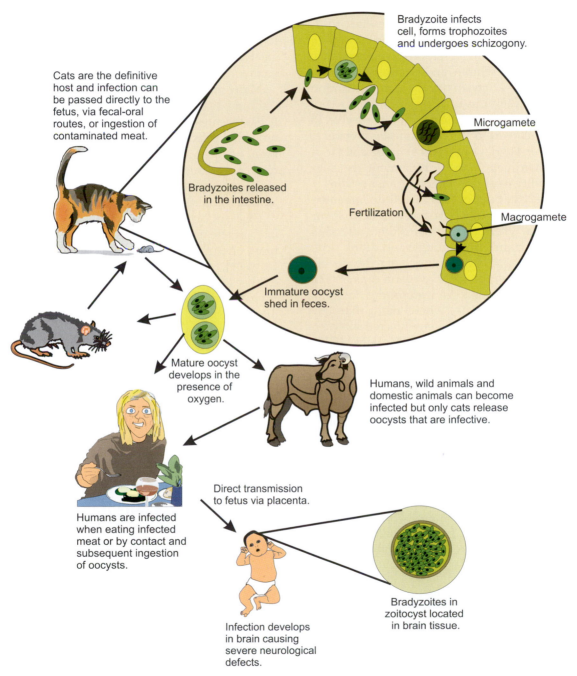

Cats are the definitive host and infection can be passed directly to the fetus, via fecal-oral routes, or ingestion of contaminated meat.

Bradyzoites released in the intestine.

Bradyzoite infects cell, forms trophozoites and undergoes schizogony.

Microgamete

Fertilization

Macrogamete

Immature oocyst shed in feces.

Mature oocyst develops in the presence of oxygen.

Humans, wild animals and domestic animals can become infected but only cats release oocysts that are infective.

Humans are infected when eating infected meat or by contact and subsequent ingestion of oocysts.

Direct transmission to fetus via placenta.

Infection develops in brain causing severe neurological defects.

Bradyzoites in zoitocyst located in brain tissue.

FIGURE 19.14   Life cycle of *Toxoplasma gondii.*

infected, tachyzoites can cross the placenta and infect the fetus. Of the fetuses that become infected, 15% have severe complications. In the United States alone, 3500 infants are born with congenital birth defects related to *T. gondii,* which costs $300 million in care and treatment (Hughes, 1985). The organism is very problematic for immunocompromised individuals. It has been estimated that 5-10% of AIDS patients in the United States develop complications such as encephalitis from *T. gondii* (Ho-Yen and Jos, 1992).

The oocysts can persist in soil and water, which can serve as a route of transmission. One study in England found that 100% of wild cats had antibodies to *T. gondii* (McOrist *et al.,* 1991). Various studies of cats in the United States have found a seropositive prevalence of approximately 40% (Dubey and Beattie, 1988). The high prevalence of *T. gondii* in the cat population demonstrates that they can be a significant source of oocysts in the environment. The oocyst can survive 18 months in the soil (-20 to $33°C$) in Kansas (Frenkel *et al.,* 1975) and over 410 days in water (Yilmaz and Hopkins, 1972). Its hardiness in the environment also suggests the possibility of environmental transmission.

To date, there have been only two well-documented outbreaks of waterborne toxoplasmosis. The first occurred in Panama, where U.S. Army soldiers attended a 3-week training course in the jungle and drank from a stagnant pond. The symptoms were headache, malaise, and flulike symptoms (Benenson *et al.,* 1982). More recently, *T. gondii* caused a waterborne outbreak in British Colombia, Canada, where more than 110 people including 12 newborns were infected after exposure to an unfiltered water supply (Mullens, 1996). In a later study of the area, four of seven domestic cats found in the watershed had antibodies to *T. gondii,* potentially being the initial source of contamination (Stephen *et al.,* 1996).

To date, no one has studied the resistance of *T. gondii* to disinfectants, other than iodine. It survived for 3 hours in 2% iodine (Dubey and Beattie, 1988). The spherical oocysts, responsible for environmental transmission of the organism, are 10 to 13 μm by 9 to 11 μm, relatively large compared with *Giardia* cysts (8–16 μm) and *Cryptosporidium* oocysts (3–6 μm). The cysts can be inactivated in meat by either freezing $-12°C$ or cooking to an internal temperature of $67°C$ (Dubey, 1996).

19.4 VIRUSES

19.4.1 Enteric Viruses

Although many waterborne outbreaks of gastroenteritis are documented every year, no etiology has been found in approximately half of them (Sobsey *et al.,* 1993). It has been suspected, however, that many

are of viral origin. During the past 25 years, it has become evident that viruses are a leading cause of gastroenteritis, in particular in infants and young children, in which they are a major cause of mortality worldwide. Since then, five major groups of human gastroenteritis virus have been identified: rotavirus, enteric adenovirus, Norwalk virus, calicivirus, and astrovirus (Table 19.11). Among these, rotavirus, astrovirus, and enteric adenovirus are the most important causes of viral gastroenteritis in children. Caliciviruses have been the enteric viruses most commonly associated with water and foodborne illness. Although endemic viral gastroenteritis can be transmitted person-to-person by the oral–fecal route, outbreaks of viral gastroenteritis may be triggered by contamination of a common source, such as water or food.

Many human viruses may infect the gastrointestinal tract and be excreted in the feces into the environment. It has been estimated that an individual with an enteric viral infection may excrete 10^{11} viral particles per gram of feces. Once in the environment, viruses can reach water supplies, recreational waters, crops, and shellfish through sewage, land runoff, solid waste landfills, and septic tanks.

Diseases caused by enteric viruses range from trivial to severe or even fatal. The viruses that are detected most often in polluted water are the enteroviruses. However, several studies have shown that adenoviruses (Hurst *et al.,* 1988), rotaviruses and hepatitis A virus (HAV) (Moore *et al.,* 1993) can be found in polluted water as well.

Waterborne outbreaks caused by enteroviruses and other enteric viruses are difficult to document because many infections by these agents are subclinical; i.e., the virus may replicate in an individual, resulting in

TABLE 19.11   Human Enteric Viruses

Enteroviruses
Poliovirus
Coxsackievirus
Echoviruses
Enteroviruses (types 68–71)
Hepatitis A virus
Reoviruses
Rotaviruses
Adenoviruses
Astroviruses
Torovirus
Caliciviruses
Hepatitis E virus
Norwalk virus
Snow Mountain agent
Small round structured viruses (SRSV)

virus shedding but without overt disease. Therefore, an individual with waterborne infection but without overt disease may infect others, who in turn may become ill, spreading the infection throughout the community. In addition, epidemiological techniques lack the sensitivity to detect low-level transmission of viruses through water. Recreational activities in swimming pools have sometimes resulted in waterborne outbreaks caused by Norwalk virus, HAV, coxsackievirus, and adenoviruses. Enteric viruses from infected individuals may contaminate recreational waters by direct contact or by fecal release. Norwalk virus is the virus most commonly associated with waterborne viral disease in the United States.

19.4.1.1 Astroviruses

Astroviruses were first observed by electron microscopy in diarrheal stools in 1975. These agents are icosahedral viruses with a starlike appearance and with a diameter of approximately 28 nm. Astrovirus have a single-stranded RNA (ssRNA) genome. The sequencing of the astrovirus genome allowed the establishment of its own virus family, the *Astroviridae*, with *Astrovirus* as the only genus. Serology assays and sequence analysis have resulted in the identification of seven distinct serotypes (Lee and Kurtz, 1994).

Astrovirus type 1 seems to be the most prevalent strain in children. Type 4 has been associated with severe gastroenteritis in young adults. Astrovirus-like particles have been found in feces of a number of animals suffering from a mild self-limiting diarrheal infection, but no antigenic cross-reactivity has been found between these agents and human astroviruses. Occasionally these viruses can be found in infants without overt disease. Astrovirus infections occur throughout the year, with a peak during the winter–spring seasons in temperate zones. In warm climates, however, the highest incidence of astrovirus infection has been observed in May (Cruz *et al.*, 1992).

Astroviruses cause a mild gastroenteritis after an incubation period of 3 to 4 days. Overt disease is common in 1 to 3-year-old children. However, adults and young children are also affected. Astrovirus infection has also been observed in immunocompromised individuals and the elderly. Outbreaks of astrovirus infection have been associated with oysters and drinking water (Kurtz and Lee, 1987).

Although astroviruses are common in developing countries, they have also been identified as a significant cause of infantile gastroenteritis in developed countries. A 10-year study of astrovirus prevalence in Japan reported that 6 to 10% of viral gastroenteritis cases in that country are caused by astroviruses. In England, these agents have been found in 7% of diarrheal samples examined by electron microscopy.

Gastroenteritis caused by astrovirus decreases steadily with age. This observation suggests that some group-specific immunity may be generated after a primary infection with this virus. Nevertheless, reinfection seems to be fairly common. Astrovirus has been identified as one of the most common gastroenteritis agents in immunocompromised individuals such as those with AIDS and bone marrow patients, in whom prolonged shedding of virus has been documented.

Astroviruses are relatively difficult to propagate in cell cultures. In 1980 it was found that a few astrovirus strains could be propagated in a continuous cell line by growing them first in primary human embryonic kidney cells, followed by a passage in a continuous cell line of rhesus monkey kidney cells (LLCMK2). As human embryo tissue is not readily available, this procedure was used to adapt a limited number of human astrovirus strains. Successful propagation of human astroviruses in a continuous cell line was finally achieved in 1990, using human continuous colonic carcinoma cells.

19.4.1.2 Adenoviruses

Adenoviruses are double-stranded DNA (dsDNA) icosahedral viruses approximately 70 nm in diameter (Fig. 19.15). At least 49 human adenovirus types have been identified. Although there are many avian and mammalian adenovirus types, they are species specific. Adenoviruses can replicate in the respiratory tract, the eye mucosa, the intestinal tract, the urinary bladder, and the liver.

Most adenovirus human disease is associated only with one third of adenovirus types. Although many adenovirus infections are subclinical, these viruses may cause acute respiratory disease (types 1–7, 14, and

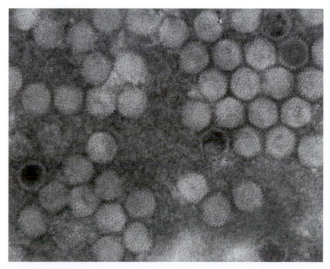

FIGURE 19.15 An enteric virus. (Photo courtesy P. Payment and R. Alain, INRS–Institut Armand-Frappier, Université du Québec.)

21), conjunctivitis (types 3, 7, 8, 11, 14, 19, and 37), acute hemorrhagic cystitis (11 and 21), acute respiratory disease (ARD) of military recruits (types 3, 4, 7, 14, and 21), and gastroenteritis (types 31, 40, and 41) (Table 19.12).

Adenoviruses gain access to susceptible individuals through the mouth, the nasopharynx, or the conjunctiva . Although initial infection may occur by the respiratory route, fecal–oral transmission accounts for most adenovirus infections in young children because of the prolonged shedding of viruses in feces (Horwitz, 1996).

Most, if not all, waterborne outbreaks of viral conjunctivitis are caused by adenoviruses. Transmission of waterborne conjunctivitis by adenovirus types 3 and 4 has been established. A large outbreak was documented in North Carolina. In this outbreak, conjunctivitis, pharyngitis, and fever were the most common manifestations among 595 affected individuals (McMillan et al., 1992). Intact conjunctival mucosa seems to be resistent to infection by adenoviruses. Attempts to induce adenovirus conjuntivitis in volunteers are unsuccessful unless the conjunctival mucosa is mildly irritated by swabbing. Therefore, chlorinated swimming pool water, dust, or optical instruments may provide the necessary conjunctival irritation and, in some cases, carry the virus as well.

The enteric adenoviruses 40 and 41 have been recognized as the second most important etiological agents of viral gastroenteritis in children. These viruses, in contrast to other adenoviruses, are not shed in respiratory secretions; thus their transmission is limited to the oral–fecal route. The enteric nature of adenoviruses 40 and 41, their presence only in the gastroenteric tract, and their extensive distribution in the environment suggest that water may play a role in the transmission of these agents. Although to a lesser degree, adenovirus type 31 has also been identified as a cause of infant gastroenteritis.

Adenoviruses are common in primary sewage sludge, where they have been found in concentrations 10 times greater than that of the enteroviruses (Williams and Hurst, 1988). In addition, a greater number of adenoviruses than enteroviruses has been consistently found in raw sewage around the world. Results of a comparative study of cytopathogenicity, immunofluorescence, and in situ DNA hybridization as methods for the detection of adenoviruses from water suggest that 80% of infectious adenoviruses in raw sewage may be enteric adenoviruses (Hurst et al., 1988).

Enteric adenoviruses propagate poorly or not at all in conventional human cell lines in which other adenoviruses can be propagated effectively. Grabow et al.

TABLE 19.12   Classification of Human Adenoviruses

Subgenus	Serotype	Human illness
A	12	Meningoencephalitis
	18, 31	Diarrhea
B	3	Acute febrile pharyngitis; adenopharyngo-conjunctival fever; pneumonia; follicular conjunctivitis; fatal infection in neonates
	7	Acute febrile pharyngitis; adenopharyngo-conjunctival fever; acute respiratory disease with pneumonia; fatal infection in neonates; meningoencephalitis
	11	Follicular conjunctivitis; hemorrhagic cystitis in children
	21	Hemorrhagic cystitis in children; fatal infection in neonates
	14, 16	Acute respiratory disease with pneumonia
	34, 35	Acute and chronic infection in patients with immunosuppression and AIDS
C	1, 2, 6	Acute febrile pharyngitis; pneumonia in children
	5	Acute febrile pharyngitis; pertussis-like syndrome; acute and chronic infection in patients with immunosuppression and AIDS
D	8, 19, 37	Epidemic keratoconjunctivitis
	9, 10, 13, 15, 17, 42	
	19, 20, 22–29	
	30	Fatal infection in neonates
	32, 33, 36, 38	Asymptomatic
	39, 42–47	Acute and chronic infection in patients with immunosuppression and AIDS
E	4	Respiratory infection
F	40, 41	Diarrhea

(1992) reported the successful propagation of both Ead40 and Ead41 in the PLC/PRF/5 human liver cell line. Pinto *et al.* (1994) were able to propagate enteric adenoviruses 40 and 41 successfully in a cell line derived from a human colon carcinoma (CaCo-2 cells).

Contaminated inanimate surfaces may play a significant role in adenovirus transmission because of its stability to drying. At room temperature, adenovirus 2 survives for 8 and12 weeks at low (7%) and high (96%) relative humidity, respectively, being more resistant than poliovirus 2, vaccinia virus, coxsackievirus B3, and herpes virus (Mahl and Sadler, 1975). The longer survival of adenoviruses has been observed in water as well. Adenovirus type 5 survives longer in tap water, at either 4 or 18°C, than either poliovirus 1 or echovirus 7 (Bagdasar'yan and Abieva, 1971), and enteric adenoviruses 40 and 41 survive longer than poliovirus 1 and HAV in tapwater and seawater (Enriquez *et al.*, 1995b). In contrast to survival patterns observed with enteroviruses, the enteric adenoviruses do not survive significantly longer in wastewater. The increased survival of the enteric adenoviruses in tapwater and seawater, and the faster inactivation in sewage, may indicate that these viruses are inactivated by different mechanisms than those affecting the enteroviruses. The enteric adenoviruses are more thermally stable than polio 1, which is inactivated faster at temperatures above 50°C (Enriquez *et al.*, 1995a). In addition, enteric adenoviruses 40 and 41 are more resistant to ultraviolet (UV) light disinfection than poliovirus type 1 and coliphage MS-2, which has been suggested as a model for enteric virus disinfection (Meng and Gerba, 1996).

The increased resistance showed by enteric adenoviruses, compared with other enteric viruses, may be associated with the double-stranded nature of their DNA, which, if damaged, may be repaired by the host cell DNA repair mechanisms. This mechanism would not be effective with ssRNA genome viruses such as polio 1 or HAV. It has been suggested that the longer survival of the enteric adenoviruses in tapwater and seawater may be associated with DNA damage, and the faster inactivation in sewage may result from protein capsid damage.

19.4.1.3 Enteroviruses

The enteroviruses are members of the family Picornaviridae and are represented by poliovirus (3 serotypes), coxsackievirus A (23 types), coxsackievirus B (6 types), echovirus (32 types), and enteroviruses (4 types). Enteroviruses are icosahedral viruses approximately 27 to 32 nm in diameter. The nucleic acid of this type of virus consists of ssRNA. These are the viruses most often detected in polluted water. However, their apparent higher prevalence may be associated, in part, with available cell lines for their propagation, because many pathogenic enteric viruses such as HAV, enteric adenoviruses, rotavirus, Norwalk virus, and other small round viruses are difficult to grow in conventional cell lines.

Although viruses belonging to the *Enterovirus* genus are capable of causing a wide variety of clinical conditions, from asymptomatic to disabling, or even fatal infections, under most circumstances they do not cause overt disease (Table 19.13).

Although there are bovine, porcine, simian, and murine enteroviruses, it is believed that humans are the only natural host of human enteroviruses. Enteroviruses replicate primarily in the gastrointestinal tract and may be shed in large numbers (approximately 10^{10}/g feces). The most common forms of transmission include the fecal–oral and respiratory routes. Waterborne transmission may be considered a form of fecal–oral transmission in which the responsible vehi-

TABLE 19.13   The Human Enteroviruses

Virus group	Serotypes	Major illnesses
Poliovirus	1, 2, and 3	Paralytic poliomyelitis, meningitis, febrile illness
Echoviruses	1–9, 11–27, 29–34[a]	Meningitis, encephalitis, paralysis, otitis, common cold, pneumonia, diarrhea, exanthema
Coxsackie A	1–22, 24[b]	Meningitis, paralysis; herpangina; striated muscle damage; common cold; pneumonia; hand, foot, and mouth disease; hepatitis; diarrhea; hemorrhagic conjunctivitis
Coxsackie B	1–6	Meningitis, myocarditis, congenital heart defects, pleurodynia, otitis, common cold, pneumonia, bronchiolitis, asthma, hepatitis, diabetes, post-viral fatigue syndrome
Enteroviruses	68–71[c]	Meningitis, paralysis, conjunctivitis, pneumonia, bronchiolitis, common cold

[a] Echovirus 10 was classified as reovirus 1 and echovirus 28 as rhinovirus 1.
[b] Coxsackievirus A23 was classified as echovirus 9.
[c] Since 1969, new enteroviruses have been assigned enterovirus type numbers.

cle is water instead of hands or fomites. Although enteroviruses are readily found in fecally contaminated drinking or recreational waters, waterborne enterovirus infection has been only occasionally documented. Waterborne outbreaks related to enteroviruses are difficult to document because many infections by these agents are subclinical. Therefore, an individual with waterborne infection without overt disease may infect others, who in turn may become ill, spreading the infection further. Attack rates of enteroviruses vary depending on the virus and the host age. Asymptomatic infections by poliovirus outnumber symptomatic disease (10:1), whereas symptomatic infection by echovirus 9 is relatively high (10:7) (Morens and Pallansch, 1995).

Both disease and infection caused by polioviruses are age related. Generally, infection is more common in infants, but adults and older children are more severely affected. However, some exceptions exist. Coxackievirus B virus infection is usually more severe in newborns than in older children and adults, often causing fulminant myocarditis, encephalitis, hepatitis, and death. Coxsackieviruses are the most prevalent nonpolio enteroviruses (Morens and Pallansch, 1995). Coxsackieviruses are also the most common nonpolio enteroviruses isolated from water and wastewater. These viruses have been associated with several serious illnesses (Table 19.12). Coxsackievirus B5 infection has been associated with recreational water. In an outbreak at a boys' summer camp, this virus was isolated from lake water; however, person to person contact appeared to be the main form of transmission (Hawley et al., 1973).

It is believed that almost all enteroviruses can be transmitted by the fecal–oral route; however, it is not clear if all of them can be transmitted by the respiratory route as well. Airborne transmission of enterovirus might include aerosol spread or direct exposure to respiratory secretions. Fecal–oral transmission may predominate in areas with poor sanitary conditions, whereas respiratory transmission may occur more often with better sanitation. In temperate climates, enteroviruses are more common during the summer season. In the United States, most enterovirus isolations (82%) occur from June to October. However, vaccine strains of poliovirus are isolated year-round because of routine vaccination of children. In contrast, in tropical and semitropical areas, enteroviruses do not show seasonality. Transmission of enteroviruses within a household is usually started by young children; then the infection spreads quickly to other family members, especially in larger families living under crowded conditions with poor hygiene. Ironically, paralytic poliomyelitis and perhaps some non-

polio enteroviral diseases are more often observed in developing countries, where sanitary conditions are improving. With poor hygiene, most individuals are infected at a very early age, when infection rarely results in overt disease and maternal immunity limits infection. This early exposure to the virus elicits a protective immune response on reexposure to these viruses later in life. In contrast, when early exposure is prevented or delayed as a result of better sanitation, an initial poliovirus infection is likely to occur at an older age, when maternal immunity has waned and the possibility of developing a more severe clinical condition is greater (see Case Study 1).

19.4.1.4 Hepatitis A Virus (HAV)

HAV is a picornavirus morphologically indistinguishable from other members of the same family. This agent was formerly classified as member of the *Enterovirus* genus (enterovirus 72). However, differences in nucleotide and amino acid sequences resulted in its classification as the only member of the hepatovirus group.

Hepatitis A was the most frequent viral waterborne disease in the United States from 1946 to 1994, with 68 outbreaks and 2297 cases. From 1989 to 1990, two waterborne outbreaks of hepatitis A occurred in Pennsylvania. Both outbreaks originated from well water that was untreated or insufficiently treated (Herwaldt *et al.*,1992).

The average HAV incubation is approximately 30 days, but it may vary from 10 to 50 days, a variation associated with the dose. Infection with very few particles results in longer incubation periods and vice versa. The period of communicability extends from early in the incubation period to about a week after the development of jaundice. The greatest danger of spreading the disease to others occurs during the middle of the incubation period, well before the first presentation of symptoms. During this period the patient remains asymptomatic; however, active shedding of the virus is the norm. Therefore, it is during the incubation time that the infected individual has the highest potential for spreading HAV.

Hepatitis A is usually a mild illness, which almost always results in complete recovery. Severity and disease manifestation are age related. An estimated 80 to 95% of infected children younger than 5 years of age do not develop overt disease, whereas clinical manifestations are observed in approximately 75 to 90% of infected adults. The mortality rate in children 14 years old or younger is 0.1%; this rate rises to 0.3% in individuals between the ages of 15 and 39 years and 2.1% in those older than 40 years.

CASE STUDY 1

A Waterborne Outbreak of Paralytic Poliomyelitis

Before the generalized use of poliovirus vaccine in the early 1950s in the United States, the annual average number of cases of paralytic poliomyelitis was approximately 21,000. Although polioviruses are readily isolated from sewage-contaminated water, their association with waterborne transmission has been elusive. Nevertheless, in the summer of 1952, a waterborne outbreak of poliomyelitis was documented in Lincoln, Nebraska. This outbreak occurred in a community of World War II veterans, who lived in apartments (formerly a military hospital) arranged in four separate sections (I, II, III and IV). On June 24, a 3-year-old boy developed an abrupt onset of fever, headache, vomiting, and meningitis. This case heralded the beginning of an outbreak of paralytic poliomyelitis in this community. By July 12 there were 11 more cases in the same apartment complex, and by mid-August the number of children suffering from paralytic poliomyelitis had increased to 20. Among these children, two developed bulbar poliomyelitis and died. Although children from the four sections shared the same playground, all cases of paralytic poliomyelitis, except one, occurred in sections I, II, and III.

In most private homes, water tanks are used for toilet flushing. This system provides an effective barrier between the drinking water source and the wastewater in the toilet bowl. In contrast, in the housing complex where the outbreak occurred, pressure in the water main was used for toilet flushing. As long as positive pressure is maintained in the water main in this type of system, no hazard exists, but if there is a failure and negative pressure develops, water from the toilet bowl may be siphoned back into the main. Normally, these toilets have a safety valve that prevents this from happening. Unfortunately, during maintenance, these safety devices had been removed from 13 toilets (12 from sections I, II, and III). A break in one of the water mains in Lincoln 8 days before the epidemic resulted in negative pressure in some points of the system, including the housing complex. This resulted in back siphoning from toilet bowls into the main, contaminating the drinking water, in particular that of sections I, II, and III. After the outbreak, the sanitary code of Lincoln made flush valve toilets illegal unless installed with a safety valve that admits air into the main in the event of negative pressure.

The poliomyelitis outbreak in Lincoln makes it evident that safe drinking water sources may be contaminated at some point in the distribution system, including private homes. Therefore, efforts to provide safe drinking water to the population should not stop at the drinking-water treatment plant but should include proper conditions at the end of the system where the water is to be used.

Hepatitis A is characterized by sudden onset of fever, malaise, nausea, anorexia, and abdominal discomfort, followed in several days by jaundice. In contrast to hepatitis B, HAV infection is not chronic. HAV is excreted in feces of infected people and can produce clinical disease when susceptible individuals consume contaminated water or foods. Cold cuts and sandwiches, fruits and fruit juices, milk and milk products, vegetables, salads, shellfish, and iced drinks are commonly implicated in outbreaks. Water, shellfish, and salads are the most frequent sources. Contamination of foods by infected workers in food processing plants and restaurants is not uncommon.

HAV is relatively stable in the environment. It is more resistant to high temperatures than poliovirus. HAV can survive at 60°C for 1 hour, and temperatures up to 85 to 95°C are needed to inactivate it in shellfish.

In addition, HAV is relatively resistant to chlorine disinfection, in particular if associated with organic matter (Margolis et al., 1997).

Poor sanitation and crowding facilitate HAV transmission. Therefore, outbreaks of hepatitis A are common in institutions, crowded house projects, prisons, and military forces. In developing countries, the incidence of disease in adults is relatively low because of exposure to the virus in childhood. Most individuals 18 years and older possess an immunity that provides lifelong protection against reinfection. In the United States, the percentage of adults with immunity increases with age (10% for those 18–19 years of age to 65% for those over 50).

No satisfactory method is presently available for routine analysis of food, but sensitive molecular methods that can detect HAV in water and clinical specimens should prove useful for detecting virus in foods. Among those, the polymerase chain reaction (PCR) amplification method seems particularly promising.

The survival of HAV on hands and its transfer to hands or inanimate surfaces were studied by Mbithi *et al.* (1992). They found that approximately 20% of the initial HAV inoculated on hands remained infectious for at least 4 hours, and that inoculated onto a stainless steel surface survived for 2 hours. They also determined that exerting higher pressure and friction between HAV-contaminated hands or fomites resulted in more efficient transfer of this virus to clean hands. These researchers indicated that the control and prevention of HAV infection in high-risk places such as day care centers, hospitals, restaurants, and other similar facilities must place a greater emphasis on both human hands and environmental surfaces as potential sources of virus transfer.

Table 19.14 summarizes the most common patterns of HAV transmission. These patterns are influenced by socioeconomic and hygienic conditions. In areas with poor sanitation, nearly all children up to 9 years of age have been infected by HAV. In these areas, outbreaks rarely occur, and clinical disease related to HAV infection is uncommon. With better sanitation, HAV infection shifts to older individuals, and the incidence of overt disease increases. In more developed countries, low levels of HAV transmission occur. However, disease outbreaks are relatively common in most of these countries as a significant segment of the population is susceptible to HAV infection. Finally, in a few industrialized countries, hepatitis A outbreaks are uncommon, and nearly all HAV infections occur among individuals who have visited areas where HAV is highly prevalent (see Case Study 2).

19.4.1.5 Hepatitis E Virus (HEV)

The hepatitis E virus, formerly known as enterically transmitted non-A, non-B hepatitis virus, is the leading cause of acute viral hepatitis among young and middle-aged adults in developing countries. HEV has a diameter of 32–34 nm and an ssRNA genome. Serologically related smaller (27–30 nm) particles are often found in feces of patients with hepatitis E and are presumed to represent degraded viral particles. Based on its physicochemical properties, HEV is thought to be related to the calicivirus family. It has been suggested that the higher prevalence of HEV in adults may be due to a silent infection early in life, with subsequent waning of immunity after 10 to 20 years, when they again become susceptible to infection by HEV (Bradley, 1992). Important epidemiological features of HEV infection are the frequent occurrence of outbreaks associated with consumption of sewage-polluted water and its severity, particularly among pregnant women, in whom the case-fatality rate may be as high as 25% (Balayan, 1993). To date, no outbreak has occurred in the United States.

Hepatitis E is clinically indistinguishable from hepatitis A. It is characterized by jaundice, malaise, anorexia, abdominal pain, arthralgia, and fever. The incubation period for hepatitis E varies from 2 to 9 weeks. The disease is most often seen in young to middle-aged adults (15–40 years old). The disease is usually mild and resolves in 2 weeks, leaving no long-term effects, and with a relatively low fatality rate (0.1–1%). There is no evidence of immunity against HEV in the population that has been exposed to this virus (Margolis *et al.*, 1997).

Diagnosis of HEV is based on the epidemiological characteristics of the outbreak and exclusion of hepatitis A and B viruses by serological tests. Confirmation requires identification of the 27–34 nm virus-like particles by immune electron microscopy in feces of acutely ill patients.

19.4.1.6 Rotavirus

Rotaviruses are classified within the family Reoviridae. These viruses have a characteristic genome consisting of 11 dsRNA segments surrounded by a distinctive two-layered protein capsid. Particles are approximately 70 nm in diameter. Six serological groups (A–F) have been identified; groups A, B, and C infect both humans and animals, and D, E, and F have been detected only in animals. Group A rotaviruses have been associated with the majority of infantile

TABLE 19.14 Patterns of Hepatitis A Virus Transmission

HAV Prevalence in relevant population	Symptomatic disease rate of infected individuals	Peak age of infection	Transmission pattern
High	Low	Early childhood	Person-to-person, outbreaks uncommon
Moderate	High	Late childhood and young adults	Person-to-person, food- and waterborne outbreaks
Low	High	Young adults	Person-to-person, food- and waterborne outbreaks
Very low	High	Adults	Travelers, outbreaks uncommon

CASE STUDY 2

A Massive Waterborne Outbreak of Infectious Hepatitis

Approximately 35,000 young adults in New Delhi, India came down with hepatitis in the fall of 1955. There were 73 deaths recorded among those affected, one of the largest viral waterborne outbreaks ever documented. New Delhi obtains its drinking water from the Jamuna River. Because of flooding following the monsoon season, water drawn from the river for treatment and distribution became heavily contaminated with sewage from a nearby creek. From November 11 to November 16, raw wastewater freely entered the drinking water treatment plant. The staff at the plant, aware of the contamination, increased the dose of alum coagulant and chlorine in an effort to eliminate contaminating pathogens. About 3 weeks later, the first hepatitis cases were identified. The observed incubation period ranged from 22 to 60 days, with an average of 40 days. Surprisingly, no increase in other types of viral, bacterial or parasitic infections was observed during the hepatitis epidemic, possibly because of the partial success of emergency disinfection using high levels of chlorine.

In an attempt to isolate the epidemic's causative agent, researchers at the Virus Research Center in Poona inoculated eight different cell lines with a variety of clinical specimens obtained from hepatitis patients. Unfortunately, no viruses were isolated from any of the samples tested. Although the epidemic's etiologic agent was not identified, it was believed that the outbreak was caused by the hepatitis A virus (HAV). A retrospective study conducted in 1980 showed that the outbreak in New Delhi was not caused by HAV or hepatitis B virus (HBV) but by the so-called *enterically transmitted non-A, non-B* (ENANB) hepatitis virus. This agent was later renamed hepatitis E virus (HEV). Although waterborne outbreaks of hepatitis E have occurred in many regions of the world, outbreaks have almost exclusively been in developing countries. HEV waterborne outbreaks primarily affect young adults, and fatalities among pregnant women are not uncommon.

The raw sewage contamination of drinking water that led to the large HEV outbreak in New Delhi was thought to be controlled by the use of high doses of alum coagulant and chlorine. Although this measure prevented an outbreak of bacterial, parasitic, and other enteric viral diseases, it failed to eliminate HEV. Therefore, prevention of contamination, rather than extensive treatment, offers the best alternative for safe drinking water.

acute gastroenteritis cases, group B with severe diarrhea epidemics in adults in China, and group C with sporadic cases of diarrhea in children, but their clinical importance has not been determined. Within each group, rotaviruses are classified into serotypes.

Rotaviruses are the most important agents of infantile gastroenteritis around the world. Group A rotavirus is endemic worldwide. It is the leading cause of severe diarrhea among infants and children and accounts for about half of the cases requiring hospitalization. In temperate areas, it occurs primarily in the winter, but in the tropics it occurs throughout the year. Group B rotavirus, also called adult diarrhea rotavirus, has caused major epidemics of severe diarrhea affecting thousands of persons of all ages in China. Group C rotavirus has been associated with rare and sporadic cases of diarrhea in children in many countries.

Rotaviruses are shed in large numbers, up to 10^{10} viral particles per gram of feces (White and Fenner, 1994). Names applied to the infection caused by the most common and widespread group A rotavirus include acute gastroenteritis, infantile diarrhea, winter diarrhea, acute nonbacterial infectious gastroenteritis, and acute viral gastroenteritis.

Rotaviruses are transmitted by the fecal–oral route. Person-to-person spread by contaminated hands is probably one of the most important routes by which rotaviruses are transmitted. Institutions or close communities such as pediatric wards, day care centers, and family homes are usually most affected by outbreaks of gastroenteritis caused by rotaviruses. Because of the low infectious dose, rotavirus-infected food handlers may contaminate foods that require handling but do not require further cooking, such as cakes, salads, and fruits. Among adults, multiple foods served at banquets were implicated in two outbreaks, and an outbreak related to contaminated municipal water occurred in Colorado in 1981. Several large outbreaks of group B rotavirus involving millions of persons as a result of sewage contamination of drinking water supplies have occurred in China since 1982. Although to date outbreaks caused by group B rotavirus have been confined to mainland China, seroepidemiological surveys have indicated lack of immunity to this group of viruses in the United States. Rotaviruses are quite stable in the environment and have been found in estuary samples at levels as high as one to five infectious particles per gallon (FDA, 1992). Sanitary measures adequate for bacteria and parasites seem to be ineffective in endemic control of rotavirus, as similar incidences of rotavirus infection are observed in countries with both high and low health standards.

Rotavirus gastroenteritis is a self-limiting disease, which can be mild to severe. It is characterized by vomiting, watery diarrhea, and mild fever. Asymptomatic rotavirus excretion has been well documented and may play a role in perpetuating endemic disease. The incubation period ranges from 1 to 3 days. Symptoms often start with vomiting followed by 4–8 days of diarrhea. As with other viral gastroenteritis, rotavirus gastroenteritis treatment consists of fluid and electrolyte replacement; if it is untreated, severe diarrhea with dehydration and death may occur. Individuals of all ages are susceptible to rotavirus infection. Premature infants, children 6 months to 2 years of age, the elderly, and the immunocompromised are particularly prone to develop more severe symptoms. In adults, rotaviral infection is usually subclinical. However, rotavirus gastroenteritis outbreaks have occurred in army recruits, geriatric patients, and hospital staff. Over 3 million cases of rotavirus gastroenteritis occur annually in the United States. It has been estimated that rotaviruses, in the United States alone, cause more than 1 million cases of severe diarrhea and up to 150 deaths per year. Worldwide, close to 1 million infants and young children die from rotavirus infection each year. Rotavirus infection does not result in an efficient or long-lasting immunity. Therefore, rotavirus infection in the same child often occurs up to six times during childhood.

Some animal rotaviruses, such as SA-11, are readily propagated in cell culture; the human rotaviruses, however, are rather fastidious. These viruses have not been grown efficiently in any conventional tissue culture system.

19.5.1.7 Small Round Viruses and Norwalk Virus

The use of electron microscopy (EM) since the early 1970s for the examination of fecal specimens from individuals suffering from nonbacterial gastroenteritis has shown many previously unknown viruses. Among these, viruses such as adenoviruses and rotaviruses are larger and well defined; thus, they are relatively easy to identify by EM. However, many smaller (20–40 nm) viruses without a distinctive morphology are often present in fecal samples of patients suffering from gastroenteritis. Most of these agents cannot be propagated in conventional cell culture. These viruses were lumped together under the loose term small round viruses (SRV). Norwalk virus was the first to be described (Kapikian et al., 1972). Subsequently, other small round viruses have been observed in diarrheal stools, namely Montgomery County, Hawaii, Wollan, Ditchling, the Parramata, the Cockle agents, and minirotavirus. Cubitt (1987) reported that Norwalk-like viruses were observed by electron microscopy in 4% of stool samples from individuals suffering from gastroenteritis in England. Human calicivirus and Norwalk-like viruses have emerged as a major cause of food- and waterborne disease. In a survey from 1976 to 1980, Norwalk virus was the most common agent associated with waterborne viral gastroenteritis. Of 17 documented outbreaks, 2 were traced to municipal waters, 7 to semipublic waters, 2 to stored water on cruise ships, and 2 more to recreational waters (Kaplan et al., 1982).

The inability to propagate these agents in conventional cell lines has limited their study and classification. The small round viruses have been classified as either featureless or structured viruses. The former include the Wollan, Ditchling, Parramata, and Cockle agents, and the latter consist of the small round structured viruses (SRSV), which include Norwalk, Montgomery County, Hawaii, Taunton, Amulree, Otofuke, Saporo, and Snow Mountain agent.

The Norwalk virus was thought to be a calicivirus because it shares several characteristics with members of this virus family. Initial studies showed that calicivirus and Norwalk virus are similar in size, both are resistant to organic solvents and detergents, and they have similar sedimentation coefficients in sucrose and cesium chloride. Furthermore, the two types of viruses have a single major structural protein, believed to be unique among caliciviruses (Cubitt, 1987). However, the identity of the Norwalk virus was not defined until Jiang et al. (1990) sequenced the Norwalk virus genome, allowing its classification in the family Caliciviridae. Norwalk virus is a nonenveloped virus with a diameter of approximately 26 to 35 nm and a positive-sense ssRNA genome.

In addition, Lew et al. (1994) demonstrated by nucleic acid sequencing that the minirotavirus is a Norwalk-like virus also belonging to the family Caliciviridae. It was suggested that the minirotavirus should be named Toronto virus, to be consistent with the designation of other Norwalk-like viruses that are named according to the location of the original outbreak. Norwalk virus appears to cause gastroenteritis in adults, whereas the caliciviruses affect young children, and infants (Roper *et al.*, 1990). The immunity to Norwalk viruses is poorly understood. However, infectivity studies with volunteers have shown that individual susceptibility is more important than acquired immunity. It has been suggested that genetically determined factors are the primary determinants of resistance to Norwalk virus infection, perhaps at the level of cellular receptor sites. In developing countries, antibodies to Norwalk virus are acquired early in life, and the peak incidence of illness may also occur among younger age groups than in developed nations (Roper *et al.*, 1990). It has been proposed that Norwalk-like viruses may circulate in low numbers in a population until an infected individual contaminates a common source of food or water, resulting in an explosive outbreak (Roper *et al.*, 1990). Norwalk and related viruses usually produce a mild and brief illness, lasting 1 to 2 days. It is characterized by nausea and abdominal cramps, followed commonly by vomiting in children and diarrhea in adults. The involvement of Norwalk virus as a pathogen of adults was further suggested by Numata *et al.* (1994), who reported a very low prevalence of antibodies to recombinant Norwalk virus capsid protein in children younger than 7 years of age but an increasing prevalence in individuals from 12 to 50 years of age in Japan.

The epidemiology of gastroenteritis associated with rotavirus and enteric adenoviruses has been well studied because reagents for their detection are available. However, epidemiologic studies of many types of SRSV have been difficult to conduct because of their small size, low numbers in feces, fastidious nature, and lack of suitable reagents for their detection.

During an abnormally rainy season in Australia, a foodborne outbreak of SRSV infection, with more than 2000 direct and indirect cases, was attributed to runoff and sewage contamination of water where the shellfish were harvested (Cubitt, 1991). In a similar outbreak in England it was shown that, in spite of a 72 hour depuration process with an acceptable reduction of indicator bacteria, consumption of only one oyster was sufficient to cause disease.

Although Norwalk virus is highly infectious, strict sanitary measures such as proper hand washing and disposal or disinfection of contaminated material may decrease the rate of transmission. Foods should not be handled by ill individuals, and only properly treated drinking water should be consumed.

19.4.2 Respiratory Viruses

Worldwide, respiratory illnesses are the most common illnesses in humans and most have a viral etiology. Respiratory disease is associated with a large number of viruses, including rhinoviruses, coronaviruses, parainfluenza viruses, respiratory syncytial virus (RSV), influenza virus, and adenovirus. These viruses, when they infect the upper respiratory tract, can cause acute viral rhinitis or pharyngitis (common cold); when the primary site of infection is the lower respiratory tract, they can cause laryngotracheitis (croup), bronchitis, or pneumonia.

Mortality related to acute respiratory disease may be especially significant in children and in the elderly. In adults, temporary disability results in important economic loss. Respiratory infection often results from self-inoculation, when virus-contaminated hands or fingers rub the eyes or when viruses are introduced into the mouth or nose. Another important form of transmission of respiratory viruses is inhalation of contaminated aerosols. The mode of transmission of viruses causing respiratory disease is influenced by the type of virus. For example, fragile enveloped viruses such as parainfluenza virus, RSV, and influenza virus are usually transmitted by close contact with infected individuals, whereas the more stable nonenveloped viruses such as adenoviruses, rhinoviruses, and coronaviruses are commonly transmitted by self-inoculation by contaminated hands.

19.4.2.1 Rhinoviruses

Rhinoviruses (Latin *rhino*, nose), belong to the family Picornaviridae. Two important characteristics differentiate these two types of viruses: stability at low pH values and temperature. Whereas the rhinoviruses are inactivated at pH below 6, the enteroviruses are stable at low pH values. In contrast, rhinoviruses are stable at temperatures (50°C) that would inactivate most enteroviruses (Couch, 1996). Human rhinoviruses have an icosahedral morphology, containing a single-stranded RNA genome. The diameter of these viruses is approximately 25 to 30 nm. Although no etiologic agent is identified in half of the acute upper respiratory illnesses, it has been estimated that 30 to 50% are caused by rhinoviruses. In fact, rhinovirus infection is probably the most common type of human acute infection (Gwaltney, 1997). The optimal temperature for rhinovirus growth is 33 to 35°C, which corresponds to the normal temperature of the nasal mucosa. This may explain why these viruses propagate most efficiently in the upper respiratory tract. These viruses can be propagated in monkey and human cell lines. There are 100 different human rhinovirus serotypes. Although these viruses are species specific, an equine rhinovirus can infect other species, including humans. Experimentally, chimpanzees and gibbons have been infected with human rhinoviruses, but these animals do not develop disease. Volunteer studies have shown that infection can be started with less than one tissue culture median infectious dose ($TCID_{50}$).

Rhinovirus infection, after 1 to 4 days of incubation, is characterized by nasal obstruction and discharge, sneezing, scratchy throat, mild cough, and malaise. The relatively high rate of rhinovirus infection may be associated with the large number of rhinovirus serotypes and the fact that the same serotype can infect an individual more than once. In addition, immunity to rhinovirus infection is short lived. It has been estimated, on the basis of culture methods, that the rhinovirus infection rate (infection per person per year) may range from 0.74 to 1.21.

Although rhinovirus infection occurs throughout the year, in temperate climates its frequency increases during colder months and in the tropics the peak incidence occurs during the rainy season. Experiments with volunteers have failed to associate exposure to low temperatures with an increased susceptibility to rhinovirus infection. Therefore, the cause of this seasonality remains unclear. Nevertheless, it has been suggested that colder temperatures or rain may increase the survivability of these viruses and/or promote crowding conditions, in which the virus may propagate more efficiently.

Children are often the source of rhinovirus infection in family units, with illness developing in approximately half of those exposed. Infection is more likely to occur when the infected person sheds large numbers of viruses and the susceptible contact is exposed for long periods. During rhinovirus infection, the highest concentration of viruses is found in the nasal and pharyngeal mucosa. In contrast, in saliva, rhinoviruses are found in low numbers and inconsistently. This suggests that aerosols may not play an important role in the transmission of these agents, as sneezing and coughing generate aerosols composed mainly of saliva. In fact, studies with volunteers have shown that rhinovirus transmission by hand contact and self-inoculation of the eye or nasal mucosa is much more efficient than through aerosols (Hendley *et al.*, 1973).

The high concentration of rhinoviruses in nasal secretions easily leads to hand contamination. Rhinoviruses have been isolated from 40 to 90% of hands of ill persons and from 6 to 15% of environmental objects such as doorknobs, dolls, coffee cups, and glasses. It is believed that contamination of objects in the environment may play a significant role in the transmission of rhinoviruses. This and the ability of rhinoviruses to survive on human skin and on inanimate objects may explain the more prevalent self-inoculation form of transmission. Both finger-to-eye and finger-to-nose contact are part of normal human behavior. Therefore, experimental rhinovirus transmission has been effectively interrupted by hand disinfection. Rhinoviruses are relatively stable on inanimate surfaces. They can survive at least 3 hours on human skin, nonporous materials such as plastic surfaces, Formica, stainless steel, and hard synthetic fabrics such as nylon and dacron. In porous materials such as paper tissue and cotton fabric, rhinoviruses can survive for 1 hour (Hendley *et al.*, 1973).

19.4.2.2 Paramyxovirus

Parainfluenza viruses and respiratory syncytial viruses (RSV) belong to the family Paramyxoviridae. These viruses are spherical enveloped particles of heterogeneous size, ranging from 125 to 250 nm. The paramyxovirus viral particle has a lipid envelope covered with spikes of about 10 nm. The nucleic acid of these agents consists of a single-stranded RNA molecule (Ginsberg, 1990).

RSV and parainfluenza viruses do not survive well in the environment. If suspended in a protein-free medium at 4°C, 90 to 99% of their infectivity is lost within 4 hours. Organic solvents and detergents rapidly inactivate these viruses by dissolving their lipid envelopes. Both RSV and parainfluenza viruses cause a variety of illnesses, primarily in infants and young children. The most important are common cold, bronchitis, bronchiolitis, laryngotracheobronchitis (croup), and pneumonia.

19.4.2.3 Parainfluenza Viruses

The parainfluenza viruses constitute two of the four genera of the family Paramyxoviridae: *Paramyxovirus* (parainfluenza virus types 1 and 3) and *Rubulavirus* (parainfluenza virus types 2 and 4). Parainfluenza viruses infect most people during childhood. Types 1 and 2 are often associated with croup in infants and type 3 with bronchiolitis and pneumonia; type 4 seldom causes illness. Although parainfluenza infections in children and infants can result in serious disease, most infections are subclinical. In adults these viruses can cause a mild cold, but, as in children, most infections are also subclinical. Because of the fragile nature of the parainfluenza viruses, these agents are mostly transmitted by direct person-to-person contact or by large-droplet spread. Survival studies have shown that parainfluenza virus type 3 remains infectious in aerosol particles for at least 1 hour (Miller and Artenstein, 1967).

Disease caused by parainfluenza viruses is observed throughout the year. Infection by parainfluenza virus types 1 and 2 occurs endemically, but small epidemics caused by these agents are observed every 2 years. Parainfluenza virus type 3 appears to spread the most efficiently. This agent is a major cause of bronchiolitis and pneumonia in infants, second in importance to RSV (Glezen and Denny, 1997). Parainfluenza virus type 3 infects approximately 60% of infants during the first 2 years of life, reaching 80% by 4 years of age, whereas infection by types 1 and 2 does not reach 80% until 10 years of age. In children, an incubation period of 2 to 4 days has been estimated for parainfluenza virus illness. Studies have shown that a child infected with parainfluenza type 3 may shed viruses an average of 8 days, but shedding viruses for up to 4 weeks has been documented (Frank et al., 1981).

19.4.2.4 Respiratory Syncytial Virus

RSV belongs to the family Paramyxoviridae, as the only member of the genus *Pneumovirus*. The characteristics of RSV have already been described. This virus is the most important respiratory pathogen during infancy and early childhood. It causes approximately half of the cases of bronchiolitis and 25% of cases of pneumonia in infants. Approximately 90,000 hospital admissions and 4500 deaths are associated each year with RSV in the United States in both infants and young children. In approximately half of infants younger than 8 months, the infection spreads to the lower respiratory tract, resulting in life-threatening bronchitis, bronchiolitis, bronchopneumonia, and croup (Ginsberg, 1990). Furthermore, RSV is the most common cause of nosocomial infections in pediatric wards and the most important cause of middle ear infection in children. In contrast to the parainfluenza viruses, which are important pathogens in adults, RSV produces serious respiratory disease mostly during the first 6 months of life. Approximately 1% of all infants suffer from a severe RSV infection that may require hospitalization, and about 1% of those die. At higher risk are infants with pulmonary hypertension, congenital heart defects, or impaired immunity, pre-

mature babies, and newborns with very low birth weight. Asymptomatic first infection with RSV is rare; almost 100% of infected children develop disease. Because of poor protective immunity, reinfection is very common, but the disease is not as severe. Second and third exposures to RSV result in reinfection rates of 75 and 65%, respectively, with approximately 50% showing respiratory illness. Interestingly, severe lower respiratory disease caused by RSV is significantly more common in males than in females (Parrot et al., 1973).

Inoculation of adult volunteers with RSV has shown that the nose and eye mucosae are the most important portals of entry of this virus. Efficient transmission of RSV by large droplets or by touching occurs when susceptible individuals are in close contact with children shedding the virus but not when they are exposed to small-particle aerosol (McIntosh, 1997).

The incubation period of RSV respiratory disease is 4 to 5 days. Replication of RSV in the upper respiratory tract can reach concentrations of 10^4 to 10^6 $TCID_{50}$/ml of secretion, with higher titers in infants. Infected individuals may shed RSV for up to 3 weeks. It has been observed that this shedding period correlates with the severity of illness but not with the age of the patient. In temperate climates RSV infection may occur throughout the year, but it peaks during winter months with few occurrences in the summer; in tropical areas outbreaks often occur during the rainy season.

19.4.2.5 Influenza Viruses

The influenza viruses belong to three genera of the family Orthomyxoviridae: influenza virus A, influenza virus B, and influenza virus C. These viruses, approximately 80 to 120 nm in size, possess a lipid envelope with a genome consisting of eight segments of single-stranded RNA. Influenza viruses are divided into types A, B, and C. The influenza virus type A is further classified into many subtypes according to host of origin, year, and geographic location of first isolation.

It is not known why several recent influenza pandemics have started in China. It has been suggested that the large pig, duck, and human population in the Canton area may facilitate co-infection of animals with influenza viruses originated from different species, leading to genetic reassortment and to the generation of viruses with novel antigenic and virulence characteristics. There is strong evidence that aquatic birds are the main reservoir of all influenza viruses in other species. For example, the catastrophic influenza pandemic of 1918, in which over 20 million people died, is believed to have been caused by an influenza A virus derived from a bird.

Influenza A virus is a highly contagious agent, causing epidemics of an acute respiratory infection known as influenza, with high mortality in the elderly. About 75% of all influenza deaths occur in individuals over 55 years of age. Influenza A virus infection may be asymptomatic, but more often it may be manifested by a wide variety of clinical conditions, ranging from an annoying flu to a fatal pneumonia. For instance, a subtype of influenza A virus that was responsible for epidemics of severe disease at the beginning of the century caused minor illness in older people during widespread epidemics in the 1980s. The reason for this wide variation is poorly understood, but it has been speculated that age, underlying illnesses, previous exposure to a similar influenza A virus subtype, and virus virulence may be associated with the type of disease presentation. In the United States in recent years more than 90% of influenza-related deaths occurred among persons of 65 years of age or older (Centers for Disease Control, 1996). Mortality due to influenza is significant and varies from season to season. During 9 of 20 influenza seasons in the United States (1972–1992), more than 20,000 people died each season; during four seasons, more than 40,000 deaths were recorded (Centers for Disease Control, 1997).

Influenza A virus can be transmitted most efficiently through aerosols. It has been demonstrated that influenza A virus remains infectious for at least 1 hour in aerosols at room temperature (Murphy and Webster, 1996). Clinical manifestations of influenza start rather suddenly, about 1 to 4 days after infection. A disabling syndrome characterized by high fever, together with muscle pain, sore throat, nasal congestion, conjunctivitis, cough, and headache, is usually the norm.

Influenza epidemics occur intermittently. In temperate climates they usually occur from early fall to late spring, but in tropical regions epidemics are observed throughout the year. Influenza epidemics spread rapidly and tend to occur worldwide. Although variable, epidemics caused by influenza A virus are observed every 2 to 4 years, whereas influenza B virus epidemics normally occur every 3 to 6 years. Influenza B virus epidemics are usually less widespread and less severe than the epidemics caused by influenza A virus. The virus source that initiates new influenza epidemics is usually unknown. It has been postulated that carriers with an apparent infection or sporadic cases may occur during periods between epidemics; however, this is unlikely, as influenza viruses are seldom isolated from individuals during interepidemic periods. A more accepted explanation is that animals may play an important role as carriers of influenza viruses during the interval between epidemics. This is supported by the fact that human strains of influenza A virus are, genetically and immunologically, related to influenza viruses of horses, birds, and pigs. There is also evidence that influenza epidemics may be caused by viruses that remain hidden for many years in a frozen state. The virus responsible for a pandemic that was initiated in northern China in 1972 was genetically indistinguishable from a virus isolated during an epidemic in Russia in 1950.

Immunity to influenza is long lived; however, it is virus subtype specific. Epidemiological studies have shown that individuals previously infected with influenza A subtype H1N1 during the 1957 pandemic were resistant to infection when the same subtype reappeared in the 1977 pandemic, but they were fully susceptible when exposed to other influenza A subtypes.

QUESTIONS AND PROBLEMS

1. What are pathogens? What is an enteric pathogen?

2. What is the difference between a waterborne and a water-based pathogen?

3. Which group of enteric pathogens survives longest in the environment and why?

4. What are some of the niches in which *Legionella* can grow to high numbers?

5. Why are *Cryptosporidium* and *Giardia* major causes of waterborne disease in the United States today?

6. What are the names of the environmentally resistant forms of waterborne protozoan parasites?

7. What group of animals are the reservoir of *Toxoplasma gondii*?

8. What is the difference between a frank pathogen and an opportunistic pathogen? Give examples of each.

9. Which groups of pathogens cannot grow in the environment outside a host? Which ones can?

10. Which virus is the leading cause of childhood gastroenteritis worldwide?

11. Which virus is most commonly associated with waterborne disease outbreaks?

12. Which bacterium is the leading cause of gastroenteritis in the United States?

13. Fomites are important in the spread of what respiratory viruses?

14. Why are respiratory infections more common at certain times of the year?

15. What virus can be transmitted by contact with the eyes?

16. What type of hepatitis has a high mortality in pregnant women?

17. What type of virus causes eye infections in persons swimming in contaminated waters?

18. What are nosocomial infections?

19. What organism is responsible for typhoid? Cholera? Winter diarrhea in infants?

20. Why are only certain strains of *E. coli* capable of causing disease in humans?

References and Recommended Readings

Adam, R. (1991) The biology of *Giardia* spp. *Microbiol. Rev.* **55**, 706–732.

Alary M., and Joly, J. R. (1991) Risk factors for contamination of domestic hot water systems by legionellae. *Appl. Environ. Microbiol.* **57**, 2360–2367.

Allen, M. J., and Geldreich, E. E. (1975) Bacteriological criteria for groundwater quality. *Ground Water* **13**, 45–52.

Bagdasar'yan, G. A., and Abieva, R. M. (1971) Survival of enteroviruses and adenoviruses in water. *Hyg. Sanit.* **36**, 333–337.

Balayan, M. S. (1993) Hepatitis E virus infection in Europe: Regional situation regarding laboratory diagnosis and epidemiology. *Clin. Diagn. Virol.* **1**, 1–9.

Baleux, B., and Trousllier, M. (1989) Optimization of a sampling design and significance of bacterial indicators: Application to the bacteriological survey of the Ardeche river, France. *Water Res.* **23**, 1183–1190.

Baumann, P. (1968) Isolation of *Acinetobacter* from soil and water. *J. Bacteriol.* **96**, 39–42.

Benenson, M. W., Takafuji, E. T., Lemon, S. M., Greenup, R. L., and Sulzer, A. J. (1982) Oocyst- transmitted toxoplasmosis associated with ingestion of contaminated water. *N. Engl. J. Med.* **307**, 666–669.

Biamon, E. J., and Hazen, T. C. (1983) Survival and distribution of *Aeromonas hydrophila* in near- shore coastal waters of Puerto Rico receiving rum distillery effluent. *Water Res.* **17**, 319–326.

Bifulco, J. M., Shirey, J. J., and Bissonnette, G. K. (1989) Detection of *Acinetobacter* spp. in rural drinking water supplies. *Appl. Environ. Microbiol.* **55**, 2214–2219.

Bingham, A. K., Jarroll, E., and Meyer, E. (1979) *Giardia* spp.: Physical factors of excystation *in vitro*, and excystation vs. eosin exclusion as determinants of viability. *Exp. Parasitol.* **47**, 284–291.

Blaser, M. J., and Smith, P. F., Wang, W. L. L., and Hoff, J. C. (1986) Inactivation of *Campylobacter* by chlorine and monochloramine. *Appl. Environ. Microbiol.* **51**, 307–311.

Bolton, F. J., Hinchliffe, P. M., Coates, D., and Robertson, L. (1982) A most probable number method for estimating small numbers of campylobacters in water. *J. Hyg. Camb.* **89**, 185–190.

Bradley, D. W. (1992) Hepatitis E: Epidemiology aetiology and molecular biology. *Rev. Med. Virol.* **2**, 19–28.

Brudastov, A. N., Lemelev, V. R., Kholnukhanedov, S. K., and Krasnos, L. N. (1971) The clinical picture of the migration phase of ascaris in self-infection. *Medskaya Parazitol.* **40**, 165–168.

Burden, D. J., Whitehead, A., Green, E. A., McFadzean, J. A., and Beer, R. (1976) The treatment of soil infected with human whipworm *Trichuris trichiura*. *J. Hyg.* **77**, 377–382.

Burns, D., Wallace, R. J., Jr., Schultz, M. E., Zhang, Y., Zubairi, S. Q., Pang, Y., Gibert, C. L., Brown, B. A., Noel, E. S., and Gordin, F. M. (1991) Nosocomial outbreak of respiratory tract colonization with *Mycobacterium fortuitum*: Demonstration of the usefulness of pulsed- field gel electrophoresis in an epidemiological investigation. *Am. Rev. Respir. Dis.* **144**, 1153–1159.

Canning, E. (1993) Microsporidia. In "Parasitic Protozoa," 2nd ed., Vol. 6. Academic Press, San Diego, pp. 299–370.

Canning, E., and Lom, J. (1986) "The Microsporidia in Vertebrates." Academic Press, London.

Carmichael, W. W. (1994) The toxins of cyanobacteria. *Sci. Am.* **273**, 64–72.

Carter, A. M., Pacha, R. E., Clark, G. W., and Williams, F. A. (1987) Seasonal occurrence of *Campylobacter* spp. in surface waters and their correlation with standard indicator bacteria. *Appl. Environ. Microbiol.* **53**, 523–526.

Casson, L., Sorber, C., Sykora, J., Gavaghan, P., Shapiro, M., and Jakubowski, J. (1990) *Giardia* in wastewater—effect of treatment. *Res. J. Water Pollut. Control Fed.* **62**, 670–675.

Centers for Disease Control. (1996) Prevention and control of influenza: Recommendations of the Advisory Committee on Immunization Practices (ACIP). *MMWR* **45** (RR-5).

Centers for Disease Control. (1997) Influenza Surveillance—United States, 1992–93 and 1993–94. *CDC Surveill. Summ.* **46** (SS-1).

Chin, D. P., Hopewell, P. C., Yajko, D. M., Vittinghoff, E., Horsburgh, C. R., Jr., Hadley, W. K., Stone, E. N., Nassos, P. S., Ostroff, S. M., Jacobson, M. A., Matkin, C. C., and Reingold, A. L. (1994) *Mycobacterium avium* complex in the respiratory or gastrointestinal tract and the risk of *M. avium* bacteremia in patients with human immunodeficiency virus infection. *J. Infect. Dis.* **169**, 289–295.

Clark, J. A., Burger, C. A., and Sabatinos, L. E. (1982) Characterization of indicator bacteria in municipal raw water, drinking water, and new main water samples. *Can. J. Microbiol.* **28**, 1002–1013.

Colbourne, J. S., Dennis, P. J., Trew, R. M., Beery, C., and Vesey, G. (1988) *Legionella* and public water supplies. *In* Proceedings for the International Conference on Water and Wastewater Microbiology, Newport Beach, CA, February 8–11, 1988, Vol. 1.

Collins, C. H., Grange, J. M., and Yates, M. D. (1984) Mycobacteria in water. *J. Bacteriol.* **57**, 193–211.

Couch, R. B. (1996) Rhinoviruses. *In* "Virology" (B. N. Fields, D. M. Knipe, P. M. Howley, *et al.*, eds.) Raven Press, New York, pp. 713–734.

Crabtree, K. D. (1996) Waterborne rotavirus and coxsackievirus: A risk assessment approach. *Health* **3**, 1–7.

Craun, G. F. (1979) Waterborne disease—a status report emphasizing outbreaks in ground-water systems. *Ground Water* **17**, 183–193.

Craun, G. (1986) "Waterborne diseases in the United States." CRC Press, Boca Raton, FL, pp. 121–168.

Crompton, D. W. T. (1988) The prevalence of ascariasis. *Parasitol. Today* **4**, 162–169.

Cruz, J. R., Caceres, P., Cano, F., Flores, J., Bartlett, A., and Torun, B. (1990) Adenovirus types 40 and 41 and rotaviruses associated with diarrhea in children from Guatemala. *J. Clin. Microbiol.* **28**, 1780–1784.

Cruz, J. R., Bartlett, A. V., Herrmann, J. E., Caceres, P., Blacklow, N. R., and Cano, F. (1992) Astrovirus-associated diarrhea among Guatemalan ambulatory rural children. *J. Clin. Microbiol.* **30,** 1140–1144.

Cubitt, W. D. (1987) The candidate caliciviruses. *In* "Novel Diarrhea Viruses." (G. Bock and J. Whelan, eds.), John Wiley & Sons, Chichester, UK, pp. 126–138.

Cubitt, W. D. (1991) A review of the epidemiology and diagnosis of waterborne viral infections. *Water Sci. Technol.* **24,** 197–203.

Curry, A., and Canning, E. (1993) Human microsporidiosis. *J. Infect.* **27,** 229–236.

Dan, T. B. B., and Koppel, F. (1992) Indicator bacteria for fecal pollution in the littoral zone of Lake Kinneret. *Water Res.* **26,** 1457–1469.

DeRegnier, D., Cole, L., Schupp, D., and Erlandsen, S. (1989) Viability of *Giardia* cysts suspended in lake, river and tap water. *Appl. Environ. Microbiol.* **55,** 1223–1229.

Doyle, M. P., and Schoeni, J. L. (1987) Isolation of *Escherichia coli* O157:H7 from retail fresh meats and poultry. *Appl. Environ. Microbiol.* **53,** 2394–2396.

Dubey J. P. (1996) Strategies to reduce transmission of *Toxoplasma gondii* to animals and humans. *Vet. Parasitol.* **64,** 65–70.

Dubey, J. P., and Beattie, C. P. (1988) "Toxoplasmosis of Animals and Man." CRC Press, Boca Raton, FL.

du Moulin, G. C., and Stottmeier, K. D. (1986) Waterborne mycobacteria: An increasing threat to health. *ASM News* **52,** 525–529.

du Moulin, G. C., Stottmeier, K. D., Pelletier, P. A., Tsang, A. Y., and Hedley-Whyte, J. (1988) Concentration of *Mycobacterium avium* by hospital water systems. *JAMA* **26,** 1599–1601.

DuPont, H., Chappell, C., Sterling, C. S., Okhuysen, P., Rose, J., and Jakubowski, W. (1995) The infectivity of *Cryptosporidium parvum* in healthy volunteers. *N. Engl. J. Med.* **332,** 855–859.

Edberg, S.C. (unpublished) Analysis of the virulence characteristics of bacteria isolated from bottled, water cooler, and tap water. Draft report 1–50, Yale University School of Medicine, New Haven, CT.

Enriquez, C. E., Garzon-Sandoval, J., and Gerba, C. P. (1995a) Survival, detection and resistance to disinfection of enteric adenoviruses. Proceedings of the 1995 Water Quality Technology Conference, American Water Works Association, New Orleans, pp. 2059–2086.

Enriquez, C. E., Hurst, C. J., and Gerba, C. P. (1995b) Survival of the enteric adenoviruses 40 and 41 in tap, sea, and wastewater. *Water Res.* **29,** 2548–2553.

Erlandsen, S. L. (1995) Biotic transmission—giardiasis a zoonosis? *In* "Giardia: From Molecules to Disease" (R. Thompson, J. Reynoldson, and A. Lymbery.) University Press, Cambridge, pp. 83–97.

Erlandsen, S., Sherlock, L., Januschka, M., Schupp, D., Schaefer, F., Jakubowski, W., and Bemrick, W. J. (1988) Cross-species transmission of *Giardia* spp.: Inoculation of beavers and muskrats with cysts of human, beaver, mouse, and muskrat origin. *Appl. Environ. Microbiol.* **54,** 2777–2785.

Falcao, D. P., Valentini, S. R., and Leite, C. Q. F. (1993). Pathogenic or potentially pathogenic bacteria as contaminants of fresh water from different sources in Araraquara, Brazil. *Water Res.* **27,** 1737–1741.

Falconer, I. R., Runnegar, M. T. C., Buckley, T., Huyn, V. L., and Bradshaw, P. (1989) Using activated charcoal to remove toxicity from drinking water containing cyanobacterial blooms. *J. Am. Water Works Assoc.* **2,** 102–105.

FDA, Anonymous. (1992) Foodborne pathogenic microorganisms and natural toxins. Washington, DC.

Fields, B. S., Shotts, E. B., Jr., Feeley, J. C., Gorman, G. W., and Mar-tin, W. T. (1984) Proliferation of *Legionella pneumophila* as an intracellular parasite of the ciliated protozoan *Tetrahymena pyriformis. Appl. Environ. Microbiol.* **47,** 467–471.

Frank, A. L., Taber, L. H., Wells, C. R., Wells, J. M., Glezen, W. P., and Paredes, A. (1981) Patterns of shedding of myxo-viruses and paramyxoviruses in children. *J. Infect. Dis.* **144,** 433–441.

Freedman, D. O. (1992) Intestinal nematodes. *In* "Infectious Diseases" (S. L. Gorbach, J. G. Bartlett, and N. Blacklow, eds.) W. B. Saunders, Philadelphia, pp. 2003–2008.

Frenkel, J. K. , Ruiz, A., and Chinchilla, M. (1975) Soil survival of *Toxoplasma* oocysts in KS and Costa Rica. *J. Trop. Med. Hyg.* **24,** 439–443.

Gerba, C. P., Wallis, C., and Melnick, J. L. (1975) Fate of wastewater bacteria and viruses in soil. *J. Irrigation Drainage Div. Proc. Am. Soc. Civ. Eng.* **101**(IR3), 145.

Ginsberg, H. S. (1990) Paramyxoviruses. *In* "Microbiology," 4th ed. (B. D. Davis, R. Dulbecco, H. N. Eisen and H. S. Ginsberg, eds.) J. B. Lippincott, Philadelphia, pp. 947-959.

Ginsberg, M., Keenan, K., Thompson, M., and Anders, B. (1994) Stool survey of asymptomatic diapered children in day care. *Pediatrics* **94S,** 1026–1027.

Glezen, W. P., and Denny, F. W. (1997) Parainfluenza viruses. *In* "Viral Infections of Humans, Epidemiology and Control," 4th ed. (A. S. Evans, and R. A. Kaslow, eds.) Plenum, New York, pp. 551–568.

Gonzalez, C., Gutierrez, C., and Grande. T. (1987) Bacterial flora in bottled uncarbonated mineral drinking water. *Can. J. Microbiol.* **33,** 1120–1125.

Goslee, S., and Wolinsky, E. (1976) Water as a source of potentially pathogenic mycobacteria. *Am. Rev. Respir. Dis.* **113,** 287–292.

Grabow, W. O. K., Putergill, D. L., and Bosch, A. (1992) Propagation of adenovirus types 40 and 41 in the PLC/PRF/5 primary liver carcinoma cell line. *J. Virol. Methods* **37,** 201–208.

Gwaltney, J. M. (1997) Rhinoviruses. *In* "Viral Infections of Humans, Epidemiology and Control," 4th ed. (A. S. Evans, and R. A. Kaslow, eds.) Plenum, New York, pp. 815–838.

Haas, H., and Fatal, B. (1990) Distribution of mycobacteria in different types of water in Israel. *Water Res.* **24,** 1233–1235.

Hawley, H. B., Morin, D. P., Geraghty, M. E., Tomkow, J., and Phillips, A. (1973) Coxsackievirus B epidemic at a boys' summer camp. JAMA **226,** 33–36.

Hazen, T. C., Fliermans, C. B., Hirsch, R. P., and Esch, G. W. (1978) Prevalence and distribution of *Aeromonas hydrophila* in the United States. *Appl. Environ. Microbiol.* **36,** 731–738.

Hendley, J. O., Wenzel, R. P., and Gwaltney, J. M. (1973) Transmission of rhinovirus colds by self-inoculation. *N. Engl. J. Med.* **288,** 1361–1363.

Herwaldt, B. L., and Ackers, M. L. (1997) An outbreak in 1996 of cyclosporiasis associated with imported raspberries. The Cyclospora Working Group. *N. Engl. J. Med.* **336,** 1548–1556.

Herwaldt, B. L., Craun, G. F., Stokes, S. L., and Juranek, D. D. (1992) Outbreaks of waterborne disease in the United States: 1989–90. *J. Am. Water Works Assoc.* **84,** 129–135.

Hibler, C., and Hancock, C. (1990) Waterborne giardiasis. *In* "Drinking Water Microbiology" (G. McFeters, ed.) Springer-Verlag, New York, pp. 271–293.

Hopkins, D. R. (1992) Homing in on helminths. *Am. J. Top. Med Hyg.* **46,** 626.

Horwitz, M. S. (1996) Adenoviruses. *In* "Fields Virology," 3rd ed. (B. N. Fields, D. M. Knipe, and P. M. Howley, eds.) Lippincott-Raven, Philadelphia, PA, pp. 2149–2171.

Hoxie, N. J., Davis, J. P., Vergeront, J. M., Nashold, R. D., and Blair, K. A. (1997) Cryptosporidiosis-associated mortality following a massive waterborne outbreak in Milwaukee, WI. *Am. J. Public Health* **87,** 2032–2035.

Ho-Yen, D. O., and Joss, A. W. L. (1992). "Human Toxoplasmosis." Oxford University Press, Oxford.

Huang, P., Wever, W., Sosin, D., Griffin, P., Long, E., Murphy, J., Kocka, F., Peters, C., and Kallick, C. (1995) The first reported outbreak of diarrheal illness associated with *Cyclospora* in the United States. *Ann. Intern. Med.* **123**, 409–414.

Hughes, H. P. A. (1985) How important is toxoplasmosis? *Parasitol. Today* **1**, 41–44.

Hurst, C. J., McClellan, K. A., and Benton, W. H. (1988) Comparison of cytopathogenicity, immunofluorescence and *in situ* DNA hybridization as methods for the detection of adenoviruses. *Water Res.* **22**, 1547–1552.

Isaac-Renton, J., Corderio, C., Sarafis, K., and Shahriar, H. (1993) Characterization of *Giardia duodenalis* isolates from a waterborne outbreak. *J. Infect. Dis.* **167**, 431–440.

Izaguirre, G. (1992) A copper-tolerant *Phormidium* species from Lake Mathews, CA, that produces 2-methylisoborneol and geosmin. *Water Sci. Technol.* **25**, 217–223.

Izaguirre, G., Wolfe, R. L., and Means, E. G. (1988) Bacterial degradation of 2-methylisoborneol. *Water Sci. Technol.* **20**, 205–210.

Jawetz, E. (1956) Antimicrobial chemotherapy. *Annu. Rev. Microbiol.* **10**, 85–102.

Jepsen, A., and Roth, H. (1952) Epizootiology of *Cysticercus bovis*-resistance of the eggs of *Taenia saginata*. Report 14. In. *Vet. Cong.* **22**, 43–50.

Jiang, X., Graham, D. Y., Wang, K., and Estes, M. K. (1990) Norwalk virus genome cloning and characterization. *Science* **250**, 1580–1583.

John, D. T. (1982) Primary amebic meningoencephalitis and the biology of *Naegleria fowleri*. *Annu. Rev. Microbiol.* **36**, 101–103.

Johnson, D. C., Reynolds, K. A., Gerba, C. P., Pepper, I. L., and Rose, J. B. (1995) Detection of *Giardia* and *Cryptosporidium* in marine waters. *Water Sci. Technol.* **31**, 439–442.

Johnson, D. C., Enriquez, C. E., Pepper, I. L., Davis, T. L., Gerba, C. P., and Rose, J. B. (1997) Survival of *Giardia, Cryptosporidium,* poliovirus and *Salmonella* in marine waters. *Water Sci Technol.* **35**, 261–268.

Jones, A. R., Bartlett, J., and McCormack, J. G. (1995) *Mycobacterium avium* complex (MAC) osteomyelitis and septic arthritis in an immunocompetent host. *J. Infect.* **30**, 59–62.

Kapikian, A. Z. (1997). Viral gastroenteritis. *In* "Viral Infections of Humans, Epidemiology and Control," 4th ed. (A. S. Evans, and R. A. Kaslow (eds.) Plenum, New York, pp. 285–343.

Kapikian, A. Z., Wyatt, R. G., Dolin, R., Thornhill, T. S., Kalica, A. R., and Chanock, R. M. (1972) Visualization by immune electron microscopy of 27 nm particle associated with acute infectious nonbacterial gastroenteritis. *J. Virol.* **10**, 1075–1081.

Kaplan, J. E., Feldman, R., Campbell, D. S., Lookabaugh, C., and Gary, G. W. (1982) The frequency of Norwalk-like pattern of illness in outbreaks of acute gastroenteritis. *Am. J. Public Health* **72**, 1329–1332.

Kemp, J. S., Wright, S. E., and Bukhari, Z. (1995) On farm detection of *Cryptosporidium parvum* in cattle, calves and environmental samples. *In* "Protozoan Parasites and Water." (W. B. Betts, D. Casemore, C. Fricker, H. Smith, and J. Watkins, eds.) The Royal Society of Chemistry, Cambridge, UK.

Kirschner, R. A., Jr., Parker, B. C., and Falkinham, J. O. III. (1992) *Mycobacterium avium, Mycobacterium intracellulare,* and *Mycobacterium scrofulaceum* in acid, brown-water swamps of the southeastern United States and their association with environmental variables. *Am. Rev. Respir. Dis.* **145**, 271–275.

Klein, P. D., Graham, D. Y., Gaillour, A., Opekun, A. R., and Smith, E. O. (1991) Water source as risk factor for *Helicobacter pylori* infection in Peruvian children. *Lancet* **337**, 1503–1506.

Knight, P. (1995) Once misidentified human parasite is a cyclosporan. *ASM News* **61**, 520–522.

Kurtz, J. B., and Lee, T. W. (1987) Astroviruses: Human and animal. *In* "Novel Diarrhea Viruses." (G. Bock and J. Whelan, eds.) John Wiley & Sons, Chichester, UK, pp. 92–107.

LeChevallier, M. W., Seidler, R. J., and Evans, T. M. (1980) Enumeration and characterization of standard plate count bacteria in chlorinated and raw water supplies. *Appl. Environ. Microbiol.* **40**, 922–930.

LeChevallier, M. W., Evans, T. M., Seidler, R. J., Daily, O. P., Merrell, B. R., Rollins, D. M., and Joseph, S. W. (1982) *Aeromonas sobria* in chlorinated drinking water supplies. *Microb. Ecol.* **8**, 325–333.

Lee, L. A., Shapiro, C. N., Hargrett-Bean, N., and Tauxe, R. V. (1991) Hyperendemic shigellosis in the United States: A review of surveillance data for 1967–1988. J. Infect. Dis. 164, 894–900.

Lee, T. W., and Kurtz, J. B. (1994) Prevalence of human astrovirus serotypes in the Oxford region 1976–92, with evidence for two new serotypes. *Epidemiol. Infect.* **112**, 187–193.

Levaditi, C., Nicolau, S., and Schoen, R. (1923) L'étiologie de l'encéphalite. *C. R. Acad. Sci. Paris* **177**, 985–988.

Lew, J. F., Petric, M., Kapikian, A. Z., Jiang, X., Estes, M. K., and Green, K. Y. (1994) Identification of minireovirus as a Norwalk-like virus in pediatric patients with gastroenteritis. *J. Virol.* **68**, 3391–3396.

Lucey, D. R., and Maguire, J. H. (1993) Schistosomiasis. *Infect. Dis. Clin. North Am.* **7**, 635–653.

MacKenzie, W., Hoxie, N., Proctor, M., Gradus, M., Blair, K., Peterson, D., Kazmierczak, J., Addiss, D., Fox, K., Rose, J., and Davis, J. (1994) A massive outbreak in Milwaukee of *Cryptosporidium* infection transmitted through the public water supply. *N. Engl. J. Med.* **331**, 161–167.

Mahl, M. C., and Sadler, C. (1975) Virus survival on inanimate surfaces. *Can. J. Microbiol.* **21**, 819–823.

Margolis, H. S., Alter, M. J., and Hadler, S. C. (1997) Viral hepatitis. *In* "Viral Infections of Humans, Epidemiology and Control," (A. S. Evans and R. A. Kaslow, eds.) 4th ed. Plenum, New York, pp. 363–418.

Marrie, T. J. (1994) Community-acquired pneumonia. *Clin. Infect. Dis.* **18**, 501–515.

Martone, W. J., Hierholzer, J. C., Keenlyside, R. A., Fraser, D. W., D'angelo, L. L., and Winkler, W. G. (1980) An outbreak of adenovirus type 3 disease at a private recreation center swimming pool. *Am. J. Epidemiol.* **111**, 229–237.

Mbithi, J. N., Springthorpe, V. S., and Sattar, S. A. (1992) Survival of hepatitis A virus on human hands and its transfer on contact with animate and inanimate surfaces. *J. Clin. Microbiol.* **30**, 757–763.

McFeters, G. A. Bissonette, G. K., and Jezeski, J. J. (1974) Comparative survival of indicator bacteria and enteric pathogens in well water. *Appl. Microbiol.* **27**, 823.

McHenry R., Bartlett, M. S., Lehman, G. A., O'Conner, K. W. (1987) The yield of routine duodenal aspiration for *Giardia lamblia* during oesphagogastroduodenoscopy. *Gastrointest. Endose.* **33**, 425–426.

McIntosh, K. (1997) Respiratory syncytial virus. *In* "Human Enterovirus Infections," (H. A. Robart, ed.) ASM Press, Washington, DC.

McMillan, N. S., Martin, S. A., Sobsey, M. D., and Wait, D. A. (1992) Outbreak of pharyngoconjunctival fever at a summer camp-North Carolina. *MMWR.* **41** 342–344.

McOrist, S., Boid, R., Jones, T., Easterbee, N., Hubbard, A., and Jarrett, O. (1991) *J. Wildlife Diseases.* **27**, 693–696.

Meenhorst, P. L., Reingold, A. L., Groothuis, D. G., Gorman, G. W., Wilkinson, H. W., McKinney, R. M., Feeley, J. C., Brenner, D. J., and Furth, R. V. (1985) Water-related nosocomial pneumonia caused by *Legionella pneumophila* serogroup 1 and 10. *J. Infect. Dis.* **152**, 356–363.

Meisel, J., Perera, and R., Meloigro, C. (1976) Overwhelming watery diarrhea associated with *Cryptosporidium* in an immunosuppressed patient. *Gastroenterology.* **70**, 1156–1160.

Meng, Q. S., and Gerba, C. P. (1996) Comparative inactivation of enteric adenovirus, polio virus, and coliphages by ultraviolet irradiation. *Water Res.* **30**, 2665–2668.

Miller, W. S., and Artenstein, M. S. (1967) Aerosol stability of three acute disease viruses. *Proc. Soc. Exp. Biol. Med.* **125**, 222–227.

Millership, S. E., and Chattopadhyay, B. (1985) *Aeromonas hydrophila* in chlorinated water supplies. *J. Hosp. Infect.* **6**, 75–80.

MMWR. (1995) Update, *Vibrio cholerae* O1 Western hemisphere, 1991–1994, and *V. cholerae* O139—Asia, 1994. *MMWR* **44**(11), 215–219.

Moffat, J. E., and L. S. Tompkins. (1992) A quantitative model of intracellular growth of *Legionella pneumophila* in *Acanthamoeba castellanii. Infect. Immun.* **60**, 292–301.

Montecalvo, M.A., Forester, G., Tsang, A. Y., du Moulin, G., and Wormser, G. P. (1994) Colonization of potable water with *Mycobacterium avium* complex in homes of HIV-infected patients. *Lancet* **343**,1639.

Moore, A. C., Herwaldt, B. L., Craun, G. F., Calderon, R. L., Highsmith, A. K., and Juranek, D. D. (1993) Surveillance for waterborne disease outbreaks—United States, 1991–1992. *MMWR* **42**, 1–22.

Morens, D. M., and Pallansch, M. A. (1995) Epidemiology. *In* "Human Enterovirus Infections." (H. A. Rotbart, ed.) ASM Press, Washington, DC, pp. 3–24.

Morgan, D. R., Johnson, P. C., DuPont, H. L., Satterwhite, T. K., and Wood, L. V. (1985) Lack of correlation between known virulence properties of *Aeromonas hydrophila* and enteropathogenicity for humans. *Infect. Immun.* **50**, 62–65.

Mosso, M.A., del Carmen de la Rosa, M., Vivar, C., and del Rosario Medina, M. (1994) Heterotrophic bacterial populations in the mineral waters of thermal springs in Spain. *J. Appl. Bacteriol.* **77**, 370–381.

Mroz, R. C., Jr., and Pillai, S. D. (1994) Bacterial populations in the groundwater on the US–Mexico border in El Paso County, Texas. *South. Med. J.* **87**, 1214–1217.

Mullens, A. (1996) I think we have a problem in Victoria: MD's respond quickly to toxoplasmosis outbreak in BC. *Can. Med. Assoc. J.* **154**, 1721–1724.

Murphy, B. R., and Webster, R. G. (1996) Orthomyxoviruses. *In* "Fields Virology," 3rd ed. (B. N. Fields, D. M. Knipe, P. M. Howley, *et al.*, eds.) Lippincott-Raven, Philadelphia, pp. 1397–1445.

Nägeli, C. 1857. Über die neue Krankheit der Seienraupe und verwante Organismen. *Bot. Ztg.* **15**, 760–761.

Narayan, L. V., and Nunez, W. J., III. (1974) Biological control: Isolation and bacterial oxidation of the taste-and-odor compound geosmin. *J. Am. Water Works Assoc.* **66**, 532–536.

Nguyen, A. M., Engstrand, L., Genta, R. M., Graham, D. Y., and el-Zaatari, F. A. (1993) Detection of *Helicobacter pylori* in dental plaque by reverse-transcriptase–polymerase chain reaction. *J. Clin. Microbiol.* **31**, 783–787.

Nishikawa, Y., and Kishi, T. (1988) Isolation and characterization of motile Aeromonas from human, food and environmental specimens. *Epidemiol. Infect.* **101**, 213–223.

Numata, K., Nakata, S., Jiang, X., Estes, M. K., and Chiba, S. (1994) Epidemiological study of Norwalk virus infections in Japan and Southeast Asia by enzyme-linked immunosorbent assays with Norwalk virus capsid protein produced by the baculovirus expression system. *J. Clin. Microbiol.* **32**, 121–126.

Ooi, W., Zimmerman, S., and Needham, C. (1995) *Cyclospora* species as a gastrointestinal pathogen in immunocompetent hosts. *J. Clin. Microbiol.* **33**, 1267–1269.

Ortega, Y., Sterling, C., Gilman, R., Cama, V., and Diaz, F. (1993) *Cyclospora* species—a new protozoan pathogen of humans. *N. Engl. J. Med.* **328**, 1308–1312.

Parrot, R. H., Kimi, H. W., Arrobio, J. O., Hodes, D. S., Murphy, B. R., Brandt, C. D., Camargo, E., and Chanock, R. (1973) Epidemiology of respiratory syncytial virus infection in Washington, DC: II. Infection and disease with respect to age, immunologic status, race and sex. *Am. J. Epidemiol.* **98**, 289–300.

Payment, P., Gamache, F., and Paquette, G. (1988) Microbiological and virological analysis of water from two water filtration plants and their distribution systems. *Can. J. Microbiol.* **34**, 1304–1309.

Payment, P., Coffin, E., and Paquette, G. (1994) Blood agar to detect virulence factors in tap water heterotrophic bacteria. *Appl. Environ. Microbiol.* **60**, 1179–1183.

Phillip, R., Brown, M., Bell, R., and Francis, F. (1992) Health risks associated with recreational exposure to blue–green algae (cyanobacteria) when windsurfing and fishing. *Health Hyg.* **13**, 115–119.

Pintó, R. M., Diez, J. M., and Bosch, A. (1994) Use of the colonic carcinoma cell line CaCO$_2$ for in vivo amplification and detection of enteric viruses. *J. Med. Virol.* **44**, 310–315.

Rao, V. C., and Melnick, J. L. (1986) "Environmental Virology." American Society for Microbiology. Washington, DC.

Rebold, J., Hoge, C., and Shlim, D. (1994) *Cyclospora* outbreak associated with chlorinated drinking water. *Lancet* **344**, 1360–1361.

Rippey, S. R., and Cabelli, V. J. (1979) Membrane filter procedure for enumeration of *Aeromonas hydrophila* in fresh waters. *Appl. Environ. Microbiol.* **38**, 108–113.

Rippey S. R., and Cabelli V. J. (1989) Use of thermotolerant *Aeromonas* group for the trophic state classification of freshwaters. *Water Res.* **23**, 1107–1114.

Rollins, D. M., and Colwell, R. R. (1986) Viable but nonculturable stage of *Campylobacter jejuni* and its role in survival in the natural aquatic environment. *Appl. Environ. Microbiol.* **52**, 531–538.

Roper, W. L., Murphy, F. A., Mahy, B. W., Anderson, L. S., and Glass, R. I. (1990) Viral agents of gastroenteritis: Public health importance and outbreak management. *MMWR* **39**, 1–24.

Savioli, L., Renganathan, E., Montresor, A., Davis, A., and Behbehani, K. (1997) Control of schistosomiasis—a global picture. *Parasitol. Today* **13**, 444–448.

Schmidt, G., and Roberts, L., ed. (1989) *In* "Foundations of Parasitology," 4th ed. Times Mirror/Mosby, St. Louis.

Schultz, M. G. (1974) The surveillance of parasitic diseases in the United States. *Am J. Trop. Med.* **23**, 744–751.

Schwartz, D. R., Bryan, R. T., Weber, R., and Visvesvara, G. (1994) Microsporidiosis in HIV positive patients: Current methods for diagnosis using biopsy, cytologic, ultrastructural, immunological, and tissue culture techniques. *Folia Parasitol.* **41**, 101–109.

Seidler, R. J., Allen, D. A., Lockman, H., Colwell, R. R., Joseph, S. W., and Daily, P. O. (1980) Isolation, enumeration, and characterization of *Aeromonas* from polluted waters encountered in diving operations. *Appl. Environ. Microbiol.* **39**, 1010–1018.

Shahamat, M., Mai, U., Paszko-Kolva, C., Kessel, M., and Colwell, R. R. (1993) Use of autoradiography to assess viability of *Helicobacter pylori* in water. *Appl. Environ. Microbiol.* **59**, 1231–1235.

Skeels, M. R., Sokolow, R., Hubbard, C. V., Foster, L. R. (1986) Screening for co-infection with *Cryptosporidium* and *Giardia* in Oregon public health clinic patients. *Am. J. Public Health* **76**, 270–273.

Soave, R. N., and Johnson, W. (1995) *Cyclospora*: Conquest of an emerging pathogen. *Lancet* **345**, 667–668.

Sobsey, M. D., Dufour, C. P., Gerba, C. P., LeChevallier, M. W., and Payment, P. (1993) Using a conceptual framework for assessing risks to health from microbes in drinking water. *J. Am. Water Works Assoc.* **85**, 44–48.

States, S. J., Kuchta, J. M., Conley, L. F., Wolford, R. S., Wadowsky, R. M., and Yee, R. B. (1989) Factors affecting the occurrence of legionnaires' disease bacterium in public water supplies. *In* "Biohazards of Drinking Water Treatment" (R. A. Larson, ed.) Lewis, Chelsea, MI, pp. 67–83.

Stephen, C., Haines, D., Bollinger, T., Atkinson, K., Schwantje, H. (1996) Serological evidence of toxoplasma infection in cougars on Vancouver Island, British Columbia. *Can. Vet. J.* **37**, 241.

Stoll, N. R. (1947) This wormy world. *J. Parasitol.* **33**, 1–18.

Stout, J. E., Yu, V. L., Yee, Y. C., Vaccarello, S., Diven, W., and Lee, T. C. (1992) *Legionella pneumophila* in residential water supplies: Environmental surveillance with clinical assessment for legionnaires' disease. *Epidemiol. Infect. Dis.* **109**, 49–57.

Tacket, C. O., Brenner, F., and Blake, P. A. (1984) Clinical features and an epidemiological study of *Vibrio vulnificus* infections. *J. Infect. Dis.* **149**, 558–561.

Thomas, J. E., Gibson, G. R., Darboe, M. K., Dale, A., and Weaver, L. T. (1992) Isolation of *Helicobacter pylori* from human feces. *Lancet* **340**, 1194–1195.

Tison, D. L., and Seidler, J. (1983) *Legionella* incidence and density in potable drinking water supplies. *Appl. Environ. Microbiol.* **45**, 337–339.

Turner, P. C., Gammie, A. J., Hollinrake, K., and Codd, G. A. (1990) Pneumonia associated with contact with cyanobacteria. *Br. Med. J.* **300**, 1440–1441.

UNDP/World Bank/WHO Special Programme for Research and Training in Tropical Disease (TDR). (1991) "Progress in Research, 1989–1990." World Health Organization, Geneva, p. 41.

Ungar, B. L. P. (1990) Cryptosporidiosis in humans (*Homo sapiens*). *In* "Cryptosporidiosis of Man and Animals" (J. P. Dubey, C. A. Speer, and R. Fayer, eds.) CRC Press. Boca Raton, FL, pp. 59–82.

Velasquez, O., Stetler, H. C., Avila, C., Ornelas, G., Alvarez, C., Hadler, S. C., Bradley, D. W., and Sepulveda, J. (1990) Epidemic transmission of enterically transmitted non-A, non-B hepatitis in Mexico, 1986–1987. *JAMA* **263**, 3261–3285.

von Reyn, C. F., Barber, T. W., Arbeit, R. D., Sox, C. H., O'Connor, G. T., Brindle, R. J., Gilks, C. F., Hakkarainen, K., Ranki, A., Bartholomew, C., Edwards, J., Tosteson, A. N. A., and Magnusson, M. (1993a) Evidence of previous infection with *Mycobacterium avium–Mycobacterium intracellulare* complex among healthy subjects: An international study of dominant mycobacterial skin test reactions. *J. Infect. Dis.* **168**, 1553–1558.

von Reyn, C. F., Waddell, R. D., Eaton, T., Arbeit, R. D., Maslow, J. N., Barber, T. W., Brindle, R. J., Gilks, C. F., Lumio, J., Lahde-

virta, J., Ranke, A., Dawson, D., and Falkinham. J. O., III. (1993b) Isolation of *Mycobacterium avium* complex from water in the United States, Finland, Zaire, and Kenya. *J. Clin Microbiol* **32**, 27–30.

von Reyn, C. F., Maslow, J. N., Barber, T. W., Falkinham, J. O., III, and Arbeit, R. D. (1994) Persistent colonization of potable water as a source of *Mycobacterium avium* infection in AIDS. *Lancet* **343**, 1137–1141.

Waller, T. (1979) Sensitivity of *Encephalitozoon cuniculi* to various temperatures, disinfectants and drugs. *Lab. Anim.* **13**, 227–230.

Weber, R., and Bryan, R. (1994) Microsporidial infections in immunodeficient and immunocompetent patients. *Clin. Infect. Dis.* **19**, 517–521

Weber, R., Bryan, R., Schwartz, D., and Owen, R. (1994) Human microsporidial infections. *Clin. Microbiol. Rev.* **7**, 426–461.

Webster, R. G., Laver, W. G., Air, G. M., and Schild, G. C. (1982) Molecular mechanisms of variation in influenza viruses. *Nature* **296**, 115–121.

West, P. M., and Olds, F. R. (1992) Clinical schistosomiasis. *R. I. Med J.* **75**, 179.

White, D. O., and Fenner, F. J. (1994) "Medical Virology," 4th ed. Academic Press, San Diego.

White, G. F., and Bradley, D. J. (1972) "Drawers of Waters." University of Chicago.

Williams, F. P., and Hurst, C. J. (1988) Detection of environmental viruses in sludge: Enhancement of enterovirus plaque assay titres with 5-iodo-2'-deoxyuridine and comparison to adenovirus and coliphage titres. *Water Res.* **22**, 847–851.

Wu, J. T., and Juttner, F. (1988) Effect of environmental factors on geosmin production by *Fischerella muscicola*. *Water Sci. Technol.* **20**, 143–148.

Yilmaz, S. M., and Hopkins, S. H. (1972) Effects of different conditions on duration of infectivity of *Toxoplasma gondii* oocysts. *J. Parasitol.* **58**, 938–939.

Young, W. F., Horth, H., Crane, R., Ogden, T., and Arnott, M. (1996) Taste and odor threshold concentrations of potential potable water contaminants. *Water Res.* **30**, 331–340.

Zacheus, O. M., and Martikainen, P. J. (1994) Occurrence of legionellae in hot water distribution systems of Finnish apartment buildings. *Can. J. Microbiol.* **40**, 993–999.

Zeman, D., and Baskin, G. (1985) Encephalitozoonosis in squirrel monkeys (*Saimiri sciureus*). *Vet. Pathol.* **22**, 24–31.

20

Indicator Microorganisms

CHARLES P. GERBA

20.1 THE CONCEPT OF INDICATOR ORGANISMS

The routine examination of environmental samples for the presence of intestinal pathogens is often a tedious, difficult, and time-consuming task. Thus, it has been customary to tackle such examinations by looking first for certain indicator microorganisms whose presence indicates that pathogenic microorganisms may also be present. Developed at the turn of the last century for assessing fecal contamination, the indicator concept depends on the fact that certain non-pathogenic bacteria occur in the feces of all warm-blooded animals. These bacteria can easily be isolated and quantified by simple bacteriological methods. Detection of these bacteria in water means that fecal contamination has occurred and suggests that enteric pathogens may also be present.

For example, **coliform bacteria,** which normally occur in the intestines of all warm-blooded animals, are excreted in great numbers in feces. In polluted water, coliform bacteria are found in densities roughly proportional to the degree of fecal pollution. Because coliform bacteria are generally hardier than disease-causing bacteria, their absence from water is an indication that the water is bacteriologically safe for human consumption. Conversely, the presence of the coliform group of bacteria is indicative that other kinds of microorganisms capable of causing disease may also be present and that the water is potentially unsafe to drink.

In 1914 the U.S. Public Health Service adopted the coliform group as an indicator of fecal contamination of drinking water. Many countries have adopted coliforms and other groups of bacteria as official standards for drinking water, recreational bathing waters, wastewater discharges, and various foods. Indicator microorganisms have also been used to assess the efficacy of food processing and water and wastewater treatment processes. As an ideal assessor of fecal contamination, it has been suggested that they meet the criteria listed in Table 20.1. Unfortunately, no one indicator meets all these criteria. Thus, various groups of microorganisms have been suggested and used as indicator organisms. Concentrations of indicator bacteria found in wastewater and feces are shown in Tables 20.2 and 20.3.

20.2 TOTAL COLIFORMS

The coliform group, which includes *Escherichia, Citrobacter, Enterobacter,* and *Klebsiella* species, is relatively easy to detect. Specifically, this group includes all aerobic and facultatively anaerobic, gram-negative, non–spore-forming, rod-shaped bacteria that produce gas upon lactose fermentation in prescribed culture media within 48 hours at 35°C.

The coliform group has been used as the standard for assessing fecal contamination of recreational and drinking waters for most of this century. Through experience it has been learned that absence of this organism in 100 ml of drinking water ensures the prevention of bacterial waterborne disease outbreaks.

TABLE 20.1   Criteria for an Ideal Indicator Organism

- The organism should be useful for all types of water.
- The organism should be present whenever enteric pathogens are present.
- The organism should have a reasonably longer survival time than the hardiest enteric pathogen.
- The organism should not grow in water.
- The testing method should be easy to perform.
- The density of the indicator organism should have some direct relationship to the degree of fecal pollution.
- The organism should be a member of the intestinal microflora of warm-blooded animals.

TABLE 20.2   Estimated Levels of Indicator Organisms in Raw Sewage

Organism	CFU per 100 ml
Coliforms	10^7–10^9
Fecal coliforms	10^6–10^7
Fecal streptococci	10^5–10^6
Enterococci	10^4–10^5
Clostridium perfringens	10^4
Staphylococcus (coagulase positive)	10^3
Pseudomonas aeruginosa	10^5
Acid-fast bacteria	10^2
Coliphages	10^2–10^3
Bacteroides	10^7–10^{10}

However, it has been learned that a number of deficiencies in the use of this indicator exist (Table 20.4).

All members of the coliform group have been observed to regrow in natural surface and drinking water distribution systems (Gleeson and Gray, 1997). The die-off rate of coliform bacteria depends on the amount and type of organic matter in the water and its temperature. If the water contains significant concentrations of organic matter and is at an elevated temperature, the bacteria may increase in numbers. This phenomenon has been observed in eutrophic tropical waters, waters receiving pulp and paper mill effluents, wastewater, aquatic sediments, and organically enriched soil (i.e., sewage sludge amended) after periods of heavy rainfall. Of greatest concern is the growth or recovery of injured coliform bacteria in a distribution system because this may give a false indication of fecal contamination (Fig. 20.1). Coliforms may colonize and grow in the biofilm found on the distribution system pipes, even in the presence of free chlorine. *Escherichia coli* is 2400 times more resistant to free chlorine when attached to a surface than as free cells in water (LeChevallier *et al.*, 1988).

Because large numbers of heterotrophic bacteria in the water may mask the growth of coliform bacteria on selective media used for their isolation, true numbers of coliforms may be underestimated. This often becomes a problem when aerobic heterotrophic bacterial numbers exceed 500/ml. Finally, the longer survival and greater resistance to disinfectants of pathogenic

TABLE 20.3   Microbial Flora of Animal Feces

Animal group	Average density per gram		
	Fecal coliforms	Fecal streptococci	*Clostridium perfringens*
Farm animals			
Cow	230,000	1,300,000	200
Pig	3,300,000	84,000,000	3,980
Sheep	16,000,000	38,000,000	199,000
Horse	12,600	6,300,000	<1
Duck	33,000,000	54,000,000	—
Chicken	1,300,000	3,400,000	250
Turkey	290,000	2,800,000	—
Animal pets			
Cat	7,900,000	27,000,000	25,100,000
Dog	23,000,000	—	—
Wild animals			
Mouse	330,000	7,700,000	<1
Rabbit	20	47,000	<1
Chipmunk	148,000	6,000,000	—
Human	13,000,000	3,000,000	1,580

Modified from Geldreich (1978).

TABLE 20.4   Deficiencies with the Use of Coliform Bacteria as Indicators of Water Quality

- Regrowth in aquatic environments
- Regrowth in distribution systems
- Suppression by high background bacterial growth
- Not indicative of a health threat
- No relationship between enteric protozoan and viral concentration

Modified from Gleeson and Gray (1997).

enteric viruses and protozoan parasites limit the use of coliform bacteria as an indicator for these organisms. Still, the coliform group of bacteria has proved its merit in assessing the bacterial quality of water. Three methods are commonly used to identify coliforms in water. These are the **most probable number (MPN)**, the **membrane filter (MF)**, and the **presence–absence (P–A) tests.**

20.2.1 The Most Probable Number (MPN) Test

The MPN test allows detection of the presence of coliforms in a sample and estimation of their numbers (see also Chapter 10.1.3). This test consists of three steps: a presumptive test, a confirmed test, and a completed test. In the **presumptive test** (Fig. 20.2a), lauryl sulfate–tryptose–lactose broth is placed in a set of test tubes with different dilutions of the water to be tested. Usually, three to five test tubes are prepared per dilution. These test tubes are incubated at 35°C for 24 to 48 hours, then examined for the presence of coliforms, which is indicated by gas and acid production. Once the positive tubes have been identified and recorded, it

is possible to estimate the total number of coliforms in the original sample by using an MPN table that gives numbers of coliforms per 100 ml. In the **confirming test** (Fig. 20.2b), the presence of coliforms is verified by inoculating selective bacteriological agars such as Levine's eosin–methylene blue (EMB) agar or Endo agar with a small amount of culture from the positive tubes. Lactose-fermenting bacteria are indicated on the medium by the production of colonies with a green sheen or colonies with a dark center. In some cases a **completed test** (not shown in Fig. 20.2) is performed in which colonies from the agar are inoculated back into lauryl sulfate–tryptose–lactose broth to demonstrate the production of acid and gas.

20.2.2 The Membrane Filter (MF) Test

The MF test also allows scientists to determine the number of coliforms in a sample, but it is easier to perform than the MPN test because it requires fewer test tubes and less labor (Fig. 20.3) (see also Chapter 10.2.1.3). In this technique, a measured amount of water (usually 100 ml for drinking water) is passed through a membrane filter (pore size 0.45 μm) that traps bacteria on its surface. This membrane is then placed on a thin absorbent pad that has been saturated with a specific medium designed to permit growth and differentiation of the organisms being sought. For example, if total coliform organisms are sought, a modified Endo medium is used. For coliform bacteria, the filter is incubated at 35°C for 18–24 hours. The success of the method depends on using effective differential or selective media that can facilitate identification of the bacterial colonies growing on the membrane filter surface (see Fig. 20.3). To determine the number of coliform bacteria in a water sample, the colonies having a green sheen are enumerated.

20.2.3 The Presence–Absence (P–A) Test

Presence–absence tests (P–A tests) are not quantitative tests; instead, they answer the simple question of whether the target organism is present in a sample or not. A single tube of lauryl sulfate–tryptose–lactose broth as used in the MPN test, but without dilutions, would be used in a P–A test. In recent years, enzymatic assays have been developed that allow the simultaneous detection of total coliform bacteria and *E. coli* in drinking water. The assay can be a simple P–A test or an MPN assay. The Colilert system (Fig. 20.4) is one such assay: It is based on the fact that total coliform bacteria produce the enzyme β-galactosidase, which hydrolyzes the substrate *o*-nitrophenyl-β-

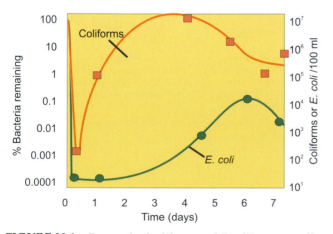

FIGURE 20.1   Regrowth of coliforms and *E. coli* in sewage effluent after inactivation with 5 mg/l chlorine. (From Shuval *et al.*, 1973.)

a) Presumptive Test

Transfer the specified volumes of sample to each tube.
Incubate 24 h at 35°C.

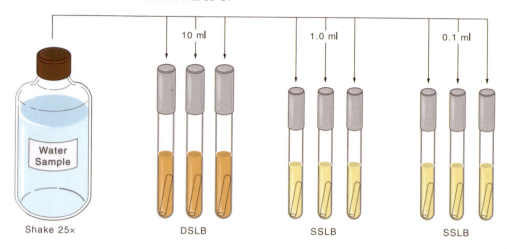

Tubes that have 10% gas or more are considered positive. The number of
positive tubes in each dilution is used to calculate the MPN of bacteria.

b) Confirmed Test

One of the positive tubes is selected, as indicated by the presence of gas trapped
in the inner tube, and used to inoculate a streak plate of Levine's EMB agar and
Endo agar. The plates are incubated 24 h at 35°C and observed for typical
coliform colonies.

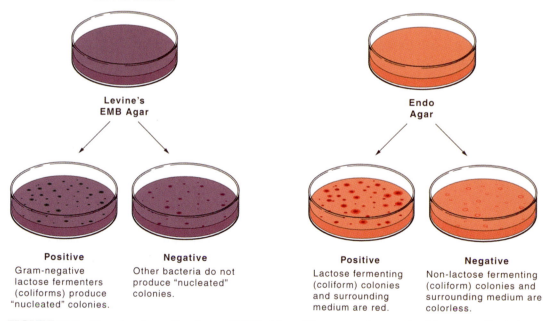

Levine's EMB Agar

Positive	Negative
Gram-negative lactose fermenters (coliforms) produce "nucleated" colonies.	Other bacteria do not produce "nucleated" colonies.

Endo Agar

Positive	Negative
Lactose fermenting (coliform) colonies and surrounding medium are red.	Non-lactose fermenting (coliform) colonies and surrounding medium are colorless.

FIGURE 20.2 Procedure for performing an MPN test for coliforms on water samples: (a) presumptive test and (b) confirmed test. (From *Pollution Science* © 1996, Academic Press, San Diego, CA.)

D-galactopyranoside (ONPG) to yellow nitrophenol. *E. coli* can be detected at the same time by incorporation of a fluorogenic substrate, 4-methylumbelliferone glucuronide (MUG) (Fig. 20.5), which produces a fluorescent end product after interaction with the enzyme β-glucuronidase found in *E. coli* but not in other coliforms. The end product is detected with a long-wave ultraviolet (UV) lamp. The Colilert test is performed

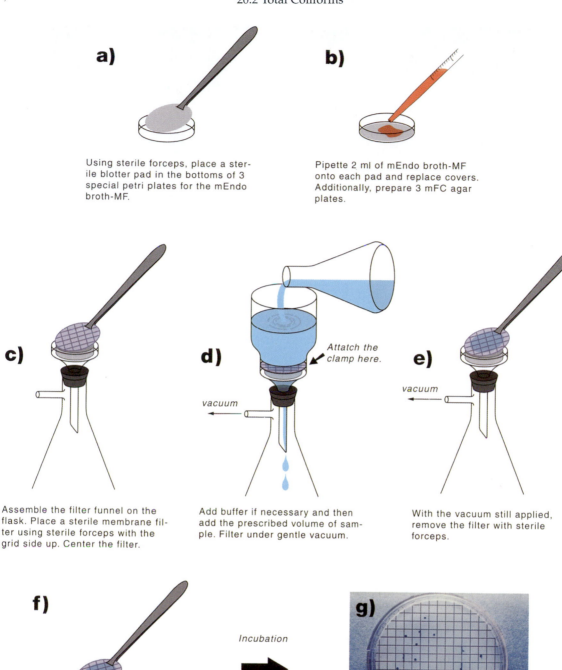

a) Using sterile forceps, place a sterile blotter pad in the bottoms of 3 special petri plates for the mEndo broth-MF.

b) Pipette 2 ml of mEndo broth-MF onto each pad and replace covers. Additionally, prepare 3 mFC agar plates.

c) Assemble the filter funnel on the flask. Place a sterile membrane filter using sterile forceps with the grid side up. Center the filter.

d) *Attach the clamp here.* *vacuum* Add buffer if necessary and then add the prescribed volume of sample. Filter under gentle vacuum.

e) *vacuum* With the vacuum still applied, remove the filter with sterile forceps.

f) *Incubation* Place the filter on the appropriate medium prepared in steps (a) and (b).

g) After incubation, count the colonies to determine the concentration of organisms in the original water sample.

FIGURE 20.3   Membrane filtration for determining the coliform count in a water sample using vacuum filtration. (From *Pollution Science* © 1996, Academic Press, San Diego, CA.)

by adding the sample to a single bottle (P–A test) or MPN tubes that contain powdered ingredients consisting of salts or specific enzyme substrates that serve as the only carbon source for the organisms (Fig. 20.4a).

After 24 hours of incubation, samples positive for total coliforms turn yellow (Fig. 20.4b), whereas *E. coli*–positive samples fluoresce under long-wave UV illumination in the dark (Fig. 20.4c).

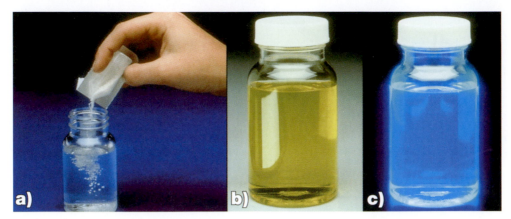

FIGURE 20.4 Detection of indicator bacteria with Colilert. (a) Addition of salts and enzyme substrates to water sample; (b) yellow color indicating the presence of coliform bacteria; (c) fluorescence under long-wave ultraviolet light indicating the presence of *E. coli*. Photographs courtesy of IDEXX, Westbrook, ME.

20.3 FECAL COLIFORMS

Although the total coliform group has served as the main indicator of water pollution for many years, many of the organisms in this group are not limited to fecal sources. Thus, methods have been developed to restrict the enumeration to coliforms that are more clearly of fecal origin—that is, the **fecal coliforms.** These organisms, which include the genera *Escherichia* and *Klebsiella,* are differentiated in the laboratory by their ability to ferment lactose with the production of acid and gas at 44.5°C within 24 hours. In general, then, this test indicates fecal coliforms; it does not, however, distinguish between human and animal contamination. The frequent occurrence of coliform and fecal coliform bacteria in unpolluted tropical waters, and their ability to survive for considerable periods of time outside the intestine in these waters, have suggested that these organisms occur naturally in tropical waters (Toranzos, 1991) and that new indicators for these waters need to be developed.

Some have suggested the use of *E. coli* as an indicator, because it can easily be distinguished from other members of the fecal coliform group (e.g., absence of urease and presence of β-glucuronidase). Fecal coliforms also have some of the same limitations in use as the coliform bacteria, i.e., regrowth and less resistant to water treatment than viruses and protozoa.

Fecal coliforms may be detected by methods similar to those used for coliform bacteria. For the MPN method EC broth is used, and for the membrane filter method m-FC agar is used for water analysis. A medium known as m-T7 agar has been proposed for use in the recovery of injured fecal coliforms from water (LeChevallier *et al.,* 1983) and results in greater re-

covery from water. The Colilert test has the advantage of detecting coliforms and *E. coli*, the principal fecal coliform, simultaneously within 24 hours.

20.4 FECAL STREPTOCOCCI

The fecal streptococci are a group of gram-positive Lancefield group D streptococci (Fig. 20.6). The fecal streptococci belong to the genera *Enterococcus* and *Streptococcus* (Gleeson and Gray, 1997). The genus *Enterococcus* includes all streptococci that share certain biochemical properties and have a wide range of tolerance of adverse growth conditions. They are differentiated from other streptococci by their ability to grow in 6.5% sodium chloride, pH 9.6, and 45°C and include *Ent. avium, Ent. faecium, Ent. durans, Ent. facculis,* and *Ent. gallinarium.* In the water industry the genus is often given as *Streptococcus* for this group. Of the genus

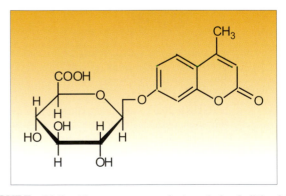

FIGURE 20.5 The structure of 4-methylumbelliferyl-β-d-glucuronide (MUG).

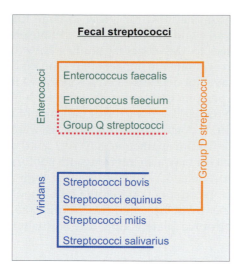

FIGURE 20.6 Definition of the terms "enterococci," "group D streptococci," and "fecal streptococci" based on *Streptococcus* species belong to each group.

Streptococcus, only *S. bovis* and *S. equinus* are considered to be true fecal streptococci. These two species of *Streptococcus* are predominately found in animals; *Ent. faecalis* and *Ent. faecium* are more specific to the human gut. It has been suggested that a fecal coliform/fecal streptococci (FC/FS) ratio of 4 or more indicates a contamination of human origin, whereas a ratio below 0.7 is indicative of animal pollution (Geldreich and Kenner, 1969) (Table 20.5). However, the validity of the FC/FS ratio has been questioned. Further, this ratio is valid only for recent (24 hours) fecal pollution.

Both the membrane filtration method and MPN method may also be used for the isolation of fecal streptococci. The membrane filter method uses fecal *Streptococcus* agar with incubation at 37°C for 24 hours. All red, maroon, and pink colonies (due to reduction of 2,4,5-triphenyltetrazolium chloride to formazan, a red dye) are counted as presumptive fecal streptococci. Confirmation of fecal streptococci is by subculture on bile aesulin agar and incubation for 18 hours at 44°C. Fecal streptococci form discrete colonies surrounded by a brown or black halo due to aesculin hydrolysis,

TABLE 20.5 The FC/FS Ratio

FC/FS ratio	Source of pollution
>4.0	Strong evidence that pollution is of human origin
2.0–4.0	Good evidence of the predominance of human wastes in mixed pollution
0.7–2.0	Good evidence of the predominance of domestic animal wastes in mixed pollution
<0.7	Strong evidence that pollution is of animal origin

and *Ent. faecium* are considered to be more specific to the human gut. Fecal streptococci are considered to have certain advantages over the coliform and fecal coliform bacteria as indicators.

- They rarely multiply in water.
- They are more resistant to environmental stress and chlorination than coliforms.
- They generally persist longer in the environment (Gleeson and Gray, 1997).

The enterococci have been suggested as useful indicators of risk of gastroenteritis for recreational bathers and standards have been recommended (Cabelli, 1989). They have been suggested as useful indicators of the presence of enteric viruses in the environment.

20.5 *Clostridium perfringens*

Clostridium perfringens is a sulfite-reducing anaerobic spore former; it is gram positive, rod shaped, and exclusively of fecal origin. The spores are very heat resistant (75°C for 15 minutes), persist for long periods in the environment, and are very resistant to disinfectants. The hardy spores of this organism limit its usefulness as an indicator. However, it has been suggested that it could be an indicator of past pollution, a tracer of less hardy indicators, and an indicator of removal of protozoan parasites or viruses during drinking water and wastewater treatment (Payment and Franco, 1993).

Other anaerobic bacteria such as *Bifidobacterium* and *Bacteroides* have been suggested as potential indicators. Because some of the *Bifidobacterium* are primarily associated with humans, they could potentially help distinguish between human and animal contamination. However, better and more standard methods are needed for detection of all of the anaerobic bacteria in the environment before they can be adequately monitored in a routine fashion.

20.6 HETEROTROPHIC PLATE COUNT

An assessment of the numbers of aerobic and facultatively anaerobic bacteria in water that derive their carbon and energy from organic compounds is conducted via the heterotrophic plate count or HPC. This group includes gram-negative bacteria belonging to the following genera: *Pseudomonas, Aeromonas, Klebsiella, Flavobacterium, Enterobacter, Citrobacter, Serratia, Acinetobacter, Proteus, Alcaligenes, Enterobacter,* and *Moraxella.* The heterotrophic plate counts of microorganisms found in untreated drinking water and chlorinated distribution water are shown in Table 20.6 (LeChevallier *et al.*, 1980). These bacteria are commonly isolated from surface waters and groundwater, and are widespread in soil and vegetation (including many vegetables eaten raw). Some members of this group are opportunistic pathogens (e.g., *Aeromonas, Pseudomonas*), but no conclusive evidence is available to demonstrate their transmission by drinking water. In drinking water, the number of HPC bacteria may vary from less than 1 to more than 10^4 CFU/ml, and they are influenced mainly by temperature, presence of residual chlorine, and level of assimilable organic matter. In reality, these counts themselves have no or little health significance. However, there has been con-

cern because the HPC can grow to large numbers in bottled water and charcoal filters on household taps. In response to this concern, studies have been performed to evaluate the impact of HPC on illness. These studies have not demonstrated a conclusive impact on illness in persons who consume water with high HPC. Although the HPC is not a direct indicator of fecal contamination, it does indicate variation in water quality and potential for pathogen survival and regrowth. These bacteria may also interfere with coliform and fecal coliform detection when present in high numbers. It has been recommended that the HPC should not exceed 500 per ml in tap water (LeChevallier *et al.*, 1980).

Heterotrophic plate counts are normally done by

TABLE 20.6   Identification of HPC Bacteria in Untreated Drinking Water and in the Chlorinated Distribution System

Organism	Distribution water % of the total number of organisms identified	Untreated drinking water % of the total number of organisms identified
Acinomycetes	10.7	0
Arthrobacter spp.	2.3	1.3
Bacillus spp.	4.9	0.6
Corynebacterium spp.	8.9	1.9
Micrococcus luteus	3.5	3.2
Staphylococcus aureus	0.6	0
S. epidermidis	5.2	5.1
Acinetobacter spp.	5.5	10.8
Alcaligenes spp.	3.7	0.6
Flavobacterium meningosepticum	2.0	0
Moraxella spp.	0.3	0.6
Pseudomonas alcaligenes	6.9	2.5
P. cepacia	1.2	0
P. fluorescens	0.6	0
P. mallei	1.4	0
P. maltophilia	1.2	5.7
Pseudomonas spp.	2.9	0
Aeromonas spp.	9.5	15.9
Citrobacter freundii	1.7	5.1
Enterobacter agglomerans	1.2	11.5
Escherichia coli	0.3	0
Yersinia enterocolitica	0.9	6.4
Hafnia alvei	0	5.7
Enterobacter aerogenes	0	0.6
Enterobacter cloacae	0	0.6
Klebsiella pneumoniae	0	0
Serratia liquefaciens	0	0.6
Unidentified	18.7	17.8

Modified from LeChevallier *et al.* (1980).

the spread plate method using yeast extract agar incubated at 35°C for 48 hours. A low-nutrient medium, R₂A (Reasoner and Geldreich, 1985), has seen widespread use and is recommended for disinfectant-damaged bacteria. This medium is recommended for use with an incubation period of 5–7 days at 28°C. HPC numbers can vary greatly depending on the incubation temperature, growth medium, and length of incubation.

20.7 BACTERIOPHAGE

Because of their constant presence in sewage and polluted waters, the use of bacteriophage (or bacterial viruses) as appropriate indicators of fecal pollution has been proposed. These organisms have also been suggested as indicators of viral pollution. This is because the structure, morphology, and size, as well as the behavior in the aquatic environment of many bacteriophage closely resemble those of enteric viruses. For these reasons, they have also been used extensively to evaluate virus resistance to disinfectants, to evaluate virus fate during water and wastewater treatment, and as surface and groundwater tracers. The use of bacteriophage as indicators of fecal pollution is based on the assumption that their presence in water samples denotes the presence of bacteria capable of supporting the replication of the phage. Two groups of phage in particular have been studied: the **somatic coliphage,** which infect *E. coli* host strains through cell wall receptors, and the **F-specific RNA coliphage,** which infect strains of *E. coli* and related bacteria through the F⁺ or sex pili. A significant advantage of using coliphage is that they can be detected by simple and inexpensive techniques that yield results in 8–18 hours. Both a plating method (the agar overlay method) and the MPN method can be used to detect coliphage (Fig. 20.7) in volumes ranging from 1 to 100 ml. The F-specific coliphage (male-specific phage) have received the greatest amount of attention because they are similar in size and shape to many of the pathogenic human enteric viruses. Coliphage f2, φ174, MS-2, and PRD-1 are the ones most commonly used as tracers and for evaluation of disinfectants. Because F-specific phage are infrequently detected in human fecal matter and show no direct relationship to the fecal pollution level, they cannot be considered indicators of fecal pollution (Havelaar *et al.*, 1990). However, their presence in high numbers in wastewaters and their relatively high resistance to chlorination contribute to their consideration as an index of waste-

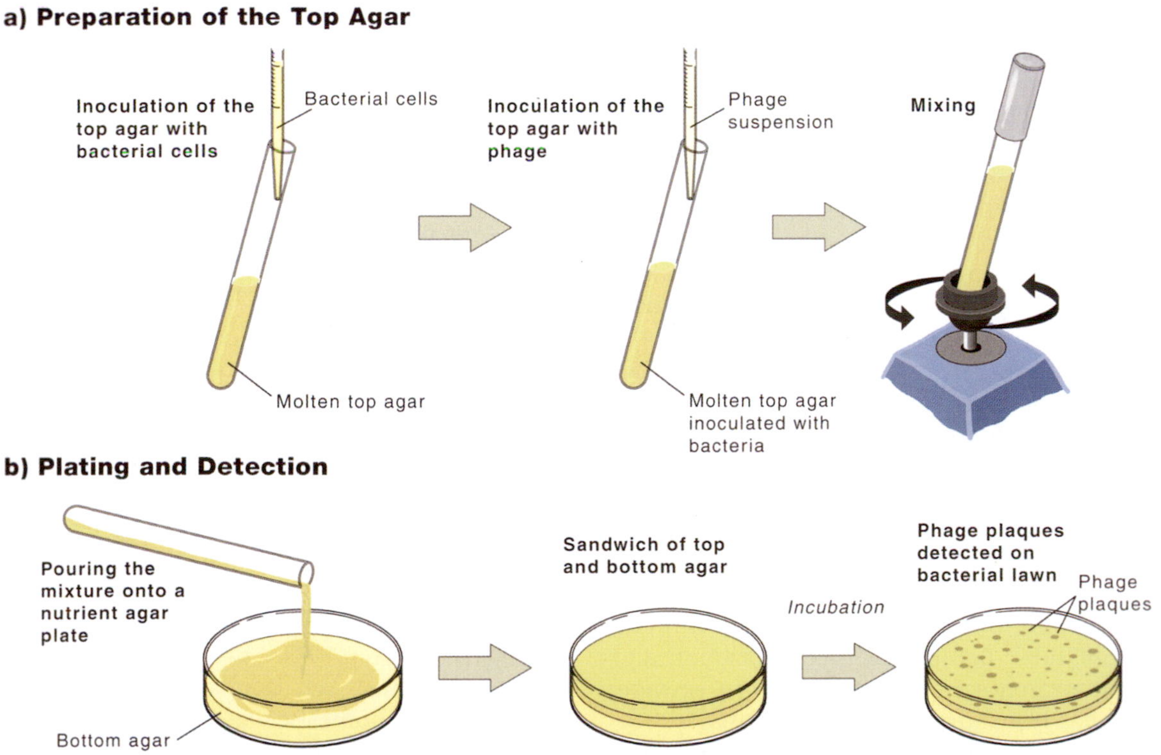

a) Preparation of the Top Agar

Inoculation of the top agar with bacterial cells — Bacterial cells — Molten top agar

Inoculation of the top agar with phage — Phage suspension — Molten top agar inoculated with bacteria

Mixing

b) Plating and Detection

Pouring the mixture onto a nutrient agar plate — Bottom agar

Sandwich of top and bottom agar

Incubation

Phage plaques detected on bacterial lawn — Phage plaques

FIGURE 20.7   Technique for performing a bacteriophage assay. (From Pepper *et al.*, 1995.)

water contamination and as potential indicators of enteric viruses.

Bacteriophage of *Bacteroides fragilis* have also been suggested as potential indicators of human viruses in the environment (Tartera and Jofre, 1987). *Bacteroides* spp. are strict anaerobes and are a major component of human feces, so bacteriophage active against these organisms have the potential to be suitable indicators of viral contamination.

Bacteriophage that infect *B. fragilis* appear to be exclusively human in origin (Tartera and Jofre, 1987) and appear to be present only in environmental samples contaminated with human fecal pollution. This may help to differentiate human from animal contamination. They are absent from natural habitats, which is a considerable advantage over coliphages, which are found in habitats other than the human gut. They are unable to multiply in the environment (Tartera *et al.*, 1989), and their decay rate in the environment appears similar to that of human enteric viruses. However, their host is an anaerobic bacterium that involves a complicated and tedious methodology, which limits their suitability as a routine indicator organism.

20.8 OTHER INDICATOR ORGANISMS

A number of other organisms have also been considered to have potential as alternative indicator organisms or for use in certain applications (e.g., recreational waters). These include *Pseudomonas* spp., yeasts, acid-fast mycobacteria (*Mycobacterium fortuitum* and *M. phlei*), *Aeromonas*, and *Staphylococcus*.

Within the genus *Pseudomonas*, the species of significant public health concern is *P. aeruginosa* a gramnegative, nonsporulating, rod-shaped bacterium. The most common diseases associated with this organism are eye, ear, nose, and throat infections. It is also the most common opportunistic pathogen causing lifethreatening infections in burn patients and immunocompromised individuals. A characteristic of the pseudomonad is that it can produce the blue–green pigment pyocyanin or the fluorescent pigment fluorescein or both. Numerous cases of folliculitis, dermatitis, and ear (swimmer's ear) and urinary tract infections are due to *P. aeruginosa* associated with swimming in contaminated water or poorly maintained swimming pools and hot tubs. Because of this association and its consistent presence in high numbers in sewage, *P. aeruginosa* has been suggested as a potential indicator for water in swimming pools, hot tubs, and other recreational waters (Cabelli, 1978). However, as this organism is known to be ubiquitous in nature and can

multiply under natural conditions (it can even grow in distilled water), it is believed to be of little value for fecal contamination studies.

Coliforms have been used for many years to assess the safety of swimming pool water, yet contamination is often not of fecal origin with infections associated primarily with the respiratory tract, skin, and eyes. For this reason *Staphylococcus aureus* and *Candida albicans*, a gram-positive bacterium and a yeast, respectively, have been proposed as better indicators of this type of infection associated with swimming. Recreational waters may serve as a vehicle for skin infections caused by *S. aureus*, and some observers have recommended that this organism be used as an additional indicator of the sanitary quality of recreational waters, because its presence is associated with human activity in recreational waters (Charoenca and Fujioka, 1993).

The genus *Aeromonas* includes straight gram-negative rods, facultatively anaerobic, that are included in the family Vibrionaceae. Only *Aeromonas hydrophila* has received attention as an organism of potential sanitary significance. Aeromonas occur in uncontaminated waters as well as in sewage and sewage-contaminated waters. The organism can be pathogenic for humans, other warm-blooded animals, and cold-blooded animals including fish. Foodborne outbreaks associated with *A. hydrophila* have been documented and it is considered an opportunistic pathogen in humans. Because of its association with nutrient-rich conditions, it has been suggested as an indicator of the nutrient status of natural waters.

Table 20.7 summarizes potential applications of indicator organisms in the assessment of water quality.

20.9 STANDARDS AND CRITERIA FOR INDICATORS

Bacterial indicators such as coliforms have been used for the development of **water quality standards**. For example, the U.S. Environmental Protection Agency (U.S. EPA) has set a standard of no detectable coliforms per 100 ml of drinking water. A drinking water standard is legally enforceable in the United States. If these standards are violated by water suppliers, they are required to take corrective action or they may be fined by the state or federal government. Authority for setting drinking water standards was given to the U.S. EPA in 1974 when Congress passed the Safe Drinking Water Act. Similarly, authority for setting standards for domestic wastewater discharges is given under the Clean Water Act (see Table 16.1). In contrast, standards for recreational waters and wastewater reuse are determined by the individual states. Micro-

TABLE 20.7 Water Quality Indicators, Their Significant Sources, and Potential Uses

Indicator	Significant source[a]	Potential use[b]
Coliforms	F S I R A	S
Fecal coliforms	F S I R A	F S
Enterococcus	F S	F S A D
Clostridium perfringens	F S	F S D
Candida albicans	F S	P F S
Bifidobacteria	F S	F S A D
Coliphage	S	S
Pseudomonas aeruginosa	S I R A	P S N
Aeromonas hydrophila	S I R A	P S N

Modified from Cabelli (1978).

[a] Relative to other sources: F, feces of warm-blooded animals; S, sewage; I, industrial wastes; R, runoff from uncontaminated soils; A, fresh and marine waters.

[b] Potential use: P, pathogen; F, fecal indicators; S, sewage indicator; A, separation of human from lower animal sources; D, proximity to fecal source; N, indicator of nutrient pollution.

bial standards set by various government bodies in the United States are shown in Table 20.8. Standards used by the European Union are given in Table 20.9.

Criteria and **guidelines** are terms used to describe recommendations for acceptable levels of indicator microorganisms. They are not legally enforceable but serve as guidance indicating that a potential water quality problem exists. Ideally, all standards would indicate that an unacceptable public health threat exists or that some relationship exists between the amount of illness and the level of indicator organisms. Such in-

TABLE 20.8 U.S. Federal and State Standards for Microorganisms

Authority	Standards
U.S. EPA	
Safe Drinking Water Act	0 coliforms/100 ml
Clean Water Act	
Wastewater discharges	200 fecal coliforms/100 ml
Sewage sludge	<1000 fecal coliforms/4 g
	<3 *Salmonella*/4 g
	<1 enteric virus/4 g
	<1 helminth oval/4 g
California	
Wastewater reclamation for irrigation	≤2.2 MPN coliforms
Arizona	
Wastewater reclamation for irrigation of golf courses	25 fecal coliforms/100 ml
	125 enteric virus/40 liters
	No detectable *Giardia*/40 liters

TABLE 20.9 Drinking Water Criteria of the European Union

Tap water	
Escherichia coli	0/100 ml
Fecal streptococci	0/100 ml
Sulfite-reducing clostrida	0/20 ml
Bottled water	
Escherichia coli	0/250 ml
Fecal streptococci	0/250 ml
Sulfite-reducing clostrida	0/50 ml
Pseudomonas aeruginosa	0/250 ml

From European Union (1995).

formation is difficult to acquire because of the involvement of costly epidemiological studies that are often difficult to interpret because of confounding factors (see Chapter 24). An area where epidemiology has been used to develop criteria is that of recreational swimming. Epidemiological studies in the United States have demonstrated a relationship between swimming-associated gastroenteritis and the densities of enterococci (Fig. 20.8) and fecal coliforms. No relationship was found for coliform bacteria (Cabelli, 1989). It was suggested that a standard geometric average of 35 enterococci per 100 ml be used for marine bathing waters. This would mean accepting a risk of 1.9% of the bathers developing gastroenteritis (Kay and Wyer, 1992). Numerous other epidemiological studies of bathing-acquired illness have been conducted. These studies have shown slightly different re-

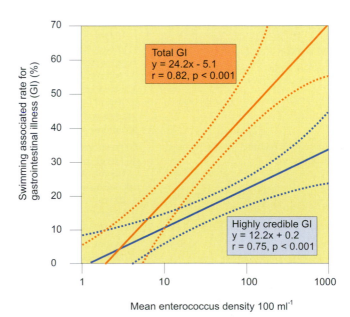

FIGURE 20.8 Dose–response relationships produced by the work of Cabelli *et al.* (1982).

lationships to illness and that other bacterial indicators were more predictive of illness rates (Kay and Wyer, 1992). These differences probably arise because of the different sources of contamination (raw versus disinfected wastewater), types of recreational water (marine versus fresh), types of illness (gastroenteritis, eye infections, skin complaints), immune status of the population, length of observation, etc. Various guidelines for acceptable numbers of indicator organisms have been in use (Table 20.10), but there is no general agreement on standards.

The use of microbial standards also requires the development of standard methods and quality assurance or quality control plans for the laboratories that will do the monitoring. Knowledge of how to sample and how often to sample is also important. All of this information is usually defined in the regulations when a standard is set. For example, frequency of sampling may be determined by the size (number of customers) of the utility providing the water. Sampling must proceed in some random fashion so that the entire system is characterized. For drinking water, no detectable coliforms are allowed in the United States (Table 20.8). However, in other countries some level of coliform bacteria is allowed. Because of the wide variability in numbers of indicators in water, some positive samples

may be allowed or tolerance levels or averages may be allowed. Usually, **geometric averages** are used in standard setting because of the often skewed distribution of bacterial numbers. This prevents one or two high values from giving overestimates of high levels of contamination, which would appear to be the case with **arithmetric averages** (see Table 20.11).

Geometric averages are determined as follows.

$$\log \bar{x} = \frac{\Sigma \, (\log x)}{N} \qquad \text{(Eq. 20.1)}$$

$$\bar{x} = \text{antilog} \, (\log \bar{x} \,) \qquad \text{(Eq. 20.2)}$$

where N is the number of samples $\bar{x}$ is and the geometric average, and x is the number of organisms per sample volume.

As can be seen, standard setting and the development of criteria is a difficult process and there is no ideal standard. A great deal of judgment by scientists, public health officials, and the regulating agency is required.

QUESTIONS AND PROBLEMS

1. What are some of the criteria for indicator bacteria?

2. What is the difference between standards and criteria?

3. Why are geometric means used to report average concentrations of indicator organisms?

4. Calculate the arithmetic and geometric averages for the following data set: Fecal coliforms/100 ml on different days on a bathing beach were reported as 2, 3, 1000, 15, 150, and 3000.

TABLE 20.10   Guidelines for Recreational Water Quality Standards

Country or agency	Regime (samples/time)	Criteria or standard[a]
U.S. EPA	5/30 days	200 fecal coliforms/ 100 ml <10% to exceed 400/ml Fresh water[b] 33 enterococci/100 ml 126 fecal coliforms/ 100 ml Marine waters[b] 35 enterococci/100 ml
European Economic Community	2/30 days[c]	500 coliforms/100 ml 100 fecal coliforms/ 100 ml 100 fecal streptococci/ 100 ml 0 *Salmonella*/liter 0 Enteroviruses/10 liters
Ontario, Canada	10/30 days	≤1000 coliforms/100 ml ≤100 fecal coliforms/ 100 ml

From Saliba, 1993; U.S. EPA, 1986.
[a] All bacterial numbers in geometric means.
[b] Proposed, 1986.
[c] Coliforms and fecal coliforms only.

TABLE 20.11   Arithmetic and Geometric Averages of Bacterial Numbers in Water

MPN[a]	Log
2	0.30
110	2.04
4	0.60
150	2.18
1100	3.04
10	1.00
12	1.08
198 = arithmetic average	1.46 = log $\bar{x}$ antilog $\bar{x}$ = 29 29 = geometric average

[a] MPN, most probable number.

5. Define coliform and fecal coliform bacteria. Why are they not ideal indicators?

6. Why have coliphage been suggested as indicator organisms?

7. What are two methods that can be used to detect indicator bacteria in water?

References and Recommended Readings

Cabelli, V. (1978) New standards for enteric bacteria. *In* "Water Pollution Microbiology," Vol. 2 (R. Mitchell, ed.) Wiley-Interscience, New York, pp. 233–273.

Cabelli, V. J. (1989) Swimming-associated illness and recreational water quality criteria. *Water Sci. Technol.* **21**, 13–21.

Cabelli, V. J., Dufour, A. P., McCabe, L. J., and Levin, M. A. (1982) Swimming associated gastroenteritis and water quality. *Am. J. Epidemiol.* **115**, 606–616.

Charoenca, N., and Fujioka, R. S. (1993) Assessment of *Staphylococcus* bacteria in Hawaii recreational waters. *Water Sci. Technol.* **27**, 283–289.

European Union (EU). (1995) Proposed for a Council Directive concerning the quality of water intended for human consumption. Com (94) 612 Final. *Offic. J. Eur. Union* **131**, 5–24.

Geldreich, E. E. (1978) Bacterial populations and indicator concepts in feces, sewage, storm water and solid wastes. *In* "Indicators of Viruses in Water and Food" (G. Berg, ed.) Ann Arbor Science, Ann Arbor, MI, pp. 51–97.

Geldreich, E. E., and Kenner, B. A. (1969) Comments on fecal streptococci in stream pollution. *J. Water Pollut. Control Fed.* **41**, R336–R341.

Gleeson, C., and Gray, N. (1997) "The Coliform Index and Waterborne Disease." E and FN Spon, London.

Havelaar, A. H., Hogeboon, W. M., Furuse, K., Pot, R., and Horman, M. P. (1990) F-specific RNA bacteriophages and sensitive host strains in faeces and wastewater of human and animal origin. *J. Appl. Bacteriol.* **69**, 30–37.

Kay, D. and Wyer, M. (1992) Recent epidemiological research leading to standards. *In* "Recreational Water Quality Management," Vol. 1. "Coastal Waters" (D. Kay, ed.) Ellis Horwood, Chichester, UK, pp. 129–156.

LeChevallier, M. W., Seidler, R. J., and Evans, T. M. (1980) Enumeration and characterization of standard plate count bacteria in chlorinated and raw water supplies. *Appl. Environ. Microbiol.* **40**, 922–930.

LeChevallier, M. W., Cameron, S. C., and McFeters, G. A. (1983) New medium for improved recovery of coliform bacteria from drinking water. *Appl. Environ. Microbiol.* **45**, 484–492.

LeChevallier, M. W., Cawthen, C. P., and Lee, R. G. (1988) Factors promoting survival of bacteria in chlorinated water supplies. *Appl. Environ. Microbiol.* **54**, 649–654.

Payment, P., and Franco, E. (1993) *Clostridium perfringens* and somatic coliphages as indicators of the efficiency of drinking water treatment for viruses and protozoan cysts. *Appl. Environ. Microbiol.* **59**, 2418–2424.

Pepper, I. L., Gerba, C. P., and Brendecke, J. W. (1995) "Environmental Microbiology: A Laboratory Manual." Academic Press, San Diego.

Pepper, I. L., Gerba, C. P., Brusseau, M. L. (1996) "Pollution Science." Academic Press, San Diego.

Reasoner, D. J., and Geldreich, E. E. (1985) A new medium for enumeration and subculture of bacteria from potable water. *Appl. Environ. Microbiol.* **49**, 1–7.

Saliba, L. (1993) Legal and economic implication in developing criteria and standards. *In* "Recreational Water Quality Management" (D. Kay and R. Hanbury, eds.) Ellis Horwood, Chichester, UK, pp. 57–73.

Shuval, H. I., Cohen, J., and Kolodney, R. (1973) Regrowth of coliforms and faecal coliforms in chlorinated wastewater effluent. *Water Res.* **7**, 537–546.

Tartera, C., and Jofre, J. (1987) Bacteriophage active against *Bacteroides fragilis* bacteriophage as indicators of the virological quality of water. *Water Sci. Technol.* **18**, 1623–1637.

Tartera, C., Lucena, F., and Jofre, J. (1989) Human origin of *Bacteroides fragilis* bacteriophage present in the environment. *Appl. Environ. Microbiol.* **55**, 2696–2701.

Toranzos, G. A. (1991) Current and possible alternative indicators of fecal contamination in tropical waters: A short review. *Environ. Toxicol. Water Qual.* **6**, 121–130.

USEPA. (1986) United States Environmental Protection Agency. Ambient water quality. Criteria—1986. EPA440/5–84–002. Washington, DC.

21

Domestic Wastes and Waste Treatment

CHARLES P. GERBA

21.1 DOMESTIC WASTEWATER

Sewage treatment is a relatively modern practice. Collection systems for human wastes date back to Roman times, but these wastes were simply discharged into the nearest body of water, i.e., lake, stream, or ocean. As the size of the human population grew, the degradation of these waters became acute as fish and other wildlife were affected by the depletion of oxygen and the threat of waterborne disease increased. Concern about the control of domestic wastewater began toward the middle of the 19th century, when waterborne diseases such as cholera were becoming common. This prompted the proper collection and disposal of wastes away from populated areas. Modern sewage treatment practices began at the turn of the 20th century. Processes were developed to treat the organic matter in domestic wastes before disposal in natural bodies of water. In more recent decades, with increasing demands on limited water supplies and the need to reuse domestic wastes, the focus of treatment has also included the reduction of pathogenic microorganisms and the removal of toxic substances (metals and certain types of organics). Today, more than 15,000 wastewater treatment plants treat approximately 150 billion liters of wastewater per day in the United States alone. In addition, septic tanks, which were also introduced at the end of the 19th century, serve approximately 25% of the U.S. population, largely in rural areas.

Domestic wastewater is primarily a combination of human feces, urine, and "graywater." **Graywater** results from washing, bathing, and meal preparation. Water from various industries and businesses may also enter the system. People excrete 100–500 grams wet weight of feces and 1–1.3 liters of urine per person per day (Bitton, 1994). Major organic and inorganic constituents of untreated domestic sewage are shown in Table 21.1.

The amount of organic matter in domestic wastes determines the degree of biological treatment required. Three tests are used to assess the amount of organic matter: **total organic carbon (TOC), biochemical oxygen demand (BOD), and chemical oxygen demand (COD).**

The major objective of domestic waste treatment is the reduction of BOD, which may be either in the form of solids (suspended matter) or soluble. BOD is the amount of dissolved oxygen consumed by microorganisms during the biochemical oxidation of organic (carbonaceous BOD) and inorganic (ammonia) matter. The methodology for measuring BOD has changed little since it was developed in the 1930s. The 5-day BOD test (written BOD_5) is a measure of the amount of oxygen consumed by a mixed population of heterotrophic bacteria in the dark at 20°C over a period of 5 days.

Aliquots of wastewater are placed in a 300-ml BOD bottle (Fig. 21.1) and diluted in phosphate buffer (pH 7.2) containing other inorganic elements (N, Ca, Mg, Fe) and saturated with oxygen. Sometimes acclimated microorganisms or dehydrated cultures of microorganisms, sold in capsule form, are added to municipal and industrial wastewaters, which may not have a sufficient microflora to carry out the BOD test.

A nitrification inhibitor is sometimes added to the sample to determine only the carbonaceous BOD. Dissolved oxygen concentration is determined at time 0

TABLE 21.1   Typical Composition of Untreated Domestic Wastewater

Contaminants	Concentration (mg/l)		
	Low	Moderate	High
Solids, total	350	720	1200
Dissolved, total	250	500	850
Volatile	105	200	325
Suspended solids	100	220	350
Volatile	80	165	275
Settleable solids	5	10	20
Biochemical oxygen demand[a]	110	220	400
Total organic carbon	80	160	290
Chemical oxygen demand	250	500	1000
Nitrogen (total as N)	20	40	85
Organic	8	15	35
Free ammonia	12	25	50
Nitrites	0	0	0
Nitrates	0	0	0
Phosphorus (total as P)	4	8	15
Organic	1	3	5
Inorganic	3	5	10

Modified from Metcalf and Eddy (1991).
[a] 5-day, 20°C (BOD, 20°C).

and after a 5-day incubation by means of an oxygen electrode, chemical procedures (e.g., Winkler test), or a manometric BOD apparatus (see Chapter 11.2.2). The BOD test is carried out on a series of dilutions of the sample, the dilution depending on the source of the sample. When dilution water is not seeded, the BOD value is expressed in milligrams per liter, according to the following equation (APHA, 1995).

FIGURE 21.1 BOD bottles used for the determination of BOD. (Photo courtesy R. M. Maier.)

$$BOD \ (mg/l) = \frac{D_1 - D_5}{P} \qquad (Eq.\ 21.1)$$

where D_1 = initial dissolved oxygen (DO), D_5 = DO at day 5, and P = decimal volumetric fraction of wastewater utilized.

If the dilution water is seeded,

$$BOD \ (mg/l) = \frac{(D_1 - D_2) - (B_1 - B_2)f}{P} \qquad (Eq.\ 21.2)$$

where D_1 = initial DO of the sample dilution (mg/l), D_2 = final DO of the sample dilution (mg/l), P = decimal volumetric fraction of sample used, B_1 = initial DO of seed control (mg/l), B_2 = final DO of seed control (mg/l), and f = ratio of seed in sample to seed in control = (% seed in D_1)/(% seed in B_1) (see Information Box 1).

Because of depletion of the carbon source, the carbonaceous BOD reaches a plateau called the ultimate carbonaceous BOD (Fig. 21.2). The BOD$_5$ test is commonly used for several reasons:

- To determine the amount of oxygen that will be required for biological treatment of the organic matter present in a wastewater
- To determine the size of waste treatment facility needed
- To assess the efficiency of treatment processes
- To determine compliance with wastewater discharge permits.

The typical BOD$_5$ of raw sewage ranges from 110 to 440 mg/l. Conventional sewage treatment will reduce this by 95%.

Chemical oxygen demand is the amount of oxygen necessary to oxidize all of the organic carbon completely to CO_2 and H_2O. COD is measured by oxidation with potassium dichromate ($K_2Cr_2O_7$) in the presence of sulfuric acid and silver and is expressed in

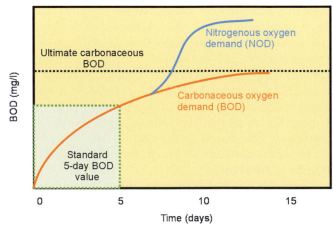

FIGURE 21.2 Carbonaceous and nitrogenous BOD. (Modified from Hammer, 1986.)

**Information Box 1
Calculation of BOD**

Determine the 5-day BOD for a wastewater sample when a 15-ml sample of the wastewater is added to a 300-ml BOD bottle containing dilution water and the dissolved oxygen is 8 mg/l. Five days later the dissolved oxygen concentration is 2 mg/l.

Using Eq. 21.1,

$$D_1 = 8 \text{ mg/L}$$

$$D_2 = 2 \text{ mg/L}$$

$$p = 15 \text{ ml} = 2\% = 0.05$$

$$\text{BOD}_5 \text{ (mg/l)} = \frac{8-2}{0.05} = 120$$

---

milligrams per liter. In general, 1 g of carbohydrate or 1 g of protein is approximately equivalent to 1 g of COD. Normally, the ratio BOD/COD is approximately 0.5. When this ratio falls below 0.3, it means that the sample contains large amounts of organic compounds that are not easily biodegraded.

Another method of measuring organic matter in water is the **TOC or total organic carbon test.** TOC is determined by oxidation of the organic matter with heat and oxygen, followed by measurement of the CO_2 liberated with an infrared analyzer. Both TOC and COD represent the concentration of both biodegradable and nonbiodegradable organics in water.

Pathogenic microorganisms are almost always present in domestic wastewater (Table 21.2). This is because large numbers of pathogenic microorganisms may be excreted by infected individuals. Both symptomatic and asymptomatic individuals may excrete pathogens. For example, the concentration of rotavirus may be as high as 10^{10} virions per gram of stool or 10^{12} in 100 g of stool (Table 21.3). Infected individuals may excrete enteric pathogens for several weeks to months. The concentration of enteric pathogens in raw wastewater varies depending on

- The incidence of the infection in the community
- The socioeconomic status of the population
- The time of year
- The per-capita water consumption.

The peak incidence of many enteric infections is seasonal in temperate climates. Thus, the highest incidence of enterovirus infection is during the late summer and early fall, rotavirus infections tend to peak in the early winter, and *Cryptosporidium* infections peak in the early spring and fall. The reason for the seasonality of enteric infections is not completely under-

stood, but several factors may play a role. It may be associated with the survival of different agents in the environment during the different seasons: *Giardia,* for example, can survive winter temperatures very well. Alternatively, excretion differences among animal reservoirs may be involved, as is the case with *Cryptosporidium.* Or it may well be that greater exposure to contaminated water, as in swimming, is the explanation for increased incidence in the summer months.

Certain populations and subpopulations are also more susceptible to infection. For example, enteric infection is more common in children because they usually lack previous protective immunity. Thus, the incidence of enteric virus and protozoa infections in day care centers, where young children are in close proxim-

TABLE 21.2 Types and Numbers of Microorganisms Typically Found in Untreated Domestic Wastewater

Organism	Concentration (per ml)
Total coliform	10^5–10^6
Fecal coliform	10^4–10^5
Fecal streptococci	10^3–10^4
Enterococci	10^2–10^3
Shigella	Present
Salmonella	10^0–10^2
Clostridium perfringens	10^1–10^3
Giardia cysts	10^{-1}–10^2
Cryptosporidium cysts	10^{-1}–10^1
Helminth ova	10^{-2}–10^1
Enteric virus	10^1–10^2

Modified from Metcalf and Eddy (1991).

TABLE 21.3   Incidence and Concentration of Enteric Viruses and Protozoa in Feces in the United States

Pathogen	Incidence (%)	Concentration in stool (per gram)
Enterovirus	10–40	10^3–10^8
Hepatitis A	0.1	10^8
Rotavirus	10–29	10^{10}–10^{12}
Giardia	3.8	10^6
	18–54[a]	10^6
Cryptosporidium	0.6–20	10^6–10^7
	27–50[a]	10^6–10^7

[a] Children in day care centers.

ity, is usually much higher than that in the general community. A greater incidence of enteric infections is also evident in lower socioeconomic groups, particularly where lower standards of sanitary conditions prevail. Concentrations of enteric pathogens are much greater in sewage in the developing world than the industrialized world. For example, the average concentration of enteric viruses in sewage in the United States has been estimated to be 10^3 per liter (Table 21.4), and concentrations as high as 10^5 per liter have been observed in Africa and Asia.

21.2 MODERN WASTEWATER TREATMENT

As previously stated, the primary goal of wastewater treatment is the removal and degradation of organic matter under controlled conditions. Complete sewage treatment comprises three major steps, as shown in Fig. 21.3.

- **Primary treatment** is a physical process that involves the separation of large debris, followed by sedimentation.

TABLE 21.4   Estimated Levels of Enteric Organisms in Sewage and Polluted Surface Water in the United States

	Concentration (per 100 ml)	
Organism	Raw sewage	Polluted stream water
Coliforms	10^9	10^5
Enteric viruses	10^2	1–10
Giardia	10–10^2	0.1–1
Cryptosporidium	1–10	0.1–10^2

Modified from U.S. EPA (1988).

- **Secondary treatment** is a biological oxidation process that is carried out by microorganisms.
- **Tertiary treatment** is usually a physicochemical process that removes turbidity caused by the presence of nutrients (e.g., nitrogen), dissolved organic matter, metals, or pathogens.

21.2.1 Primary Treatment

Primary treatment is the first step in municipal sewage treatment. It physically separates large solids from the waste stream. As raw sewage enters the treatment plant, it passes through a metal grating that removes large debris, such as branches, tires, and the like. A moving screen then filters out smaller items such as diapers and bottles, after which a brief residence in a grit tank allows sand and gravel to settle out. The waste stream is then pumped into the primary settling tank (also known as a sedimentation tank or clarifier), where about half the suspended organic solids settle to the bottom as **sludge** or **biosolids.** The resulting sludge is referred to as **primary sludge.** Microbial pathogens are not effectively removed in the primary process, although some removal of adsorbed pathogens occurs.

21.2.2 Secondary Treatment

Secondary treatment consists of biological degradation, in which the remaining suspended solids are decomposed and the number of pathogens is reduced. In this stage, the effluent from primary treatment may be pumped into a trickling filter bed, an aeration tank, or a sewage lagoon. A disinfection step is generally included at the end of the treatment.

21.2.2.1 Trickling Filters

The **trickling filter bed** is simply a bed of stones or corrugated plastic sheets through which water drips (see Fig. 21.4). It is one of the earliest systems introduced for biological waste treatment. The effluent is pumped through a system overhead sprayer onto this bed, where bacteria and other microorganisms reside. These microorganisms intercept the organic material as it trickles past and decompose it aerobically.

The media used in trickling filters may be stones, ceramic material, hard coal, or plastic media. Plastic media of polyvinyl chloride (PVC) or polypropylene are commonly used in modern high-rate trickling filters. Because of the light weight of the plastic, they can be stacked in towers 6 to 10 m high referred to as **biotowers.** As the organic matter passes through the trickling filter, it is converted to microbial biomass which forms a biofilm on the filter medium surfaces.

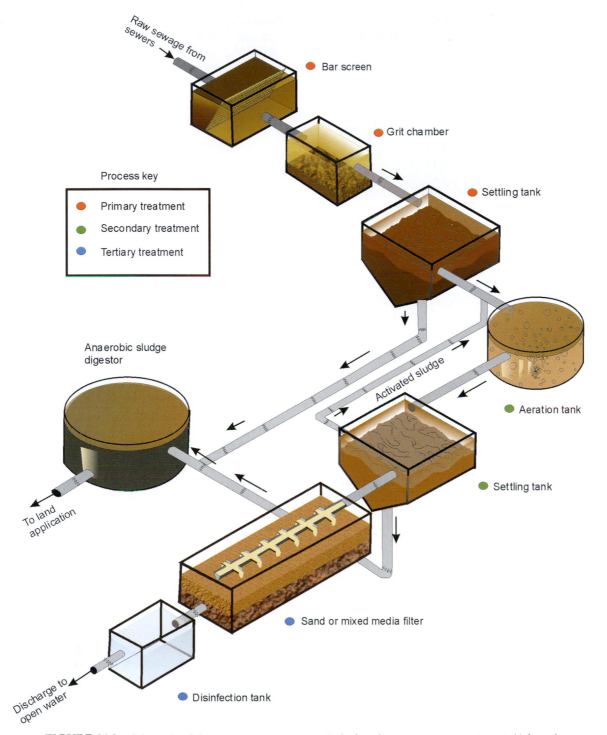

FIGURE 21.3   Schematic of the treatment processes typical of modern wastewater treatment. (Adapted from Pepper *et al.*, 1996.)

The biofilm that forms on the surface of the filter medium is called the **zooleal film.** It is composed of bacteria, fungi, algae, and protozoa. Over time, the increase in biofilm thickness leads to limited oxygen diffusion to the deeper layers of the biofilm, creating an anaerobic environment near the filter medium surface. As a result, the organisms eventually slough from the surface and a new biofilm is formed (Bitton, 1994). BOD removal by trickling filters is approximately 85% for low-rate filters (U.S. EPA, 1977).

FIGURE 21.4   A trickling filter bed. Here, rocks provide a matrix supporting the growth of a microbial biofilm that actively degrades the organic material in the wastewater under aerobic conditions. (From *Pollution Science*, 1996 Academic Press, San Diego, CA.)

The removal of enteric pathogens by trickling filters has generally been found to be low and erratic. Filtration rates have been shown to affect the removal of enteric viruses and probably other pathogenic microorganisms (Moore *et al.*, 1981).

21.2.2.2 Conventional Activated Sludge

Aeration-tank digestion is also known as the activated sludge process. In the United States, wastewater is most commonly treated by this process (this is the process illustrated in Fig. 21.3). Effluent from primary treatment is pumped into a tank and mixed with a bacteria-rich slurry known as **activated sludge.** Air or pure oxygen pumped through the mixture promotes bacterial growth and decomposition of the organic material. It then goes to a secondary settling tank, where water is siphoned off the top of the tank and sludge is removed from the bottom. Some of the sludge is used as an inoculum for the incoming primary effluent (activated sludge). The remainder of the sludge, known as **secondary sludge,** is removed. The concentration of pathogens is reduced in the activated sludge process by antagonistic microorganisms as well as adsorption to or incorporation into the secondary sludge.

An important characteristic of the activated sludge process is the recycling of a large proportion of the biomass. This results in a large number of microorganisms that oxidize organic matter in a relatively short time (Bitton, 1994). The detention time for sewage in the aeration basin varies from 4 to 8 hours. The content of the aeration tank is referred to as **mixed-liquor suspended solids (MLSS).** The organic part of the MLSS is called the **mixed-liquor volatile suspended solids (MLVSS),** which includes the nonmicrobial organic

matter as well as dead and living microorganisms and cell debris. The activated sludge process must be controlled to maintain a proper ratio of **food-to-microorganisms (F/M)** (Bitton, 1994). This is expressed as BOD per kilogram per day. It is expressed as

$$F/M = \frac{Q \times BOD}{MLSS \times V} \qquad \text{(Eq. 21.3)}$$

where Q = flow rate of sewage in million gallons per day (MGD), BOD = 5-day biochemical oxygen demand (mg/l), MLSS = mixed-liquor suspended solids (mg/l), and V = volume of aeration tank (gallons).

F/M is controlled by the rate of activated sludge wasting. **Wastage or secondary sludge** is sludge that is not returned as activated sludge. The higher the wasting rate, the higher the F/M ratio. For conventional aeration tanks the F/M ratio is 0.2–0.5 lb BOD_5/day/lb MLSS, but it can be higher (up to 1.5) for activated sludge when high-purity oxygen is used (Hammer, 1986). A low F/M ratio means that the microorganisms in the aeration tank are starved, leading to more efficient wastewater treatment.

The important parameters controlling the operation of an activated sludge process are organic loading rates, oxygen supply, and control and operation of the final settling tank. This tank has two functions: clarification and thickening. For routine operation, sludge settleability is determined by use of the **sludge volume index (SVI)** (Bitton, 1994).

SVI is determined by measuring the sludge volume after it has settled for 30 minutes and is given by the following formula:

$$SVI = \frac{V \times 1000}{MLSS} \qquad \text{(Eq. 21.4)}$$

where V = volume of settled sludge after 30 minutes (ml/l).

The microbial biomass produced in the aeration tank must settle properly from suspension so that it may be wasted or returned to the aeration tank. Good settling occurs when the sludge microorganisms are in the endogenous phase, which occurs when carbon and energy sources are limited and the microbial specific growth rate is local (Bitton, 1994). A mean cell residence time of 3–4 days is necessary for effective settling (Metcalf and Eddy, 1991). Poor settling may also be caused by sudden changes in temperature, pH, absence of nutrients, and presence of toxic metals and organics. A common problem in the activated sludge process is **filamentous bulking,** which consists of slow settling and poor compaction of solids in the clarifier. Filamentous bulking is usually caused by the excessive growth of filamentous microorganisms. The filaments produced by these microorganisms interfere with sludge settling and compaction. A high SVI (>150 ml/g) indicates bulking conditions. Filamentous microorganisms are able to predominate under conditions of low dissolved oxygen, low F/M, low nutrient, and high sulfide levels. Filamentous bacteria can be controlled by treating the return sludge with chlorine or hydrogen peroxide which kill filamentous microorganisms selectively.

21.2.2.3 Nitrogen Removal by the Activated Sludge Process

Activated sludge processes can be modified for nitrogen removal to encourage nitrification followed by denitrification. The establishment of a nitrifying population in activated sludge depends on the wasting rate of the sludge and therefore on the BOD load, MLSS, and retention time. The growth rate of nitrifying bacteria (μ_n) must be higher than the growth rate (μ_n) of heterotrophs in the system. In reality, the growth rate of nitrifiers is lower than that of heterotrophs in sewage; therefore a long sludge retention time is necessary for the conversion of ammonia to nitrate. Nitrification can occur with sludge retention times above 4 days (Bitton, 1994).

Nitrification must be followed by denitrification to remove nitrogen from wastewater. The conventional activated sludge system can be modified to encourage denitrification. Three such processes are

- *Single sludge system* (Fig. 21.5a). This system comprises a series of aerobic and anaerobic tanks in lieu of a single aeration tank. Methanol or settled sewage serves as the source of carbon for denitrifiers.
- *Multisludge system* (Fig. 21.5b). Carbonaceous oxidation, nitrification, and denitrification are carried out in three separate systems. Methanol or settled sewage serves as the source of carbon for denitrifiers.
- *Bardenpho process* (Fig. 21.6). The process consists of two aerobic and two anoxic tanks followed by a sludge settling tank. Tank 1 is anoxic and is used for denitrification, with wastewater used as a carbon source. Tank 2 is an aerobic tank utilized for both carbonaceous oxidation and nitrification. The mixed liquor from this tank, which contains nitrate, is returned to tank 1. The anoxic tank 3 removes the nitrate remaining in the effluent by denitrification. Finally, tank 4 is an aerobic tank used to strip the nitrogen gas that results from

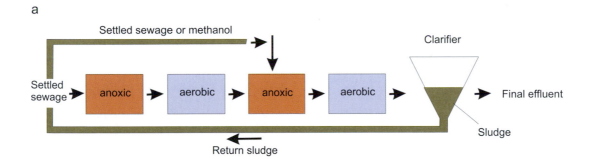

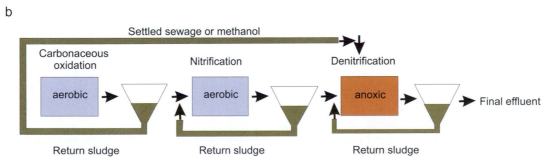

FIGURE 21.5 Denitrification systems: (a) Single-sludge system. (b) Multisludge system. (Modified from Curds and Hawkes, 1983.)

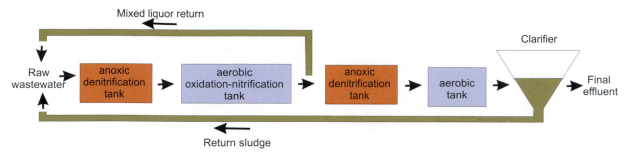

FIGURE 21.6   Denitrification system: Bardenpho process. (From U.S. EPA, 1975.)

denitrification, thus improving mixed liquor settling.

21.2.2.4 Phosphorus Removal by Activated Sludge Process

Phosphorus can also be reduced by the activity of microorganisms in modified activated sludge processes. The process depends on the uptake of phosphorus by the microbes during the aerobic stage and subsequent release during the anaerobic stage. Two of several systems in use are the

- *A/O (aerobic/oxid) process* (Fig. 21.7). The A/O process consists of a modified activated sludge system that includes an anaerobic zone (detention time 0.5–1 hour) upstream of the conventional aeration tank (detention time 1–3 hours). Figure 21.8 illustrates the microbiology of the A/O process: Under anaerobic conditions, microbes release stored phosphorus to generate energy. The energy liberated is used for the uptake of BOD from wastewater. Removal efficiency is high when the BOD/phosphorus ratio exceeds 10 (Metcalf and Eddy, 1991). When aerobic conditions are restored, microbes exhibit phosphorus uptake levels above those normally required to support the cell maintenance, synthesis, and transport reactions required for BOD oxidation. Excess phosphorus is stored as polyphosphates within the cell. The

sludge containing the excess phosphorus is then wasted (Fig. 21.8).
- *Bardenpho process*. This system also removes nitrogen as well as phosphorus by a nitrification–denitrification process.

21.2.3 Tertiary Treatment

Tertiary treatment involves a series of additional steps after secondary treatment to further reduce organics, turbidity, nitrogen, phosphorus, metals, and pathogens. Most processes involve some type of physicochemical treatment such as coagulation, filtration, activated carbon adsorption of organics, and additional disinfection. Tertiary treatment of wastewater is practiced for additional protection of wildlife after discharge into rivers or lakes or when the wastewater is to be reused for irrigation (e.g., food crops, golf courses), for recreational purposes (e.g., lakes, estuaries), or for drinking water.

21.2.4 Removal of Pathogens by Sewage Treatment Processes

There have been a number of reviews on the removal of pathogenic microorganisms by activated sludge and other wastewater treatment processes (Leong, 1983). This information suggests that significant removal especially of enteric bacterial patho-

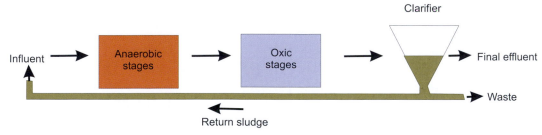

FIGURE 21.7   A/O process for phosphorus removal. (Modified from *Wastewater Microbiology*, G. Bitton, © 1994 by Wiley-Liss, with permission of Wiley-Liss, Inc., a subsidiary of John Wiley & Sons, Inc.)

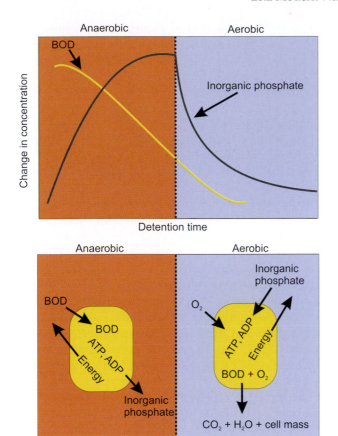

FIGURE 21.8 Microbiology of the A/O process. (Modified from Deakyne *et al.*, 1984.)

gens can be achieved by these processes (Table 21.5). However, disinfection and/or advanced tertiary treatment is necessary for many reuse applications to ensure pathogen reduction. Current issues related to pathogen reduction are treatment plant reliability, removal of new and emerging enteric pathogens of concern, and the ability of new technologies to effect pathogen reduction. Although the ability of conventional treatment processes to remove pathogens has been demonstrated in pilot and full-scale systems, how often this can be achieved over time given the variation in quality of raw wastewater and dynamics of plant processes is not well documented. Wide variation in wastewater composition can result in significant numbers of pathogens passing through the process for various time periods. The issue of reliability is of major importance if the reclaimed water is intended for recreational or potable reuse, where short-term exposures to high levels of pathogens could result in significant risk to the exposed population (Tanaka *et al.*, 1998). Recognition of the importance of water and food in the transmission of new emerging enteric pathogens have created a need for information on the ability of treatment processes to remove these pathogens.

Compared with other biological treatment methods (i.e., trickling filters), activated sludge is relatively efficient in reducing the numbers of pathogens in raw wastewater. Both sedimentation and aeration play a role in pathogen reduction. Primary sedimentation is more effective in the removal of the larger pathogens such as helminth eggs, but solid-associated bacteria

TABLE 21.5   Pathogen Removal during Sewage Treatment

	Enteric viruses	*Salmonella*	*Giardia*	*Cryptosporidium*
Concentration in raw sewage (number per liter)	10^5–10^6	5,000–80,000	9,000–200,000	1–3,960
Removal during				
Primary treatment [a]				
% removal	50–98.3	95.8–99.8	27–64	0.7
No. remaining/l	1,700–500,000	160–3,360	72,000–146,000	
Secondary treatment [b]				
% removal	53–99.92	98.65–99.996	45–96.7	
No. remaining/l	80–470,000	3–1075	6,480–109,500	
Secondary treatment [c]				
% removal	99.983–99.9999998	99.99–99.999999995	98.5–99.99995	2.7[d]
No. remaining/l	0.007–170	0.000004–7	0.099–2,951	

Data from Yates (1994); Robertson *et al.* (1995); Enriquez *et al.* (1995); Modore *et al.* (1987).

[a] Primary sedimentation and disinfection.
[b] Primary sedimentation, trickling filter or activated sludge, and disinfection.
[c] Primary sedimentation, trickling filter or activated sludge, disinfection, coagulation, filtration, and disinfection.
[d] Filtration only.

and even viruses are also removed. During aeration, pathogens are inactivated by antagonistic microorganisms and by environmental factors such as temperature. The greatest removal probably occurs by adsorption or entrapment of the organisms within the biological floc that forms. Thus, the ability of activated sludge to remove viruses is related to the ability to remove solids. This is because viruses tend to be solid associated and are removed along with the floc. Activated sludge typically removes 90% of the enteric bacteria and 80 to 90–99% percent of the enteroviruses and rotaviruses (Rao *et al.*, 1986). Ninety percent of *Giardia* and *Cryptosporidium* can also be removed (Rose and Carnahan, 1992; Casson *et al.*, 1990), being largely concentrated in the sludge. Because of their large size, helminth eggs are effectively removed by sedimentation and are rarely found in sewage effluent in the United States, although they may be detectable in the sludge. However, although the removal of the enteric pathogens may seem large, it is important to remember that initial concentrations are also large (i.e., the concentration of all enteric viruses in 1 liter of raw sewage may be as high as 100,000 in some parts of the world).

Tertiary treatment processes involving physical–chemical processes can be effective in further reducing the concentration of pathogens and enhancing the effectiveness of disinfection processes by the removal of soluble and particulate organic matter (Table 21.6). Filtration is probably the most common tertiary treatment process. Mixed-media filtration is most ef-

fective in the reduction of protozoan parasites. Usually, greater removal of *Giardia* cysts occurs than of *Cryptosporidium* oocysts because of the larger size of the cysts (Rose and Carnahan, 1992). Removal of enteroviruses and indicator bacteria is usually 90% or less. Addition of a coagulant can increase the removal of poliovirus to 99% (U.S. EPA, 1992).

Coagulation, particularly with lime, can result in significant reductions of pathogens. The high-pH conditions (pH 11–12) that can be achieved with lime can result in significant inactivation of enteric viruses. To achieve removals of 90% or greater, the pH should be maintained above 11 for at least an hour (Leong, 1983). Inactivation of the viruses occurs by denaturation of the viral protein coat. The use of iron and aluminum salts for coagulation can also result in 90% or greater reductions in enteric viruses. The degree of effectiveness of these processes, as in other solids separating processes, is highly dependent on the hydraulic design and operation of the coagulation and flocculation operation. Thus, the degree of removal observed in bench-scale tests may not approach those seen in full-scale plants where the process is more dynamic.

The removal of enteric viruses by granular activated carbon has been found to be highly variable and not very effective. Viruses are believed to be adsorbed to the activated carbon, but it appears that sites available for adsorption are quickly exhausted (Gerba *et al.*, 1975).

Reverse osmosis and ultrafiltration are also believed to result in significant reductions in enteric pathogens,

TABLE 21.6 Average Removal of Pathogen and Indicator Microorganisms in a Wastewater Treatment Plant, St. Petersburg, Florida

	Raw wastewater to secondary wastewater		Secondary wastewater to postfiltration		Postfiltration to postdisinfection		Postdisinfection to poststorage		Raw wastewater to poststorage	
	Percentage	log$_{10}$	Percentage	log$_{10}$	Percentage	log$_{10}$	Percentage	log$_{10}$	Percentage	log$_{10}$
Total coliforms (CFU/100 ml)	98.3	1.75	69.3	0.51	99.99	4.23	75.4	0.61	99.999992	7.10
Fecal coliforms (CFU/100 ml)	99.1	2.06	10.5	0.05	99.998	4.95	56.8	0.36	99.999996	7.42
Coliphage 15597 (PFU/ml)	82.1	0.75	99.98	3.81	90.05	1.03	90.3	1.03	99.999997	6.61
Enterovirus (PFU/100 l)	98.0	1.71	84.0	0.81	96.5	1.45	90.9	1.04	99.999	5.01
Giardia (cysts/100 l)	93.0	1.19	99.0	2.00	78.0	0.65	49.5	0.30	99.993	4.13
Crystosporidium (oocysts/100 l)	92.8	1.14	97.9	1.68	61.1	0.41	8.5	0.04	99.95	3.26

From Rose and Carnahan (1992).

although few studies have been done in full-scale facilities. Removal of enteric viruses in excess of 99.9% can be achieved (Leong, 1983). Although the pore sizes of the membranes used in this process are even smaller than viruses, the smallest waterborne pathogens, they should not be considered absolute barriers. In an evaluation of an on-site wastewater recycling system, it was found that although poliovirus was reduced by 99.9998% by the ultrafiltration of the wastewater, coliphage MS-2 (similar in size and shape to the poliovirus) was reduced only 99.93% (Naranjo *et al.*, 1993). This may be due to differences between the two viruses in the degree of attachment to solids in the wastewater. Whereas poliovirus readily adsorbs to solids, MS-2 coliphage does not (Hurst *et al.*, 1980). Nanofiltration was also observed to achieve a 99% reduction of MS-2 coliphage (Yahya *et al.*, 1993). It is possible that viruses may pass through a few openings in the membranes or around the sealing of the elements. Because such "leakage" occurs at a very low level, it is not observed by other measures of water quality used to judge performance of the membranes (e.g., conductivity).

21.2.5 Sludge Processing

The sludge or biosolids resulting from the various stages of sewage processing (both the primary sludge and the activated sludge generated by secondary processes) also require treatment, in this case to stabilize their organic matter and reduce their water content. Primary sludge from the primary clarifier contains 3–8% solids, and secondary sludge contains 0.5 to 2% solids. Treating the organic matter prevents the formation of odors and decreases the number of pathogens, such as enteric parasites and viruses, which are often concentrated in the sludge. Reducing the water content reduces the weight of the sludge, making it more economical to transport it to its final disposal site.

Sludge treatment typically involves several steps, as shown in Fig. 21.9. **Thickening** reduces the volume of the sludge and can be accomplished by allowing the solids to settle in a tank or by centrifugation. **Digestion** is a microbial process that results in stabilization of the organic matter and some destruction of pathogens because of the higher temperatures generated. Sludge can be digested both anaerobically and aerobically. **Anaerobic digestion** is the most common type of treatment and typically takes place over a period of 2–3 weeks in large covered tanks at the sewage treatment facility (Fig. 21.10). This process has the advantage of producing methane gas, which can be recovered as an energy source.

The complex organic matter is degraded in a series of microbial processes to methane (see Chapter 3.4).

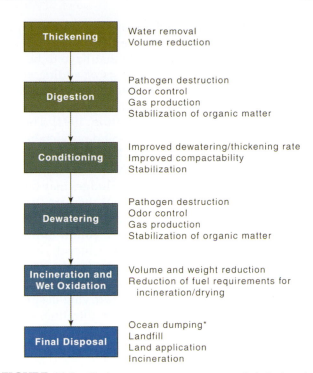

FIGURE 21.9   Sludge treatment processes and their function. *Ocean dumping is no longer allowed in the United States. (From *Pollution Science* © 1996, Academic Press, San Diego, CA.)

Four groups of bacteria are involved in this transformation and operate in a synergistic reaction (Bitton, 1994). Anaerobic sludge digestion is affected by temperature, retention time, pH, chemical composition of wastewater, competition with sulfate-reducing bacteria, and the presence of toxic substances (e.g., heavy metals).

Aerobic digestion also stabilizes sludge. In this process, air or oxygen is passed through the sludge in open tanks. The advantages of aerobic digestion are

FIGURE 21.10   Anaerobic sludge digestor tank. (Photo courtesy P. E. Gerba.)

low capital costs, easy operation, and production of odorless, stabilized sludge. However, it also produces greater amounts of waste sludge that have to be disposed of. Aerobic digestion consists of adding air or oxygen to sludge in a 10- to 20-feet deep open tank. Oxygen concentration is maintained above 1 mg/l to avoid production of foul odors (Bitton, 1994). The detention time in the digestor is 12-30 days, depending on the prevailing temperature. Microorganisms degrade the available organic matter aerobically, resulting in a reduction of sludge solids. Sludges may be disposed of after they are digested, but they are usually treated further to reduce the volume of water. This is accomplished by a process called **conditioning** in which chemicals such as alum, ferric chloride, or lime are added to aggregate suspended particles. This is followed by **dewatering,** which can be accomplished by a number of methods, such as air drying in spreading basins, centrifugation, or vacuum filtration.

Coagulation with alum, lime, or polyelectrolytes is effective in removing suspended matter and phosphorus from sludges. It is also very effective in removing viruses. The precipitate that forms is removed and the residual effluent is usually passed through sand or mixed-media filters. Mixed-media filters are composed of granular coal, garnet, and sand, which together enhance filtration performance.

21.2.6 Pathogen Occurrence and Fate in Biosolids

Although treatment by anaerobic or aerobic digestion and/or dewatering reduces the numerical population of disease agents in these biosolids, significant numbers of the pathogens present in raw sewage often remain in sewage biosolids. On a volume basis, the concentration of pathogens in sewage biosolids can be fairly high because of settling (of the large organisms, especially helminths) and adsorption (especially viruses). Moreover, most microbial species found in raw sewage are concentrated in sludge during primary sedimentation. In addition, although enteric viruses are too low in mass to settle alone, they are also concentrated in sludge because of their strong binding affinity to particles.

The densities of pathogenic and indicator organisms in primary sludge (biosolids) shown in Table 21.7 represent typical, average values detected by various investigators. Note that the indicator organisms are normally present in fairly constant amounts. But bear in mind that different biosolids may contain significantly greater or smaller numbers of any organism, depending on the kind of sewage from which the

TABLE 21.7   Densities of Microbial Pathogens and Indicators in Primary Biosolids

Type	Organism	Density (number per gram of dry weight)
Viruses	Various enteric viruses	10^2–10^4
	Bacteriophage	10^5
Bacteria	Total coliforms	10^8–10^9
	Fecal coliforms	10^7–10^8
	Fecal streptococci	10^6–10^7
	Salmonella sp.	10^2–10^3
	Clostridium sp.	10^6
	Mycobacterium tuberculosis	10^6
Protozoa	*Giardia* sp.	10^2–10^3
Helminths	*Ascaris* sp.	10^2–10^3
	Trichuris vulpis	10^2
	Toxocara sp.	10^1–10^2

Modified from Straub *et al.* (1993).

biosolid was derived. Similarly, the quantities of pathogenic species are especially variable because these figures depend on which kind are present in a specific community at a particular time. Finally, note that concentrations determined in any study are dependent on assays for each microbial species; thus, these concentrations are only as accurate as the assay itself, which may be compromised by such factors as inefficient recovery of pathogens from the sample.

Secondary sludges (biosolids) are produced following the biological treatment of wastewater. Microbial populations in biosolids after these treatments depend on the initial concentrations in the wastewater, die-off or growth during treatments, and the association of these organisms with the biosolids. Some treatment processes, such as the activated sludge process, may limit or destroy certain enteric microbial species. Viral and bacterial pathogens, for example, are reduced in concentration by activated sludge treatment. Even so, the ranges of pathogen concentration in secondary biosolids obtained from this and most other secondary treatments may not be significantly different from those of primary biosolids, as shown in Table 21.8.

The reduction of pathogenic microorganisms by anaerobic digestion is both time and temperature dependent. Thermophilic digestion (50–60°C) and longer detention times favor greater reduction of pathogens (Straub *et al.*, 1993). In general, a plant using mesophilic digestion (30–38°C) with a mean retention time of 14–15 days can expect a 1–2 $\log_{10}$ removal of total coliforms, fecal coliforms, and fecal streptococcus.

TABLE 21.8   Densities of Pathogenic and Indicator Microbial Species in Secondary Sludge Biosolids

Type	Organism	Density (number per gram of dry weight)
Viruses	Various enteric viruses	3×10^2
Bacteria	Total coliforms	7×10^8
	Fecal coliforms	8×10^6
	Fecal streptococci	2×10^2
	Salmonella sp.	9×10^2
Protozoa	*Giardia* sp.	10^2–10^3
Helminths	*Ascaris* sp.	1×10^3
	Trichuris vulpis	$<10^2$
	Toxocara sp.	3×10^2

Modified from Straub *et al.* (1993).

Helminth ova apparently survive anaerobic digestion with little reduction in viable numbers.

Temperatures for aerobic digestion are usually mesophilic (37°C) with a mean retention time of 10–20 days. Conversion of organic matter into carbon dioxide and water leads to decreased carbon sources for bacteria; hence, the numbers of bacteria are most likely reduced because of nutrient deprivation.

After digestion, sludges may be air dried or treated with lime. Both processes reduce the numbers of pathogens. In lime stabilization, liquid sludge is mixed with a sufficient amount of lime to raise the pH to 12.0 for at least 2 hours. At this pH, the NH_4^+ ion is deprotonated, resulting in the production of ammonia gas. The combination of high pH and ammonia can reduce

enteroviruses by four orders of magnitude (Satter *et al.*, 1976) and coliform indicator bacteria by two to seven orders of magnitude; however, no reduction of *Acaris* ova may occur. Composting is another method of sludge treatment that is in current use, and this is discussed in Section 21.7.2. Other nonconventional treatment or disinfection processes such as heat drying, pasteurization, heat treatment, and gamma irradiation also act to reduce the numbers of pathogens before disposal.

21.2.6.1 Land Disposal of Biosolids

Landfarming is the practice of disposing biosolids produced by wastewater plants on agricultural land. These biosolids may be added to the soil as either solids or liquids. Such practices have the benefits of adding nutrients and water to the soil. Liquid sludges can be injected into the soil from tanker trucks as shown in Fig. 21.11.

The U.S. Environmental Protection Agency has established two categories of sewage biosolids before land applications (Outwater, 1994), class A and class B. Sewage sludge that is applied to lawns and home gardens and is sold or given away in bags or other containers must meet the class A criteria. All sewage sludge that is land applied must meet class B requirements. Various microbial criteria and treatment requirements are given as options (Tables 21.9 and 21.10). Class B sludges may be treated by **"processes which significantly reduce pathogens"** or PSRP (Table 21.11). Sludges treated by PSRP may be land applied if certain restrictions are met with regard to crop production (no food crops should be grown

a) Injecting Sludge

b) Sludge-Treated Soil

FIGURE 21.11   Sewage biosolids are injected (a) beneath the soil surface for subsequent growth of furrow-irrigated cotton (b). (Photographs courtesy J. F. Artiola.)

TABLE 21.9   Class A Sewage Sludge

Class A sludge must meet one of the following criteria:

1. A fecal coliform density less than 1000 MPN/g total dry solids (TS);

or

2. A *Salmonella* sp. density less than 3 MPN/4 g TS;

or

the requirements of **one** of the following alternatives must be met:

1. **Time/temperature**—An increased sewage sludge temperature should be maintained for a prescribed period of time according to the guidelines.

2. **Alkaline treatment**—The pH of the sewage sludge is raised to greater than 12 for at least 72 h. During this time, the temperature of the sewage sludge shall be greater than 52°C for at least 12 h. In addition, after the 72-h period, the sewage sludge is to be air dried to at least 50% TS.

or

3. **Prior testing for enteric viruses/viable helminth ova**—If the sludge is analyzed *before* the pathogen reduction process and found to have densities of enteric viruses <1 plaque forming unit (pfu) per 4 g TS and <1 viable helminth ova per 4 g TS, the sludge is class A with respect to enteric viruses and viable helminth ova until the next monitoring session. If the sludge is monitored *before* the pathogen reduction process and found to have densities of enteric virus ≥1 pfu/4 g TS or ≥1 viable helminth ova per 4 g TS and is tested again *after* processing and found to meet the enteric virus and viable helminth ova levels listed under point 4 below, then the sludge will be class A with respect to enteric viruses and viable helminth ova when the operating parameters for the pathogen reduction process are monitored and shown to be consistent with the values or ranges of values documented.

4. **No prior testing for enteric viruses/viable helminth ova**—If the sludge is not analyzed *before* pathogen reduction processing for enteric viruses and viable helminth ova, the sludge must meet the enteric virus and viable helminth ova levels noted below to be class A at the time the sludge is used or disposed of, prepared for sale, or given away in a bag or container, or when the sewage sludge or derived material meets "exceptional quality" requirements—pollutant concentration limits, class A pathogen reduction, and vector attraction reduction requirements:

   • The density of enteric viruses must be <1 pfu/4 g TS.
   • The density of viable helminth ova must be <1 per 4 g TS.

within 18 months after application), animal grazing (prevention of grazing for at least 1 month by animals that provide products consumed by humans), and public access to the treated site (public access must be controlled for at least 12 months). There are no restrictions if the sludge is treated by processes referred to as **"processes to further reduce pathogens" or PFRP** (Table 21.12). PFRP are required if sludge is applied to edible crops.

TABLE 21.10   Class B Sewage Sludge

Class B sludge must meet one of the following pathogen reduction requirements:

1. The sewage sludge must be treated with a PSRP or a PSRP-equivalent process, as shown in Table 21.12;

or

2. At least seven sewage sludge samples should be collected at the time of use or disposal and analyzed for fecal coliforms during each monitoring period. The geometric mean of the densities of these samples will be calculated and should meet the following criteria:

   • <2 million MPN/g TS
   or
   • <2 million cfu/g TS

21.2.7 An Example of a Modern Sewage Treatment Plant

Roger Road Wastewater Treatment Plant, serving Pima County and the Tucson, Arizona metropolitan area, is an example of a modern sewage treatment plant that combines both activated sludge and biotowers (Fig. 21.12).

Raw sewage enters the plant at the headworks, which houses the bar screen and grit chamber (compare with Fig. 21.3). Once out of the headworks, the raw sewage flows into the primary settling tanks, which perform an initial separation of the solids from the wastewater. The primary sewage flows on to the biotowers, in which the effluent is trickled over a plastic latticework exhibiting a high surface-to-volume ratio. In this case, the biotowers take the place of the simpler trickling filters shown in Fig. 21.4, which are

TABLE 21.11   Pathogen Treatment Processes: Processes to Significantly Reduce Pathogens (PSRP)

1. *Aerobic digestion*—Sewage sludge is agitated with air and oxygen to maintain aerobic conditions for a mean cell residence time and temperature between 40 days at 20°C and 60 days at 15°C.

2. *Air drying*—Sludge is dried on sand beds or on paved or unpaved basins for a minimum of 3 months; during 2 of the 3 months, the ambient average daily temperature is above 0°C.

3. *Anaerobic digestion*—Sludge is treated in the absence of air for a mean cell residence time and temperature of between 15 days at 35 to 55°C and 60 days at 20°C.

4. *Composting*—Using either in-vessel, static aerated pile, or windrow composting methods, the temperature of the sludge is raised to 40°C or higher for 5 days. For 4 h during the 5 days, the pile temperature must exceed 55°C.

5. *Lime stabilization*—Sufficient lime is added to the sludge to raise the pH of the sludge to 12 after 2 h of contact.

TABLE 21.12   Pathogen Treatment Processes: Processes to Further Reduce Pathogens (PFRP)

1. *Composting*—Using either in-vessel or static aerated pile composting, the temperature of the sewage sludge is maintained at 55°C or higher for 3 days. Using windrow composting, the temperature of the sewage sludge is maintained at 55°C or higher for 15 days or longer. During this period, a minimum of five windrow turnings are required.

2. *Heat drying*—Sewage sludge is dried by indirect or direct contact with hot gases to reduce the moisture content of the sludge to 10% or lower. Either the temperature of the gas in contact with the sludge exceeds 80°C or the wet bulb temperature of the gas in contact with the sludge as the sludge leaves the dryer exceeds 80°C.

3. *Heat treatment*—Liquid sludge is heated to a temperature of 180°C or higher for 30 min.

4. *Thermophilic aerobic digestion*—Liquid dewatered sludge is agitated with air or oxygen to maintain aerobic conditions, and the mean cell residence time for the sewage sludge is 10 days at 55 to 60°C.

5. *Beta-ray irradiation*—Sewage sludge is irradiated with beta rays from an accelerator at dosage of at least 1.0 Mrad at room temperature (about 20°C).

6. *Gamma-ray irradiation*—Sewage sludge is irradiated with gamma rays from certain isotopes such as ^{60}Co and ^{137}Ce, at dosages of at least 1.0 Mrad at room temperature (about 20°C).

7. *Pasteurization*—The temperature of the sludge is maintained at 70°C or higher for at least 30 min.

used in some communities. In the cooler winter months, when levels of microbial activity are reduced, additional treatment of the effluent occurs in the adjoining aeration basins under forced aeration. The aerobically treated effluent proceeds to the final or secondary settling tanks, where further suspended solids are removed from the effluent (Fig. 21.13). The secondary effluent is then chlorinated before discharge into a dry steam bed. Part of the wastewater receives tertiary treatment in which it passes through mixed-media filters to reduce turbidity, and is further disinfected. This effluent is used to irrigate area golf courses or further treated by rapid infiltration extraction (RIX) (see Section 21.5) through application to dry constructed infiltration basins.

The sludges collected from the primary and final settling tanks are combined and further processed in the anaerobic digesters. The anaerobically treated sludge is piped to a neighboring sewage treatment plant where the combined output from both plants is centrifuged. The treated sludge (now 8–9% solids) then resides in a holding pond, awaiting land application to local agricultural fields by a contracted company. Currently, all of Tucson's municipal sludge is land applied.

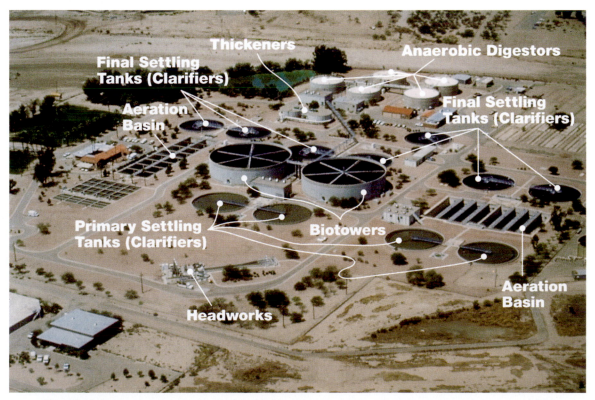

FIGURE 21.12   Roger Road Wastewater Treatment Plant, Tucson, Arizona. (Photo courtesy C. P. Gerba.) (From *Pollution Science* © 1996, Academic Press, San Diego, CA.)

FIGURE 21.13 *Secondary sewage settling tank. (Photograph courtesy C. P. Gerba.)*

21.3 OXIDATION PONDS

Sewage lagoons are often referred to as **oxidation** or **stabilization ponds** and are the oldest of the wastewater treatment systems. Usually no more than a hectare in area and just a few meters deep, oxidation ponds are natural "stewpots" where wastewater is detained while organic matter is degraded. A period of time ranging from 1 to 4 weeks (and sometimes longer) is necessary to complete the decomposition of organic matter. Light, heat, and settling of the solids can also effectively reduce the number of pathogens present in the wastewater.

There are four categories of oxidation ponds, which are often used in series: aerobic ponds, aerated ponds, anaerobic ponds, and facultative ponds:

- **Aerobic ponds** (Fig. 21.14a), which are naturally mixed, must be shallow because they depend on penetration of light to stimulate algal growth that promotes subsequent oxygen generation. The detention time of wastewater is generally 3 to 5 days.

- **Anaerobic ponds** (Fig. 21.14b) may be 1 to 10 m deep and require a relatively long detention time of 20–50 days. These ponds, which do not require expensive mechanical aeration, generate small amounts of sludge. Often, anaerobic ponds serve as a pretreatment step for high-BOD organic wastes rich in protein and fat (e.g., meat wastes) with a heavy concentration of suspended solids.

- **Facultative ponds** (Fig. 21.15) are most common for domestic waste treatment. Waste treatment is provided by both aerobic and anaerobic processes. These ponds range in depth from 1 to 2.5 m and are subdivided in three layers: an upper aerated zone, a middle facultative zone, and a lower

anaerobic zone. The detention time varies between 5 and 30 days.

- **Aerated ponds** (Fig. 21.16), which are mechanically aerated, may be 1–2 m deep and have a detention time of less than 10 days. In general, treatment depends on the aeration time and temperature as well as the type of wastewater. For example, at 20°C an aeration period of 5 days results in 85% BOD removal.

Because sewage lagoons require a minimum of technology and are relatively low in cost, they are most common in developing countries and in small communities in the United States, where land is available at reasonable prices. However, biodegradable organic matter and turbidity are not as effectively reduced as they are in activated sludge treatment.

Given sufficient retention times, oxidation ponds can cause significant reductions in the concentrations of enteric pathogens, especially in helminth eggs. For this reason they have been promoted widely in the developing world as a low-cost method of pathogen reduction for wastewater reuse for irrigation. However, a major drawback of ponds is the potential for inadequate mixing or short-circuiting because of thermal gradients even in multipond systems designed for long retention times (i.e., 90 days). Even though the amount of short-circuiting may be small, detectable levels of pathogens can often be found in the effluent from oxidation ponds.

Inactivation and/or removal of pathogens in oxidation ponds is controlled by a number of factors including temperature, sunlight, pH, bacteriophage, predation by other microorganisms, and adsorption to or entrapment by settleable solids. Indicator bacteria and pathogenic bacteria may be reduced by 90–99% or more, depending on retention times. The die-off of indicator bacteria in waste stabilization ponds has been studied extensively and several models have been proposed, largely based on retention time, solar radiation, and temperature (Marais, 1974; Mills *et al.*, 1992). The more sunlight and the higher the temperature, the higher the rate of die-off of enteric pathogens. Enteric viruses are also affected by the same factors but to a lesser degree. Because these factors vary greatly by location and seasonally, the removal to be expected is somewhat site specific. For example, in one study it was found that it took 5 days for 99% inactivation of poliovirus type 1 in a model oxidation pond, whereas in the winter 25 days was required (Funderburg *et al.*, 1978). In a study of waste stabilization ponds in Africa, it was found that a retention time of at least 37.3 days would be required to ensure at least 99.9% reduction of *Giardia* and *Cryptosporidium* (Grimason *et al.*, 1993).

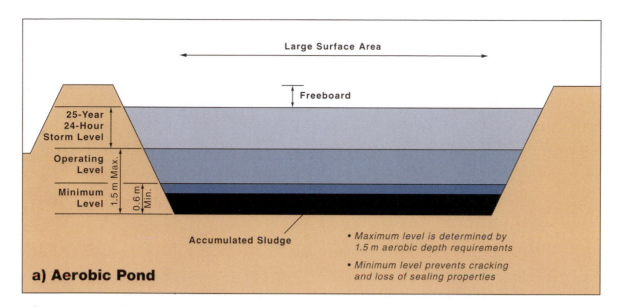

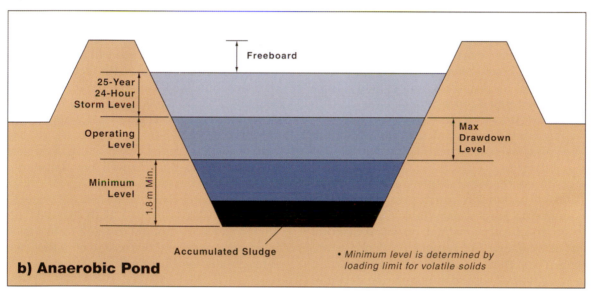

FIGURE 21.14 Pond profiles: (a) Aerobic waste pond profile, and (b) Anaerobic waste pond profile. (Adapted with permission from The Guidebook: Nitrogen Best Management Practices for Concentrated Animal Feeding Operations in Arizona, by Robert J. Freitas and Thomas A. Doerge; copyright © 1996 by the Arizona Board of Regents all rights reserved.)

Helminth eggs can be removed below detection within 10 days of detention (Schwartzbrod *et al.*, 1989).

21.4 SEPTIC TANKS

Until the middle of the 20th century in the United States, many rural families and quite a few residents of towns and small cities depended on pit toilets or "outhouses" for waste disposal. These pit toilets, however, often allowed untreated wastes to seep into the groundwater, allowing pathogens to contaminate drinking water supplies. This risk to public health led to the development of septic tanks and properly constructed drain fields. Primarily, septic tanks serve as repositories where solids are separated from incoming wastewater and biological digestion of the waste organic matter can take place under anaerobic conditions. In a typical septic tank system (Fig. 21.17), the wastewater and sewage enter a tank made of concrete, metal, or fiberglass. There, grease and oils rise to the top as scum, and solids settle to the bottom. The wastewater and sewage then undergo anaerobic bacterial decomposition, resulting in the production of a

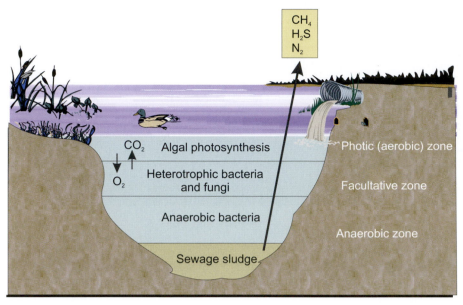

FIGURE 21.15   Microbiology of facultative ponds. (Modified from Bitton, 1980.)

sludge. The wastewater usually remains in the septic tank for just 24–72 hours, after which it is channeled out to a drainfield. This drainfield or leachfield is composed of small perforated pipes that are embedded in gravel below the surface of the soil. Periodically, the residual septage in the septic tank is pumped out into a tank truck and taken to a treatment plant for disposal.

Although the concentration of contaminants in septic tank septage is typically much greater than that found in domestic wastewater (Table 21.13), septic tanks can be an effective method of waste disposal where land is available and population densities are not too high. Thus, they are widely used in rural and suburban areas. But as suburban population densities increase, groundwater and surface water pollution may arise, indicating a need to shift to a commercial municipal sewage system. (In fact, private septic systems are sometimes banned in many suburban areas.) Moreover, septic tanks are not appropriate for every area of the country. They do not work well, for example, in cold, rainy climates, where the drainfield may be too wet for proper evaporation, or in areas where the water table is shallow. High densities of septic tanks can also be responsible for nitrate contamination of groundwater. Finally, most of the waterborne disease outbreaks associated with groundwater in the United States are thought to result from contamination by septic tanks. For a discussion of pathogen transport through soil, see Chapter 7.

21.5 LAND APPLICATION OF WASTEWATER

Although treated domestic wastewater is usually discharged into bodies of water, it may also be disposed of via land application—sometimes for crop irrigation and sometimes as a means of additional treatment or disposal. The three basic methods used in the application of sewage effluents to land include low-rate irrigation; overland flow; and high-rate infiltration, whose leading characteristics are listed in Table 21.14. The choice of a given method depends on the conditions prevailing at the site under consideration (loading rates, methods of irrigation, crops, and expected treatment).

With **low-rate irrigation** (Fig. 21.18a), sewage effluents are applied by sprinkling or by surface applica-

FIGURE 21.16   Aerated lagoon. (Photo courtesy C. P. Gerba.)

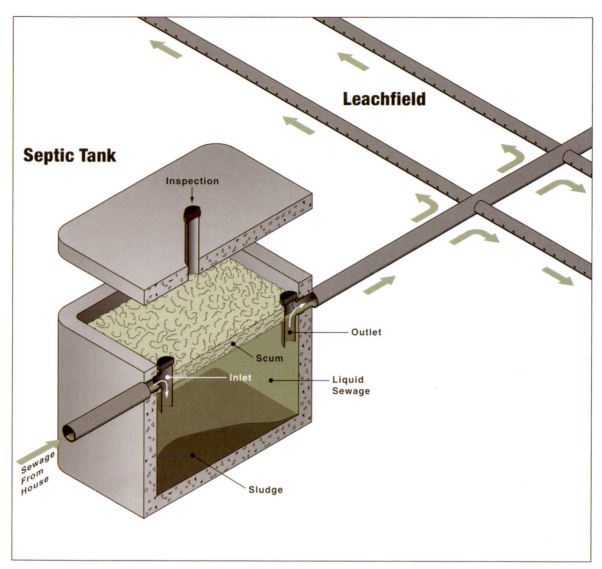

FIGURE 21.17   Septic tank system. (From *Pollution Science* © 1996, Academic Press, San Diego, CA.)

TABLE 21.13   Typical Characteristics of Septage

Constituent	Concentration (mg/l)	
	Range	Typical
Total solids	5,000–100,000	40,000
Suspended solids	4,000–100,000	15,000
Volatile suspended solids	1,200–14,000	7,000
5-day, 20°C BOD	2,000–30,000	6,000
Chemical oxygen demand	5,000–80,000	30,000
Total Kjeldahl nitrogen (as N)	100–1,600	700
Ammonia, NH_3 as N	100–800	400
Total phosphorus as P	50–800	250
Heavy metals[a]	100–1,000	300

Modified from Metcalf and Eddy (1991).
[a] Primarily iron (Fe), zinc (Zn), and aluminum (Al).

tion at a rate of 1.5 to 10 cm per week. Two thirds of the water is taken up by crops or lost by evaporation, and the remainder percolates through the soil matrix. The system must be designed to maximize denitrification in order to avoid pollution of groundwater by nitrates. Phosphorus is immobilized within the soil matrix by fixation or precipitation. The irrigation method is used primarily by small communities and requires large areas, generally on the order of 5–6 hectares per 1000 people.

In the **overland flow** method (Fig. 21.18b), wastewater effluents are allowed to flow for a distance of 50–100 m along a 2–8% vegetated slope and are collected in a ditch. The loading rate of wastewater ranges from 5 to 14 cm a week. Only about 10% of the

**TABLE 21.14   General Characteristics of the Three Methods Used for Land Application
of Sewage Effluent**

	Application method		
Factor	Low-rate irrigation	Overland flow	High-rate infiltration
Main objectives	Reuse of nutrients and water Wastewater treatment	Wastewater treatment	Wastewater treatment Groundwater recharge
Soil permeability	Moderate (sandy to clay soils)	Slow (clay soils)	Rapid (sandy soils)
Need for vegetation	Required	Required	Optional
Loading rate	1.3–10 cm/week	5–14 cm/week	>50 cm
Application technique	Spray, surface	Usually spray	Surface flooding
Land required for flow of 10^6 l/day	8–66 hectares	5–16 hectares	0.25–7 hectares
Needed depth to groundwater	About 2 m	Undetermined	5 m or more
BOD and suspended solid removal	90–99%	90–99%	90–99%
N removal	85–90%	70–90%	0–80%
P removal	80–90%	50–60%	75–90%

From Pepper *et al.* (1996)

water percolates through the soil, compared with 60% that runs off into the ditch. The remainder is lost as evapotranspiration. This system requires clay soils with low permeability and infiltration.

The primary objective of **high-rate infiltration** (Fig. 21.18c)—also referred to as **soil aquifer treatment (SAT)** or **rapid infiltration extraction (RIX)**—is the treatment of wastewater at loading rates exceeding 50 cm per week. The treated water, most of which has percolated through coarse-textured soil, is used for groundwater recharge and may be recovered for irrigation (see Case Study). This system requires less land than irrigation or overland flow methods. Drying periods are often necessary to aerate the soil system and avoid problems with clogging. The selection of a site for land application is based on many factors, including soil types, drainability and depth, distance to groundwater, groundwater movement, slope, underground formations, and degree of isolation of the site from the public.

Inherent in land application of wastewater are risks of transmission of enteric waterborne pathogens. The degree of risk is associated with the concentration of pathogens in the wastewater and the degree of contact with humans. Land application of wastewater is usually considered an intentional form of reuse and is regulated by most states. Because of limited water resources in the western United States, reuse is considered essential. Usually, stricter treatment and microbial standards must be met before land application. The highest degree of treatment is required when wastewater will be used for food crop irrigation, with

lesser treatment for landscape irrigation or fiber crops. For example, the state of California requires no disinfection of wastewater for irrigation for fiber crops and no limits on coliform bacteria. However, if the reclaimed wastewater is used for surface irrigation of food crops and open landscaped areas, chemical coagulation (to precipitate suspended matter), followed by filtration and disinfection to reduce the coliform concentration to 2.2/100 ml is required.

Because high-rate infiltration may be practiced to recharge aquifers, addition treatments of secondary wastewater may be required. However, as some removal of pathogens can be expected from the infiltration process itself, treatment requirements may be less. The degree of treatment needed may be influenced by the amount of time it takes the reclaimed water to travel from the infiltration site to the point of extraction and the width and depth of the unsaturated zone. The greatest concern has been with the transport of viruses, which, because of their small size, have the greatest chance of traveling long distances in the subsurface (see Chapter 7). Generally, several meters of moderately fine-textured, continuous soil layer are necessary for virus reductions of 99.9% or more (Yates and Gerba, 1998).

21.6 WETLANDS AND AQUACULTURE SYSTEMS

Wetlands, which are typically less than 1 m in depth, are areas that support aquatic vegetation and

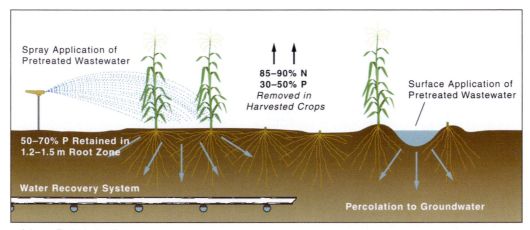

Spray Application of
Pretreated Wastewater

**85–90% N
30–50% P**
*Removed in
Harvested Crops*

Surface Application of
Pretreated Wastewater

**50–70% P Retained in
1.2–1.5 m Root Zone**

Water Recovery System

Percolation to Groundwater

a) Low-Rate Irrigation

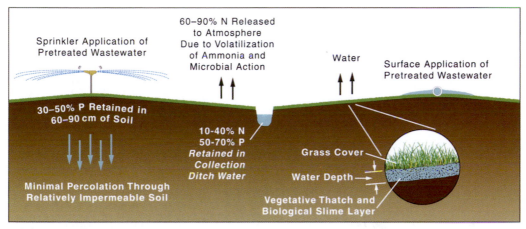

Sprinkler Application of
Pretreated Wastewater

**60–90% N Released
to Atmosphere
Due to Volatilization
of Ammonia and
Microbial Action**

Water

Surface Application of
Pretreated Wastewater

**30–50% P Retained in
60–90 cm of Soil**

**10-40% N
50–70% P
*Retained in
Collection
Ditch Water***

Grass Cover

Water Depth

**Vegetative Thatch and
Biological Slime Layer**

**Minimal Percolation Through
Relatively Impermeable Soil**

b) Overland Flow

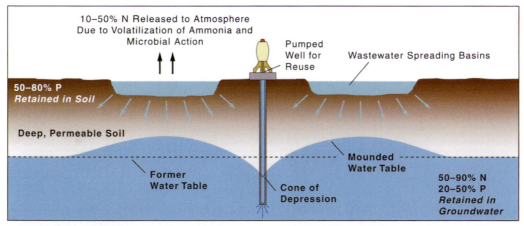

**10–50% N Released to Atmosphere
Due to Volatilization of Ammonia and
Microbial Action**

Pumped
Well for
Reuse

Wastewater Spreading Basins

**50–80% P
*Retained in Soil***

Deep, Permeable Soil

Former
Water Table

Cone of
Depression

Mounded
Water Table

**50–90% N
20–50% P
*Retained in
Groundwater***

c) High-Rate Infiltration

FIGURE 21.18   Three basic methods of land application of wastewater. (From *Pollution Science* © 1996, Academic Press, San Diego, CA.)

CASE STUDY

Rapid Infiltration–Extraction Project in Colton, California

The cities of San Bernardino and Colton, California, are required to filter and disinfect their secondary effluent prior to discharge to the Santa Ana River, which is a source of drinking water and is used for body contact recreation. The two cities joined under the auspices of the Santa Ana Watershed Project Authority (SAWPA) to seek a regional solution to their wastewater treatment requirements. They hoped to develop a cost-effective alternative to conventional tertiary treatment (chemical coagulation, filtration, and disinfection) that would still result in an effluent that was essentially free of measurable levels of pathogens. SAWPA conducted a 1-year demonstration project to examine the feasibility of a rapid infiltration–extraction (RIX) process to treat unchlorinated secondary effluent prior to discharge to the river, and to determine whether or not the RIX process is equivalent—in terms of treatment reliability and quality of the water produced—to the conventional tertiary treatment processes specified in the California Department of Health Services Wastewater Reclamation Criteria.

The demonstration was conducted on a site in Colton, California, adjacent to the Santa Ana River bed. Infiltration basins were built at two sites on the property to allow testing the RIX system under a variety of operating conditions. The soils were coarse sands with clean water infiltration rates of about 15 m/day (50 ft/day). Forty-four monitoring wells were sampled for a number of organic, inorganic, microbiological, and physical measurements.

The study indicted that the optimal filtration rate was 2 m/day (6.6 ft/day) with a wet-to-dry ratio of 1 : 1 (1 day of flooding to 1 day of drying). Mounding beneath the infiltration basins ranged from 0.6 to 0.9 m (2 to 3 ft), and infiltrated wastewater migrated up to 24.4 m (80 ft) vertically and over 30.5 m (100 ft) laterally before it was extracted and had an aquifer residence time of 20 to 45 days, depending on the recharge site. Extraction of 110% of the volume of infiltrated effluent by downgradient extraction wells effectively contained the effluent on the RIX site and minimized mixing with regional groundwater outside the project area.

The soil–aquifer treatment reduced the concentration of total coliform organisms 99 to 99.9% in samples collected approximately 7.6 m (25 ft) below the infiltration basins, and, generally, water from the extraction wells prior to disinfection contained less than 2.2 total coliforms per 100 ml. Although viral levels were as high as 316 viruses per 200 L in the unchlorinated secondary effluent applied to the infiltration basins, only one sample of extracted water prior to disinfection was found to contain detectable levels of viruses. In addition, the turbidity of the extracted water generally averaged less than 0.04 NTU.

Although the RIX process greatly reduced microorganism levels in the wastewater, disinfection of the extracted wastewater still proved to be necessary prior to its discharge to the Santa Ana River. Ultraviolet radiation was evaluated as an alternative to chlorination for disinfection. Due in part to the high quality of the extracted water, ultraviolet radiation was shown to be effective for the destruction of both bacteria and viruses and will probably be used instead of chlorine in the full-scale project (National Research Council, 1994).

foster the growth of emergent plants such as cattails, bulrushes, reeds, sedges, and trees. They also provide important wetland habitat for many animal species. Recently, wetland areas have been receiving increasing attention as a means of additional treatment for secondary effluents. The vegetation provides surfaces for the attachment of bacteria and aids in the filtration and removal of wastewater contaminants. Although both natural and constructed wetlands have been used for wastewater treatment, recent work has focused on constructed wetlands because of regulatory requirements. Two types of constructed wetland systems are in general use: (1) **free water surface (FWS)** systems and (2) **subsurface flow systems (SFS).** An FWS wetland is similar to a natural marsh because the water surface is exposed to the atmosphere. Floating and submerged plants, such as those shown in Fig. 21.19a, may be present. SFSs consist of channels or trenches with relatively impermeable bottoms filled with sand or rock media to support emergent vegetation.

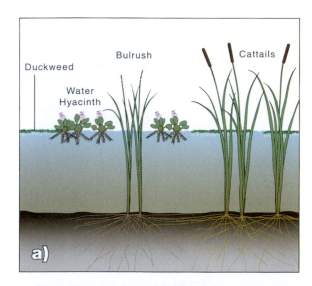

FIGURE 21.20 Sweetwater Site, Tucson, Arizona. This is an example of a subsurface flow wetland used to treat secondary treated wastewater. (Photo courtesy C. P. Gerba.)

FIGURE 21.19 (a) Common aquatic plants used in constructed wetlands. (b) An artificial wetland system in San Diego, California, utilizing water hyacinths. (Photo courtesy C. P. Gerba.)

During wetland treatment, the wastewater is usable. It can, for instance, be used to grow aquatic plants such as water hyacinths (Figs. 21.19b and 21.20) and/or to raise fish for human consumption. The growth of such aquatic plants provides not only additional treatment for the water but also a food source for fish and other animals. Such aquaculture systems, however, tend to require a great deal of land area: moreover, the health risk associated with the production of aquatic animals for human consumption in this manner must be better defined.

There has been increasing interest in the use of natural systems for the treatment of municipal wastewater as a form of tertiary treatment (Kadlec and Knight, 1996). Artificial or constructed wetlands have a higher degree of biological activity than most ecosystems; thus transformation of pollutants into harmless byproducts or essential nutrients for plant growth can take place at a rate that is useful for application to the treatment of municipal wastewater. Most artificial

wetlands in the United States use reeds or bulrushes, although floating aquatic plants such as water hyacinths and duckweed have been used. To reduce potential problems with flying insects, subsurface flow wetlands have also been built (Fig. 21.20). In these types of wetlands all of the flow of the wastewater is below the surface of a gravel bed containing plants tolerant of water-saturated soils. Most of the existing information on the performance of these wetlands concerns coliform and fecal coliform bacteria. Kadlec and Knight (1996) have summarized the existing literature on this topic. They point out that natural sources of indicators in treatment wetlands never reach zero because wetlands are open to wildlife. Reductions in fecal coliforms are generally greater than 99%, but there is a great deal of variation, probably depending on the season, type of wetland, numbers and type of wildlife, and retention time in the wetland. Volume-based and area-based bacterial die-off models have been used to estimate bacterial die-off in surface flow wetlands (Kadlec and Knight, 1996).

Virus removal has been studied in several types of wetlands. Ginsberg *et al.*, (1989) found that coliphage MS-2 was closely related to the removal of suspended solids in a surface flow wetland. In a study of a mixed-species surface flow wetland with a detention time of approximately 4 days, *Cryptosporidium* was reduced by 53%, *Giardia* by 58%, enteric viruses by 98%, and fecal coliform reduction averaged 98% (Karpiscak et al., 1996). In the same study, greater removal of the protozoan parasites (*Giardia* 98% and *Cryptosporidium* 87%) than fecal coliform bacteria (57%) was seen in a duckweed pond with a similar retention time. These differences may reflect greater opportunity for settling of the large parasites in the duckweed pond or input of the parasites from animals in the mixed-species wetland. Additional research is needed to understand the influence of wetland design on its potential for pathogen reduction.

21.7 SOLID WASTE

All major cities in the United States generate two types of solid waste that must be either disposed of or treated and reused: **municipal solid waste (MSW)** and sludge (biosolids). Municipal solid waste originates in households as garbage, in commercial establishments, or at construction or demolition sites. Sludge (biosolids), as just discussed, comes from municipal sewage treatment plants, either as activated sludge or as nonhazardous industrial waste.

21.7.1 Municipal Solid Waste

The Environmental Protection Agency (EPA) estimates that the United States currently produces more than 200 million metric tons of MSW each year, or about 2 kg of trash per person per day. Developed nations, such as the United States and Canada, produce much more discarded material than developing countries. Currently, about 17% of the MSW is recycled; the other 83% is mostly disposed of in landfills or incinerated. It was once hoped that much of the landfilled material would biodegrade. In reality, however, most landfilled waste remains 30–40 years later. Surveys of landfills have shown that conditions at those sites are often not favorable for biodegradation. Such conditions include various factors: low moisture, low oxygen concentration, and high heterogeneity of materials. Further, many waste components are nondegradable or very slow to degrade. Thus, many old landfills serve as "waste repositories," releasing pollutants to the groundwater and the atmosphere.

Despite the fact that many landfills are nearly full, the number of landfills in the United States has been decreasing, from 18,500 in 1979 to 6500 in 1988. In fact, the EPA predicts that by the year 2000, there will only be 3250 active landfills. This decrease is partially attributable to the fact that many landfills have been filled and closed.

Ninety-five percent of landfilling in the United States involves MSW. Municipal solid waste consists of food, animal, and plant residues, as well as miscellaneous household nontoxic wastes such as paper, cans, plastic, metals, and bottles (Fig. 21.21). Until the mid-1980s, these materials were buried or dumped haphazardly within soil or the vadose zone, then simply covered with a layer of soil as a cap (Fig. 21.22a).

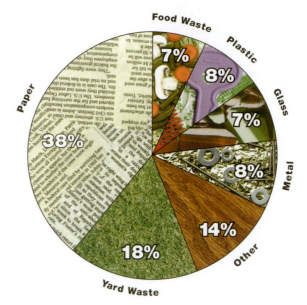

FIGURE 21.21 The nature of household trash (municipal solid waste, MSW) in the United States (on a volume basis). (Data from U.S. EPA, 1994. Reprinted from *Pollution Science* © 1996, Academic Press, San Diego, CA.)

There was no regulation of these landfills, nor were protective mechanisms in place to prevent or minimize releases of pollutants from the landfills.

21.7.2 Modern Sanitary Landfills

Old landfills were usually located in old quarries, mines, natural depressions, or excavated holes in abandoned land. Conditions in these landfills are conducive to pollution, as these sites are often hydrologically connected to surface streams or groundwater water sources. Water pollution is caused by the formation and movement of **leachate,** which is produced by the infiltration of water through the waste material. As the water moves through the waste, it dissolves components of the waste; thus, landfill leachate consists of water containing dissolved chemicals such as salts, heavy metals, and often synthetic organic compounds. Such water–solute transport processes are typically slow, so the effects of leachate migration from old landfills may take years to affect surface and groundwater supplies. In addition to contaminating water, landfills release pollutants into the air. Anaerobic microbial processes generate greenhouse gases, such as nitrous oxide (N_2O), methane (CH_4), and carbon dioxide (CO_2) (see Chapter 14.2.3.3).

A modern **sanitary landfill** is designed to meet exact standards with respect to containment of all materials, including leachates and gases (Fig. 21.22b). New design specifications are predicated on the desire for

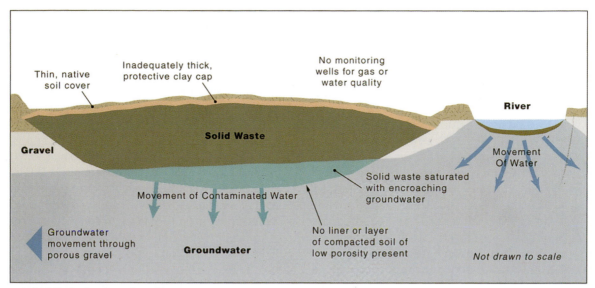

a) Old-Style Sanitary Landfill

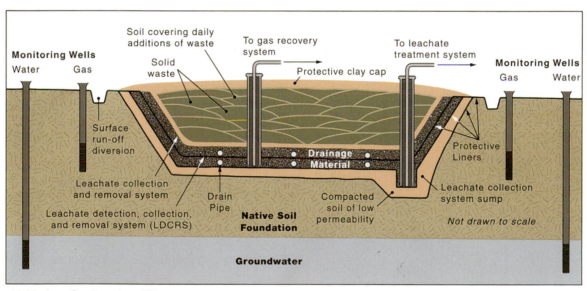

b) Modern Sanitary Landfill

FIGURE 21.22 Landfills. (a) Old-style sanitary landfill, where location was chosen more for of convenience or budgetary concerns than for any environmental considerations. Here, an abandoned gravel pit located near a river exemplifies an all too common siting arrangement. (b) Cross-section of a modern sanitary landfill showing pollutant monitoring wells, a leachate management system, and leachate barriers. In contrast to the old-style site in (a), the modern sanitary landfill emphasizes long-term environmental protection. In addition, whereas landfills were formerly abandoned when full, a modern landfill is monitored long after closure. (From *Pollution Science* © 1996, Academic Press, San Diego, CA.)

minimal impact on the environment, in both the short and long term, with particular emphasis on groundwater protection. Landfill site selection is based on geology and soil type, together with such groundwater considerations as depth of the water table and use. Normally, a new landfill is located in an excavated depression, and fresh garbage is covered daily with a layer of soil. The bottom of the landfill is lined with a low-permeability liner made out of high-density plastic or clay. In addition, provisions are made to collect and analyze leachate and gases that emanate from the landfill.

Municipal solid waste may contain a variety of enteric pathogens originating from pet feces, food waste, garden waste, and disposable diapers. The greatest concentration of fecal indicator bacteria has been observed in paper, garden waste, and food (Pahren, 1987). Pet wastes (1–4% of the solid waste) and disposal baby diapers (0.53 to 1.82% of the solid waste) are also known to contribute enteric pathogens to the waste stream. The concentration of fecal coliforms averages 4.7×10^8 and fecal streptococci 2.5×10^9 per gram of dry MSW (Pahren, 1987), and as much as 10% of the locally soiled disposable diapers may contain enteroviruses. The high temperature (50°C) generated during the decomposition of buried or composted solid waste is probably responsible for the inactivation of most enteric pathogens (Gerba, 1996), although some growth of indicator bacteria may occur because of the abundance of nutrients. Enteric viruses have been observed only in leachates in improperly lined landfills (Sobsey, 1978), not in properly constructed landfills.

21.7.3 Composting of Biosolids and Domestic Solid Waste

One additional way to deal with solid waste is **composting,** in which the organic component of solid waste is biologically decomposed under controlled aerobic conditions. (See Chapter 15.8 for more detail on the microbiology of composting.) Heat produced during this decomposition destroys human pathogens, including many that survive other treatment methods. The result is a product that can be safely handled, easily stored, and readily applied to the land without adversely affecting the environment.

Composting systems are generally divided into three categories: windrow, static pile, and in-vessel. In the **windrow** approach, the MSW mixtures are composted in long rows (called windrows) and aerated by convective air movement and diffusion. In this case, the mixtures are turned periodically by mechanical means to expose the organic matter to ambient oxygen. An overview of a compost facility using the windrow approach is shown in Fig. 21.23 In the **static pile** (or forced-aeration) approach, piles of MSW mixture are aerated by using a forced-aeration system, which is installed under the piles to maintain a minimum oxygen level throughout the compost mass (Fig. 21.24). **In-vessel** composting (also known as mechanical or enclosed-reactor composting) takes place in a partially or completely enclosed container in which environmental conditions can be controlled. In-vessel systems may incorporate the features of windrow and/or static pile methods of composting.

Table 21.15 lists the advantages of each composting system. At present, an estimated 90% of the operational facilities in the United States use static pile composting; the remainder use windrow methods. These two methods both yield stable products, and they are relatively low in cost. However, the more costly in-vessel systems are currently being evaluated in Europe and the United States for large-scale MSW treatment than can meet stringent regulations with respect to pathogens and metals.

Compost—the end product of the composting process—is a stable, humus-like substance with valuable properties as a soil conditioner. It also contains several macro- and micronutrients favorable to plant growth but not enough nitrogen to be considered a fertilizer. Although it is usually pathogen free, the product is not completely stabilized; that is, the organic mater is not 100% degraded. Rather, compost is sufficiently stabilized to reduce the potential for odor generation, thereby allowing the product to be stored and marketed.

For the composting process, large debris and easily recycled materials, such as metals (e.g., aluminum cans, batteries), plastics, and the like, must first be removed. Then the remaining solids are ground and mixed. During the actual composting, the mix is aerated and biological processes decompose the organic matter, thus generating temperatures (above 55°C) high enough to destroy any pathogenic microorganisms that may be present. The oxygen required to fuel the biological processes can be supplied in two ways:

- By mechanically turning the mixture so that the compost is periodically exposed to oxygen in the atmosphere. (In the windrow approach, for example, convective air movement and diffusion move oxygen; thus, turning is necessary to increase porosity, which facilitates air movement, and to distribute anaerobic areas to the aerobic zone.)
- By using a blower to force or draw air through the mix, as in the static pile approach.

Composting usually takes about 3–4 weeks, after which the material is frequently cured for about 30 days. During this phase, further decomposition, stabilization, and degassing take place. Some systems have an additional drying stage, which can vary from a few days to several months. During composting, fungi may grow to large numbers, especially members of the genus *Aspergillus*.

Aspergillus fumigatus is commonly found in composting vegetation, wood chip piles, MSW compost, refuse sludge compost, and moldy hay. It grows over a range of 20 to about 50°C. Spores of this fungus can cause bronchopulmonary hypersensitivity, marked by asthmatic spasm, fever, and malaise.

Most of the information about the occurrence of *A.*

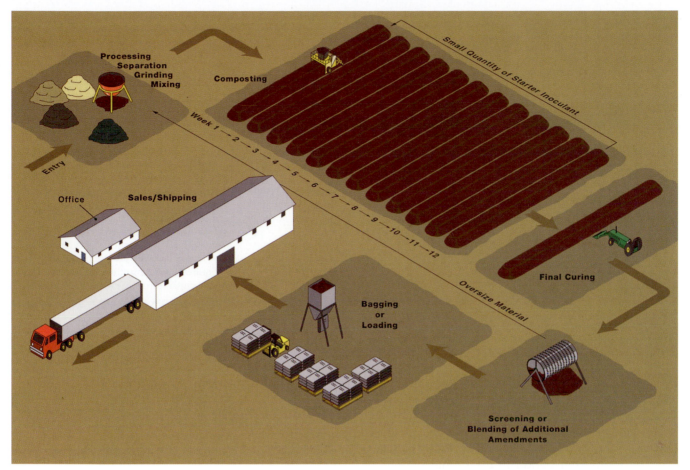

FIGURE 21.23   Overview of a compost facility using the windrow process. Such a system is appropriate for both municipal and agricultural waste. See Section 16.10. Reprinted by permission from "The Guidebook: Nitrogen Best Management Practices for Concentrated Animal Feeding Operations in Arizona," by Robert J. Frietas and Thomas A. Doerge; © 1996 by the Arizona Board of Regents. All rights reserved.

fumigatus in compost comes from studies of sewage sludge composting. Millner *et al.* (1977) were the first to study the occurrence of *A. fumigatus* in various stages of aerated static composting. The composting mixture consisted of raw sewage sludge mixed with wood chips as a bulking agent. Wood chips appeared to be the major source of the fungus. Samples of old wood chips contained as many as 6.1×10^7 CFU/g dry weight. Semiquantitative studies of air at the composting site indicated that *A. fumigatus* constituted 75% of the total viable microflora. This decreased to about 2% at a distance of 320 m to 8 km from the site. In a later study, the same authors (Millner *et al.*, 1980) demonstrated that aerosol concentrations related to wind-blown losses from stationary compost piles are relatively small in comparison with those generated during mechanical movement of the piles. After mixing of the compost, *A. fumigatus* concentrations at 3 and 30 m downwind of the static piles were about 33 to 1800 times less than those measured during pile

movement and not significantly above background levels. The authors estimated a maximum emission of 4.6×10^6 *A. fumigatus* particles per second during pile moving operations.

An epidemiological study of workers at composting facilities showed that the workers did not experience any ill effects in comparison with a control group (Clark, *et al.*, 1984). Minor effects were observed, however, consisting of skin irritation and inflammation of the nose and eyes that resulted from exposure to dust containing *A. fumigatus*. To reduce potential risks for susceptible individuals, it has been recommended that personnel working in areas of high agitation such as mixing and screening buildings be evaluated for the development of sensitivity to *Aspergillus* (Haug, 1993).

Composting is a thermophilic process, and the temperatures achieved during this treatment (55 to 70°C) are believed to be adequate to kill enteric pathogens. However, for the process to be successful in pathogen elimination, it is important that all of the material un-

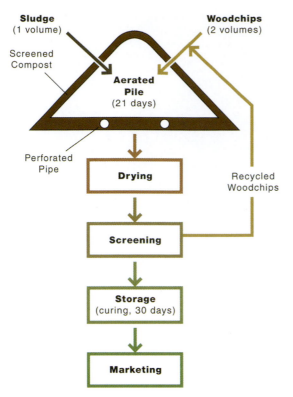

FIGURE 21.24 Static aerated pile composting method. (Adapted from Bitton, 1980.)

TABLE 21.15   Advantages of Three Composting Systems

Windrow systems	Rapid drying of the compost because moisture is released as the piles are turned over.
	Drier compost material, which results in easier separation of bulking agent from the compost during screening and relatively high rates of recovery for bulking materials
	High volume of material can be utilized.
	Good product stabilization.
	Relatively low capital investment: materials required are a pad for the piles, a windrow machine, and generally a front-end loader.
Static pile systems	Low capital costs. The capital equipment required consists of a paved surface, some front-end loaders, a screen, relatively inexpensive blowers, and a water trap.
	A high degree of pathogen destruction. The insulation over the pile and uniform aeration throughout the pile help maintain pile temperatures that destroy pathogens.
In-vessel systems	Better odor control than windrow composting. The pile is kept aerobic at all times. Also, with blowers in the suction mode, the odors can be treated as a point source.
	Good product stabilization. Oxygen and temperature can be maintained at optimal levels.
	Space efficiency.
	Better process control than outdoor operations.
	Protection from adverse climatic conditions.
	Good odor control should be possible.
	Potential heat recovery depending on system design.

Modified from U.S. EPA (1985).

dergoing composting be exposed to these temperatures for a sufficient period of time. Cold pockets or zones in nonreactor systems or inadequate resident times in reactor systems may allow pathogens to escape thermal destruction.

Most of the research on pathogen fate has been on composted sewage sludge. Sewage sludge is likely to contain higher concentrations of enteric pathogens than MSW. In a year-long study of the occurrence of enteric bacteria, viruses, and *Ascaris* eggs in both windrow and static pile composting of sewage sludge, only viruses were detected (Gerba, 1996). It was concluded that viruses could survive 25 days of composting if the composting mass did not achieve adequately high temperatures. Fecal coliforms were also found in high concentrations if maximum composting temperatures were less than 50°C. Research suggests that temperatures maintained above 53°C for 3 days are sufficient to eliminate enteric pathogens including *Ascaris* eggs (Haug, 1993). It has been found, however, that enteric bacteria can regrow in composted material when temperatures are reduced to sublethal levels. This phenomenon has been reported for total and fecal coliforms, *Salmonella,* and fecal streptococci in liquid sewage sludges. Regrowth may occur through bacteria surviving at undetectable levels or through recontamination from an outside source. Regrowth of *Salmonella*

in composted sewage sludges has been noted in several studies (Russ and Yanko, 1981). The active, indigenous flora of composts act as a barrier to colonization by *Salmonella*. Only in the absence of competing flora, such as with sterilization, were reinoculated *Salmonella* able to regrow.

QUESTIONS AND PROBLEMS

1. What are the three major steps in modern wastewater treatment?

2. Why is it important to reduce the amount of biodegradable organic matter and nutrients during sewage treatment?

3. When would tertiary treatment of wastewater be necessary?

4. What are some types of tertiary treatment?

5. What are some methods of sludge (biosolids) reuse?

6. What is the major component of biosolids?

7. What conditions are necessary for nitrogen removal during the activated sludge process?

8. What are the three types of land application of wastewater? Which one is most likely to contaminate the groundwater with enteric viruses? Why? What factors determine how far viruses will be transported in groundwater?

9. What are the major sources of enteric pathogens in solid waste?

10. What factors may determine the concentration of enteric pathogens in domestic raw sewage?

11. What is the major factor that causes destruction of enteric pathogens during composting? Which enteric bacterial pathogen is capable of regrowth?

12. Five milliliters of a wastewater sample is added to dilution water in a 300-ml BOD bottle. If the following results are obtained, what is the BOD after 3 days and 5 days?

Time (days)	Dissolved oxygen (mg/l)
0	9.55
1	4.57
2	4.00
3	3.20
4	2.60
5	2.40
6	2.10

13. Are pathogen concentrations greater in treated wastewater or undigested sludge? Why?

14. How does removal of pathogens occur during activated sludge treatment?

References and Recommended Readings

APHA. (1995) "Standard Methods for Water and Wastewater." American Public Health Association, Washington, DC.

Bitton, G. (1980) "Introduction to Environmental Virology." John Wiley & Sons, New York.

Bitton, G. (1994) "Wastewater Microbiology." Wiley-Liss, New York.

Casson. L. W., Sorber, C. A., Sykora, J. L., Cavaghan, P. D., Shapiro, M. A., and Jakubowski, W. (1990) Giardia in wastewater—Effect of treatment. J. Water Pollut. Control Fed. **62,** 670–675.

Clark, C. S., Bjornson, H. S., Schwartz-Fulton, J., Holland, J. W., and Garside, P. S. (1984) Biological health risks associated with the composting of wastewater treatment plant sludge. J. Water Poll. Control Fed. **56,** 1269–1276.

Curds, C. R., and Hawkes, H. A., eds. (1983) "Ecological Aspects of Used-Water Treatment," Vol. 2. Academic Press, London.

Deakyne, C. W., Patel, M. A., and Krichten, D. J. (1984) Pilot plant demonstration of biological phosphorus removal. J. Water Pollut. Control Fed. **56,** 867–873.

Enriquez, V., Rose, J. B., Enriquez, C. E., and Gerba, C. P. (1995) Occurrence of Cryptosporidium and Giardia in secondary and tertiary wastewater effluents. In "Protozoan Parasites and Water" (W. B. Betts, D. Casemore, C. Fricker, H. Smith, and J. Watkins, eds.) Royal Society of Chemistry, Cambridge, UK, pp. 84–86.

Freitas, R. J., and Doerge, T. A. (1996) "The Guidebook: Nitrogen Best Management Practices for Concentrated Animal Feeding Operations in Arizona." Arizona Board of Regents, Tucson, AZ.

Funderburg, S. W., Moore, B. C., Sorber, C. A., and Sagik, B. P. (1978) Survival of poliovirus in model wastewater hold pond. Prog. Water Technol. **10,** 619–629.

Gerba, C. P. (1996) Microbial pathogens in municipal solid waste. In "Microbiology of Solid Waste" (A. C. Palmisano, and M. A. Berlaz, eds.) CRC Press, Boca Raton, FL, pp. 155–174.

Gerba, C. P., Sobsey, M. D., Wallis, C., and Melnick, J. L. (1975) Factors influencing the adsorption of poliovirus onto activated carbon in wastewater. Environ. Sci. Technol. **9,** 727–731.

Ginsburg, R. M., Gearheat, R. A., and Ives, M. (1989) Pathogen removal in constructed wetlands. In "Constructed Wetlands for Wastewater Treatment: Municipal, Industrial, and Agricultural" (D. A. Hammer, ed.) Lewis, Chelsea, MI, pp. 431–445.

Grimason, A. M., Smith, H. V., Thitai, W. N., Smith, P. G., Jackson, M. H., and Girdwood, R. W. A. (1993) Occurrence and removal of Cryptosporidium spp. oocysts and Giardia cysts in Kenyan waste stabilization ponds. Water Sci. Technol. **27,** 97–104.

Hammer, M. J. (1986) "Water and Wastewater Technology." John Wiley & Sons, New York.

Haug, R. T. (1993) "The Practical Handbook of Compost Engineering." Lewis, Chelsea, MI.

Hurst, C. J., Gerba, C. P., and Cech, I. (1980) Effects of environmental variables and soil characteristics on virus survival in soil. Appl. Environ. Microbiol. **40,** 1067–1079.

Kadlec, R. H., and Knight, R. L. (1996) "Treatment Wetlands." Lewis, Boca Raton, FL.

Karpiscak, M. M., Gerba, C. P., Watt, P. M., Foster, K. E., and Falabi, J. A. (1996) Multi-species plant systems for wastewater quality improvements and habitat enhancement. Water Sci. Technol. **33,** 231–236.

Leong, L. Y. C. (1983) Removal and inactivation of viruses by treatment processes for portable water and wastewater—A review. Water Sci. Technol. **15,** 91–114.

Marais, G. V. R. (1974) Fecal bacteria kinetics in stabilization ponds. J. Sanit. Eng. Div. Am. Soc. Civ. Eng. **100,** 119–139.

Metcalf and Eddy, Inc. (1991) "Wastewater Engineering." McGraw-Hill, New York.

Millner, P. D., Bassett, D. A., and Marsh, P. B. (1980) Dispersal of Aspergillus-fumigatis from sewage-sludge compost piles subjected to mechanical agitation in open air. Appl. Environ. Microbiol. **39,** 1000–1009.

Millner, P. D., Marsh, P. B., Snowden, R. B., and Parr, J. F. (1977) Occurrence of Aspergillus fumigatus during composting sewage sludge. Appl. Environ. Microbiol. **34,** 765–772.

Mills, S. W., Alabaster, G. P., Mara, D. D., Pearson, H. W., and Thitai, W. N. (1992) Efficiency of fecal bacterial removal in waste stabilization ponds in Kenya. Water Sci. Technol. **26,** 1739–1748.

Modore, M. S., Rose, J. B., Gerba, C. P., Arrowood, M. J., and Sterling, C. R. (1987) Occurrence of Cryptosporidium in sewage effluents and selected surface waters. J. Parasitol. **73,** 702–705.

Moore, B. E., Sagik, B. P., and Sorber, C. A. (1981) Viral transport to groundwater at a wastewater land application site. J. Water Pollut. Control Fed. **53,** 1492–1502.

Naranjo, J. E., Gerba, C. P., Bradford, S. M., and Irwin, J. (1993) Virus removal by an on-site wastewater treatment and recycling system. Water Sci. Technol. **27,** 441–444.

National Research Council (NRC). (1994) "Ground Water Recharge Using Waters of Impaired Quality." National Academy Press, Washington, DC.

Outwater, A. B. (1994) "Reuse of Sludge and Minor Wastewater Residuals." Lewis, Boca Raton, FL.

Pahren, H. R. (1987) Microorganisms in municipal solid waste and public health implications. *CRC Crit. Rev. Environ. Control* **17**, 187–228.

Pepper, I. L., Gerba, C. P., and Brusseau, M. L. (1996) "Pollution Science." Academic Press, San Diego.

Rao, V. C., Metcalf, T. G., and Melnick, J. L. (1986) Removal of pathogens during wastewater treatment. *In* "Biotechnology," Vol. 8 (H. J. Rehm and G. Reed, eds.) VCH, Berlin, pp. 531–554.

Robertson, L. J., Smith, H. V., and Paton, C. A. (1995) Occurrence of *Giardia* cysts and *Cryptosporidium* oocysts in sewage influent and six sewage treatment plants in Scotland and prevalence of cryptosporidiosis and giardiasis diagnosed in the communities by those plants. *In* "Protozoan Parasites and Water" (W. B. Betts, D. Casemore, C. Fricker, H. Smith, and J. Watkins, eds.) Royal Society of Chemistry, Cambridge, UK, pp. 47–49.

Rose, J. B., and Carnahan, R. P. (1992) Pathogen removal by full scale wastewater treatment. Report to Florida Department of Environmental Regulation, Tallahassee, FL.

Russ, C. F., and Yanko, W. A. (1981) Factors affecting *Salmonella* repopulation in composted sludges. *Appl. Environ. Microbiol.* **41**, 597–602.

Sattar, S. A., Ramia, S., and Westwood, J. C. N. (1976) Calcium hydroxide (lime) and the elimination of human pathogenic viruses from sewage: Studies and experimentally-contaminated (poliovirus type 1), Sabin, and pilot plant studies. *Can. J. Microbiol.* **67**, 221–226.

Schwartzbrod, J., Stein, J. L., Bouhoum, K., and Baleux, G. (1989) Impact of wastewater treatment on helminth eggs. *Water Sci. Technol.* **21**, 295–297.

Sobsey, M. D. (1978) Field survey of enteric viruses in solid waste landfill leachates. *Am. J. Public Health* **68**, 858–864.

Straub, T. M., Pepper, I. L., and Gerba, C. P. (1993) Hazards of pathogenic microorganisms in land disposed sewage sludge. *Rev. Environ. Contam. Toxicol.* **132**, 55–91.

Tanaka, H., Asano, T., Schroeder, E. D., and Tchobanglous, G. (1998) Estimating the reliability of wastewater reclamation and reuse using enteric virus monitoring data. *Water Environ. Res.* **70**, 39–51.

U.S. EPA. (1974) Process design manual for sludge treatment and disposal. Technology transfer. EPA 625/1–74–006, Washington, DC.

U.S. EPA. (1975) Process design manual for nitrogen control. Office of Technology Transfer, Washington, DC.

U.S. EPA. (1977) Wastewater treatment facilities for sewered small communities. EPA–62511–77–009, Washington, DC.

U.S. EPA. (1985) Composting of municipal wastewater sludges. EPA 62514–85-014, Washington, DC.

U.S. EPA. (1988) Comparative health risks effects assessment of drinking water. U. S. Environmental Protection Agency, Washington, DC.

U.S. EPA. (1992) United States Environmental Protection Agency. Guidelines for water reuse. EPA/625/R–92/004, Washington, DC.

Yahya, M. T., Cluff, C. B., and Gerba, C. P. (1993) Virus removal by slow sand filtration and nanofiltration. *Water Sci. Technol.* **24**, 445–448.

Yates, M. V. (1994) Monitoring concerns and procedures for human health effects. *In* "Wastewater Reuse for Golf Course Irrigation" CRC Press, Boca Raton, FL, pp. 143–171.

Yates, M. V., and C. P. Gerba. (1998) Microbial considerations in wastewater reclamation and reuse. *In* "Wastewater Reclamation and Reuse" (T. Asano, ed.) Technomic, Lancaster, PA, pp. 437–488.

22

Drinking Water Treatment and Distribution

CHARLES P. GERBA

Rivers, streams, lakes, and underground aquifers are all potential sources of potable water. In the United States, all water obtained from surface sources must be filtered and disinfected to protect against the threat of microbiological contaminants. (Note: Such treatment of surface waters also improves esthetic values such as taste, color, and odors.) In addition, groundwater under the direct influence of surface waters (e.g., nearby rivers) must be treated as if it were a surface supply. In many cases, however, groundwater needs either no treatment or only disinfection before use as drinking water. This is because soil itself acts as a filter to remove pathogenic microorganisms, decreasing their chances of contaminating drinking supplies.

At first, slow sand filtration was the only means employed for purifying public water supplies. Then, when Louis Pasteur and Robert Koch developed the germ theory of disease in the 1870s, things began to move more quickly. In 1881, Koch demonstrated in the laboratory that chlorine could kill bacteria. Following an outbreak of typhoid fever in London, continuous chlorination of a public water supply was used for the first time in 1905 (Montgomery, 1985). The regular use of disinfection in the United States began in Chicago in 1908. The application of modern water treatment processes had a major impact on water-transmitted diseases such as typhoid in the United States (Fig. 22.1).

22.1 WATER TREATMENT PROCESSES

Modern water treatment processes provide barriers, or lines of defense, between the consumer and waterborne disease. These barriers, when implemented as a succession of treatment processes, are known collectively as a **treatment process train** (Fig. 22.2). The simplest treatment process train, known as **chlorination,** consists of a single treatment process, disinfection by chlorination (Fig. 22.2a). The treatment process train known as **filtration,** entails chlorination followed by filtration through sand or coal, which removes particulate matter from the water and reduces turbidity (Fig. 22.2b). At the next level of treatment, **in-line filtration,** a coagulant is added prior to filtration (Fig. 22.2c). Coagulation alters the physical and chemical state of dissolved and suspended solids and facilitates their removal by filtration. More conservative water treatment plants add a flocculation (stirring) step before filtration, which enhances the agglomeration of particles and further improves the removal efficiency in a treatment process train called **direct filtration** (Fig. 22.2d). In direct filtration, disinfection is enhanced by adding chlorine (or an alternative disinfectant, such as chlorine dioxide or ozone) at both the beginning and end of the process train. The most common treatment process train for surface water supplies, known as conventional treatment, consists of disinfection, coagulation, flocculation, sedimentation, filtration, and disinfection (Fig. 22.2e).

As already mentioned, **coagulation** involves the addition of chemicals to facilitate the removal of dissolved and suspended solids by sedimentation and filtration. The most common primary coagulants are hydrolyzing metal salts, most notably alum $[Al_2(SO_4)_3 - 14H_2O]$, ferric sulfate $[Fe_2(SO_4)_3]$, and ferric chloride $(FeCl_3)$. Additional chemicals that may be added to enhance coagulation are charged organic

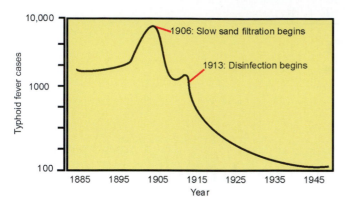

FIGURE 22.1 Impact of water filtration and chlorination on typhoid fever death rate in Albany, New York. (From Logsdon and Lippy, 1982.)

molecules called polyelectrolytes; these include high-molecular-weight polyacrylamides, dimethyldiallyl-ammonium chloride, polyamines, and starch. These chemicals ensure the aggregation of the suspended solids during the next treatment step, flocculation.

Sometimes polyelectrolytes (usually polyacrylamides) are added after flocculation and sedimentation as an aid in the filtration step.

Coagulation can also remove dissolved organic and inorganic compounds. Hydrolyzing metal salts added to the water may react with the organic matter to form a precipitate, or they may form aluminum hydroxide or ferric hydroxide floc particles on which the organic molecules adsorb. The organic substances are then removed by sedimentation and filtration, or filtration alone if direct filtration or in-line filtration is used. Adsorption and precipitation also remove inorganic substances.

Flocculation is a purely physical process in which the treated water is gently stirred to increase interparticle collisions, thus promoting the formation of large particles. After adequate flocculation, most of the aggregates settle out during the 1 to 2 hours of sedimentation. Microorganisms are entrapped or adsorbed to the suspended particles and removed during sedimentation (Fig. 22.3).

Sedimentation is another purely physical process,

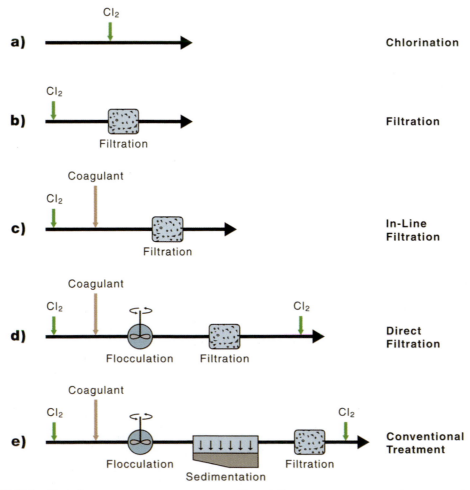

FIGURE 22.2 Typical water treatment process trains. (From *Pollution Science* © 1996, Academic Press, San Diego, CA.)

FIGURE 22.3 Drinking water treatment plant showing sand filter beds in the foreground and tanks containing alum flocculant in the background. (Photo courtesy C. P. Gerba.)

TABLE 22.2 Coagulation, Sedimentation, Filtration: Typical Removal Efficiencies and Effluent Quality

Organisms	Coagulation and sedimentation (% removal)	Rapid filtration (% removal)	Slow sand filtration (% removal)
Total coliforms	74–97	50–98	>99.999
Fecal coliforms	76–83	50–98	>99.999
Enteric viruses	88–95	10–99	>99.999
Giardia	58–99	97–99.9	>99
Cryptosporidium	90	99–99	99

From U.S. EPA (1988).

involving the gravitational settling of suspended particles that are denser than water. The resulting effluent is then subjected to **rapid filtration** to separate out solids that are still suspended in the water. Rapid filters typically consist of 50–75 cm of sand and/or anthracite having a diameter between 0.5 and 1.0 mm (Fig 22.3). Particles are removed as water is filtered through the medium at rates of 4–24/min/10 dm². Filters need to be backwashed on a regular basis to remove the buildup of suspended matter. This backwash water may also contain significant concentrations of pathogens removed by the filtration process. Rapid filtration is commonly used in the United States. Another method, **slow sand filtration,** is also used. Employed primarily in the United Kingdom and Europe, this method operates at low filtration rates without the use of coagulation. Slow sand filters contain a layer of sand (60–120 cm deep) supported by a gravel layer (30–50 cm deep). The hydraulic loading rate is between 0.04 and 0.4 m/hr. The buildup of a biologically active layer, called a *schmutzdecke*, occurs during the operation of a slow sand filter. This eventually leads to head loss across the filter, requiring removing or scraping the top layer of sand. Factors that influence pathogen removal by filtration are shown in Table 22.1.

Taken together, coagulation, flocculation, sedimentation, and filtration effectively remove many contaminants as shown in Tables 22.2 and 22.3. Equally important, they reduce turbidity, yielding water of good clarity and hence enhanced disinfection efficiency. If not removed by such methods, particles may harbor microorganisms and make final disinfection more difficult. Filtration is an especially important barrier in the removal of the protozoan parasites *Giardia lamblia* and *Cryptosporidium* (Fig. 22.4). The cysts and oocysts of these organisms are very resistant to inactivation by disinfectants, so disinfection alone cannot be relied on to prevent waterborne illness (see Case Study). However, because of their smaller size, viruses and bacteria can pass through the filtration process. Removal of viruses by filtration and coagulation depends on their attachment to particles (adsorption), which is dependent on the surface charge of the virus. This is related to the isoelectric point (the pH at which the virus has no charge) and is both strain and type dependent (see also Chapter 7.2.4). The variations in surface properties have been used to explain why different types of viruses are removed with different efficiencies by coagulation and filtration. Thus, dis-infection remains the ultimate barrier to these microorganisms.

TABLE 22.1 Factors Effecting the Removal of Pathogens by Slow Sand Filters

Temperature

Sand grain size

Filter depth

Flow rate

Well-developed biofilm layer

TABLE 22.3 Removal of Virus by Coagulation–Settling–Sand Filtration

Virus	Viral assays, total PFU/200 l (percentage removed)		
	Input	Settled water	Filtered water
Poliovirus	5.2×10^7	1.0×10^6 (98)	8.7×10^4 (99.84)
Rotavirus	9.3×10^7	4.6×10^6 (95)	1.3×10^4 (99.987)
Hepatitis A virus	4.9×10^{10}	1.6×10^9 (97)	7.0×10^8 (98.6)

Adapted from Rao *et al.* (1988).

CASE STUDY

Cryptosporidiosis in Milwaukee

Early in the spring of 1993, heavy rains flooded the rich agricultural plains of Wisconsin. These rains produced an abnormal runoff into a river that drains into Lake Michigan, from which the city of Milwaukee obtains its drinking water. The city's water treatment plant seemed able to handle the extra load: it had never failed before, and all existing water quality standards for drinking water were properly met. Nevertheless, by April 1, thousands of Milwaukee residents came down with acute watery diarrhea, often accompanied by abdominal cramping, nausea, vomiting, and fever. In a short period of time, more than 400,000 people developed gastroenteritis, and as many as 100—mostly immunocompromised individuals—ultimately died. Finally, after much testing, it was discovered that *Cryptosporidium* oocysts were present in the finished drinking water after treatment. These findings pointed to the water supply as the likely source of infection; and on the evening of April 7, the city put out an urgent advisory for residents to boil their water. This measure effectively ended the outbreak. All told, direct costs and loss of life are believed to have exceeded $150 million dollars.

The Milwaukee episode was the largest waterborne outbreak of disease ever documented in the United States. But what happened? How could such a massive outbreak occur in a modern U.S. city in the 1990s? And how could so many people die? Apparently, high concentrations of suspended matter and oocysts in the raw water resulted in failure of the water treatment processes-a failure in which *Cryptosporidium* oocysts passed right through the filtration system in one of the city's water treatment plants, thereby affecting a large segment of the population. Among this general population were many whose systems could not withstand the resulting illness. In immunocompetent people, cryptosporidiosis is a self-limiting illness; it is very uncomfortable, but it goes away of its own accord. However, in the immunocompromised, cryptosporidiosis can be unrelenting and fatal.

Generally, disinfection is accomplished through the addition of an oxidant. Chlorine is by far the most common disinfectant used to treat drinking water; but other oxidants, such as chloramines, chlorine dioxide, and even ozone, are also used (Table 22.4).

FIGURE 22.4 Cattle are potential source of *Cryptosporidium* that have been associated with waterborne outbreaks. (Photo courtesy P. E. Gerba.)

22.2 WATER DISTRIBUTION SYSTEMS

Once drinking water is treated, it must often travel through many miles of pipe or be held in storage reservoirs before it reaches the consumer. The presence of dissolved organic compounds in this water can cause problems, such as taste and odors, enhanced chlorine demand, and bacterial colonization of water distribution systems (Bitton, 1994). Biofilms of microorganisms in the distribution are of concern because of the potential for protection of pathogens from the action of

TABLE 22.4   Methods of Disinfection

Drinking water systems serving a population > 10,000	
Chlorination	70%
Chloramination	25%
Chlorine dioxide	5%
Ozonation	1%
Drinking water systems serving a population < 10,000	
Chlorination	>95%

From Craun, 1993.

residual disinfectant in the water and the regrowth of indicator bacteria such as coliforms (Table 22.5).

Biofilms may appear as a patchy mass in some pipe sections or as uniform layers (see Chapter 6.2.4.1). They may consist of a monolayer of cells in a microcolony or can be as thick as 10 to 40 mm, as in algal mats at the bottom of a reservoir (Geldreich, 1996). These biofilms often provide a variety of microenvironments for growth that include aerobic and anaerobic zones because of oxygen limitations within the biofilm. Growth of biofilms proceeds up to a critical thickness, at which nutrient diffusion across the biofilm becomes limiting. Biofilm microorganisms are held together by an extracellular polymeric matrix called a **glycocalyx.** The glycocalyx is composed of glucans, uronic acids, glycoproteins, and mannans (see Chapter 2.2.4.2). The glycocalyx helps protects microorganisms from predation and adverse conditions (e.g., disinfectants) (see Chapter 15.2.2).

The occurrence of even low levels of organic matter in the distribution system allows the growth of biofilm microorganisms. Factors controlling the growth of these organisms are temperature, water hardness, pH, redox potential, dissolved carbon, and residual disinfectant .

It has been demonstrated that biofilms can coexist with chlorine residuals in distribution systems (Geldreich, 1996). *Escherichia coli* is 2400 times more resistant to chlorine when attached to surfaces than as free cells in the water, leading to high survival rates within the distribution system (Le Chevallier *et al.*, 1988a). The health significance of coliform growth in distribution systems is an important consideration for water utilities, because the presence of these bacteria may mask the presence of indicator bacteria in water supplies resulting from a breakdown in treatment barriers. Total numbers of heterotrophic bacteria growing in the water or biofilm may interfere with the detection of coliform bacteria. High heterotrophic plate counts (HPC) are also indicative of a deterioration of water quality in the distribution system. It has been recommended that HPC numbers not exceed 500 organisms per milliliter.

Biofilms in distribution systems are difficult to inactivate. Chlorine levels commonly used in water treatment are inadequate to control biofilms. Free chlorine levels as high as 4.3 mg/l have proved inefficient in eliminating coliform occurrence. It has been suggested that monochloramine may be more effective in controlling biofilms because of its ability to penetrate the biofilm (LeChevallier *et al.*, 1988b).

Water-based pathogens such as *Legionella* may also have the ability to grow in biofilms and the distribution system. *Legionella* is known to occur in distribution systems and colonize the plumbing, faucet fixtures, and taps in homes (Colbourne *et al.*, 1988). *Legionella* are more resistant to chlorine than *E. coli,* and small numbers may survive in the distribution system. Hot-water tanks in homes and hospitals favor their growth. *Legionella* survives well at 50°C, and it is capable of growth at 42°C (Yee and Wadowsky, 1982). It has been found that sediments in water distribution systems and the natural microflora favor their survival. *Legionella* associated with hot-water systems in hospitals have caused numerous outbreaks of disease in immunocompromised patients. Sporadic cases in communities show a strong association with the colonization of household taps. Forty percent of sporidic cases of this illness may be related to household taps or showers.

22.3 ASSIMILABLE ORGANIC CARBON

Bacterial growth in distribution systems is influenced by the concentration of biodegradable organic matter, water temperatures, nature of the pipes, disinfectant residual concentration, and detention time within the distribution system (Bitton, 1994). Bacteria such as *Pseudomonas aeruginosa* and *P. fluorescens* are able to grow in tap water at relatively low concentrations (μg/l) of low-molecular-weight organic substrates such as acetate, lactate, succinate, and amino acids. The amount of biodegradable organic matter available to microorganisms is difficult to determine from data from dissolved organic carbon (DOC) or total organic carbon (TOC) measurements. These measurements capture only the bulk water portion of organic matter, of which only some is biodegradable.

Several bioassay tests have been proposed using either pure cultures of selected bacteria or mixed flora from the source water for assessment of the **assimilable organic carbon (AOC)** in water (Geldreich, 1996). Measurements of bacterial action in the test sample

TABLE 22.5 Problems Caused by Biofilms in Distribution Systems

Frictional resistance of fluids

Photoreduction of H_2S because of anaerobic conditions

Taste and odor problems

Colored water (red, black) from activity of iron- and manganese-oxidizing bacteria

Resistance to disinfection

Regrowth of coliform bacteria

Growth of pathogenic bacteria (i.e., *Legionella*)

TABLE 22.6 Concentrations of Assimilable Organic Carbon (AOC) in Various Water Samples

Source of water	Dissolved organic carbon (DOC), mg C/l[a]	Assimilable organic carbon (AOC), mg C/l
River Lek	6.8	0.062–0.085
River Meuse	4.7	0.118–0.128
Brabantse Diesbosch	4.0	0.08–0.103
Lake Yssel, after open storage	5.6	0.48–0.53
River Lek, after bank filtration	1.6	0.7–1.2
Aerobic groundwater	0.3	<0.15

Adapted from van der Kooij *et al.* (1982).
[a] DOC, dissolved organic carbon; C/l, carbon/liter.

over time are determined by plate counts, direct cell count, ATP, turbidity, etc. It is estimated that the assimilable organic carbon in tap water is between 0.1 and 9% of the total organic carbon (van der Kooij *et al.*, 1982), although this fraction may be higher if the treatment train involves ozonation, which breaks down complex organic molecules and makes them more available to microorganisms (Table 22.6 and Fig. 22.5). In fact, rapid regrowth of heterotrophic plate counts usually occurs after ozonation of tapwater. Addition of a secondary disinfectant is required to control this regrowth.

Because determination of AOC levels based on *P. fluorescens* does not always appear to be a good indicator of the growth potential for coliforms, a **coliform growth response (CGR)** test has been developed (Geldreich, 1996) (Fig. 22.5). This procedure uses *Enterobacter cloacae* as the bioassay organism. Changes in viable densities of this organism in the test over a 5-day period at 20°C are used to develop an index of nutrients available to support coliform biofilm growth. The CGR result is calculated by log transformation of the ratio of the colony density achieved at the end of the incubation period to the initial cell concentration. Thus:

$$CGR = \log(N_5/N_0) \qquad \text{(Eq. 22.1)}$$

where N_5 = number of CFU per milliliter at day 5 and N_0 = number of CFU per milliliter at day 0. Any sample that demonstrates a 1-log or greater increase is interpreted as supporting coliform growth. Calculated values between 0.51 and 0.99 are considered to be moderately growth supportive, and those less than 0.5 are regarded as not supportive of coliform growth.

It is important to note that the CGR test responds only to the concentrations of assimilable organic materials that support growth of coliforms characteristic of regrowth in biofilms. In fact, parallel assays comparing *E. coli* response with that of *Ent. cloacae* indicate a significant difference in growth response between these two coliforms. *Ent. cloacae* growth can occur in nutrient concentrations far below those required by *E. coli*.

Detecting any changes in the dissolved organic concentrations is another approach to obtaining a measure of assimilable organic carbon. Rather than using pure cultures, these procedures utilize the indigenous microflora of the raw surface-water source of the biomass washed from sand used in the sand filter during drinking water treatment.

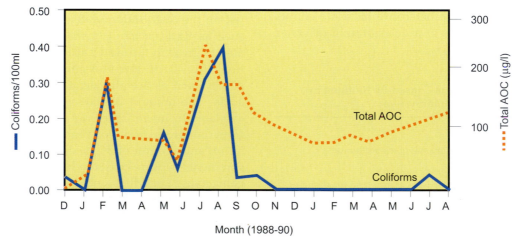

FIGURE 22.5 Relationship between mean coliform densities and total assimilable organic carbon (AOC) levels. (From LeChevallier *et al.*, 1992.)

The biodegradable dissolved organic carbon (BDOC) is given by the following formula:

$$BDOC \ (mg/l) = initial \ DOC - final \ DOC \qquad (Eq. \ 22.2)$$

The general approach is as follows. A water sample is sterilized by filtration through a 0.2-μm pore size filter, inoculated with indigenous microorganisms, and incubated in the dark at $20°C$ for 10–30 days until DOC reaches a constant level. BDOC is the difference between the initial and final DOC values (Servais *et al.*, 1987). The advantage of using the consortium of heterotrophic microorganisms that occurs in these aquatic habitats is in their acquired proficiency to degrade a diverse spectrum of dissolved organics that may be in test samples. Whether the inoculum is from the raw source water or the sand filter biofilm has not been found to be critical for optimal test performance.

QUESTIONS AND PROBLEMS

1. Why is it important to reduce the amount of biodegradable organic matter and nutrients during water treatment?

2. Describe the major steps in the conventional treatment of drinking water.

3. Why is growth of bacteria in the distribution system a concern?

4. What methods can be used to assess the growth of bacteria in water?

5. Which pathogenic microorganisms are the most difficult to remove by conventional water treatment and why?

6. What are biofilms and why are they important in the drinking water industry?

7. Why does the HPC increase after ozonation of drinking water?

References and Recommended Readings

Bitton, G. (1994) "Wastewater Microbiology." Wiley-Liss, New York.

Colbourne, J. S., Dennis, P. J., Trew, C., Berry, C., and Vesey, G. (1988) *Legionella* and public water supplies. *Water Sci. Technol.* **20**, 5–10.

Craun, G. F. (1993) "Safety of Water Disinfection: Balancing Chemical and Microbial Risks" ILSl Press, Washington, DC.

Geldreich, E. E. (1996) "Microbial Quality of Water Supply in Distribution Systems." Lewis , Boca Raton, FL.

LeChevallier, M. W., Cawthon, C. P., and Lee, R. G. (1988a) Factors promoting survival of bacteria in chlorinated water supplies. *Appl. Environ. Microbiol.* **54**, 649–654.

LeChevallier, M. W., Cawthon, C. P., and Lee, R. G. (1988b) Inactivation of biofilm bacteria. *Appl. Environ. Microbiol.* **54**, 2492–2 499.

LeChevalier, M. W., Becker, W. C., Schorr, P., and Lee, R. G. (1992) Evaluating the performance of biologically active rapid filters. *J. Am. Water Works Assoc.* **84**, 136–146.

Logsdon, G. S., and Lippy, E. C. (1982) The role of filtration in preventing waterborne disease. *J. Am. Water Works Assoc.* **74**, 649–655.

Montgomery, J. M. (1985) "Water Treatment Principles and Design." John Wiley & Sons, NewYork.

Servais, P., Billen, G., and Hascoet, M.-C. (1987) Determination of the biodegradable fraction of dissolved organic matter in waters. *Water Res.* **21**, 445–452.

U.S. EPA (1988) "Comparative Health Effects Assessment of Drinking Water Treatment" Technologies. Office of Drinking Water, U.S. Environmental Protection Agency, Washington, DC.

van der Kooij, D., Visser, A., and Hijnen, W. A. M. (1982) Determining the concentration of easily assimilable organic carbon in drinking water. *J. Am. Water Works Assoc.* **74**, 540–545.

Yee, R. B., and Wadowsky, R. M. (1982) Multiplication of *Legionella pneumophila* in unsterilized tap water. *Appl. Environ. Microbiol.* **43**, 1130–1134.

23

Disinfection

CHARLES P. GERBA

The destruction or prevention of growth of microorganisms is essential in the control of infectious disease transmission and preservation of foodstuffs and biodegradable materials. This is most commonly accomplished by heat, chemicals, filtration, or radiation. Heat acts to kill or **inactivate** by denaturation of essential proteins (enzymes, viral capsids) and nucleic acid. Chemicals may act by many different means to kill organisms or prevent their growth, including destruction of membranes, cell wall, enzymes, and replication of the nucleic acids (Table 23.1). Filtration is a process that acts to remove the organisms physically by size exclusion and does not result in destruction of the organism. Ultraviolet light and gamma radiation act directly on nucleic acids.

Sterilization is a process, physical or chemical, that destroys or eliminates all organisms. A **sanitizer** is an agent that reduces the number of bacterial contaminants to safe levels as judged by public health requirements. According to the official sanitizer test used in the United States, a sanitizer is a chemical that kills 99.999% of the specific test bacteria within 30 seconds under the conditions of the test (Block, 1991). A **disinfectant** is a physical or chemical agent that destroys disease-causing or other harmful microorganisms, but does not necessarily kill all microorganisms. Disinfectants are usually applied to water and inanimate objects **(fomites)** to control the spread of pathogenic microorganisms. They can also be used to treat foods and aerosols. A **bacteriostat** is usually a chemical agent that prevents the growth of bacteria but does not necessarily kill them. For example, silver is often added to activated carbon to prevent the growth of bacteria in home faucet-mounted water treatment devices.

23.1 THERMAL DESTRUCTION

The thermal destruction of microorganisms has been studied in great detail by the food industry because of the importance of this process in killing foodborne spoilage and pathogenic bacteria. However, less well studied are the waterborne pathogens. The thermal death of microorganisms is generally considered a first-order relationship, i.e., linear with time. The time necessary to kill a given number of organisms at a specific temperature is called the **thermal death time (TDT).** The general procedure for determining TDT by these methods is to place a known number of organisms in a sufficient number of sealed containers to get the desired number of survivors for the test period. At the end of the heating period, the containers are quickly removed and cooled quickly in cold water. Viability of the organism is assessed on standard culture media. The TDTs of some foodborne and waterborne pathogens are shown in Table 23.2. The **D value** or decimal reduction time is the time required to destroy 90% of the organisms. This value is numerically equal to the number of minutes required for the survivor as a function of time curve to traverse one log (Fig. 23.1). This is equal to the reciprocal of the slope of the survivor curve and is a measure of the death rate of an organism. The temperature at which the D value is determined is given as a subscript. For example, the D value for *Clostridium perfringens* at 250°F is $D_{250} = 0.1$ to 0.2 (Jay, 1996). The **z value** refers to the degrees Fahrenheit required for the thermal destruction as a function of temperature curve to traverse one log. This value is equal to the reciprocal of the slope of the TDT curve (Fig. 23.2). Whereas D reflects the resistance of an organism to a specific temperature, z pro-

TABLE 23.1   Mechanisms of Inactivation of Common Disinfectants

Target	Agent	Effect
Cell wall	Aldehydes Anionic surfactants	Interaction with $-NH_2$ groups Lysis
Cytoplasmic membrane	Quaternary ammonium compounds, biguanides, hexachlorophene	Leakage of low-molecular-weight material
Nucleic acids	Dyes, alkylating agents, ionizing and ultraviolet radiation	Breakage of bonds, cross-linking, binding of agents to nucleic acids
Enzymes or proteins	Metal ions (Ag, Cu) Alkylating agents Oxidizing agents (chlorine, hydrogen peroxide)	Bind to –SH groups of enzymes Combine with DNA or RNA

From Block (1991)

vides information on the relative resistance of an organism to different destructive temperatures; it allows the calculation of equivalent thermal processes at different temperatures. If, for example, 3.5 minutes at 140°F is considered to be an adequate process and $z = 8.0$, either 0.35 minutes at 148°F or 35 minutes at 132°F would be considered equivalent processes.

Spore-forming bacteria such as *Bacillus* and *Clostridium* are the most resistant to heat inactivation. Of the non–spore-forming waterborne and foodborne enteric pathogens, enteric viruses are the most heat resistant, followed by the bacteria and protozoa (Table 23.2). In addition to the type of microorganism, factors that influence TDT in foods include water, fat, salts, sugars,

TABLE 23.2   Thermal Death Times of Water- and Foodborne Pathogenic Organisms

Organism	Temperature (°C)/time (min)	Reference
Campylobacter spp.	75/1	Bandres *et al.*, 1988
Escherichia coli	65/1	Bandres *et al.*, 1988
Legionella	66/0.45[a]	Sanden *et al.*, 1989
Mycobacterium spp.	70/2	Robbecke and Buchholtz, 1992
M. avium	70/2.3[a]	
Salmonella spp.	65/1	Bandres *et al.*, 1988
Shigella spp.	65/1	Bandres *et al.*, 1988
Vibro cholerae	55/1[a]	Roberts and Gilbert, 1979
Cryptosporidium parvum	72.4/1	Fayer, 1994
Giardia lamblia	50/1[a]	Cerva, 1955
Poliovirus	60/25	Larkin and Fassolitis, 1979
Hepatitis A virus	70/10	Siegl *et al.*, 1984
Rotavirus	50/30	Estes *et al.*, 1979

[a] In buffered distilled water.

pH, and other substances. The heat resistance of microbial cells increases with decreasing humidity or moisture. Dried microbial cells are considerably more heat resistant than moist cells of the same type. Because protein denaturation occurs at a higher rate with heating in water than in air, it is likely that protein denaturation is closely associated with thermal death. The presence of fats and salts increases the heat resistance of some microorganisms. The effect of salt is variable and dependent on the kind of salt, concentration, and cation. Cationic salts at molar concentrations greatly increase the thermal resistance of enteric viruses. For this reason, $MgCl_2$ is added to poliovirus vaccine to aid in extending its useful life. The presence of sugars causes an increase in the heat resistance of microorganisms, in part because of decreased water activity. Microorganisms are most resistant to heat at their optimal pH of growth, which is generally about 7.0. As the pH is lowered or raised from the optimal value, there is an increase in heat sensitivity. Thus, acid and alkaline foods require less heat processing than neutral foods. In water, suspended solids or organic matter increase heat resistance (Liew and Gerba, 1980).

23.2 KINETICS OF DISINFECTION

Inactivation of microorganisms is a gradual process that involves a series of physical–chemical and biochemical steps. In an effort to predict the outcome of disinfection, various models have been developed on the basis of experimental data. The principal disinfection theory used today is still the Chick–Watson model, which expresses the rate of inactivation of microorganisms by a first-order chemical reaction.

$$Nt/No = e^{-kt} \qquad \text{(Eq. 23.1)}$$

or

$$\ln N_t/No = -kt \qquad \text{(Eq. 23.2)}$$

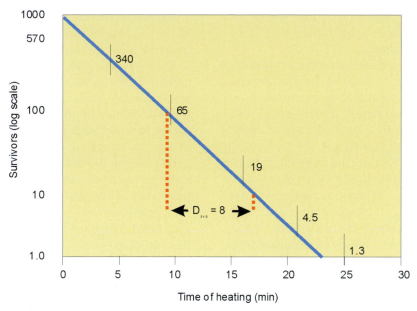

FIGURE 23.1 Thermal inactivation curve for a microorganism. The D value is the time required for inactivation of 90% of the organisms at a given temperature. In this case it required 8 minutes to kill 90% of the organisms at 240°F or $D_{240} = 8$ minutes.

where N_0 = number of microorganisms at time 0, N_t = number of microorganisms at time t, k = decay constant (1/time), and t = time. The logarithm of the survival rate (N_t/N_0) plots as a straight line versus time (Fig. 23.3). Unfortunately, laboratory and field data often deviate from first-order kinetics. Shoulder

curves may result from clumps of organisms or multiple hits of critical sites before inactivation. Curves of this type are common in disinfection of coliform bacteria by chloramines (Montgomery, 1988). The tailing-off curve, often seen with many disinfectants, may be explained by the survival of a resistant sub-

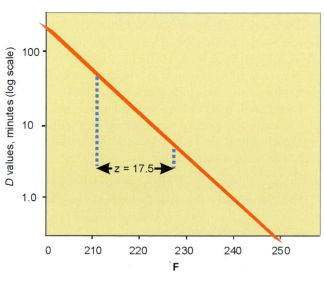

FIGURE 23.2 Thermal death time curve. The z value is equal to the degrees Fahrenheit required for the thermal destruction curve to traverse one log cycle.

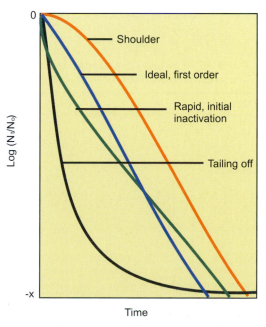

FIGURE 23.3 Types of inactivation curves observed for microorganisms.

population as a result of protection by interfering substances (suspended matter in water), clumping, or genetically conferred resistance.

In water applications, disinfectant effectiveness can be expressed as $C \cdot t$, where C is the disinfectant concentration and t the time required to inactivate a certain percentage of the population under specific conditions (pH and temperature). Typically, a level of 99% inactivation is used when comparing $C \cdot t$ values. In general, the lower the $C \cdot t$ value, the more effective the disinfectant. The $C \cdot t$ method allows a general comparison of the effectiveness of various disinfectants on different microbial agents (Tables 23.3 through 23.6). It is used by the drinking water industry to determine how much disinfectant must be applied during treatment to achieve a given reduction in pathogenic microorganisms. $C \cdot t$ values for chlorine for a variety of pathogenic microorganisms are shown in Table 23.3. The order of resistance to chlorine and most other disinfectants used to treat water is protozoan cysts > viruses > vegetative bacteria.

23.3 FACTORS AFFECTING DISINFECTANTS

Numerous factors determine the effectiveness and/or rate of kill of a given microorganism. Temperature has a major effect as it controls the rate of chemical reactions. Thus, as temperature increases, the rate of kill with a chemical disinfectant increases. The pH can affect the ionization of the disinfectant and the viability of the organism. Most waterborne organisms

are adversely affected by pH levels below 3 and above 10. In the case of halogens such as chlorine, pH controls the amount of HOCl (hypochlorous acid) and ⁻OCl (hypochlorite) in solution (Fig. 23.4). HOCl is more effective than ⁻OCl in the disinfection of microorganisms. With chlorine, the $C \cdot t$ increases with pH. Attachment of organisms to surfaces or particulate matter in water such as clays and organic detritus aids in the resistance of microorganisms to disinfection. Particulate matter may interfere by either acting chemically to react with the disinfectant, thus neutralizing the action of the disinfectant, or physically shielding the organism from the disinfectant (Stewart and Olson, 1996). The particulate–microbial complex may be thought of as

- Adsorption of microbes to larger particles
- Adsorption of small particles to the surface of the microbe
- Encasement of the microbe by one or more large particles or many associated small particles

Disinfectant protection is enhanced with decreasing size of the organism and increasing particle availability. Therefore, viruses are afforded greater protection than bacteria. For these reasons, particulate or turbidity removal in drinking water treatment is necessary to ensure the effectiveness of disinfection in the destruction of waterborne pathogens. Dissolved chemical substances that interfere with chemical disinfection include organic compounds, inorganic and organic nitrogenous compounds, iron, manganese, and hydrogen sulfide.

Studies have demonstrated that pathogenic and indicator bacteria occurring in the natural environment may be more resistant to disinfectants than laboratory-grown bacteria. This resistance is cell mediated and physiological in nature, requiring that the organism develop adaptive features to survive under adverse environmental conditions (Stewart and Olson, 1996). Cell-mediated mechanisms of resistance to disinfectant agents are poorly understood compared with physicochemical protective effects. Examples of cell-mediated resistance include

- Polymer or capsule production, which may act to limit diffusion of the disinfectant into the cell
- Cellular aggregation, providing physical protection to internal cells
- Cell wall and/or cell membrane alterations that result in reduced permeability to disinfectants
- Modification of sensitive sites, i.e., enzymes (Stewart and Olson, 1996)

TABLE 23.3 $C \cdot t$ Values for Chlorine Inactivation of Microorganisms in Water (99% Inactivation)[a]

Organism	°C	pH	$C \cdot t$
Bacteria			
E. coli	5	6.0	0.04
E. coli	23	10.0	0.6
L. pneumophila	20	7.7	1.1
Viruses			
Polio 1	5	6.0	1.7
Protozoa			
G. lamblia cysts	5	6.0	54–87
G. lamblia cysts	5	7.0	83–133
G. lamblia cysts	5	8.0	119–192
G. muris cysts	5	6.0	250
Cryptosporidium oocysts	25	7.0	>7200

From Sobsey (1989); Rose et al. (1997).
[a] In buffered distilled water.

TABLE 23.4 $C \cdot t$ Values for Chlorine Dioxide in Water

Microbe	ClO$_2$ residual (mg/l)	Temperature (°C)	pH	% reduction	$C \cdot t$
Bacteria					
E. coli	0.3–0.8	5	7.0	99	0.48
L. pneumophila	0.5–0.35	23		99.9–99.99	ND
Viruses					
Polio 1	0.4–14.3	5	7.0	99	0.2–6.7
Rotavirus SA11					
Dispersed	0.5–1.0	5	6.0	99	0.2–0.3
Cell-associated	0.45–1.0	5	6.0	99	1.0–2.1
Hepatitis A	0.14–0.23	5	6.0	99	1.7
Coliphage MS-2	0.15	5	6.0	99	5.1
Protozoa					
G. muris	0.1–5.55	5	7.0	99	10.7
G. muris	0.26–1.2	25	5.0	99	5.8
G. mursi	0.21–1.12	25	7.0	99	5.1
G. muris	0.15–0.81	25	9.0	99	2.7
Cryptosporidium	4.03	10	7.0	95.8	6.0

Adapted from Sobsey (1989); Rose *et al.* (1997).

It has been speculated that many of these physiological events are a function of adaptation to low- nutrient conditions in the environment.

Repeated exposure of bacteria and viruses to chlorine appears to result in selection for greater resistance (Bates *et al.*, 1977; Haas and Morrison, 1981). However, the enhanced resistance has not been great enough to overcome concentrations of chlorine applied in practice.

TABLE 23.5 $C \cdot t$ Values for Chloramines in Water (99% Inactivation)[a]

Microbe	°C	pH	$C \cdot t$
Bacteria			
E. coli	5	9.0	113
M. fortuitum	20	7.0	2667
M. avium	17	7.0	ND**
M. intracellulare	17	7.0	ND
Viruses			
Polio 1	5	9.0	1420
Hepatitis A	5	8.0	592
Coliphage MS2	5	8.0	2100
Rotavirus SA11			
Dispersed	5	8.0	4034
Cell-associated	5	8.0	6124
Protozoa			
G. muris	3	6.5–7.5	430–580
G. muris	5	7.0	1400
Cryptosporidium	25	7.0	>7200

Adapted from Sobsey (1989); Rose *et al.* (1997).
[a] In buffered distilled water.
** ND = no data.

23.4 HALOGENS

23.4.1 Chlorine

Chlorine and its compounds are the most commonly used disinfectants for treating drinking and wastewater. Chlorine is a strong oxidizing agent which, when added as a gas to water, forms a mixture of hypochlorous acid (HOCl) and hydrochloric acids.

$$Cl_2 + H_2O \rightleftharpoons HOCl + HCl \qquad (Eq. 23.3)$$

In dilute solutions, little Cl$_2$ exists in solution. The dis-

TABLE 23.6 $C \cdot t$ Values for Ozone Inactivation of Microorganisms in Water (99% Inactivation)

Organism	°C	pH	$C \cdot t$
Bacteria			
E. coli	1	7.2	0.006–0.02
Viruses			
Polio 1	5	7.2	0.2
Polio 2	25	7.2	0.72
Rota SA11	4	6.0–8.0	0.019–0.064
Protozoa			
G. muris	5	7.0	1.94
G. lamblia	5	7.0	0.53
Cryptosporidium	7	—	7.0
Cryptosporidium	22	—	3.5

From Sobsey (1989); Rose *et al.* (1997).

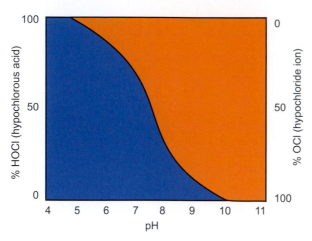

FIGURE 23.4 Distribution of HOCl and OCl⁻ in water as a function of pH. (From Bitton, 1994.)

infectant's action is associated with the HOCl formed. Hypochlorous acid dissociates as follows:

$$HOCl \rightleftharpoons H^+ \ OCl^- \qquad (Eq.\ 23.4)$$

The preparation of hypochlorous acid and OCl⁻ (hypochorite ion) depends on the pH of the water (Fig. 23.4). The amount of HOCl is greater at neutral and lower pH levels, resulting in greater disinfection ability of chlorine at these pH levels. Chlorine as HOCl or OCl⁻ is defined as **free available chlorine.** HOCl combines with ammonia and organic compounds to form what is referred to as **combined chlorine.** The reactions of chlorine with ammonia and nitrogen-containing organic substances are of great importance in water disinfection. These reactions result in the formation of monochloramine, dichloramine, trichloramine, etc.

$$NH_3 + HOCl \rightarrow NH_2Cl + H_2O$$
$$\text{monochloramine} \qquad (Eq.\ 23.5)$$

$$NH_2Cl + HOCl \rightarrow NHCl_2 + H_2O$$
$$\text{dichloramine} \qquad (Eq.\ 23.6)$$

$$NHCl_2 + HOCl \rightarrow NCl_3 + H_2O$$
$$\text{trichloramine} \qquad (Eq.\ 23.7)$$

Such products retain some disinfecting power of hypochlorous acid but are much less effective at a given concentration than chlorine.

Free chlorine is quite efficient in inactivating pathogenic microorganisms. In drinking water treatment, 1 mg/l or less for about 30 minutes is generally sufficient to reduce significantly bacterial numbers. The presence of interfering substances in wastewater reduces the disinfection efficacy of chlorine, and relatively high concentrations of chlorine (20–40 mg/l) are required (Bitton, 1994). Enteric viruses and protozoan parasites are more resistant to chlorine than bacteria (Table 7.3) and can be found in secondary wastewater

effluents after normal disinfection practices. *Cryptosporidium* is extremely resistant to chlorine. A chlorine concentration of 80 mg/l is necessary to cause 90% inactivation following a 90-minute contact time (Korich *et al.*, 1990). Chloramines are much less efficient than free chlorine (about 50 times less efficient) in inactivation of viruses.

Bacterial inactivation by chlorine may result from (Stewart and Olson, 1996)

- Altered permeability of the outer cellular membrane, resulting in leakage of critical cell components
- Interference with cell-associated membrane functions (e.g., phosphorylation of high-energy compounds)
- Impairment of enzyme and protein function as a result of irreversible binding of the sulfyhydryl groups
- Nucleic acid denaturation

The actual mechanism of chlorine inactivation may involve a combination of these actions or merely the effect of chlorine on a few critical sites. Currently, it appears that bacterial inactivation by chlorine is primarily caused by impairment of physiological functions associated with the bacterial cell membrane.

Chlorine may inactivate viruses by interaction with either the viral capsid proteins or the nucleic acid (Thurman and Gerba, 1988). The site of action may also depend on the concentration of chlorine and the type of virus. It has been found that at free chlorine concentrations of less than 0.8 mg/l inactivation of poliovirus RNA occurs without major structural changes, whereas chlorine concentrations in excess of 0.8 mg/l result in damage to the viral RNA and protein capsid (Alvarez and O'Brien, 1982) (Fig. 23.5). The protein coat appears to be the target for the double-stranded RNA rotaviruses (Vaughn and Novotny, 1991).

23.4.2 Chloramines

Inorganic chloramines are produced by combining chlorine and ammonia (NH₄) for drinking water disinfection. The species of chloramines formed (see Eqs. 23.5 through 23.7) depend on a number of factors, including the ratio of chlorine to ammonia-nitrogen, chlorine dose, temperature, and pH. Up to a chlorine-to-ammonia mass ratio of 5, the predominant product formed is monochloramine, which demonstrates greater disinfection capability than other forms, i.e., dichloramine and trichloramine. Chloramines are used to disinfect drinking water by some utilities in the United States, but because they are slow acting, they have mainly been used as secondary disinfectants when a residual in the distribution system is desired.

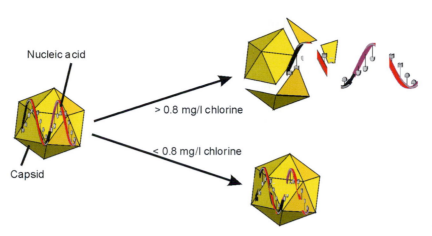

FIGURE 23.5   Virus inactivation by chlorine.

For example, when ozone is used to treat drinking water, no residual disinfectant remains. Because bacterial growth may occur after ozonation of tap water, chloramines are added to prevent regrowth in the distribution system. In addition, chloramines have been found to be more effective in controlling biofilm microorganisms on the surfaces of pipes in drinking water distribution systems because they interact poorly with capsular polysaccharides (LeChevallier *et al.*, 1990).

Because of the occurrence of ammonia in sewage effluents, most of the chlorine added is converted to chloramines. This demand on the chlorine must be met before free chorine is available for disinfection. As chlorine is added, the residual reaches a peak (formation of mostly monochloramine) and then decreases to a minimum called the breakpoint (Fig. 23.6). At the breakpoint, the chloramine is oxidized to nitrogen gas in a complex series of reactions summarized in Eq. 23.8.

$$2NH_3 + 3HOCl \rightarrow N_2 + 3H_2O + 3HCl \qquad \text{(Eq. 23.8)}$$

Addition of chlorine beyond the breakpoint ensures the existence of a free available chlorine residual.

Although numerous studies have been conducted to determine the mode of microbial inactivation by free chlorine, there have been fewer studies concerning chloramine inactivation mechanisms. It should be noted, however, that because of the poorly controlled experimental conditions employed by early investigators, many of the postulated chlorine inactivation mechanisms may have involved the action of chloramines rather than free chlorine. Research to date indicates that chloramines primarily inactivate microorganisms by irreversible denaturation of proteins (Stewart and Olson, 1996). Chloramine inactivation of bacteria is caused primarily by the oxidation of sulfyhydryl-containing enzymes and, to a lesser extent, a reaction with the nucleic acid. In contrast with chlorine, there are no existing data to suggest that chloramines can modify the permeability state of the cell. Viral inactivation by chloramines is similar to the mechanism of inactivation by chlorine, in which primary targets consist of both capsid proteins and nucleic acid.

23.4.3 Chlorine Dioxide

Chlorine dioxide is an oxidizing agent that is extremely soluble in water (five times more than chlorine) and, unlike chlorine, does not react with ammonia or organic compounds to form trihalomethane, which is potentially carcinogenic. Therefore it has received attention for use as a drinking water disinfectant. Chlorine dioxide must be generated on site be-

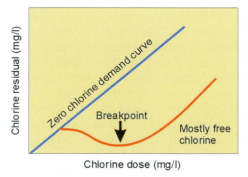

FIGURE 23.6   Dose–demand curve for chlorine.

cause it cannot be stored. It is generated from the reaction of chlorine gas with sodium chlorite:

$$2NaClO_2 + Cl_2 \rightarrow 2\, ClO_2 + 2NaCl \qquad \text{(Eq. 23.9)}$$

Chlorine dioxide does not hydrolyze in water but exists as a dissolved gas.

Studies have demonstrated that chlorine dioxide is as effective as or more effective in inactivating bacteria and viruses in water than chlorine (Table 23.4). As is the case with chlorine, chlorine dioxide inactivates microorganisms by denaturation of the sulfyhydryl groups contained in proteins (Stewart and Olson, 1996); inhibition of protein synthesis (Bernarde *et al.*, 1967); denaturation of nucleic acid; and impairment of permeability control.

Studies with bacteriophage have suggested that the protein in the capsids is irreversibly damaged by chlorine dioxide. However, studies with poliovirus have suggested that the viral RNA is separated from the capsid during treatment (Vaughn and Novotny, 1991). The viricidal efficiency of chlorine dioxide increases as the pH is increased from 4.5 to 9.0 (Chen and Vaughn, 1990).

23.4.4 Bromine and Iodine

Bromine undergoes reactions in water similar to those of chlorine. However, its disinfecting capacity and mode of action differ from those of chlorine. The primary use of bromine is limited to hot tubs or spas and certain industrial applications (cooling towers). It is not as fast acting as chlorine but is effective against bacteria (*Legionella*), viruses and protozoan parasites (*Entamoeba histolytica*). Bromine appears primarily to attach to the protein of viruses without causing structural damage (Keswick *et al.*, 1981). It does not appear to be able to penetrate the protein coat to inactivate the viral RNA.

Iodine has been used as a disinfectant primarily for small-scale water treatment needs such as campers, the space shuttle, and small water treatment systems. On a comparative mg/l basis, more iodine than chlorine is required for a comparative bacterial kill. Iodine reacts in water as follows:

$$I + H_2O \rightleftharpoons HOI + H^+ + I^- \\ \text{(iodine hydrolysis)} \qquad \text{(Eq. 23.10)}$$

$$3I_2 + 3H_2O \rightleftharpoons IO_3^- + 5I^- + 6H^+ \\ \text{(iodate formation)} \qquad \text{(Eq. 23.11)}$$

Iodide and iodate are primarily formed above pH 8.0 and have no viricidal action and little action against bacteria. Thus, at pH 9.0 HOI is the dominant form, whereas at pH 5.0, I_2 is dominant. At low pH iodine is more effective against some protozoan cysts because they are more sensitive to I_2 than to HOCl (Gottardi, 1991). This behavior is explained by the higher diffusibility of molecular iodine through the cell walls of cysts.

Iodine is not effective for all protozoa. For example, while *Giardia* cysts can be inactivated by iodine, *Cryptosporidium* oocysts are very resistant (Gerba *et al.*, 1997). In contrast to protozoa, viruses are more readily inactivated at pH levels above 7.0 because of the stronger oxidizing power of HOI. Iodine displays first-order inactivation kinetics, indicating single-site inactivation. Iodine oxidizes sulfyhydryl groups and tryptophan and, perhaps more important, substitutes tyrosyl on histidyl moieties at neutral pH and room temperature. Structural changes in viral integrity have been noted by electron microscopy after treatment with iodine, and thus infectious RNA could be released into the environment.

23.5 OZONE

Ozone (O_3), a powerful oxidizing agent, can be produced by passing an electric discharge through a stream of air or oxygen. Ozone is more expensive than chlorination to apply to drinking water, but it has increased in popularity as a disinfectant because it does not produce trihalomethanes or other chlorinated by-products, which are suspected carcinogens. However, aldehydes and bromates may be produced by ozonation and may have adverse health effects. Because ozone does not leave any residual in water, ozone treatment is usually followed by chlorination or addition of chloramines. This is necessary to prevent regrowth of bacteria because ozone breaks down complex organic compounds present in water into simpler ones that serve as substrates for growth in the water distribution system. The effectiveness of ozone as a disinfectant is not influenced by pH and ammonia.

Ozone is a much more powerful oxidant than chlorine (Tables 23.3 and 23.6). The $C \cdot t$ values for 99% inactivation are only 0.0011–0.2 for enteric bacteria and 0.04–0.42 for enteric viruses (Bitton, 1994). Ozone appears to inactivate bacteria by the same mechanisms as chlorine-based disinfection: by disruption of membrane permeability (Stewart and Olson, 1996); impairment of enzyme function and/or protein integrity by oxidation of sulfyhydryl groups; and nucleic acid denaturation. *Cryptosporidium* oocysts can be inactivated by ozone, but a $C \cdot t$ of 1–3 is required. Viral inactivation may proceed by breakup of the capsid proteins into subunits, resulting in release of the RNA, which may then be damaged (Fig. 23.7).

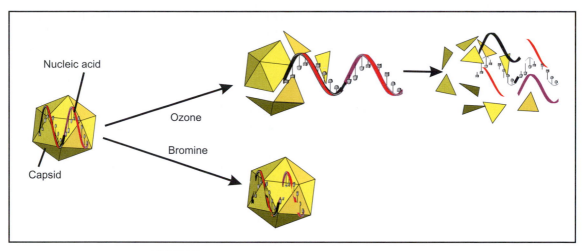

FIGURE 23.7   Virus inactivation by ozone and bromine.

23.6 METAL IONS

Heavy metals such as copper, silver, zinc, lead, cadmium, nickel, and cobalt all exhibit antimicrobial activity; however, because of toxicity to animals, only copper and silver have seen application to water disinfection. Copper and silver have seen use as swimming pool and hot tub disinfectants. Copper has been used to control the growth of *Legionella* in hospital distribution systems. Silver has been used as a bacteriostat added to the activated carbon used in faucet-mounted water treatment devices for home use. Concentrations of copper used range from 200 to 400 μg/l. Silver exhibits greater antimicrobial action

and concentrations of 40–90 μg/l give the same effectiveness. The effectiveness of metal ions is influenced by pH, presence of anions, and soluble organic matter. Unlike halogens and other oxidizing disinfectants, metals remain active for long periods of time in water. The rate of inactivation is very slow compared with oxidizing agents (Fig. 23.8); however, their action is enhanced in the presence of low concentrations of oxidizing agents such as chloramines (Straub *et al.*, 1995). The enhanced rate of inactivation is due to a synergistic interaction of both disinfectants.

Metal ions may inactivate bacteria or viruses by reacting outside or inside the cell or virus either directly or indirectly. It has been suggested that the inactivat-

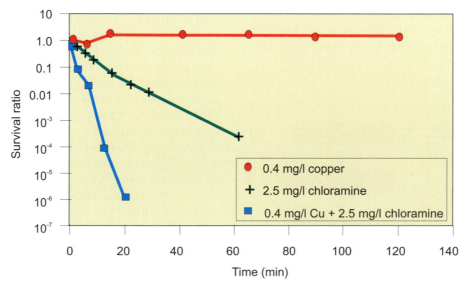

FIGURE 23.8   Synergistic inactivation of *Escherichia coli* by chloramines and copper. (From Straub *et al.*, 1995.)

ing capacity of heavy-metal ions is due to their oxidation power and that a functional relationship exists between the inactivation rate and the oxidation potential of the ion (Thurman and Gerba, 1989). Inactivation of the macromolecules (proteins or nucleic acids) is thought to involve site-specific Fenton mechanisms. It is assumed that the metal ion binds to a biological target and is reduced by superoxide radicals or other reductants and subsequently reoxidized by H_2O_2, generating hydroxide radicals (Fig. 23.9). Repeated cyclic redox reactions may result in multi-hit damage as radical formation occurs near the target site. Copper and silver may bind to proteins, interfering with the normal function of enzymes, resulting in cell death. Silver readily reacts with sulfhydryl groups in proteins. Metals may also bind to the nucleic acids, forming complexes that interfere with replication.

The action of metals is slow and may be reversed by addition of chelating agents. For example, assay of samples containing silver for bacteria will give lower counts if the silver is not first neutralized by addition of sodium thiosulfate-sodium thioglycolate to inhibit the bacteristatic effect of silver (Chambers *et al.*, 1962).

23.7 ULTRAVIOLET DISINFECTION

The use of ultraviolet disinfection of water and wastewater has seen increased popularity because it is not known to produce carcinogenic or toxic byproducts or taste and odor problem, and there is no need to handle or store toxic chemicals. Unfortunately, it has several disadvantages including higher costs than halogens, no disinfectant residual, difficulty in determining the UV dose, maintenance and cleaning of UV lamps, and potential photoreactivation of some enteric bacteria (Bitton, 1994). However, advances in UV technology are providing lower cost, more effi-

cient lamps and more reliable equipment. These advances have aided in the commercial application of UV for water treatment in the pharmaceutical, cosmetic, beverage, and electronic industries in addition to municipal water and wastewater application.

Microbial inactivation is proportional to the UV dose, which is expressed in microwatt-seconds per square centimeter (μW-s/cm^2) or

$$\text{UV dose} = I \cdot t \qquad \text{(Eq. 23.12)}$$

where $I = \mu$W/cm^2 and t = exposure time.

In most disinfection studies, it has been observed that the logarithm of the surviving fraction of organisms is nearly linear when it is plotted against the dose, where dose is the product of concentration and time ($C \cdot t$) for chemical disinfectants or intensity and time ($I \cdot t$) for UV. A further observation is that constant dose yields constant inactivation. This is expressed mathematically in Eq. 23.13.

$$\log \frac{N_S}{N_i} = \text{function}(I_i t) \qquad \text{(Eq. 23.13)}$$

where N_S is the density of surviving organisms (number/cm^3) and N_i is the initial density of organisms before exposure (number/cm^3). Because of the logarithmic relationship of microbial inactivation versus UV dose, it is common to describe inactivation in terms of log survival, as expressed in Eq. 23.14. For example, if one organism in 1000 survived exposure to UV, the result would be a -3 log survival, or a 3 log reduction.

$$\log \text{survival} = \log \frac{N_S}{N_i} \qquad \text{(Eq. 23.14)}$$

Determining the UV susceptibility of various indicator and pathogenic waterborne microorganisms is fundamental in quantifying the UV dose required for adequate water disinfection. Factors that may affect UV dose include cell clumping and shadowing, suspended solids, turbidity, and UV absorption. UV susceptibility experiments described in the literature are often based on the exposure of microorganisms under conditions optimized for UV disinfection. Such conditions include filtration of the microorganisms to yield monodispersed, uniform cell suspensions and the use of buffered water with low turbidity and high transmission at 254 nm. Thus, in reality, higher doses are required to achieve the same amount of microbial inactivation in full-scale flow through operating systems.

The effectiveness of UV light is decreased in wastewater effluents by substances that affect UV transmission in water. These include humic substances, phenolic compounds, lignin sulfonates, and

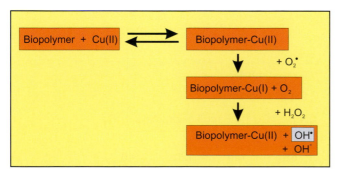

FIGURE 23.9 Modified site-specific Fenton mechanism. (From Thurman and Gerba, 1989.)

ferric iron. Suspended matter may protect microorganisms from the action of UV light; thus filtration of wastewater is usually necessary for effective UV light disinfection.

UV inactivation data are usually collected by placing a suspension of organisms in a stirred, flat, thin-layer dish in water with low UV light absorbance. In UV batch reactors there are uniform UV intensities and contact time can be controlled. To deliver UV to these reactors, a collimating beam apparatus should be used (Fig. 23.10). The light emitted at the end of the collimating beam is perpendicular to the batch reactor surface, thus creating a uniform, constant irradiation field that can be accurately quantified by means of a radiometer and photodetector calibrated for detecting 254-nm light. In general, the resistance of microorganisms to UV light follows the same pattern as the resistance to chemical disinfectants, i.e., double-stranded DNA viruses > MS-2 coliphage > bacterial spores > double-stranded RNA enteric viruses > single-stranded RNA enteric viruses > vegetative bacteria (Table 23.5).

Ultraviolet radiation damages microbial DNA or RNA at a wavelength of approximately 260 nm. It causes thymine dimerization (Fig. 23.11), which blocks nucleic acid replication and effectively inactivates microorganisms. The initial site of UV damage in viruses is the genome, followed by structural damage to the virus protein coat. Viruses with high molecular weight, double-stranded DNA or RNA are easier to inactivate than those with low-molecular-weight, double-stranded genomes. Likewise, viruses with single-stranded nucleic acids of high molecular weight are easier to inactivate than those with single-stranded nucleic acids of low molecular weight. This is presumably because the target density is higher in larger genomes. However, viruses with double-stranded genomes are less susceptible than those with single-stranded genomes because of the ability of the naturally occurring enzymes within the host cell to repair damaged sections of the double-stranded genome, using the nondamaged strand as a template (Roessler and Severin, 1996).

A phenomenon known as **photoreactivation** occurs in some UV light-damaged bacteria when exposed to visible wavelengths between 300 and 500 nm. The UV light damage is repaired by activation of a photoreactivating enzyme, which binds and then splits the thymine dimers. DNA damage can also be repaired in the dark by a mechanism that excises dimerized pyrimidine base pairs and allows the reinsertion of undimerized bases by other enzymes. The regenerative capacity of any organism is dependent on the type of organism. Total and fecal coliforms are capable of photoreactivation, but fecal streptococci are not. To prevent photoreactivation, sufficient doses must be applied or exposure to direct sunlight prevented.

A minimum dose of 16,000 μW s/cm^2 has been recommended for treating drinking water, as this results in a 99.9% reduction in coliforms. However, this level is not enough to inactivate enteric viruses or protozoan cysts (Table 23.7) (Abbaszadegan et al., 1997). Filtration can be applied to remove protozoan cysts before UV light disinfection to improve performance.

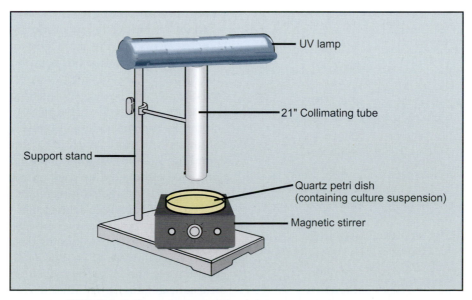

FIGURE 23.10   Collimating tube apparatus for UV dose application.

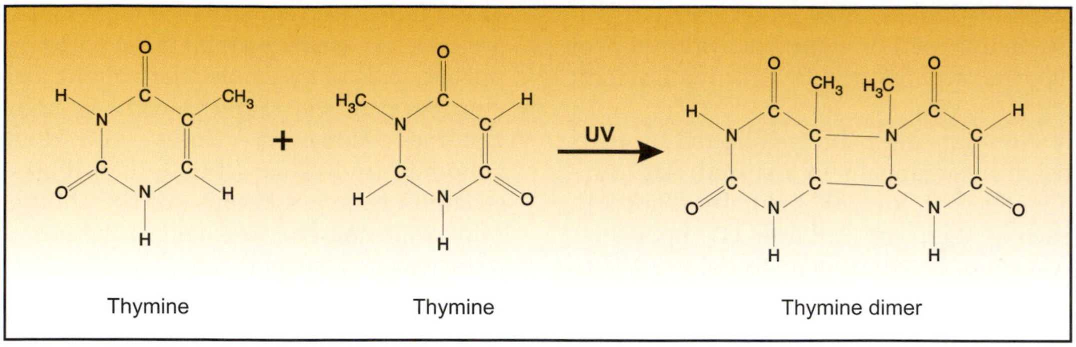

FIGURE 23.11   Formation of thymine dimers in the DNA of irradiated nonsporulating bacteria.

23.8 PHOTODYNAMIC INACTIVATION

The usefulness of photoreactive dyes for inactivating microorganisms and oxidizing toxic compounds and organic matter in wastewater has been demonstrated. **Photodynamic action** may be defined as the sensitization of microorganisms to inactivation by visible light through the action of certain dyes (e.g., methylene blue). The dye combines with the nucleic acid or another critical site, and the complex absorbs light energy and attains an excited energy state. The excited complex then combines with oxygen as the energy is released in a reaction that results in disruption of chemical bonds and loss of infectivity of the organism. Solid material such as titanium dioxide has a similar effect in the presence of UV light in sunlight, causing strong oxidizing reactions at the surface of the metal oxide (Watts *et al.*, 1995).

Poliovirus is inactivated in tap water, sewage, or seawater by methylene blue in the presence of sunlight, with a 2.5 logarithmic drop in virus titer within 5 minutes in solutions containing 1 mg/l of methylene blue. Various factors such as temperature, pH (affects binding of the dye), oxygen concentration, and light energy affect the rate and sensitivity of microbial inactivation (Thurman and Gerba, 1988).

23.9 GAMMA AND HIGH-ENERGY IRRADIATION

Ionizing radiation generated by radioactive materials such as cesium 127 or cobalt 60 and high-energy electron beams can inactivate microorganisms either directly or indirectly by production of free radicals. Nucleic acids are the main targets of ionizing radiation. Ionizing radiation has been studied in great detail for preservation of foods and more recently for wastewater and sewage sludge treatment. Factors that influence the effectiveness of ionizing radiation include the type of organism (generally, the smaller the organism the more resistant); composition of the suspending medium (organic material offers protection); presence of oxygen (greater resistance in the absence of oxygen); and moisture (greater resistance of dried cells and radiolysis of water). The unit of dose is the **rad,** which is equivalent to the absorption of 100 ergs per gram of matter. A kilorad (krad) is equal to 1000 rads. Typical doses to produce a D value of 90% inactivation are shown in Table 23.8. Viruses are the most resistant to ionizing irradiation in water and sludge.

Sludge irradiators have been built in Europe and experimental electron beam irradiators in the

TABLE 23.7   UV Dose to Kill Enteric Microorganisms

Organism	Ultraviolet dose (μW-s/cm^2) required for 90% reduction
Campylobacter jejuni	1,100
Escherichia coli	1,300–3,000
Klebsiella terrigena	3,900
Legionella pneumophila	920–2,500
Salmonella typhi	2,100–2,500
Shigella dysenteriae	890–2,200
Vibro cholerae	650–3,400
Yersinia enterocolitica	1,100
Adenovirus	23,600–30,000
Coxsackievirus	11,900–15,600
Echovirus	10,800–12,100
Poliovirus	5,000–12,000
Hepatitis A	3,700–7,300
Rotavirus SA11	8,000–9,900
Coliphage MS2	18,600

From Roessler and Severin (1996).

TABLE 23.8 Sludge Irradiation: *D* Values for Selected Pathogens and Parasites

Organism	*D* value (krad)
Bacteria	
E. coli	<22–36
Klebsiella spp.	36–92
Enterobacter spp.	34–62
Salmonella typhimurium	<50–140
Streptococcus faecalis	110–250
Viruses	
Poliovirus	350
Coxsackievirus	200
Echovirus	170
Reovirus	165
Adenovirus	150
Parasites	
Ascaris spp.	<66

Modified from Ahlstrom and Lessel (1986).

United States. The electron beams are generated by a 750-kV electron accelerator. The unit treats a thin layer (≈ 2 mm) of liquid sludge spread on a rotating drum. Such systems are costly for waste treatment and require thick concrete shielding.

QUESTIONS AND PROBLEMS

1. Of the non–spore-forming bacteria, which microbial group is the most resistant to thermal inactivation in water?

2. What is thermal reduction time? *D* value?

3. Why are all microorganisms not inactivated according to first-order kinetics?

4. How long would you have to maintain a residual of 1.0 mg/l of free chlorine to obtain a $C \cdot t$ of 15? A $C \cdot t$ of 0.1?

5. Why is chlorine more effective against microorganisms at pH 5.0 than at pH 9.0?

6. Which chlorine compound is most effective against biofilms? Why?

7. What factors interfere with chlorine disinfection? Ultraviolet disinfection?

8. What is the main site of UV light inactivation in microorganisms? What group of microorganisms are the most resistant to UV light? Why?

9. At what pH is iodine most effective against protozoan parasites? Why?

10. What is photoreactivation? Are all microorganisms capable of photoreaction? If not, why?

11. What are two sources of ionizing radiation? How does ionizing radiation kill microorganisms?

12. Why does suspended matter interfere with the disinfection of microorganisms?

References and Recommended Readings

Abbaszadegan, M., Hasan, M. N., Gerba, C. P., Roessler, P. F., Wilson, B. R., Kuennen, R., and Van Dellen, E. (1997) The disinfection efficacy of a point-of-use water treatment system against bacterial, viral and protozoan waterborne pathogens. *Water Res.* **31**, 574–582.

Ahlstrom, S. B., and Lessel, T. (1986) Irradiation of municipal sludge for pathogen control. *In* "Control of Sludge Pathogens" (C. A. Sorber, ed.) *Water Pollution Control Federation*, Washington, DC.

Alvarez, M. E., and O'Brien, R. T. (1982) Effects of chlorine concentration on the structure of poliovirus. *Appl. Environ. Microbiol.* **43**, 237–239.

Bandres, J. C., Mawhewson, J. J., and Du Pont, H. L. (1988). Heat susceptibility of bacterial enteropathogens—implications for the prevention of travelers diarrhea. *Arch. Intern. Med.* **148**, 2261–2263.

Bates, R. C., Shaffer, P. T. B., and Sutherland, S. M. (1977) Development of poliovirus having increased resistance to chlorine inactivation. *Appl. Environ. Microbiol.* **33**, 849–853.

Bernarde, M. A., Snow, N. B., Olivieri, V. P., and Davidson, B. (1967) Kinetics and mechanism of bacterial disinfection by chloride dioxide. *Appl. Microbiol.* **15**, 257–265.

Bitton, G. (1994) "Wastewater Microbiology." Wiley-Liss, New York.

Block, S. S. (1991) "Disinfection, Sterilization, and Preservation," 4th ed. Lea & Febiger, Philadelphia.

Cerva, L. (1955) The effect of disinfectants on the cyst of *Giardia intestinalis*. *Cesk. Parasitol.* **2**, 17–21.

Chambers, C. W., Procter, C. M., and Kabler, P. W. (1962) Bactericidal effect of low concentrations of silver. *J. Am. Water Works Assoc.* **54**, 208–216.

Chen, Y.-S., and Vaughn, J. (1990) Inactivation of human and simian rotaviruses by chlorine dioxide. *Appl. Environ. Microbiol.* **56**, 1363–1366.

Estes, M. K., Graham, D. Y., Smith, E. R., and Gerba, C. P. (1979) Rotavirus stability and inactivation. *J. Gen. Virol.* **43**, 403–409.

Fayer, R. (1994) Effect of high temperature on infectivity of *Cryptosporidium parvum* oocysts in water. *Appl. Environ. Microbiol.* **60**, 2732–2735.

Gerba, C. P., Johnson, D. C., and Hasan, M. N. (1997) Efficacy of iodine water purification tablets against *Cryptosporidium* oocysts and *Giardia* cysts. *Wilder. Environ. Med.* **8**, 96–100.

Gottardi, W. (1991) Iodine and iodine compounds. *In* "Disinfection, Sterilization, and Preservation," 4th ed. (S. S. Block, ed.) Lea and Febiger, Philadelphia, pp. 152–166.

Haas, C. N., and Morrison, E. C. (1981) Repeated exposure of *Escherichia coli* to free chlorine: Production of strains possessing altered sensitivity. *Water Air Soil Pollut.* **16**, 233–242.

Hoff, J. C. (1986) Inactivation of microbial agents by chemical disinfectants. U.S. Environmental Protection Agency Report 600/2, 86/067, Cincinnati.

Jay, J. M. (1996) "Modern Food Microbiology," 5th ed. Chapman & Hall, New York.

Keswick, B. H., Fujioka, R. S., and Loy, P. C. (1981) Mechanism of poliovirus inactivation by bromine chloride. *Appl. Environ. Microbiol.* **42,** 824–829.

Korich, D. G., Mead, J. R., Madore, M. S., Sinclair, N. A., and Sterling, C. R. (1990) Effect of zone, chlorine dioxide, chlorine, and monochloramine on *Cryptosporidium parvum* oocyst viability. *Appl. Environ. Microbiol.* **56,** 1423–1428.

Larkin, E. P., and Fassolitis, A. C. (1979) Viral heat resistance and infectious ribonucleic acid. *Appl. Environ. Microbiol.* **38,** 650–655.

LeChevallier, M. W., Lowry, C. H., and Lee, R. G. (1990) Disinfecting biofilm in a model distribution system. *J. Am. Water Works Assoc.* **82,** 85–99.

Liew, P., and Gerba, C. P. (1980) Thermostabilization of enteroviruses on estuarine sediment. *Appl. Environ. Microbiol.* **40,** 305–308.

Montgomery, (1988) "Water Treatment and Design." John Wiley & Sons, New York.

Robbecke, R. S., and Buchholtz, K. (1992) Heat susceptibility of aquatic Mycobacteria. *Appl. Environ. Microbiol.* **58,** 1869–1873.

Roberts, D., and Gilbert, R. J. (1979) Survival and growth of noncholera vibrios in various foods. *J. Hyg.* Cambridge **82,** 123–131.

Roessler, P. F., and Severin, B. F. (1996) Ultraviolet light disinfection of water and wasteater. *In* "Modeling Disease Transmission and Its Prevention by Disinfection" (C. J. Hurst, ed.) Cambridge University Press, Cambridge, UK, pp. 313–368.

Rose, J. B., Lisle, J. T., and LeChevallier, M. (1997) Waterborne *Cryptosporidiosis:* Incidence, outbreaks, and treatment strategies. *In* "*Cryptosporidium* and *Cryptosporidiosis*" (R. Fayer, ed.) CRC Press, Boca Raton, FL, pp. 93–109.

Sanden, G. N., Filds, B. S., Barbaree, J. M., and Feeley, J. C. (1989) Viability of *Legionella pneumophila* in chlorine-free water at elevated temperatures. *Curr. Microbiol.* **18,** 61–65.

Siegl, G., Weitz, M., and Kronauer, G. (1984) Stability of hepatitis A virus. *Intervirology* **22,** 218–226.

Sobsey, M. D. (1989) Inactivation of health-related microorganisms in water by disinfection processes. *Water Sci. Technol.* **21,** 179–195.

Straub, T. M., Gerba, C. P., Zhou, X., Price, R., and Yahya, M. T. (1995) Synergistic inactivation of *Escherichia coli* and MS-2 coliphage by chloramine and cupric chloride. *Water Res.* **24,** 811–818.

Stewart, M. H., and Olson, B. H. (1996) Bacterial resistance to potable water disinfectants. *In* "Modeling Disease Transmission and Its Prevention by Disinfection." Cambridge University Press, Cambridge, UK, pp. 140–192.

Thurman, R. B., and Gerba, C. P. (1988) Molecular mechanisms of viral inactivation by water disinfectants. *Adv. Appl. Environ. Microbiol.* **33,** 75–105.

Thurman, R. B., and Gerba, C. P. (1989) The molecular mechanisms of copper and silver ion disinfection of bacteria and viruses. *CRC Crit. Rev. Environ. Control* **18,** 295–315.

Vaughn, J. M., and Novotny, J. F. (1991) Virus inactivation by disinfectants. *In* "Modeling the Environmental Fate of Microorganisms" (C. J. Hurst, ed.) American Society for Microbiology, Washington, DC, pp. 217–241.

Watts, R. J., Kong, S., Orr, M. P., Miller, G. C., and Henry, B. E. (1995) Photcatalytic inactivation of coliform bacteria and viruses in secondary wastewater effluent. *Water Res.* **29,** 95–100.

24

Risk Assessment

CHARLES P. GERBA

The task of interpreting data on the occurrence of pathogens in the environment is often critical in making decisions about potential health risks and corrective actions. In the past this has largely been done in a qualitative or subjective fashion. **Quantitative risk assessment (QRA)** is an approach that allows the expression of risks in a quantitative fashion in terms of risk of infection, illness, or mortality from microbial pathogens. In this format, such information can be better utilized by decision makers to determine the magnitude of such risks and weigh the costs and benefits of corrective action. The purpose of this chapter is to provide a general background to the topic of risk analysis and how it can be used in the problem-solving processes.

24.1 THE CONCEPT OF RISK ASSESSMENT

Risk, which is common to all life, is an inherent property of everyday human existence. It is therefore a key factor in all decision making. Risk assessment or analysis, however, means different things to different people: Wall Street analysts assess financial risks and insurance companies calculate actuarial risks, while regulatory agencies estimate the risks of fatalities from nuclear plant accidents, the incidence of cancer resulting from industrial emissions, and habitat loss associated with increases in human populations. All these seemingly disparate activities have in common the concept of a measurable phenomenon called risk that can be expressed in terms of probability. Thus, we can define risk assessment as the process of estimating both the probability that an event will occur and the probable magnitude of its adverse effects—economic, health or safety related, or ecological—over a specified time period. For example, one might estimate the probable incidence of cancer in the community where a chemical was spilled over a period of years. Or one might calculate the health risks associated with the presence of pathogens in drinking water or in food.

There are, of course, several varieties of risk assessment. Risk assessment as a formal discipline emerged in the 1940s and 1950s, paralleling the rise of the nuclear industry. Safety hazard analyses have been used since at least the 1950s in the nuclear, petroleum refining, and chemical processing industries, as well as in aerospace. Health risk assessments, however, had their beginnings in 1986 with the publication of the "Guidelines for carcinogenic risk assessment" by the Environmental Protection Agency (EPA). Microbial risk assessment is relatively new, beginning in the mid-1980s, but has already been used in the development of government regulations (Regli *et al.*, 1991).

24.2 ELEMENTS OF RISK ASSESSMENT

Chemical risk assessment has been used to judge the safety of our food and water supply. Such assessments are important in setting standards for chemical contaminants in the environment. Whether chemical or microbial, risk assessment consists of four basic steps:

- **Hazard identification**—Defining the hazard and nature of the harm: for example, identifying a contaminant (e.g., a chemical, such as lead or carbon tetrachloride, or a microbial pathogen, such as *Legionella*), and documenting its toxic effects on humans.

- **Exposure assessment**—Determining the concentration of a contaminating agent in the environment and estimating its rate of intake. For example, determining the concentration of *Salmonella* in a meat product and determining the average dose a person would ingest.
- **Dose–response assessment**—Quantitating the adverse effects arising from exposure to a hazardous agent based on the degree of exposure. This assessment is usually expressed mathematically as a plot showing the response in living organisms to increasing doses of the agent (e.g., rotavirus).
- **Risk characterization**—Estimating the potential impact (e.g., human illness or death) of a microorganism or chemical based on the severity of its effects and the amount of exposure.

Once the risks are characterized, various regulatory options are evaluated in a process called **risk management,** which includes consideration of social, political, and economic issues as well as the engineering problems inherent in a proposed solution. One important component of risk management is **risk communication,** which is the interactive process of information and opinion exchange among individuals, groups, and institutions. Risk communication includes the transfer of risk information from expert to nonexpert audiences. In order to be effective, risk communication must provide a forum for balanced discussions of the nature of the risk, lending a perspective that allows the benefits of reducing the risk to be weighed against the costs.

In the United States, the passage of federal and state laws to protect public health and the environment has expanded the application of risk assessment. Major federal agencies that routinely use risk analysis include the Food and Drug Administration (FDA), the Environmental Protection Agency (EPA), and the Occupational Safety and Health Administration (OSHA). Together with state agencies, these regulatory agencies use risk assessment in a variety of situations:

- Setting standards for concentrations of toxic chemicals or pathogenic microorganisms in water or food
- Assessing the risk from the release of genetically altered organisms
- Conducting baseline analyses of contaminated sites or facilities to determine the need for remediation and the extent of cleanup required

- Performing cost–benefit analyses of contaminated-site cleanup or treatment options (including treatment processes to reduce exposure to pathogens)
- Developing cleanup goals for contaminants for which no federal or state authorities have promulgated numerical standards: evaluating acceptable variance from promulgated standards and guidelines (e.g., approving alternative concentration limits)
- Constructing "what if" scenarios to compare the potential impact of remedial or treatment alternatives and to set priorities for corrective action
- Evaluating existing and new technologies for effective prevention, control, or mitigation of hazards and risks (e.g., new drinking water treatment technologies)
- Articulating community public health concerns and developing consistent public health expectations among different localities

Risk assessment provides an effective framework for determining the relative urgency of problems and the allocation of resources to reduce risks. Using the results of risk analyses, we can target prevention, remediation, or control efforts toward areas, sources, or situations in which the greatest risk reductions can be achieved with the resources available. However, risk assessment is not an absolute procedure carried out in a vacuum; rather, it is an evaluative, multifaceted, comparative process. Thus, to evaluate risk, we must inevitably compare one risk with a host of others. In fact, the comparison of potential risks associated with several problems or issues has developed into a subset of risk assessment called **comparative risk assessment.** Some commonplace risks are shown in Table 24.1. Here we see, for example, that risks from chemical exposure are fairly small compared with those associated with driving a car or smoking cigarettes.

Comparing different risks allows us to comprehend the uncommon magnitudes involved and to understand the level, or magnitude, of risk associated with a particular hazard. But comparison with other risks cannot itself establish the *acceptability* of a risk. Thus, the fact that the chance of death from a previously unknown risk is about the same as that from a known risk does not necessarily imply that the two risks are equally acceptable. Generally, comparing risks along a single dimension is not helpful when the risks are widely perceived as qualitatively different. Rather, we must take account of certain qualitative factors that affect risk perception and evaluation when selecting risks to be compared. Some of these qualifying factors

TABLE 24.1   Examples of Some Commonplace Risks in the United States

Risk	Lifetime risk of mortality
Cancer from cigarette smoking (one pack per day)	$1:4$
Death in a motor vehicle accident	$2:100$
Homicide	$1:100$
Home accident deaths	$1:100$
Cancer from exposure to radon in homes	$3:1000$
Death from hepatitis A	$3:1000$
Exposure to the pesticide aflatoxin in peanut butter	$6:10,000$
Diarrhea from rotavirus	$1:10,000$
Exposure to typical EPA maximum chemical contaminant levels	$1:10,000$–$1:10,000,000$

Based on data in Wilson and Crouch (1987) and Gerba and Rose (1992).

are listed in Table 24.2. We must also understand the underlying premise that *voluntary risk is always more acceptable than involuntary risk.* For example, the same people who cheerfully drive their cars every day—thus incurring a $2:100$ lifetime risk of death by automobile—are quite capable of refusing to accept the $6:10,000$ involuntary risk of eating peanut butter contaminated with aflatoxin.

In considering risk, then, we must also understand another principle,—the *de minimis* principle, which means that there are some levels of risk so trivial that they are not worth bothering about. However attractive, this concept is hard to define, especially if we are trying to find a *de minimis* level acceptable to an entire society. Understandably, regulatory authorities are reluctant to be explicit about an "acceptable" risk. (How much aflatoxin would you consider acceptable in your peanut butter and jelly sandwich? How many dead insect parts?) Some prefer the term "tolerable risk," i.e., the level of risks we can accept given the economic

TABLE 24.2   Factors Affecting Risk Perception and Risk Analysis

Factor	Conditions associated with increased public concern	Conditions associated with decreased public concern
Catastrophic potential	Fatalities and injuries grouped in time and space	Fatalities and injuries scattered and random
Familiarity	Unfamiliar	Familiar
Understanding	Mechanisms or process not understood	Mechanisms or process understood
Controllability (personal)	Uncontrollable	Controllable
Voluntariness of exposure	Involuntary	Voluntary
Effects on children	Children specifically at risk	Children not specifically at risk
Effects manifestation	Delayed effects	Immediate effects
Effects on future generations	Risk to future generations	No risk to future generations
Victim identity	Identifiable victims	Statistical victims
Dread	Effects dreaded	Effects not dreaded
Trust in institutions	Lack of trust in responsible institutions	Trust in responsible institutions
Media attention	Much media attention	Little media attention
Accident history	Major and sometimes minor accidents	No major or minor accidents
Equity	Inequitable distribution of risks and benefits	Equitable distribution of risks and benefits
Benefits	Unclear benefits	Clear benefits
Reversibility	Effects irreversible	Effects reversible
Origin	Caused by human actions or failures	Caused by acts of nature

From Covello *et al.* (1988).

costs, and social and scientific constraints. But it is generally agreed that a lifetime risk on the order of one in a million (or 10^{-6}) is trivial enough to be acceptable for the general public. Although the origins and precise meaning of a one-in-a-million acceptable risk remain obscure, its impact on product choices, operations, and costs is very real—running, for example, into hundreds of billions of dollars in hazardous waste site cleanup decisions alone. The levels of acceptable risk can vary within this range. Levels of risk at the higher end of the range (10^{-4} rather than 10^{-6}) may be acceptable if just a few people are exposed rather than the entire populace. For example, workers dealing with production of solvents can often tolerate higher levels of risk than can the public at large. These higher levels are justified because workers tend to be a relatively homogeneous, healthy group and because employment is voluntary; however, the sum level of risks would not be acceptable for the same solvents for the general population.

24.3 THE PROCESS OF RISK ASSESSMENT

24.3.1 Hazard Identification

The first step in risk assessment is to determine the nature of the hazard. For pollution-related problems, the hazard in question is usually a specific chemical, a physical agent (such as irradiation), or a microorganism identified with a specific illness or disease. Thus the hazard identification component of a pollution risk assessment consists of a review of all relevant biological and chemical information bearing on whether or not an agent poses a specific threat.

As Fig. 24.1 shows, clinical studies of disease can be used to identify very large risks (between 1:10 and 1:100), most epidemiological studies can detect risks down to 1:1000, and very large epidemiological studies can examine risks in the 1:10,000 range. However, risks lower than 1:10,000 cannot be studied with much certainty using epidemiological approaches. Because regulatory policy objectives generally strive to limit risks below 1:100,000 for life-threatening diseases such as cancer, these lower risks are often estimated by extrapolating from the effects of high doses given to animals.

24.3.2 Exposure Assessment

Exposure assessment is the process of measuring or estimating the intensity, frequency, and duration of human exposures to an environmental agent. Exposure to contaminants can occur via inhalation, ingestion of water or food, or the skin. Contaminant sources, release mechanisms, transport, and transfor-

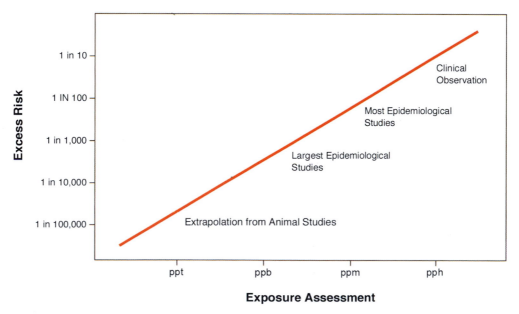

FIGURE 24.1   Sensitivity of epidemiology in detecting risks of regulatory concern. The generalized units of ppt (parts per thousand), ppb (parts per billion), ppm (parts per million), and pph (parts per hundred) are used here for comparative purposes. (Modified from National Research Council, 1993.)

mation characteristics are all important aspects of exposure assessment, as are the nature, location, and activity patterns of the exposed population. (This explains why it is critical to understand the factors and processes influencing the transport and fate of a contaminant.)

An **exposure pathway** is the course that a hazardous agent takes from a source to a receptor (e.g., human or animal) via environmental carriers or media—generally, air (volatile compounds, particles) or water (soluble compounds). An exception is electromagnetic radiation, which needs no medium. The **exposure route,** or intake pathway, is the mechanism by which the transfer occurs-usually by inhalation, ingestion, and/or dermal contact. Direct contact can result in a local effect at the point of entry and/or in a systemic effect.

The quantitation of exposure, intake, or potential dose can involve equations with three sets of variables:

- Concentrations of chemicals or microbes in the media
- Exposure rates (magnitude, frequency, duration)
- Quantified biological characteristics of receptors (e.g., body weight, absorption capacity for chemicals, level of immunity to microbial pathogens)

Exposure concentrations are derived from measured, monitored, and/or modeled data. Ideally, exposure concentrations should be measured at the points of contact between the environmental media and current or potential receptors. It is usually possible to identify potential receptors and exposure points from field observations and other information. However, it is seldom possible to anticipate all potential exposure points and measure all environmental concentrations under all conditions. In practice, a combination of monitoring and modeling data, together with a great deal of professional judgment, is required to estimate exposure concentrations.

In order to assess exposure rates via different pathways, one has to consider and weigh many factors. For example, in estimating exposure to a substance via drinking water, one first has to determine the average daily consumption of that water. But this is not as easy as it sounds. Studies have shown that daily fluid intake varies greatly from individual to individual. Moreover, total water intake depends on how much fluid is consumed as tap water and how much is ingested in the form of soft drinks and other non-tap water sources. Tap water intake also changes significantly with age, body weight, diet, and climate. Because these factors are so variable, the EPA has suggested a number of very conservative "default" exposure values that can be used when assessing contaminants in tap water, vegetables, soil, and the like (see Table 24.3).

One also has to consider how much of the population to include in the exposure. For example, the average consumption of tap water is 1.5 liters/day; some persons drink less than this and some significantly more. Approximately 95% of the population consumes 2 liters or less. If we assume the exposure to be 1.5

TABLE 24.3 EPA Standard Default Exposure Factors

Land use	Exposure pathway	Daily intake	Exposure frequency (days/year)	Exposure duration (years)
Residential	Ingestion of potable water	2 l/day	350	30
	Ingestion of soil and dust	200 mg (child)	350	6
		100 mg (adult)		24
	Inhalation of contaminants	20 m^3 (total)	350	30
		15 m^3 (indoor)		
Industrial and commercial	Ingestion of potable water	1 liter	250	25
	Ingestion of soil and dust	50 mg	250	25
	Inhalation of contaminants	20 m^3 (workday)	250	25
Agricultural	Consumption of homegrown produce	42 g (fruit)	350	30
		80 g (vegetable)		
Recreational	Consumption of locally caught fish	54 g	350	30
Swimming		10–100 ml[a]	1–10	—

Modified from Kolluru (1993).
[a] Per event.

liters/day, then we are excluding half the population, which consumes more than this amount. Using the 2 liter/day value allows inclusion of 95% of the population.

24.3.3 Dose–Response Assessment

Chemical or microbial contaminants are not equal in their capacity to cause adverse effects. To determine the capacity of agents to cause harm, we need quantitative toxicity or infectivity data. These data can sometimes be derived from occupational, clinical, and epidemiological studies. Most toxicity data, however, come from animal experiments in which researchers expose laboratory animals, mostly mice and rats, to increasingly higher concentrations or doses and observe their corresponding effects. The result of these experiments is the **dose–response relationship**—a quantitative relationship that indicates the agent's degree of toxicity to exposed species. Dose is normalized as milligrams of substance or pathogen ingested, inhaled, or absorbed (in the case of chemicals) through the skin per kilogram of body weight per day (mg/kg/day). Responses or effects can vary widely—from no observable effect, to temporary and reversible effects (e.g., enzyme depression caused by some pesticides or diarrhea caused by viruses), to permanent organ injury (e.g., liver and kidney damage caused by chlorinated solvents, heavy metals, or viruses), to chronic functional impairment (e.g., bronchitis or emphysema arising from smoke damage), to death.

The goal of a dose–response assessment is to obtain a mathematical relationship between the amount (concentration) of a toxicant or microorganism to which a human is exposed and the risk of an adverse outcome from that dose. The data resulting from experimental studies are presented as a dose–response curve, as shown in Fig. 24.2. The abscissa represents the dose, and the ordinate represents the risk that some adverse health effect will occur. In the case of a pathogen, for instance, the ordinate may represent the risk of infection and not necessarily illness.

However, dose–response curves derived from animal studies must be interpreted with care. The data for these curves are necessarily obtained by examining the effects of large doses on test animals. Because of the costs involved, researchers are limited in the numbers of test animals they can use—it is both impractical and cost prohibitive to use thousands (even millions) of animals to observe just a few individuals that show adverse effects at low doses (e.g., risks of 1 : 1000 or 1 : 10,000). Researchers must therefore extrapolate low-dose responses from their high-dose data. And therein lies the rub: Dose–response curves are subject to con-

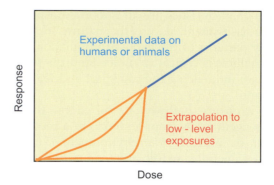

FIGURE 24.2 Extrapolation of dose–response curves. (Adapted from U.S. EPA, 1990.)

troversy because their results change depending on the method chosen to extrapolate from the high doses actually administered to laboratory test subjects to the low doses humans are likely to receive in the course of everyday living.

This controversy revolves around the choice of several mathematical models that have been proposed for extrapolation to low doses. Unfortunately, no model can be proved or disproved from the data, so there is no way to know which model is the most accurate. The choice of models is therefore strictly a policy decision, which is usually based on understandably conservative assumptions. Thus, for noncarcinogenic chemical responses, the assumption is that some *threshold* exists below which there is no toxic response; that is, no adverse effects will occur below some very low dose (say, one in a million). Carcinogens, however, are considered *nonthreshold*—that is, the conservative assumption is that exposure to any amount of carcinogen creates some likelihood of cancer. This means that the only "safe" amount of carcinogen is zero, so the dose–response plot is required to go through the origin (0), as shown in Fig. 24.2.

In the microbiological literature the term **"minimum infectious dose"** is used frequently, implying that a threshold dose exists for microorganisms. In reality, the term used usually refers to the ID50 or the dose at which 50% of the animals or humans exposed became infected or exhibit any symptoms of an illness. Existing infectious dose data are compatible with nonthreshold responses, and the term "infectivity" is probably more appropriate when referring to differences in the likelihood of an organism causing an infection. For example, the probability of a given number of ingested rotaviruses causing diarrhea is greater than that for *Salmonella*. Thus, the infectivity of rotavirus is greater than that of *Salmonella*.

There are many mathematical models to choose

from, including the *one-hit model*, the *multistage model*, the *multihit model*, and the *probit model*. The characteristics of these models for nonthreshold effects are listed in Tables 24.4 and 24.5.

24.3.4 Risk Characterization

24.3.4.1 Uncertainty Analysis

Uncertainty is inherent in every step of the risk assessment process. Thus, before we can begin to characterize any risk, we need some idea of the nature and magnitude of uncertainty in the risk estimate. Sources of uncertainty include

- Extrapolation from high to low doses
- Extrapolation from animal to human responses
- Extrapolation from one route of exposure to another
- Limitations of analytical methods
- Estimates of exposure

Although the uncertainties are generally much larger in estimates of exposure and the relationships between dose and response (e.g., the percent mortality), it is important to include the uncertainties originating from all steps in a risk assessment as part of risk characterization.

Two approaches commonly used to characterize uncertainty are sensitivity analyses and Monte Carlo simulations. In **sensitivity analyses,** the uncertain quantities of each parameter (e.g., average values, high and low estimates) are varied, usually one at a

TABLE 24.5   Lifetime Risks of Cancer Derived from Different Extrapolation Models[a]

Model applied	Lifetime risk (mg/kg/day) of toxic chemical
One-hit	6.0×10^{-5} (1 in 17,000)
Multistage	6.0×10^{-6} (1 in 167,000)
Multihit	4.4×10^{-7} (1 in 2.3 million)
Probit	1.9×10^{-10} (1 in 5.3 billion)

From U.S. EPA (1990).
[a] All risks are for a full lifetime of daily exposure. The lifetime is used as the unit of risk measurement because the experimental data reflect the risk experienced by animals over their full lifetimes. The values shown are upper confidence limits on risks.

time, to find out how changes in these quantities affect the final risk estimate. This procedure gives a range of possible values for the overall risk and provides information on which parameters are most crucial in determining the size of the risk. In a **Monte Carlo simulation,** however, it is assumed that all parameters are random or uncertain. Thus, instead of varying one parameter at a time, a computer program is used to select distributions randomly every time the model equations are solved, the procedure being repeated many times. The resulting output can be used to identify values of exposure or risk corresponding to a specified probability, say the 50th percentile or 95th percentile. Details of these methods of dealing with uncertainty can be found in "Quantitative Microbial Risk Assessment" (Haas *et al.*, 1999).

24.3.4.2 Risk Projection and Management

The final phase of the risk assessment process is risk characterization. In this phase, exposure and dose-response assessments are integrated to yield probabilities of effects occurring in humans under specific exposure conditions. Quantitative risks are calculated for appropriate media and pathways. For example, the risks of lead in water are estimated over a lifetime assuming

1. That the exposure is 2 liters of water ingested over a 70-year lifetime
2. That different concentrations of lead occur in the drinking water

This information can be used by risk managers to develop standards or guidelines for specific toxic chemicals or infectious microorganisms in different media, such as the drinking water or food supply.

TABLE 24.4   Primary Models Used for Assessment of Nonthreshold Effects[a]

Model	Comments
One-hit	Assumes (1) single-stage for cancer, (2) malignant change induced by one molecular or radiation interaction *Very conservative*
Linear multistage	Assumes multiple stages for cancer *Fits curve to the experimental data*
Multihit	Assumes several interactions needed before cells becomes transformed *Least conservative model*
Probit	Assumes probit (lognormal) distribution for tolerances of exposed population *Appropriate for acute toxicity; questionable for cancer*

Modified from Cockerham and Shane (1994).
[a] All these models assume that exposure to the pollutant will always produce an effect, regardless of dose.

In the case of a microorganism, a treatment strategy may be used. For example, 99.9% removal of *Giardia* cysts is required for drinking water treatment plants in the United States to ensure that the yearly risk of infection in a community is no greater than 1:10,000. The assumption is made that this amount of removal will guarantee a concentration of *Giardia* cysts in the finish water that would not result in a risk greater than 1:10,000 (Gerba *et al.*, 1997; Teunis *et al.*, 1997).

24.4 MICROBIAL RISK ASSESSMENT

Outbreaks of waterborne disease caused by microorganisms usually occur when the water supply has been obviously and significantly contaminated. In such high-level cases, the exposure is manifest and cause and effect are relatively easy to determine. However, exposure to low-level microbial contamination is difficult or impossible to determine epidemiologically. We know, for example, that long-term exposure to microbes can have a significant impact on the health of individuals within a community, but we need a way to measure that impact.

For some time, methods have been available to detect the presence of low levels (one organism per 1000 liters) of pathogenic organisms in water, including enteric viruses and protozoan parasites. The trouble is that the risks posed to the community by these low levels of pathogens in a water supply over time are not like those posed by low levels of chemical toxins or carcinogens. For example, it takes just one amoeba in the wrong place at the wrong time to infect one individual, whereas the same individual would have to consume some quantity of a toxic chemical to be comparably harmed. Microbial risk assessment is, therefore, a process that allows us to estimate responses in terms of the *risk of infection* in a quantitative fashion. Although no accepted formal framework for microbial risk assessment exists, it generally follows the steps used in other health-based risk assessments—hazard identification, exposure assessment, dose-response, and risk characterization. The differences are in the specific assumptions, models, and extrapolation methods used.

Hazard identification in the case of pathogens is complicated because several outcomes—from asymptomatic infection to death (Fig. 24.3)—are possible, and these outcomes depend on the complex interaction between the pathogenic agent (the "infector") and the host (the "infectee"). This interaction, in turn, depends on the characteristics of the host as well as the nature of the pathogen. Host factors, for example, in-

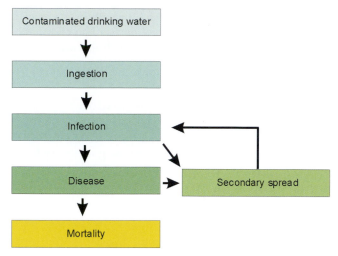

FIGURE 24.3   Outcomes of enteric viral exposure.

clude preexisting immunity, age, nutrition, ability to mount an immune response, and other nonspecific host factors. Agent factors include type and strain of the organism as well as its capacity to elicit an immune response.

Among the various outcomes of infection is the possibility of **subclinical illness.** Subclinical (**asymptomatic**) infections are those in which the infection (growth of the microorganism within the human body) results in no obvious illness such as fever, headache, or diarrhea That is, individuals can host a pathogen microorganism—and transmit it to others—without ever getting sick themselves. The ratio of clinical to subclinical infection varies from pathogen to pathogen, especially for viruses, as shown in Table 24.6. Poliovirus infections, for instance, seldom result in obvious clinical symptoms; in fact, the proportion of individuals developing clinical illness may be less than 1%. However, other enteroviruses, such as the coxsackieviruses, may exhibit a greater proportion. In many cases, as in that of rotaviruses, the probability of developing clinical illness appears to be completely unrelated to the dose an individual receives via ingestion (Ward *et al.*, 1986). Rather, the likelihood of developing clinical illness depends on the type and strain of the virus as well as host age, nonspecific host factors, and possibly preexisting immunity. The incidence of clinical infection can also vary from year to year for the same virus, depending on the emergence of new strains.

Another outcome of infection is the development of clinical illness. Several host factors play a major role in this outcome. The age of the host is often a determin-

TABLE 24.6   Ratio of Clinical to Subclinical Infections with Enteric Viruses

Virus	Frequency of clinical illness[a] (%)
Poliovirus 1	0.1–1
Coxsackie	
A16	50
B2	11–50
B3	29–96
B4	30–70
B5	5–40
Echovirus	
Overall	50
9	15–60
18	Rare–20
20	33
25	30
30	50
Hepatitis A (adults)	75
Rotavirus	
(adults)	56–60
(children)	28
Astrovirus (adults)	12.5

From Gerba and Rose (1992).

[a] The percentage of the individuals infected who develop clinical illness.

ing factor. In the case of hepatitis A, for example, clinical illness can vary from about 5% in children younger than 5 years of age to 75% in adults. Similarly, children are more likely to develop rotaviral gastroenteritis than are adults. Immunity is also an important factor, albeit a variable one. That is, immunity may or may not provide long-term protection from reinfection, depending on the enteric pathogen. It does not, for example, provide long-term protection against the development of clinical illness in the case of the Norwalk virus or *Giardia*. However, for most enteroviruses and for the hepatitis A virus, immunity to reinfection is believed to be lifelong. Other undefined host factors may also control the odds of developing illness. For example, in experiments with the Norwalk virus, human volunteers who did not become infected upon an initial exposure to the virus also did not respond to a second exposure. In contrast, volunteers who developed gastroenteritis upon the first exposure also developed illness after the second exposure.

The ultimate outcome of infection—mortality—can be caused by nearly all enteric organisms. The factors that control the prospect of mortality are largely the same factors that control the development of clinical illness. Host age, for example, is significant. Thus, mortality for hepatitis A and poliovirus is greater in adults than in children. In general, however, one can say that the very young, the elderly, and the immunocompromised are at the greatest risk of a fatal outcome of most illnesses (Gerba et al., 1996). For example, the case–fatality rate (%) for *Salmonella* in the general population is 0.1%, but it has been observed to be as high as 3.8% in nursing homes (Table 24.7). In North America and Europe, the reported case–fatality rates (i.e., the ratio of cases to fatalities reported as a percentage of persons who die) for enterovirus infections range from less than 0.1 to 0.94%, as shown in Table 24.8. The case–fatal rate for common enteric bacteria ranges from 0.1 to 0.2% in the general population. Enteric bacterial diseases can be treated with antibiotics, but no treatment is available for enteric viruses.

Recognizing that microbial risk involves a myriad of pathogenic organisms capable of producing a variety of outcomes that depend on a number of factors—many of which are undefined—one must now face the problem of exposure assessment, which has complications of its own. Unlike chemical-contaminated water, microorganism-contaminated water does not have to be consumed to cause harm. That is, individuals who do not actually drink, or even touch, contaminated water also risk infection because pathogens—partic-

TABLE 24.7   Case Fatality Observed for Enteric Pathogens in Nursing Homes versus General Population

Organism	Case fatality (%) in general population	Case fatality (%) in nursing homes
Campylobacter jejuni	0.1	1.1
Escherichia coli 0157:H7	0.2	11.8
Salmonella	0.1	3.8
Rotavirus	0.01	1.0

Modified from Gerba et al. (1996).

TABLE 24.8   Case–Fatality Rates for Enteric Viruses and Bacteria

Organism	Case–fatality rate (%)
Viruses	
Poliovirus 1	0.90
Coxsackie	
A2	0.50
A4	0.50
A9	0.26
A16	0.12
Coxsackie B	0.59–0.94
Echovirus	
6	0.29
9	0.27
Hepatitis A	0.30
Rotavirus	
(total)	0.01
(hospitalized	0.12
Norwalk	0.0001
Bacteria	
Shigella	0.2
Salmonella	0.1
Escherichia coli 0157 : H7	0.2
Campylobacter jejuni	0.1
Astrovirus	0.01

From Gerba and Rose (1992) and Gerba *et al.* (1995).

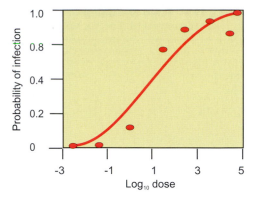

FIGURE 24.4   Dose–response for human rotavirus by oral ingestion.

ularly viruses—may be spread by person-to-person contact or subsequent contact with contaminated inanimate objects (such as toys). This phenomenon is described as the **secondary attack rate,** which is reported as a percentage. For example, one person infected with poliovirus can transmit it to 90% of the persons with whom he or she associates. This secondary spread of viruses has been well documented for waterborne outbreaks of several diseases, including that caused by Norwalk virus, whose secondary attack rate is about 30%.

The question of dose is another problem in exposure assessment. How does one define "dose" in this context? To answer this question, researchers have conducted a number of studies to determine the infectious dose of enteric microorganisms in human volunteers. Such human experimentation is necessary because determination of the infectious dose in animals and extrapolation to humans is often impossible. In some cases, for example, humans are the primary or only known host. In other cases, such as that of *Shigella* or Norwalk virus, infection can be induced in laboratory-held primates, but it is not known whether the infectious dose data can be extrapolated

to humans. Much of the existing data on infectious doses of viruses has been obtained with attenuated vaccine viruses or with avirulent laboratory-grown strains, so that the likelihood of serious illness is minimized. An example of a dose–response curve for a human feeding study with rotavirus is shown in Fig. 24.4.

Next, one must choose a dose–response model, whose abscissa is the dose and whose ordinate is the risk of infection (see Fig. 24.4). The choice of model is critical so that risks are not greatly over or underestimated. A modified exponential (beta–Poisson distribution) or a log-probit (simple lognormal, or exponential, distribution) model may be used to describe the probability of infection in human subjects for many enteric microorganisms (Haas, 1983). These models have been

TABLE 24.9   Best-Fit Dose–Response Parameters for Enteric Pathogen Ingestion Studies

Microorganism	Best model	Model parameters
Echovirus 12	Beta–Poisson	$\alpha = 0.374$ $\beta = 186.69$
Rotavirus	Beta–Poisson	$\alpha = 0.26$ $\beta = 0.42$
Poliovirus 1	Exponential	$r = 0.009102$
Poliovirus 1	Beta–Poisson	$\alpha = 0.1097$ $\beta = 1524$
Poliovirus 3	Beta–Poisson	$\alpha = 0.409$ $\beta = 0.788$
Cryptosporidium	Exponential	$r = 0.004191$
Giardia lamblia	Exponential	$r = 0.02$
Salmonella	Exponential	$r = 0.00752$
Escherichia coli	Beta-Poisson	$\alpha = 0.1705$ $\beta = 1.61 \times 10^6$

Modified from Regli *et al.* (1991).

found to best fit the experimental data. For the beta-Poisson model the probability of infection from a single exposure, P, can be described as follows:

$$P = 1 - (1 + N/\beta)^{-\alpha} \qquad \text{(Eq. 24.1)}$$

where N is the number of organisms ingested per exposure and α and β represent parameters characterizing the host–virus interaction (dose–response curve). Some values for α and β for several enteric waterborne pathogens are shown in Table 24.9; these values were determined from human studies. For some microorganisms, an **exponential model** may better represent the probability of infection.

$$P = 1 - \exp(-rN) \qquad \text{(Eq. 24.2)}$$

In this equation, r is the fraction of the ingested microorganisms that survive to initiate infections (host–microorganism interaction probability). Table 24.9 shows examples of results of both models for several organisms.

These models define the probability of the microorganism overcoming the host defenses (stomach pH, finding a susceptible cell, nonspecific immunity, etc.) to establish an infection in the host. When one uses these models, one estimates the probability of becoming infected after ingestion of various concentrations of pathogen. For example, Case Study 1 shows how to calculate the risk of acquiring a viral infection from consumption of contaminated drinking water containing echovirus 12 using Eq. 24.1.

Annual and lifetime risks can also be determined, again assuming a Poisson distribution of the virus in the water consumed (assuming daily exposure to a constant concentration of viral contamination), as follows:

$$P_A = 1 - (1 - P)^{365} \qquad \text{(Eq. 24.3)}$$

where P_A is the annual risk (365 days) of contracting one or more infections, and

$$P_L = 1 - (1 - P)^{25,550} \qquad \text{(Eq. 24.4)}$$

where P_L is the lifetime risk (assuming a lifetime of 70 years = 25,550 days) of contracting one or more infections.

Risks of clinical illness and mortality can then be determined by incorporating terms for the percentage of clinical illness and mortality associated with each particular virus:

CASE STUDY 1

Application of a Virus Risk Model to Characterize Risks from Consuming Shellfish

It is well known that infectious hepatitis and viral gastroenteritis are caused by consumption of raw or, in some cases, cooked clams and oysters. The concentration of echovirus 12 was found to be 8 plaque-forming units per 100 g in oysters collected from coastal New England waters. What are the risks of becoming infected and ill from echovirus 12 if the oysters are consumed? Assume that a person usually consumes 60 g of oyster meat in a single serving:

$$\frac{8\,\text{PFU}}{100\,\text{g}} = \frac{N}{60\,\text{g}} \qquad N = 4.8\,\text{PFU consumed}$$

From Table 24.9, $\alpha = 0.374$, $\beta = 186.64$. The probability of infection from Eq. 24.1 is then

$$P = 1 - \left(1 + \frac{4.8}{186.69}\right)^{0.374} = 9.4 \times 10^{-3}$$

If the percent of infections that result in risk of clinical illness is 50%, then from Eq. 24.5 one can calculate the risk of clinical illness:

$$\text{Risk of clinical illness} = (9.4 \times 10^{-3})\,(0.50) = 4.7 \times 10^{-3}$$

If the case–fatality rate is 0.001%, then from Eq. 24.6:

$$\text{Risk of mortality} = (9.4 \times 10^{-3})\,(0.50)\,(0.001) = 4.7 \times 10^{-6}$$

If a person consumes oysters 10 times a year with 4.8 PFU per serving, then one can calculate the risk of infection in one year from Eq. 24.3:

$$\text{Annual risk} = P_A = 1 - (1 - 9.4 \times 10^{-3})^{365} = 9.7 \times 10^{-1}$$

TABLE 24.10   Risk of Infection, Disease, and Mortality
for Rotavirus

Virus concentration per 100 liters	Risk	
	Daily	Annual
Infection		
100	9.6×10^{-2}	1.0
1	1.2×10^{-3}	3.6×10^{-1}
0.1	1.2×10^{-4}	4.4×10^{-2}
Disease		
100	5.3×10^{-2}	5.3×10^{-1}
1	6.6×10^{-4}	2.0×10^{-1}
0.1	6.6×10^{-5}	2.5×10^{-2}
Mortality		
100	5.3×10^{-6}	5.3×10^{-5}
1	6.6×10^{-8}	2.0×10^{-5}
0.1	6.6×10^{-9}	2.5×10^{-6}

Modified from Gerba and Rose (1992).

$$\text{Risk of clinical illness} = PI \qquad \text{(Eq. 24.5)}$$

$$\text{Risk of mortality} = PIM \qquad \text{(Eq. 24.6)}$$

where I is the percentage of infections that result in clinical illness and M is the percentage of clinical cases that result in mortality.

Application of this model allows estimation of the risks of infection, development of clinical illness, and mortality for different levels of exposure. As shown in Table 24.10, for example, the estimated risk of infection from one rotavirus in 100 liters of drinking water (assuming ingestion of 2 liters per day) is 1.2×10^{-3}, or almost one in a thousand for a single-day exposure. This risk would increase to 3.6×10^{-1}, or approximately one in three, on an annual basis. As can be seen from this table, the risk of developing of clinical illness also appears to be significant for exposure to low levels of rotavirus in drinking water.

The EPA has recently recommended that any drinking water treatment process should be designed to ensure that human populations are not subjected to risk of infection greater than 1 : 10,000 for a yearly exposure. To achieve this goal, it would appear from the data shown in Table 24.10 that the virus concentration in drinking water would have to be less than one per 1000 liters. Thus, if the average concentration of enteric viruses in untreated water is 1400/1000 liters, treatment plants should be designed to remove at least 99.99% of the virus present in the raw water. A further application of this approach is to define the required treatment of a water source in terms of the concentration of a disease-causing organism in that supply. Thus, the more contaminated the raw water source, the more treatment is required to reduce the risk to an acceptable level. An example of this application is shown in Fig. 24.5. The plausibility of validation of microbial risk assessment models has been examined by using data from foodborne outbreaks in which information has been available on exposure and outcomes (Rose et al., 1995; Crockett et al., 1996). These studies suggest that microbial risk assessment can give reasonable estimates of illness from exposure to contaminated foods (Table 24.11).

In summary, risk assessment is a major tool for decision making in the regulatory arena. This approach is used to explain chemical and microbial risks as well as ecosystem impacts. The results of such assessments can be used to inform risk managers of the probability and extent of environmental impacts resulting from exposure to different levels of stress (contaminants). Moreover, this process, which allows the quantitation and comparison of diverse risks, lets risk managers utilize the maximum amount of complex information in the decision-making process. This information can also be used to weigh the cost and benefits of control options and to develop standards or treatment options (see Case Study 2).

TABLE 24.11   Comparison of Outbreak Data to Model Predictions for Assessment of
Risks Associated with Exposure to *Salmonella*

Food	Dose CFU	Amount consumed	Attack rate (%)	Predicted P (%)
Water	17	1 liter	12	12
Pancretin	200	7 doses	100	77
Ice cream	102	1 portion	52	54
Cheese	100–500	28 g	28–36	53–98
Cheese	10^5	100 g	100	>99.99
Ham	10^6	50–100 g	100	>99.99

Modified from Rose *et al.* (1995)

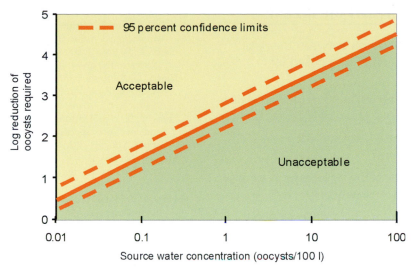

FIGURE 24.5 Relationship of influent *Cryptosporidium* concentration and log reduction by treatment necessary to produce acceptable water. (From Haas *et al.*, 1996).

CASE STUDY 2

How Do We Set Standards for Pathogens in Drinking Water?

In 1974 the U. S. Congress passed the Safe Drinking Water Act giving the U.S. Environmental Protection Agency the authority to establish standards for contaminants in drinking water. Through a risk analysis approach, standards have been set for many chemical contaminants in drinking water. Setting standards for microbial contaminants proved more difficult because (1) methods for the detection of many pathogens are not available, (2) days to weeks are sometime required to obtain results, and (3) costly and time-consuming methods are required. To overcome these difficulties, coliform bacteria had been used historically to assess the microbial quality of drinking water. However, by the 1980s it had become quite clear that coliform bacteria did not indicate the presence of pathogenic waterborne *Giardia* or enteric viruses. Numerous outbreaks had occurred in which coliform standards were met, because of the greater resistance of viruses and *Giardia* to disinfection. A new approach was needed to ensure the microbial safety of drinking water.

To achieve this goal a new treatment approach was developed called the Surface Treatment Rule (STR). As part of the STR, all water utilities that use surface waters as their source of potable water would be required to provide filtration to remove *Giardia* and enough disinfection to kill viruses. The problem facing the EPA was how much removal should be required. To deal with this issue, the EPA for the first time used a microbial risk assessment approach. The STR established that the goal of treatment was to ensure that microbial illness from *Giardia lamblia* infection should not be any greater than 1 per 10,000 exposed persons annually (10^{-4} per year). This value is close to the annual risk of infection from waterborne disease outbreaks in the United States (4×10^{-3}). Based on the estimated concentration of *Giardia* and enteric viruses in surface waters in the United States from the data available at the time, it was required that all drinking water treatment plants be capable of removing 99.9% of the *Giardia* and 99.99% of the viruses. In this manner it was hoped that the risk of infection of 10^{-4} per year would be achieved. The STR went into effect in 1991.

To better assess whether the degree of treatment required is adequate, the EPA developed the Information Collection Rule, which requires major drinking water utilities that use surface waters to analyze these surface water for the presence of *Giardia, Cryptosporidium,* and enteric viruses for a period of almost 2 years. From this information the EPA hopes to assess whether the treatment currently required is enough to ensure the 10^{-4} yearly risk. Potentially, utilities that have heavily contaminated source water may require greater levels of treatment in the future (see Fig. 24.5).

QUESTIONS AND PROBLEMS

1. What are some differences between the risks posed by chemicals and those posed by microorganisms?

2. What are some of the potential applications of risk assessment?

3. What is the difference between risk assessment and risk management?

4. Why is the selection of the dose–response curve so important in risk assessment?

5. What is meant by a threshold dose–response curve? Give arguments for a threshold and a nonthreshold dose–response curve for microorganisms.

6. What is the difference between a voluntary and an involuntary risk? Give examples of both.

7. List the four steps in a formal health risk assessment.

8. Does infection always lead to illness with enteric pathogens? What are the factors that determine morbidity and mortality outcomes with microbial infections?

9. What types of dose–response curve best reflect pathogen exposure?

10. What are some potential applications of microbial risk assessment?

11. Calculate the risk of infection from rotavirus during swimming in polluted water. Assume that 100 ml of water is ingested during swimming and the concentration of virus was one per 10 liters. What would the risk be in a year if a person went swimming 10 times in the same water with the same concentration of rotavirus?

References and Recommended Readings

Cockerham, L. G., and Shane, B. S. (1994) "Basic Environmental Toxicology." CRC Press, Boca Raton, FL.

Covello, V., von Winterfieldt, D., and Slovic, P. (1988) Risk communication: A review of the literature. *Risk Anal.* **3,** 171–182.

Crockett, C. S., Haas, C. N., Fazil, A., Rose, J. B., and Gerba, C. P. (1996) Prevalence of shillgellosis: Consistency with dose–response information. *Int. J. Food Prot.* **30,** 87–99.

Gerba, C. P., and Rose, J. B. (1992) Estimating viral disease risk from drinking water. *In* "Comparative Environmental Risks" (C. R. Cothern, ed.) Lewis, Boca Raton, FL.

Gerba, C. P., Rose, J. B, and Haas, C. N. (1995) Waterborne disease—Who is at risk? Water Quality Technology Proceedings, American Water Works Association, Denver, CO, pp. 231–254.

Gerba, C. P., Rose, J. B., and Haas, C. N. (1996) Sensitive populations: Who is at the greatest risk? *Int. J. Food Microbiol.* **30,** 113–123.

Gerba, C. P., Rose, J. B., Haas, C. N., and Crabtree, K. D. (1997) Waterborne rotavirus: A risk assessment. *Water Res.* **12,** 2929–2940.

Haas, C. N. (1983) Estimation of risk due to low levels of microorganisms: A comparison of alternative methodologies. *Am. J. Epidemiol.* **118,** 573–582.

Haas, C. N., Crockett, C. S., Rose, J. B., Gerba, C. P., and Fazil, A. M. (1996) Assessing the risk posed by oocysts in drinking water. *J. Am. Water Works Assoc.* **88,** 113–123.

Haas, C. N., Rose, J. B., and Gerba, C. P. (1999) "Quantitative Microbial Risk Assessment." John Wiley & Sons, New York.

ILSI Risk Science Institute Pathogen Risk Assessment Working Group. (1996) A conceptual framework to assess the risks of human disease following exposure to pathogens. *Risk Anal.* **16,** 841–848.

Kolluru, R.V. (1993) "Environmental Strategies Handbook." McGraw-Hill, New York.

National Research Council. (1983) "Risk Assessment in the Federal Government: Managing the Process." National Academy Press, Washington, DC.

National Research Council. (1989) "Improving Risk Communication." National Academy Press, Washington, DC.

National Research Council. (1991) "Frontiers in Assessing Human Exposure." National Academy Press, Washington, DC.

National Research Council. (1993) "Managing Wastewater Coastal Urban Areas." National Academy Press, Washington, DC.

Regli, S., Rose, J. B., Haas, C. N., and Gerba, C. P. (1991) Modeling the risk from *Giardia* and viruses in drinking water. *J. Am. Water Works Assoc.* **83,** 76–84.

Ricci, P. F., and Molton, L. S. (1985) Regulating cancer risks. *Environ. Sci. Technol.* **19,** 473–479.

Rodricks, J. V. (1992) "Calculated Risks. Understanding the Toxicity and Human Health Risks of Chemicals in Our Environment." Cambridge University Press, Cambridge, UK.

Rose, J. B., Haas, C. N., and Gerba, C. P. (1995) Linking microbiological criteria for foods with quantitative risk assessment. *J. Food Safety* **15,** 11–132.

Teunis, P. F. M., Medema, G. J., Kruidenier, L., and Havelaar, A. H. (1997) Assessment of the risk of infection of *Cryptosporidium* and *Giardia* in drinking water from a surface water source. *Water Res.* **31,** 1333–1346.

U.S. Environmental Protection Agency (U.S. EPA). (1990) Risk Assessment, Management and Communication of Drinking Water Contamination. EPA/625/4–89/024, Washington, DC.

U.S. Environmental Protection Agency. (1992) Dermal Exposure Assessment: Principles and Applications. EPA 600/8–91/011B, Washington, DC.

U.S. Environmental Protection Agency. (1986) Guidelines for carcinogen risk assessment. Fed. Regist. September 24.

Ward, R. L., Berstein, D. I., and Young, E. C. (1986) Human rotavirus studies in volunteers of infectious dose and serological response to infection. *J. Infect. Dis.* **154,** 871–877.

Wilson, R., and Crouch, E. A. C. (1987) Risk assessment and comparisons: An introduction. *Science* **236,** 267–270.

Index